NASA SP-4015

ASTRONAUTICS AND AERONAUTICS, 1970

Chronology on Science, Technology, and Policy

Text by
Science and Technology Division
Library of Congress

Sponsored by
NASA Historical Office

Scientific and Technical Information Office   1972
NATIONAL AERONAUTICS AND SPACE ADMINISTRATION
Washington, D.C.

Foreword

This volume provides an immediate reference to aerospace-related events of 1970, enabling all of us to broaden or to refresh our crowded memories. As part of NASA's chronology series it is intended to help historians and other analysts in preserving historical accuracy and precision.

Perhaps the *Apollo 13* problems that resulted in using the lunar lander *Aquarius* as a "lifeboat" will be the most readily recalled of the 1970 events. Yet, as this volume clearly indicates, there were many significant milestones in space and aeronautics. The lunar samples returned by *Apollo 11* and *12* were intensively studied in the United States and 16 other countries; the last two Apollo missions scheduled were canceled; the feasibility of reusable space shuttles as a means of reducing the cost and expanding the versatility of space operations was examined in greater detail; the Skylab program to evaluate man's adaptation to the space environment proceeded toward its launchings in early 1973. Dividends continued to be returned from our scientific satellites already in orbit: *Pioneer VI* completed six orbits of the sun, providing data on solar weather on the far side of the sun; *Oao II* began transmitting ultraviolet spectral data on some 25 000 stars; *Ogo III, IV, V,* and *VI* made sky surveys which could correlate geophysical data from other satellites, sounding rockets, and ground-based instruments. Among the 38 new payloads launched by the United States were the first prototype satellite for an operational weather system, ITOS, and three Intelsat communications satellites. Japan and China both launched their first satellites during 1970. In aeronautics, research and development on the supercritical wing culminated in the first flight tests. And the X-24 "lifting body" made its first flight early in the year and moved into supersonic flight in October.

Other kinds of events are also included in this chronology. Dr. Cyril A. Ponnamperuma of the Ames Research Center reported the first positive identification of amino acids of extraterrestrial origins, found in the meteorite that fell near Murchison, Australia, in 1969. Numerous entries evidence increasing international cooperation in and mutual benefits from space activities.

The overall space program was being evaluated during all of 1970—at perhaps the very peak of our technological and scientific achievements in exploration, in basic science, and in application to earth-bound needs. Questions of the basic value of and the need for space activities in relation to needs for solutions of problems on earth were being explored in all sectors of our national life. The space frontier has been crossed and men, their thoughts and their machines, are no longer bound to our home planet Earth. The accessibility of space will have a place in man's life and thinking for all his future. Even against a backdrop of the broad range of earth-bound problems—poverty, pollution, world peace, and disruptive changes besetting most social institutions—man's new perspective on his place in a dynamic universe is a basic challenge. We appear to have been in an escalating scientific and technological revolution during the 20th century, its intellectual stimulus and constructive values perhaps best symbolized by the intense space venture of the past decade.

The space program as a human enterprise, gaining new knowledge and skills and applying them to practical purpose, seems a symbol much needed in a world beset by doubts and fears and frustrations and with many difficult problems yet to solve.

Homer E. Newell
Associate Administrator
National Aeronautics and Space Administration

September 17, 1971

Contents

	PAGE
Foreword	iii
NASA Associate Administrator Homer E. Newell	
Illustrations	vii
Preface	ix
January	1
February	35
March	71
April	107
May	161
June	189
July	223
August	255
September	283
October	323
November	357
December	383
Appendix A: Satellites, Space Probes, and Manned Space Flights, a Chronicle for 1970	415
Appendix B: Chronology of Major NASA Launches, 1970	439
Appendix C: Chronology of Manned Space Flight, 1970	443
Appendix D: Abbreviations of References	447
Index and List of Abbreviations and Acronyms	451

Illustrations

	PAGE
Itos I Improved Tiros Operational Satellite prepared for mating to its launch vehicle and later waiting atop the first "super-six" Thor-Delta before Jan. 23 launch	24
Japan's first satellite, *Ohsumi*, prepared for Feb. 11 launch	48
Three lifting bodies active in NASA-USAF research program: HL-10, M2-F3, and X-24A	56
Totality of March 7 eclipse of the sun, above Wallops Station, passing over tracking antenna, and during Nike-Cajun sounding rocket launch	80
Moon's shadow moving up across the Eastern Seaboard from the Gulf of Mexico during March 7 eclipse; photographed by *Ats III*	81
Dr. Wernher von Braun sworn in March 13 as NASA Deputy Associate Administrator for Planning by Administrator Thomas O. Paine	89
Feb. 17, 1958, visit by Vice President Nixon to Jet Propulsion Laboratory, celebrating *Explorer I* launch with JPL and Cal Tech officials Dr. Clark B. Millikan, Dr. Lee A. DuBridge, Dr. William H. Pickering, Dr. Jack E. Froelich, V. C. Larsen, and Robert J. Parks	104
NASA Deputy Administrator George M. Low conferring April 10 with Associate Administrator for Manned Space Flight Dale D. Myers on launch status of *Apollo 13*	117
Launch of *Apollo 13* toward the moon from Kennedy Space Center April 11 and the service module that ruptured 56 hours later, endangering the flight	120
Officials consulting in Manned Spacecraft Center's Mission Operations Control Room as *Apollo 13* astronauts worked to bring the spacecraft home	122
Apollo 13 LM *Aquarius* photographed from the CM just after jettison April 17	123
Crew of *Apollo 13*—Fred W. Haise, Jr., James A. Lovell, Jr., and John L. Swigert, Jr.—arriving safely onboard recovery ship April 17	124
Mission officials congratulating each other in MSC Mission Operations Control Room after safe *Apollo 13 splashdown*	124
Apollo 11 Astronauts Michael Collins and Edwin E. Aldrin, Jr., and NASA Administrator Paine, representing Neil A. Armstrong, standing with the Gen. Thomas D. White Space Trophy presented by Secretary of the Air Force Robert C. Seamans April 29	155
City of Chicago welcoming *Apollo 13* astronauts with ticker tape and cheers May 1	163

	PAGE
Demonstrations of down-to-earth uses of Fluorel fireproofing material developed for spacecraft, which were presented at May 6-7 MSC conference	168
Space Orbiter Shuttlecraft model after air-drop test at White Sands Missile Range May 27	182
The speakers' table at NASA's June 3-5 space station briefing for ESRO officials in Paris	193
Rep. George P. Miller examining an Apollo oxygen tank with NASA Administrator Paine and Langley Research Center Director Edgar M. Cortright June 16 before the Senate Space Committee hearing on the *Apollo 13* accident	205
Full-scale model of the lunar roving vehicle in June practice exercises at Marshall Space Flight Center, with Astronauts Charles M. Duke, Jr., and John W. Young	206
Experimental thermal reactor in tests (announced by NASA Aug. 9) on V-8 engine at Lewis Research Center to reduce pollution from automobile exhaust, with technician Warren A. Moore	261
Some of the lunar craters named at IAU's Aug. 18-21 triennial assembly, shown on a photograph of the moon's far side made by *Lunar Orbiter V*	268
Cartoonist Paul Conrad's reaction to NASA cancellation of lunar landing missions, announced Sept. 2	288
Last Saturn V 1st stage (S-IC-15) captive-fired at Mississippi Test Facility Sept. 30, with photo of the 14th S-IC stage on the test stand earlier	319
Supersonic Planetary Entry Decelerator (SPED) experiment's parasol deployed in June systems check-out before Oct. 9 launch	332
Soviet Cosmonauts Andrian G. Nikolayev and Vitaly I. Sevastyanov welcomed at National Airport by Astronauts Edwin E. Aldrin, Jr., and Neil A. Armstrong at start of U.S. tour Oct. 18	339
Full-scale hypersonic research engine (HRE) poised before Oct. 20 test in LaRC's 8-Foot High-Temperature Structures Tunnel	344
Principal investigator Dr. Torquato Gualitierotti for OFO Orbiting Frog Otolith satellite standing beside part of spacecraft before its Nov. 9 launch, with a cutway diagram of the experiment package inserted	363
USN T2-C aircraft on its first flight with NASA's supercritical test wing Nov. 24 and the Nov. 3 delivery to Flight Research Center of the thin version of the new airfoil mounted on an F-8 fighter	378
Dr. Cyril A. Ponnamperuma of Ames Research Center displaying a meteorite that he announced Dec. 5 had been found to contain amino acids, constitutents of living cells	388
Uhuru (Explorer XLII) shown in preflight testing with Mrs. Marjorie Townsend, project manager, and during Dec. 12 launch from the San Marco sea platform	396
USN F-14 Tomcat fighter just before takeoff on its successful maiden flight Dec. 21	404

Preface

As with its seven annual ancestors and several precursors, this volume is intended to be as complete and accurate a chronicle of known events and commentary in 1970 as could be documented by immediately available sources. Essentially it provides a running start on aeronautical and astronautical data of interest to future historians. But it also serves as a general reference for those interested in science, technology, and policy aspects of aerospace affairs during 1970.

For the first time, this *Astronautics and Aeronautics* chronology gives all measurements in the metric system according to the International System of Units (SI). Heretofore, the English system of units was used. In this volume they are retained in parentheses following SI units. NASA Policy Directive 2220.4 of September 14, 1970, required conversion to the International System of Units, previously used in GSFC's *Satellite Situation Report* and most European publications. NASA was the first Federal agency to adopt this policy.

General editor of this volume was the Deputy NASA Historian, Frank W. Anderson, Jr., and the technical editor was Mrs. Carrie E. Karegeannes. The entire NASA Historical Office participated in source selection, review, and publication. Archivist Lee D. Saegesser collects current documentation. The Science and Technology Division of the Library of Congress, under an exchange of funds agreement, drafts monthly segments in comment edition form, which are circulated for corrections and use. At the Library Mrs. Patricia D. Davis, Mrs. Carmen B. Brock-Smith, and Mrs. Shirley M. Singleton carry principal responsibility. At the end of the year, the entire manuscript is reworked to include comments received and recent additional information. Arthur G. Renstrom of the Library of Congress prepared the extensive index, which is indispensable to the usefulness of a chronology. The index serves also as a glossary of abbreviations and acronyms used in the chronology.

Appendix A, "Satellites, Space Probes, and Manned Space Flights, 1970," Appendix B, "Chronology of Major NASA Launches, 1970," and Appendix C, "Chronology of Manned Space Flight, 1970," were prepared by Leonard C. Bruno of the Library staff. Appendix D, "Abbreviations of References," was prepared by Mrs. Brock-Smith.

Without the validation throughout NASA and other Federal agencies, the content of this volume would be less reliable and less nearly complete. Comments, additions, and criticisms are always welcomed by the NASA Historical Office.

Eugene M. Emme
NASA Historian

January 1970

January 1: Apollo 11 lunar landing had been voted biggest news story of 1969 in poll of Associated Press member newspapers and radio and TV stations, AP announced. Astronaut Neil A. Armstrong, *Apollo 11* commander, had been voted top newsmaker of year. AP poll ranked Vietnam war second biggest story and "story of Sen. Edward M. Kennedy's political fortunes" third. (Rock, AP, W *Star*, 1/1/70, A3)
- President Nixon in San Clemente, Calif., signed National Environmental Policy Act of 1969. Act declared Government policy "to create and maintain conditions under which man and nature can exist in productive harmony and fulfill the social, economic, and other requirements of present and future generations of Americans." It established within Presidential Executive Office full-time, three-member Council on Environmental Quality and required President to submit environmental quality reports to Congress annually, beginning July 1, 1970. President said that "the 1970's absolutely must be the years when America pays its debt to the past by reclaiming the purity of its air, its waters, and our living environment. It is literally now or never." (*PD*, 1/5/70, 11-3)
- Dr. Charles H. Townes, Nobel Prize winning Univ. of California physicist and member of NAS Council, was appointed Chairman of NAS-NRC Space Science Board. Dr. Townes had headed President Nixon's task force to make recommendations on national space program in 1969. (NAS PIO; NAS-NRC-NAE *News Rpt*, 3/70; *A&A 68*)

January 2: Nike-Apache sounding rocket was launched by NASA from TERLS carrying Indian payload to study ionosphere. Rocket and instruments functioned satisfactorily. (SR list)
- *Apollo 10* and *11* photos of lunar far side had revealed number of probable igneous intrusions, including three probable dikes that crosscut wall and floor of 75-km (46.6-mi) crater, Farouk El-Baz of Bellcomm, Inc., reported in *Science*. Intrusions were distinguished by their settings, textures, structures, and brightness. Recognition of probable igneous intrusions in lunar highlands supported indications of heterogeneity of lunar materials and plausibility of intrusive igneous activity on moon, in addition to extrusive volcanism. (*Science*, 1/2/70, 49-50)
- Results of study of infrared spectra recorded by *Mariner VI* and *VII* were reported in *Science* by Univ. of California at Berkeley chemists Kenneth C. Herr and Dr. George C. Pimentel. Reflections at 4.3μ suggested presence of solid carbon dioxide in upper atmosphere of Mars. (*Science*, 1/2/70, 47-9)
- MSC announced appointment of Sigurd A. Sjoberg as Director of Flight Operations, replacing Dr. Christopher C. Kraft, Jr., who had been appointed MSC Deputy Director. Sjoberg had been Deputy

Director of Flight Operations since 1963. Appointment was effective Dec. 28, 1969. (MSC Release 70-1)

- With advent of SST, information on meteorological conditions at higher levels of stratosphere would be needed for routine civil-aviation operations, Frederick G. Finger and Raymond M. McInturff of ESSA's Weather Bureau said in *Science*. Forecast system would be extended from 12-km (7.5-mi) to 20-km (12.4-mi) altitudes or higher. Forecasting techniques to support SST might be facilitated by exploiting similarities between stratosphere and troposphere weather. Much remained to be done on phenomenon forecasting. Wind effects should be smaller for SST than for subsonic aircraft but wind forecasts might be necessary during winter, when variations were high enough to effect economics of operation. Knowledge of temperature fields and of ozone distribution could prove valuable. Possible solar cosmic radiation hazards would be minimized by ground-based forecasting system and onboard detection equipment but sonic boom problem at ground level might be less tractable. Amplification and attenuation of boom by meteorological conditions could be predicted. World Weather Watch of U.N. and ESSA's satellites and spacecraft with ability to depict continuous three-dimensional temperature distribution and wind pressure over entire globe would play major role in providing SST meteorological information, as would SST itself. (*Science*, 1/2/70, 16-24)

January 3: Photos of falling meteorite were taken by automatic cameras at Hominy, Okla., and Pleasanton, Kans., field stations of Prairie Network operated for NASA by Smithsonian Astrophysical Observatory (SAO). Fireball was observed over large area of midwestern U.S. Observatory scientist Dr. Richard E. McCrosky later predicted from film records probable impact point near Lost City, Okla. Meteorite entered earth's atmosphere at 56 300 km per hr (35 000 mph), creating trail that was visible for nine seconds. (SAO Release 70-01; McElheny, *Boston Globe*, 1/20/70)

- Accuracy of *Apollo 12* LM lunar landing was described by Henry S. F. Cooper, Jr., in post-mission report from KSC published by the *New Yorker*. After successful completion of mission, Cooper had talked with Flight Dynamics Officer David Reed, who had been in charge of LM trajectory during lunar landing. Cooper reported that Reed and other flight engineers had been "too accurate in plotting their trajectory, for Commander (now Captain) Charles Conrad, Jr., who was at the controls, had to change the LM's landing point at the last minute. 'He was afraid he'd land right on top of the Surveyor if he kept on going according to plan, and later, when we fed the data from the LM into a computer, we found that if he'd gone on in the way he was, he *would* have hit it,' Reed told me." (*New Yorker*, 1/3/70, 46-56)

January 4: Lunar Science Institute near MSC in Houston, Tex., was dedicated at ceremonies attended by Dr. Thomas O. Paine, NASA Administrator, other NASA officials, and some 70 space scientists who had gathered early for Jan. 5 opening of *Apollo 11* Lunar Science Conference at MSC. Dr. Paine said, "This is the institution to help us bring all the expertise we need together to solve the riddles of the moon." LSI, created by 49-university Universities Space Research Assn., had one-year contract with NASA, renewable annually. Facili-

ties would be open to all scientists, including those from Iron Curtain countries. Dr. William W. Rubey, Univ. of California and Rice Univ. geologist, would be Director. (*H Post*, 1/5/70)
- Budget limits had caused cancellation of one lunar landing and stretchout of remaining seven Apollo missions into 1974, Dr. George M. Low, NASA Deputy Administrator, said in interview following LSI dedication in Houston, Tex. Eighth lunar landing had been dropped to permit use of Saturn V rocket to launch first U.S. space station in July 1972. (UPI, *NYT*, 1/5/70, 1)
- Capt. W. J. Bright of Ascot, England, was named provisional winner of 1970 London-Sydney Air Race after flying Britten-Norman Islander aircraft 19 300 km (12 000 mi) from Gatwick Airport, London, to Bankstown Airport, Sydney, Australia, in 76 hrs elapsed time. Second and third prizes in $112 000 competition to commemorate 1919 England-to-Australia flight of Ross and Keith Smith were won by Australian pilot Jack Masling in Cessna 310 and British pilot Brian Holland in Piper Twin Comanche B. (Trumbull, *NYT*, 1/5/70, 61)

January 5: NASA launched two Nike-Cajun sounding rockets carrying GSFC grenade experiments. One was launched from Churchill Research Range; other was launched from Wallops Station. Rockets and instruments functioned satisfactorily. (SR list)
- ERC announced development of silicon carbide diode that had survived and functioned over temperature range of 889 K (1600°F)—from 978 K (+1300°F) to 89 K (−300°F). Diode was believed to be first of its kind to withstand extremely high temperatures while performing effectively as detector of high-frequency radiation at frequencies to 10 billion cps and would be operable at Venus surface temperatures—about 700 K (800°F). Diode was developed by ERC's Richard Farrell. (ERC Release 70-1)
- *Time* magazine named "the Middle Americans" its Men and Women of the Year: "They sent Richard Nixon to the White House and two teams of astronauts to the moon. They were both exalted and afraid. The mysteries of space were nothing...compared with the menacing confusions of their own society." Middle Americans admired "men like Neil Armstrong and, to some extent, Spiro Agnew." *Apollo 11* and *12* were "a quintessential adventure of American technology and daring." Astronauts were "paragons of Middle American aspiration. Redolent of charcoal cookouts, their vocabularies an engaging mix of space jargon and 'gee whiz,' the space explorers gave back to Middle America a victory of its own values. It was little noted, except in Middle America where such things still matter, that among Neil Armstrong's extraterrestrial baggage was a special badge of his college fraternity, Phi Delta Theta. He used it symbolically to establish Moon Alpha Chapter." Some liberals had "grumbled that the Apollo program's $26 billion would have been better spent on curing hunger or the urban malaise." Yet Middle Americans "reveled in the lunar landings precisely because they were victories purely accomplished; in Vietnam, in the carious slums, in the polluted environment, no clear victories seemed possible any longer." (*Time*, 1/5/70, 10-2)
- *Chicago Tribune* article blamed Apollo program for demise of USAF's Project Blue Book UFO study: "Wonderful as the saucers were, the

moon shots probably killed them. The idea of little men with bulbous heads and wagging antennae landing...wasn't a bit more amazing than the spectacle of human beings traversing space and walking the surface of a dead world. And, in addition, the latter was undoubtedly true." (*C Trib*, 1/5/70)

- Dr. Max Born, nuclear physicist and corecipient of 1954 Nobel Prize in physics, died at age 87. (*NYT*, 1/6/70, 1)

January 5-9: Apollo 11 Lunar Science Conference at MSC presented results of first systematic studies of lunar samples by more than 500 scientists from nine countries. Samples from Tranquility Base consisted of basaltic igneous rocks; microbreccias, which were mechanical mixture of soil and small rock fragments compacted into coherent rock; and lunar soil. Soil was mixture of crystalline and glassy fragments with small fragments of iron meteoroids. Most rock fragments derived from larger igneous rocks—probably once part of underlying bedrock. Some crystalline fragments might have come from Highlands near Tranquility site. Rocks and fragments in soil showed evidence of surface erosion from hypervelocity impacts accompanied by local melting, splashing, evaporation, and condensation. Crystalline rocks ranged from fine-grained vesicular to vuggy, medium-grained equigranular.

Most common minerals found were pyroxene, plagioclase, ilmenite, olivine, and cristobalite. Three new minerals occurring in igneous rocks were pyroxmanganite, ferropseudobrookite, and chromium-titanium spinel. Free metallic iron and troilite, rare on earth, were common accessory minerals in igneous rocks. Silicate minerals were unusually transparent because of complete absence of hydrothermal alteration. Tests had shown that, at time of crystallization, observed phases could have coexisted only in very dry, highly reducing system in which oxygen pressure was estimated to be $10-13$ atmospheres—more than five orders of magnitude lower than for typical terrestrial basaltic magmas.

Melting experiments indicated 98% of primary igneous liquid crystallized in 1483-1333 K (1210-1060°C) temperature range with minor interstitial liquids continuing to crystallize down to 1223 K (950°C). Microscopic and microprobe examinations had produced evidence for existence of interstitial liquid rich in potassium and aluminum, which probably was immiscible with main liquid. Viscosity of lunar magmas was calculated an order of magnitude below that of terrestrial basaltic magmas. This characteristic might be significant in explaining textural features, differentiation mechanisms that produced chemical composition, and morphological features of lunar seas.

Chemical compositions of all igneous rocks were similar except for concentration of potassium, rubidium, cesium, uranium, thorium, and barium. These elements distinguished two groups of igneous rocks, with fine-grained rocks containing more than coarse-grained rocks. All rocks had high concentrations of titanium, scandium, zirconium, hafnium, yttrium, and trivalent rare earth elements, and low concentrations of sodium. Low abundance of europium was striking feature of igneous rocks. Composition of soils and breccias was similar to but distinguishable from igneous rocks. Igneous rocks contained at least one other "rock" component distinct from lunar basalts sampled. Soil was rich in nickel and volatile elements—cadmium, zinc, silver, gold, copper, and thallium—that occurred in carbonaceous chondrites. Enrichment was consistent with observed occurrence of meteorite material in soil.

Many elements that occurred in low abundance in lunar igneous rocks were strongly enriched in terrestrial crustal rocks that were product of igneous differentiation. It had been suggested that low abundance of these elements derived from residual lunar liquids was indication that whole moon was depleted in number of volatile elements. If inference was correct, it could be inferred that lunar material separated from high-temperature, dispersed nebula at 1273 K (1000°C) or higher.

Coincidence of high abundance of titanium, separation of europium from other rare earth elements, and separation of barium and strontium suggested igneous liquids were end product of fractional crystallization process or, less likely, that they were produced by partial melting and subsequent segregation of liquid in lunar interior.

High-pressure and temperature experiments had shown that materials with chemical composition of Tranquility Base basalts would, at conditions inferred for lunar interior, have densities exceeding moon's average density. Therefore, basalts could not represent moon's bulk composition.

Age of basaltic crystalline rock was determined at 3.7 billion yrs. Igneous rocks were melted and crystallized about 10 billion yrs after moon's formation. Single exotic rock fragment had yielded age of 4.4 billion yrs and indicated variability in age between different areas on lunar surface. Relatively young basalts showed moon had not been completely dead planet from its formation, but that it had undergone significant differentiation, at least locally, in thin lunar crust. Period before 3.7-billion-yr event recorded in older Highland areas was interval in which earth's record had been obliterated. Therefore much of lunar surface was important in understanding early evolution and differentiation of planets.

Observations showed lunar materials were product of considerable geochemical and petrological evolution. Optical properties of soil measured in laboratories agreed with those inferred from telescopic studies and indicated thin layer of soil must have covered most of moon. Correlation of spectral features determined in laboratory with telescopic studies suggested that iron content of lunar surface might vary. Measured dielectric constant of soil agreed with known radar reflectivity of moon. Penetration depth suggested upper layers were electrically transparent. Measured thermal conductivity suggested that diurnal variations in temperature should extend to less than one meter in soil regions. Thermoluminescence studies on core samples showed substantial diurnal variation in temperature 12 cm (4.7 in) below surface.

Seismic study showed that attenuation of lunar sound waves decreased with pressure. High-pressure velocities were consistent with those observed in *Apollo 12* seismic experiments, suggesting possibility of seismic wave guide at shallow depths on moon. Natural remanent magnetization found in crystalline rocks and breccias suggested ancient moon might have had magnetic field with strength of few percent of earth's field. Field existing 3.7 billion yrs ago might have been result of fluid motions in lunar interior—effect of earth's field when moon and earth were closer together or result of processes not yet understood.

Lunar samples were valuable for study of low-energy, weakly penetrating radiations and had provided sample of gases blown off sun in solar wind, making it possible to determine isotopic composition and abundance of elements in sun by direct measurement. Results would lead to better understanding of sun's evolution. Studies of stored nuclear tracks and induced radioactivities had shown effect of bombardment by solar flare particles and suggested activity had persisted for long periods. Abundance of cosmic-ray-produced nuclides in samples showed some rocks had been on or within several centimeters of lunar surface for at least 10 million yrs and within one or two meters (three to six and a half feet) for at least 500 million yrs.

In search for life origins on moon, micropaleontological examination of sample by optical microscopy and by electron and scanning electron microscopy had produced uniformly negative results. Intensive search for viable organisms using many environmental and media combinations had produced negative results, as had the one quarantine study.

Reports given at conference were published in Jan. 30 issue of *Science* under NASA contract. (*Science*, 1/30/70)

Cambridge Univ. astronomer Dr. Fred Hoyle was principal speaker at banquet given by National Space Hall of Fame and City of Houston during conference. He said it was significant that in year when men first set foot on another world, desire to save their own planet had suddenly blossomed. "Quite suddenly everyone worldwide has become seriously interested in protecting the environment. Something new has happened to create this feeling of awareness about our planet." He warned that man had to face probability that he could no longer discover important physical laws solely by observation in earth laboratories. "We are in a transition period, and in the future we must rely more on outside laboratories than astronomers did in the past." (Burgess, *CSM*, 1/8/70)

January 6: Dr. Gary V. Latham, Columbia Univ. seismologist and principal seismic investigator for Apollo program, during Houston, Tex., press conference withdrew proposal that atomic bomb be detonated on moon. Instead, he said, NASA should find way to hit moon hard enough to create strong internal reverberations. Pressure from scientists and public had led to retraction of original proposal. (W *Star*, 1/8/70, B4)

- Former Astronaut Michael Collins was sworn in as Assistant Secretary of State for Public Affairs in Washington, D.C. He would be responsible for State Dept. relations with U.S. public, particularly youth. (State Dept Release 3)

- NASA announced it was publishing new quarterly, *Computer Program Abstracts*, listing documented computer programs developed by and for NASA and DOD. First issue listed 500 programs available to domestic purchasers for nominal fee. Journal, published by Office of Technology Utilization, was available by subscription or from Superintendent of Documents, GPO. (NASA Release 70-1)

- MSC announced contract awards: General Motors Corp. AC Electronics Div. received $3-million letter contract for primary navigation, guidance, and control system support for Apollo CM and LM; Boeing Co. received $3-million letter contract for systems engineering and assessment for Apollo spacecraft program; and Raytheon Co. received $1-million letter contract for 250 fixed-memory modules for CM and LM guidance computers in support of Apollo and Apollo Applications programs. (MSC Releases 70-2, 70-3, 70-4)

January 7: Interagency Committee on Back Contamination notified NASA that *Apollo 12* lunar samples and Surveyor parts could be released from LRL quarantine to principal investigators. (MSC Release 70-7)

- MSC announced NASA had awarded United Aircraft Corp. Hamilton Standard Div. $5 461 203 contract supplement for changes in Apollo portable-life-support-system (PLSS) contract. Agreement formally incorporated changes permitting PLSS to accept an improved communications system supplied by NASA. (MSC Release 70-6)
- President Nixon announced in San Clemente, Calif., redesignation of Secor D. Browne as Chairman and of Whitney Gillilland as Vice Chairman of Civil Aeronautics Board for period ending Dec. 31, 1970. (*PD*, 1/12/70, 35)

January 8: NASA announced that Apollo 13 manned lunar landing mission, scheduled for launch March 12, had been rescheduled for April 11 to allow "additional time for more detailed analysis of specific plans." Launches after Apollo 13 were being studied to determine best interval between them to obtain maximum operational and scientific returns. Apollo 14, scheduled for July would be rescheduled for autumn 1970. (NASA Release 70-5)

- Col. Douglas H. Frost (USAF) set A-7D Corsair II flight-endurance record with 10-hr flight from Edwards AFB, Calif., consisting of two roundtrips to New Mexico covering 8047 km (5000 mi), with one aerial refueling. Purpose of flight was to test aircraft for pilot comfort, cruise capability, and deployment characteristics during long flight. Col. Frost pronounced flight comfortable and did not realize he had set new time record for Corsair. (AFSC Release 9.70)
- NASA announced appointment of Dale D. Myers, Vice President and General Manager of North American Rockwell Corp.'s Space Shuttle Program, as NASA Associate Administrator for Manned Space Flight, effective Jan. 12. Myers would succeed Dr. George E. Mueller, who left NASA Dec. 10, 1969. (NASA Release 70-4)
- U.S. and Australia would begin construction of top-secret communications station at Woomera rocket range, Australia, AP reported. Australian Defense Minister Malcolm Fraser had announced $4-million contract had been let to Australian firm. U.S. would pay about three fourths of construction bill. Facility would be used for space and defense programs. (*P Bull*, 1/8/70, 3)
- Highly sensitive magnetometer that could measure earth's total magnetic field 10 times more accurately than best instruments in use had been developed by ESSA scientists Joe H. Allen and Dale L. Vance and by Dr. Peter L. Bender of Joint Institute for Laboratory Astrophysics, ESSA said. Instrument was expected to disclose whether earthquakes were preceded by local changes in earth's magnetic field. (ESSA Research Lab Release 69-125-357)

- Establishment of Charles S. Draper Research Center and appointment of MIT scientist Dr. C. S. Draper as its President were announced by Dr. Jerome P. Keuper, President of Florida Institute of Technology, site of new center. Dr. Draper would continue as Vice Director of MIT Instrumentation Laboratory, which had been renamed Charles S. Draper Laboratory, with cognizance of DOD projects and responsibility for advanced technology in control navigation guidance. (FIT Release)

January 9: U.S.S.R. launched *Cosmos CCCXVIII* from Baykonur into orbit with 276-km (171.5-mi) apogee, 203-km (126.1-mi) perigee, 89.2-min period, and 64.9° inclination. Satellite reentered Jan. 21. (GSFC *SSR*, 1/15/70, 1/30/70; *SBD*, 1/12/70, 38)

- NASA announced appointment of Philip E. Culbertson, Director of Project Integration for Apollo Applications Program, as NASA Director of Advanced Manned Missions Program, OMSF. (NASA Release 70-6)

- GSFC announced award of $2 831 670 contract to Hughes Aircraft Co. to develop millimeter-wave communications experiment for NASA's ATS-F satellite, scheduled for launch in 1972. Millimeter-wave frequencies would offer relief for saturated lower frequency bands and provide large bandwidth channels for transmission of TV and high-density data-relay traffic. (GSFC Release 1-70)

- Bronzite chondrite meteorite weighing 10 kg (22 lbs) was found by Gunther Schwartz, field manager of Smithsonian Astrophysical Observatory's Prairie Network, near Lost City, Okla., 73 km (45 mi) east of Tulsa, six days after it had been photographed by network's automatic cameras [see Jan. 3]. Schwartz found meteorite less than three fourths kilometer (half mile) from impact point predicted by Smithsonian Observatory scientist Dr. Richard E. McCrosky. (SAO Release 70-01)

- Importance of studying asteroids was cited in *Science* by Dr. Hannes O. Alfven of Royal Institute of Technology in Stockholm and Univ. of California at San Diego and Dr. Gustaf O. S. Arrhenius of Scripps Institution of Oceanography: "There are now good reasons to believe that the asteroidal belt represents an intermediate stage in the formation of the planets. This links the present conditions in the asteroidal region with the epoch in which the earth and the other planets were accreting from interplanetary grains. Hence, in order to understand how the solar system originated it may be essential to explore the asteroids." Since manned Mars landing would take place after 1980, "it is of interest to discuss whether a sample of an asteroid may be obtained in an easier way, at an earlier time, and as a technologically intermediate step." Asteroid sample could be obtained by landing spacecraft on asteroid or by capturing small asteroid and landing it on earth's surface or storing it in earth orbit for later investigation. (*Science*, 1/9/70, 139-41)

Apollo 11 Astronaut Neil A. Armstrong received honorary Doctor of Engineering degree from Purdue Univ. Later he told students and press he had no political ambitions but did not criticize astronauts who did. "After all, politics should have something more than lawyers representing the people." (AP, *W Post*, 1/10/70; *KC Times*, 1/10/70).

- DOD announced allotment of $100 000 000 to Lockheed-Georgia Co. under USAF contract for production of C-5A aircraft, bringing total obligated to $2 026 229 928. Contract would be managed by Aeronautical Systems Div. (DOD Release 24-70)

January 10: ComSatCorp's *Intelsat-III F-6*—first Intelsat comsat insured against launch failure—misfired but remained on launch pad, apparently undamaged. Satellite was insured for $4.5 million with Lloyd's of London and Associated Aviation Underwriters. Launch by NASA, originally scheduled for Jan. 7, had been postponed because of bad weather and difficulties with Delta booster. (UPI, *NYT*, 1/11/70, 52; *Newsweek*, 1/26/70)

- NASA launched two Nike-Cajun sounding rockets carrying GSFC grenade experiments—one from Point Barrow, Alaska, and one from Wallops Station. Rockets and instruments functioned satisfactorily. (SR list)

- *The Economist* commented on NASA Lunar Science Conference: "It has probably been a long time since so many scientists enjoyed themselves so visibly at a meeting or laughed so much or had reflected in their eyes the sheer pleasure of discovery that only children have. For the moment they are up and Nasa is down." NASA was "between space programmes and is running short of friends and also short of cash as its budget is daily cut. The rate of moon shots has now been cut to two a year. This really ought not to please the scientists but it pleases them just the same since they have complained bitterly that they cannot conduct research at Nasa's hectic pace and their work is suffering as a result. This is not exactly the happiest time to be a Nasa administrator waiting daily for the President's next word but it is a wonderful time to be a moon man." (*Economist*, 1/10/70, 15-6)

- Cosmonaut Pavel Belyayev, commander of U.S.S.R.'s *Voskhod II* mission March 18, 1965, died from complications following operation for stomach ulcers. (UPI, W *Star*, 1/11/70, D6)

January 11: Development of Fluorel, which could be applied by spray to fireproof material against temperatures to 1500 K (2200°F) was described by Jerome F. Lederer, Director of Safety, OMSF, in interview released by AP. Compound was developed by Minnesota Mining and Manufacturing Co. under direction of MSC engineers. Fluorinated copolymer paste developed to reduce flammability hazards in spacecraft consisted of 20% of Fluorel solids homogenized to form methylethylketone solution. It had been designated Refset L-3203-6 by manufacturer, Raybestos-Manhattan, Inc. Fluorel would be among nonflammable and fire-resistant materials developed for U.S. space program and reviewed by NASA for industry and Government agencies at safety conference at MSC in May. (AP, *NYT*, 1/12/70, 10; NASA Release 70-7)

- Optical Doppler system powered by carbon dioxide laser was being developed by MSFC physicist Robert M. Huffnaker to detect and measure clear-air turbulence, MSFC announced. System, which measured atmospheric wind velocity directly, operated on Doppler radar principle except that coherent laser pulse was used instead of radio frequency energy. Huffnaker hoped to have airborne version of system operational in year. (MSFC Release 70-5)

- Nixon Administration had decided against big space program for 1970s, Thomas O'Toole wrote in *Washington Post*. There would be "no set goal for landing men on Mars, no plan to colonize the moon and no sudden push to orbit a permanent manned space station above the earth in the next five years." While space program would "not be allowed to shrink much" in 1970, Administration would ask Congress for no more than $3.7 billion for space in FY 1971, "the lowest request for space funds in almost a decade." Low funding level meant space spending would not rise above $4 billion annually for at least several years and new programs would not be started in near future. (*W Post*, 1/11/70, A1)

January 11-18: Washington *Evening Star* published series of seven articles on environmental pollution. Many scientists feared "that the 1970s will be the dawn of Doomsday." Mail to White House on environment had doubled after first lunar landing. "To many, the flights raised the question of where technological priorities should be directed—into space or back on the earth?" Some astronauts had said that from space, air pollution was so visible it cut into the joy of seeing earth from hundreds or thousands of miles away. Articles examined "what man has done to his environment, and what he can do to save it." Observers saw "great promise" in National Environmental Policy Act of 1969 signed Jan. 1. "To Congress, the President, and so many others who will become embroiled in this recently dramatized issue, the task at hand can be summed up in the phrase 'quality of life.' It will be a task of cleaning up, of making the air and the water and the land healthier and more enjoyable. Over the decades, as Americans have built a richer economic standard, they have run up a huge bill to the natural world around them. The bill is overdue." Total cost of cleaning up environment had been estimated at $100 billion to $125 billion. (Hornig, Welsh, *W Star*, 1/11-18/70, A1)

January 12: AEC Chairman, Dr. Glenn T. Seaborg, in *New York Times* article predicted new frontiers for nuclear energy during 1970s. Major advances would include development of large commercial fast-breeder reactor which should lead to "breeder economy in the 1980's that will inaugurate vast savings in fuel resources and costs"; use of abundant cheap power for desalting seawater and introducing new industrial processes; use of nuclear fission as power source; and melding of nuclear and space ages "as nuclear energy becomes a major source of auxiliary and propulsive power for space activities." SNAP systems for nuclear auxiliary power would become "principle source of electricity in space." Seventies also would see expanded use of nuclear explosives for peaceful purposes, such as power source for navaids, for underground excavation in stimulating production of natural gas, and in extraction of minerals from low-grade ores and oil from oil shales. Advances were expected in use of radioisotopes in medicine, agriculture, and industry and in basic and applied research in nuclear fields. (*NYT*, 1/12/70, 81)

- Structural failure of left wing stemming from "preexisting defect in the steel used in a pivot fitting where the wheel joins the fuselage" was cause of Dec. 22, 1969, crash of USAF F-111, Dr. Robert C. Seamans, Jr., Secretary of the Air Force, announced. USAF would

inspect remaining F-111s for similar defects. (Wieghart, *NYT*, 1/13/70, 16)
- Mathematics had become "key tool in almost all of the physical and social sciences," Harry Schwartz said in *New York Times*. "Theories and working hypotheses in the most diverse fields are now expressed in terms of equations using not only the language of ordinary algebra, but also derivatives, matrices, and even more advanced concepts." It was "stress on numerical results and quantitative, not qualitative, predictions, that gives the computer its present central and increasing role in scientific research." (*NYT*, 1/12/70, 81)
- Editorial by Robert Hotz in *Aviation Week & Space Technology* outlined space outlook for 1970: "Space will...see some titanic struggles in the market-place as the key programs in the space transportation system of the mid-1970s come up for contractor selection. Many observers will be misled by the relatively small dollar increases in Fiscal 1971 space budgets. Many new programs can be initiated with the same level of funding that was required for the Apollo era, plus some heavy reprogramming of existing projects that are approaching technical obsolescence much faster than any planners imagined. By judiciously paring its present programs and getting only a modest boost for Fiscal 1971, NASA can lay the foundations for the next decade's major space program. Military space, temporarily deflated by the manned orbiting laboratory cancellation, will trend toward a broader program involving several major new fields. Commercial and non-NASA space will provide a modestly expanding market, and the big battle between celestial and terrestrial communications is shaping into a major clash." (*Av Wk*, 1/12/70, 11)

January 13: Dr. Thomas O. Paine, NASA Administrator, said at NASA Hq. news conference on future plans he was directing portions of space program to bring operations in line with FY 1971 budget. "We recognize the many important needs and urgent problems we face here on earth. America's space achievements in the 1960's have rightly raised hopes that this country and all mankind can do more to overcome pressing problems of society. The space program should inspire bolder solutions and suggest new approaches. It has already provided many direct and indirect benefits and is creating new wealth and capabilities." Strong space program continued as one of Nation's major priorities. "However, we recognize that under current fiscal restraints NASA must find new ways to stretch out current programs and reduce our present operational base." NASA would "press forward in 1971 at a reduced level, but in the right direction with the basic ingredients we need for major achievements in the 1970s and beyond....we will not dissipate the strong teams that sent men to explore the moon and automated spacecraft to observe the planets."

NASA would suspend Saturn V production indefinitely after Saturn V 515 completion, stretch Apollo lunar missions to six-month launch intervals, defer lunar expeditions during AAP space station flights in 1972, and postpone Viking/Mars 1973 unmanned lander launch to next Mars opportunity in 1975. Number of employees in NASA program was expected to decline from 190 000 at end of FY 1970 to 140 000 at close of FY 1971.

Positive aspects of space program for 1970s included Apollo missions through Apollo 19, use of one Saturn V to launch first experimental AAP space station into earth orbit in 1972, Mars orbit by two unmanned spacecraft in 1971, launch of first Jupiter probe in 1972, launch of spacecraft past Venus to Mercury in 1973, and continuation of scientific satellite launches.

Design of reusable space shuttle would be started and first experimental space station would be launched, as planned, in 1972. FY 1971 budget would permit proceeding with design of advanced space station as permanent base in earth orbit, continued work on NASA-AEC NERVA project, practical earth applications of space technology, and advances in applications of space technology in fields such as meteorology and communications. Although "austere," space program for 1971 was "forward looking" and contained "basic ingredients needed for an effective space program in the 1970s." It would "extend our space capabilities, expand our scientific knowledge, and make available new applications of benefit to people here on earth."

In response to question, Dr. Paine said he foresaw no shutdown of KSC and had no plans for closing other centers. Intended actions "represent a decision to proceed through the year 1971 and see what our studies turn up. I think it is important to realize these are very advanced, perhaps, and we have a lot to learn about the technological feasibility, the cost, schedules—and I wouldn't want to say we now have a complete free road into the future. We don't." Asked if he thought NASA was close to point where "team started falling to pieces," Dr. Paine said that "you can carry out space programs for the United States at a large number of different levels. I think the level we have now is a level we can take great pride in and I think the country in the 1970's will be very proud of the achievements NASA will turn in." (Transcript)

- At NASA Hq. news conference on future NASA plans Julian Scheer, Assistant Administrator for Public Affairs, announced that 10-kg (22-lb) meteorite recovered Jan. 9 near Lost City, Okla., had come from orbit which took it outside Mars' orbit at planet's farthest point from sun. (Transcript)
- NASA launched four sounding rockets from Churchill Research Range. Two Nike-Cajuns carried GSFC grenade experiments. Nike-Apache carried Univ. of Michigan experiment to study atmospheric structure. Nike-Tomahawk carried Univ. of Maryland payload through northern edge of bright aurora to 259-km (160.9-mi) altitude to observe visual aurora. Observations included measurement of energy spectra of auroral electrons and auroral lights, measurement of electric fields, and measurement of visual aurora and aurora at three wavelengths. (SR list; NASA Rpt SRL)
- Robert J. Perchard of Bourne, Mass., had obtained exclusive royalty-free patent license from NASA to manufacture and sell inflatable life raft developed for manned space flight program, NASA announced. Patent had been issued to NASA Nov. 11, 1964. Nontippable raft with radar-reflective surface had been part of survival equipment aboard all Mercury, Gemini, and Apollo spacecraft. Under NASA patent policy exclusive royalty-free license could be granted if commercial use of invention had not occurred within two years after

patent's issuance. Perchard had become interested in developing new life raft after death of his son at sea during Coast Guard air rescue mission. (NASA Release 70-3)

- Massachusetts Gov. Francis W. Sargent visited President Nixon at White House to discuss other uses for ERC complex in Cambridge, Mass., which NASA was closing. Later governor told press he had received "commitment" from President to seek other uses for the Center and he hoped Administration would consider other kinds of Federal research. (Doyle, W *Star*, 1/14/70, A15)
- Pan American World Airways' Boeing 747 arrived at London Airport after first passenger transatlantic crossing. Aircraft, world's largest passenger plane, carried 361 passengers—mostly staff—on 6-hr 30-min flight from John F. Kennedy International Airport in New York. First regularly scheduled passenger flight had been set for Jan. 21. (*NYT*, 1/13/70, 73)
- In FAA precertification test of Boeing 747 superjet airliner at Roswell, N. Mex., 423 Roswell residents proved passengers could be evacuated through four hatches in less than 86 secs. Five hatches were to have been used, but one evacuation chute failed and "passengers" had to scramble out of remaining exits. (UPI, W *Star*, 1/14/70, A6)
- TV commentator Walter Cronkite was named by Boston Museum, Boston, Mass., as 1970 recipient of Bradford Washburn Award for his "superlative presentation" of U.S. space program. Gold medal and $5000 honorarium were awarded annually in spring for "outstanding contribution toward public understanding of science, appreciation of its fascination, and the vital role it plays in all our lives." (UPI, *NYT*, 1/16/70)
- Boeing Co. would lay off some 18 000 persons in Puget Sound area during 1970, company Vice President Lowell P. Michelwait said in Seattle. Company employed 101 500 in peak year 1968. Nationwide force declined by 25 500 during 1969. Boeing employment in Seattle area was 80 000 and would drop to 62 000. Figures indicated 40% drop in work force in 2 1/2 yrs. Decline was attributed to slower commercial aircraft sales, failure to win new Government contracts, reduced Government spending, completion of current Government contracts, and increased productivity in 747 jet airliner assembly lines. (AP, *NYT*, 1/14/70)
- Blanche S. Scott, first woman pilot in U.S., died at age 84. Mrs. Scott had made solo flight in Hammondsport, N.Y., Sept. 6, 1910, in Curtiss Pusher aircraft and had become first woman to ride in jet aircraft in 1948. (UPI, *W Post*, 1/14/70, C6)

January 13-14: NASA successfully launched one Nike-Cajun and six Nike-Apache sounding rockets carrying chemical cloud experiments from Wallops Station between 5:36 pm and 6:35 am EST. Rockets ejected vapor trails between 40- and 217-km (25- and 135-mi) altitudes to measure wind velocities and directions.

Nike-Apache launched at dusk was unsuccessful because 2nd stage failed to ignite. Nike-Apache launched at dawn carried sodium experiment that created reddish-orange cloud visible for hundreds of miles along East Coast. Nike-Cajun and four Nike-Apaches carried trimethylaluminum (TMA) payloads that formed pale white clouds. Data were obtained by photographing continuously motions of trails from five ground-based camera sites. Similar tests had been conducted Feb. 13-14, 1969.

In conjunction with vapor series two Nike-Apaches and one Nike-Cajun were launched at 6:47 am, 12:23 pm, and 12:55 pm EST Jan. 14, carrying acoustic grenade experiments to obtain correlative data on atmosphere between 48- and 121-km (30- and 75-mi) altitudes by detonating grenades and recording their sound arrivals on ground. Arcas was launched at 1:27 pm EST Jan. 14 to obtain additional meteorological data. (WS Release 70-1)

January 14-16: *Intelsat-III F-6* was successfully launched by NASA for ComSatCorp on behalf of International Telecommunications Satellite Consortium, after series of delays including misfiring Jan. 10. The 293.9-kg (648-lb) cylindrical satellite, launched from ETR at 7:16 pm EST by four-stage long-tank, thrust-augmented Thor-Delta booster, entered elliptical transfer orbit with 35 871.2-km (22 289.3-mi) apogee, 289.5-km (179.9-mi) perigee, 634.5-min period, and 28.05° inclination. All systems were functioning normally.

Apogee motor was fired at 8:29 am EST Jan. 16 to place satellite in near-synchronous orbit of 35 900-km (22 300-mi) altitude over Atlantic and antenna was deployed. Full-time commercial service was scheduled to begin Feb. 1, with 955 circuits carrying telephone and telegraph traffic between 17 earth stations in Atlantic region. Satellite could also carry TV traffic at same time and had 1200-circuit capacity.

Intelsat-III F-6 was fourth successful launch in Intelsat III series. Last successful mission, *Intelsat-III F-4* (May 21-23, 1969), had been followed by *Intelsat-III F-5* (July 25, 1969), which failed to reach planned orbit. (GSFC Dir Wkly Rpt; ComSatCorp Releases 70-1, 70-4, 70-6; Proj Off)

January 14: USAF launched unidentified satellite from Vandenberg AFB by Titan IIIB-Agena D booster. Satellite entered orbit with 407.2-km (253-mi) apogee, 125.5-km (78-mi) perigee, 89.8-min period, and 109.9° inclination and reentered Feb. 1. (Pres Rpt 71: GSFC *SSR*, 1/15/70; 2/15/70)

- At NASA Hq. briefing, NASA and Smithsonian Astrophysical Observatory officials displayed and described 10-kg (22-lb) meteorite found by Prairie Network [see Jan. 9]. Observatory's Dr. Richard Mc-Crosky said meteorite was "not sucked into earth by gravity." It had flamed through atmosphere on direct collision course and was "almost a direct hit. Earth's gravity had a very little role." Eight billion meteors struck earth's atmosphere daily, but most burned out and only few remaining ones were discovered. NASA official said Lost City, Okla., meteorite was "freshest" ever examined. (NASA PAO; *W Post*, 1/14/70, A7)

- Cancellation of 8 743 070 first day covers of moon landing postage stamp—nearly trebling previous top of three million covers canceled—was announced by Postmaster General Winton M. Blount. Requests for souvenir envelope commemorating July 20, 1969, *Apollo 11* lunar landing had been received from 100 different countries and reached 60 000-80 000 daily within days after stamp was announced July 9, 1969. Stamp sales totaled 9 614 685, with value of $961 468.50. Previous record had been set in 1962 for stamp com-

memorating first U.S. earth-orbiting flight by Astronaut John H. Glenn, Jr. (PO Dept Philatelic Release 3)
- Former President Lyndon B. Johnson was briefed on *Apollo 12* mission by Astronauts Alan L. Bean, Richard F. Gordon, Jr., and Charles Conrad, Jr.; at LBJ ranch near Stonewall, Tex. (P *Bull,* 1/15/70)
- FAA Administrator John H. Shaffer said in speech before Aviation/ Space Writers Assn. in New York that FAA was not satisfied with Boeing 747 jumbo jet emergency evacuation system after Jan. 13 tests at Roswell, N. Mex. If problem of chute failure was not solved by Jan. 21, scheduled start of Pan American World Airways' 747 service between New York and London might be delayed. (Lindsey, *NYT,* 1/15/70, 77)
- U.S.S.R. had agreed in December to build comsat station in Cuba, *Philadelphia Inquirer* reported. Experts had said purpose could be to monitor and track U.S. missiles. Officially station was to permit exchange of radio and TV between Moscow and Havana. (*P Inq,* 1/14/70)

January 15: U.S.S.R. launched *Cosmos CCCXIX* from Plesetsk into orbit with 1490-km (925.8-mi) apogee, 195-km (121.2-mi) perigee, 101.9-min period, and 81.8° inclination. Satellite reentered July 1. (GSFC *SSR,* 1/15/70; 7/31/70; *SF,* 7/70, 282)
- Interagency Committee on Back Contamination suggested elimination of requirement for quarantine after lunar exploration because no evidence of back contamination had been found after *Apollo 11* and *12* missions. Committee report released to press said, "There is no need to impose quarantine on the crews, the lunar samples or equipment on subsequent missions." (AP, *W Post,* 1/16/70, A5)
- Mrs. Richard M. Nixon christened *Clipper Young America,* Pan American World Airways' first operational Boeing 747, in Dulles International Airport ceremony. Aircraft was parked alongside Boeing 707 and 40-yr old Ford Trimotor, first aircraft flown by Pan Am. (AP, *W Post,* 1/16/70, 14)
- Advertisement offering "genuine" moon rock fragments at $1 per carat had been denounced as fraud by West German police in Frankfurt. Check with U.S. authorities had revealed fragments could not possibly have come from *Apollo 11* samples, AP reported. (*W Post,* 1/16/70, A12)
- USN announced support aircraft carrier U.S.S. *Hornet* would be inactivated by June 30. Ship, which had served as recovery vessel in Apollo missions, would be mothballed after 26 yrs of service. (DOD Release 38-70)
- Secretary of Transportation John A. Volpe and Under Secretary of Transportation James M. Beggs met with West German Minister of Transport and of Postal Affairs and Communications George Leber in Washington, D.C. They agreed to give priority to continuing collaboration on R&D in transportation technology. (DOT Release 2170)
- Czechoslovakian press agency CTK had said in Prague that astronomical dictionary in seven languages, issued by Czechoslovakia's Akademia publishing house, was first work of its kind in world, AP reported. (*NYT,* 1/15/70, 12)
- *Denver Post* editorial commented: "The earth problems seem to us more important at the moment than the space problems, and human

survival will be in jeopardy unless the earth problems are solved. Instead of allowing some of NASA's talents and resources to leave the public service, we believe they could be put to work developing scientific and technological solutions to the problems of environment." (*D Post*, 1/15/70, 22)

January 16: U.S.S.R. launched *Cosmos CCCXX* from Kapustin Yar into orbit with 297-km (184.6-mi) apogee, 239-km (148.5-mi) perigee, 89.8-min period, and 48.4° inclination. Satellite reentered Feb. 10. (GSFC *SSR*, 1/31/70; 2/15/70; *SF*, 7/70, 282)

- U.K.'s *Skynet A* (IDCSP-A) comsat, launched by NASA for USAF Nov. 21, 1969, was adjudged successful by NASA. Satellite was adjusted into final orbit Jan. 7 and USAF reported all systems were operating satisfactorily. (NASA Proj Off)
- MSC released *Surveyor III* spacecraft parts, returned from moon to earth by *Apollo 12* crew, from LRL quarantine for examination by principal investigators. (MSC Release 70-10)
- NASA launched two Nike-Tomahawk sounding rockets from Churchill Research Range. One carried GSFC payload to 210-km (130.5-mi) altitude to measure extreme UV auroral emissions. Rocket and instruments functioned satisfactorily, but door covering photometer did not deploy. Second rocket carried Univ. of Maryland payload into bright aurora to 252-km (156.6-mi) altitude to make detailed observations of visual aurora. Rocket and instruments functioned satisfactorily and good data were obtained. (NASA Rpts SRL)
- Bullpup-Cajun sounding rocket, launched by NASA from Wallops Station, carried GSFC payload to 41-km (25.5-mi) altitude to measure ozone distribution in mesosphere and stratosphere and ozone in 20- to 65-km (12- to 40-mi) region. Rocket underperformed because of pitch-roll lock-in during 2nd-stage burn, which increased coning angle. Instrument performance was satisfactory. (NASA Rpt SRL)
- USAF grounded all 10 C-5A transports, world's largest aircraft, after discovery of wing crack in aircraft undergoing modifications at Lockheed-Georgia Co. plant. Crack was in same area where weakness had been reported during earlier static load tests, USAF said. (AP, *W Post*, 1/17/70, A1)
- Use of range and Doppler tracking data to determine earth-moon mass ratio and mass of Mars was described in *Science* by JPL astronomers John D. Anderson, Leonard Efron, and S. Kuen Wong. Accurate navigation of *Mariner VI* and *VII* to Mars had required use of two-way, phase-coherent range and Doppler tracking data from NASA-JPL Deep Space Network. Coverage of both Mariners from launch to encounter with Mars, closest approach to Mars, and beyond had provided new information on ratio of mass of earth to that of moon; Mars gravity field, particularly planet's mass; and ephemerides of Mars and earth. Data complemented Doppler data from *Mariner IV* with exception of inconsistency of about 0.0004% in values for mass of moon determined from lunar trajectories. Determinations of mass of Mars by *Mariner IV* and *VI* were in agreement. (*Science*, 1/16/70, 277-9)
- Gen. James Ferguson, AFSC Commander, delivered graduation address before Aerospace Research Pilot School at Edwards AFB, Calif.: "To the man in the street, we may have made it look too easy to prevent nuclear wars for more than 20 years, to deter or contain lim-

ited wars, and even to put men on the moon. The average American may well be bored by the third moon landing, but the Soviet planners—who know exactly what magnitude of effort is involved—do not get bored. They see us researching and developing new strategic and tactical weapons; they see us producing the largest airplane in the world and developing a new superiority fighter and advanced bomber; they see us committed to a larger and faster supersonic transport. And, knowing full well that they had a significant head start, they see us on the moon—in living color." They "know in the Kremlin—far better than most of our own laymen know—the time, the effort, the skill and dedication, and the vast commitment of resources all this requires. To them, what we do does not look easy—and in any contest of endurance, it must by now have become obvious that our far greater productivity must inevitably prevail." (Text)

- William T. Piper, founder of Piper Aircraft Corp.—one of world's largest light-aircraft corporations—died at age 89. Piper, known as "the Henry Ford of aviation," had built more aircraft than anyone else in the world, including durable, inexpensive Piper Cub. (Krebs, NYT, 1/17/70, 31)

January 17: Washington *Evening Star* editorial commented on Nixon Administration's cut in space funding: "The lowering of the sights in space is regrettable. It is unfortunate that, with the new frontier now open to us, we cannot move out to the full extent of our technical ability. But the decision to cut down on expenditures and to spread out the enormous cost of space exploration was no mistake.... The planets and the ocean of space beyond have endured for quite some time. They will wait undisturbed for man's intrusions. The economy is not so accommodating." (W *Star*, 1/17/70, A4)

- Bob P. Helgeson, former NASA Director of Safety, died of cancer at age 49. Helgeson had retired from NASA in September 1969 because of illness. He had been Deputy Manager of AEC's Hanford Atomic Works and Director of AEC-NASA nuclear rocket development station in Jackass Flats, Nev., before becoming Director of Safety in January 1968. (*W Post*, 1/19/70, D4)

January 18: Lockheed-Georgia Co. said 8 of 10 C-5As grounded by USAF Jan. 16 because of wing crack had been cleared to fly again. (AP, NYT, 1/20/70, 33)

- Mt. Palomar Observatory and Mt. Wilson Observatory were collectively renamed Hale Observatories, in honor of founder George E. Hale. As California facilities operated jointly by Cal Tech and Carnegie Institution of Washington, Mt. Palomar housed 5080-mm (200-in) Hale telescope, world's largest optical telescope, and Mt. Wilson, 1524-mm (60-in) reflection telescope. (UPI, W *News*, 1/19/70, 14)

- In Washington *Sunday Star* William Hines advocated abandonment of SST program "and substitution of a longer range objective: a hypersonic transport airplane (HST) to become operational around 1990." R&D for 500-passenger, 8000-km-per-hr (5000-mph) global-range HST was "going to be bought and paid for anyway" as part of NASA's space shuttle program. This would be a classic case of 'spinoff'—a byproduct of the space program useful to the civilian economy."

There seemed little doubt "that an aerospace-plane capable of shuttling Earth-to-orbit could be adapted to serve long-range world markets as a pure airplane if travel demands could be shown to justify its construction." SST was "not a spinoff but a drainoff." Its "most dubious aspect" was "direct federal subsidy of R&D, which in effect amounts to a $1.5 billion loan to private enterprise at zero interest," Hines said. (W *Star*, 1/18/70, C4)

During Week of January 18: India successfully test-fired four Rohini rockets at Thumba Equatorial Rocket Launching Station. Additional launches would be conducted following week. (AP, *W Post*, 1/23/70, A21)

January 19: NASA's HL-10 lifting-body vehicle, piloted by Maj. Peter C. Hoag (USAF), reached 25 600-m (84 000-ft) altitude and mach 1.3 during powered flight after air-launch from B-52 aircraft at 13 700-m (45 000-ft) altitude west of Rosamond, Calif. Purpose of flight, 32nd in series, was to obtain stability and control data. (NASA Proj Off)

• Modularized equipment transport system (METS) to be used by lunar explorers starting with October 1971 Apollo 14 mission was described by MSC Chief of Crew Systems Div. Robert E. Smylie in interview at AIAA's annual Aerospace Sciences meeting in New York. Two-wheel, rickshaw-like cart would double walking range on moon and increase sample collection. Other plans to improve astronauts' mobility included more flexible spacesuits, modified LM that could stay more than two days on moon, and improved backpack to allow astronauts to work outside spacecraft for more than six hours at a time. Cart, 7.6 cm (30 in) high, weighing 9 kg (20 lbs), had been designed with attachments for cameras, tools, and scientific equipment. It would be carried to moon folded in compartment outside LM. (Wilford, *NYT*, 1/20/70, 46)

• *Apollo 12* commander Charles Conrad, Jr., received award of excellence from Governor's Committee of 100 000 Pennsylvanians at ceremonies in Pittsburgh. Conrad, cited for excellence in science and technology, was born in Philadelphia. Ceremony climaxed day during which Conrad and *Apollo 12* Astronauts Alan L. Bean and Richard F. Gordon, Jr., toured Pennsylvania. (P *Bull*, 1/19/70, 1/20/70)

• Vice President Spiro T. Agnew completed 12-nation Asiatic goodwill tour which began Dec. 26, 1969. On return to Washington, D.C., he was accompanied by Astronaut Thomas P. Stafford and wife, who had joined tour at Bangkok, Thailand, Jan. 3. Astronaut Eugene A. Cernan and wife had accompanied Agnew party as far as Afghanistan and had returned to U.S. Jan. 9. During tour Vice President distributed chips from *Apollo 11* moon rock to government representatives and passed out foreign flags that had been carried to moon aboard *Apollo 11* spacecraft. (Off of VP; Naughton, *NYT*, 1/13/70, C3)

• Second fragment of Lost City, Okla., meteorite weighing 277.8 g (9.8 oz) had been found by farmer Philip Halpain in Lost City pasture, Smithsonian Astrophysical Observatory announced. It had been flown to Richland, Wash., for analysis of short-lived radioactive byproducts of radiation in outer space. (McElheny, *Boston Globe*, 1/20/70)

- USAF ordered that C-5As, world's largest aircraft, be limited to carrying 50% of intended capacity until each could be fitted with wing-strengthening braces. Senate Armed Services Committee announced it would investigate C-5A structural soundness before authorizing further funds for aircraft's production. (Homan, *W Post*, 1/20/70, A3)
- FAA Deputy Administrator David D. Thomas announced his resignation and retirement, effective Feb. 15. He had joined FAA as air traffic controller in 1938 and had worked his way up to second-ranking job in agency. Following resignation of Gen. William F. McKee in 1968, Thomas had served as acting FAA administrator for eight months. (*NYT*, 1/19/70, 71; UPI, *W Post*, 1/20/70, A12)
- *Kansas City Times* editorial commented on effect of budget cuts on future of NASA programs: Cutback of 50 000 employees, many in production jobs, would not alone be critical. "The real damage would come with the loss of key scientific personnel and teams engaged in shaping the medium and long-term future of the U.S. space effort. Lucrative opportunities in education and industry await such people, and they are not about to fritter away their talents and careers in trying to hold together an underfunded program whose goals are either limited or uncertain." (*KC Times*, 1/19/70, 28)
- *Aviation Week & Space Technology* quoted from article by Soviet Academician Boris N. Petrov that had appeared in July 1969 *Space World*. Communist Party First Secretary Leonid I. Brezhnev had praised manned orbital platform as possible "cosmodrome in space, a launch platform for flights to other planets. There will emerge great scientific laboratories for the study of space technology and biology, medicine and geophysics, astronomy, and astrophysics." Platform was "the main highway into space for man." (*Av Wk*, 1/19/70, 18)
- Pace of U.S. space exploration program had been "critically slowed and its direction is in serious doubt," William J. Normyle said in *Aviation Week & Space Technology*. "Manned and unmanned programs are being severely curtailed by the Nixon Administration's refusal to commit itself to a firm post-Apollo plan while beset with social and economic problems." (*Av Wk*, 1/19/70, 16)

January 19-21: AIAA held eighth Aerospace Sciences meeting in New York.

At banquet Jan. 20, $10 000 Goddard Award was presented to Gerhard Neumann, Vice President and Group Executive of GE Aircraft Engine Group, for "pioneering achievements in the development of high performance aircraft gas turbine engines, including the invention of the variable stator compressor system that is making possible an engine generation of both subsonic and supersonic engines, and for leadership in fostering the development of the high bypass turbofan."

AIAA Sylvanus Albert Reed Award of $500 was presented to Dr. Richard T. Whitcomb, Jr., Head of 8-Foot Tunnels Branch at LaRC, for "the imaginative use of wind tunnel experimentation for the solution of aerodynamic problems for the improvement of aerodynamic performances of transonic aircraft." AIAA Space Science Award of $500 was given to Dr. Carl E. McIlwan, Univ. of California at San Diego physicist, for ordering radiations trapped in earth's geomagnetic field and for outstanding investigations of these radiations. Newton A. Lieurance, ESSA Director of Aviation Affairs, received $500 Robert M. Losey Award for outstanding contributions to the science of meteorology as applied to aeronautics. Dr. Wilmot N. Hess, ESSA Director of Research Laboratories, received $500 G. Edward Pendray Award "for an outstanding contribution to the literature in the field of space physics by synthesizing and reflecting in a single text the results of a decade of research on 'The Radiation Belt and Magnetosphere.'" (AIAA Release; *Langley Researcher*, 1/23/70, 1)

AIAA Board established Committee for International Cooperation in Space with Dean Francis H. Clauser of Cal Tech as chairman. Committee would give AIAA contact point with aerospace men and societies in countries around the world through IAF and other international groups. (*A&A*, 8/70, 25)

January 20: U.S.S.R. launched *Cosmos CCCXXI* from Plesetsk into orbit with 473-km (293.9-mi) apogee, 271-km (168.4-mi) perigee, 91.9-min period, and 70.9° inclination. Satellite reentered March 23. (GSFC *SSR*, 1/31/70; 3/31/70; *SF*, 7/70, 282)

• FAA approved Boeing 747 jumbo jet for commercial service. (*W Post*, 1/21/70, D9)

• Results of study by ARC geologist Dr. William L. Quaide contradicted 4.5-billion-yr age reported for oldest sample at Jan. 5-9 Lunar Science Conference in Houston, ARC announced. Most of thick layer of lunar soil and rocks at Tranquility Base appeared to have been formed from local volcanic bedrock underlying *Apollo 11* landing site. Bedrock had been dated by three methods at approximately 3.65 billion yrs old. Same dating methods had shown one sample to be almost 1 billion yrs older, or 4.5 billion yrs old. Dr. Quaide had found when individual particles of rock were taken from soil and dated they again measured at 3.65-billion-yr age. (ARC Release 70-1)

• President Nixon had asked Dr. Lee A. DuBridge, Science Adviser, to explore future use of ERC, Office of Science and Technology announced. (OST PIO)

• Richard C. McCurdy, immediate past president and chief executive officer of Shell Oil Co., was sworn in as NASA consultant. He would serve on Management Advisory Panel. (NASA Release 70-9)

• MSFC announced appointment of Robert E. Lindstrom as Deputy Director of Manufacturing Engineering Laboratory, Science and Engineering Directorate. Lindstrom had worked in aerospace industry since his resignation from NASA in 1963. (MSFC Release 70-10)

• Astronauts Alan L. Bean, R. Walter Cunningham, and Joseph P. Kerwin, grounded Dec. 23, 1969, for minor infractions of aircraft flight rules, resumed training in T-38 jet aircraft at MSC. (UPI, *W Star*, 1/21/70, A7)

• Many Turkish peasants were convinced that flu that had been ravaging Turkey came from germs brought to earth by Apollo astronauts, *San Francisco Chronicle* said. Peasants thought astronauts had been imprisoned in room for bringing microbes back from the moon and that wind had carried microbes from space capsule to Europe and Turkey. Charges were appearing in Turkish press. (Arnow, *SF Chron*, 1/20/70)

- Agreement to reduce jet aircraft air pollution by 1972 was reached in Washington, D.C., by Nixon Administration and representatives from 31 scheduled and charter U.S. airlines. Airlines agreed to install pollution control devices on 1000 Boeing 727 and 737 and DC-9 aircraft during regular maintenance overhauls. Devices, called burner cans, would eliminate 70%-80% of solid particles spewed into air by jets. (Curry, *W Post*, 1/21/70, A2)
- *Washington Post* editorial commented on Boeing 747 jumbo jet debut: "We have a sneaking suspicion that the 747, as the new behemoth of the skies is called, has appeared on the scene three or four years prematurely. It is, or so everyone who has flown in it says, a magnificent airplane that brings a jump in passenger comfort equal to or surpassing that which occurred when the smaller jets went into service. But passenger comfort in the air is only part of the business of air travel and it looks to us as if other parts of that business—the airports and the airlines themselves, to be exact—are not really ready." (*W Post*, 1/20/70, A14)

January 21: U.S.S.R. launched *Cosmos CCCXXII* from Plesetsk into orbit with 311-km (193.3-mi) apogee, 195-km (121.2-mi) perigee, 89.5-min period, and 65.4° inclination. Satellite reentered Jan. 29. (GSFC *SSR*, 1/31/70; *SF*, 7/70, 282)
- Scientists at Royal Observatory in Edinburgh unveiled Galaxy, machine conceived by Dr. P. N. Fellgett of Reading Univ., Scotland, to measure star images at rate of 10 000 per hr. Machine had found 1103 stars less than 28 million yrs old in constellation Perseus, where only 15 had been found previously. Galaxy contained Elliot 4130 computer linked to 406-mm (16-in) Schmidt telescope and could measure accuracy of star's position to one micron. British scientists said they believed machine established world lead in astronomy for U.K. (*W Post*, 1/21/70, A3)
- AIAA announced reelection of Dr. Ronald Smelt as AIAA President and of Dr. John C. Houbolt as Vice President-Technical and election of eight new Directors: Dr. Mac C. Adams, Vice President and Deputy Group Executive of Avco Government Products Group and former NASA Associate Administrator for Advanced Research and Technology; A. Scott Crossfield, Division Vice President for Flight Research and Development for Eastern Air Lines; Christopher C. Kraft, Jr., MSC Deputy Director; Carlos C. Wood, Division Vice President of United Aircraft Corp.'s Sikorsky Aircraft Div.; Edward N. Hall, Assistant Chief Scientist for United Aircraft Corp.'s Corporate Technical Staff; Dr. Walter T. Olsen, LeRC Assistant Director; Alan Y. Pope, Director of Aerothermodynamics for Sandia Corp.; and Charles W. Duffy, Jr., Director of Personnel for Boeing Co.'s Commercial Airplane Div., SST Branch. (AIAA Release)
- Officials from Massachusetts, City of Cambridge, and ERC met at White House with Dr. Lee A. DuBridge, President's Science Adviser, to discuss prospects for new uses for ERC facilities. (OST PIO)
- Portuguese Government announced plans to establish permanent study committee on space exploration. Committee would collect, collate, and study information on space travel, promote training of scientists and technicians, and maintain contact with world organizations on space exploration. (Reuters, *W Post*, 1/22/70, A7)

- FAA reported it had executed 513 grant agreements committing $107.5 million for airport development during FY 1969 and raising 23-yr total of grants provided under Federal-aid Airport Program to almost $1.2 billion. (FAA Release 70-6)
- Consumer adviser and safety advocate Ralph Nader released report in Washington, D.C., to Government agencies and industry spokesmen. It charged that travel by light aircraft was most dangerous form of transportation in U.S. and that general-aviation industry had neglected simple remedies which could save hundreds of lives annually. (AP, W *Star,* 1/22/70, A4)
- DOD announced USAF contract awards: $1 800 000 supplemental agreement to MIT for basic research on properties of matter in intense magnetic fields, managed by Office of Scientific Research; and $354 000 initial increment to $1 062 000 cost-reimbursement contract to Trustees of Princeton Univ. for basic research on hypersonic flow, managed by Aeronautical Systems Div. (DOD Release 54-70)

January 21-22: Pan American World Airways Boeing 747 jumbo jet completed first commercial transatlantic flight in 7-hr 8-min flight from New York to London with 196 passengers and 18 crew members. (Witkin, *NYT,* 1/23/70, 93)

January 22: Discovery by *Oao II* of great hydrogen cloud surrounding comet Tago-Sato-Kosaka was announced by NASA. Instruments in Univ. of Wisconsin payload aboard *Oao II,* launched Dec. 7, 1968, had first studied comet Jan. 14 as it moved from sun. Data gathered to date had shown hydrogen cloud surrounding comet's head was large as sun. Data and ground-based observations should facilitate more accurate determination of amount of mass ejected from comet and provide better knowledge of its composition, NASA said. (NASA Release 70-10)

- President Nixon delivered State of the Union Message before Joint Session of Congress: "The moment has arrived to harness the vast energies and abundance of this land to the creation of a new American experience...more truly a reflection of the goodness and grace of the human spirit. The seventies will be a time of new beginnings, a time of exploring both on the earth and in the heavens, a time of discovery."

  GNP would increase by $500 billion in next 10 yrs—increase greater than entire growth of U.S. economy from 1790 to 1950. "Our purpose in this period should not be simply better management of the programs of the past. The time has come for a new quest—a quest not for a greater quantity of what we have had but for a new quality of life in America."

  President proposed total reform of welfare system; assessment and reform of institutions of government at Federal, state, and local level; and "reforms which will expand the range of opportunities for all Americans." Greatest question of seventies was "shall we surrender to our surroundings or shall we make our peace with nature and begin to make reparations for the damage we have done to our air, to our land, and to our water?"

  President would propose to Congress $10 billion nationwide clean waters program "to put modern municipal waste treatment plants in every place in America where they are needed to make our waters clean again...now." He called for further advances in engine design and fuel composition for automobile, "our worst polluter of the air," and new rural environment to reverse migration to urban centers. He also planned balanced FY 1971 budget to help stop rise of cost of living.

". . . I see a new America as we celebrate our 200th anniversary 6 years from now. I see an America in which we have abolished hunger, provided the means for every family in the nation to obtain a minimum income, made enormous progress in providing better housing, faster transportation, improved health, and superior education. . . . I see an America in which we have made great strides in stopping pollution of our air, cleaning up our water, opening up our parks, continuing to explore in space. Most important, I see an America at peace with all the nations of the world.

"This is not an impossible dream. The goals are within our reach. . . . Let it not be recorded that we were the first American generation that had the means but not the vision to make this dream come true." (*PD*, 1/26/70, 58-66)

- Univ. of Southern California conferred degree of M.S. in aerospace engineering on *Apollo 11* Astronaut Neil A. Armstrong. Astronaut had pursued graduate studies under USC-Edwards AFB program from 1955 to 1960 and had nearly completed degree requirements when he was transferred. Lecture delivered Jan. 22 by Armstrong on "Lunar Landing: Techniques and Procedures" completed requirements. (AP, W *Star*, 1/23/70)
- MSFC announced selection of Chrysler Corp., General Electric Co., and Boeing Co. for competitive negotiations leading to three-year, $25-million, cost-plus-incentive-award-fee contract for management, sustaining engineering, and logistics support for Saturn ground support equipment and for operations and maintenance of Saturn systems development facilities. (MSFC Release 70-12)

January 23: NASA successfully launched *Itos I* (Tiros-M) Improved Tiros Operational Satellite from WTR at 3:31 am PST by two-stage, long-tank Thor-Delta (DSV-3N-6) booster with six solid-fuel, strap-on Thiokol rockets. Satellite entered polar, sun-synchronous orbit with 1478.2-km (918.7-mi) apogee, 1432.3-km (890.2-mi) perigee, 115-min period, 102° inclination, and 0.02° per day orbital drift.

Australis Oscar V (Oscar A) Orbiting Satellite Carrying Amateur Radio, carried pickaback on 2nd stage, was successfully ejected and entered orbit with 1482-km (920.9-mi) apogee, 1434-km (891.1-mi) perigee, 115-min period, and 101.9° inclination. *Oscar V* was fifth in series of Oscar launches; previous spacecraft had been launched in conjunction with DOD missions. Satellite had been constructed by amateur radio operators at Melbourne Univ. in Australia and prepared for launch by U.S. Radio Amateur Satellite Corp.; 17.7-kg (39-lb) satellite would transmit at 29.45 mhz in 10-m band and 144.5 mhz in 2-m band.

Primary objective of *Itos I* was to flight-qualify prototype spacecraft for Improved Tiros Operational System and obtain engineering data for evaluation of single-momentum, wheel-stabilized system for earth-oriented stabilized platform. Secondary objective was to evaluate use of stabilized platform for operational meteorology by observing cloud cover by direct readout.

January 23: NASA launched Itos I Improved Tiros Operational Satellite into orbit to photograph the earth day and night and flight-qualify the spacecraft for use in the National Operational Meteorological Satellite System. Itos I would take night infrared pictures and transmit cloud-top and surface temperatures. TV photos could be received by automatic picture transmission system stations. "Super-six" version of two-stage, long-tank Thor-Delta launch vehicle (at right)—thrust augmented for the first time by six solid-fuel strap-on Castor rockets instead of three—made its debut to orbit the spacecraft weighing more than twice as much as previous TOS satellites. Before the launch, above, the spacecraft was photographed as it was prepared for mating to the Delta-N at the Western Test Range. Oscar V was launched pickaback for use by amateur radio operators at Melbourne University, Australia.

'Itos I was first in new series of operational meteorological satellites to be launched in joint NASA-ESSA program to replace TOS satellites operating in National Operational Meteorological Satellite System. Itos I weighed 306 kg (675 lbs)—more than twice as much as previous TOS satellites—and carried both automatic picture transmission (APT) and advanced vidicon camera system (AVCS) cameras, previously carried on separate missions. Satellite was box-shaped with three-panel solar array and had new "stabilite" three-axis stabilization system. Onboard experiments included two redundant AVCS cameras for stored global cloud-cover data for remote readout; two redundant APT cameras for local readout of global cloud-cover data; redundant scanning radiometers for global cloud-cover data for remote and local readout, both day and night; flat-plate radiometer for data on global heat balance; and solar proton monitor for routine observation of solar flares. Spacecraft would more than double daily weather coverage possible from ESSA satellites already in orbit and give more effective coverage at less expense for longer time. Nighttime photo capability would provide cloud-cover photos night and day every 12 hrs; ESSA series provided one picture per day.

Itos I was 19th spacecraft to be launched in NOMSS. First, Tiros I, had been launched April 1, 1960; last, Essa IX, had been launched Feb. 26, 1969. Itos I was funded by NASA except for meteorological sensors; KSC was responsible for launch operations and GSFC was responsible for Delta booster. Future ITOS spacecraft would be funded by ESSA and would be turned over to ESSA by NASA after postlaunch checkout. (NASA Proj Off; NASA Release 70-2)

- NASA successfully conducted first test firing of experimental solid-propellant rocket motor with 1.07-m (42-in) diameter and built-in thrust stopper at Eniwetok Atoll in Pacific. Motor burned 34 secs at 53 400-newton (12 000-lb) thrust before it was stopped by internal water-quench system. Motor had been constructed by Hercules Inc., for JPL. (NASA Release 70-12)
- NASA's ALSEP 1, placed on lunar surface by Apollo 12 crew Nov. 19, 1969, was functioning satisfactorily after more than 64 days of uninterrupted operation, MSC announced. Some 1977 commands had been transmitted to and carried out by ALSEP 1 since deployment. Scientific and engineering data were being transmitted continuously and radioisotope power remained constant at 74 w. Passive seismic experiment had recorded episode of large tilts on long-period horizontal sensors that lasted one hour and coincided with third lunar sunrise Jan. 16. (MSC Release 70-12)
- NASA released photo of emblem selected by Astronauts James A. Lovell, Jr., Thomas K. Mattingly II, and Fred W. Haise, Jr., for Apollo 13 mission. Emblem, designed by Lumen Martin Winter of New York, depicted three horses pulling sun chariot from earth to moon, with Latin phrase "Ex luna scientia" (From the moon, knowledge). (NASA PAO; B News, 1/25/70)
- President Nixon announced Administration's recommendations on use of comsats for domestic telecommunications. White House press office released memorandum to FCC Chairman Dean Burch outlining recommendations: Government policy "should encourage and facilitate the development" of commercial domestic comsat

systems "to the extent that private enterprise finds them economically and operationally feasible." Subject to appropriate conditions, "any financially qualified public or private entity, including Government corporations, should be permitted to establish and operate domestic satellite facilities for its own needs; join with related entities in common-user, cooperative facilities; establish facilities for lease to prospective users; or establish facilities to be used in providing specialized carrier services on a competitive basis." Common carriers "should be free to establish facilities for either switched public message or specialized services, or both. The number of classes of potential offerers...should not be limited arbitrarily. Nor should there be any a priori ranking of potential types of systems...." Guidelines were set forth for establishment and operation of domestic comsats. (PD, 1/26/70, 66; Text)

- ComSatCorp announced it would file promptly with FCC for authority to proceed with domestic satellite services. (ComSatCorp Release 70-5)
- Dr. Thomas O. Paine, NASA Administrator, named Oran W. Nicks, Deputy Associate Administrator for Space Science and Applications, as Acting Associate Administrator for Advanced Research and Technology, effective immediately. Nicks succeeded Bruce T. Lundin, who had been named LeRC Director. (NASA Release 70-11; NASA Ann, 1/23/70)
- Results of acquisition measurements of round-trip travel time of light from McDonald Observatory, Tex., to laser ranging retroreflector deployed on moon by *Apollo 11* astronauts—made Aug. 20 and Sept. 3, 4, and 22, 1969—were reported in *Science* by astronomers from Univ. of Maryland, GSFC, and Univ. of Texas. Uncertainty in round-trip travel time was ± 15 nanoseconds, with pulsed ruby laser and timing system used for acquisition. Uncertainty in later measurements of planned long-term sequence from McDonald Observatory was expected to be order of magnitude smaller. Succesful performance of retroreflector at several angles of solar illumination, as well as during and after a lunar night, confirmed prediction of thermal design analyses. (*Science*, 1/23/70, 368-70)
- Analytical results for lunar material in Sinus Medii derived from alpha-scattering experiment on *Surveyor VI* (launched by NASA Nov. 7, 1967; softlanded on moon Nov. 9, 1967) were reported in *Science* by JPL, Argonne National Laboratory, and Univ of Chicago chemists. Amounts of principal constituents at Sinus Medii mare were approximately same as those at Sea of Tranquility. Sodium contents of both maria were lower than those of terrestrial basalts. Titanium content at Sinus Medii was lower than at Sea of Tranquility. Results suggested important differences in detailed chemical composition at different mare areas on moon. (*Science*, 1/23/70, 376-9)
- USAF announced it was flight-testing dual sidearm controller developed by Hughes Aircraft Co. that might replace pilot's control stick or wheel in future aircraft. Built into copilot's seat frame, device was designed for electrical flight control using power augumentation system similar to power steering in automobile. (AFSC Release 215.69)

- *New York Times* editorial commented on President Nixon's State of the Union address: "Where the President was at his best was in his call for an environmental program not only the most comprehensive in the nation's history but the most costly as well. . . . The proposal is thoroughly commendable as far as it goes, but unless it is combined with effective legislation to curb private industrial polluters as well, it is bound to fall considerably short of the objective." (*NYT*, 1/23/70, 46)
- Situation at Lockheed-Georgia Co. aircraft plant in Marietta, Ga., typified "impact of the changed national priorities outlined by President Nixon in his State of the Union message today," George C. Wilson said in *Washington Post*. "Workers wonder when—not if—they will be laid off as they see the end of the C-5A airplane program." Work force was expected to drop from 29 000 to 24 000 during 1970, "with longer range employment picture much bleaker." (*W Post*, 1/23/70, 3)

January 24: Aerobee 150MI sounding rocket, launched by NASA from WSMR with VAM-20 booster, carried Princeton Univ. payload to 168.9-km (105.0-mi) altitude to obtain UV spectra of hot atmospheres of stars. Rocket and instruments functioned satisfactorily and unique UV photo of comet Tago-Sato-Kosaka was recorded. (NASA Rpt SRL)
- World's largest tuna fishing boat was launched in San Diego, Calif. It was insulated with polyurethane foam which had been developed for use as insulation in Saturn V 2nd-stage fuel tank. North American Rockwell Corp. had used foam in 61-m (200-ft) clipper. (MSFC Release 70-23)
- USN launched DSRV-1 (Deep Submergence Rescue Vehicle) in San Diego, Calif. It was first of fleet of six planned, at total $480-million cost, to rescue survivors of submarine disasters. (Lannan, W *Star*, 1/24/70, A2)

January 25: U.S. Ambassador to Guyana Spencer King said in Guyana that Apollo moon rocks would go on exhibition there during February observance of Guyana's becoming a republic. (*NYT*, 1/27/70)
- "Heavily censored testimony of defense officials before congressional committees" indicated laser development field was making "dramatic advances" in race to develop "death ray," Orr Kelly said in Washington *Sunday Star*. But most intriguing developments had come in unclassified research. Reports indicated U.S.S.R. also was seeking to exploit same basic scientific knowledge. (W *Star*, 1/25/70)

January 26: NASA's HL-10 lifting-body vehicle, piloted by NASA test pilot William H. Dana, reached 26 800-m (88 000-ft) altitude and mach 1.4 during powered flight after air launch from B-52 aircraft at 13 700-m (45 000-ft) altitude west of Rosamond, Calif. Purpose of flight, 33rd in series, was to obtain stability and control data at various mach numbers and angles of attack. (NASA Proj Off)
- LeRC scientists and engineers were "applying their special talents in off hours to studying the dirty air they breathe," Cleveland *Plain Dealer* said. Several had attended Cleveland hearing on air quality and "had a great impact on the state air pollution control board." More than 100 LeRC volunteers had formed clean air committee to offer talents as concerned citizens. (McCann, Cleveland *Plain Dealer*, 1/26/70)

January 26 —

- Secretary of Transportation John A. Volpe and Canadian Minister of Transport Donald C. Jamieson agreed in Washington, D.C., to exchange information on problems of automobile and aviation safety and pollution common to both countries. (DOT Release 2670)
- President Nixon submitted to Senate nomination of Charles D. Baker, Deputy Under Secretary of Transportation, as Assistant Secretary of Transportation for Policy and International Affairs. He would succeed Paul Cherington who had resigned. (PD, 2/2/70, 74, 99)
- MSC announced award of $5.1-million NASA contract to Itek Corp. for six high-resolution cameras to be used for lunar photography on Apollo 16, 17, and 18 missions. (MSC Release 70-11)
- MSFC announced award of $4 360 260 modification to Saturn V contract with Boeing Co. for installation of accumulators—small gas reservoirs in liquid-oxygen prevalves—in 1st stage (S-IC) to change frequency of oscillation in propulsion system. Installation had been made after second Saturn V had oscillated severely during unmanned *Apollo 6* flight April 4, 1968. (MSFC Release 70-14)
- Feasibility of replacing existing primary and secondary radar in 1970s with satellite system to perform air surveillance would be studied by new ad hoc air traffic control panel of President's Science Advisory Committee, *Aviation Week & Space Technology* reported. Panel's report was due in June. (*Av Wk*, 1/26/70, 25)
- Washington *Evening Star* editorial praised Nixon Administration's Jan. 23 recommendation that FCC allow open competition in domestic comsat services. President was "relying on innovation and competition—rather than government regulation of a single, monopolistic organization—to provide the best service at the lowest possible price. This would be vastly different from the regulated monopoly approach on which this country has come to rely more and more in such public utility areas as electricity and telephone service. The White House has chosen wisely to abandon such an approach for the nation's first step into space communication for domestic use." (W *Star*, 1/26/70)
- AT&T announced it would apply to FCC for permission to use satellites in its domestic operations. (Samuelson, *W Post*, 1/27/70)

January 27: Dr. Thomas O. Paine, NASA Administrator, announced appointment of Dr. Wernher von Braun, MSFC Director, as NASA Deputy Associate Administrator for Planning, effective March 1. Dr. von Braun would be succeeded by his deputy, Dr. Eberhard F. M. Rees. Dr. Paine said: "It is essential that we bring NASA's best talents to bear on our future space planning." NASA must select its "new space ventures with the best critical judgment and make every dollar count. Dr. Wernher von Braun has an unmatched record of looking to the future to choose the most promising avenues of technical advance. He brings to his new assignment sound vision, insight, and technical competence...." (MSFC Release 70-19; NASA Release 70-13)

- Charles W. Harper, NASA Deputy Associate Administrator for Advanced Research and Technology (Aeronautics), discussed NASA's role in development of anticollision systems for aircraft before House Committee on Government Operations' Subcommittee on Government Activities. It appeared "impractical" to force small general-aviation aircraft "to avoid controlled airspace or to carry the expensive equipment and undergo the intensive training necessary to place them under positive control of the ATC [air traffic control] system. For these reasons, FAA has asked NASA to investigate the possibility of developing inexpensive, lightweight equipment for small aircraft which would augment significantly the collision avoidance probability when the pilot is flying on the see-and-be-seen principle. . . ."

NASA program was guided by FAA's Collision Prevention Advisory Group, of which NASA was member, and by NASA Research and Technology Advisory Committee's Subcommittee on Aircraft Operating Problems, of which FAA was member. Both groups included representatives of industry and other Government agencies. (Testimony)

- MSFC announced award of $3.2-million extension to contract with Mason-Rust for institutional support services at Michoud Assembly Facility. Contract covered Jan. 1 through June 30. (MSFC Release 70-16)

January 28: Rep. Seymour Halpern (R-N.Y.) introduced H.R. 15609 "to establish a sonic boom and aircraft noise damage fund to provide for the payment of damages caused by sonic boom and other aircraft noise." Legislation was referred to House Judiciary Committee. (*CR*, 1/28/70, H472)

- Rep. F. Bradford Morse (R-Mass.) introduced H.R. 15605 "to provide for a National Laboratory for Environmental Sciences." (*CR*, 1/28/70, H478)

January 28-29: Review meeting at MSFC on design of lunar roving vehicle (LRV) was attended by nearly 100 representatives of NASA, prime contractor Boeing Co., and Boeing subcontractor, General Motors AC Electronics Div. (MSFC Release 70-17; *Marshall Star*, 1/28/70, 1)

January 29: NASA launched two Nike-Cajun sounding rockets from Point Barrow, Alaska, carrying GSFC grenade experiments. Rockets and instruments functioned satisfactorily. (SR list)

- *Apollo 8* Astronaut Frank Borman would leave active duty with NASA July 1, NASA announced. He would become Vice President of Electronic Data Systems, Inc., assist in establishment of American Horizons Foundation, and continue to serve NASA as consultant on earth-orbiting space stations. Borman, assigned to NASA by USAF since 1962 and NASA Field Director of Space Station Studies since May 1969, would retire from USAF after 20 yrs active duty. He had received NASA Distinguished and Exceptional Service Medals, USAF Comand Astronaut Wings and Distinguished Flying Cross, American Astronautical Flight Achievement Award, Air Force Assn.'s David C. Schilling Flight Trophy, Cal Tech's Distinguished Alumni Service Award, and New York State Medal for Valor.

Experience in space program, Borman said, "has been the most rewarding of my life. This new opportunity will enable me to work with a major industrial firm, continue an association with the space effort and to devote a substantial amount of my time through the Foundation to many issues which have interested me for a long time." (NASA Release 70-14)

- MSC announced award of one-year, $11-million, cost-plus-award-fee contract extension to Service Technology Corp. for facility support services. Award was for third year of five-year program initially

awarded to LTV Aerospace Corp. and brought total value of contract since December 1967 to $34.8 million. (MSC Release 70-17)
- President Nixon announced appointment of Russell E. Train, Robert Cahn, and Gordon J. F. MacDonald as members of new Council on Environmental Quality and outlined Council's chief responsibilities: to study condition of Nation's environment, develop new environmental programs and policies, coordinate Federal programs, ensure that all activities of Federal Government took environmental considerations into account, and assist President in preparing annual report on Environmental Quality. (*PD*, 2/2/70, 90-1)
- At annual meeting of American Physical Society in Chicago, symposium on cosmic physics suggested stellar explosion could have precipitated condensation of dust and gas to form sun and planets. Analysis of meteorite discovered several years earlier at St. Severin, France, had led to suggestion by Cal Tech physicist Dr. Gerald J. Wasserburg that great stellar explosion had accompanied or immediately preceded formation of solar system, Walter Sullivan said in *New York Times*. Meteorite had contained calcium phosphate crystals in tiny tracks of particles ejected by radioactive atoms. Suspected source was plutonium 244—element manufactured in laboratory but not found in nature. Because plutonium 244's mean lifetime was only 118 million yrs, any plutonium in earth rocks or meteorites, both more than 4.5 billion yrs old, would have decayed radioactively unless something occurred just before or during solar system's formation that synthesized this substance. According to Wasserburg hypothesis, material was incorporated into meteorite where, during solar system's infancy, it decayed possibly into neutron-rich forms, or isotopes, of xenon found in meteorite, leaving radiation tracks. (*NYT*, 1/30/70, 23)

January 29-30: U.N. Committee on Peaceful Uses of Outer Space met at MSC for review of Earth Resources Satellite Program. Representatives from 21 of 28 nations on Committee were shown conceptual design of satellite system to survey global atmospheric patterns by mid-1970s. Dr. John M. DeNoyer, Director of NASA Earth Observations Programs, described Global Atmospheric Research Project: It would be "significant step toward adequate modeling of the atmosphere and will take us much closer to a reliable prediction capability for the weather." Space technology could also contribute to management of world's food supply, quality and availability of water, mineral sources, land use, urban development, and pollution. "Degradation of our planet may be inevitable. If so, we must work as a community of all nations to make the rate of degradation as slow as possible." (MSC Release 70-16; Hill, *H Chron*, 1/30/70)

January 30: AEC announced portable atomic camera using californium 252 to generate neutrons to penetrate most metals and other opaque materials and detect weaknesses or foreign substances had been developed by Battelle-Northwest at AEC's Pacific Northwest Laboratory, Richland, Wash. Other applications of camera included safety checks of radioactive heat sources and reactor components and possible use to examine bone marrow and flesh for tumors. (AEC Release N-7)

January 31: Library of Congress Legislative Reference Service published *United States and Soviet Progress in Space: How do the Na-*

tions Compare? Report by Dr. Charles S. Sheldon II, Chief of Science Policy Div., summarized 12-yr efforts of major space powers. U.S. expected to have spent by June 30, 1970, just over $56 billion on combined military and civilian space programs in which some half million persons had been employed. It was likely U.S.S.R. had "committed a similar amount of resources." U.S. space employment had peaked; Soviet had not, "as judged by their rising number of launchings." There was no evidence that total Soviet aerospace industry was as fully equipped as U.S., but missiles of differing design philosophies seen in Moscow parades suggested "more than one design and development team for space work just as in Soviet aviation." Each country seemed adequately equipped with launch pads. U.S.S.R. claimed nationwide tracking system but relied on ships to fill gaps.

U.S.S.R. was still using ICBM with 1400-kg (3000-lb) thrust and improved staging to boost 4700-kg (10 400-lb) Vostok and 6600-kg (14 500-lb) Soyuz. Proton launch vehicle had carried 16 800 kg (37 000 lbs) to earth orbit and might have potential to lift 27 200 kg (60 000 lbs) with circumlunar capacity of 6000 kg (14 500 lbs). "We have not yet seen the very large Soviet vehicle confidently predicted by NASA officials the last three years. Lesser Soviet launch vehicles account for unmanned programs." There was no sign of Soviet breakthrough in fuels "but their engines are run at relatively high pressure, increasing efficiency." There was no sign that solid-fuel rockets had been used by U.S.S.R. for space purposes and question of Soviet development of nuclear rocket was "highly speculative." Soviet work on electric rockets had reached only "early proof-of-principle" stage. U.S. Saturn V was "clearly the best operational weight lifter in the world."

Odds were "overwhelming" that Vladimir Komarov had been only Soviet Cosmonaut killed in space so far. "This is because the Russians have followed a conservative approach to manned flight with heavy vehicles allowing much redundancy in equipment."

There seemed to be "little difference between the Soviet and U.S. space programs as to general purpose and direction." About three fifths of U.S. flights were conducted for DOD; similar heavy military emphasis existed in U.S.S.R. "Soviet Union has a large and regular program of military support flights including use of satellites for photographic observation, electronic listening, weather reporting, communications relaying, and ship navigation." U.S. did not have FOB program, "because it does not regard them as necessary or desirable. The Russians obviously do, for reasons not wholly clear." While assessment could not be conclusive, "suspicion remains that a capability to inspect and destroy satellites has been created" by U.S.S.R.

In space applications, U.S. "has held a clear lead from the earlier days of the space program, and this experience has permitted us to move ahead to more advanced systems while the Soviet Union was still working hard to catch up." U.S.S.R. had made greater effort in number of lunar and planetary flights, "but so far has not gained as good results as we have from a more modest program." Some 14% of Soviet flights had been in escape mission category and 8% of U.S. flights. Reasonable assumption was "that manned flights by the

Russians to the Moon will come as soon as they solve their present problems of unreliability of hardware." Question of U.S. and U.S.S.R. going to moon together "has been asked more in rhetorical sense than as a concrete offer by either side." No assessment of prospects could be made "without forecasting the future political climate which is beyond our present capabilities." (Text)

- Twelfth anniversary of *Explorer I,* first U.S. satellite. Launched Jan. 31, 1958, stovepipe-shaped 14-kg (30.8-lb) satellite was in orbit with 585.7-km (364-mi) apogee and 294-km (182.7-mi) perigee and was expected to reenter and disintegrate in May. (MSFC Release 70-11)
- Aerospace Defense Command reported that U.S. had 290 satellites in orbit around earth at end of January, with 18 deep-space probes still active. U.S.S.R. had 75 satellites and 14 deep-space probes. U.K. had 3, France 5, Canada 3, West Germany 1, Australia 1, and ESRO 3. (*SD Union,* 2/8/70)

During January: Space/Aeronautics reviewed status of aerospace and aeronautical projects at commencement of 1970s. Commercial transport industry was "microcosm of the capitalistic system. The health of any constituent, SST or airbus or Stol, is on trial in a harshly profit-motivated atmosphere that says... 'shape up or ship out.' Nowhere is this test more evident than in the troubled, hesitant posture of the U.S. SST program." With retrenchment to fixed-wing design "and a continual dropback in the urgency level of our SST effort," SST's competitive advantages were "being dissipated by an inexorable calendar."

Apollo 11 lunar landing had alleviated Congressional opposition to further expensive long-term space commitments. "Suddenly, it was practical to think of going ahead with an expanded lunar exploration program."

Space science and applications program had been marking time, "with most programs stretched out to match the diminished funding made smaller still by the nasty bite of inflation." But "light is beginning to show at the end of the long, dark tunnel of fiscal retrenchment." Success of satellite infrared spectrometer (SIRS), "welcome and long overdue test of ATS-3" for coast-to-coast relay of TV programs, and climb in "Comsat's '69 operating profiles" above $1-million mark had led space science program steadily to shift "its dollar focus away from the earth."

Civil systems activity by aerospace community was only token. "Yet 1969 may prove to have been a landmark year for the industry's relations with this strange and often self-contradictory market. New initiatives by the federal government indicate a growing awareness in Washington of what aerospace may be able to contribute to the solution of complex civil problems." (*S/A,* 1/70, 27-69, 81-132)

- *Fortune* editorial commented on end of decade: "Before the 1960's are too hastily buried, it is well worth recalling...that the decade just ended was, for U.S. business, far and away the most profitable and productive ever. Indeed, in the long view of history, the Sixties will be remembered not just for tragic episodes, the war, riots, and assassinations, but also for the greatest floodtide of material abundance ever seen by human eyes. It is, after all, rare in human affairs that resources, talent, and institutions coincide to release ful-

ly man's remarkable productive capacities—and this happened in the U.S. in the Sixties." (*Fortune*, 1/70, 69-70)
- Lightweight double-wall meteoroid shield developed for enlarged propulsion system of Mariner Mars 1971 was described in *Journal of Spacecraft and Rockets* by JPL scientist John R. Howard. Outer sheet of Teflon-impregnated glass fabric and multilayer thermal insulation was more effective than aluminum sheet of same weight because of impact phenomena modification by heterogeneous first sheet and impact energy absorption by thermal insulation. Series of tests over range of impact velocities with projectiles of glass, nylon, and syntactic paste had been conducted. Linear extrapolation had shown that composite outer sheet was sufficient for specified meteoroid threat. (*JSR*, 1/70, 69-72)
- J. W. Chamberlain, Associate Director for Planetary Sciences at Kitt Peak National Observatory, Ariz., reviewed cosmological theory for systematical exploration of outer planets in *Astronautics & Aeronautics*. Jovian planets—Jupiter, Saturn, Uranus, and Neptune—had high intrinsic interest because their characteristics were similar to one another's but "vastly different" from those of terrestrial planets. Atmospheric distinctions which did exist among Jovian planets had probably "arisen from relatively minor differences in the prevailing conditions at the times of their birth."

  Major planets constituted "completely different kind of object." Nearest and biggest—Jupiter—had over 1000 times earth's volume and 300 times earth's mass for mean density of only 1.3 times that of water. It was "guardian (if not the parent) of a large family of at least 12 satellites, two of which are as large as a small planet. It has a daily rotation period of only 10 hr; and its atmosphere is composed primarily of hydrogen and helium, the most abundant elements in the Sun (and the universe itself...)." Neptune was "most distant of the giants" and Pluto was "nonconformist of the solar system" whose orbit "is far from circular, the solar distance ranging between about 30 and 50 A.U. [30 and 50 times distance from earth to sun] during its 248-year revolution, at a 17-deg inclination to the ecliptic." Many astronomers suspected larger differences between terrestrial and major planets held clues to formation of solar system. (*A&A*, 1/70, 20-2)
- ARC engineer Gerald J. Miatech and Aerojet-General engineer Clarence A. Lysdale reported application of multiple-satellite concept to particles and field research in *Journal of Spacecraft and Rockets*. Emphasis had been on experiments to resolve temporal and spatial interreactions of solar wind and earth's magnetosphere. Study had revealed feasibility, using proved technology, of establishing four-satellite cluster for particles and field research in acceptable orbital array using single Thor-Delta launch vehicle. (*JRS*, 1/70, 60-8)
- Lord Ritchie-Calder wrote in *Foreign Affairs:* "Past civilizations are buried in the graveyards of their own mistakes, but as each died of its greed, its carelessness or its effeteness another took its place. That was because such civilizations took their character from a locality or region. Today ours is a global civilization; it is not bounded by the Tigris and the Euphrates nor even the Hellespont and the Indus; it is the whole world. Its planet has shrunk to a neighborhood

round which a man-made satellite can patrol sixteen times a day, riding the gravitational fences of Man's family estate. It is a community so interdependent that our mistakes are exaggerated on a world scale.... Modern Man can outboast the Ancients, who in the arrogance of their material achievements built pyramids as the gravestones of their civilizations. We can blast our pyramids into space to orbit through all eternity round a planet which perished by our neglect." (*Foreign Affairs*, 1/70, 207-20)

February 1970

February 2: President Nixon in message transmitting FY 1971 budget to Congress said: "Man has ventured to the moon and returned—an awesome achievement. In determining the proper pace for future space activities, we must carefully weigh the potential benefits of: scientific research by unmanned spacecraft; continued exploration of the solar system, including manned exploration of the planets; and the application of space and aeronautics technology to the direct benefit of mankind.

"I have reviewed many exciting alternatives for the future. Consistent with other national priorities, we shall seek to extend our capabilities in space—both manned and unmanned. I intend to do this within total space outlays 12% smaller than in 1970. In our current efforts, we will continue to stress additional uses of space technology. Our actions will make it possible to begin plans for a manned expedition to Mars."

President said national budget, first for which he bore full responsibility, fulfilled pledge to submit balanced FY 1971 budget. "This anti-inflationary budget begins the necessary process of reordering our national priorities. For the first time in two full decades, the Federal Government will spend more money on human resource programs than on national defense." About 41% of estimated outlays would be devoted to spending for education and manpower, health, income security, and veterans benefits and services. Pollution control, crime reduction, transportation, and housing programs were "planned to grow substantially in the years ahead." (*PD,* 2/9/70, 106-19)

- President Nixon sent $200.8-billion FY 1971 budget request to Congress, including total space budget of $4.954 billion. Of this sum, NASA would receive $3.148 billion (plus $185.5 million for aircraft technology); DOD would receive $1.674 billion; AEC, $99 million; ESSA, $26 million; Dept. of Interior, $4 million; NSF, $2 million; and Dept. of Agriculture, $1 million.

  Total NASA new obligational authority (NOA) requested, $3.333 billion, was $404.9 million less than FY 1970 NOA of $3.737 billion. NASA expenditures were budgeted to decline $486.4 million, from FY 1970 level of $3.889 billion to $3.403 billion. Reductions placed proposed NASA FY 1971 funding at lowest level since 1962. Of budget request, $2.606 billion would go for R&D, $34.6 million for construction of facilities, and $692.3 million for research and program management.

  Apollo funding decrease of $729.6 million, to $956.50 million in FY 1971, reflecting successful completion of manned lunar landing goal, would be partially offset by increase to $515.2 million for manned space flight operations. The $515.2 million would include $364.3 million for Apollo Applications, $110 million for space shut-

35

tle and station, and $40.9 million for operations. Advanced missions would receive $2.5 million, same as in FY 1970, to bring total for manned space flight to $1.474 billion, down $557.5 million from FY 1970.

Funding for space science and applications programs would increase $45.9 million, from $519.7 million in FY 1970 to $565.7 million. Increases would go to physics and astronomy program, for airborne research ($3 million), solar observatories ($16.1 million), and Explorer satellites ($25.6 million). Funding for Pioneer program of planetary exploration would increase from $20 million in FY 1970 to $32.9 million. Funding for biosatellites would decrease from $6 million in FY 1970 to $1.5 million. Earth resources survey would receive more than 50% increase in funding, from $26 million to $52.5 million, with $41.5 million allocated for Earth Resources Technology Satellite. New projects funded were navigation and traffic control studies (with DOT), $3 million, and Global Atmospheric Research Program (GARP) study, $1 million.

Total for advanced research and technology programs would be reduced from $272 million in FY 1970 to $264 million. NERVA rocket funding would increase from $27 million to $32 million. Increases also were proposed for V/STOL aircraft technology ($15 million in FY 1971, from $9.5 million in FY 1970), subsonic aircraft technology ($11.9 million, from $9.4 million in FY 1970), and supersonic aircraft technology ($21.9 million, from $18.9 million in FY 1970), within total $185.5 million requested for aeronautics.

Sustaining university program was being phased out, with no funds requested for FY 1971.

DOD space funding, down $82 million, reflected cancellation of MOL program. FY 1971 funding would support satellite development, portions of missile development, and range operations.

AEC's requested $99 million reflected slight decreases for development of NERVA rocket and space nuclear electric power sources for space applications.

Funds for Depts. of Interior and Agriculture would support participation in ERTS project. NSF's $2-million space funding was for research rockets and satellite and observation instruments.

FAA 1971 budget reflected restoration of $275 million in Federal funds for development of two prototype SSTs. (NASA Release, 2/2/70; US Budget for FY 1971; BOB Special Analysis)

• NASA Hq. briefing on FY 1971 budget, held Jan. 31, was released. Dr. Thomas O. Paine, NASA Administrator, had said: "We are putting into motion with this budget a program which does not advance toward a single climactic event. It follows the recommendations of the Space Task Group Report's Option II, although at a reduced pace. The budget to implement the recommendations, however, is austere and, as a result, does not take full advantage of our total capability. But the new program has the flexibility to accommodate this reduced budget in FY 1971."

Lowest NASA budget since FY 1962 "means substantial cutbacks and deferrals in a number of programs. Our best current estimate is that the total nationwide employment on NASA work—which was once 420,000 people—will decline from current level by another 45,000, to about 144,000 people by the end of fiscal year 1971. The impact will be felt in many sections of the country. This is regrettable, but the Administration has faced the hard fact that these steps must be taken to achieve a fiscally responsible balanced budget in this time of inflation. . . .this budget level, which we hope will be NASA's lowest level, is strongly oriented to the future. The necessary reductions. . .have not been made 'across the board.' We have made careful selections and decisions."

NASA would preserve strong future capability centered on development of economical, reusable space transportation system including shuttle, space station, and reusable nuclear rocket; continue efforts on practical applications of space technology for early practical benefits on earth; continue lunar exploration with Apollo vehicles and Apollo Applications earth orbital workshop flights with hardware previously under procurement; continue, "but with stretchouts and deferrals," balanced program of planetary exploration and scientific investigations in space focused on potentially exciting discoveries; continue increasing emphasis on aeronautics; and maintain "essential cadre of NASA's technical, operational, and management capabilities on which the continuing and future position of the United States in space depends." [See Jan. 13.]

Assistant Administrator for Administration William E. Lilly during briefing had replied to question that planned fund request for space station and shuttle had been for $250 million to $260 million—supporting Task Group's Option II, to provide operational systems in 1977—but request had been cut to final $110 million. Associate Administrator for Manned Space Flight Dale D. Myers had said program should give CY 1978 operational date. Dates for Mars mission were not yet planned.

Michoud Assembly Facility and Mississippi Test Facility were among NASA installations to be affected by budgetary action, Lilly had said. Plans on Michoud would be to "complete the last of 15 Saturn 5's in 1970; then we will be carrying on activities for the storage, repairing and refurbishing of the 1-B and the Saturn 5 stages, providing the engineering and support necessary for the flights through 1974, and maintaining the capability to restart or reinitiate the production of Saturn 5's; maintaining the capability to accept new work. . .and. . .attempting to find other users of that facility in New Orleans." At MTF, Saturn V static test firings would be completed. "Then, the plans in our budget are to mothball the facilities. Personnel will be phased down. . .to essentially a mothball or caretaker mode." NASA was working with other agencies to seek other uses for that location. (Transcript)

- At Washington Heart Assn. meeting in Washington, D.C., Dr. Charles A. Berry, Director of Medical Research and Operations at MSC, received Spacemedic Award "for outstanding contributions to the study of hearts in space."

  He also reported evidence from *Apollo 12* mission that prolonged space flight might intensify action of some bacteria normally found on human skin. Astronauts Alan L. Bean and Charles Conrad, Jr., had sustained temporary post-flight infections from staphylococcus bacteria carried from earth, which had increased in virulence during flight. (AP, *H Post*, 2/3/70; *W Post*, 2/3/70)

- ALSEP 1 had operated without interruption for over 1948 hrs since being deployed on lunar surface Nov. 19, 1969, by *Apollo 12* crew, MSC announced. To date ALSEP 1 had received and implemented 2034 commands. (MSC Release 70-19)

- Between June 30, 1970, and June 30, 1971, there would be reduction of from 1000 to 2000 in number of aerospace contractor personnel at MSC, NASA Deputy Administrator George M. Low said during meetings with press and staff at MSC. Civil service employees would be reduced by 49. (Maloney, *H Post*, 2/3/70)

February 3: NASA successfully launched 1500-kg (3300-lb) *Sert II* (Space Electric Rocket Test) from WTR at 6:49 pm PST by Thorad-Agena booster. Spacecraft entered orbit with 1008.9-km (626.9-mi) apogee, 998.4-km (620.4-mi) perigee, 105.2-min period, and 99.1° inclination. Deployment of solar panels, disabling and dumping of Agena attitude control system, and transfer of Agena horizon sensor to spacecraft were accomplished satisfactorily. Spacecraft attitude was being maintained by gravity gradient forces and control moment gyros to well within required limits.

Primary mission objective was to operate electric ion thruster system in space satisfactorily for six months. Secondary objectives were to determine variation of thruster power efficiency with mission time; measure RF noise from ion beam in frequency bands associated with interplanetary communications systems; measure extent of coupling between ion beam and space plasma; measure magnitude of ion engine's thrust by electrostatically suspended accelerometer, orbit change, or thruster measurements; measure long-term effects of ion thruster efflux on silicon-cell solar array; and measure change in equilibrium temperature of optically reflective coating in space environment.

Sert II system had large solar array, cylindrical spacecraft support unit (SSU), and spacecraft ring housing two ion engines, electric propulsion experiments, power conditioning, power switching and signaling equipment, and reacquisition control system. SSU housed two telemetry transmitters, two command systems, two tape recorders for data storage, power distribution system, battery for reacquisition control, and four control moment gyros.

Sert II was second mission in SERT project. SERT I, launched on suborbital mission July 20, 1964, had proved that ion engines could produce thrust in space. SERT project was managed by LeRC under OART direction. Tracking and data acquisition were conducted by GSFC. (NASA Proj Off; NASA Release 70-8; LeRC Release 70-2)

- NASA launched three sounding rockets. Two Nike-Cajuns launched from Point Barrow, Alaska, carried GSFC grenade experiment. Nike-Tomahawk launched from Andoeya, Norway, carried Norwegian payload to conduct auroral studies. (SR list)
- NASA announced it had granted Alaska permission to use *Ats I* (launched Dec. 7, 1966) to transmit educational radio programs to remote areas in Alaska and instructional and public TV from Fairbanks to Kodiak, Nome, and Fort Yukon, Alaska. Experiment—intended to help Alaska develop operational experience with which to plan follow-on operational satellite system—would operate seven hours per day through spring 1971. Radio transmissions would begin in March 1970 and TV in October 1970. ComSatCorp would provide at least three ground stations and RCA Global Communications Inc.

would provide one. Alaska would finance operation of experiment, provide programming, and evaluate results, reimbursing NASA for any costs over those for normal experimental operations.

Potential users of operational satellite systems willing to invest in necessary facilities, provide programming, and cover ground costs were being permitted to present proposals for use of *Ats I* and *Ats III* (launched Nov. 6, 1967) for worthwhile experiments as long as satellites continued to operate. (NASA Release 70-16)

- President Nixon transmitted to Congress *Aeronautics and Space Report of the President* for 1969 (dated January 1970). He said: "The year 1969 was truly a turning point in the story of space exploration —the most significant of any year in that still brief history." U.S. had achieved "most prominent of our goals in space—one which had long been a focus for our energies. As we enter a new decade, we must now set new goals which make sense for the Seventies. The space budget that I am submitting to Congress reflects my view of a balanced space program, one which will build on the progress we have already made. Our space and aeronautics program has benefited this Nation in many ways. It has contributed to our national security, to our educational, transportation, and commercial strength, to our scientific and medical knowledge, to our international position and to our sense of the dignity and the capacity of man. And the story is only beginning. We have made long strides into the future during the past year; now we must build on those accomplishments in the coming years and decades." (*Pres Rpt* 70[69]; *PD*,2/9/70; 146)

- Subcommittee on Space Science and Applications submitted to House Committee on Science and Astronautics *Assessment of Space Communications Technology:* NASA program of R&D in comsats and associated equipment during past decade had been "remarkably successful." ComSatCorp also had made "significant strides." As result, U.S. was "undisputed leader in this new technology." State of the art was sufficiently advanced to "make important contributions to the improvement of the American domestic telecommunications system, particularly in Alaska...largest and most remote State of the Union."

  Committee urged expansion of NASA space applications program. It should be "principle purpose of the U.S. Government...to bring to the American people the best, the most effective, the least expensive communications services. It is the clear responsibility of the Government to develop and implement public policies, and improve the effectiveness of its regulatory functions, so as to promote the improvement of the Nation's telecommunications at a rate which reflects the progress of technology. To the extent that governmental indecision has delayed the efficient integration of innovation and invention, the Government must be considered to have failed in this responsibility." (Text)

- Sen. Clinton P. Anderson (D-N. Mex.) introduced S. 3374, $3.333-billion FY 1971 NASA authorization bill. [See Feb. 2.] (Text)

- NASA budget cutbacks would not substantially affect JPL, Dr. Thomas O. Paine, NASA Administrator, told press following conference with JPL leadership teams in Pasadena, Calif. "Our cutbacks will fall most heavily upon the manned space flight program." Some 16 000 aerospace employees would be cut from Governmental payroll throughout California, Dr. Paine said. (Pottage, Glendale *Ledger*, 2/5/70)

- In paper at AIAA Launch Operations Meeting at Cocoa Beach, Fla., Dr. Adolf H. Knothe, Chief of KSC Range Safety Staff, said teletype received July 17, 1969, from German ship *Vegesack* reported "numerous pieces of Stage 1 [S-IC stage of Saturn V] of Apollo 11 were sighted dropping into sea around the vessel." Ship had been 604 km (375 mi) east northeast of Cape Kennedy when *Apollo 11* had lifted off toward moon July 16, 1969. (*W News*, 2/3/70; AIAA PIO)
- Photos of central Alabama taken by *Apollo 9* astronauts had been of major significance for geological exploration in the state, AP reported. Alabama State Geologist Philip LaMoreaux had said photos were "new aid and approach in locating stable areas for construction, exploration for minerals and studying geologic structures as an aid in oil and gas exploration." (*Huntsville Times*, 2/3/70)
- President Nixon submitted to Senate nomination of Robert H. Cannon, Jr., as Assistant Secretary for Systems Development and Technology. He would succeed Secor D. Browne, appointed to CAB Oct. 6, 1969. (*PD*, 2/9/70, 125, 147; DOT PIO)
- Japan became 94th nation to sign nuclear nonproliferation treaty. (Roberts, *W Post*, 2/4/70, A14)

February 4: Two Nike-Tomahawk sounding rockets, launched by NASA from Churchill Research Range, carried Univ. of Alaska payloads to 310-km (192.6-mi) and 340-km (211.3-mi) altitudes to obtain data on spatial variation of auroral light emissions and on relationship between their intensities and volume emission rates. Rockets and instruments functioned satisfactorily; good data were obtained. (NASA Rpts SRL)

- Dr. Thomas O. Paine, NASA Administrator, told Cleveland, Ohio, press conference LeRC would receive $22-million increase for FY 1971. "We expect Lewis to play a very significant role in the space station-space base evolution. The challenge in electric power alone is one that will keep Lewis people going indefinitely." Dr. Paine was visiting NASA centers to explain effects of FY 1971 budget cuts. (*Lewis News*, 2/13/70, 1)
- Astronaut James A. Lovell, Jr., told press at KSC that Apollo 13 mission would be his last space flight. "There are a lot of people standing in line for flights. I've had more than my share. So I'll step aside after this one." Apollo 13 would be Lovell's fourth space flight. (AP, *W Star*, 2/5/70, A2)
- MSC Director of Flight Crew Operations Donald K. Slayton, on behalf of himself and MSC Flight Crew Operations staff, received 1969 IEEE Reliability Award in Los Angeles for exceptional contributions to field of reliability. (NASA Hq Reliability & Quality Assurance *Newsletter*, 3/70, 2)
- ARC biochemists were developing process to convert principal body wastes—carbon dioxide and water—into palatable food for long-duration space missions, ARC announced. Methods to turn carbon dioxide and water vapor exhaled and water recovered from urine into sugars and glycerol in series of chemical reactions might also be applied to solve food problems on earth. Work was under way by ARC, Worcester Polytechnic Institute, and two industrial contractors to develop system to produce chemical foods under space condi-

tions, including weightlessness, and reduce food deliveries to orbiting space station. (ARC Release 70-3)
- *Apollo 12* commander Charles Conrad, Jr., accepted Silver Quill Award of American Business Press on NASA's behalf at dinner in Wahington, D.C. Award was given for "greatest exploration expedition in the history of mankind," ABP Board Chairman Donald V. Buttenheim said. It also recognized "tremendous feat of organization of human and material resources" and translation of discoveries and development of manned space flight into "down to earth benefits of the new technology for the good of industry, America, and all mankind." (ABP Release, 1/1/70; ABP Off)
- MSC announced award of $3 270 897 one-year extension to contract with Zia Co. for maintenance and operations support at White Sands Test Facility. Extension brought total value of cost-plus-award-fee contract to $17.9-million since November 1966. (MSC Release 70-20)
- Rep. George P. Miller (D-Calif.) introduced H.R. 15695, $3.333-billion FY 1971 NASA authorization bill. (Text)

February 4-6: AIAA held Advanced Space Transportation Meeting at Patrick AFB and Cocoa Beach, Fla. In keynote address Charles W. Mathews, NASA Deputy Associate Administrator for Manned Space Flight, discussed capabilities of NASA reusable space shuttle. "In the area of multi-mission and multi-agency use, we are talking and thinking in the broadest sense. Certainly, the Department of Defense, in addition to NASA, has many requirements to carry payloads into space. Other agencies, the industrial community, and the academic community also have been involved in like fashion but have been...inhibited by the limited capabilities and costs of present launch systems and payloads. However, in addition to a higher level of participation within our own country, a much greater involvement is expected internationally because of the flexibility and utility of the space shuttle. Other nations might utilize a United States shuttle to carry and deploy their payloads or carry their personnel to a space station, but ultimately, nations or a consortium of nations may desire to operate their own shuttles just as foreign airlines operate U.S. developed commercial aircraft."

Design baselines at start of Phase B studies of shuttle included 1.6 million kg (3.5 million lbs) at liftoff and payload compartment volume of 4.6-m (15-ft) dia by 18.3 m (60 ft) long.

Pan American World Airways President Najeeb E. Halaby said NASA reusable space shuttle was logical step toward ship for commercial travel to orbit and back and for point-to-point earth transportation. (Mathews Text; UPI, W *Star*, 2/6/70, A10)

February 5: NASA launched Aerobee 170—new 12.5-m (41-ft)-tall sounding rocket configuration consisting of liquid-fuel Aerobee 150 sustainer and solid-fuel Nike booster—from Wallops Station at 10:51 am EST, carrying 95.7-kg (211-lb) instrumented payload. Primary objectives were to flight-test first Aerobee 170 production vehicle, flight-test new water recovery system and conduct water recovery operation like one to be conducted during solar eclipse March 7, and flight-qualify new telemetry instrumentation. Rocket and instruments functioned satisfactorily; payload and Aerobee sustainer were recovered. (WS Release 70-2)

February 5

- *Apollo 12* Astronauts Charles Conrad, Jr., Richard F. Gordon, Jr., and Alan L. Bean were presented Distinguished Service Medals by Adm. Thomas H. Moorer, Chief of Naval Operations, in Pentagon ceremony. Conrad also received Navy Command Insignia for his role as *Apollo 12* commander. Bean, for whom *Apollo 12* was first space flight, received USN astronaut wings. (DOD Release 97-70)
- Rep. James G. Fulton (R-Pa.) introduced H.R. 15747, $3.333-billion FY 1971 NASA authorization bill. (Text)
- "American and British space experts believe that the Soviet Union has built and successfully tested a satellite capable of intercepting and destroying other orbiting spacecraft," Richard D. Lyons reported in *New York Times*. Tracking data and intelligence reports suggested that U.S.S.R.'s *Cosmos CCXLVIII* (launched Oct. 19, 1968) had homed in on *Cosmos CCXLIX* (launched Oct. 20, 1968) and *Cosmos CCLII* (launched Nov. 1, 1968) and destroyed them. Air Force Aerospace Defense Command radar had monitored launches and orbits and later detected about 25 pieces of the two spacecraft, indicating explosions had occurred while they were near *Cosmos CCXLVIII*.

  USAF plans for orbital attack system code-named Saint (satellite inspection and interception satellite) prepared 11 yrs ago had been shelved, but were apparently being revived under new code name, Lyons said. (*NYT*, 2/6/70, 1)
- MSFC announced award of separate $150 000 three-month contracts to three firms to study possible conversion of 12 existing rocket engine test stands for captive firings of propulsion system for new space shuttle engine. New engine, not yet under development, would be used in clusters to power reusable space shuttle which could be flown up to 100 times. Four test positions at MSFC would be studied by Aerojet-General Corp.; four at NASA Rocket Engine Test Site at Edwards, Calif., by NAR Rocketdyne Div.; and four at MTF, by United Aircraft Corp.'s Pratt & Whitney Div. (MSFC Release 70-27)
- NASA announced delay in selection of contractor to develop, fabricate, test, and operate ATS-F and ATS-G Applications Technology Satellites because of "recent decisions in connection with the NASA Fiscal Year 1971 budget." Fairchild Hiller Corp. and General Electric Co., final competitors, had been asked to revise their proposals. (NASA Release 70-18)
- Members of President's Council on Environmental Quality indicated reservations about proceeding with SST development during meeting with press in Washington, D.C. Chairman Russell E. Train said, "The environmental problems posed by the SST are exceedingly serious and have not been solved yet." Dr. Gordon J. F. MacDonald said he shared Train's views and mentioned large quantities of water vapor introduced into atmosphere by SST in flight. Robert Cahn said that, while SST would fly at subsonic speeds over populated areas, "we don't know the effects [of supersonic speeds] on wildlife in nonpopulated areas." (Kenworth, *NYT*, 2/6/70, 73)

February 6: At NASA Hq. industry briefing on space program for 1970s and beyond, Dr. Thomas O. Paine, NASA Administrator, said: "I think that when we talk about reducing the overall aerospace program by 12 percent, we have to look at the other side of the point too...we will still be spending $3.3 billion in the 71 budget, and I think by any possible measure that puts the space program very, very high in the nation's priorities." Major reduction would be in manned spaceflight "from a little over $2 billion in FY 70 to $1,474,000,000 in 71." With "the manned lunar landing behind us, with almost all the initial buy of 15 vehicles now completed, we are able to take a major reduction in the Apollo account as that program reaches a logical termination. We had hoped to continue the Saturn V at limited production of several a year. The decision was made under these austere conditions that we would terminate that production and take the savings."

Peak NASA expenditure of $6 billion in 1965 included outlay for buildup of total NASA base. Fact "that we are now down between the three and four billion mark by no means means that we will be operating at a substantially lower space activity than we had at the peak. It reflects more the fact that the base is now completed; we can put a much higher percentage of our expenditures into program."

However, FY 1971 budget would delay Apollo Applications flight until end of 1972; stretch Apollo lunar flights to two a year, with Apollo 16 occurring before AA program launch in 1972-1973 and Apollo 17 and 18 after AA program missions; delay ATS-F and ATS-G by one year, and Viking-Mars lander from 1973 to 1975; and reduce NASA operational base by shutting down ERC and possibly mothballing Mississippi Test Facility. Dr. Paine said NASA aeronautics effort would increase. "We feel that much of the work that will be going forward in the shuttle program will probably have some spin-off into the aeronautics area." Emphasis would be placed on V/STOL and quiet engines.

Orbiting Solar Observatory launch recommended by Space Task Group for 1975 would be delayed to 1976. Interplanetary Monitoring Probes originally scheduled for 1971 and 1972 would be launched in 1972 and 1973 and ISIS would be moved from 1970 and 1972 to 1971 and 1973. Synchronous Meteorological Satellites each would slip one year, to 1972 and 1973, and Applications Technology Satellites, one year, to 1973 and 1975. lunar roving vehicle would be carried on Apollo 16, 17, 18, and 19. Decision on whether to fund two or three space shuttle definition studies would be made after proposals were received. In space station area experimental requirements would be emphasized. Two Phase-B definition studies were under way. Summer study of utilization of space stations would be international conference "with, we hope, strong participation from the European space research organization." NERVA studies would continue.

Dale D. Myers, Associate Administrator for Manned Space Flight, told conference that "with the technology we already have in hand" antipollution program could be initiated "to produce a space reflector—maybe it will have to be a mile [1.6 km] in diameter—that can direct the sun's energy to create holes over our cities which would act as chimneys to evacuate smog." (Transcript)

- Value of space program to national defense was discussed by SAMSO Commander, L/G Samuel C. Phillips (USAF), before Executives' Club in Chicago. Program had produced "greatly improved military communications, strategical and tactical; the ability to detect and monitor nuclear detonations anywhere in the world; new capabilities for precise navigation of fast-moving aircraft, ships, submarines or land vehicles; and a variety of capabilities for early warning and other defensive space systems." First defense comsat system had been transmitting "substantial load of voice communications between Washington and Vietnam since 1967. It can also handle high-speed digital data such as tactical reconnaissance photographs. Improved adaptations of this 26-satellite initial system have been developed for the United Kingdom and NATO."

In year's time "we will be launching a still further improved Phase II United States system with greater traffic handling capacity, higher reliability and longer life expectancy." First experimental satellite had also been launched "in development of a new tactical communications capability that will enable troops...to communicate with each other in spite of intervening obstacles like mountain ranges or enemy forces. Another new defense capability was built into our VELA satellites...launched in October of 1963 to monitor the nuclear test ban treaty." Velas had become "oldest man-made functioning systems in space; they still respond to command, although their duties have largely been taken over now by more advanced versions." Velas had performed scientific chores like "monitoring the level of solar radiation during manned space flights, to ensure against danger to the astronauts."

USAF navigation satellites under study would permit pinpoint accuracy in navigation of supersonic aircraft and surface and subsurface vehicles. They held promise for prevention of midair collisions. "We're also doing extensive studies on a space shuttle system that would be...a reusable space booster capable of supporting both military and civilian space operations." It could put satellites into orbit, service satellites operating in space, supply manned space station, or assist in rescue of astronauts in space emergencies. (Text)

- Saturn V 1st stage (S-IC), scheduled to become part of SA-513 launch vehicle, was successfully static-fired at MTF. (MSFC Release 70-7; MSFC PIO)
- NASA awarded Service Technology Corp. one-year $3.5-million cost-plus-award-fee contract to provide technical services for development and verification testing of space propulsion systems and related subsystems for follow-on MSC missions. (MSC Release 70-22)
- R&D aspects of Nixon Administration's FY 1971 budget were discussed in *Science*. Total Federal obligations in support of R&D would continue to decline—from $16.4 billion in FY 1970 to $15.8 billion in FY 1971. Obligations for basic and applied research would increase by $260 million, from $5.54 billion to $5.80 billion. Obligations for R&D facilities would drop from $727 million to $585 million, "the decrease chiefly reflecting a big drop in atomic energy spending." (Boffey, Carter, Hamilton, *Science*, 2/6/70, 845-8)
- Sen. Winston L. Prouty (R-Vt.) introduced S. 3412, authorizing appropriations of $498 million for National Science Foundation during FY 1971. (*CR*, 2/6/70, S1343)
- Sen. Howard H. Baker, Jr. (R-Tenn.), introduced for himself and Sen. Edmund S. Muskie (D-Me.) S. 3410, "to establish a structure that will provide integrated knowledge and understanding of the ecological, social and technological problems associated with air pollution,

water pollution, solid waste disposal, general pollution and degradation of the environment." (CR, 2/6/70, S1341-3)

February 7: *Intelsat-III F-6*, launched by NASA for ComSatCorp Jan. 14, began full-time commercial service with 955 circuits carrying telephone and telegraph messages between 17 earth stations in Atlantic area. (ComSatCorp PIO; ComSatCorp Release 70-6)

• Imaging photopolarimeter for Jupiter probe was being developed by Santa Barbara Research Center for ARC, *Armed Forces Journal* reported. Instrument for Pioneer F and G spacecraft to be launched in 1972 and 1973 would map density and distribution of "asteroidal debris," measure gas above Jupiter's cloud layers, and transmit two-color spin-scan images of planet. (*AFJ*, 2/7/70, 2)

• Aerobee 170 sounding rocket was launched by NASA from WSMR carrying American Science and Engineering, Inc., experiment to study x-ray spectra. Mission did not meet minimum scientific requirements. (SR list)

February 8: NSF released results of survey which showed growth rate of R&D expenditures by universities and colleges had slowed with leveling off of Federal obligations for R&D. Trend had begun in 1960s. R&D expenditures by institutions of higher learning had grown at annual rate of 17.0% during academic years 1958-1966 and at 11.6%, 1966-1968. Federal Government financed 61% of total 1966-1968.

Analysis of President Nixon's FY 1971 budget had shown Federal support of R&D at universities and colleges decreased by 2% between FY 1968 and 1969 and increased by 2% between FY 1969 and 1970. FY 1971 budget requested increase of 2% between 1970 and 1971. (NSF Release 70-108)

• George J. Vecchietti, NASA Director of Procurement, assumed additional duties as Deputy Assistant Administrator for Industry Affairs. (NASA Release 70-38)

• Boom "was busted" at Cape Kennedy, *New York Times* article said. Engineers and technicians "who helped put American astronauts on the moon" had moved away by thousands, leaving "a sagging economy and perhaps the best house buys in the United States." Employment in Cape Kennedy area had dropped from 26 000 to 18 900 and was still falling. Business in Brevard County had dropped 15%; total bank deposits were down by $1.7 million from $241 million on Dec. 31, 1968; and unemployment rate had risen to almost 4%. Motel business was off estimated $400 000 per month, with half of county's 1500 units vacant most of time. In December 1969 county had become eligible for Federal food commodities and 1327 families had qualified. Real estate was only business booming, with local residents buying homes sold by engineers at low prices and interest rates. (Waldron, *NYT*, 2/8/70, 80)

February 9: Dr. Thomas O. Paine, NASA Administrator, received British Interplanetary Society awards commemorating man's first lunar landing at special meeting of Society in London. Awards were silver replica of *Apollo 11* LM presented to NASA and gold medals for *Apollo 11* Astronauts Neil A. Armstrong, Michael Collins, and Edwin E. Aldrin, Jr. (*Spaceflight*, 5/70, 191)

• President Nixon transmitted to Congress Reorganization Plan 1 of 1970, "to establish an Office of Telecommunications Policy in the Executive Office of the President." Unit would be headed by Director and Deputy Director appointed by President. Existing office of Director of Telecommunications Management would be abolished. New office would serve as President's principal adviser on telecommunications policy, would help formulate policies and coordinate operations for Federal Government's communications systems and carry out responsibilities of President under Communications Satellite Act, and would "enable the executive branch to speak with a clearer voice and to act as a more effective partner in discussions of communications policy with both the Congress and the Federal Communications Commission." (PD, 2/16/70, 156-8)

- Dr. Charles A. Berry, MSC Director of Medical Research and Operations, described medical phenomena encountered by *Apollo 12* astronauts in keynote address at Air Force School of Aerospace Medicine seminar in San Antonio, Tex. Astronauts Charles Conrad, Jr., Richard F. Gordon, Jr., and Alan L. Bean had observed brilliant flashes of light on closing eyes in darkened portions of spacecraft. They were believed to be caused by charged particles or rays from sun or from space. Astronauts also experienced sensitivity to lunar dust, body changes including face fullness, reddened eyes, and flushed faces during early part of flight, and increase in skin bacteria beyond amount normal on earth. Dr. Berry blamed body changes on pooling of fluids in head from lack of gravity. Behavior of bacteria might make doctors consider altering normal skin surface bacteria content by special baths for astronauts before flight.

   Dr. Berry also said that woman might share in planetary exploration but tests were needed to ensure "she doesn't create more problems than she solves. I think it would definitely be more comfortable if we took women along." (*Huntsville Times*, 2/10/70; *Huntsville News*, 2/10/70)

- SAMSO had successfully tested thrust vector control system (TVC), improved steering mechanism for Titan IIIC booster, AFSC announced. TVC was lighter and less complicated than present system and would permit orbiting of heavier payloads by Titan IIIC. Mechanism had ring of 24 electrical valves that injected liquid nitrogen tetroxide from pressurized tank into flaming exhaust to deflect hot gases and steer booster. Operational system already in use had hydraulic valves with spaghetti-like collection of hydraulic lines, heavy pumps, and separate tanks for liquid and pressurizing gas. (AFSC Release 17.70)

- Hundredth anniversary of U.S. Weather Bureau. (Latham, *W Post*, 2/10/69, 20)

- *New Haven Register* editorial commented on intention of *Apollo 8* Astronaut James A. Lovell, Jr., to end active participation in space after commanding Apollo 13 mission: "The space agency can ill afford to let go of Jim Lovell, whose 24 days in space give him a world's record. When he concludes the Apollo 13 mission he will be, without doubt, the best qualified man in the agency. A top administrative post should be found for him." (*New Haven Register*, 2/9/70)

February 10: U.S.S.R. launched *Cosmos CCCXXIII* into orbit with 314-km (195.1-mi) apogee, 200-km (124.3-mi) perigee, 89.6-min period, and 65.3° inclination. Satellite reentered Feb. 18. (GSFC *SSR*, 2/15/70; 2/28/70)

February 10

- NASA Hq. briefing explained to university administrators effects of NASA budget cuts on funds for educating space scientists. Termination of sustaining university program in FY 1971 would eliminate more than 200 predoctoral training grants which had been funded at $4.18 million. Funding for multidisciplinary research grants also would end. Students would not be cut off immediately, since grants had been step-funded and would be reduced over three years. NASA Assistant Administrator for University Affairs F. B. Smith said other agencies might assume funding of some grants, since university R&D funding in total Federal budget would increase by $114 million during FY 1971. Presidential Science Adviser, Dr. Lee A. DuBridge, said increase would not compensate for leveling off of R&D budgets in past few years nor for effects of inflation. He noted Government support for graduate students was declining generally and said Government did not need as many scientists as previously.

  Total $21 million would be cut by NASA in contracts and grants to universities. In OSSA, university R&D grants in bioscience would be cut by 30%. Cuts would eliminate a high-energy astronomical observatory. OART would experience 30% cut in university R&D, chiefly affecting electronics and space vehicles research. OMSF contracts with MIT for Apollo program would be reduced by $2 million. University grants and contracts which survived would be reoriented in FY 1971, with emphasis on what universities could do for NASA. (NASA Off of Univ Aff.; Gruchow, Science, 2/20/70, 1107)
- President Nixon recommended to Congress 37-point administrative and legislative program on environmental quality. It included five-year, $10-billion Clean Waters Act; more rigorous standards to limit pollution by automobiles; establishment of nationwide air quality standards and power to enforce compliance; extension of Solid Waste Disposal Act; use of Federal land near metropolitan areas for recreational purposes; and enlistment of volunteer aid in restoring healthy environment. (PD, 2/16/70, 158-73)
- U.S. Patent Office granted patent No. 3 495 260 to GSFC engineers Charles R. Laughlin and Roger C. Hollenbaugh for North Atlantic air traffic control system using satellite. System, in which aircraft would transmit positions to ground stations and data would be relayed to all aircraft via satellite, would be tested with ATS in 1972 or 1973. Position of SSTs flying at three times speed of subsonic aircraft would be calculated by computers and supplied to traffic controllers. (NYT, 2/14/70)
- NASA issued RFPs to 14 firms for engineering study on space environment in which average person could work, sleep, eat, and relax comfortably and efficiently for long periods. Objective of study was to prepare handbooks to be used for basic criteria for design and to set standards. (MSC Release 70-23)
- Inter-Academy Exchange program of NAS and Soviet Academy of Sciences, renewed and broadened for 1970 and 1971, was signed. Agreement, negotiated in Washington, D.C., was latest in series of two-year agreements begun in 1959. It was retroactive to Jan. 1, 1970. Pact provided for 180 man-months of study in U.S. by scientists designated by Soviet Academy and for 180 man-months of research in U.S.S.R. by scientists selected by NAS. It also contained

February 11: *Japan launched its first satellite, Ohsumi, becoming the fourth nation to orbit a spacecraft on its own booster. The satellite was photographed during preparation for launch from Kagoshima Space Center, Uchinora, in the Ohsumi district.*

provision for joint research. (NAS-NRC-NAE News Rpt, 3/70, 1; NAS PIO)

- NASA launched two sounding rockets from Churchill Research Range. Nike-Tomahawk carried Univ. of Alaska payload to examine spatial distribution of ionospheric currents near visual auroral forms, examine relationship between electron and proton precipitation and usual auroral distributions produced by precipitation, and track flashing light on payload with image orthicon TV system while simultaneously observing aurora. Payload penetrated intense auroral arc forms near northern edge of auroral breakup. All experiments operated successfully and TV system tracked rocket through aurora.

    Nike-Apache carried Univ. of Texas at Dallas experiment to study ionosphere. Mission did not meet minimum scientific requirements. (NASA Rpt SRL ; SR list)

February 11: Japan successfully launched 38-kg (84-lb) *Ohsumi* (Lambda 4S-5) satellite carrying 10.9-kg (24-lb) instrumented payload from Kagoshima Space Center at Uchinora at 1:25 pm local time, becoming fourth nation to orbit earth satellite on own launch vehicle. Orbital parameters: apogee, 5136 km (3191.4 mi); perigee, 525 km (326.2 mi); period, 116.1 min; and inclination, 31.4°. Total cost of satellite and 16.5-m (54-ft) four-stage Lambda 4S booster was 118 million yen. Onboard equipment included 4th-stage spherical motor 48 cm (18.9 in) in diameter with small radio transmitter, battery, thermometer, and accelerometer in nosecone and four antenna spikes projecting from cone. Launch was fifth attempt by Japanese Institute of Space and Aeronautical Science to orbit satellite. Spacecraft was named for district from which it was launched. (GSFC SSR, 2/15/70; SF, 5/70, 189; NASA Int Aff)

- USAF launched unidentified satellite from Vandenberg AFB by Thor-Burner II booster. Satellite entered orbit with 872.3-km (542-mi) apogee, 772.5-km (480-mi) perigee, 101.3-min period, and 98.6° inclination. (Pres Rpt 71; GSFC SSR, 2/15/70)
- *Apollo 11* spacecraft, first manned vehicle to land on moon, would start year-long tour of state capitals April 17, NASA announced. (AP, P Bull, 2/11/70)
- NASA announced award of $38-million, cost-plus-incentive-award-fee contract with multiple incentives to TRW Systems Group to design, develop, fabricate, assemble, test and deliver Pioneer F and Pioneer G. Spacecraft would be launched toward Jupiter in 1972 and 1973 to collect data on interplanetary medium beyond Mars, observe asteroid belt, explore environment and atmosphere of Jupiter, and develop technology for long-duration flights to outer planets. Contract was follow-on to previous TRW contract for *Pioneer VI, VII, VIII,* and *IX.* (NASA Release 70-20)
- President Nixon established Advisory Council on Management Improvement to recommend improved methods and procedures to sharpen efficiency of Government operations. He appointed Gen. Bernard A. Schriever (USAF, Ret.) Chairman. (PD, 2/16/70, 176-7)

February 11-14: Symposium, "A Century of Weather Progress," was held in Washington, D.C., to honor 100th anniversary of U.S. weather services and 50th anniversary of American Meteorological Society. Dr. Frederick Sargent II of College of Environmental Science in Washington said current method of "ecological adaptation" was not appropriate strategy for managing quality of environment. He proposed environmental planning and control motivated by "new humanism" that recognized all men as sharers of environment.

Dr. Walter Orr Roberts, President of University Corp. for Atmospheric Research in Colorado, said man and his technology were changing weather and environment. They were creating megalopolises with effect on atmospheric and surface radiation, permitting heat dumping by air conditioners, creating worldwide buildup of carbon dioxide from increased use of oil and coal, fostering increased storm activity from industrial smoke particles in air, and producing sun-reflecting cloud layers by jet aircraft use. (*NYT*, 2/15/70, L36)

February 12: ESRO's *Boreas* satellite (launched by NASA Oct. 1, 1969) was adjudged successful by NASA. Satellite had entered lower than planned orbit and had reentered Nov. 23, 1969, but all experiments

functioned satisfactorily. ESRO had obtained large quantity of scientific data and considered mission a success. (NASA Proj Off)
- Japanese scientists reported loss of signals from *Ohsumi* satellite (launched Feb. 11) during eighth orbit. Scientists discontinued tracking on ninth orbit; they said mission had provided valuable data for development of Japan's Mu rocket, scheduled for launch in autumn of 1970. (*SBD*, 2/13/79, 201)
- House Committee on Science and Astronautics' Subcommittee on NASA Oversight released *Manned Space Flight: Present and Future.* Staff study concluded that growth of national space program in decade had "provided major national technological and scientific resource of personnel and facilities." Decline in resource could increase risk in accomplishing current programs and was causing "dislocations in affected communities throughout the United States."

  With termination of Apollo fabrication, 8 Saturn Vs and 6 Saturn IBs, 12 CMs and SMs, and 1 or 2 LMs would be available for manned flight. After 1974, U.S. would have no capability for lifting manned payloads over 27 200 kg (60 000 lbs) into space and would have only three vehicles that could lift 27 200-kg payloads unless Saturn V production was resumed or substitute was developed. Launch rate of less than two manned vehicles per year could materially increase risk of manned launch and spaceflight operations; production rate of two or more Apollo spacecraft and launch vehicles per year would minimize annual cost without inordinate increases in total vehicle cost over current costs, while maintaining critical skills of major subcontractors.

  U.S. would have no long-duration manned earth-orbital capability after 1973 without extension of orbital workshop program or initiation of development of space station. Early development of low-cost earth-to-orbit transportation system appeared to offer greatest opportunity for reducing space flight costs while increasing flexibility and variety of missions in near-earth orbit. For increased flexibility in low-cost earth-to-orbit recoverable transportation system, development of chemical space tug was advised. Report noted that NASA maintained the only U.S. manned space flight capability and advised major decisions on manned space flight program to prolong capability beyond 1974. (Text)
- USAF Chief of Staff, Gen. John D. Ryan, approved flights of seven heavily instrumented F-111s needed in R&D programs. Remaining 226 F-111s continued to be grounded pending final determination of test and inspection procedures to ensure against repetition of manufacturing defect that led to fatal Dec. 22, 1969, crash. (AP, *NYT*, 2/12/70, C9)
- *Christian Science Monitor* editorial said NASA had postponed manned Mars mission because of budget cutback but "it is keeping its space shuttle program pretty much on schedule." Reusable space vehicle was "guarantee that space progress won't be wholly shunted aside by budget stringencies." It would "help to keep alive the interest of the big rocket and airframe contractors, and the enthusiasm of the thousands of researchers and scientists who have hoped to make space exploration their life work." (*CSM*, 2/12/70)

February 13: MSC announced establishment of Space Shuttle Program

Office and appointment of Robert F. Thompson as Manager. Kenneth S. Kleinknecht, Manager for Command and Service Modules, would succeed Thompson as Manager of Apollo Applications Program Office. Clifford E. Charlesworth, a flight director, was appointed to new position of Apollo Applications Program Deputy Manager. (MSC Release 70-25)

- Distribution of 13 kg (28.6 lbs) of *Apollo 12* lunar material to 139 U.S. scientists and 54 scientists from 16 foreign countries had begun at MSC, NASA announced. Domestic analysis would be performed in 139 university, industrial, and government laboratories in 25 states and District of Columbia. Foreign investigators represented Australia, Belgium, Canada, Czechoslovakia, Finland, West Germany, Japan, Korea, Spain, Switzerland, U.K., South Africa, Italy, France, Norway, and India. Preliminary investigation at LRL had revealed *Apollo 12* samples were similar in characteristics to *Apollo 11* samples. Principal investigators would report findings at MSC Lunar Science Conference in January 1971. (NASA Release 70-19)
- Aerobee 170 sounding rocket, launched by NASA from WSMR, carried Columbia Radiation Laboratories payload to 230.9-km (143.5-mi) altitude to study soft x-ray sources and diffuse background. Rocket and instruments functioned satisfactorily. (NASA Rpt SRL)
- White House announced in Key Biscayne, Fla., that President Nixon had asked *Apollo 12* Astronauts Charles Conrad, Jr., Richard F. Gordon, Jr., and Alan L. Bean to visit Latin America, Europe, Africa, and Asia as his personal representatives. Tour, to take 40 days, would begin Feb. 16. Astronauts would be accompanied by wives. (*PD*, 2/16/70, 178-9)
- NASA awarded $5.7-million contract extension to Aerojet-General Corp. for continued R&D on SNAP-8 nuclear power system. (LeRC Hist Monitor; Pasadena *Star-News*, 2/21/70)

February 14: Electron-bombardment ion engine onboard NASA's *Sert II* spacecraft, launched Feb. 3, was advanced to full thrust to begin endurance test. During six-month test NASA would evaluate engine performance and engine's 27-millinewton (six-millipound) thrust would increase spacecraft's apogee by about 97 km (60 mi). (NASA Release 70-34)

- OSSA Science Steering Committee's Planetary Biology Subcommittee, concerned about validity of Lunar Receiving Laboratory tests on lunar samples and unexplained biological effect of lunar soil on some plants, had recommended continuation of quarantine procedures through Apollo 13 mission, Victor Cohn reported in *Washington Post*. Subcommittee had strongly opposed Jan. 15 suggestion by Interagency Committee on Back Contamination that quarantine be abandoned and had prompted NASA to ask NAS for reevaluation by a special committee. (*W Post*, 2/14/70, A5; NASA Exobiology Off)
- Prince Philip of Great Britain inspected Apollo 13 rocket during tour of KSC. (UPI, *W Post*, 2/15/70, A2)
- Scientists at Japanese astronomical laboratory discovered new star of seventh magnitude in constellation of Serpens. Later it was upgraded to fifth magnitude as light brightened. (AP, *W Post*, 2/17/70, A15)

February 15: NASA announced selection of payload and principal investigators for OSO-I (Eye) Orbiting Solar Observatory, scheduled for launch in 1973 to investigate sun's chromosphere-corona interface.

OSO-I experiments would study unexplained temperature change between solar disc and corona over 16 000-km (10 000-mi) altitude and cosmic x-radiation, including both general x-ray background of universe and discrete x-ray sources. (NASA Release 70-17)
- NASA was placing new emphasis on development of profitable manufacturing facilities orbiting in space and operated by private industry, because of budget cuts, *New York Times* article said. Industry participation could lead to operation and ownership of facilities by organizations like ComSatCorp—"prototype of such a joint undertaking." NASA would make available facilities like space shuttle transportation system, space workshops, and station at mutually agreeable price. Some manufacturing might be more economical in space because of high vacuum and zero g, but before program could become practical NASA would have to move into "more highly developed operational stage...more applicable to research and development and to manufacturing." (Tomaszewski, *NYT*, 2/15/70, C1)
- West German government was ready to accept partnership arrangement with U.S. in seven-year space shuttle program, William Hines reported in Washington *Sunday Star*. Nixon Administration had "gone openly shopping" for foreign assistance in project. There was dispute over how far negotiations had gone, "but the West Germans believe they have been offered a 10 percent piece of the action in exchange for about $570 million in...Deutschemarks, payable 300 million marks annually for seven years." NASA had said discussions had been going on with eight foreign countries. (W *Star*, 2/15/70)

February 16: Wild, 227-kg (500-lb) elk on National Elk Refuge near Jackson Hole, Wyo., would be fitted with 10-kg (23-lb) electronic collar to transmit data to orbiting *Nimbus III* (launched April 14, 1969), NASA announced. Experiment—sponsored by NASA, Dept. of Interior, and Smithsonian Institution—was attempt to learn more about migratory habits of large animals. Collar containing small antenna, transmitter, receiver, batteries, and solar cells would transmit data on atmospheric temperature, elk's skin temperature, altitude above sea level, light intensity, and location daily at noon and midnight for six months. Data would be received by interrogation, recording, and location system (IRLS) onboard *Nimbus III* and stored in spacecraft's memory system to be matched with another interrogation two or three minutes later. With time and range of two consecutive interrogations, elk's location could be determined to within one mile.

Electronic collar had been tested on semitame elk at National Bison Range in Montana with no ill effect on the elk. Scientists would attempt to place electronic collar on one of two wild female elk which had been wearing 10-kg dummy collars. Elk used in experiment would be named "Monique." (NASA Release 70-24)
- Vice President Spiro T. Agnew presented National Geographic Society's Hubbard Medal to *Apollo 11* Astronauts Neil A. Armstrong, Edwin E. Aldrin, Jr., and Michael Collins at ceremonies in Washington, D.C. Medal honored exceptional research, exploration, and discovery. (Program; UPI, W *News*, 2/16/70)
- Atheist Mrs. Madalyn Murray O'Hair carried her suit to ban broadcast and telecast of prayers by astronauts in space to U.S. Supreme

Court. She appealed Dec. 2, 1969, decision of U.S. District Court that prayers were individual decisions of astronauts. (AP, *NYT*, 2/17/70, C3)
- INTELSAT international planetary conference convened in Washington, D.C. U.S. Under Secretary of State for Political Affairs U. Alexis Johnson welcomed 100 delegates to meeting, which was to work out definitive arrangements for establishment of single global commercial comsat system with control apportioned among members. (UPI, *W Post*, 2/17/70)
- In message transmitting to Congress 19th annual report of National Science Foundation President Nixon said, "As we go forward into the decade of the 70s, the role of science will surely become more and more important in the search for solutions to our problems and in the effort to enhance our environment." (*CR*, 2/16/70, H812)

February 17: Agreement to ensure that space transportation system (STS) would be of maximum utility to both NASA and DOD was signed in Washington, D.C., by NASA Administrator, Dr. Thomas O. Paine, and Secretary of the Air Force, Dr. Robert C. Seamans, Jr. STS was to provide U.S. with "economical capability for delivering payloads of men, equipment, supplies, and other spacecraft to and from space by reducing operating costs an order of magnitude below those of present systems." Program would include international participation and use. NASA would manage STS development, with project generally unclassified. STS would consist of "earth-to-orbit space shuttle." Agreement established NASA-USAF STS Committee to conduct continuing review of STS program and recommend steps to achieve "system that meets DOD and NASA requirements." Recommendations would include, but not be limited to, "development and operational aspects, technology status and needs, resource considerations, and interagency relationships." (Text)
- Dr. Thomas O. Paine, NASA Administrator, opened testimony before House Committee on Science and Astronautics on NASA FY 1971 authorization: "Apollo 11 and 12 demonstrated that we can fly across a quarter of a million miles [400 000 km] of space, land with precision, carry out research and exploration, and return safely to earth. We met our commitments for the decade of the sixties. Now we must chart a new course for the next decade in space. The Congress again shares with the Administration the responsibility for continuing progress and leadership in space and aeronautics. At the same time, NASA is going through a difficult period of redirection, and austere budget goals must be met. This is therefore a time of great challenge. But I firmly believe that with your support we can continue to forge ahead in space exploration, in science, in applications, and in aerospace technology. The program which we will lay before you, although austere, provides for balanced progress toward the challenging goals for the 1970's and the decades to follow."

NASA budget as it came from President included "major reduction in space program," to achieve balanced U.S. budget in "time of inflation." By reducing operating base, making selective cut-backs, and deferring new starts NASA had developed program permitting progress toward Space Task Group's Option II goals, but at reduced pace. [see Jan. 13 and Feb. 2]. Dr. Paine noted it was significant that, "in a sharply reduced total budget, we have been able to increase our space applications efforts. Our aeronautics effort will also continue to be strengthened."

He urged Committee to keep in mind that proposed program "could not be conducted in the face of further cuts." NASA had no plans to fly astronauts in space between last Apollo flight in 1974 and the time space shuttle would be operative in late 1970s. "Any significant reduction in our FY 1971 request will further extend this gap in American manned space flight, and reduce other important programs." (Testimony)

- *Apollo 12* LM in final moments of descent to lunar surface had caused dust shower that pocked *Surveyor III* and sand-blasted camera, NASA announced. Burton G. Cour-Palais, Chief of Meteoroid Sciences Branch at MSC, reported that *Surveyor III* TV camera experienced little meteoroid damage during 950 days of exposure to space environment, but had numerous shallow craters of recent origin—apparently caused by LM 183 m (600 ft) away. LM-caused craters were "shallow. . .mostly of recent origin, as indicated by their whiteness against the sandy brown color of the TV camera housing painted surface." White craters were concentrated on arc of TV camera facing LM, 10 to 100 times as many as on side away from LM. Preliminary assessment of meteoroid impacts indicated more than five impacts on camera housing and four on polished tube. (NASA Release 70-23)

- ESSA announced it had received NASA contract to investigate lightning hazards to rockets as part of NASA effort to develop lightning-warning system and techniques for lightning suppression. Dr. Heinz W. Kasemir, head of ESSA research project, said that "launching a rocket through thunder or rain clouds can be hazardous, not so much because the rocket will be hit accidently by natural lightning, but because the rocket itself may trigger a lightning stroke in the storm." NASA's *Apollo 12* had suffered temporary power failure from lightning during launch through storm at KSC Nov. 14, 1969.

    Suspected triggering effect of rockets in thunderheads had been demonstrated when specially equipped ESSA DC-6 research aircraft operated by ESSA's Research Flight Facility was struck by lightning three separate times under similar meteorological conditions during thunderstorm research project near Flagstaff, Ariz. Each time lightning struck, aircraft was in dissipating thunderhead near freezing level and in area of cloud that contained both ice and water. Corona discharge had been audibly detected on aircraft communications system for two seconds before each lightning strike. Observations of electrical field within the storm before and after lightning bolts suggested that aircraft itself, moving through thunderhead, might have triggered discharges. (ESSA Release 70-8-30)

- First public TV programming via satellite was inaugurated by transmission of South Carolina ETV Network program from Columbia, S.C., via NASA's *Ats III* to public TV network station in Los Angeles. Program featured John W. Macy, Jr., President of Corp. for Public Broadcasting, addressing special "Satellite Dinner" in Columbia honoring South Carolina General Assembly. CPB would reimburse NASA for costs of experiment. (NASA Release 70-22)

- NASA announced award of $3 300 000 cost-plus-fixed-fee contract to

Boeing Co. Aerospace Systems Div. for engineering support of Apollo Lunar Exploration program. Contract covered engineering evaluation, flight readiness reviews, and mission evaluations Jan. 1 through Dec. 31. (NASA Release 70-25)
- Abolition of 21-day quarantine for astronauts returning from moon was recommended to NAS panel by NASA officials and representatives of Interagency Committee on Back Contamination. Panel Chairman, Dr. Allan H. Brown of Univ. of Pennsylvania, said his group would relay decision to NASA Administrator, Dr. Thomas O. Paine. (UPI, *NYT*, 2/18/70, C17)
- *Apollo 12* astronauts, on 40-day tour, arrived in Lima, Peru, from Caracas, Venezuela. They were greeted by about 5000 persons. (AP, *W Post*, 2/19/70, C7)
- Sen. Barry M. Goldwater (R-Ariz.) urged construction of air and space museum at Smithsonian Institution to house aeronautical and astronautical displays. (*CR*, 2/17/70, S1757)
- President Nixon presented National Science Medals in White House ceremony. Recipients had been announced Dec. 31, 1969; Purdue Univ. chemist Herbert C. Brown, Princeton Univ. mathematician William Feller, Jack S. C. Kilby of Texas Instruments Inc., and Wolfgang K. H. Panofsky, Director of Stanford Univ. Linear Accelerator Center. (*PD*, 2/23/70, 190-1; 1/5/70, 10-1)
- U.S.S.R. was constructing astrophysical observatory atop 5633-m (18 481-ft) Mt. Elbrus in Caucasus mountains to house 599-cm (236-in) altazimuthal telescope, *Los Angeles Times* said. Instrument would be largest optical telescope in the world. Its mirror would be tested in autumn 1971, with observations to begin in autumn 1972. Telescope would be used for extragalactic exploration; study of nebulae and stars, particularly faint stars; and planetary research. (Bengelsdorf, *LA Times*, 2/17/70)

February 18: NASA's HL-10 lifting-body vehicle, piloted by Maj. Peter C. Hoag (USAF), reached 19 800-m (65 000-ft) altitude and mach 1.86 —highest speed to date—during powered flight after air-launch from B-52 aircraft at 14 300-m (47 000-ft) altitude at FRC. Primary objectives were to investigate suspected roll reversal at high speed and to operate vehicle for extended time with stability augmentation system off. Preliminary indications were that no roll reversal existed and that handling qualities were better than anticipated. Pilot said vehicle flew "beautifully" in spite of moderate air turbulence. (NASA Proj Off; LA *Her-Exam*, 2/20/70)
- NASA issued RFPs for preliminary definition and planning studies for main propulsion system of reusable space shuttle. Preliminary concepts called for cluster of throttleable engines with 1780-kilonewton (400 000-lb) thrust each at sea level. NR Rocketdyne Div., United Aircraft Corp. Pratt & Whitney Div., Aerojet-General Corp., TRW Inc., Bell Aerospace Systems, and Marquardt Corp. would submit proposals by March 20. Three firms would be awarded fixed-price contracts for 11-mo parallel studies. (NASA Release 70-26)
- Only medical side effect to astronauts found thus far from long periods spent in space was "some cardiovascular deterioration," Dr. Charles A. Berry, MSC Director of Medical Research and Operations, told Los Angeles press conference. This was "normal response of the body to a less demanding environment." But medical

February 18: NASA's *HL-10* lifting body (above right) reached a lifting-body speed record of mach 1.86 and an altitude of 19 800 meters (65 000 feet) during powered flight after air launch from a B-52 aircraft. Experimental vehicles in the joint NASA-USAF Lifting Body Flight Research Program to develop the design of future space shuttle craft also included the *X-24A* (above left), which made its first supersonic flight Oct. 14, and the *M2-F3* (center), which made its first glide flight June 2.

research was still too flimsy to risk two-year manned Mars flight. (Krimsky, Long Beach [Calif] *Independent*, 2/19/70)
- Javelin sounding rocket, launched by NASA from Churchill Research Range, carried Univ. of California at Berkeley payload to 494-km (307-mi) altitude. Objectives were to observe auroral protons and electrons over energy spectrum from thermal levels to several hundred kilovolts and to study electric fields and their relation to particle fluxes; density, temperature, and motion of ionospheric plasma currents; alpha particle and neutral fluxes in aurora; and auroral radiation. All experiments functioned satisfactorily and useful data were obtained. (NASA Rpt SRL)
- Lend-lease of quarters and equipment aboard U.S. earth-orbiting space station of 1970s to help foreign scientists study terrain of underdeveloped countries was advocated by Dr. Kraft A. Ehricke, Chief Scientific Adviser for Advanced Programs at North American Rockwell Corp. Space Div., before press in Los Angeles. (Miles, *LA Times*, 2/19/70)
- DOD announced USAF award of $3 000 000 supplement to previously awarded contract with General Dynamics Corp. for F-111 aircraft. (DOD Release 134-70)

February 19: U.S.S.R. launched *Molniya I-13* comsat from Plesetsk into orbit with 39 309-km (24 425.4-mi) apogee, 335-km (208.2-mi) perigee, 703.4-min period, and 65.3° inclination. (GSFC *SSR*, 2/28/70; *SF*, 5/70, 189)
- Nike-Apache sounding rocket was launched by NASA from Churchill

Research Range carrying GSFC experiment to conduct auroral studies. Rocket and instruments functioned satisfactorily. (SR list)
- In message read at opening of Warsaw exhibit of *Apollo 11* lunar landing materials, Dr. Thomas O. Paine, NASA Administrator, said, "It is our sincere hope and belief that Poland, with its rich scientific heritage, will play an important role in these [space] ventures." (*C Trib*, 2/20/70)
- Arthur S. Flemming Award was presented by Downtown Jaycees of Washington, D.C., to 10 young men for outstanding and meritorious service in Federal Government, including *Apollo 11* Astronaut Neil A. Armstrong and Eugene F. Kranz, Chief of MSC's Flight Control Div. (*W Post*, 2/16/70, B6)
- Scientists accidently tranquilized wrong elk for *Nimbus III* tracking experiment [see Feb. 16]; since electronic collar fit, they attached it to the 227-kg (500-lb) elk and named her "Monique." Dart containing tranquilizer had missed intended target, one of two elk which had been wearing dummy collars in preparation for experiment. (AP, *KC Star*, 2/20/70)
- President Nixon transmitted to Congress *The Physical Sciences*, second annual report of National Science Board. Board expressed desire to participate in preparation of Government-wide plan to realize excellence in science as national goal. "The United States scientific effort is currently threatened with possible mediocrity. Funding limitations currently imposed by the Federal Government on scientific research should be lifted before the present vitality of the physical sciences, which is essential to the progress of all science, is lost." (*PD*, 2/13/70, 246; Text)
- Chairman Willard F. Rockwell, Jr., was named Chief Executive Officer and Robert Anderson was elected President and Chief Operating Officer of North American Rockwell Corp. by board of directors at annual meeting. (*NAR News*, 2/20/70, 1)

February 19-20: Medical behavioral experiments to be conducted on Apollo Applications missions were described by NASA Associate Administrator for Manned Space Flight Dale D. Myers in authorization hearings before House Committee on Science and Astronautics: Experiments would seek "information required to understand man's capability for long-duration flight and to provide the confidence that man can participate in an optimal manner in these...activities." Four experiments to determine skeletal and muscular alterations during space flight and evaluate biochemical changes and nutritive requirements would measure input and output of calcium, nitrogen, and other biochemical constituents; make x-ray studies of bone demineralization; and assess hormonal and electrolyte constituents of blood and body waste products. Cardiovascular study would use lower body negative pressure—preflight, inflight, and postflight—to test cardiovascular system reflexes regulating regional perfusion pressure in distribution of blood throughout body as man changed posture on earth. Inflight measurements would allow establishment of onset, rate of progression, and severity of adverse changes in protective reflex responses. Investigations of hematology and immunology would be made.

Two neurophysiological experiments would evaluate central nervous system responses. First, human vestibular experiment would extend studies initiated during Gemini program to investigate effects of weightlessness and subgravity states on perception and organization of personal and extrapersonal space and establish integrity of vestibular apparatus during prolonged weightlessness. In second, electroencephalogram (EEG) would determine effects of prolonged space flight on patterns of sleep and wakefulness and assess attention levels in task performance and work-rest cycles. Energy expenditure in space would be measured for first time by comparison of metabolic rate during rest, during calibrated exercise, and while performing operational tasks. (Testimony)

February 20: Dr. Thomas O. Paine, NASA Administrator, began NASA testimony on FY 1971 authorization before Senate Committee on Aeronautical and Space Sciences [See Feb. 17]. Among FY 1969 aerospace accomplishments he noted that "the promising NASA supercritical wing development moved from conceptual and wind tunnel work into preparations for full-scale tests at our Flight Research Center on a modified jet aircraft acquired from the Navy. These flight tests will begin in November 1970. Because of its potential for enhancing both the cruise performance and the operations economics of subsonic jet aircraft, this new NASA concept has generated widespread interest within the aircraft industry." (Testimony)

- ComSatCorp reported net income had increased from $6 841 000 (68 cents per share) for 1968 to $7 129 000 (71 cents per share) for 1969. Net operating income had increased from $988 000 for 1968 to $1 832 000 for 1969. Revenues had increased from $30 495 000 for 1968 to $47 034 000 for 1969. On Dec. 31, 1969, ComSatCorp was leasing full-time to its customers equivalent of 1435 half circuits, 494 more than 941 being leased on Dec. 31, 1968. (ComSatCorp Release 70-8)
- South African surgeon Dr. Christian N. Barnard, heart transplant pioneer, toured MSFC and addressed NASA audience. He asked for closer cooperation between heart transplant research and space program so that studies of effects of weightlessness could be more closely observed. (*Marshall Star*, 2/25/70, 1)
- DOD announced USAF contract awards: $1 600 000 definitive contract to Aerojet-General Corp. for procurement of spare parts for Titan IIIC booster and $1 841 000 cost-plus-incentive-fee contract to Lockheed Aircraft Corp. for R&D on Air Force Satellite Control Facility. (DOD Release 144-70)

February 21: Instruments placed on moon by *Apollo 12* astronauts recorded 306 K (91°F) drop in temperature and 40% decline in light intensity as moon went through partial eclipse. MSC scientists said no other significant readings were recorded. (AP, *W Post*, 2/21/70, A3; *W Post*, 2/22/70, A2)

- Second attempt to locate Monique, instrumented elk, by satellite [see Feb. 16] was unsuccessful. Feb. 19 attempt had failed. NASA attributed failures to interference or failure of collar transmitter. NASA observer would be sent to track Monique on foot. (AP, *NYT*, 2/22/70)
- Article by Dr. Ralph E. Lapp in *New Republic* criticized space transportation system and NASA participation in Feb. 4-6 classified conference at Patrick AFB at which system was discussed. System, al-

though NASA budget item, was "Air Force finesse to pave the way for a space-bomber or interceptor," and "could rival the Safeguard ABM as a waste of money," Lapp declared. (*New Republic*, 2/21/70)

February 22: Scientists located elk Monique, using ground tracking system in animal's collar, but were unable to track her with orbiting *Nimbus III.* Monique had been fitted with electronic collar Feb. 19 in experiment being conducted by NASA, Dept. of Interior, and Smithsonian Institution to obtain data on elk migration. (AP, P *Bull*, 2/23/70)

- Univ. of Maryland astronomer Dr. Joseph Weber had located source of gravity waves that permeated universe in concentrated area near constellation Sagittarius at center of Milky Way, *New York Times* reported. (*NYT*, 2/22/70, F4)

February 23: At annual meeting of American Chemical Society in Houston, Tex., Univ. of Chicago lunar chemist Dr. Anthony Turkevich described lunar highlands as "comparable to the continents of earth, floating like icebergs on a more dense medium of deeper rocks." Only chemical basis of comparison between lunar plains and highlands had been provided by Turkevich-designed experiment flown on Surveyor mission to highland region near crater Tycho. Compared with *Apollo 11* lunar sample data, experiment data proved plains were darker and denser than highlands because they contained twice the abundance of elements heavier than calcium. Moon's chemical content also showed "evidence for a crust on the moon."

Dr. John A. Wood of Smithsonian Astophysical Observatory presented corroborating evidence based on traces of highland material in *Apollo 11* samples. He explained meteoroid impact at nearby highland area might have showered lighter-colored material—anorthosite—on plain where *Apollo 11* landed. Anorthositic crust, formed by melting, probably covered moon at one time. Then giant craters, which became plains, were blasted into crust by early impacts and lavas from beneath surface welled into holes and solidified. Dr. Wood estimated depth of crust at 24 km (15 mi), but other scientists were skeptical.

In interview following meeting, Cal Tech chemist Dr. Gerald J. Wasserburg said crust was thin "like the skin of an onion." (Wilford, *NYT*, 2/24/70, 25)

- Orbiting *Nimbus III* successfully located elk Monique, fitted Feb. 19 with electronic collar containing transmitter. (*SBD*, 2/26/70, 254)
- *Apollo 12* astronauts began goodwill tour of Europe with one-day visit to Lisbon, Portugal. (*NYT*, 2/24/70)
- Japan's space goals were described in *Aviation Week & Space Technology:* "Japan is redirecting the orientation of its national space program with the ambitious goal of placing a series of advanced satellites into orbit by the mid-1970s." Plans provided for variety of payloads, including comsats, navigation, and geodetic satellite packages. "If successful, and if the funding to support it is forthcoming, [Japan's] satellite and booster technology will outstrip that of the older programs of western Europe, placing Japan third in line behind the U.S. and the Soviet Union." (Brownlow, *Av Wk*, 2/23/70, 75-85)

February 24: NASA's X-24A lifting-body vehicle, piloted by Maj. Jerauld R. Gentry (USAF), successfully completed ninth glide flight over South Rogers Lake Bed, Calif. Objectives were to obtain stability and control data with flap and rudder settings that would be used on powered flights, obtain expected angle of attack and mach number for powered flights, and determine effects of modifying nose-gear door to reduce pitch-down tendency at gear deployment. Pilot flew X-24A through turbulence without difficulty and landing gear door modification apparently alleviated pitch-down. Next X-24A flight would be powered. (NASA Proj Off)

- Challenge and objectives of space science and applications program for next decade were outlined by Dr. John E. Naugle, NASA Associate Administrator for Space Science and Applications, before House Committee on Science and Astronautics during NASA FY 1971 authorization hearings: "In the allocation of its scarce resources NASA is emphasizing those projects which help solve immediate problems and deferring to a later year those major new starts whose objectives are primarily to gain the new knowledge. Obviously, we cannot continue to defer such promising scientific projects indefinitely without mortgaging the economic health, the technical progress, and the security of this Nation.

"In the immediate future these postponements are tolerable because of our active ongoing program. Future Aeronomy and Interplanetary Monitoring Platform (IMP) Explorer missions will be directed at the earth's environment: improved OSO's will study the sun and provide data for correlation with ecology studies; the Small Astronomy Satellites (SAS) will start giving us continued data on galactic x-rays and gamma-rays; and future missions of the OAO will continue the successful observation program begun with OAO-2. Before the middle of the decade, we will visit two additional planets, Jupiter and Mercury. We will orbit Mars in 1971 and land in 1976. Under our supporting programs, we are analyzing space data, developing theories to explain the findings; we are conducting relatively less expensive but very productive investigations with sounding rockets and airplanes; and are developing concepts and investigations for future missions." Establishing scientific priorities, allocating proper resources to research, and deciding between various expensive scientific projects would continue to be "a most difficult problem." NASA would work with in-house scientists, advisory groups, and NAS and NAE "to be sure that the scientific projects we are undertaking are sound and worthy of support." (Testimony)

Donald P. Hearth, Director of Planetary Programs, OSSA, testified on continuation of Mars program before Committee: "The results of the Mariner 6 and 7 missions have affected very little the probability that life, in some form, may exist and they have not dampened the enthusiasm of the biologist to search for life on Mars. The apparent low nitrogen content in the atmosphere can be tolerated by many organisms. In fact, biota can be readily conceived which do not directly utilize gaseous nitrogen at all. The results have increased the geological interest in Mars primarily due to the discovery of the chaotic and featureless terrains. An understanding of these geological differences could shed new light on the origin and evolution of Mars and, thereby, on the origin and evolution of the solar system as a whole. We are, therefore, continuing the Mars program. . . . However because of the need to minimize expenditures. . .it has been necessary to delay the Viking missions from 1973 to 1975." (Testimony)

R. B. Marsten, Director of Communications Programs, OSSA, testified on broadcast service of NASA advanced flight experimentation in satellite communications: "Broadcast services come in three classes—the distribution class, which, through a relatively small satellite, would broadcast program material to large receivers, perhaps co-located with local broadcast transmitters for rebroadcast in the conventional mode; a somewhat larger satellite in the community mode, which would broadcast to moderate size receivers located in village squares, or in remote areas and connected to a cable distribution system for redistribution of the program material to viewing sets; and. . .direct broadcast mode, in which multi-kilowatt power level satellites broadcast directly to the user, who may have a conventional TV receiver. This application has the potential of providing instant national service where there is no ground infrastructure, and is much talked about now for educational purposes." (Testimony)

Jesse L. Mitchell, Director of Physics and Astronomy Programs, OSSA, said in submitted statement that research with instrumented aircraft was integral part of NASA program. "This research helps to bridge the gap between groundbased experiments and those that require the more expensive rockets and space vehicles. We have conducted auroral and airglow research observations of the solar corona during total eclipses and infrared observations. The ability to investigate interstellar dust and other celestial objects by observing their infrared emissions is opening up a promising line of research. A small jet aircraft has been used to take a 12-inch [305-mm] infrared telescope to 50 000 feet [15 200 m] above the earth. At this altitude, much of the infrared radiation from the universe is observable; in contrast, at the surface of the earth it is obscured by water vapor. Investigators from the University of Arizona have used a Lear jet from ARC for observations of planets, stars, and galaxies at wavelengths from 30 to 300 microns." They had discovered centers of galaxies were bright in infrared. Because of importance of this area of astronomy, "36-inch [914-mm] telescope is presently being constructed to be placed in a C-141 jet airplane. This plane and the larger telescope will constitute a unique and effective airborne astronomical observatory." (Testimony)

Joseph B. Mahon, Director of Launch Vehicle and Propulsion Programs, OSSA, submitted statement on NASA plans for use of Titan IIIC and Titan-Centaur booster configuration. Titan IIIC would be used to launch ATS-F and ATS-G into synchronous orbit. Introduction of Titan-Centaur configuration in time scale to support Viking mission would represent "vital portion of the effort associated with the Centaur Program in Calendar Years 1971, 1972, and 1973." NASA and USAF had concluded basic management agreements for operation of Titan IIIC and Titan-Centaur systems. USAF would act as Titan IIIC launch vehicle systems manager, responsible to NASA mission direction. NASA would be Titan-Centaur launch vehicle systems manager but would use selected elements of existing USAF contractor organi-

zations. Titan III complex at ETR would be modified to provide single launch pad capability for supporting planetary opportunities with two launches from the same pad. Plans for Titan III-Centaur included engineering proof flight of all-up Viking configuration late in 1973. First operational use was planned for 1974 to support Helios-A solar probe, subject to discussions with West Germany. During FY 1970, preliminary design and program definition effort for Titan III-Centaur would be completed and final design and qualification effort initiated, as well as action for procurement of proof-flight vehicle and supporting hardware. (Testimony)
- Dr. Thomas O. Paine, NASA Administrator, began six-day visit to Australia to discuss space program cooperation. (Reuters, *W Post*, 2/13/70)
- Monique, elk fitted with electronic collar Feb. 19 to transmit data on elk migration to orbiting *Nimbus III*, died of pneumonia-like disease. NASA would select another elk and continue the tracking experiment. (*SBD*, 2/26/70, 254)
- NASA announced designation of Apollo Applications program as "Skylab." (NASA Release 70-30)
- Arcas sounding rocket was launched by NASA from Churchill Research Range carrying Univ. of Houston experiment to study ionosphere. Mission was partially successful and met minimum scientific requirements. (SR list)
- *Apollo 12* astronauts arrived in Luxembourg and met with Grand Duke during 40-day, 21-nation tour. (AP, W *Star*, 2/25/70, A5)
- Boeing 747 had emerged "as the most thoroughly tested transport in history," Charles Yarborough said in Washington *Evening Star*. For year Boeing Co. had been subjecting complete 747 minus operating systems to total destruction, "flexing its great wings upward 29 feet [9 m] and applying wing-airframe stresses of 116 percent of the ultimate design load." Test figures had shown aircraft withstood 174% of maximum load anticipated in normal operations. Aircraft, in operation, was encountering problem of wake turbulence. French government had threatened to ban inauguration of 747 service to Paris on March 1 because of "troubled air its massive engines leave behind." U.K. had warned all aircraft to follow FAA rule calling for 16 km (10 mi) horizontal separation between 747 and other aircraft. Normal separation was three to five miles (five to eight kilometers). FAA edict, applicable only in U.S., also called for 600-m (2000-ft) vertical separation. (W *Star*, 2/24/70, A9)

February 25: NASA announced plans to place White Sands, N. Mex., Test Facility on standby status in June 1971 because of reduced FY 1971 budget requests and reduction of Apollo program work. Move could reduce employment from 641 to caretaker force of 50 following completion of 1970-1971 tests of Apollo LM and Skylab program CMs and SMs. Facility would be kept in condition for resumption of work in 60-90 days. In announcing decision Dr. Thomas O. Paine, NASA Administrator, said, "The NASA facilities at White Sands are unique and valuable. We plan to consider them carefully for possible use in future NASA programs, and to undertake a survey of possible uses by other government agencies." (NASA Release 70-31)
- Apollo 13 Astronaut Fred W. Haise, Jr., described plans for EVA on lunar surface at KSC news conference. Haise said he and Astronaut

James A. Lovell, Jr., would explore hilly region, collect samples, and climb 180-220 m (250-400 ft) up Cone Crater. "We hope to get back to at least the base. . . . From a sampling standpoint there really is a very definite ejecta blanket of rock fragments that's thrown out of the hole at the base of the cone and it would certainly be of interest to get up to the rim. . .and even if you just got pictures of that it would be of extreme interest. Another benefit. . .high ground would give you a magnificient panorama. You could see a good deal more of the countryside that you can see from. . .walking around. . . ."

Haise would drill three holes 3 m (10 ft) deep in lunar surface—two to hold heat sensors and one for sample collection—using battery-powered, rotary-percussion drill. Depth of 3 m (10 ft) "compared to the center of the moon may not be very deep, but it does give a look-see through the surface structure at least in place that in time represents several million years." (Transcript)

- NASA announced selection of 15 investigators and experiments for Atmosphere Explorer (AE) flights C, D, and E scheduled for launch in 1973, 1974, and 1975. Spacecraft would investigate atmosphere at altitudes between 120 and 150 km (75 and 95 mi). Region had not been investigated by satellites because air drag slowed satellites' periods and caused premature reentry. New AE spacecraft would have onboard hydrazine propulsion systems that would overcome air drag and permit automated variable orbits. (NASA Release 70-29)
- Oran W. Nicks, NASA Acting Associate Administrator for Advanced Research and Technology, testified before House Committee on Science and Astronautics on OART's FY 1971 program. Primary objective of V/STOL program was "research needed for the development of advanced rotorcraft which will offer hovering capability and low-noise operations for inner-city operations plus the cruise flight efficiency necessary for inner-city to outlying airport service. Experience with conventional helicopters shows that this service is desired but that it falls short of economic success. If the gap can be closed by improved technology, then the way will be clear for rapid development of this type of travel service. To this end, we are working on several advanced rotor concepts." LaRC V/STOL Transition Research Wind Tunnel would be operational in late 1970. Facility would study effect of down flow from rotors on fuselage and tail, high-lift flaps for STOL air transport applications, and effect of jet-lift engine exhaust on aerodynamic and propulsion characteristics of VTOL fighter configurations. (Testimony)

Dr. Walton L. Jones, Director of Biotechnology and Human Research Div., OART, testifying on human factors program, told Committee: "No seriously limiting effects have been noted in cardiovascular function during the relatively brief manned space flight to date. However, longer duration exposure to weightlessness and the absence of hydrostatic stresses of earth gravity may cause deterioration in cardiovascular reactivity." (Testimony)

Charles W. Harper, Deputy Associate Administrator for Aeronautics, OART, testified before House Committee on NASA's V/STOL technology: "In the past few years NASA has focused its research increasingly toward the objective of enabling development of several types of V/STOL aircraft, each directed at a single important mission. These include an advanced rotor type with airplane-like cruise capability for those missions where vertical takeoff and landing is mandatory, a small and agile STOL aircraft with modest cruise performance and capable of operating easily out of 1500' [460-m] runways having minimum site preparation and a large STOL aircraft having jet transport type cruise performance while also operating easily out of well prepared 1500' [460-m] STOL ports. In addition to this more focused activity, the V/STOL program contains broader research objectives leading eventually to multimission V/STOL aircraft satisfying this growing transportation mode most effectively." (Testimony)

Dr. Hermann H. Kurzweg, Director of Research, OART, testified on NASA basic research programs: NASA-supported research had developed new approach permitting for first time step-by-step numerical calculation of growth of disturbances in atmospheric shear flows, which should contribute to detection and avoidance of clear-air turbulence by jet aircraft. First steps had been taken to apply fluid physics knowledge to atmospheric pollution from airborne and ground sources, testing pollutant action in models of city streets. New research programs would apply knowledge of fluid dynamics to noise abatement studies. Establishment of direct relationship between surface structure and electrical characteristics of germanium surface during oxidation had increased knowledge of oxygen's role in device performance and should increase efficiency in electronic devices. Special attention was being given to evaluation of alloys at the high temperatures of supersonic speeds and to evaluation of rigid polymeric foams as lightweight structural materials for aircraft. (Testimony)

- Milton Klein, Manager of Space Nuclear Propulsion Office, OART, testified before House Committee on Science and Astronautics' Subcommittee on Advanced Research and Technology on role of nuclear rockets in future space transportation. Potential for reuse of nuclear stage provided additional prospects for spaceflight economy. Nuclear rocket would be in-space link between earth-to-orbit transportation system and operations at moon and in distant earth orbits. NASA budget request for $38 million for program in FY 1971, however, could "be viewed as maintaining progress in only the most important longest-lead-time development activities in NERVA." Progress made in "early, critical years, will permit later development activities. . .to proceed smoothly." Effort was essential to meet major milestones: first ground test of series of NERVA flight reactors in 1973 and first ground test of series of NERVA flight engines in 1974. (Testimony)

- *Mission Agency Support of Basic Research*, report by Research Management Advisory Panel of Subcommittee on Science, Research, and Development was submitted to House Committee on Science and Astronautics: "The scientists who work in fundamental research have always proved to be the best reservoir of manpower for tasks requiring objectivity, innovative ideas, and imaginative approaches. We must not let the source of this reservoir dry up." Report recommended Congressional reaffirmation of policy that mission agencies fund their proportionate share of basic research, Federal funding of research to ensure against U.S. research gap,

and Federal support of younger scientists as well as established investigators. (Text)
- Executive nomination of Charles D. Baker as Assistant Secretary of Transportation, submitted Jan. 26, was withdrawn from Senate. (CR, 2/25/70, S2394)
- Harry Mitchell, NASA Deputy Director of Office of Facilities, died at age 47 after heart attack at NASA Hq. (W Star, 2/27/70, B5)

February 26: President Nixon sent message to Congress setting forth Federal Economy Act of 1970: "I propose reduction, termination or restructuring of 57 programs which are obsolete, low priority or in need of basic reform. These program changes would save a total of $2.5 billion in the fiscal year 1971." Among "most significant" items in total savings effort was space program. "After the recent successful Apollo missions, scientific needs for more manned lunar explorations were reassessed. We concluded that fewer manned expeditions to the moon were needed, and production of additional Saturn V launch vehicles and spacecraft has been suspended. Eight Saturn Vs remain in our inventory for manned flights during the early 70s. Savings as a result of these and related space research decisions total $417 million in fiscal year 1971 appropriations." (PD, 3/2/70, 271-5)

- President Nixon sent to Congress *Annual Report on Activities and Accomplishments under the Communications Satellite Act of 1962.* He said: "Communications between Earth and the Moon, while certainly the most dramatic use, is only one of many ways in which satellite communications can now be employed. The Intelsat Consortium of more than 70 nations has been highly successful in bringing the benefits of communications satellite technology to the people of many nations. This report reflects the steady progress being made toward an improved global communications network. Already we see major improvements in international telecommunications capabilities—improvements that will ultimately benefit all of the world's people."

  Report said 1969 "saw the largest annual increase in the number of operational earth stations in the history of INTELSAT." At year's end, 41 earth station antennas were in operation in 24 countries. "Nine other stations or additional antennas are expected to be in service within the first six months of 1970." (PD, 3/2/70, 275; H Doc 91-264)

- Dr. George M. Low, NASA Deputy Administrator, testified in NASA FY 1971 authorization hearings before House Committee on Science and Astronautics: "Our most serious problem in recent years has been our inability to hire newly graduated scientists, engineers and administrative professionals in sufficient numbers. The average age of NASA employees is now increasing 8/10 of a year each year. While we were able to hire 965 graduating college students in 1966, the number dropped to 738 in 1967, 567 in 1968, 253 in 1969, and an estimated 180 in 1970."

  NASA Civil Service staffing would decline from 31 350 at end of FY 1970 to 30 550 at end of FY 1971. Of this reduction, 600 would result from phasing out of ERC employees. Total of 200 would be reduced at manned space flight centers and NASA Hq. (Testimony)

- NASA Associate Administrator for Tracking and Data Acquisition

February 26

Gerald M. Truszynski testified before House Committee: Tracking networks had supported 101 manned and unmanned missions in addition to *Apollo 11* in 1969. Meeting basic requirements of flight projects when budget was at "minimum level" had required significant reductions in overall network capability. Planned improvements had been deferred in FY 1969 for second year and, for first time, sustaining equipment items had been deferred.

FY 1971 request of $299.3 million—$298.0 million for R&D and $1.3 million for construction—included MSFN equipment to support Apollo Applications program while continuing ALSEP support. Satellite Network reductions in FY 1969 had terminated or greatly reduced support of several in-orbit spacecraft still capable of transmitting valuable scientific data. Remaining workload of some 45 satellites continued to require around-the-clock operations at most network stations. Many satellites already aloft would require support in FY 1971, as would 10 new spacecraft scheduled for launch. FY 1971 equipment replacement program was critical. Network could support only portion of spacecraft passes and reliability was essential.

Continued support from DSN overseas facilities would be needed to support four on-going Pioneer missions and *Mariner VI* and *VII*. Major increase in workload in late FY 1971 with launch of dual Mariner '71 mission would require substantial preparatory engineering effort, as would multimission capability for future programs of Pioneer F and G, Mariner/Mercury '73, and Helios. (Testimony)

- Dr. William H. Pickering, JPL Director, announced formation of Mariner Venus-Mercury Office at JPL, headed by project manager Walker E. Giberson. (JPL Release 543)

February 27: U.S.S.R. launched *Cosmos CCCXXIV* into orbit with 465-km (288.9-mi) apogee, 264-km (164.0-mi) perigee, 91.9-min period, and 71.0° inclination. Satellite reentered May 23. (GSFC SSR, 2/28/70; 5/31/70)

- NASA's HL-10 lifting-body vehicle, piloted by NASA test pilot William H. Dana, successfully completed 35th flight after air launch from B-52 aircraft at FRC. Purposes of powered flight were to obtain data with speed brakes open at supersonic speed and to obtain additional stability and control data at supersonic speeds. If data were satisfactory XLR-11 engine would be replaced by hydrogen-peroxide landing engines for powered approach and landing phase of program. (NASA Proj Off)

- Dale D. Myers, NASA Associate Administrator for Manned Space Flight, testified on NASA FY 1971 authorization before Senate Committee on Aeronautical and Space Sciences: Manned space expenditures for FY 1971 would be about 25% less than in FY 1970 and about half FY 1966 peak. "We are able to make these rapid changes . . . as the result of two fundamental policies . . . . One of these policies is that the great bulk of the NASA program is carried out under contracts with industry, university laboratories and other institutions in the private sector of our economy. The other . . . is that NASA must maintain a civil service staff sufficient in quality and numbers to properly supervise the work performed under contract . . . . Essentially all the reductions being implemented in the manned space flight programs in Fiscal Year 1971 are in those efforts carried out under contract."

In FY 1971 NASA proposed to complete definition of space shuttle and space station and continue studies of systems, missions, and payloads for further economies in future years. In Apollo program, Myers pointed out, scheduled stretchout of manned lunar landings would "add to operational costs of the lunar exploration program, although not to our costs this year."

In Skylab program, design, development, and ground testing of subsystems and assemblies of Saturn Workshop, airlock module, multiple docking adapter, and ATM were well advanced. Fabrication had begun on first flight articles, with delivery scheduled for late 1971.

In space shuttle program, shuttle would be "designed so that it can be maintained in a state of launch readiness for lengthy periods and yet be launched within two hours notice into an orbit of any inclination. In the event of an emergency in space, it would be able to carry engineering or repair specialists to overcome the problem or rescue an endangered crew." (Testimony)

- Secretary of Defense Melvin R. Laird urged repeal of Congressional restriction that forced DOD to prove military usefulness of its basic research, during testimony before House Committee on Armed Services hearings on military posture: If "we reduce our support of research projects on the theory that they do not seem to be related directly to military programs, I believe it is essential that the support of high-quality basic research projects of broad national interest be provided immediately by some other agency of the government."

  Later Laird told House Appropriations Committee's Subcommittee on Defense Appropriations that driving pace at which U.S.S.R. was expanding missile forces could reduce U.S. to second-rate strategic position by mid-1970s. (Testimony; *CR*, 4/1/70, S4794)

- President Nixon, in White House ceremony, presented Atomic Pioneers Award to Dr. Vannevar Bush for his "exceptional contributions to the national security as Director of the Office of Scientific Research and Development in marshalling the resources of American science for national defense during World War II and for his pioneering leadership as a Presidential advisor in fostering the establishment of new Federal agencies, including the National Science Foundation, and the Atomic Energy Commission, which have made possible the unprecedented growth of scientific research and development in the last two decades."

  President also presented Award to Dr. James B. Conant for his "exceptional contributions to the national security as Chairman of the National Defense Research Committee in overseeing the successful development of weapons systems, including the atomic bomb, during World War II and for his pioneering leadership in the nation's atomic energy program after the war as Chairman of the Committee on Atomic Energy of the Joint Research and Development Board and as a member of General Advisory Committee to the Atomic Energy Commission."

  President presented third Atomic Pioneers Award to L/G Leslie R. Groves (USA-Ret) for his "exceptional contributions to the national security as Commanding General of the Manhattan Engineer District, United States Army, in developing the world's first nuclear weapons during World War II, and for his pioneering efforts in es-

tablishing administrative patterns adopted by the Atomic Energy Commission in effecting the use of atomic energy for military and peaceful purposes." (PD, 3/2/70, 279-80)
- *Christian Science Monitor* published interview with Dr. Charles S. Sheldon II, Chief of Library of Congress's Science Policy Research Div. U.S.S.R. had not closed door to manned lunar landings but since successful *Apollo 11* lunar landing, "I think . . . they have succeeded in taking the heat off themselves psychologically." Soviet space program "definitely seems to have been delayed in so far as use of any large vehicle is concerned." Still, program was "probably larger than ours." It was "about equivalent to our program at its peak." U.S.S.R. was "putting close to two percent of their GNP into their space program." There was "possibility" that U.S.S.R. was testing FOBS. (Stanford, *CSM*, 2/27/70)
- At meeting of American College of Cardiology in New Orleans, La., ARC-Stanford Univ. team of scientists reported successful use of sonar to provide data on heartbeat and blood circulation formerly unobtainable from cardiac patients without cardiac catherization, blood sampling, or x-rays. Stanford work—supported by ARC, NIH and American Heart Assn.—used commercially available machine which emitted and received high-frequency sound waves. As sonic impulses were bounced against walls of heart, they were recorded and converted into electrical signals which were displayed on TV screen. Sonar also was being used at Stanford to detect heart changes that led to early rejection in heart transplant patients. (NASA Release 70-32)
- Common digitizer, new radar processing device developed by AFSC's Electronic Systems Div. and FAA, would replace radar signal processors used by DOD, USAF announced. It would allow USAF and FAA to share radar information for air traffic control and military air defense and would reduce costs by relaying radar information over telephone lines. (AFSC Release 253.69)
- John H. Chaffee, Secretary of the Navy, and Dr. Robert C. Seamans, Jr., Secretary of the Air Force, announced selection of United Aircraft Corp. Pratt & Whitney Div. to develop and produce high-performance, afterburning turbofan engines for USN F-14B and USAF F-15 fighter aircraft. Target price for first 90 engines was $448 162 600; first increment to be funded was $47 450 000. (DOD Release 162-70)

February 28: Apollo 12 astronauts and wives arrived in Bucharest, Romania, only Communist capital on their world tour itinerary. Astronaut Charles Conrad, Jr., said it was "warmest welcome we've had yet," as crowd of estimated 15 000 persons braved icy winds along 16-km (10-mi) motorcade route from airport. In rare exception to strict rules, Romanian authorities allowed vehicles in motorcade to fly U.S. flag, practice usually reserved for official or state occasions. After stopover in Vienna, astronauts and wives would continue to Africa. (UPI, W *Star*, 3/1/70, A17)
- Vial containing 2.3 g (0.07 oz) of *Apollo 11* lunar material disappeared from display at charity dinner in Los Angeles, Calif. Julian Scheer, NASA Assistant Administrator for Public Affairs, later said sample, one of 100 issued to scientists in U.S., had been lent to Cal Tech sci-

entist for research. Use of sample at fund-raising affair was violation of NASA regulations. (Rosenbaum, *NYT*, 3/2/70, 26)
- Nike-Apache sounding rocket was launched by NASA from Churchill Research Range carrying GSFC experiment to conduct auroral studies. Mission did not meet minimum scientific requirements. (SR list)

During February: In AFSC *Newsreview* editorial Gen. James Ferguson (USAF), AFSC Commander, praised command's two decades of "distinguished progress" in science and technology: "When the F-15 air-superiority fighter outclasses everything in the air—when the B-1 advanced bomber is operational—when a new generation of ICBMs helps to maintain peace—when a million-pound airplane can remain on station indefinitely because of nuclear power—and when military men maneuver into and out of space orbits, at will, from conventional airfields—all these, and more, will have been possible not only because of what we learn today and tomorrow, but because of all the things we learned in our first two decades." (AFSC *Newsreview*, 2/70, 2)

- In *Space/Aeronautics* Frank Leary commented on NASA plans for manned Mars mission: "A commitment to placing human footprints on the red planet will spark an arduous and enormously expensive effort. And it will rankle those critics who want the money spent on earthly ills. Technologically and politically, the manned Mars mission is no piece of cake. But, like Apollo, it is virtually inevitable." (*S/A*, 2/70, 28-38)
- RAF took delivery of Belfast aircraft equipped with Smith MK 29 fully automatic landing (autoland) system. Built by Short Brothers & Harland, aircraft was world's first military transport cleared for automatic landings under civil safety standards and fully operational conditions. (*AF/SD*, 3/70, 12)
- Progress report on plans for National Air and Space Museum in Washington, D.C., was published in *Astronautics and Aeronautics*. During hearings on H.R. 6125, signed by President Johnson in July 1966, Smithsonian Institution had been enjoined from asking Congress for construction funding until after settlement of Vietnam War. Assistant Secretary for Science Sidney R. Galler was requesting FY 1971 funds of $2 million to reappraise architectural plans for museum and "listening for a 'signal from Congress' " before asking for construction funds. In interim, small air and space display was housed in Smithsonian's Arts and Industries Building and in small, temporary Air and Space Building. "The bulk of the 200 plus airframes and the 300-400 engines is stored at the Silver Hill (Maryland) Navy facility. About 80% of the stored collection is either in sheds or crates." Museum plans were reviewed periodically to reflect latest in design. Until 747 aircraft and Apollo spacecraft, Smithsonian could house aerospace history displays.

    "The total Smithsonian annual budget runs about $30 million. The Air and Space portion amounted to $538 000 in 1969 or about 2% of the total. . . . S. Paul Johnston retired in September as Director of the National Air and Space Museum. Galler is still looking for a replacement." (*A&A*, 2/70, 16-7)
- Univ. of Arizona astronomer Dale P. Cruikshank in *Sky and Telescope* reported on his 10-mo visit to U.S.S.R. in 1968-1969 under exchange program between NAS and Soviet equivalent, Akademiya

Nauk. He had left U.S.S.R. "with the feeling that further Soviet-American cooperation in ground-based astronomy is possible and desirable. This is not the case for research from spacecraft, because of the vast difference between the space policies of the two countries." (*Sky and Telescope*, 2/70, 76-9)

March 1970

March 1: NASA launched two Nike-Tomahawk sounding rockets simultaneously—one from Dew Line Station, Bar Main, Barter Island, Alaska, and one from Alaska Rocket Range at Fairbanks—carrying GSFC payloads. Primary objective was to make simultaneous measurements of electric fields over wide range of latitudes crossing auroral zone and extending into polar cap region. Four barium clouds were released from each rocket and were photographed by cameras at six sites in Alaska and one in Canada. (NASA Rpts SRL)
- Arcas sounding rocket was launched by NASA from Churchill Research Range carrying Univ. of Houston experiment to study ionosphere. Rocket and instruments functioned satisfactorily. (SR list)

March 2: Apollo 13 Mission Director's briefing was held at NASA Hq.: Mission would be launched from KSC April 11 on hybrid trajectory to moon's Fra Mauro, rugged highland area. Saturn V 3rd stage (S-IVB) would be ejected onto trajectory that would crash it onto lunar surface, providing signals that could be recorded by seismometer left on moon by *Apollo 12* crew. Astronauts James A. Lovell, Jr., and Fred W. Haise, Jr., would land LM on moon and conduct two five-hour EVA periods on lunar surface while Astronaut Thomas K. Mattingly II orbited moon in CSM, photographing candidate landing sites. Lunar surface EVA would be covered with color TV camera; black-and-white camera would be carried as backup.

After return to earth crew, lunar samples, and spacecraft would be quarantined under procedures similar to those for *Apollo 11* and *Apollo 12*. Apollo Mission Director, Capt. Chester M. Lee (USN, Ret.), said continuation of quarantine—thought unnecessary by many scientists—was based "solely upon the fact that a segment of the scientific community considered the hilly upland site to constitute a new lunar environment.... The bio tests... physical science, and organic chemistry analyses, have found absolutely no evidence of any life forms or precursors of life."

To reduce risk of lightning from high-potential near thunderstorms which had affected *Apollo 12*, NASA had tightened rules for launching in bad weather. Spacecraft would not be launched if nominal flight path carried vehicle through cumulonimbus clouds, within 8 km (5 mi) of thunderstorm cloud, or within 5 km (3 mi) of overhang from a cumulonimbus cloud; through a cold front or squall lying clouds extending 3000 m (10 000 ft); through middle cloud layers 1800 m (6000 ft) or more deep; or through cumulus clouds with tops 3000 m (10 000 ft) or higher. Capt. Lee said although Apollo vehicle was well protected, lightning discharge did occur and NASA officials were "doing everything we possibly can to avoid it, without restricting ourselves unduly" to never launching at all. (Transcript)
- NSF published *Federal Funds for Research, Development, and Other Scientific Activities, Fiscal Years 1968, 1969, and 1970:* Federal ob-

ligations for R&D were $15.9 billion in FY 1968 and were expected to total $15.8 billion in FY 1969 and $16.5 billion in FY 1970. Cycle of rapid advancement in Federal R&D funding that began in 1950s had ended. Basic research obligations totaled $2.1 billion in FY 1968 and were expected to be same in FY 1969. In FY 1970 research obligations were expected to rise to $2.4 billion. During decade, basic research had grown more rapidly than either applied research or development. Applied research obligations totaled $3.3 billion in FY 1968 and were expected to be same in FY 1969. For FY 1970, obligations were expected to reach $3.7 billion. Applied research growth rate since FY 1967 had slowed to less than 1% annually. Development obligations totaled $10.6 billion in FY 1968 and were expected to decrease to $10.4 billion in FYs 1969 and 1970. From FYs 1959 to 1969 development portion of Federal R&D total dropped from 79% to 66%.

In FY 1969, 77% of Federal R&D funds was expected to go to extramural performers and remainder to intramural at Federal facilities.

DOD, NASA, and AEC continued to provide bulk of Federal R&D dollars (83%) in FY 1969, with 75% of their obligations allocated to development. Industrial firms performed 67% of DOD, NASA, and AEC R&D work in FY 1969. Engineering was scheduled to receive most research funds in FY 1969 (29% of total), followed by life sciences (29%), environmental sciences (11%), social sciences (4%), mathematics (2%), and psychology (2%). (Text)

- In letter to Deputy Secretary of Defense David M. Packard, Lockheed Aircraft Corp. Chairman Daniel J. Haughton asked DOD for up to $655 million in assistance funds. Haughton blamed "unprecedented magnitude" of Lockheed's disputes with USAF, USN, and USA for corporation's serious financial plight and said Lockheed wanted advance on $770 million it claimed Government owed on defense contracts. Unless money was forthcoming, Haughton said, it would be "financially impossible" for Lockheed to continue C-5A cargo aircraft production and to fulfill contracts for Cheyenne helicopter, destroyer escorts and amphibious floating docks, and short-range attack missile (SRAM). (Text)

March 2-4: AIAA Earth Resources Conference to consider technical, economic, and political aspects of global satellite system was held in Annapolis, Md. In keynote address Under Secretary of State U. Alexis Johnson said surveying earth's resources from space could change face of international relations. He warned nations must organize international survey program to avoid serious diplomatic problems when system became operational in few years. "We should start now to consider with other countries how best to capitalize and coordinate earth resource surveying systems as they come to serve the interests of many countries. People value what they pay for and they will not contribute to a U.S. program." Problem of how to handle data gathered from country not participating or not wanting information known was major problem.

Rep. Joseph E. Karth (D-Minn.), Chairman of Subcommittee on NASA Oversight of House Committee on Science and Astronautics, said NASA's projected new program costs were "unrealistically low." He predicted, "There will be pressure on Congress to reduce the space budget still further unless the future orientation of the program is based less on space spectaculars and more on the production of tangible benefits."

ERC and MIT research teams reported that camera and sensor systems for survey satellites had shown in aircraft and balloon tests along northeastern coast of U.S. that air and water pollution could be monitored effectively and traced to sources by satellites. (AP, *Huntsville News*, 3/30/70; Lannan, W *Star*, 3/4/70, A4; AP, *C Trib*, 3/4/70)

March 3: Nike-Tomahawk sounding rocket was launched by NASA from Alaska Rocket Range at Fairbanks, carrying GSFC payload to make simultaneous measurement of electric fields over wide range of latitudes crossing auroral zone and extending into polar cap region. Four barium clouds were released during disturbed magnetic conditions and good photographic coverage was obtained. Rocket was launched in conjunction with series begun March 1. (NASA Rpt SRL)

March 4: USAF launched two unidentified satellites on one Thor-Agena booster from Vandenberg AFB. One entered orbit with 226.9-km (141.0-mi) apogee, 224.7-km (139.7-mi) perigee, 88.4-min period, and 88.4° inclination and reentered March 26. Second entered orbit with 506.9-km (315.0-mi) apogee, 439.4-km (273.0-mi) perigee, 94.0-min period, and 88.1° inclination. (GSFC *SSR*, 4/31/70; Pres Rpt 71)

- U.S.S.R. launched *Cosmos CCCXXV* into orbit with 321-km (199.5-mi) apogee, 199-km (123.7-mi) perigee, 89.7-min period, and 65.4° inclination. Satellite reentered March 12. (GSFC *SSR*, 3/31/70)

- NASA launched two Nike-Tomahawk sounding rockets 11 min apart—one from Dew Line Station, Bar Main, Barter Island, Alaska, and one from Alaska Rocket Range at Fairbanks—carrying GSFC payloads. Primary objective was to make simultaneous measurement of electric fields over wide range of latitudes crossing auroral zone and extending into polar cap region. Four barium clouds were released from each rocket and good photographic coverage was obtained from ground stations. Similar launches had been conducted March 1 and 3. (NASA Rpts SRL)

- NASA announced appointment of Donald P. Hearth, Director of Planetary Programs in OSSA since 1967, as GSFC Deputy Director, effective April 6. Hearth had joined NASA in 1962 as Manager of Advanced Programs and Technology. In 1966 he became Assistant Director for Planetary Flight Programs and was Deputy Director of Voyager Program in 1967. (NASA Release 70-35)

- Oran W. Nicks, NASA Acting Associate Administrator for Advanced Research and Technology, testified before Senate Committee on Aeronautical and Space Sciences during NASA FY 1971 authorization hearings: Jet aircraft noise-reduction goals of NASA's acoustically treated nacelle program had been achieved. McDonnell Douglas design modifying nacelles of DC-8 had reduced noise by 10.5 epndb with estimated 4.2% direct-operating-cost increase. Boeing design for 707 aircraft had reduced noise by 15.5 epndb with estimated 9.2% direct-operating-cost increase. Long-range approach to reducing jet transport noise was under way in quiet engine program, with quietness as design requirement for first time in aviation history. Quiet engine incorporating high bypass ratio, low-noise fan design, and sound attenuating ducts would be built by General Electric Co.

Three lifting-body vehicles—USAF X-24A, NASA M2-F3, and NASA HL-10—were investigating low-speed flight problems to be encountered by space shuttle orbiter and booster in future and providing information on subsonic and transonic flying qualities.

In human factors program, NASA was preparing life support systems for extended space missions. "Continuous testing of candidate sub-systems integrated into functional life support systems has been accomplished for up to 60 days. The space vehicle simulation chambers employed...have accommodated four-man teams. A 90-day test with the four-man simulator will be conducted this year." Test would include onboard oxygen recovery system with water electrolysis unit—major component in subsystem for oxygen recovery from carbon dioxide. During 90-day run, advanced subsystems for water recovery employing vacuum distillation and vapor pyrolysis with isotope power source and water vapor electrolysis unit for humidity control would be validated. (Testimony)

Gerald M. Truszynski, NASA Associate Administrator for Tracking and Data Acquisition, testified on communications satellites before Committee: "NASA has recently completed negotiations and signed an agreement with the Communications Satellite Corporation (COMSAT) for a continuation of satellite services.... New contracts have been negotiated for continuing this service with Cable and Wireless, Ltd., England; Australian Overseas Telecommunications Commission; and Compañia Telefónica Nacional de España, Spain." (Testimony)

- NASA was reviewing policy of allowing astronauts to sell publications exclusive personal stories, John N. Wilford said in *New York Times*. Review had been initiated because current contract with *Life* magazine would expire July 24. Though magazine denied it had made decision to renew or to drop contract, astronauts' attorney had said he would entertain contract discussion with other publications. NASA committee, which included three astronauts, was expected to announce policy decision within few weeks. While details of *Life* contract had not been made public, Wilford said it was understood magazine had paid astronauuts $200 000 a year and provided $100 000 worth of life insurance. (*NYT*, 3/4/70, C13)
- Sen. Charles H. Percy (R-Ill.) introduced for himself and several co-sponsors S.R. 56, requesting President to call conference on international exploration of space. (*CR*, 3/4/70, S3020)
- Australian government would review urgently its commitment to purchase 24 F-111 fighter-bombers from U.S., Defense Minister Malcolm Fraser told Parliament in Canberra. Fraser said, "Since the last crash [Dec. 22, 1969], the government has been pressing for sufficient information to enable the situation concerning the F-111 to be reassessed." (Reuters, *W Post*, 3/5/70, A9)
- Steady increase of particulate matter in atmosphere might ultimately create eternal winter on earth, ESSA scientist Dr. Earl W. Barrett told International Solar Energy Conference in Melbourne, Australia. His calculations from data gathered in relatively clean environment of Boulder, Colo., indicated current loss of one percent of

available solar energy in summer and slightly higher loss in winter. (ESSA Research Labs Release 70/17/54)
* Permanent injunction to bar sale of Cornell Aeronautical Laboratory by Cornell Univ. to EDP Technology, Inc., private research company, was issued by Justice Harold P. Kelly of New York State Supreme Court. He ruled Cornell should continue to operate laboratory and its work should be restricted to public research projects. (*NYT*, 3/5/70, 40C)
* DOD announced it would close or scale down activities at 371 military bases by June 30, 1971, to trim defense spending by $914 million. (*NYT*, 3/5/70, 1)

March 5: Nike-Tomahawk sounding rocket was launched by NASA from Dew Line Station, Bar Main, Barter Island, Alaska, carrying GSFC payload to measure electric fields over wide range of latitudes crossing auroral zone and extending into polar cap region. Rocket, last in Alaska series [see March 4], released four barium clouds that were photographed from ground. (NASA Rpt SRL)
* Jerome F. Lederer, director of OMSF safety programs, was appointed NASA Director of Safety, succeeding Bob P. Helgeson who died Jan. 17. Lederer would continue in charge of manned flight safety, with staff and activities merged into safety director's office. Philip H. Bolger would become Assistant Director for Safety Programs and Research and Daniel F. Hayes would become Assistant Director for Industrial Safety. Assistant Director for Aviation Safety would be appointed shortly. (NASA Release 70-36)
* Milton Klein, Manager of AEC-NASA Space Nuclear Propulsion Office and Director of AEC Space Nuclear Systems, testified before Senate Committee on Aeronautical and Space Sciences: "During the past year, the AEC program to provide technology for space nuclear power has furnished operational systems for an earth satellite and the surface of the Moon, and in the near future will provide nuclear power for a new navigational satellite and probes of Jupiter and Mars. Moreover, nuclear power will be imperative for the deep space unmanned planetary missions, the space station/space base and a variety of other missions planned for the new decade and beyond." Basic technology program had demonstrated "nuclear rocket technology is ready to fulfill this vital role in space and forms the basis for the development of the flight rated NERVA." (Testimony)
* Nuclear nonproliferation treaty went into effect. President Nixon in Washington, D.C., and Soviet Premier Aleksey N. Kosygin in Moscow presided in ceremonies at which last of 47 nations deposited instruments of ratification. (Finney, *NYT*, 3/6/70, 1)
* President Nixon signed E.O. 11514, putting into effect policies and responsibilities of National Environmental Policy Act of 1969 (P.L. No. 91-190, approved Jan. 1, 1970). (*PD*, 3/9/70, 320-1)
* Wallops Station announced selection of General Electric Co. and Computer Sciences Corp. for competitive negotiations leading to $400 000 cost-plus-fixed-fee contract to provide engineering and technical services at Wallops through March 1971. (WS Release 70-3)
* Secretary of Defense Melvin R. Laird told Washington, D.C., news conference he was limiting F-111 procurement to four wings. "In

order to have four wings of the F-111, it would be necessary to purchase an additional 40 aircraft. The decision as to whether those 40 aircraft will be purchased or not is dependent upon the scientific review that we are making now and the Air Force has set up a special committee of outstanding scientists and technical advisors" to make review. (Transcript)

- FAA released report by Air Traffic Control Advisory Committee on capabilities and needs of FAA air traffic control system. Report said semiautomated third-generation air traffic control system being implemented nationwide by FAA "must be substantially upgraded if it is even to accommodate the aviation growth of the 1970s." With major modifications, useful life of system could be extended into the 1990s, at which time fourth-generation system might be needed. (FAA Release 5770)
- Pawel Elsztein's *Rakiety Sondujace Atmosfere,* published in Warsaw, Poland, listed site of U.S.S.R.'s Baykonur Cosmodrome at 47° 22' north latitude and 65° 45' east longitude, *Space Business Daily* reported. U.S.S.R. had said launch site was in south, east of Aral Sea, but had never revealed exact location. (*SBD*, 3/5/70, 24)

March 6: Sert II spacecraft with experimental electric rocket engine (launched Feb. 3) was already setting endurance records and returning valuable technical data, NASA announced. Endurance test begun Feb. 14 was continuing satisfactorily. Only problem was failure of miniature electrostatic accelerometer, one of three instruments for measuring thrust. (NASA Release 70-34)

- *Science* published report of *Apollo 12* Lunar Sample Preliminary Examination Team. Samples from Ocean of Storms were about one billion years younger than *Apollo 11* samples from Tranquility Base. While most conclusions reached on *Apollo 11* rocks also applied to *Apollo 12* rocks, *Apollo 11* crystalline rocks displayed one texture (lath-shaped ilmenite and plagioclase with interstitial pyroxene) and similar modes (50% pyroxene, 30% plagioclase, 20% opaque, 0 to 5% olivine), while *Apollo 12* crystalline rocks showed wide range in texture and mode. Most *Apollo 12* igneous rocks fitted fractional crystallization sequence, indicating that they represented either parts of single intrusive sequence or samples of number of similar sequences. Breccias were of lower abundance at Ocean of Storms site than at Tranquility Base, "presumably because the regolith at the Ocean of Storms is less mature and not as thick as at Tranquility Base." Complex stratification existed in lunar regolith, "presumably due mainly to the superposition of ejecta blankets." *Apollo 12* sample 12033 indicated presence of layer of volcanic ash. Greater carbon content in *Apollo 12* breccias and fines than in crystalline rocks was "presumably due largely to contributions of meteoritic material and solar wind." Amount of indigenous organic material capable of volatilization or pyrolysis, or both, appeared extremely low in *Apollo 12* samples. Content of noble gas of solar wind origin was less in fines and breccias of *Apollo 12* rocks than in fines and breccias from *Apollo 11*'s Tranquility Base site. Breccias contained less solar wind contribution than did fines, indicating that breccias were formed from fines lower in solar wind noble gases than were fines at surface. Presence of nuclides produced by cosmic rays showed *Apollo 12* rocks had been within 1 m (3.3 ft) of surface for

10 million to 200 million yrs. Igneous rocks had crystalized 1.7 billion to 2.7 billion yrs ago. (*Science*, 3/6/70, 1325-39)

- Dr. John E. Naugle, NASA Associate Administrator for Space Science and Applications, testified before Senate Committee on Aeronautical and Space Sciences during FY 1971 NASA authorization hearings: "One of the immediate challenges to NASA, and particularly to the Office of Space Science and Applications, for this decade will be to understand the problems of the ecology of the earth, to determine and develop the space systems which will be required to solve these problems, and to understand the long-range implications of their use so that we do not, by their use, create new problems for mankind."

  In FY 1971 NASA budget had almost doubled resources allocated to applications in FY 1969. Agency was proceeding with ERTS, cooperating with ESSA in Global Atmospheric Research Program, and studying with FAA and ESRO "feasibility of establishing a joint navigation and traffic control satellite system over the North Atlantic within the next decade." (Testimony)

- Dr. Donald G. Rea, former NASA Deputy Director for Planetary Programs in OSSA, would become JPL Assistant Laboratory Director for Science March 16, NASA announced. In newly created post Dr. Rea would be responsible for managing JPL's total science program and would act as JPL's chief scientist. (NASA Release 70-37)

- In *Science* editorial, Preston Cloud of UCLA Dept. of Geology said: "Planet Earth, seen from space, appears as a fragile and limited life-supporting system in a vast emptiness. That such a view is, in fact, an accurate one is now more widely accepted than it was not long ago, before Earth's problems of overpopulation, overconsumption, limited resources, and environmental degradation attained the limelight. The danger is that a government and public now deluged by doomsday prophecies and bemused by equally confident reiterations that technology will cure all may hesitate in coming to grips with the problem, in the hope that, given time, it will go away." (*Science*, 3/6/70, 1323)

March 6-8: NASA launched barrage of 31 sounding rockets—Arcas, Nike-Apache, Nike-Cajun, Nike-Tomahawk, Nike-Iroquois, Aerobee 150, Aerobee 170, and Javelin—from Wallops Station to collect data on solar eclipse. Meteorology, ionospheric physics, and solar physics experiments were provided by four universities and seven other research organizations. All 31 rockets were launched on time, all vehicles performed satisfactorily, and excellent radar traces and telemetry records were received. Only one experiment, ozone and water vapor experiment on a Nike-Cajun, failed to return any scientific data. Three other experiments—solar flare spectrum on an Aerobee 170, airglow experiment on a Nike-Tomahawk, and Pitot probe on a Nike-Apache—were partial successes.

Of three supporting Aerobee 150 sounding rocket launches from WSMR two experiments were successful and were recovered. Third was recovered but pointing-control system had not trained experiment on sun. (NASA Releases 70-39, 70-28)

March 7: President Nixon issued statement from Key Biscayne, Fla., on future of U.S. space program: "Having completed that long stride into the future which has been our objective for the past decade, we must now define new goals which make sense for the seventies. We must build on the successes of the past, always reaching out for new achievements. But we must also recognize that many critical problems here on this planet make high priority demands on our attention and our resources. By no means should we allow our space program to stagnate. But—with the entire future and the entire universe before us—we should not try to do everything at once. Our approach to space must continue to be bold—but it must also be balanced."

General purposes of space program should be exploration, acquisition of scientific knowledge, and use of practical applications to benefit life on earth. Six objectives of space program should be lunar exploration; planetary exploration, including eventual manned Mars exploration; reduction in cost of space operations; extension of man's capability to live and work in space; expansion of practical applications of space technology; and encouragement of greater international cooperation in space. "A program which achieves these goals will be a balanced space program...which will extend our capabilities and knowledge...and put our new learning to work for the immediate benefit of all people. As we enter a new decade, we are conscious...that man is also entering a new historic era. For the first time, he has reached beyond his planet; for the rest of time, we will think of ourselves as men from the planet Earth."

President Nixon said NASA Administrator, Dr. Thomas O. Paine, had met recently with space authorities of Western Europe, Canada, Japan, and Australia to "find ways in which we can cooperate more effectively in space." (PD, 3/9/70, 328-31)

- Dr. Thomas O. Paine, NASA Administrator, at Key Biscayne press conference following release of President's space message outlined NASA plans for "Grand Tour" missions in 1970s and space developments for next decade: "A strong effort in exploration, in science and in the practical application of space for the benefit of man here on Earth is laid out, with increasing emphasis on international cooperation." Space shuttle and space station proposals contained reduction in cost of space activities and increase in man's capability to travel back and forth and to work in space for long periods. Space shuttle would take off vertically; at 64-km (40-mi) altitude and 11 300 km per hr (7000 mph), 2nd stage would separate and carry 22.5-metric-ton (25-short-ton) payload of men, equipment, and supplies into earth orbit, with 1st stage returning to earth for horizontal landing. After mission of up to two weeks, 2nd stage would reenter atmosphere and land horizontally for reuse.

President's program of planetary exploration would "provide man with a closehand look of every planet in the solar system."

Administration placed "great emphasis...on the practical applications of space techniques, not only in communication satellites and weather satellites, but in new areas such as Earth resources satellites."

Dr. Paine saw as "most statesmanlike portion" of President's statement "increasing recognition of the international participation of other nations in both the exploration and the utilization of space." It was too early to report specific results of Dr. Paine's visit to world's space capitals but "there certainly is a very substantial interest overseas in utilizing more fully the American program as a

means of closing the technology gap on the part of other nations and, at the same time, allowing them to participate in both the exploration and the utilization of space in the decades to come." (Transcript)
- Photo of eclipse taken by National Center for Atmospheric Research scientist Dr. Gordon A. Newkirk, Jr., at 2680-m (8800-ft) altitude in mountains 48 km (30 mi) southwest of San Carlos Yautepec, Mexico, showed more detail of solar corona than any picture ever taken, Center said. Dr. Newkirk used special filter installed on white-light telescope centered precisely on sun during brief period of totality, in experiment to compare structure of corona to measurements of sun's magnetic fields. (AP, B *Sun*, 5/21/70)
- In NASA-sponsored solar eclipse project scientists used spacecraft in deep space, earth-orbiting satellites, instruments on ground, and sounding rockets to make intense study, concentrating on eclipse's effect on earth's atmosphere and ionosphere.

In-space observations were conducted with four NASA spacecraft and three U.S.-Canada satellites. Radio signals from *Mariner VI* on far side of sun were measured to determine how eclipse affected signals and to study changes in charged particles in earth's atmosphere. *Oso V* and *Oso VI*, in earth orbit, pointed instruments at sun to gather data on sun and solar atmosphere, and *Ats III* photographed earth and its cloud-cover to show path of eclipse across northern hemisphere. American-Canadian satellites *Alouette I*, *Alouette II*, and *Isis I* investigated eclipse-caused changes in ionosphere.

Ground observations were conducted by observers from three universities and four NASA centers at sites in Virginia and Mexico. At site near Sandbridge, Va., Univ. of Michigan team photographed solar corona and LaRC experimenter photographed solar flash spectrum that was too weak to be seen when sun was not obscured. LaRC, GSFC, and SAO scientists used four Super Schmidt Meteor Cameras at Wallops Island, Eastville, and Sandbridge, Va., to photograph faint comets believed to be near sun. At LaRC, Univ. of Colorado scientists used optical assembly for two Wallaston prisms on LaRC's 241-mm (9.5-in) Cassegrain telescope to investigate difference between color of tangentially and radially polarized radiation from solar K-corona. ARC scientists photographed full intermediate-scale coronal structure, using special radial transmission filter in front of LaRC telescope's prime focus.

In Mexico LaRC scientists used NASA's mobile Satellite Photometric Observatory in path of totality at Miahuatlan to determine intensity of coronal radiation to 1% accuracy. Simultaneous wide-band monitor measurements of intensity and polarization of coronal radiation were made out to 3.2 million km (2 million mi) from sun to determine for first time fine structure of coronal electron density and temperature. Team of MSC and Univ. of Houston scientists made variety of observations near Miahuatlan in attempt to separate F- and K-coronas, determine structure of coronas, and study interplanetary dust and particles by measuring coronal intensity and radiation, recording flash chromosphere spectrum, and using photographic and photoelectric techniques to study shadow bands.

Data from satellite and ground observations would be compared

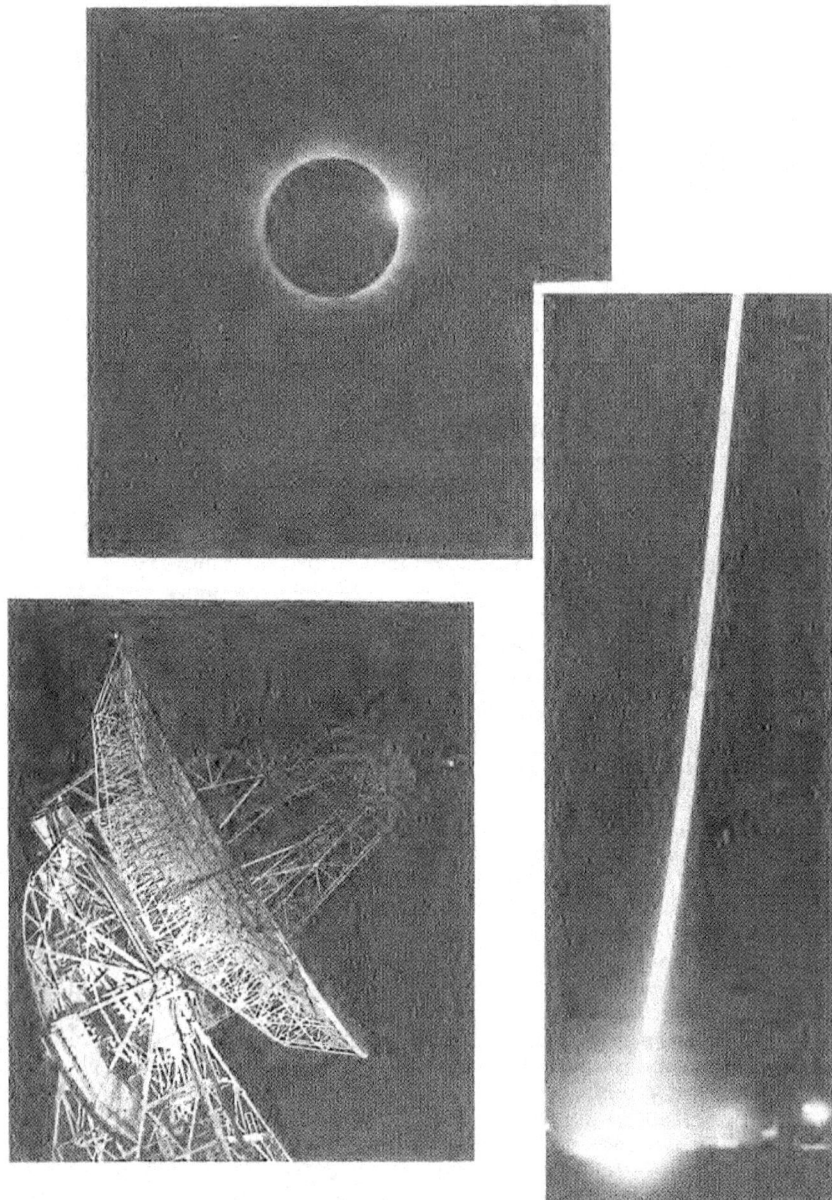

March 7: Spacecraft in deep space, earth-orbiting satellites, barage of sounding rockets, and instruments on the ground were used to make intense study of the 1970 solar eclipse and its effect on the earth's atmosphere and ionosphere. Totality of the eclipse (at top) occurred at Wallops Station at 1:38 pm EST. passing over tracking antenna (lower left) as it began to sweep the Eastern Seaboard. A Nike-Cajun (above right), one of 31 sounding rockets launched from Wallops March 6-8, lifted off during totality. Five photos (left to right on facing page) taken by Ats III satellite from synchronous altitude 35 900 kilometers (22 300 miles) above the earth show the moon's shadow (dark spots among clouds) moving northeast from the Gulf of Mexico and out into the Atlantic.

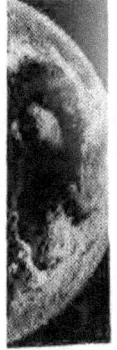

with data from barrage of sounding rockets launched March 6-8. (NASA Release 70-28; MSC *Roundup*, 3/27/70, 1)

- AFCRL KC-135 research aircraft, carrying new infrared sensing system and multichannel tape recorders above moon's shadow, collected data on solar eclipse. Primary objective was to develop techniques for predicting wavelengths at which atmosphere would be most transparent under different conditions. Although eclipse lasted only 3 min 30 secs on earth, by chasing lunar shadow scientists prolonged effect to 5 min 38 secs. Abrupt switching off and switching on of sunlight provided ideal means of learning nature of airglow reaction, particularly in 80- to 100-km (50- to 60-mi) region. Aircraft flew from Kelly AFB, Tex., down eclipse path over Pacific four times—before eclipse, as sun became increasingly covered, during eclipse totality, and during final stages of partial eclipse. (Sullivan, *NYT*, 3/8/70, 60)

- First total solar eclipse seen over heavily populated areas of U.S. since 1925 was greeted by millions of viewers who crowded beaches, towns, and islands where viewing was most favorable. Eclipse was total along path 137-161 km (85-100 mi) wide through Mexico, Florida, Georgia, the Carolinas, Virginia, and Nantucket Island, Mass. It was 96% total in New York City and 95% in Washington, D.C. (Wilford, *NYT*, 3/8/70, 1)
- Ivory Coast President Felix Houphouet-Boigny presented gold plaques to *Apollo 12* astronauts, who had arrived in Abidjan March 6. They would leave for Tanzania March 8 during world tour. (Reuters, *W Post*, 3/8/70, A7)
- *Pravda* published policy statement which said U.S.S.R. was not seeking nuclear superiority over U.S. *Pravda* said there were possibilities for reaching agreement on limiting strategic weapons at U.S.-U.S.S.R. talks scheduled for April 16 in Vienna but questioned U.S. sincerity in wanting accord. (Gwertzman, *NYT*, 3/8/70, 1)

March 8: Nike-Tomahawk sounding rocket was launched by NASA from Andoeya, Norway, carrying Norwegian experiment to conduct auroral studies. Rocket and instruments functioned satisfactorily. (SR list)
- Cleveland *Plain Dealer* editorial on President's space goal message: "Underlying the message seemed to be one word—flexibility. And because the space program deals with the unknown, it is reasonable to accept flexibility as a necessary factor." (C *Plain Dealer*, 3/8/70)
- *New York Times* editorial commented on solar eclipse: "Today an eclipse is practically a scientific event, predictable as high noon, right down to the fraction of a second. The big mystery is gone, though it is an occasion for scientific measurement and further calculation. But there is another dimension to such an occasion. It puts the earth into perspective as few other occasions ever do." (*NYT*, 3/8/70, 4:10)
- U.S. submitted to U.N. Seabed Committee 12-point guide for international regulation of deep seabed to provide equal share of benefits of exploitation for all nations. (Brewer, *NYT*, 3/8/70, 24)

March 9: ComSatCorp sent annual report to shareholders on 1969 activities. Financial report had been released Feb. 20. In letter to shareholders ComSatCorp President Joseph V. Charyk and Chairman James McCormack said: "Outstanding progress has been made in the establishment of the global communications satellite system...a goal toward which COMSAT and the International Telecommunications Satellite Consortium (INTELSAT) have worked since inception. This basic goal was achieved when a satellite was placed over the Indian Ocean in the summer of 1969.... In addition, the pace of earth station development by national entities in many countries has been significant. Investment worldwide of about $200 million in earth stations and more than $100 million in satellites reflects a widespread confidence in the present and future benefits of satellite communications."

Highlights of 1969 included achievement of major INTELSAT goal, global coverage by comsat; full commerical operation of 22 new antennas in 18 countries; start of construction on Intelsat IV series satellites and selection of Atlas-Centaur as series launch vehicle;

completion of ComSat Laboratories, major R&D facility; and reaffirmation by ComSatCorp of its readiness to finance, establish, and operate domestic comsat system. (Text)
- Press editorials commented on President Nixon's statement on Nation's space program.

New York Times: "If the President's program is fulfilled on schedule, the economics of space activity—both its costs and its dividends for nonscientists here on earth—will be revolutionized. The reconnaissance activities of earth resources satellites have enormous potentials for aiding agriculture, forestry, the search for new minerals, and the exploitation of the seas. The prospect of communications satellites that can broadcast radio and television programs directly to homes everywhere offers exciting new perspectives for further unifying all peoples, while simultaneously raising delicate political problems. . . . When it becomes available, the reusable space shuttle will. . .facilitate construction of a permanent manned orbiting space station that will open up new areas of scientific and economic activity in the near neighborhood of earth." (*NYT*, 3/9/70)

Washington *Evening Star:* "The 1970's promise to be an exciting decade in space, not just a comedown from the exhilaration of the initial moon landings. The next 10 years will be even more satisfying if another of the President's goals—increased international cooperation in space—is realized. To that end, Soviet space scientists have been told about the U.S. plans. The Russian leaders would do well to consider the economies and other benefits that could come from pooling skills and resources as man reaches farther and farther beyond his planet." (W *Star*, 3/9/70)

Washington Post: "It was appropriate that President Nixon should announce his proposed plans for exploration in space on the last day in this century when a total eclipse of the sun was visible in the United States. Opportunities to make strides in learning about our universe do not come often and must be seized when they arrive. The scientists. . .did the best that could be done in Saturday's brief moment of darkness to learn more about the forces in the sun that dominate our solar system. And the President seems to have done about the best that can be done with the resources now available to keep our space program on the right track.

"The emphasis the Administration is giving to unmanned flight activities—the commitments to launch vehicles in 1977 and 1979 for grand tours of our outermost planets are the most exciting—will test the willingness of the nation to spend huge sums on basic research. The future programs of NASA, at least for the next decade, will have little of the appeal that seized the public during the race to land men on the moon first. And the worth of these programs, in terms of knowledge they may yield about the solar system and the usefulness of that knowledge, is difficult to appreciate fully. The test in the future will be whether we are willing to pay for these probes for knowledge without the exhilarating experience of watching the first man step onto the moon's surface, and of beating the Russians." (*W Post*, 3/9/70)
- MSC announced award of $57 264 989, cost-plus-fixed-fee NASA con-

tract to General Electric Co. for spacecraft checkout, reliability and quality assurance engineering, and systems engineering in support of Apollo and Skylab programs. Contract, which definitized letter contract, covered work from October 1969 through December 1972. (MSC Release 70-31)

- "Grand Tour" of outer solar system was discussed by William Hines in *Chicago Sun-Times:* "The grand tour idea came out of Caltech several years ago after a graduate student in Prof. Homer J. Stewart's aeronautical engineering department calculated the added energy that could be imparted to a spacecraft by the massive gravitational fields of various planets. This information, together with the unusual configuration of outer planets in the 1980s, was patched together by Stewart into the grand tour concept. Theoretically, Stewart calculated, it should be possible to fire a spacecraft into a near-miss trajectory toward Jupiter and then let that planet, biggest in the solar system, take over. Jupiter would add energy, and hence velocity, to the spacecraft.... This would send the craft toward Saturn, where more energy would be borrowed from the ringed planet's gravitational field, and then on to Uranus for more of the same, Neptune and finally Pluto." (*C Sun-Times*, 3/9/70)

March 9-10: NASA launched series of three Nike-Tomahawk sounding rockets from TERLS, carrying GSFC payloads to investigate relationship between ion composition, airglow emissions, and vertical drift velocities in F region of ionosphere. Rockets and instruments functioned satisfactorily and good data were obtained. (NASA Rpts SRL)

March 10: French Centre National d'Études Spatiales (CNES) successfully launched French Diamant-B rocket carrying Franco-German scientific satellite *Dial* (Diamant pour l'Allemagne) from Guiana Space Center (GSC), Kourou, French Guiana. Mission, under joint agreement with West Germany, was France's fifth successful launch. The 115-kg (253-lb) payload included German scientific "WI Kapsule" WIKA and French technological "Mini Kapsule" MIKA which was damaged by vibration between 2nd and 3rd stages. French capsule was to have checked Diamant-B stages during flight. Satellite entered orbit with 1613-km (1002.3-mi) apogee, 308-km (191.4-mi) perigee, 104.3-min period, and 5.4° inclination. (GSFC SSR, 3/31/70; *C Trib*, 3/11/70; *Av Wk*, 4/6/70; NASA Int Aff; *SBD*, 3/12/70, 57; *SF*, 5/70, 189)

- Subsystem test bed (STB), experimental, low-cost spacecraft module, began 10-day vacuum-chamber test at MSC to determine how STB reacted to temperature and vacuum extremes. It also would evaluate heat pipe, new device that might be useful in controlling temperature of future space vehicles. Test would subject STB, which had 4.6-m (15-ft) diameter and 2.7-m (9-ft) height, to temperatures from 200 K to 383 K (−100°F to +230°F) and to vacuum equivalent to that at 160 km (100 mi) above earth. (MSC Release 70-32)

- Quincy (Mass.) *Patriot Ledger* editorial noted President's space goal message had not included Mars mission as major goal at this point. "More important is the consolidation and mastery of the space environment closer to earth, as well as greater application of space research to human problems. Fortunately, this is the emphasis of Mr. Nixon's new program." (*Q Patriot Ledger*, 3/10/70)

- U.S. intention to begin deployment of MIRV missiles in June was disclosed by Secretary of the Air Force, Dr. Robert C. Seamans, Jr., at closed session of Senate Armed Services Committee, AP reported. (*C Trib*, 3/11/70, 18)

March 11: MSC scientists reported three kinds of earth bacteria had died when exposed to lunar soil gathered by *Apollo 11*: staphylococcus aureus scraped from *Apollo 11* astronauts when they returned to earth; azobacter vinlandii, solid bacterium; and pseuodomonas aeroginosa, used in laboratory studies. LRL chief of preventative medicine Dr. William Kemmerer said conjecture was that all were killed by poison in *Apollo 11* core-tube material. Tube held lunar soil gathered from five to eight inches beneath lunar surface. There had been no ill effects in similar tests with *Apollo 11* soil or *Apollo 12* surface or underground samples. (Cohn, *W Post*, 3/12/70, A1)

- Dr. Thomas O. Paine, NASA Administrator, reported on international cooperation in NASA programs before Senate Committee on Aeronautical and Space Sciences: NASA had established "sound base of extensive and successful cooperation" with over 70 countries in 1960s, entered into almost 250 agreements for international space projects, orbited foreign satellites and flown foreign experiments on U.S. spacecraft, participated in more than 500 cooperative scientific rocket soundings from sites throughout world, involved more than 50 foreign scientists in analysis of lunar samples, and provided for direct daily reception by some 50 countries of data from weather satellites. "Third" countries together spent "roughly $300 million a year on space."

"On the aeronautics side of NASA's responsibilities, cooperative projects with Canadian, French, German, and British agencies are contributing importantly to the development and testing of a variety of V/STOL aircraft configurations."

In experimental meteorological satellite and balloon program, Project Eole, NASA would launch French satellite to track several hundred balloons to derive global circulation of winds. Cooperative earth resources survey projects had been undertaken with Brazil and Mexico to acquaint foreign scientists and policy makers with potential of remote sensing. NASA had added earth resources disciplines to international graduate fellowship program and planned international workshop in 1971. NASA was providing technical assistance to India in aircraft survey of coconut palm blight in Kerala State so that its spread could be controlled. Agency was considering proposal to launch French synchronous meteorological satellite as joint contribution to Global Atmospheric Research Program and had met informally with ESRO and European aviation officials to consider mission requirements for possible preoperational air-traffic-control satellite system for North Atlantic.

Dr. Paine had written new letters to Soviet Academy of Sciences President Mstislav V. Keldysh and to Academician Anatoly A. Blagonravov "inviting new initiatives in space cooperation." Keldysh had accepted suggestion for meeting, but had deferred further discussion for "three or four months" from Dec. 12, 1969. Keldysh had declined invitation of Soviet proposals for experiments on NASA planetary probes, "advocating instead a relationship in which NASA and the Soviet Academy would coordinate 'planetary goals' and

'exchange results' of unmanned planetary investigations.'' (Testimony)
- Saturn V 2nd stage (S-II), scheduled to boost Apollo 17 toward moon in 1972, was successfully static-fired for 378 secs at MTF. (*Marshall Star*, 3/11/70)
- *Apollo 12* lunar samples were presented in Bonn to Science and Education Ministry's Parliamentary State Secretary Klaus von Dohnanyi by U.S. Ambassador to West Germany Kenneth Rush. Samples would be investigated at Max-Planck Institute in Heidelberg. (Leyendecker, *Bonner Rundschau*, 3/12/70)
- James M. Beggs, Under Secretary of Transportation, addressed Western Governor's Conference on application of science and technology to problems of pollution, transportation, and employment in Salt Lake City, Utah: "We have left the Age of technology, where a technical innovation...was sufficient cause for the formation of a great new industry or the initiation of a large complex project. We seem to be shifting now to a period of more complexity in decision-making, where the longer term effects of our technological advances may be the true determinant of the decision to proceed or not to proceed...the 'Technosocial Age.'"

  In aviation, "we find a sector of transport supported by burgeoning technology, effective salesmanship, and extremely rapid expansion of traffic demand. In 1964, there were 83 million air carrier passengers. This figure grew to nearly 153 million in 1968. The number of air carrier operations...grew accordingly from 7.4 to 9.9 million." Of total daily operations handled by FAA terminal facilities, four fifths were general aviation. (Text)
- *Kansas City Times* editorial commented on President Nixon's space recommendations: "The hang-the-cost generosity of Congress that marked the early years of the Apollo moon program seems not to have been rekindled by last summer's landing on the Sea of Tranquility, for all the justified national pride that feat stimulated. And Mr. Nixon, judging from his last budget message, is disinclined to buck this trend. The President's conviction seems to be that important and dramatic goals await the U.S. in space, but that the reality of competing demands on the national resources prohibits the setting of binding deadlines for their achievement. His caution is reasonable, even if it does not lend itself to flights of public imagination." (*KC Times*, 3/11/70)
- Status of C-5 Galaxy aircraft program funding was described by Dr. Robert C. Seamans, Jr., Secretary of the Air Force, before Senate Armed Services Committee during hearings on USAF FY 1971 budget request: In 1969 it had appeared "that the 120 aircraft program would cost about $4,831 million, including initial spares. As we interpreted the contract, the contractor would lose some $285 million on 120 aircraft, and any smaller buy the contractor's loss would be higher. The contractor disagreed. By his interpretation...he expected to make a small profit on the total program. As a consequence of budget restraints, rising costs, and an overall reappraisal of defense requirements, we have now limited the program to 81 aircraft. The program's final cost to the Government, however, depends on resolution of the issues between the Air Force and contractor over the contract. We are still trying to work these out with Lockheed

but...litigation may be required. If the Air Force wins all the arguments, the 81 aircraft program will cost about $4 billion. If the contractor wins all his points, the cost will be...on the order of $500 million more." It might be "several years before we know the exact outcome of this program." (Testimony)

March 12: House Committee on Science and Astronautics ordered favorably reported H.R. 15695, FY 1971 NASA authorization with amendments that added $297.8 million to President Nixon's budget request of $3.333 billion. New authorization total of $3.631 billion included addition of $144.9 million to Apollo budget request, for total Apollo allocation of $1.1 billion, and $155 million for space flight operations, making total of $670.2 million in that category. (H Rpt 91-929; *CR*, 3/12/70, D225; *W Post*, 3/13/70, A10)

- NASA announced award of $22-million, 45-mo, cost-plus-incentive-award-fee contract to General Electric Co. for management, sustaining engineering, logistics, and support for Saturn ground-support equipment and for operations and maintenance of Saturn systems development facility. Contract, to be managed by MSFC under OMSF direction, was expected to achieve substantial savings by consolidating work previously performed under six separate contracts. (NASA Release 70-40)

- AFSC announced acceptance by SAMSO's 6555th Aerospace Test Wing of new, $1.6-million spacecraft environmental enclosure for Titan IIIC Launch Complex 40 at Cape Kennedy. Enclosure, 25 m (83 ft) tall, extending from 9th to 14th level of 73-m (240-ft) mobile service tower, would allow crews to adjust and repair spacecraft in place instead of removing them from MST and transporting them to "clean rooms." (AFSC Release 44.70)

- NAE President Eric A. Walker announced selection of Dr. C. Stark Draper, vice-director of MIT's Charles Stark Draper Laboratory (formerly Instrumentation Laboratory), to receive NAE's fifth Founders Medal. Dr. Draper would receive medal and deliver Founders Lecture at luncheon April 30 in conjunction with NAE's annual meeting. (NAE Release)

March 13: U.S.S.R. launched *Cosmos CCCXXVI* into orbit with 239-km (148.5-mi) apogee, 209-km (129.9-mi) perigee, 90.1-min period, and 81.3° inclination. Satellite reentered March 21. (GSFC *SSR*, 3/31/70)

- *Itos I* (Tiros-M), launched Jan. 23, was adjudged successful by NASA. Primary objective of obtaining inflight engineering data for evaluation of single-momentum wheel stabilization system and secondary objective of evaluating stabilized platform for operational meteorology had been achieved. New stabilization system had proved very effective, maintaining spacecraft attitude within specified $\pm 1°$. After completion of engineering tests and evaluation of spacecraft subsystems, spacecraft would be turned over to ESSA for use in meteorological operations. (NASA Proj Off)

- Office of Science and Technology sent to Congress *The Next Decade in Space*, report of Space Science and Technology Panel of President's Science Advisory Committee: National program for next decade in space should focus on using "space capabilities for the welfare, security, and enlightenment of all people." Recommended goals were to contribute to Nation's economic strength and security and expand earth-oriented research and applications of space science and technology; explore solar system with phased program of lunar exploration, search for extraterrestrial life, and diversified planetary exploration; and use space platforms to expand knowledge of universe and basic physical laws. Panel urged that U.S. develop technology to expand capability for automated equipment controlled by man in remote location for space exploration and civil applications, reduce cost of space operations by cost reduction for expendable vehicles and payload development, strengthen biomedical basis for possible long-duration manned space flights, and encourage international cooperative programs.

Panel recommended NASA and DOD continue joint study of reusable space transportation system "in the expectation that a decision can be made for a FY 1972 budget decision on an STS development program." Design target should be replacement "of the largest possible part of the national launch vehicle stable," including Saturn V. "We urge the deferral of a development decision on space stations more elaborate than the AAP space station...concept until the prior availability of the STS can be confidently anticipated." It was "likely that some form of multiply reusable space transportation system will indeed become the most attractive major new technology development for the decade."

Other recommendations were basic biomedical research program supplemented by 12- to 18-mo Apollo Applications flight program, to provide human factors information for more ambitious space station; possible second-generation space station to utilize STS; and earth-orbital manned space flight activity with systems using STS. (Text)

- Dr. Wernher von Braun, former MSFC Director, was sworn in as NASA Deputy Associate Administrator for Planning by Dr. Thomas O. Paine, NASA Administrator. (Off NASA Dep Assoc Admin)
- Dr. Thomas O. Paine, NASA Administrator, and Prof. Giampietro Puppi of Italian National Research Council (CNR) signed agreement specifying conditions under which NASA would provide reimbursable launchings for Italian scientific spacecraft and responsibilities of each party. CNR would provide spacecraft and experiments, inform NASA of spacecraft's design and mission, and reimburse NASA for launch vehicle and associated services. NASA would furnish booster that would meet mission requirements and provide launch, tracking and data acquisition, and other services.

  First launch under agreement would be Italy's Sirio, scheduled for launch by Delta booster into 35 400-km (22 000-mi) stationary orbit in 1972. Satellite would carry telecommunications, technology, and scientific experiments to be conducted with ground stations in Europe. (NASA Release 70-42)
- NASA held Hq. briefing on space station and shuttle programs for representatives of 17 nations, ELDO, ESC, and ESRO. Dr. Thomas O. Paine, NASA Administrator, said foreign participation was step in NASA's continuing efforts to inform other nations of post-Apollo program planning so that they might determine extent to which they wished to participate in it. (NASA Release 70-41)
- Vice President Spiro T. Agnew presented Harmon International Aviation and Space Trophy for 1969 to Maj. Jerauld R. Gentry (USAF) for "extraordinary and brilliant piloting feat" while testing NASA HL-10

March 13: *Dr. Wernher von Braun (left), former MSFC Director, was sworn in as NASA Deputy Associate Administrator for Planning by Dr. Thomas O. Paine, Administrator.*

lifting-body vehicle and to *Apollo 8* Astronauts Frank Borman, William A. Anders, and James A. Lovell, Jr., for first manned flight to moon. (Program; AFSC *Newsreview*, 4/70)
- Dr. Charles A. Berry, MSC Director of Medical Research and Operations, participated in international medical conference by two-way satellite TV relay between MSC and 11 European cities. Conference was part of program of 18th International Post-Graduate Congress of German Federal Medical Council. Central meeting was at Davos, Switzerland, with other medical groups participating through TV projection from nine German cities and Vienna, Austria. (MSC Release 70-29; NASA PAO)
- Models of Viking Mars lander had been tested at speeds from mach 0.6 to mach 0.3 in Propulsion Wind Tunnel at AFSC's Arnold Engineering Development Center at Tullahoma, Tenn., AFSC announced. Atmosphere on Mars had been simulated in tunnels for tests conducted for NASA and Martin Marietta Corp. on aeroshell and base cover that would protect lander during entry through Martian atmosphere. (AFSC Release 67.70)
- President Nixon announced appointment of new members of President's Science Advisory Committee: Dr. Solomon J. Buchsbaum, Vice President of Research, Sandia Laboratories; Dr. Theodore L. Cairns, Assistant Director, Central Research Dept., E. I. du Pont de Nemours & Co.; Dr. James S. Coleman, Dept. of Social Relations, Johns Hopkins Univ.; Dr. Val L. Fitch, Princeton Univ. physicist;

and Dr. Lloyd H. Smith, Jr., Chairman of Dept. of Medicine, Univ. of California, San Francisco Medical Center. New members would serve until Dec. 31, 1973. They succeeded William R. Hewlett, who had resigned, and Dr. Ivan L. Bennett, Jr., Dr. Sidney D. Drell, Dr. Charles P. Slichter, and Dr. Charles H. Townes, whose terms had expired. (*PD*, 3/16/70, 357-8)

- Dr. Donald F. Hornig—Vice President and Director of Eastman Kodak Co., professor at Univ. of Rochester, and former Presidential Science Adviser—was named President of Brown Univ. in Providence, R.I., effective June 1. (W *Star*, 3/14/70, A2; *Science*, 3/20/70, 1597)

- *Chicago Daily News* editorial commented on President Nixon's space recommendations: "Compared to the feverish reach for the moon in the Sixties, the plans...seem almost pedestrian. Not until men take off for Mars—at the earliest in 1983—will there be anything like the excitement of the first trip to the moon. Yet it is a practical timetable NASA has developed, one that promises the maximum in scientific research within a budget reflecting a lowered priority for space. Exploration of space is by no means being abandoned, but urgent needs on Earth clearly come first." (*C Daily News*, 3/13/70)

- *Science* article commented on suggestions of President's Science Advisory Committee panel that NASA devote more time to biomedical research: "There appear to be practical limits other than the budgetary ones on the expansion of basic studies in the manned space program. It will be a long time before astronauts on missions operate in conditions anything like those of the laboratory." MSC Director of Medical Research and Operations, Dr. Charles A. Berry, and his staff "have been responsible for establishing medical protocols for the missions, in a way that balances the need for biomedical data with operational demands and the willingness and ability of the astronauts to cooperate. Some medical procedures are tedious, unpleasant, and even humiliating, and when the work load on astronauts is heavy, data gathering has to be limited. For this reason in part, then, tension between biomedical insiders and outsiders is likely to continue. At the same time the interests of the two groups are nearer to converging than ever before." (Walsh, *Science*, 3/13/70, 1469-71)

March 14: Apollo 13 Astronauts James A. Lovell, Jr., Thomas K. Mattingly II, and Fred W. Haise, Jr., held press conference at MSC. Lovell announced final code names for LM and CM had been selected according to theme of mythology. LM would be called "Aquarius" after mythical Egyptian water carrier who brought fertility and knowledge to Nile Valley. Code name originally selected for CM—"Auriga," for constellation Auriga—had been dropped in favor of "Odyssey," epic Greek poem about long adventures and wanderings of Odysseus in Greek mythology. Lovell said Apollo 13 crew had "already had quite an odyssey just getting trained" for mission.

Landing of Aquarius on moon would be different from previous landings, Lovell said. Key to landing was Cone Crater on Fra Mauro. LM was "being targeted for a spot that we don't want to land at," uprange about 1/2 km from desired site; "we're not sure of the actual height of this ridge [of Cone Crater]...so we're targeting uprange to make sure we get over it."

Reason for going to Fra Mauro was "its peculiar geologic formations there and...a place called the Mare Imbrium...a large circular basin that was formed early in the moon's history...by a large impact. In the formation of this impact it made hummocky ridges, radiating outward from the center of the impact down into...Fra Mauro.... Now, by going to Fra Mauro, we hope to...bring back some of the basin type material that was thrown out by this catastrophe." (Transcript; UPI, *C Trib*, 3/15/70)
- Aerobee 150 sounding rocket was launched by NASA from WSMR carrying NRL experiment for stellar studies. Mission did not meet minimum scientific requirements, but instruments on board rocket detected molecular hydrogen in interstellar space for first time. Although much indirect evidence for presence of large amounts of molecular hydrogen in space had been accumulated, detection by Aerobee 150 was first direct measurement. (SR list; *Naval Research Reviews*, 7/70, 31)
- Philadelphia *Evening Bulletin* editorial commented on President's space goal message: "President Nixon's 'low profile' space program represents a realistic appraisal of the American temper at this time. Without the Russians nipping at our heels, apparently, there is little public pressure to keep space exploration and development going at better than a jog. It's worth noting, however, that although Mr. Nixon has accepted implicit funding limitations reflecting both his own anti-inflation drive and public opinion, he hasn't really slammed any doors." (P *Bull*, 3/14/70)

The *Economist* commented: "Nasa's future plans may not brighten the back of cereal boxes but they are solid, varied and balanced. The dangers from now on are not likely to be financial but technical: the space shuttle will be hard to make." (*Economist*, 3/14/70, 53)

March 15: Dr. Thomas O. Paine, NASA Administrator, and Dr. Wernher von Braun, NASA Deputy Associate Administrator, discussed future of U.S. space program on ABC program "Issues and Answers." Dr. Paine said he did not anticipate further cutbacks in NASA personnel and facilities. "In my opinion we have slimmed down the NASA base to the point where we now have something which I would expect to see extended throughout the seventies and eighties." He was "quite optimistic about the future of the space agency."

Dr. von Braun said "in the long haul" U.S. could not compete with U.S.S.R. at $3.333-billion budget level. It was obvious "that we are ahead in some areas," but U.S.S.R. was probably ahead in space station field. (NASA Special Ann, 3/11/70; *SBD*, 3/17/70, 76)
- U.S., Japanese Government, and U.S.S.R. pavilions attracted largest crowds on opening day of Expo '70 in Osaka, Japan. *Apollo 12* moon rock was primary interest of 8000 visitors per hour who passed through U.S. exhibit. (Shabecoff, *NYT*, 3/16/70)

March 16: West Germany's *Azur* research satellite, launched by NASA Nov. 7, 1969, was adjudged successful by NASA. Satellite had achieved nominal orbit and had obtained very good data. Spacecraft and instruments were functioning satisfactorily except for tape recorder that had failed 30 days after launch. (NASA Proj Off)
- Custer Channel Wing Corp. demonstrated takeoff and climbing ability of Custer CCW-5 channel-wing aircraft at Teterboro Airport in New

Jersey. Prototype of possible STOL aircraft for serving small airports in downtown urban areas had half-circle channel in each wing, weighed 2100 kg (4600 lbs), and flew at minimum speed of 64 km per hr (40 mph)—almost half that of conventional aircraft of same size, weight, and power. Its maximum speed was 258 km per hr (160 mph). (Hudson, *NYT*, 3/17/70, 69)

- Senate Committee on Aeronautical and Space Sciences conducted hearing on S.R. 49, providing for Congressional recognition of Goddard Rocket and Space Museum at Roswell, N. Mex., as fitting memorial to Dr. Robert H. Goddard, U.S. rocket pioneer. Dr. Thomas O. Paine, NASA Administrator, testified: "Americans can ill afford to ignore the contributions of Robert H. Goddard in the history of the coming of the space age. Young Americans need to appreciate that what appear as unsoluble problems today are not new to man's experience. Dr. Robert H. Goddard's rocket artifacts and his lifelong labors should be an inspiration to each generation of youth as they grapple with their concerns of their day and their dreams of a better world of their tomorrows in a dynamic universe. This nation cannot afford not to have Robert H. Goddards in the future."

  Dr. Wernher von Braun, NASA Deputy Associate Administrator, testified: "It seems very important that we and our children after us never lose awareness of what key individuals such as Robert Goddard provided to a fuller understanding of the progress of mankind. He, like Konstantine Tsiolkovsky of Russia and Hermann Oberth of Germany, correctly theorized on the use of liquid hydrogen as a fuel with liquid oxygen for high-energy thrust. Based upon the liquid-hydrogen work at the Lewis Research Center, NASA developed the upper stages for the Saturn V which enabled us to achieve the Apollo missions to the moon."

  Hearing took place on 44th anniversary of first successful launching of liquid-fuel rocket by Dr. Goddard. (Testimony; *CR*, 3/17/70, S3828-30; *Huntsville News*, 3/17/70)

- Harold G. Miller, Chief of Mission Simulation Branch in MSC's Flight Control Div., received AIAA's DeFlorez Training Award in Cape Kennedy for "contributing significantly to the United States space flight programs through development of flight controller training techniques and capabilities." (AIAA Release, 3/11/70)

- Hospital in Sydney, Australia, was using blankets coated with aluminum foil and lined with polyester film to keep infants warm, *Chicago Tribune* said. Both substances, used in astronauts' spacesuits, had high-heat-retention properties. (*C Trib*, 3/16/70)

- Dual-input transponder using two transmitting-receiving antennas to achieve uninterrupted line of sight to ATC ground interrogators within range was described in one of eight technical reports announced by FAA. (FAA Release T 70-16)

March 17: U.S.S.R. launched *Meteor III* weather satellite from Plesetsk into orbit with 633-km (393.3-mi) apogee, 537-km (333.7-mi) perigee, 96.3-min period, and 81.1° inclination. (GSFC *SSR*, 3/31/70; *SBD*, 3/19/70, 94)

- Rep. George P. Miller (D-Calif.), for himself and others, introduced H.R. 16516, $3.630-billion NASA FY 1971 authorization bill which reflected March 12 committee action, superceded H.R. 15695 introduced Feb. 4, and added $297.8 million to President's budget request of $3.333 billion. New bill would allocate $2.903 billion for R&D, including: Apollo, $1.101 billion; space flight operations, $670.2 million; advanced missions, $1 million; physics and astronomy, $110.4 million; lunar and planetary exploration, $144.9 million; bioscience, $12.9 million; space applications, $172.6 million; launch vehicle procurement, $124.9 million; space vehicle systems, $30 million; electronics systems, $23.9 million; human factor systems, $18.3 million; and basic research, $18 million. Space power and electric propulsion systems would remain at $30.9 million; nuclear rockets, at $38 million; chemical propulsion, at $20.3 million; and aeronautical vehicles, at $87.1 million. Tracking and data acquisition would drop to $293.8 million; technology utilization would increase to $4.5 million.

Construction of facilities would be cut to $33.9 million. Research and program management funds would be increased to $693.7 million. (Text)

- Gen. John D. Ryan, USAF Chief of Staff, presented USAF FY 1971 budget request to House Committee on Armed Services: U.S.S.R.'s operational ICBM force outnumbered that of U.S. and, "by mid-1971, will probably exceed ours by several hundred launchers." Soviets had technology to develop hard-target MIRV and "Fractional Orbit Bombardment (FOBS) or depressed trajectory ICBM (DICBM) may already be operational. Their new 'swing-wing' bomber prototype could become operational, and their submarine-launched ballistic missile force is expanding rapidly."

U.S. continued to reduce bomber strength of Strategic Air Command. "We have phased out the B-58s, and severely cut back the FB-111s. At end FY-71, we will have fewer B-52s and 66 FB-111s. We hope to offset the reduced strength by increasing the capability of our bombers with the Short Range Attack Missile (SRAM) and the Subsonic Cruise Armed Decoy (SCAD) and providing for increased prelaunch survivability through satellite basing and improved warning." In last five years, "we have cut our active interceptor strength by two-thirds and our radars by one-half. The Soviets during this period have maintained their Long Range Aviation forces at about 200 heavy bombers and over 700 medium bombers and tankers...." (Testimony)

In testimony before Senate Armed Services Committee, L/G Otto J. Glasser (USAF) said development of short-range attack missile (SRAM) would be completed in September. USAF would run 12 tests of advanced ballistic reentry system (ABRES), push studies of new midcourse surveillance system to warn of attacking missiles, operate four large-payload flight tests using Atlas E and F boosters in ABRES program, and continue advanced development "to apply new technology to the mid-course tracking and discrimination problem." Satellite system looked "promising" but USAF was investigating alternate sensor/platform combinations "to assure that no promising concept is overlooked." USAF advanced avionics program was pursuing electronically scanned, phased-array radar antenna for strike aircraft and completion of all-weather close-support weapon-delivery system. (*Aero Daily*, 3/18/70, 20)

- Alabama Space and Rocket Center at Huntsville opened with $10-million exhibit of hardware from MSFC, Army Missile Command, and U.S. aerospace-missile companies. Speakers at dedication ceremonies included Dr. Wernher von Braun—NASA Deputy Associate Administrator and former MSFC Director—and Astronaut Frank Borman. Center, managed by Alabama Space Science Exhibit Commission, was financed by $1.9-million state bond issue and built on 14 hectares (35 acres) transferred by USA from Redstone Arsenal property. (*Marshall Star*, 3/11/70, 1)

- Secretary of Transportation John A. Volpe announced award of $3-million DOT contract to Grumman Aerospace Corp. to design 480-km-per-hr (300-mph) tracked air-cushion research vehicle (TACRV). (DOT Release 6370)

March 17-19: Manned Flight Awareness Workshop at MSFC was attended by 51 representatives of nearly 20 industrial firms and several Government agencies, to plan promotion of space flight safety and hardware reliability among work forces during 1970s. (MSFC Release 70-49; NASA PAO)

March 18: U.S.S.R. launched *Cosmos CCCXXVII* from Plesetsk into orbit with 823-km (511.4-mi) apogee, 268-km (166.5-mi) perigee, 95.5-min period, and 70.9° inclination. (GSFC *SSR*, 3/31/70; *SBD*, 3/20/70, 99)

- National Space Club's Robert H. Goddard Memorial Trophy was presented to *Apollo 11* Astronauts Neil A. Armstrong, Edwin E. Aldrin, Jr., and Michael Collins at 13th Annual Goddard Memorial Dinner in Washington, D.C. Astronauts were cited for successfully accomplishing lunar landing that "climaxed a decade of dedication and technological advancement, and gave millions of Americans pride in their country and its achievements."

  Other awards included Goddard Historical Essay Award to Prof. John M. Logsdon of Catholic Univ. and Frank H. Winter of National Air and Space Museum and NSC Press Award to Thomas O'Toole of *Washington Post*. Astronautics Engineer Award was given to Dr. George M. Low, NASA Deputy Administrator; and Nelson P. Jackson Aerospace Award, to Grumman Aerospace Corp. for Apollo LM. (Program)

- *Apollo 12* astronauts arrived in Djakarta, Indonesia, during world tour. They were welcomed by several thousand persons. (AP, *W Post*, 3/19/70)

- Rep. George P. Miller (D-Calif.) introduced for himself and Rep. James G. Fulton (R-Pa.) and Rep. Ken Hechler (D-W. Va.) H.R. 16539, "to amend the National Aeronautics and Space Act of 1958 to provide that the Secretary of Transportation shall be a member of the National Aeronautics and Space Council." (*CR*, 3/18/70, H1891)

- Dr. Glenn T. Seaborg, AEC Commissioner, announced selection of five scientists to receive 1970 Ernest Orlando Lawrence Memorial Award for meritorious work in atomic energy: Dr. William J. Blair, Pacific Northwest Laboratory; Dr. James W. Cobble, Purdue Univ.; Dr. Joseph M. Hendrie, Brookhaven National Laboratory; and Dr. Michael M. May and Dr. Andrew M. Sessler, Lawrence Radiation Laboratory. Awards would be presented May 11 at Univ. of California at Berkeley. (AEC Release N-39)

March 19: USAF's X-24A lifting-body vehicle, piloted by Maj. Gerauld R. Gentry (USAF), successfully completed first powered flight after air launch from B-52 aircraft at 12 200-km (40 000-ft) altitude. XLR-11 rocket engine burned 160 secs, boosting vehicle to 13 700-km (45 000-ft) altitude and mach 0.8 before it glided to landing on Rogers Dry Lake, Calif. Objective of flight, 10th in X-24A series, was to check operation of rocket propulsion system, stability and control system, and handling characteristics during powered flight. (NASA Proj Off; AFFTC Release 70-3)

- NASA launched series of three Nike-Apache sounding rockets from TERLS to measure positive ion composition in daytime equatorial D and E region and to measure variation of ion composition and density and electron density with altitude. Rockets and instruments functioned satisfactorily and data were expected to yield valuable information. (NASA Rpts SRL)
- House Committee on Science and Astronautics favorably reported H.R. 16516, FY 1971 NASA authorization bill totaling $3.630 billion, as introduced March 17. (CR, 3/19/70, D262; H Rpt 91-929)
- *Apollo 11* commander Neil A. Armstrong was one of 10 persons named to receive National Civil Service League's 15th annual career service awards. Each honoree would receive $1000, inscribed watch, and citation at banquet May 8. (W *Star*, 3/19/70, A2)
- MSFC was launching "low-key campaign to encourage almost 1000 of its federal civil servants to retire," *Huntsville Times* said. Move was "to get 'old timers'...to voluntarily move out of their jobs, making room for younger workers who conceivably could lose their jobs in the space agency's belt-tightening." (*Huntsville Times*, 3/19/70)
- Dr. Robert C. Seamans, Jr., Secretary of the Air Force, had told closed session of House Armed Services Committee that USAF planned "continued rapid retrenchment" to lowest expenditure level since Korean War buildup, *Washington Daily News* reported. Planned purchase of 390 aircraft was "smallest number...procured in any one year" in USAF history (as a separate service). (*W News*, 3/19/70, 20)

March 20-25: North Atlantic Treaty Organization's *Nato I* (NATO-A) military comsat was successfully launched from ETR at 6:52 pm EST by NASA for USAF and NATO by long-tank, thrust-augmented Thor-Delta (DSV-3M) booster. Spacecraft entered transfer orbit with 36 926.6-km (22 950.0-mi) apogee, 290.0-km (180.7-mi) perigee, 653-min period, and 25.6° inclination. NASA objective was to place spacecraft into transfer orbit accurate enough for apogee motor to place spacecraft into synchronous equatorial orbit.

Apogee motor was fired March 23 and spacecraft entered circular orbit with 36 619.2-km (22 759.0-mi) apogee, 34 421.3-km (21 393.0-mi) perigee, 23.5-hr period, 0.3° inclination, and 5.7°-per-day eastward orbital drift. Spacecraft would arrive on station over Atlantic at 18° west longitude June 10. All systems had operated normally since March 25 turn-on.

Nato I was first of two NATO military comsats scheduled to be launched into orbit over Atlantic under DOD-NATO agreement. USAF managed project for DOD and would reimburse NASA for launch services. NATO would reimburse USAF. Spacecraft weighed 242.7 kg (535 lbs) and consisted of two concentric cylinders containing redundant X-band communication systems.

Initial operation of spacecraft telemetry and command functions were performed from USAF satellite control facility. Control of orbital operations would be transferred to NATO after spacecraft reached station. (NASA Proj Off; *SBD*, 3/24/70, 111)

March 20: Nike-Cajun sounding rocket, launched by NASA from WSMR, carried MSC payload to 56-km (34.8-mi) altitude to measure, simultaneously, hydroxyl airglow and ozone distributions during day and night. Results would be used to calculate atomic hydrogen distributions in 75-90-km (47-56-mi) region. Rocket trajectory was low and clamshell nosecone separated at liftoff. Data were not useable, but payload was recovered in reusable condition. (NASA Rpt SRL)

- Nixon Administration was reviewing need for Government program to stockpile helium, *Science* reported. Program, "which provides for extracting helium from streams of natural gas and storing it underground for future use, has recently been running at a huge deficit." Scientists were worried that Administration might allow squandering of "priceless natural resource that may be desperately needed by future generations." Helium had unique properties "for which there is no real substitute in high-technology applications." Largest use was "as a purging and pressurizing agent in liquid-fueled rockets." Helium also was expected to play key role in development of nuclear reactors, lasers and masers, magnetohydrodamics, and superconducting cables for electric power.

  Threat to supply lay in use of natural gas as domestic fuel. Unless helium was extracted first, it was passed into atmosphere when natural gas was burned. Federal Government had begun conservation program in 1960. (Boffey, *Science*, 3/20/70, 1593-6; *Budget of the U.S. Government, FY 1971*, 357)

- *Apollo 8* Astronaut Frank Borman appeared as host on ABC TV program "Mission Possible—They Care for a City." Program outlined attempts to improve environment of San Francisco. (W *Star*, 3/17/70)

- *Science* published letter on environment from John Caffrey of American Council on Education: "I date my own reawakening of interest in man's environment to the Apollo 8 mission and to the first clear photographs of the earth from that mission. My theory is that the views of the earth from that expedition and from the subsequent Apollo flights have made many of us see the earth as a whole, in a curious way—as a single environment in which hundreds of millions of human beings have a stake. . . . I suspect that the greatest lasting benefit of the Apollo missions may be. . .this sudden rush of inspiration to try to save this fragile environment—the *whole* one—if we still can." (*Science*, 3/20/70, 1561)

- International conference on global communications satellite system adjourned until autumn without reaching agreement on permanent operating arrangements. Conference established Intersessional Working Group to negotiate and draft intergovernmental and operating agreement reflecting compromise proposals offered by Australia and Japan. (Rpt of US Del to Sec State 4-3-70)

- *Science* article discussed "puzzling observations and bizarre theories" accrued from studies of galaxies and quasars: "Starting with the discovery of radio galaxies in the mid-1950s astronomers have shown that many types of galaxies and (since 1963) quasars release more energy than can be accounted for by known physical processes." New theories discussed seemed "to bring two trends to the surface. First is the idea that a collection of discrete sources helps get around the severe limitations placed on the size of the radiating object and may help account for the random variability observed in the emission of many extragalactic objects. Second, the idea that many collections of matter may be expanding seems to be making a small dent in the long-held view that all objects are condensing from a homogeneous universe." (Holcomb, *Science,* 3/20/70, 1601-3)

March 21: Theory that helped to explain source of uncontrolled malignant growth and indicated short cuts to development of chemical countermeasures against cancer was described by LaRC biophysicist Clarence D. Cone, Jr., at 12th Annual Science Writers Seminar of American Cancer Society in San Antonio, Tex. Theory proposed that division of body cells was controlled precisely by pattern of ion concentrations on surface tissue of cells. Pattern was formed by electrical voltage that normally existed across cellular surfaces and varied from one part of body to another. Theory provided explanation of functional connection between two major pathological features of cancer—uncontrolled growth of cells and spread of disease in body—and implied that basic functional aberrancy producing both conditions lay in alteration of molecular structure of cell surface. Theory proposed central mechanism for control of body cell division, which, if proved valid, would provide new basis for research progress on key biomedical problems, such as human conception, birth defects, growth, aging, and cancer. (Text; NASA Release 70-43)

• Laser weapons were likely to enter U.S. armory within five years, *Manchester Guardian* said. Since defense expenditure on lasers began in 1959, U.S. had become prime research agency and had allocated $8.4 million for high-energy laser research in 1971. U.S. plans were reported as including "possibility of destroying low flying missiles, the disabling and destruction of aircraft, probably through ignition of fuel, and the disabling of fuse mechanisms and of infra-red homing or detection devices." (*Manchester Guardian,* 3/21/70)

• U.S.S.R. began series of carrier rocket tests in the Pacific about 725 km (450 mi) north northeast of Midway Island and warned ships and aircraft to stay away from area with 177-km (110-mi) diameter until completion of tests April 10. (AP, *NYT,* 3/22/70, 7)

March 21-26: IEEE held annual meeting in New York. Successful use of laser beams to transmit long-distance telephone calls was described by Prof. Raphael A. Kazaryan of Armenian Academy of Sciences. He said 24-km (15-mi), 24-channel telephone link-up in Soviet Armenia was world's first "operational" application of lasers in telephoning. System had been in service between Yerevan, capital of Armenia, and Burakan Astrophysical Observatory since July 1969. Another six-mile system was being tested in Moscow. In interview Prof. Kazaryan said system, using helium-neon laser, had been 80 percent reliable during most difficult transmission times when reflected sunlight interfered with communications light beam. More powerful laser, using carbon dioxide to generate beam, would be employed soon to improve system. (Wilford, *NYT,* 3/29/70, 53)

March 22: NASA's intention to assist Texas with problems of pollution and preservation of natural resources and environment had been announced by Texas Gov. Preston Smith, *Houston Chronicle* reported. Governor had asked NASA Administrator, Dr. Thomas O. Paine, for access to "vast wealth of scientific information" in Earth Resources data bank at MSC. Data would include remote sensor material from aircraft and satellites over Texas and its coastline. (*H Chron*, 3/22/70)

• Lawrence W. Vogel, former Special Assistant to NASA Acting Associate Administrator for Organization and Management, became Director of NASA Hq. Administration. He succeeded Charles G. Haynes, who became Deputy Auditor General at Agency for International Development. (NASA Ann, 3/16/70)

March 23: NSF released *Federal Support for Research and Development at Universities and Colleges and Selected Nonprofit Institutions, FY 1968: A Report to the President and Congress* (NSF 69-33). Federal Government had obligated $1.4 billion for R&D in FY 1968—5% increase over FY 1967. Average annual increase of Federal obligations to universities and colleges for R&D 1955-1967 had been over 21%. Dept. of Agriculture and DOD reported cutbacks in academic R&D funding in FY 1968, including $21 million in defense R&D. Number of academic institutions receiving Federal support rose 8%, from 573 in FY 1967 to 620 in FY 1968. One hundred top-ranking institutions received 93% of R&D totals obligated by NASA and AEC and 88% of DOD R&D total in 1968. MIT was largest recipient again, with $80 million, "three-fourths of which was obligated by DOD and NASA." (Text)

• North American Rockwell Corp. would lay off 8000 employees between April 1 and Nov. 1, *Wall Street Journal* quoted NR Chairman Willard F. Rockwell as saying in interview. Most reductions would be in aerospace operations at Southern California facilities. About 30% of workers to be laid off were engineers and 10% were management employees. (James, *WSJ*, 3/23/70)

• First volume of third edition of Bolshaya *Sovetskaya Entsklopediya*, new Soviet encyclopedia, was released in U.S.S.R. It credited Wright Brothers with first airplane flight. Second edition, which appeared from 1950 to 1958, had contended that Russian, A. F. Mozhaisky, was designer of first airplane, which had been flown first during summer of 1882. This version had appeared in U.S.S.R. as late as 1968, in history of Russian aviation and astronautics published by Ministry of Defense. New edition of encyclopedia said Mozhaisky's airplane was built in 1885 and "crashed on take-off." (Gwertzman, *NYT*, 3/24/70, 6)

• House Committee on Science and Astronautics submitted to Congress *Issues and Directions for Aeronautical Research and Development*, report of Subcommittee on Advanced Research and Technology. Report recommended that U.S. establish national aeronautics and aviation policy emphasizing aeronautical R&D, Government-industry roles be clarified, and DOT be statutory member of NASC. Report found reason for concern over maintenance of long-term U.S. leadership in aeronautics and aviation. Technological R&D base needed to be rebuilt and existing technology should be used more fruitfully. (Text)

March 24: FCC announced it would consider "applications to establish and operate domestic communication satellite facilities for multiple or specialized common carrier services, for lease to other common carriers, for private use, joint cooperative use, or any combination of such services." FCC had been unable to "determine on the basis of information it now has whether a domestic satellite program could best be developed by authorization of a multi-purpose or specialized or a combination of systems or through an 'open entry' policy." Commission would permit "potential applicants to take the initiative in submitting concrete system proposals for . . . consideration." (FCC Public Notice Rpt 5862)

- Senate investigators disclosed F-111 range in very-low-level flight at supersonic speed was only 56 km (35 mi) rather than 386 km (240 mi) called for in production contract. *New York Times* later said USAF sources privately acknowledged high-dash range near tree-top level had been prime goal in F-111 development. Sources had suggested deficiency could be mitigated by reduction in speed or by use of extra, droppable fuel tanks. (Witkin, *NYT,* 3/25/70, 11)

- *Apollo 12* Astronauts Alan L. Bean, Richard F. Gordon, Jr., and Charles Conrad, Jr., visited Expo '70 in Osaka, Japan, during world tour. They climbed out of mock spaceship in U.S. Pavilion and onto simulated lunar surface. (AP, W *Star,* 3/24/70, A2)

- President Nixon submitted to Senate nomination of Kenneth M. Smith, President and Chief Executive of Windecker Research Inc., to be FAA Deputy Administrator. He would replace David D. Thomas, who had resigned. (*PD,* 3/30/70, 421, 447)

- VTOL Dynastat, helium-filled airship that could carry 100 passengers 300-800 km (200-500 mi) in 30- by 9-m (100- by 30-ft) car, was described at symposium of American Helium Society in Washington, D.C. Goodyear Aerospace Corp. scientists Dr. Robert S. Ross and Philip F. Myers said airship would be powered by four to six engines driving propellers with forward and vertical thrust. (Goodyear *Profile,* 5/70, 9-10; *Huntsville Times,* 3/29/70)

- House Committee on Science and Astronautics ordered favorably reported H.R. 16595, NSF FY 1971 authorization bill, after adding $27.6 million to President Nixon's budget request of $500 million. (*CR,* 3/24/70, D283; Boffey, *Science,* 4/3/70, 95-7)

March 25: President Nixon announced transfer of Electronics Research Center in Cambridge, Mass., from NASA to DOT, effective July 1. James C. Elms would continue as Director and Center would be renamed Transportation Development Center. Later, Secretary of Transportation John A. Volpe announced DOT would use center for R&D efforts on systems and equipment to automate air traffic control system more highly, economically feasible collision avoidance system for commercial and private aircraft, sensors to measure and monitor air pollution caused by transportation elements, automatic landing systems for aircraft, system analysis on urban transit systems and highway traffic control, ocean-data buoy systems to transmit weather and oceanographic data, and auto-driver simulation systems to study accident causes and prevention. (*PD,* 3/30/70, 446; DOT Release 6870)

- NASA launched two sounding rockets from Churchill Research Range. Aerobee 150MI launched with VAM-17 booster carried Johns Hopkins Univ. and Univ. of Pittsburgh payload to 164.4-km (102.2-mi) altitude. Objective was to measure neutral and ion composition in auroral discharge, visible emission features of aurora, primary and secondary electron energy distributions, and auroral height profile. Rocket and instruments functioned satisfactorily.

Nike-Tomahawk carried Univ. of Colorado payload to 235-km (146.0-mi) altitude to measure Vegard-Kaplan and Lyman-Birge-Hopfield bands of nitrogen in UV auroral emissions. Rocket and instruments functioned satisfactorily and good data were obtained. Prototype parachute recovery system, being tested for first time, failed and payload was not recovered. (NASA Rpts SRL)

- Three security automobiles burst into flames in fog-like oxygen vapor 335 m (1100 ft) from Apollo 13's Saturn V booster at KSC Launch Complex 39, Pad A, during rocket fueling for countdown demonstration test. Patrolmen escaped injury and booster was not damaged. Test was delayed nearly four hours, but engineers successfully completed countdown with simulated ignition. Vented oxygen vapor apparently mixed with gasoline in carburetors and was ignited by spark plugs. Dr. Kurt H. Debus, KSC Director, appointed board to investigate accident. (UPI, *W Post*, 3/26/70, A3)

- U.S. and Australia signed 10-yr extension to agreement under which Australia managed NASA tracking activities. Original 10-yr agreement effective Feb. 26, 1960, would extend to Feb. 26, 1980. Australian Dept. of Supply would continue to manage NASA tracking, data acquisition, and communications facilities near Canberra, Woomera, Carnarvon, and Cooby Creek. New antenna 64 m (210 ft) in diameter for deep space tracking would be installed at Tidbinbilla. Australian facilities—largest concentration of NASA facilities outside U.S.—represented $77-million investment; employed more than 700 Australian engineers, technicians, and support staff; and operated at annual cost of $14 million. (NASA Release 70-33)

- L/G John W. O'Neill (USAF), AFSC Vice Commander, testified on status of F-111 aircraft before Senate Committee on Government Operations' Permanent Subcommittee on Investigations. USAF investigation of Dec. 22, 1969, crash that had killed two pilots near Nellis AFB, Nev., had shown "structural failure of the left wing occurred at very low altitude, during a normal dive recovery after a routine delivery of air-to-ground ordnance. Although the aircrew escape module performed as it was designed, time between ejection and ground impact did not allow the recovery chute to open." Metallurgical flaw, "somewhat like an ultra-thin hairline crack, existed in the lower plate of the wing pivot fitting." In addition to USAF investigations, special ad hoc committee of Air Force Scientific Advisory Board had reviewed findings and technical data. "We have concluded that the flaw was introduced at a subcontractor facility at some point during the manufacture of the lower plate. . . ."

No modifications to present structure were planned but structural inspections of all F-111s would "utilize improved ultrasonic inspection techniques and procedures developed as a direct result of this problem." Program would cost estimated $31.2 million. First aircraft were to be completed and returned to flight status in June 1970. All aircraft should be restored by shortly after first of year. (Testimony)

- Lowell Observatory astronomer Dr. Peter Boyce had reported "blue smog" of Mars was caused by physical changes on planet's surface rather than by looking at Mars through earth's atmosphere, *Philadelphia Inquirer* said. Dr. Boyce had used special area-scanning photometer attached to Lowell Observatory's 610-mm (24-in) reflector at Cerro Tololo in Chilean Andes to obtain accurate photometric measurements of Mars. (*P Inq*, 3/25/70)

March 26-27: Princeton Univ. astronomers successfully launched Stratoscope II, 91-cm-dia (36-in-dia) balloon-borne optical telescope, from National Scientific Balloon Flight Station at Palestine, Tex. Telescope remained aloft during night at 24 400-m (80 000-ft) altitude where it photographed Uranus, Jupiter, nebulae, and nuclei of Seyfert galaxy with resolution three times greater than ground-based photos. Project was supported by NASA and NSF. (NASA Release 70-44; NASA PAO)

March 26: Apollo 13 Astronauts James A. Lovell, Jr., Fred W. Haise, Jr., and Thomas K. Mattingly II underwent brief medical examination, donned spacesuits, and practiced countdown procedures at KSC for 2 1/2 hrs in preparation for launch toward moon April 11. (AP, *NYT*, 3/27/70, 23)

- NASA named flight and support crews for Apollo 15 lunar landing mission scheduled for autumn of 1970. Prime crewmen were Astronauts David R. Scott, commander; Alfred M. Worden, CM pilot; and James B. Irwin, LM pilot. Backup crew consisted of Astronauts Richard F. Gordon, Jr., Vance D. Brand, and Harrison H. Schmitt. Support crewmen were Scientist-Astronauts Karl G. Henize and Robert A. Parker. (NASA Release 70-46)

- Two photographs taken from *Apollo 9* spacecraft in March 1969 had been used as backdrop for experimental topographic map of Phoenix, Ariz., area, *New York Times* reported. U.S. Geological Survey scientist Dr. A.P. Colvocoresses had said that "on ordinary maps the lines and symbols used to represent geographic features in such terrain leave many areas essentially blank. On this map, however, there are no blank areas.... The map is easy to 'read' because the photo shows the actual land surface, enhanced by superposition of cartographic details." (*NYT*, 3/26/70)

- Responding to question on astronaut pay, *Washington Daily News* columnist Hy Gardner quoted NASA as saying astronauts received no supplemental allowance for flying space missions. During last weeks before mission launch, they were on temporary duty assignment at KSC where they resided at Government expense. Status continued until they returned home to Houston after mission. During temporary duty period, military astronauts received allotment of $2 per day and civilian astronauts, $3.20 per day. (*W News*, 3/26/70, 40)

- Sun shone less brightly than had been supposed, studies at GSFC and by Cal Tech team and Dr. Andrew J. Drummond of Eppley Laboratory had determined, *New York Times* reported. GSFC had used NASA's Convair 990 jet aircraft to collect data. Combined results had been evaluated by committee of Government and academic scientists headed by Dr. Matthew P. Teokaekara, who led GSFC experiments. Intensity of visible sunlight was now believed to be eight percent lower than generally accepted figure of two calories per square centimeter per minute. Estimate for total flow of energy from sun to earth—solar constant—had been revised downward three percent. (Sullivan, *NYT*, 3/26/70, 26)

- Four particles of moon rock no bigger than tea leaf went on display at Science Museum in London. (Reuters, *W Post*, 3/27/70, C5)
- New reconnaissance aircraft, Q-Star, was successfully demonstrated by Lockheed Missiles & Space Co. in Palo Alto, Calif. Modified glider featured quiet flight achieved through special Curtiss-Wright Corp. engine and slow-turning wooden propeller. Observers said they could not tell aircraft was overhead without watching sky. (UPI, *W Post*, 3/27/70)

March 27: U.S.S.R. launched *Cosmos CCCXXVIII* from Plesetsk into orbit with 316-km (196.4-mi) apogee, 206-km (128.0-mi) perigee, 89.7-min period, and 72.8° inclination. Satellite reentered April 9. (GSFC *SSR*, 3/31/70, 4/30/70; *SBD*, 3/31/70, 142)

- NASA launched series of four Boosted Arcas sounding rockets from TERLS to measure positive and negative ion densities with Gerdien condenser and make simultaneous radio propagation measurements of electron density. NASA also launched Nike-Apache from TERLS carrying Indian experiment to study magnetic field and ionosphere. Rockets and instruments functioned satisfactorily. (NASA Rpts SRL; SR list)
- Scientific data on March 7 solar eclipse collected by NASA Aerobee 170 sounding rocket at 173.8-km (108-mi) altitude had been recovered from 1783 m (5850 ft) under water by USN cable-controlled underwater recovery vehicle (CURV III), Wallops Station announced. NRL payload carrying four cameras had photographed solar flash spectrum in eclipsed sun's chromosphere, but parachute failed to open and payload sank 120 km (75 mi) off coast of Wallops Island, Va. After recovery, water-soaked film was maintained at 275 K (35°F) while scientists devised special processing techniques to restore film. Photos containing prime UV spectrum during total eclipse were salvaged. (WS Release 70-5)
- President Nixon announced intention to nominate four assistant directors of NSF: Edward C. Creutz to be Assistant Director for Research; Lloyd E. Humphreys to be Assistant Director for Education; Lewis Levin to be Assistant Director for Institutional Programs; and Thomas B. Owen to be Assistant Director for National and International Programs. Nominations were submitted to Senate March 31, 1970. (*PD*, 3/30/70, 444; 4/6/70, 475)
- FAA issuance of advance notice of proposed aircraft engine emission standards, soliciting comments, was announced by Secretary of Transportation John A. Volpe. Studies had shown that one percent of total U.S. air pollution might be attributed to aircraft emissions. (FAA Release 70-28)

March 27-28: NASA launched two Nike-Cajun sounding rockets—Rehbar 21 and Rehbar 22—from Karachi, Pakistan, carrying Pakistan-U.K. grenade experiments. Rockets and instruments functioned satisfactorily. (SR list)

March 28: X-15 rocket research aircraft, holder of 7274-km per hr (4520-mph) speed record for manned aircraft, went on display at Alabama Space and Rocket Center near MSFC. Once piloted by *Apollo 11* Astronaut Neil A. Armstrong and one of three manufactured for NASA,

the 22 700-kg (50 000-lb) $50-million X-15 was on loan from Air Force Museum at Wright-Patterson AFB, Ohio. Second X-15 was on display at Smithsonian Institution in Washington, D.C. Third had been destroyed in 1967 crash. (*Birmingham Post-Herald*, 3/28/70)
- Retrorocket parachute—with rocket to provide upthrust to descending parachute carrying heavy and delicate equipment just before it hit ground—was among items shown in Soviet film purchased for U.S. market, AP reported. Half-hour movie made to show Soviet people power of their armed forces, depicted what it called "awesome arsenal" including small rocket that skittered just off ground in search of enemy tanks. (*C Trib*, 3/29/70, 1A)

March 28-29: U.S.S.R. tested two SS-9 missiles with triple warheads. Missiles traveled more than 9260 km (5754 mi) before dropping warheads into Pacific south of Aleutian Islands, according to Jerry W. Friedheim, Principal Deputy Assistant Secretary of Defense for Public Affairs. (Kelly, *W Star*, 3/30/70, A4)

March 29: NASA invited world scientific community to submit by June 1 proposals for scientific research on lunar samples to be returned from Apollo missions 14 through 19. Specific RFPs were being sent to nearly 200 teams of scientists in U.S. and 16 foreign countries that were analyzing *Apollo 11* and *Apollo 12* samples and would analyze samples from Apollo 13. (NASA Release 70-45)
- Dr. William H. Pickering, JPL Director, announced establishment of Civil Systems Projects Office at JPL to "develop applications of the laboratory's capabilities for solution of problems in medical engineering, public safety support, urban land use and transportation." Office would be headed by Howard H. Haglund, former Surveyor Project Manager. (JPL Release 548)

March 30: Japanese government had asked U.S.S.R. to suspend series of carrier rocket tests begun in Pacific March 21, *Space Business Daily* reported. Japan said tests so close to Japanese islands of Kuriles, Shikoku, and Niigaia would disrupt shipping, air service, and fishing and warned that continuing tests could damage Soviet-Japanese relations. (*SBD*, 3/30/70, 138)
- DOD announced award of $51 081 576 to General Dynamics Corp. under existing USAF contract for production of F-111 aircraft. (DOD Release 239-70)

March 31: First U.S. satellite—14-kg (30.8-lb) *Explorer I*, launched from Cape Canaveral, Fla., by Jupiter C booster Jan. 31, 1958—reentered atmosphere over South Pacific. Satellite had discovered Van Allen radiation belts, had completed 58 408 revolutions of the earth, and traveled 2.67 billion km (1.66 billion mi).

Dr. Wernher von Braun, NASA Deputy Associate Administrator for Planning, who was Director of Army Ballistic Missile Agency at time of launch, said: "By today's standards Explorer 1 was a feeble, first step in space. But in its day it was an outstanding accomplishment, done on short notice to place the free world in the space race. . . . We have come a long way from that tiny Explorer, demonstrated by the fact that the Saturn V we are now flying can place in Earth orbit about 10,000 times as much payload as the little Jupiter-C that launched Explorer 1." (NASA Release 70-49; GSFC Mission & Analysis Div)
- In first public appearance since assuming duties at NASA Hq. as Depu-

March 31: Explorer I, first U.S. satellite (launched Jan. 31, 1958), reentered the atmosphere over the South Pacific after completing 58 408 revolutions and traveling 2.67 billion kilometers (1.66 billion miles) during its 13-year lifetime. To celebrate the launch, Vice President Richard M. Nixon visited Jet Propulsion Laboratory Feb. 17, 1958. Examining the spacecraft model with him (left to right above) were Dr. Clark B. Millikan, Director of the Guggenheim Aeronautical Laboratory at Cal Tech; Dr. Lee A. DuBridge, President of Cal Tech; Dr. William H. Pickering, Director of JPL; Dr. Jack E. Froehlich, Explorer Project Director at JPL; V. C. Larsen, Jr., Laboratory Administrator at JPL; and Robert J. Parks, Chief of JPL Guidance Control Division.

ty Associate Administrator for Planning, Dr. Wernher von Braun held press conference on NASA planning procedures: Skylab program would consist of four flights. First would be an unmanned flight boosted by two-stage Saturn V—only first and second stage, with third stage replaced by orbital workshop, now called Skylab.

"It will be this country's first orbital space station. Attached to this Skylab is an airlock module, and a multiple docking adapter, to which arriving command and service modules can dock for crew exchange. And also attached to it is the ATM...a manned solar observatory that will be serviced by the people living in the Skylab. One day after this Skylab has been launched into orbit and has deployed its solar panels, both on the ATM and the Skylab itself, it will be visited by a Saturn 1B-launched command and service module, bringing the first crew complement of three. The command module will dock to the multiple docking adapter, the crew will slip through the docking adapter and the airlock, into the workshop. . . . And the men will stay up there for 28 days." Crew would then "crawl back into their command module, detach...use the service module to deboost themselves, and the command module will reenter the ocean and make a normal Apollo landing.

"About three months after the first flight, another flight will go up, and this time the crew stay...will be extended to 56 days.... And finally, there will be a third revisit, again of 56 days, after which the third crew will descend. The Skylab will then go into storage and can potentially be reactivated any time thereafter if more revisits are planned."

Even without budget restraints, it would be "something like 1977 before we could have a [space] shuttle flying."

NASA was "in continuous touch with the Air Force" in space shuttle and space station field. "The armed forces...have the duty to look into everything that may have a defense aspect; and so it is entirely proper that they want to understand the ramifications of space." (Transcript)

- First powered launch of USN's solid-propellant Condor air-to-surface missile was successfully conducted at China Lake, Calif. Missile flew preprogrammed trajectory to impact after launch from F-4A Phantom jet at high altitude and mach 0.58. All test objectives were met. (NR News, 4/3/70, 1)
- *Christian Science Monitor* editorial commented on transfer of ERC to DOT for use as Transportation Systems Center: "However glamorous it may be, travel in space is of little immediate importance to the average man. But travel on earth becomes daily more important—and often more difficult.... This is an extraordinarily important and welcome move. It means that, perhaps for the first time in the United States, there will be continuous, large-scale research into all aspects of land, sea, and air transportation by a large and highly competent staff." (*CSM*, 3/31/70)
- U.S.S.R. announced successful completion of carrier rocket tests begun in Pacific March 21. Tests, scheduled for completion April 10, had been protested by Japanese government March 30. (AP, *W Post*, 4/1/70, A17)
- DOD announced USAF contract awards for operation and maintenance of Air Force Satellite Control Facility tracking stations: $4 300 000 initial increment to Lockheed Aircraft Corp. under $7 772 050 contract and $3 078 000 initial increment under $4 578 444 contract; and $4 500 000 initial increment to Philco-Ford Corp. under $8 424 250 fixed-price contract. (DOD Release 241-70)

During March: Dr. James A. Van Allen, Univ. of Iowa Director of Physics and Astronomy Depts., discussed significance of lunar landings, in university lecture series: "The Apollo missions, to me, are straightforward though immensely difficult tasks. They do, however, yield relatively little in the way of fundamental understanding of nature. They are not scientific in that sense. There is a longstanding controversy as to whether a manned spacecraft is a better way to conduct science in space than an automated, commandable spacecraft. I'm sure there is no simple answer to that question. As the general romance and entertainment value of manned flight tend to wear off a little, I think this question will be attacked in a thoroughly pragmatic way." *Apollo 11* lunar landing "might properly be compared to the explorations of Amundsen and Perry and Byrd in the Arctic and Antarctic, or perhaps Lindbergh flying the Atlantic. These are great achievements, heroic achievements, but the general potential of the Moon in its relationship to human life on a large scale is by no means obvious to me. I don't think any competent person has found a significant, economic, human use for the Moon." There was general feeling, mostly from aeronautics and allied industries, that development of space flight was important to U.S. future in same sense that aircraft development had been. "I think this is fundamentally a false analogy." Space shuttle and space station were feasible. "Whether or not it is sensible to pursue them, I have a great difficulty in judging." (Univ of Iowa *Spectator*, 3/70, 6)

- AIA President Karl G. Harr, Jr., said in *Aerospace Technology:* "In terms of its scope and novelty no single enterprise matches man's landing on the moon—the Apollo program. Nor has any other enterprise in history broken so much new technical ground. The technical and managerial ingredients of these space accomplishments, as well as of comparable national security programs, are available for broader applications in the nation's service. The problem is how to harness them effectively for such purposes. The Apollo program centered on a clearly stated goal with adequate funding for its attainment. Because it represented achievement of a *national* goal, lines of authority and responsibility were clear, simple and strong. Our nation's environmental problems, however, often do not enjoy such clear delineations. . . . The greatest need is for the development of innovative means to overcome such constraints so that technology's revolutionary new tools can be more effectively applied to the pressing needs of our society. . . . The transfer of aerospace technology to domestic areas is underway and is already making significant contributions."

    Magazine cited computer technology used by NASA to evaluate spacecraft systems and by USN to study pilot behavior as space age spinoff which might "soon be used to conduct highway safety research or to design new houses and urban renewal projects." (*Aero Tech*, 3/70, 3)

- Soviet Cosmonaut Konstantin P. Feoktistov's comments on October-November 1969 visit to U.S. were published in *Space World:* "During the talks with the astronauts and specialists from NASA we learned many interesting things about the preparation for, and realization of, the American Space project. We visited the enterprises of the North American Rockwell Company which manufactures the command module of Apollo. I managed to sit in Apollo 14 which was being tested then. I saw quite a few good design solutions." (*Space World*, 3/70, 38-9)

- AIAA and AAS jointly established Technology Utilization Awards to stimulate aerospace engineers, scientists, and technicians to apply aerospace technology to commercial uses. Categories for consideration were pollution control, transportation, earth resources, communications, data handling, agriculture, medicine, housing, education, and earth environmental sciences. Abstracts on technology utilization ideas were due May 31 and full papers, Aug. 31. Plaques and $150-$800 awards would be presented in October. (NR *News*, 3/20/70, 1; AAS Release)

April 1970

April 1: Tenth anniversary of *Tiros I* (Television Infrared Observation Satellite), world's first weather satellite, launched by NASA from Cape Canaveral, Fla. Satellite had transmitted 22 952 pictures during 89-day lifetime. Since *Tiros I,* 23 meteorological satellites—10 Tiros, 3 Nimbus, 9 ESSA, and 1 ITOS—had been orbited, returning more than one million pictures of earth's cloud cover and other data for weather forecasting and research. (ESSA Release 70-18)

• Scientists tranquilized 227-kg (500-lb) elk near Jackson Hole, Wyo., and fitted her with 10-kg (23-lb) electronic collar to transmit data to orbiting *Nimbus III* (launched April 14, 1969). Elk, which had worn dummy collar for 44 days, rejoined herd and showed no ill effects. Experiment was sponsored by NASA, Dept. of Interior, and Smithsonian Institution and was second attempt to study migratory habits of large animals. First attempt had ended prematurely when Monique, first elk fitted with collar, died of pneumonia Feb. 24.

Nimbus III, orbiting at 1100-km (700-mi) altitude, successfully contacted second elk four times over 10-min period and would contact her twice daily as she migrated to summer grazing area. (NASA Release 70-53)

• Space shuttle mission capabilities were described by LeRoy E. Day, Manager of Space Shuttle Task Group in NASA OMSF, in paper presented at AIAA Conference on Test Effectiveness in the 70s at Palo Alto, Calif.: "It is envisioned that the shuttle will eventually replace essentially all of the present day launch vehicles or their derivatives except for very small vehicles of the Scout class and the Saturn V. This will be possible because low operational costs of the reusable shuttle will make it competitive even if it carries only a fraction of its full payload capability on particular missions. In addition to the low launch costs, we expect the benign acoustic and acceleration environment of the shuttle to allow significant reductions in the cost of payloads. Acceleration during descent and reentry will be less than 3 g's. Payload design will be further aided by allowing greater volume and weight for many payloads because of the shuttle cargo bay accommodations. Preliminary analyses indicate reductions in payload development costs of 25-30 percent may be expected." (Text)

• Approximately 155 representatives from industry, universities, and government attended pre-proposal briefing on proposed High Energy Astronomy Observatory (HEAO) at MSFC. (MSFC Release 70-58)

• Transfer of SST program from FAA to Office of Secretary of Transportation and appointment of William M. Magruder, Deputy Director of Commercial Engineering at Lockheed California Co., as Director of Supersonic Transport Development Program were announced by Secretary of Transportation John A. Volpe. (DOT Release 7670)

• AFSC announced that AFWTR at Vandenberg AFB, Calif., and SAMSO's aerospace test wings at Vandenberg and Patrick AFB, Fla., would be combined to form Space and Missile Test Center (SAMTEC) under SAMSO command. (AFSC Release 65.70)

April 2: USAF's X-24A lifting-body vehicle piloted by NASA test pilot John A. Manke reached 18 300-m (60 000-ft) altitude and mach 0.85 during second powered flight after air launch from B-52 aircraft at 12 200-m (40 000-ft) altitude from FRC. Purpose of flight—11th in X-24A series and first for a NASA pilot—was to obtain stability and control data. (NASA Proj Off)

• Javelin sounding rocket, launched by NASA from Churchill Research Range, carried Univ. of California at Berkeley payload to 557-km (346-mi) altitude. Objectives were to observe auroral protons and electrons over energy spectrum from thermal levels to several hundred kilovolts; study electric fields and their relation to particle fluxes; study density, temperature, and motion of ionospheric plasma currents; and study alpha-particle and neutral fluxes in aurora. All experiments performed satisfactorily and excellent data were obtained. (NASA Rpt SRL)

• Fédération Aéronautique Internationale (FAI) certified establishment of two space flight records by U.S.S.R.'s *Soyuz VI, VII,* and *VIII* mission (Oct. 11-18, 1969). One was for group flight of two spacecraft, with five men logging 35 hrs 19 min 33 secs over 989 242-km (614 685.3-mi) distance. Second was for group flight of three spacecraft, with seven men for 4 hrs 24 min 29 secs over 123 425 km (76 692.6 mi). (*SBD*, 4/27/70, 268; FAI)

• Ice breaker S.S. *Manhattan* left Newport News, Va., for Arctic Circle carrying experimental antenna, receiver, and decoder for NASA experiment in ranging and data transmission between vessel and orbiting *Ats V.* ERC scientist Richard M. Waetjen would test satellite's contact with ship at high latitudes of Arctic Circle through data transmission at teletype rates and obtain data on effects of Arctic weather, ocean surfaces, and reflections of signals from surrounding ice. NASA would use Mojave Ground Station in Barstow, Calif., to send signals to S.S. *Manhattan* through *Ats V.* (ERC Release 70-5)

• Enrique G. Regueiro, who broadcast Apollo missions in Latin America, estimated his audience at 100 million out of total population of 190 million. At Washington, D.C., dinner party he said he felt "people in Latin America are more excited about these moon flights than the people of the United States." (McCardle, *W Post,* 4/3/70, B3)

• NASA announced selection of NAS for negotiation of $3.5-million, cost-reimbursement contract to continue administration of NASA Resident Research Associateship Program. Contract would cover Sept. 1, 1969, through Aug. 31, 1971, during which maximum of 187 doctoral and postdoctoral researchers would work with experienced engineers. (NASA Release 70-51)

• NR Space Div. had earned incentive award fees of $1 006 080 and $643 200 under contracts for Apollo CSM and Saturn V 2nd stage (S-II), NASA announced. Awards, determined by NASA Performance Evaluation Board, covered management performance during 1969. (NASA Release 70-52)

• Reorganization of FAA engineering and development activities was announced by Secretary of Transportation John A. Volpe. Move would place functions of FAA Aircraft Development Service in Systems Research and Development Service. National Aviation Facilities Experimental Center at Atlantic City, N.J., and National Airspace Systems Program Office would come under direction of Associate Administrator for Engineering and Development. (FAA Release T 70-17)

April 3: U.S.S.R. launched *Cosmos CCCXXIX* from Plesetsk into orbit with 241-km (149.8-mi) apogee, 210-km (130.5-mi) perigee, 88.9-min period, and 81.4° inclination. Satellite reentered April 15. (GSFC *SSR*, 4/30/70; *SBD*, 4/6/70, 168)

- JPL and USAF officials commemorated 25 yrs of rocket engine testing at Edwards Test Station, Edwards AFB, Calif., by placing historical marker at approach to ETS administration building. Plaque cited April 2, 1945, date of arrival at Muroc Dry Lake of first cadre of 10 JPL engineers and technicians. (Boron, Calif *Enterprise*, 4/2/70)

- Paul E. Purser, Special Assistant to Dr. Robert R. Gilruth, MSC Director, retired after more than 30 yrs service to NASA and NACA. Purser had served as MSC Acting Director in Dr. Gilruth's absence and had been on leave of absence from MSC since September 1968 to assist Dr. Philip G. Hoffman, President of Univ. of Houston. (MSC Release 70-36; MSC *Roundup*, 4/13/70, 1)

- USA's Meritorious Civilian Service Award for Bravery was presented to Milton W. Alberry of Army Aeronautical Laboratory for exhibiting unusual courage and competence during wind-tunnel accident at ARC September 17, 1969. Cheyenne AH-56A helicopter under test broke loose and threw parts into control room. Alberry shut off fuel and electrical supplies to aircraft and rendered first aid to injured coworker "possibly saving his life." (ARC *Astrogram*, 4/23/70; ARC PAO; NASA Release 69-154)

- DOT takeover of NASA's Electronics Research Center was discussed in *Science:* Since inception in September 1965, ERC had built up to about 825-member staff including 420 professionals. "If the NASA and DOT proponents are right it may not require so wild a transition to bring space electronics research down to earth and deploy it against problems of air traffic control. There seems to be a fair amount of optimism that the technical problems of developing effective control systems will yield to electronics research, but DOT, which plans a budget of about $20 million next year for the center, is likely to encounter the practical problems of getting research funds equal to the task." (*Science*, 4/3/70, 100)

- *Science* editorial noted that President's Council of Economic Advisers had projected increase in Federal expenditures from $189 billion in 1970 to $206 billion in 1975. It was "within this budgetary situation—one so cramped as to allow little room for maneuvering—that future research funds must be considered. The President's budget for 1971 calls for $15.8 billion of R&D funds, an amount which constitutes a smaller percentage of the total federal budget than in any year since 1959. Research and development funds hit a high of 12.6 percent of the federal budget in 1965 and have been decreasing by an average rate of 0.8 percentage point a year, to 8.7 percent in 1970 and a requested 7.8 percent in 1971. Any statistical projection of R&D funds for the next several years looks bleak, and there is little that can be done to improve matters for the coming year." (Wolfle, *Science*, 4/3/70, 69)

April 4: Project Tektite II began at Charlotte Amalie, V.I., after three-day delay caused by late arrival of equipment. Participants in underwater research project were NASA, Dept. of Interior, NSF, USN, Smithsonian Institution, U.S. Coast Guard, HEW, General Electric Co., Virgin Islands Government, universities, and technical institutions. Four scientists and one engineer descended 15 m (50 ft) to habitat laboratory in which they would live and conduct research for 11 days. During ensuing seven months, 62 scientists, engineers, and doctors, including five women, would descend in teams.

Objectives were to accelerate development of ocean science and technology and provide impetus for national man-in-the-sea program, expand Tektite I successes by program of diving scientists in all marine disciplines; train scientists in saturation techniques, amplify Tektite I behavioral program and develop crew selections and performance criteria for manned undersea and space missions, conduct biomedical studies and operational procedures to extend saturated nitrogen diving to 30 m (100 ft) and evaluate diving equipment for marine undersea research. (Tektite II Press Info; Wilford, *NYT*, 4/5/70, 70)

- Phasing out of Naval Air Reserve's Floyd Bennett Field near Jamaica, N.Y., as military installation began, with official closing scheduled for end of June 1971. In almost four decades, airport had been used by Amelia Earhart, Douglas C. "Wrong Way" Corrigan, Col. Roscoe Turner, Gen. Italo Balbo of Italy, and Jacqueline Cochran. It was start and finish of July 1933 round-the-world flight of Wiley Post. Field was dedicated May 23, 1931, and named after USN warrant officer who was copilot of Adm. Richard E. Byrd on first flight over North Pole in 1926. (Silver, *NYT*, 4/5/70, 90)

- Discovery of carbon monoxide in five regions of Milky Way was made by scientists of Bell Telephone Laboratories using National Radio Astronomy Observatory telescope at Kitt Peak, Ariz. Carbon monoxide was sixth molecular substance identified in outer space. NSF later said discoveries would "stimulate further work by chemists as well as astrophysicists, looking toward better understanding of the chemical processes taking place in space, and eventually more knowledge of the possibilities of life elsewhere in the universe." (UPI, *W Star*, 4/11/70, A14)

- Article by woman cosmonaut Valentina V. Tereshkova-Nikolayava published in January UNESCO magazine *Impact* asserted equality of sexes in space, *New York Times* reported. Cosmonaut—who had orbited earth 48 times in June 16-19, 1963, mission—said that "a woman can stand all the conditions of space flight as well as a man." Women had endured silence and isolation as well as men and had adapted to weightlessness more quickly, although takeoff and landing of spacecraft should be timed to consider menstrual cycle. (*NYT*, 4/4/70, 20)

- Containerization system for airline passengers had been designed by Fried. Krupp Co. scientists in Essen, West Germany, *New York Times* reported. Units containing washrooms, galley, and luggage compartment would be filled at airline terminal and moved into waiting airliners. At destination, containers would be rolled from aircraft to terminal or directly to electric trains; smaller containers could be shifted from airliner to small aircraft or helicopters. De-

signers estimated 500 passengers and their luggage could be exchanged in 10 min. (*NYT*, 4/4/70, 58)
- Cornell Univ. team of scientists headed by Dr. Carl Sagan had sent shock waves through mixture of methane, ethane, ammonia, and water vapor—gases which simulated "air" on earth four billion years ago—*Christian Science Monitor* reported. Results, which showed amino acids in great abundance, indicated protein molecules (building blocks of organic life) might have been shocked into existence by thunder and impact of meteors in earth's primitive atmosphere. Protein molecules evolved from amino acids. (*CSM*, 4/4/70)

April 4-5: Lunar sample material was displayed at Wallops Station during two-day joint Federal activities including Assateague National Seashore Park, Chincoteague Coast Guard Station, Chincoteague National Wildlife Refuge, and ESSA exhibits and tours. (WS Release 70-6; NASA PAO)

April 5: Countdown for NASA's Apollo 13 manned lunar landing mission, scheduled for launch April 11, began at KSC despite serious problem in helium tank. NASA spokesman said super-cold helium was heating too rapidly, suggesting possibility of leak or contaminants in system. (Wilford, *NYT*, 4/6/70, 1; *W Post*, 4/6/70, A4)
- Apollo 13 launch would open "decade in which space will have a new shape...leaner shape, with spending well below the hell-bent-to-the-moon space budgets of the 1960s," Victor Cohn said in *Washington Post*. It would also be "subdued shape, lowered and Nixonian. Political and popular enthusiasm for expensive space travel is down several percentage points." New shape "might turn out to be a more sensible one for many seasons." It might place "more stress on using space flight to learn the facts of the universe and ways these facts can be applied to improve life on earth." NASA was still undergoing "wrenching change, with many controversies that are far from ended. There is still a deep split between many space scientists and many aerospace leaders and technologists over the best ways to use America's new capability." (*W Post*, 4/5/70, B1)

April 6: Apollo 13 Astronauts James A. Lovell, Jr., Fred W. Haise, Jr., and Thomas K. Mattingly II underwent comprehensive medical examination at KSC, including laboratory tests to determine their immunity to German measles (rubella). Dr. Charles A. Berry, Director of Medical Research and Operations at MSC, said crew was "in good physical condition and good shape to carry out the mission and the lunar surface activities," but expressed concern that astronauts could develop German measles during mission. They had been exposed to rubella virus when Astronaut Charles M. Duke, Jr., member of backup crew, developed disease April 5. (Wilford, *NYT*, 4/7/70, 1)
- Dr. Thomas O. Paine, NASA Administrator, in invited testimony on space benefits and potential, told Senate Committee on Aeronautical and Space Sciences that during first 12 yrs in space U.S. orbital payloads had increased 10 000 times, from *Explorer I*'s 14-kg (30.8 lbs) to Apollo 13's planned 136 000 kg (300 000 lbs); speed record 13 times, from 3000 km per hr (1900 mph) to 40 000 km per hr (25 000 mph); and flight altitude 10 000 times, from 38 400 m (126 000 ft) to 377 400 km (234 500 mi). Astronauts had logged 5843 hrs in space,

safely flying 113 400 000 km (70 472 000 mi). Twelve Americans had orbited moon and four had walked on its surface. NASA had launched 155 unmanned spacecraft to return scientific and practical data, 23 in cooperative international programs.

"Today we live in a different world because in 1958 America recognized the challenge of space and boldly made the required national investment to meet it. Since then over a billion children have been born all around the world, the first space age generation. Because of the space program they will learn a new science, a new cosmology, and a new view of man and his destiny in the universe. . . . Today's children can look ahead confidently to new opportunities and to great new strides that man will make in the twenty-first century. . . . Their generation will view the earth as a whole for the first time, and be able to deal with technology, with science, and with philosophy as a unified experience, common to all men of the blue planet earth."

Appendixes to prepared statement described NASA's technology utilization program: 1.6 million scientific and technical publications distributed in 1969, including 3 211 500 microfiche copies; contribution to U.S. schools including spacemobile project, which had reached 3 306 410 students live and 20 391 500 via TV in 1969; and space research laboratories built at 34 institutions of higher learning during 1960s, in which more than 1000 students had worked towards doctorates. NASA had received 968 830 letters in 1969; its exhibits were viewed by 37.6 million persons and its films, by 9.8 million directly and by 248 million on TV.

News media coverage of NASA activities in 1969 included 3497 newsmen from 57 countries accredited to *Apollo 11*. (Testimony; Schlstedt, B *Sun*, 4/7/70)

- Fifth anniversary of ComSatCorp's *Early Bird I*, world's first commercial comsat, launched by NASA from KSC. First commercial space link between North America and Europe, *Early Bird I* had been retired from full-time service in 1969 after receiving and transmitting more than 235 hrs of TV and operating for more than 3 1/2 yrs with 100% reliability. Satellite could still be used for emergency service to handle 300 circuits between U.S. and Europe. (ComSatCorp Release 70-16)

- Edmund J. Habib, Associate Chief of GSFC's Advanced Development Div., Tracking and Data Systems Directorate, received AIAA's 1970 Space Communications Award at Los Angeles luncheon. Habib was cited for "pioneering developments of the advanced tracking and telemetry data processing systems of NASA's Space Tracking and Data Acquisition Network, and for leadership of the development of technology for the future Tracking and Data Relay Satellite space network." (AIAA Release 4/6/70)

- MSC issued RFPs for preliminary study of reusable space tug that could be economically operated around earth, near moon, and in support of interplanetary missions. Preliminary study of highly versatile, four-module system would lead to award of nine-month, $250 000 fixed-price contract to investigate space tug missions and operations and determine if single-design vehicle could accomplish all tasks proposed. Space tug was one of new flight hardware items identified

in Space Task Group report to President in September 1969. (MSC Release 70-38)
- MSFC announced award to SPACO Inc. of 10-mo, $2.8-million, cost-plus-award-fee contract for engineering and operation support to MSFC Quality and Reliability Assurance Laboratory. (MSFC Release 70-66)

April 7: U.S.S.R. launched *Cosmos CCCXXX* from Plesetsk into orbit with 538-km (334.3-mi) apogee, 517-km (321.3-mi) perigee, 95.1-min period, and 74.0° inclination. (GSFC *SSR*, 4/30/70; *SBD*, 4/13/70, 201)
- Edward C. Polhamus of LaRC Stability and Performance Branch testified before Senate Committee on Government Operations' Permanent Subcommittee on Investigations during F-111 hearings. Extensive wind-tunnel tests of F-111 during aircraft's development stage had led LaRC engineers in 1963 to conclude F-111 would not meet USAF specification for primary mission dash requirement; aircraft would not develop maneuver capability at supersonic speeds specified by contractor; and aircraft's directional stability was extremely low at supersonic speeds. Primary mission shortcoming had been "associated with a high drag level which required significant design changes to solve." Polhamus said contractor, General Dynamics Corp., had insisted F-111 would exceed primary mission requirement "and did little in the way of airframe modification." (Testimony; *P Inq*, 4/8/70, 2)
- U.S. Supreme Court dismissed Feb. 16 appeal by atheist Mrs. Madalyn Murray O'Hair of District Court decision rejecting her suit against NASA to ban prayer in space. (Reuters, B *Sun*, 4/7/70, A6)
- Senate passed S.C.R. 49 to provide Congressional recognition of Dr. Robert H. Goddard and Goddard Rocket and Space Museum. (*CR*, 4/7/70, S5153)
- Comet Bennett, discovered Dec. 28, 1969, by J. C. Bennett of Pretoria, South Africa, had been visible to naked eye for several hours before dawn, *New York Times* said. It was expected to fade by end of April and might never be seen again. (*NYT*, 4/7/70, 11)
- FAA-DOT Certificate of Commendation was presented by FAA Administrator John H. Shaffer to Col. A. B. McMullen (USAF, Ret.) for outstanding contribution to development of general-aviation airports throughout U.S. and for his lifelong dedication to advancement of airport facilities throughout the world. Certificate was presented during ceremonies at National Aviation Club in Washington, D.C. (FAA Release 70-30)
- MSFC announced award of 10-mo contract extension at $6.4 million to Brown Engineering Co. for engineering and operation support to MSFC Astronautics Laboratory. (MSFC Release 70-70)
- James J. Kilpatrick in Washington *Evening Star* criticized Administration's intention to invest additional $290 million in SST: "Except for speed, the SST offers not a single advantage in range, comfort, passenger volume, profits, safety, ease of handling—none of these. And unlike the [Boeing] 747, which the industry developed at its own expense, the SST would cost the taxpayers a fortune in subsidies unlikely ever to be recouped." (W *Star*, 4/7/70, A7)

April 8: *Nimbus IV* (Nimbus-D) meteorological satellite was successfully launched by NASA from WTR by long-tank, thrust-augmented Thor

(THORAD)-Agena D booster. Launch was 400th for USAF's Thor booster. Satellite entered orbit with 1098.0-km (682.4-mi) apogee, 1087.2-km (675.7-mi) perigee, 107.1-min period, and 99.9° inclination.

USA's *Topo I* first in new series of topographic satellites to be used with equipment modified from USA's SECOR (Sequential Collation of Range) program, was carried as secondary payload on Agena D 2nd stage. *Topo I* was ejected into orbit with 1092-km (678.5-mi) apogee, 1085-km (674.2-mi) perigee, 107-min period, and 99.8° inclination, where it would investigate new technique for accurate, real-time determination of positions on earth's surface.

Primary objectives for *Nimbus IV* were to acquire sufficient number of global samples of atmospheric radiation measurements for comparing vertical-temperature, water-vapor, and ozone profiles and to provide basis for comparing merits of at least three of five spectrometric experiments on board. As secondary objectives spacecraft would demonstrate feasibility of determining wind-velocity fields by tracking multiple balloons and demonstrate satisfactory operation of advanced modular three-axis attitude-control system for six months.

Nimbus IV carried nine meteorological experiments: three new (backscatter UV spectrometer, filter wedge spectrometer, and selective chopper radiometer) and six carried on previous Nimbus missions (infrared interferometer spectrometer; satellite infrared spectrometer; interrogation, recording, and location system; temperature and humidity infrared radiometer; image-dissector camera system; and monitor of UV solar energy). Spacecraft's new attitude-control system stabilized *Nimbus IV* constantly pointing toward earth within 1° accuracy and permitted initial acquisition and reacquisition of earth from any attitude.

Butterfly-shaped, 675.2-kg (1488.6-lb) *Nimbus IV* was fifth in series of seven spacecraft designed to explore and collect data for understanding nature and development of atmosphere and for reduction of economic impact of adverse weather on all nations. *Nimbus III* had been launched April 14, 1969. (NASA Proj Off; NASA Release 70-47; AFSC Release 104.70; GSFC *SSR*, 4/30/70)

- USAF launched two Vela nuclear-detection satellites from ETR on one Titan IIIC booster. *Vela XI* entered orbit with 110 739.5-km (68 810.3-mi) apogee, 110 091.8-km (68 478.4-mi) perigee, 6695-min period, and 37.7° inclination. *Vela XII* entered orbit with 111 379.8-km (69 281.8-mi) apogee, 111 111.1-km (69 041.2-mi) perigee, 6699-min period, and 32.9° inclination. Each 350-kg (770-lb) satellite carried optical sensors to detect electromagnetic pulses—brief radio signals generated by atomic explosions—anywhere in atmosphere. Sensors could also detect atomic weapon tests up to 160 million km (100 million mi) into space and could distinguish between weapon tests and exploding stars. Launch was last in Vela series begun in 1963. (GSFC *SSR*, 4/30/70; *W Post*, 4/9/70, A8; AFSC Release 106.70; Pres Rpt 71)

- *Cosmos CCCXXXI* was launched by U.S.S.R. into orbit with 323-km (200.7-mi) apogee, 203-km (126.1-mi) perigee, 89.6-min period, and 64.9° inclination. Satellite reentered April 16. (GSFC *SSR*, 4/30/70)

- Astronaut John L. Swigert, Jr., Apollo 13 backup CM pilot, began last-minute training program to determine whether he could replace Astronaut Thomas K. Mattingly II on mission scheduled for launch April 11. Apollo 13 prime crewmen—James A. Lovell, Jr., Fred W. Haise, Jr., and Mattingly—had been tested for immunity to German measles April 6 after exposure to disease and Mattingly had not shown immunity. NASA said final decision on moving Swigert to prime crew depended upon demonstration that Swigert, Lovell, and Haise could work together as a team and "precisely execute the few intricate time-critical maneuvers which require rapid and close coordination." (Lannan, W *Star,* 4/9/70, A1; O'Toole, *W Post,* 4/9/70, A1)
- Interview on Apollo 13 cuisine with MSC chief of food and nutrition Dr. Malcolm Smith, Jr., was published by *New York Times.* Increased use of spoons, use of bowls, and facilities for spreading sandwich fillings on slices of fresh bread would provide Apollo 13 astronauts more nearly normal eating procedures than on previous missions. Freeze-dried pork and scalloped potatoes would contain chunks rather than be of baby food consistency but meal would still be lukewarm. Wider variety of fresh bread had been baked for Apollo 13 mission, in lidded pans to provide uniform crust as insurance against loose crumbs which had been problem with former freeze-dried cracker sandwiches. Loaves were passed through 1900 K (3000°F) flame to destroy all mold-producing spores. Two slices of bread made from irradiated flour would be carried to test potential alternate method of preserving food without refrigeration. Apollo 13's food was 70% freeze-dried, mainly because of space limitation; bread was treated with nitrogen gas to retard staling. Other "firsts" in foods to be carried were pecans to stimulate astronauts' appetites, dehydrated orange crystals modified to prevent caking, and instant rice product to be reconstituted by hot tap water.

    Dr. Smith was already planning for 120-day Skylab missions, when hot food would be welcomed. Major obstacle was difficulty in heating food at zero g. He hoped to place test heating device aboard Apollo 14. (Hewitt, *NYT,* 4/8/70, 51)
- Wristwatch calibrated on lunar time system developed by Kenneth L. Franklin, Assistant Chairman of American Museum-Hayden Planetarium, had been devised by Helbros Watches Div. of Elgin National Industries, *Wall Street Journal* reported. Watch used three hands of conventional earth watch plus "window" telling "lune" time, based on fact that time from sunrise to sunrise on moon was equal to 29.53 earth days. Franklin called this period "lunation" and divided it into 30 "lunes" each equal to about 24 earth hours. Each lune was devided into 24 "lunours" about an earth-hour long. (*WSJ,* 4/8/70)
- NASA announced selection of General Electric Co. for negotiation of $50 000 000 cost-plus-award-fee contract for development, fabrication, test, and project operations of ATS-F and ATS-G, scheduled for launch into stationary orbit over equator in 1973 and 1975. (NASA Release 70-55)
- MSC announced award of $875 000 contract to Hycon Manufacturing Co. for five lunar topographic camera systems. Cameras, each equipped with 457-mm (18-in) focal length lens and 61 m (200 ft) of

film to take more than 400 photos without reloading, would be flown on Apollo 13, 14, and 15 missions to photograph future landing sites and other targets. (MSC Release 70-31)

- *Philadelphia Inquirer* editorial commented on Apollo 13 crew's exposure to German measles: "It is an untimely reminder of how much America spends to send men to the moon and how little to combat disease on earth.... We are in favor of continued space exploration, to a point, but we cannot help thinking what the money spent on just one shot to the moon could do if devoted instead to the prevention and cure of diseases that afflict millions of earthlings." (*P Inq*, 4/8/70)

April 9: Dr. Charles A. Berry, Director of Medical Research and Operations at MSC, said chances were high that Astronaut Thomas K. Mattingly II would develop German measles next week and that, in his opinion, Mattingly's lack of immunity would not permit him to fly on Apollo 13 April 11.

NASA had not officially ruled out Mattingly for flight, but countdown was proceeding on schedule and backup CM pilot John L. Swigert, Jr., was undergoing intensive training in Apollo simulators at KSC. (O'Toole, *W Post*, 4/10/70, A2)

- Hydrogen cloud 13 million km (8 million mi) in diameter around comet Bennett had been discovered by Lyman-alpha experiment on board *Ogo V* (launched March 4, 1968), NASA announced. Measurements had been made during three-day period while Bennett was about 105 000 000 km (65 000 000 mi) from earth and *Ogo V* was orbiting earth at 23 000- to 108 000-km (14 000- to 67 000-mi) altitude, operating in special spin-scan mode. Univ. of Colorado scientists said data from *Ogo V* photometer indicated comet was composed largely of water. (NASA Release 70-56; UPI, *W Star*, 4/10/70, A2)

- MSC announced award of $4.5-million, cost-plus-fixed-fee NASA contract to Fairchild Camera and Instrument Corp. for three-inch-focal-length mapping camera, stellar reference camera, and timing device for precision mapping of moon on future Apollo missions. Mapping camera would provide 114-mm-sq (4.5-in-sq) photos with precise geometric controls. Stellar reference camera, used with timing device and laser altimeter—being produced under separate NASA contract with RCA—would provide spacecraft attitude, time, and altitude at which each photo was taken. (MSC Release 70-40)

- AEC-NASA Space Nuclear Propulsion Office had awarded Aerojet-General Corp. $50 219 497 contract extension for continued design and development work on NERVA engine, NASA announced. Extension definitized work to be performed from Oct. 1, 1969, through Sept. 30, 1970. (NASA Release 70-54)

- President Nixon signed E.O. 11523 establishing National Industrial Pollution Control Council to advise on industry programs on environmental quality. He announced appointments of Bert S. Cross, board chairman of Minnesota Mining and Manufacturing Co., and Willard F. Rockwell, board chairman of North American Rockwell Corp., as Chairman and Vice Chairman and appointments of 53 members to Council. Council would submit, through Secretary of Commerce, specific recommendations for further action on industrial environmental pollution and consider "role it can play in helping

to implement the Nation's environmental protection program." (PD, 4/13/70, 502-4)
- Senate, by vote of 72 to 6, adopted S.R. 211, which sought agreement with U.S.S.R. on limiting offensive and defensive strategic weapons and suspension of test flights of reentry vehicles. (CR, 4/9/70, S5479-508)
- Sen. Edward M. Kennedy (D-Mass.) introduced S. 3700, FY 1971 NSF authorization providing $548 million—$50 million more than President Nixon's recommendation plus additional $2 million annually in foreign currencies to be spent outside U.S. (CR, 4/9/70, S5428-9)

April 9-10: U.S.-U.S.S.R. editorial board to prepare and publish joint review of space biology and medicine met in Moscow to select authors and establish procedures and schedules. Project was established under Oct. 8, 1965, agreement between NASA and Soviet Academy of Sciences. First exchange of reference materials had been made in January. (NASA Release 70-61)

April 10: Astronaut John L. Swigert, Jr., would replace Thomas K. Mattingly II as CM pilot on Apollo 13 mission April 11, Dr. Thomas O. Paine, NASA Administrator, announced. Mattingly, exposed to German measles, had not developed immunity and was expected to show symptoms of disease next week. "The recommendation of

April 10: NASA announced that Astronaut John L. Swigert, Jr., would replace Thomas K. Mattingly II as CM pilot on Apollo 13 mission. Cancellation of the mission had been considered following Mattingly's exposure to German measles. Below, Dr. George M. Low (left), Deputy Administrator, conferred with Associate Administrator for Manned Space Flight Dale D. Myers at April 10 launch status meeting at Kennedy Space Center.

everyone...was unanimous that Apollo 13 be launched tomorrow," Dr. Paine told press at KSC. Decision was made after Swigert demonstrated in intensive training that he could replace Mattingly on short notice. "When it comes to the problems you can get into in the command module...Jack [Swigert] literally wrote the book on the malfunctions and how to overcome them," Dr. Paine said.

Director of Flight Crew Operations Donald K. Slayton said Swigert had performed critical maneuvers flawlessly in training. "Our simulator people said it couldn't be done...but we anticipated more problems than we had. We were all surprised. It went very smoothly." (Wilford, *NYT*, 4/11/70, 1; O'Toole, *W Post*, 4/11/70, A1)

- President Nixon honored visiting West German Chancellor and Mrs. Willy Brandt at White House dinner on eve of Apollo 13 launch. He said: "We are particularly happy that tomorrow you will go to see the takeoff of Apollo 13. We are very honored that tonight among our very honored guests is Wernher Von Braun, which reminds us of the debt we owe to those who have helped our space project and who are of German background."

  President said he had telephoned Astronaut James A. Lovell, Jr., and Apollo 13 colleagues as they were having dinner together at MSC. "I wished them well and told them that the Chancellor would be there to see the take off and they promised much better weather than when I saw it last year." (*PD*, 4/13/70, 508-10)

- Few people "really seem to care" that $375-million Apollo 13 mission "may never get off the ground tomorrow" because astronauts had been exposed to measles, *Wall Street Journal* said. Mission was inspiring "little excitement at KSC or elsewhere across the land." Perhaps 100 000 tourists were on hand for event, "but that's a small crowd compared with the million or so that came here for the first moon shot. . . ." There would be 700 newsmen, "far below the 2000 who covered the first shot." (Tanner, *WSJ*, 4/10/70)

- First aircraft designed and produced in Israel—twin-engine, 20-passenger Arava STOL aircraft—made inaugural flight from Lydda Airport near Tel Aviv. Built by Israel Aircraft Industries, Arava cost $400 000 and was intended for feeder lines in developed nations and more rugged operations in developing areas. Aircraft was powered by Pratt & Whitney turboprop engines from Canada. (*NYT*, 4/10/70, 2)

- *Nimbus IV* meteorological satellite, launched April 8, had contacted elk fitted with instrumented collar in Wyoming April 1 and had transmitted data on elk's temperature, NASA announced. (UPI, *W Post*, 4/12/70, 60)

- MSC announced award of $1.9-million contract to Itek Corp. for design, development, and delivery of multispectral photographic equipment for Project Skylab. Itek would provide one six-lens camera flight unit, one backup, and associated lenses and magazines. (MSC Release 70-41)

- *Washington Post* editorial commented on launch of Project Tektite II: "As the astronauts get ready to soar to the heights, it is worth mentioning that a group of aquanauts have sunk to the bottoms—in this case, the crew of Tektite II. . . . The drama of undersea exploration is considerably less than probes into outer space; but, in many ways, the world on the floor of the sea is more mysterious, more demanding—and potentially richer in what it offers mankind—than anything on the craterscape of the moon." (*W Post*, 4/10/70, A18)

April 11-17: NASA's *Apollo 13* (AS-508)—carrying three-man crew, LM-7 *Aquarius* and CSM-109 *Odyssey*—was launched on lunar landing mission, but 56 hrs into flight toward moon mission was aborted because of SM oxygen tank rupture. In most serious inflight situation experienced in manned space program, crew followed LM "lifeboat" emergency plan, transferring to LM and using its systems and supplies to swing around moon and back to earth on fastest feasible course. Emergency procedures were used to make electrical power, water, and oxygen last to end of journey. Crew returned to CM and jettisoned LM before reentry and splashed down safely in Pacific April 17.

April 11-12: Mission began with successful, on-schedule liftoff from KSC Launch Complex 39, Pad A, at 2:13 pm EST April 11, carrying Astronauts James A. Lovell, Jr. (commander), John L. Swigert, Jr. (CM pilot), and Fred W. Haise, Jr. (LM pilot). Swigert was last-minute substitution from backup crew for Thomas K. Mattingly II, who had failed to develop immunity after exposure to German measles. Launch was watched by 4500 persons in VIP stands. Vice President Spiro T. Agnew, West German Chancellor Willy Brandt, and Dr. Thomas O. Paine, NASA Administrator, watched from KSC firing Room No. 1. Weather conditions at launch were satisfactory: peak ground winds of 10 knots, overcast at 6000 m (20 000 ft), and visibility of 16 km (10 mi).

Spacecraft and S-IVB combination entered parking orbit with 185.5-km (115.3-mi) apogee and 181.5-km (112.8-mi) perigee. During 2nd-stage boost S-II center engine cut off 132 secs early and remaining four engines burned 34 secs longer than predicted. Space vehicle velocity after boost was 68 m per sec (223 fps) lower than planned, causing S-IVB orbital insertion burn to be 9 secs longer than predicted, but sufficient propellants remained for translunar insertion. Onboard TV was initiated at 1:35 GET for 5 1/2 min. Second S-IVB burn, on schedule at 2:41 GET, injected spacecraft on translunar trajectory.

CSM separated from S-IVB and LM *Aquarius*. Onboard TV initiated after separation for 72 min clearly showed CSM "hard-docking" with LM, ejection of CSM/LM from S-IVB/IU at 4:01 GET, S-IVB auxiliary propulsion system (APS) evasive maneuver, and spacecraft exterior and interior.

First S-IVB APS burn, lasting 217 secs beginning at 6:00 GET, aimed stage for lunar target point so accurately that second burn was not required. S-IVB/IU hit lunar surface at 77:57 GET (8:10 pm EST April 14) at 259 m per sec (8465 fps) and energy equivalent to 11.5 tons of TNT at 2.4° south latitude and 27.9° west longitude—137.1 km (85.2 mi) from *Apollo 12* seismometer that detected impact. Seismic signal for S-IVB/IU was similar to signal from *Apollo 12* LM *Intrepid* that had impacted moon Nov. 20, 1969, but was 20-30 times larger and lasted 3 hrs 20 mins—four times longer. Signal was so large that ground command had to reduce seismometer gain to keep recording on scale. Suprathermal ion detector experiment (SIDE) deployed by *Apollo 12* crew recorded jump in the number of ion counts after S-IVB/IU impact. Instrument, in lunar shadow,

April 11-17: Apollo 13, launched toward moon from Kennedy Space Center by Saturn V (left), was aborted 56 hours into flight by the rupture of a service module oxygen tank. Astronauts transferred to LM in an emergency "lifeboat" plan and used its systems and supplies to return to earth without landing on the moon. The severely damaged SM (above) was photographed from the LM/CM following jettison just before earth reentry. An entire panel was blown away by the explosion.

recorded no ions at time of impact; count increased from few ions 22 secs after impact to 250 and up to 2500 before decreasing to few ions and back to none. Scientists speculated ionization had been produced by 6300 K to 10 300 K (6000°C to 10 000°C) temperatures generated by impact or that particles that reached up to 60 km (37 mi) had been ionized by sunlight.

Translunar insertion maneuver was so accurate that midcourse correction No. 1 (MCC-1) was not necessary. MCC-2, for 3.4 secs at 30:14 GET, placed spacecraft on desired hybrid circumlunar trajectory with closest approach of 114.9 km (71.4 mi). MCC-3 was not necessary. Good-quality TV coverage of preparations for and performance of midcourse maneuver was transmitted for 49 min, beginning at 30:03 GET, including dumping waste water from spacecraft fuel cells with moon visible in background.

While joking with Mission Control, Swigert suddenly remembered that he had failed to file his Federal income tax return, but was assured that he could be granted an extension.

April 13-14: Lovell and Haise entered LM for housekeeping and system checks, reported that helium tank that had malfunctioned before launch was satisfactory, transmitted TV pictures of LM interior, and joked about sleeping and eating in space. Haise reported he had shaved to improve his TV image. Lovell closed telecast by wishing everyone on earth "a nice evening," saying astronauts were ready "to get back for a pleasant evening in *Odyssey.*" Minutes later, at 55:55 GET (10:08 pm EST April 13), crew reported undervoltage alarm on CSM main bus B, rapid loss of pressure in SM oxygen tank No. 2, and dropping of current in fuel cells 1 and 3 to zero.

"Hey, we've got a problem here," Swigert told ground controllers.

"This is Houston, say again please."

"Houston, we've had a problem," Lovell replied. And Haise reported "a pretty large bang associated with the caution and warning."

A few minutes later crew reported spacecraft was venting something—later determined to be oxygen—out into space. Astronauts could see blizzard of particles flying out of SM, presumably from ruptured tank. Despite desperate efforts improvised on board and suggested by Mission Control, nothing seemed to halt drop in oxygen pressure. Spacecraft dipped repeatedly, apparently because of venting gas. By 57:24 GET, Mission Control reported tank No. 2 pressure was "slowly going to zero and we're starting to think about the LM lifeboat." Swigert replied, "Yes, that's something we're thinking about, too. We've been talking it over...." Decision was made to abort mission.

Astronauts hurriedly had to salvage enough oxygen to supply CM during reentry and keep gyros operating until LM navigation equipment could be powered up. With only minutes of power remaining, crew fed oxygen into reserve tank and hooked up emergency batteries to CSM navigation gyros until LM was powered up. Crew activated LM, powered down CSM, and used LM systems for power and life support.

Astronauts remained calm as they wrestled with procedures to stabilize spacecraft and stretch consumables. All information on

April 11-17: *Officials conferred at the flight directors' console in the MSC Mission Operations Control Room as Apollo 13 astronauts attempted to bring their crippled spacecraft safely back to earth. Dr. Donald K. Slayton, Director of Flight Crew Operations, leans an elbow on the console at far left. Circling to the right in front of him are Apollo Spacecraft Program Manager James A. McDivitt; Deputy Director of Flight Operations Howard W. Tindall, Jr.; Dr. Rocco A. Petrone, Apollo Program Director at NASA Headquarters; and Dr. Christopher C. Kraft, Jr., Director of Flight Operations. Apollo Mission Director Chester M. Lee stands in front of Dr. Kraft. Seated at the console (center, left to right) are Flight Directors Gerald D. Griffin and Glynn S. Lunney. Behind the console (left to right) are Astronauts Joseph P. Kerwin and Vance D. Brand, Flight Dynamics Officer Philip C. Shaffer, Retro Officer John S. Llewellyn, Flight Dynamics Officer Charles F. Dieterich, and Lawrence S. Canin, guidance navigation and control engineer at MSC.*

mission—including details of accident and emergency procedures—were made public immediately as data were received. Public's apparent boredom at beginning of mission was transformed into deep concern for astronauts' safety. Messages and offers of aid were sent to U.S. from world leaders and prayers were offered by religious leaders on every continent.

MCC-4 was conducted with LM's descent propulsion system (DPS) at 61:30 GET, placing *Apollo 13* on free-return trajectory around moon on which CM *Odyssey* would splash down in Indian Ocean at 152:00 GET. Both LM guidance systems were powered up and primary system used for 263.4-sec DPS burn at 79:28 GET as spacecraft swung out from behind moon, two hours after closest approach. Burn produced 262-m-per-sec (860 fps) differential velocity, shortened predicted return time to 143:53 GET, and moved predicted landing point to mid-Pacific. After maneuver, passive thermal control was established and LM was powered down to conserve consumables.

"Aquarius has really been quite a winner," Haise said to Mission Control about three hours later. "When this flight's all over we'll really be able to figure out what a LM can do."

During remainder of transearth coast crew continued emergency procedures to deal with shortage of water to cool LM electronic systems, decreasing temperatures in both cabins, and buildup of carbon dioxide in LM. Fearing that CM instruments needed for reentry might freeze, crew kept spacecraft in passive thermal control—"barbecue mode" of rotation to let sun uniformly heat all sides. Astronauts met cooling water shortage by filling every plastic bag they could with water from CM and carrying it to LM. Shortage of air-cleaning lithium hydroxide in LM was met by rigging combination of CM and LM cartridges; following instructions worked out on ground, crew built adapter and ran lunar spacesuit air hoses from LM fans through tunnel to connect to CM air-scrubbing canisters. Transmission of health data was eliminated and communications held to minimum to conserve electricity. Crew tried to keep warm by wearing boots carried for walking on moon, sleeping in narrow tunnel between LM and CSM that seemed warmer, and wearing extra clothing.

April 15-17: MCC-5 at 105:19 GET, by 15-sec LM DPS burn, decreased velocity by 2.3 m per sec (7.5 fps) and raised entry-flight-path angle to -6.52°. Crew partially powered up CSM to check ther-

April 15-17: *"Farewell, Aquarius, and we thank you," Mission Control responded to crew report "LM jettison," just over an hour before Apollo 13 CM splashed down April 17. The LM, which had served as a lifeboat after an oxygen tank explosion aborted the mission, was photographed from the CM after jettison. Industry, university, and NASA ground teams had been working continuously during the last days of the return trip to develop new reentry procedures for the unusual CM/LM configuration.*

mal conditions. Because of unusual spacecraft configuration (with LM still attached) and LM's inability to survive reentry heat, officials on ground developed new reentry procedures and verified them in ground-based simulations. A 30-man team at MIT's Charles S. Draper Laboratory worked throughout night studying stressed *Apollo 13* guidance system and working out new trajectories. Simulations and programming of emergency problems were conducted by NR engineers in Downey, Calif., by Astronaut Richard F. Gordon, Jr., in LM at KSC, and by Astronauts Eugene A. Cernan and David R. Scott at MSC, so that almost every move *Apollo 13* crew made had first been proved out on ground.

MCC-7 was successfully accomplished at 137:40 GET with 22.4-sec LM RCS burn, resulting in predicted $-6.49°$ entry-flight-path angle. Crew viewed and photographed SM when it was jettisoned at 138:02 GET, reporting that one entire panel was missing near S-band high-gain antenna and that great deal of debris was hanging out.

CM was powered up and LM was jettisoned at 141:30 GET. SNAP-27 nuclear generator carrying 3.9 kg (8.6 lbs) of radioactive plutonium fell from LM into Pacific northeast of New Zealand.

"Farewell, Aquarius, and we thank you," Mission Control responded to spacecraft's report "LM jettison." "She sure was a good ship," Swigert added later.

Parachute deployment and other reentry events occurred as planned and *Odyssey* splashed down in mid-Pacific southeast of American Samoa at 142:54 GET (1:07 pm EST April 17), 6.4 km (4 mi) from recovery ship U.S.S. *Iwo Jima*. Astronauts—exhausted but in good health—were picked up by recovery helicopter and were safely aboard recovery ship less than one hour after splashdown.

Primary *Apollo 13* mission objectives—to make selenological inspection, survey, and sampling of materials in preselected Fra Mauro area; deploy and activate ALSEP; develop man's capability to work in lunar environment; and photograph candidate exploration sites—were not achieved. Dr. Thomas O. Paine, NASA Administrator, announced April 17 that *Apollo 13* Review Board, headed by LaRC Director Edgar M. Cortright, would be established to determine cause of accident.

Apollo 13 flight crew performance was outstanding throughout mission; calm, precise reaction to emergency and diligence in configuring and maintaining LM was noteworthy. Despite lack of adequate sleep and low temperature in spacecraft, neither performance nor spirits faltered. Similarly, performance of flight operations team was outstanding in planning and aiding crew to safe return.

Apollo 13 was 10th Apollo mission and third manned lunar landing attempt. Accident was first inflight failure in 22 manned flights in U.S. space program. *Apollo 12* (Nov. 14-24, 1969) and *Apollo 11*

April 17: *Safely home, the Apollo 13 crew stepped aboard the U.S.S. Iwo Jima recovery ship after splashdown in the South Pacific. Leaving the helicopter (at top, from left) are Astronauts Fred W. Haise, Jr., James A. Lovell, Jr., and John L. Swigert, Jr. In the once-tense Mission Operations Control Room at MSC (at bottom), Dr. Donald K. Slayton (left of center), Director of Flight Crew Operations, shook hands with Apollo Mission Director Chester M. Lee as arriving Astronaut Lovell appeared on the TV monitor. Dr. Rocco A. Petrone, Apollo Program Director, stands behind Lee, with Flight Control Division Chief Eugene F. Kranz sitting at the console in front.*

(July 16-24, 1969) manned missions had successfully landed on moon and returned to earth with lunar samples. (NASA Proj Off; *FonF*, 4/16-22/70, 253-5; W *Star*, 4/12-14/70; *NYT*, 4/15/70; 4/18/70; 4/19/70; W *Post*, 4/15/70; 4/18/70, A1; MSC Hist Off; Off NASA Administrator)

April 11: U.S.S.R. launched *Cosmos CCCXXXII* from Plesetsk into orbit with 760-km (472.2-mi) apogee, 755-km (469.1-mi) perigee, 99.9-min period, and 74.0° inclination. (GSFC *SSR*, 4/30/70; *SBD*, 4/20/70, 236)

- Nike-Tomahawk sounding rocket was launched by NASA from Churchill Research Range, carrying Univ. of Minnesota experiment to conduct auroral studies. Rocket and instruments functioned satisfactorily. (SR list)
- *New York Times* editorial commented on substitution of Astronaut John L. Swigert, Jr., for Astronaut Thomas K. Mattingly II on *Apollo 13* mission. "The decision adds a new note of tension and a new reason for worry in what will be—despite the brilliant success of the two earlier moon trips—an exceedingly dangerous venture whose maneuvers require the most precise possible coordination among all three crew members. In the entire history of the space program there can have been few if any more difficult or more hotly debated choices faced by the program's top leaders." (*NYT*, 4/11/70, 30)
- *Apollo 12* moon rock was guarded by three men as firemen fought fire at Stardust Hotel in Las Vegas, Nev. Lunar sample lent by NASA for display at physics symposium in hotel was kept in place during fire for security reasons. (UPI, W *Post*, 4/12/70, A3)

April 12: U.S.S.R. celebrated National Cosmonauts Day—ninth anniversary of *Vostok I*, first manned orbital space flight by Cosmonaut Yuri A. Gagarin April 12, 1961. Soviet press reviewed U.S.S.R. space triumphs and praised 18 cosmonauts.

Except for brief announcement in *Pravda* that *Apollo 13* had "departed for the moon," press ignored third U.S. manned lunar landing attempt. AP reported mood of Soviet people was not festive and quoted one Muscovite as saying, "It seems a bit silly celebrating Gagarin's first space flight while the Americans are paying their third visit to the moon." (B *Sun*, 4/13/70, A5)

- Philadelphia *Sunday Bulletin* editorial commented on *Apollo 13:* "The ease with which Mr. Swigert fitted into the prime crew lineup would suggest that the real risks in the Apollo 13 mission are still the risks the public never sees or comprehends. These involve mechanical and electronic components beyond lay understanding. They are immediately interchangeable on Earth, but they lack the adaptability of Captain Lovell and his companions at the point of no return." (P *Bull*, 4/12/70)
- Dr. William H. Pickering, JPL Director, received Society of Manufacturing Engineers Interprofessional Cooperation Award at dinner in Detroit, Mich., for giving U.S. "through his work in jet propulsion, missilery, and telemetry . . . a viable defense system while contributing substantially to her pre-eminence in the investigation of outer space." (Pasadena *Star-News*, 4/2/70)
- Financial plight of Lockheed Aircraft Corp. was "symptomatic of a far more pervasive malaise spreading in the aerospace industry,"

Robert A. Wright said in *New York Times.* Because of "intricate interwoven pattern" of subcontracting common to large defense programs, Lockheed's problems would affect other companies including 2000 that supplied or made parts for C-5A cargo aircraft. "In the current atmosphere of defense spending cuts and political criticism of the so-called military-industrial complex, the attention generated by Lockheed's problem is unlikely to enhance the poor prospects of the depressed aero-space industry." (*NYT*, 4/12/70, F1)

April 13: With *Apollo 13* proceeding "in almost textbook fashion," newsmen at MSC were discussing "minor 'glitches' " because "there had been nothing of major nature to talk about," Los Angeles *Herald-Examiner,* said. Houston TV had failed to carry launch live and only 295 U.S. and 55 foreign journalists were covering MSC. *Apollo 13* astronauts had prompted complaints from newsmen for being "too quiet." Reason might be "that few newsmen have listened to find out whether the crewmen are talking or not." (Bane, LA *Her-Exam,* 4/13/70)

- Timing of *Apollo 13* could upstage U.S.S.R. 's plans for "biggest ideological event in Soviet history, the Lenin centennial," *New York Times* reported. Mission had been launched on 1970 Leninist Subbotnik—day when loyal citizens donated full day's work to their country. Second day of Apollo 13 mission was Cosmonauts Day, anniversary of Cosmonaut Yuri A. Gagarin's triumph as first man to orbit earth. *Apollo 13* astronauts were scheduled to return to earth the day before the Lenin anniversary, "almost exactly at the hour that Secretary General Leonid I. Brezhnev is scheduled to crown his career by delivering the major address commemorating the founder of the Soviet system." (Schwartz, *NYT,* 4/13/70)
- Baltimore *Sun* editorial on *Apollo 13* launch: "If there is less national emotional interest in the moon flights now as they become almost routine, and as their goals are related to specific scientific quests, rather than to the romance of just making the flight and landing, there is still great interest and concern for a safe voyage. That America is not jaded by these brave and difficult ventures is shown in the large crowds that show up for launches in Florida and the attention televised coverage draws in the rest of the nation." (B *Sun,* 4/13/70, A12)
- Aerobee 150 sounding rocket was launched by NASA from WSMR carrying Univ. of Michigan experiment to study airglow. Mission was unsuccessful. (SR list)
- Papers of U.S. rocket pioneer Dr. Robert H. Goddard, recording his life work, were published by McGraw-Hill Book Co. of New York. (*AFJ,* 4/18/70, 6)
- President Nixon sent to Congress annual report on marine resources and engineering development. Budget request for FY 1971 would provide $533.1 million for marine science and technology activities to improve management of coastal zone, expand Arctic research, develop program for restoring damaged lakes, expand collection of data on ocean and weather conditions, and undertake other projects. Funds also would support U.S. participation in International Decade of Ocean Exploration. (*PD,* 4/13/70, 518)

April 14: As *Apollo 13* approached critical point in efforts to return safely to earth, Dr. Thomas O. Paine, NASA Administrator, told press in Houston, "The performance of the entire NASA team with the remainder of this mission will undoubtedly have an effect on a view that Congress has of the activities that we are proposing for the future." Specific problems encountered by *Apollo 13* were "things that will be reviewed by the Congress." (Wilson, *W Post*, 4/15/70, A13)

- President Nixon received 45-min briefing by NASA officials at GSFC on problems of *Apollo 13* mission. Dr. John F. Clark, GSFC Director, said President was deeply concerned about mission's difficulties.

    Later President entertained Danish Prime Minister Hilmar Baunsgaard at White House dinner. President said: "I think tonight of three men . . . in outer space coming around the moon . . . . I do know that they have the spirit of the Vikings. They are men of adventure. They are men of courage. Back home—I talked to the wives of two of the men and the mother of the one who is not married—they have women, who like the wives and the mothers of the Vikings of old waited at home with faith that their men would come back."

    President, at his request, was informed during dinner of success of *Apollo 13* maneuver which put spacecraft on homeward course. (*PD*, 4/20/70, 538; Lyons, *NYT*, 4/15/70, 1; Beale, *W Star*, 4/15/70, F1)

- Senate unanimously passed S.R. 388 "that the Senate of the United States commends the Apollo 13 astronauts . . . for their fortitude and courage; extends its support to their families, their friends and all who are involved in their mission; and urges all businesses, commercial operations, communications media, and others who wish to and can comply to pause at 9 p.m. today, April 14, 1970, in order that all persons who so desire may join in asking the help of Almighty God in assuring the safe return of the astronauts." (*CR*, 4/14/70, 85644-5)

- International reaction to *Apollo 13* crisis:

    In Moscow, many persons relied on foreign radio broadcasts for information on *Apollo 13* progress; Soviet news media continued to give only brief reports. At U.N. in New York, Dr. Anatoly A. Blagonravov, Soviet delegate to Committee on Peaceful Uses of Outer Space, said people of world were hoping for safe return of "courageous astronauts."

    Pope Paul VI told audience of 10 000 Romans and tourists in St. Peter's Basilica, "We cannot forget at this moment the lot of the astronauts of *Apollo 13*. We hope that at least their lives can be saved."

    Japanese businessman said in Tokyo: "How could it happen? Why, they almost shuttled to the moon as safely as we ride commuter trains."

    In Paris, radio stations interrupted regular programs with news bulletins and interviews with space experts. *La Monde* said, "The whole human race is participating with them in the agony of their return."

    Thousands flocked to churches in Georgetown, Guyana. They were saying mission should have been called Apollo 12B or Apollo 14, rather than *Apollo 13*.

In Australia, Sydney TV station superimposed *Apollo 13* bulletins over film of "Lost in Space."

BBC continued newscasts until 4 am local time to report critical maneuver of spacecraft.

Budapest woman said, "Oh God, I hope they return safely." Hers was "common theme," *Washington Post* said; "around the world people, whether lowly or mighty, paused to participate in the first life-and-death drama in deep space." (Johnson, *W Post*, 4/15/70, A1; Arnold, *NYT*, 4/15/70, 1)

- Families of *Apollo 13* astronauts monitored flight communications on NASA squawk boxes and watched TV in their homes. Mrs. Marilyn Lovell and Mrs. Mary Haise kept older children home from school. Parents of unmarried Astronaut John L. Swigert, Jr., remained at home in Denver, Colo. (AP, *NYT*, 4/15/70, 29)
- America's "sanguine attitude about manned space flight" had been shattered by *Apollo 13*'s technical problems, Howard Simons said in *Washington Post*. "As the first words came that the three astronauts were in peril, earthbound fears began to race along with the moonbound craft. The astronauts seemed remarkably calm, most other persons desperate." As tension heightened, "inevitable questions began to intrude on one's innate fears. Would they get back? Could they get back?" It seemed certain situation would revive debate over rescue capability. "Mysterious 'bang' in space is bound to further slow an already slowed effort to stretch man's reach beyond his immediate horizon." (*W Post*, 4/14/70, A1)
- Aerobee 350 sounding rocket, launched by NASA from WSMR, carried GSFC payload to 87.8-km (54.6-mi) altitude. Objectives of mission, first Aerobee 350 launch from WSMR, were to demonstrate suitability of complete Aerobee 350 system for scientific research and to investigate solar and stellar x-ray phenomena. Rocket performance was unsatisfactory because premature sustainer-thrust termination seriously limited apogee. Payload recovery system operated successfully and ACS turned on and tried to erect. No useful x-ray data were obtained, but good performance data were recorded. (NASA Rpt SRL)
- House Armed Services Committee, by vote of 25 to 5, authorized $544.2 million for construction of C-5A transport aircraft, including $200 million toward funds requested by manufacturer Lockheed Aircraft Corp. [see March 2]. (AP, *NYT*, 4/16/70, 31)
- Secretary of Defense Melvin R. Laird and Australian Minister for Defence Malcolm Fraser reached accord in Washington, D.C., on Australian purchase of F-111C aircraft. Delivery of aircraft would be delayed until technical difficulties were fully evaluated and corrected; Australia would be provided on lease or other reimburseable basis up to 24 F-4E Phantom aircraft in interim; alternative Australian actions to meet strike aircraft requirements remained as options. (DOD Release 301-70)
- President Nixon submitted to Senate nominations of new members to National Science Board: H. Guyford Stever, President of Carnegie-Mellon Univ.; Robert A. Charpie, President of Cabot Corp.; Lloyd M. Cooke, Director of Urban Affairs for Union Carbide Corp.; Robert H. Dicke, Princeton Univ. physicist; David M. Gates, botanist of Washington Univ. in St. Louis; and Frank Press, MIT geologist and physicist. (*PD*, 4/20/70, 521-2)

- Officials of Stardust Hotel in Las Vegas, Nev., said 22 000 persons had filed past hotel's exhibit of *Apollo 12* lunar sample in six days. About 10 000 viewed most successful Strip "show" in one week. (AP, *NYT*, 4/16/70)
- INTELSAT announced election of new officers of Interim Communications Satellite Committee, for term of office beginning July 1. John A. Johnson, ComSatCorp Vice President-International and U.S. representative to INTELSAT, would succeed Carlos Nuñez A. of Mexico as Chairman. Yves Fargette, Regional Director of French Ministry of Posts and Telecommunications, would succeed Johnson as Vice Chairman. (INTELSAT Release 70-19)

April 15: Soviet Premier Aleksey N. Kosygin sent message to President Nixon: "We follow with concern the flight of the spacecraft Apollo 13 in which failure developed. I want to inform you the Soviet government has given orders to all citizens and members of the armed forces to use all necessary means to render assistance in the rescue of the American astronauts."

Italian Defense Ministry ordered Italian armed forces to be ready to give immediate help as needed if *Apollo 13* splashed down in Mediterranean. Reuters reported that U.K. Prime Minister Harold Wilson pledged help of Royal Navy ships for possible recovery work and that French President Georges J. Pompidou made similar offer of his navy. (UPI, W *Star*, 4/15/70; Reuters, *W Post*, 4/16/70)

- USAF launched unidentified satellite from Vandenberg AFB by Titan IIIB-Agena D booster into orbit with 386.2-km (240-mi) apogee. 136.8-km (85-mi) perigee, 89.7-min period, and 110.9° inclination. Satellite reentered May 6. (GSFC *SSR*, 4/30/70; UPI, *NYT*, 4/17/70, 48; Pres Rpt 71)
- U.S.S.R. launched *Cosmos CCCXXXIII* into orbit with 226-km (140.4-mi) apogee, 211-km (131.1-mi) perigee, 88.8-min period, and 81.3° inclination. Satellite reentered April 28. (GSFC *SSR*, 4/30/70)
- Concept of using LM to help return space crew to earth in case of serious trouble aboard CM had been worked out in 1964 and was practiced during *Apollo 9*'s 10-day orbit March 3-13, 1969, LM chief designer Thomas J. Kelly told *New York Times*. "But I can't say we foresaw exactly this type of failure, with everything out." Kelly said he had flown from Massachusetts, where he had been taking course at MIT, to Grumman Aerospace Corp. plant at Bethpage, N.Y., to help answer questions about how to deal with *Apollo 13* emergency. (Witkin, *NYT*, 4/16/70)
- International reaction to *Apollo 13* emergency continued:

  Dr. Eugene Carson Blake, General Secretary of World Council of Churches, said in Geneva that people of world were praying for astronauts.

  In Santa Ninfa, Sicily, devasted by earthquake in January, Mayor Vito Bellafore said, "After the first Apollo success people did lose interest; it seemed that it was only a mechanical exercise. Now we see again the human drama, and all of our worry is for those three lonely men."

  London Times said *Apollo 13* would leave "scar on the American space program." Best outcome of accident would be "greater sense of deliberation in the exploration of space." (Lyons, *NYT*, 4/16/70, 1)

- Editorial comment as damaged *Apollo 13* attempted safe return to earth:

    New York Times: "An emergency in a space craft on the voyage between earth and moon was bound to occur sometime. The enormous hazards of the long journey through the most hostile of environments, the ever present possibility of failure by one or more of the millions of components contained in an Apollo vehicle, the ubiquitous threat of human error, all these combine in lunar flights to make them the most dangerous expeditions human beings have ever undertaken. In retrospect, the remarkable fact is not that such ominous peril has arisen in the now-aborted trip of Apollo 13, but that the actually hazardous trips to the moon and back by Apollos 8, 10, 11, and 12 were accomplished in safety." (*NYT*, 4/15/70)

    Washington Post: "Lindbergh gave up a continent, as the astronauts gave up a planet, because he had measured the risks and found them reasonable, for all the possibility of disaster lurking around the corner—it does so every day for men and nations. So if we are going to apply the glories, we must confront the dangers along the way. All we can do now, as that unlikely looking craft limps back from the moon, is to hope that the men who ride in it and the men in Houston who guide it can find what Lindbergh found. 'Somewhere in an unknown recess of my mind,' he wrote 'I've discovered that my ability rises and falls with the essential problems that confront me. What I can do depends largely on what I have to do to stay alive and on course.' " (*W Post*, 4/15/70)

    Baltimore *Sun:* "The safety record of the Apollo program has, except for one hideous ground accident, been a perfect one. Perhaps we have too easily assumed that therefore all expeditions would begin and end without serious and unforseen breakdown, without the torments of suspense when the line between splendid accomplishment and bitter tragedy is drawn very fine. The truth is, of course, that exploration, especially in a realm new to man, can never be free of deadly hazards. But the astronauts and the men associated with them in their ventures have always known it." It was clear astronauts were equal to emergency, "but they remain at the mercy of machinery which in some still unexplained way has failed them. The rest of us can only wait from desperate hour to hour, trusting that superlative skill, universal hope and fervent prayers will prevail over cruel chance." (B *Sun*, 4/15/70)

- As U.S. and U.S.S.R. opened strategic arms limitation talks (SALT) in Vienna, thoughts of delegates and local citizens were on *Apollo 13*. *Washington Post* later said mission was "a symbol of the space age that reflects the arms control problem at these talks. The rocket that put men into space also is the device that can rain death for millions by one superpower upon the other." (Roberts, *W Post*, 4/16/70, A10)

- Researchers at Ohio State Univ. headed by physiologist Dr. Harold S. Weiss had received grants totaling $452 975 to study artificial atmospheres inside space vehicles, UPI reported. They were trying to eliminate undesirable side effects on astronauts from breathing pure oxygen. (*NYT*, 4/16/70, 30)

- NASA announced award of one-year, $40 000, cost-plus-fixed-fee contract with two one-year options to Computing and Software, Inc., for support services for biomedical programs at FRC. Work would include design of life-support systems for advanced research flight vehicles. (NASA Release 70-57)

- Subcommittee on Science, Research, and Development of House Committee on Science and Astronautics transmitted to parent committee *The National Institutes of Research and Advanced Studies: A Recommendation for Centralization of Federal Science Responsibilities*. Report, result of 10-mo study by subcommittee, recommended immediate establishment of National Institutes of Research and Advanced Studies (NIRAS) to consolidate Federal responsibilities for basic research and graduate education. NIRAS would be independent agency below cabinet level with administrator appointed by President subject to Senate confirmation. NSF, whose charter was basis for NIRAS, would be distributed in new centralized agency, with education programs at undergraduate level or below transferred to HEW Office of Education. Basic research programs determined to be no longer relevant to individual agency missions would be considered for transfer to NIRAS.

  Committee also recommended increase of OST staff with director given Cabinet status and responsibilities extended. (Text)

- F-111 wing carry-through box underwent 16 000th hour of ground testing—four times estimated number of hours aircraft would actually fly in 10 yrs—at Fort Worth Div. of General Dynamics Corp. Box had to withstand four times the number of maneuver loadings anticipated during decade of service operation at rate of 400 flight hours per year. (*PD*, 4/20/70, 525-6)

April 16: Tass announced Soviet efforts to aid recovery of *Apollo 13* astronauts: fishing trauler *Chumikan* and two cargo ships were being sent to recovery area, several whaling ships in South Pacific were being diverted to Samoa, and Soviet ships throughout world had been alerted. President Nixon sent message to Soviet Premier Aleksey Kosygin expressing "deep gratitude" and said, "I will let you know fast if we need your government's help."

  Later, *New York Times* quoted "informed sources" as saying *Chumikan* was missile instrumentation ship which had already been deployed in area to gather data on *Apollo 13* mission with electronic sensing devices.

  Times said 13 countries had offered to make ships available for recovery, at least six said they would permit U.S. aircraft to fly over their territory during recovery operations, and more than 70 countries had responded to U.S. request not to use radio frequencies used by recovery forces to avoid possible conflicts on airways. (Lindsey, *NYT*, 4/18/70, 12)

- Cosmonaut Vladimir A. Shatalov, on behalf of all Soviet cosmonauts, sent message to *Apollo 13* crew: "We Soviet cosmonauts are following your flight with great attention and anxiety. We wish wholeheartedly your safe return to our mother earth." (AP, B *Sun*, 4/17/70, A9)

- *Apollo 11* Astronaut Neil A. Armstrong told press at MSC *Apollo 13* crewmen would be "considerably disappointed that they weren't able to use all that practice they've had and that wonderful opportunity to walk on the moon's surface." Possibility of not returning safely was a real one that "always exists in the back of your mind,"

he said. "I suspect that the attitude that they [astronauts] reflect over the communication loops reflects what they're really thinking. Namely, they're trying to do each and every job precisely as well as they can and not overlook anything." (AP, W *Star,* 4/17/70, A7)
- NASA launched two sounding rockets from Churchill Research Range. Arcas carried Univ. of Houston payload to study ionosphere. Rocket and instruments functioned satisfactorily. Nike-Apache carried Univ. of Texas at Dallas payload to study ionosphere, but mission was unsuccessful. (SR list)
- NASA announced tracking station at Antigua, no longer required for manned space flight support, would be closed by June 30. Station, which included unified S-band radar complex and 9-m (30-ft) dish antenna, had been established in 1967 and had played major role in tracking Apollo flights through *Apollo 11.* (NASA Release 70-58)
- FAA published notice of proposed rule to prohibit flights by civil aircraft over U.S. at speeds that would cause sonic boom to reach ground. (FAA Release 70-33)
- President Nixon sent message to Ambassador Gerard C. Smith, chief of U.S. delegation to U.S.-U.S.S.R. strategic arms limitations talks (SALT) in Vienna: "I express...the hope that an agreement can be reached on the limitation and eventual reduction of strategic arsenals with proper recognition of the legitimate security interests of the United States and the Soviet Union and of third countries." (*PD,* 4/20/70, 531)

April 17: President Nixon issued statement following safe return of *Apollo 13* astronauts: "From the beginning, man's ventures into space have been accompanied by danger. Apollo 13 reminds us how real those dangers are...of the special qualities of the men who dare to brave the perils of space. It testifies, also, to the extraordinary concert of skills, in space and on the ground, that goes into a moon mission. To the astronauts, a relieved Nation says 'Welcome home.' To them and to those on the ground who did so magnificent a job of guiding Apollo 13 safely back from the edge of eternity, a grateful Nation says 'Well done.' "

In proclamation designating April 19 as national day of prayer and thanksgiving for safe return of *Apollo 13* astronauts, President said: "The imperiled flight and safe return of the crew of Apollo 13 were events that humbled and inspired people all over the world. We were humbled by the knowledge that in this stage of man's exploration of space, a safe splashdown of an imperiled mission is in its way as successful as a landing on and safe return from the Moon. We were inspired by the courage of the crew, the devotion and skill of the members of the NASA team on the ground and by the offers of assistance from nations around the world."

President spoke with *Apollo 13* astronauts by telephone from White House to U.S.S. *Iwo Jima* in Pacific. He told them he would like to present them the Presidential Medal of Freedom in Hawaii on following day. President then announced intentions at White House press briefing. He would go "first to Houston, where we will pick up the two wives, and then go on to Hawaii. ...what these men have done has been a great inspiration to all of us. I think also what the men on the ground have done is an inspiration to us. How men react in adversity determines their true greatness, and these men have demonstrated that the American character is sound and strong and capable of taking a very difficult situation and turning it into a very successful venture. . . .in recognition, also, of the men on the ground, I am going to present the Medal of Freedom tomorrow in Houston. . . to the Apollo 13 crew on the ground who have made these very difficult decisions on the spot, decisions that had to be right." (*PD*, 4/20/70, 538-40; Text)

- NASA held series of three *Apollo 13* post-recovery press conferences at MSC hosted by NASA Administrator, manned space flight officials, and technical specialists. Dr. Thomas O. Paine, NASA Administrator, said although mission "must be recorded as a failure, there's never been a more prideful moment in the space program." Most impressive aspect of mission was "tremendous effectiveness in team work of our ground support operation . . . . From every state . . . we've had people. . . working to backstop the many different system decisions that had to be systematically and methodically faced up to and made. . . . The Flight Operations Directors . . . certainly demonstrated that flight controllers can operate under these very adverse conditions. They've been able to do things in a far shorter period of time than we'd ever thought possible before."

    Dr. George M. Low, NASA Deputy Administrator, said basic procedures used for every phase of mission were predetermined. "For example, the burn made behind the moon . . . was a standard emergency burn that would be used in case the service propulsion system would not work on a normal lunar mission going into lunar orbit. Each one of these steps had been predetermined and also simulated. But putting them all together into a sequence of steps . . . and working out some of the detailed procedures on how to save power and how to save batteries; how to save water; those things were determined during the course of the flight, both by the men . . . on the ground and by the astronauts." (Transcript)

- At White House press conference former astronauts Michael Collins, Assistant Secretary of State for Public Affairs, and William A. Anders, NASC Executive Secretary, described arrangements that had kept President Nixon informed on *Apollo 13* crisis. President had had access to direct TV, radio, and telephone contact with MSC Mission Control. He had talked with NASA Administrator, Dr. Thomas O. Paine, and Astronaut Frank Borman there. Collins and Anders had set up "squawk box" into Mission Control outside President's office and briefed President and his staff on developments. President's chief concerns had been CM-SM separation, and crew's safety and physical condition. Collins said, "The fact that we had this incident really changes very little. We have always known that such a situation could develop and he [President] does not feel that it should unduly cloud the future when it comes to space exploration." (Transcript)

- President Nixon made brief appearance at banquet for Republican leaders in Washington, D.C., after learning of *Apollo 13* astronauts' safe return. He told group that when he called mission commander James A. Lovell, Jr., after splashdown, Astronaut Lovell had said he was sorry mission had not been completed. "My reply," President said, "was: 'Captain, you and your colleagues did complete the mission.'" President told group, "After what happened this tumultuous week, let's not take American wealth and strength for granted...where machines failed, we had to turn to man, and man came through." President said it had been "proudest day of my life and in the life of the country." Vice President Spiro T. Agnew said "social levelers of the New Left" would seize on *Apollo 13* near-disaster to demand space budget be "soaked down into the nearest slum." It would be "shortsighted tragic blunder" to heed warnings of "fair weather friends" of space program who were "readying the life-boats for a quick exit." (Hope, W *Star*, 4/18/70, A2)

- U.S. reaction to *Apollo 13*'s successful splashdown:

  Sign at MSC was changed from "Our hearts are with the Apollo 13 crew," to "Sigh of relief party here tonight." Mission Control broke into applause before whipping out traditional cigars to fill huge room with blue smoke. Dr. Robert R. Gilruth, MSC Director, observed that mission had not been complete failure. Spacecraft had gone around moon—feat that would have caused sensation two years before. Dr. Thomas O. Paine, NASA Administrator, said later in day, "I'm a little tired and bedraggled but I'm feeling 10 or 20 years younger than I did yesterday."

  In New York, thousands watched recovery on giant TV screen at Grand Central Terminal. Loud applause sounded when capsule appeared in sky, at splashdown, and when astronauts appeared. Hot dog stand operator at Vanderbilt Ave. and 47th Street assured passersby, "It's all right." Long strands of multicolored ticker tape and bushels of confetti poured from skyscraper windows. At City Hall, Mayor John V. Lindsay said, "Now they are safely back and New Yorkers join with men everywhere in saying 'welcome home.'"

  Church bells pealed throughout Nation. Traffic came to standstill in Indianapolis, Los Angeles reported no crime for brief period, and dice stopped rolling in Las Vegas briefly. Richmond, Va., department store manager, watching crowd before store TV set, said everybody "had a quiet smile, but it was curious the way they stood looking at all that good news—silently and almost as if they were still praying." (*W Post*, 4/18/70, A1, All; *NYT*, 4/18/70, 1, 13)

- International reaction to *Apollo 13* spashdown:

  European Broadcasting Union said in Geneva that TV audience at splashdown might well have been biggest TV audience of all time.

  In American Samoa, girls were limbering up for siva-siva dance to greet astronauts. Pilgrims prayed their thanks in India. Press item from Johannesburg, South Africa, said witch doctor named Magomezulu had tossed bones which told him *Apollo 13* astronauts would return safely but that "these men have interfered with God and he has turned them back. This is a warning that the Americans should stop landing on the moon."

  Thousands in Lima, Peru, received permission to take early lunch to watch final moments of splashdown.

  In Canada, Prime Minister Pierre E. Trudeau expressed "great relief."

  Queen Elizabeth wired congratulations from England to President Nixon. Roman Catholic Archbishop of Westminster John Cardinal Heenan said, "Thank God." Telephone operator at U.S. Embassy in London was inundated with calls. She said, "People were sobbing with obvious relief and happiness. I just don't know what to say to them" (Johnson, *W Post*, 4/18/70, A1; Arnold, *NYT*, 4/18/70, 1; Waldron, *NYT*, 4/18/70, 13)

- *New York Times* editorial appeared before *Apollo 13* splashdown: "Here on earth the drama in space has done more than shatter any complacency that may have begun to surround lunar flight. It has drawn men and nations together in common concern over astronauts who are envoys of mankind even before they are nationals of a particular state . . . . Lovell, Swigert and Haise have already written an epic of heroism and skill that assures them immortality in the annals of space exploration." (*NYT*, 4/17/70)
- ComSatCorp reported increase in net income for first quarter of 1970 to $3 345 000 (33 cents per share), up from $1 525 000 (15 cents per share) for first quarter of 1969. First quarter net operating income was $1 850 000—up from $137 000 for first quarter of 1969 and from $1 832 000 for all of 1969. First quarter revenues totaled $15 435 000, some $5 213 000 more than for first quarter of 1969. On March 31, 1970, ComSatCorp was leasing full time 1566 half circuits—357 more than on March 31, 1969. (ComSatCorp Release 70-22)
- *Philadelphia Inquirer* editorial commented on Communist Chinese satellite: "To assume that the Red Chinese will refrain from placing hydrogen warheads in orbiting satellites, capable of being released by remote control on signal from Peking, is to engage in the most dangerous kind of head-in-the-sand complacency." (*P Inq*, 4/17/70)
- Marxist syndrome "—that R&D must be directly relevant to social problems, from national security to drug addition—is becoming the U.S. science policy rule of thumb," S. Dedijer of Univ. of Lund, Sweden, said in *Science*. During past generation, "there seemed to be no limit in the United States to the support of any project so long as it was classified science, research, development, test or evaluation." Demand for more rational R&D policy had been growing in U.S. for years. "This social reexamination is resulting in the flight of motivated young talent from the physical to the life sciences, the humanities, and ethics and religion, on the one hand, and to a growth of interest in the policy sciences, in social problems studies, and in the definition of R&D relevant to them, on the other." (*Science*, 4/17/70, 344-5)

April 17-18: Full-scale plywood-and-metal model of Apollo LM that had stood on MSC grounds for two years collapsed when MSC officials attempted to move it to serve as backdrop for President Nixon's scheduled speech. Workmen worked through night to install replacement—LM that had been used as trainer for *Apollo 11* lunar landing. (*W Post*, 4/19/70, A18)

April 18: *Apollo 13* astronauts were taken in three helicopters from U.S.S. *Iwo Jima* to Tafunda Airport in American Samoa, to set foot on solid ground for first time since mission launch. They were greeted by Samoan Governor John M. Haydon. Governor presented IRS form 1040A to Astronaut John L. Swigert, Jr., who had forgoten to file his income tax before departing on mission. Colorfully dressed crowd of more than 3000 cheered astronauts during half-hour reception before their departure in USAF C-141 jet aircraft for Hawaii to meet President Nixon. (Trumbull, *NYT*, 4/19/70, 1)

- President Nixon presented Presidential Medal of Freedom to Apollo

Mission Operations team at MSC. Citation read: "We often speak of scientific 'miracles'—forgetting...that these are the product of hard work, long hours and disciplined intelligence. The men and women of the Apollo mission operations team performed such a miracle, transforming potential tragedy into one of the most dramatic rescues of all time. Years of intense preparation made this rescue possible. The skill, coordination and performance under pressure of the mission operations team made it happen. Three brave astronauts are alive and on Earth because of their dedication, and because at the critical moments the people of that team were wise enough and self-possessed enough to make the right decisions. This extraordinary feat is a tribute to man's ingenuity, to his resourcefulness and to his courage." Award was accepted by Sigurd A. Sjoberg, Director of Flight Operations, on behalf of himself, Glynn S. Lunney, Milton L. Windler, Gerald D. Griffin, and Eugene F. Kranz. Following award ceremony President and Mrs. Nixon left Houston for Honolulu with *Apollo 13* astronauts' wives, Mrs. James A. Lovell, Jr., Mrs. Fred W. Haise, Jr., and parents of bachelor astronaut, John L. Swigert, Jr.

In Honolulu, President Nixon presented Presidential Medal of Freedom to *Apollo 13* astronauts. He said in part: "we have received over 100 messages from foreign governments—from the Soviet Union, from Poland, other countries behind the Iron Curtain, from countries in the free world. This is truly a welcome from all the people of the world to three very brave men." Citation on each medal read: "Adversity brings out the character of a man. Confronted suddenly and unexpectedly with grave peril in the far reaches of space, he demonstrated a calm courage and quiet heroism that stand as an example to men everywhere. His safe return is a triumph of the human spirit—of those special qualities of man himself we rely on when machines fail, and that we rely on also for those things that machines cannot do.

"From the start, the exploration of space has been hazardous adventure. The voyage of Apollo 13 dramatized its risks. The men of Apollo 13, by their poise and skill under the most intense kind of pressure, epitomized the character that accepts danger and surmounts it. Theirs is the spirit that built America. With gratitude and admiration, America salutes their spirit and their achievement." (*PD*, 4/27/70, 548-50)

- More than 50 engineers at MSC and at NR's Downey, Calif., plant were working through weekend to complete "life history" of ruptured *Apollo 13* oxygen tank, John N. Wilford reported in *New York Times*. "The only time in the tank's history even the slightest doubt was raised about its preflight condition was during precountdown preparations . . . . Technicians reported having trouble draining the liquid oxygen from the tank after a test. The tank seemed to empty slower than usual. But the next time it was checked out, the drainage problem did not occur." Report on tank's manufacture, installation, testing, and flight would be submitted to *Apollo 13* Review Board. (*NYT*, 4/19/70, 54)
- NBC estimated *Apollo 13* splashdown TV audience at more than 40 million viewing portions of coverage from Houston Control or pickups from sea off American Samoa. Since millions abroad watched

event, "global total that joined in the final moment of relief probably can never be measured," Jack Gould said in *New York Times*. ABC, CBS and NBC had "acquitted themselves most handsomely" during day "that added emotion to emotion before the climax that left a viewer almost drained of the powers of response." Apollo Astronauts Walter M. Schirra, Jr., Charles Conrad, Jr., and Eugene A. Cernan had assisted network commentators in reporting technical aspects of mission. "For the first time on TV, the viewer could see for himself the fruits of the imaginative training program of NASA."

Washington Post said networks "deserved the criticism they got for slow reaction to the unexpected danger...on Monday night" but "deserve praise today for the crisp, professional reporting and for some unusual restraint." (*NYT*, 4/18/70; Laurent, *W Post*, 4/18/70, A12)

- *Apollo 13* editorials:

Baltimore *Sun:* "The simple, grateful 'welcome home' spoken throughout the world yesterday not only lifted a burden of dread; it celebrated a new and unforgettable triumph in man's exploration of space and his own resources of courage and ability at their highest." (B *Sun*, 4/18/70)

New York Times: "Only in a formal sense, Apollo 13 will go into history as a failure . . . Apollo 13's booster rocket was sent crashing into the moon on schedule, and that did produce some useful and important scientific data. But in a larger sense, Apollo 13's flight was enormously productive. The emergency which so gravely endangered its crew turned this flight into a crucial—though unintended—experiment in space rescue . . . . The lessons learned from this near-disaster will help save spacefarers who are still unborn. And Apollo 13's ordeal must raise more insistently than ever the question of creating an international space rescue organization with the rockets, crews, and other necessities required to give aid to distressed travelers in space." (*NYT*, 4/18/70)

Cleveland *Plain Dealer:* "The very real triumph of *Apollo 13*, the successful return of a crippled craft, could help the United States space program. What NASA can learn from its inquiry into the cause of the accident could further advance space science and technology —and safety. Certainly further evaluation of the space program's timetable is in order. But man learns from his failures as well as from his spectacular achievements." (C *Plain Dealer*, 4/18/70)

New York *News:* "President Nixon has very fittingly proclaimed tomorrow a day of national thanksgiving. And with that attended to, the U.S. space research and exploration program should proceed full-speed ahead, with special reference to more manned moon shots, and the sooner the better. To falter or fail now would be...unworthy of Americans or America." (NY *News*, 4/18/70)

Nature: "One of the striking features of the past days has been the sheer competence of the people who have been concerned on the ground as well as in the sky. As always, NASA has put on a good show, and there is more to this than the mere capacity to keep a stiff upper lip or even to diminish real dangers." NASA had functioned "superbly" during daring exploits. "A part of the secret has been that there have been enough men on hand to think of everything, or

nearly everything. Another has been the logic of the planning." Conduct of Apollo would be "monument to the way in which pure reason can accelerate complicated technical developments. Nothing decided in the next few months should diminish the importance of this brilliant demonstration. Those who are inclined to complain that there had been an accident should, rather, wonder why it has not happened before." (*Nature*, 4/18/70, 197)
- Discovery by radar astronomers at JPL's Goldstone Tracking Station of several nearly continent-sized rough areas on Mercury's surface was reported by Dr. Richard M. Goldstein, head of JPL Communications Research Section, at meeting of U.S. National Committee of International Scientific Radio Union in Washington, D.C. Areas were similar to ones observed on Venus but "larger relative to the size of the disk," and had much less contrast to the surrounding areas. (JPL Release 550)
- Display of *Apollo 11* lunar rocks and *Apollo 12* space helmet opened at New York State Museum in Albany, N.Y. Display had been lent by General Electric Co. and NASA. Exhibit would continue through April 26. (UPI, *NYT*, 4/18/70, 12)
- Study of defense industry profits by Logistics Management Institute, DOD-subsidized research organization, had shown lower profits on defense work than on commercial sales, *Armed Forces Journal* said. Overall profit percentage on total capital investment for major defense companies was 6.8% for 1968, down 0.7% from 1967. (*AFJ*, 4/18/70, 2-3)

April 19: U.S. observed national day of prayer in gratitude for safe return of *Apollo 13* astronauts. In New York 150 persons attended sunset service at Temple Emanuel. Invocation of gratitude was added to Sunday mass at St. Patrick's Cathedral, expressing hope that space exploration would contribute to peace and welfare of mankind. Greek Orthodox Primate of North and South America, Archbishop Iakovos, delivered special message for faith's Palm Sunday observances. He said he believed "God permitted the agony of these last four days" to aid men in "overcoming the temporal with the spiritual, the temporary with the eternal."

Church and interfaith councils in St. Petersburg, Fla., made combined appeal for prayer in area churches and kept chapels open during day. In Marietta, Ohio, Jaycees completed petition containing 50 000 signatures, opposing efforts to ban prayers in broadcasts from space. Petition was later presented to space officials in Washington, D.C. Gov. Nelson D. Rockefeller proclaimed "day of prayer and thanksgiving" in New York State. Sen. Margaret Chase Smith (R-Me.) said crisis refuted claims that space exploration dealt fatal blow to religion. "The manner in which great numbers of Americans attended special prayer services... has been inspirational." J. F. Meredith, treasurer of Central State Conference of the Seventh Day Adventists, said in Kansas City, "We feel that this points up the fact that despite all of man's knowledge and ingenuity there is still a supreme being in control of things, and that we have to trust him." (*NYT*, 4/20/70, 28; 4/19/70, 54)
- President and Mrs. Nixon and party of *Apollo 13* astronauts and families attended services at Kawaiahao Church in Honolulu commemorating National Day of Prayer and Thanksgiving [see April 17]. Pres-

ident said: "There are very deep ideological differences that divide the world today. But when it was learned that these men were in danger, there poured into the White House from all over the world messages from the Communist countries, from people of various religions, saying that they wished their best, offering their assistance. When they learned they were back, there was an outpouring of relief and rejoicing from people, regardless of their political or religious differences . . . . If only we could think in that way about every individual on this earth, we could truly have world peace." Following service Presidential party flew from Hawaii to San Clemente, Calif. (*PD*, 4/27/70, 550-3)

- *Apollo 13* astronauts and families were greeted by crowd of 5000 at Ellington AFB on return to Houston from Hawaii. Astronaut James A. Lovell, Jr., said: "Jack and Fred and I are very proud and glad to be back in Texas because there were times when we weren't sure we would make it back here. We would not have been here today if it were not for the people on the ground. The people in Mission Control guided us all the way, day and night." (Pasadena *Star-News*. 4/20/70; Lannan, W *Star*, 4/20/70, A1)
- NASA released two black-and-white photos taken by *Apollo 13* crew, showing damage to SM from oxygen tank explosion. Former Astronaut James A. McDivitt, Manager of Apollo Spacecraft Program at MSC, said primary information revealed by pictures was that entire 4- by 1½-m (13- by 5½-ft) panel held by 250 bolts had been blown off and that fuel cells above tanks had remained intact. (Auerbach, W *Post*, 4/20/70, A1)
- *Apollo 13* editorials:

    Washington Post: "The footprints are still imperishably graven on the lunar dust and the flag still flies on the surface of the moon, but it all seems more distant, unreal and unattainable now. The loud bang that rocked Apollo 13 and echoed back to us from outer space last week did more than challenge our comfortable assumptions about manned space flight. It reminded us how much we have taken for granted, how much has been achieved and how far man has yet to go." (*W Post*, 4/19/70, B1)

    Washington *Sunday Star:* ". . .men, the world over, instinctively recognize the metaphysical importance of the reach into space. Lovell, Haise, and Swigert were not just three Americans on a dangerous mission. They, and those who will follow, are ambassadors of life to new, unknown regions. The success of these first halting steps into space has brought a new vision to a slightly weary world and given mankind a new pride in its humanity. It is a gift without price. It is an offering that must not be withdrawn because of a momentary setback and a dramatic demonstration of the dangers that have always been present." (W *Star*, 4/19/70)

    Philadelphia *Sunday Bulletin:* "Of all the machines man has made to plumb the depths of space, surely none has served him better or more faithfully in desperate circumstances [than the LM]. So Aquarius and the men who designed and manufactured her earned the plaque in Smithsonian, just as the astronauts and the Apollo 13 ground crew earned the Freedom Medals President Nixon has awarded them." (P *Bull*, 4/19/70)
- City council of Roermond, the Netherlands, had announced it would

name street after *Apollo 13* and include names of Astronauts James A. Lovell, Jr., Fred W. Haise, Jr., and John L. Swigert, Jr., in street signs, AP reported. (*W Post*, 4/20/70, A8)

April 20: House leaders of both parties urged that U.S. push ahead with space exploration despite *Apollo 13* abort. Speaker John W. McCormack (D-Mass.) said in floor speech, "There may be a small minority who will urge that we pull back our space efforts. This is not my position. Neither am I convinced it will be the view of the Congress or the overwhelming majority of American people." House Minority Leader Gerald R. Ford (R-Mich.) said he concurred. Chairman George P. Miller (D-Calif.) of House Committee on Science and Astronautics said committee would not conduct investigation of *Apollo 13* accident but would review findings of NASA inquiry board. (*CR*, 4/20/70, H3247, 3276; UPI, *W Post*, 4/2/70)

- Astronauts James A. Lovell, Jr., Fred W. Haise, Jr., and John L. Swigert, Jr., were debriefed on *Apollo 13* mission at MSC, tape-recording details for almost eight hours.

  NASA released 6½-min 15-mm movie of mission, including scenes of astronauts after SM explosion. Film showed Lovell rubbing his hands vigorously as if trying to warm them, Haise sleeping with his hands folded inside his spacesuit to keep them from floating free in weightlessness, and Swigert studying flight plan. (AP, *W Post*, 4/21/70, A4)

- Max Lerner described reactions to *Apollo 13* accident in Philadelphia *Evening Bulletin:* "What was I doing, glued to the TV screen like a zombie with the blood drained out me . . .? What was wrong with me—and with millions of others? . . . we had all played the game of apathy up to that moment . . . confident that Apollo 13 was just one more sure-thing deployment of what man already proved he can do. The flight reports had to compete for newspaper space and, in some newspapers, were slipping into the inside pages." After news of oxygen tank explosion "astronauts were no longer either three invincible heroes or three stooges for 'the Establishment,' but simply three human beings caught in space a quarter-million miles from home . . . . Through the plight of our proxies . . . we caught a glimpse of how perilous man's whole enterprise is, how tragic, in fact, his whole existence is in this shaky cosmos that runs by laws until you hit the unexplainable."

  Apollo 13 was "a tiny, mobile version of the earth, with a power failure and the shortage of water and a danger of polluted air and the erratic fortunes of communicating and the image of three men having to read their instruments with a flashlight in the darkness of the cabin." (*P Bull*, 4/20/70)

- *Apollo 13* editorials:

  National Observer: "At a time when Americans were becoming bored with successful flights to the moon, the ordeal of Apollo 13 reminded the nation of the dangers and difficulties of space. But because of courage, care, and uncommon resourcefulness—because, in fact, of history's most dramatic field expedient—the astronauts made their way safely back to earth. The first walk on the moon was a wonder of wonders, but the return of Apollo 13 was the gladdest moment of them all." (*National Observer*, 4/20/70)

Wall Street Journal: "Assuming all feasible precautions are taken, then, the issue of death potential should not really be a major factor in assessing the space program, and particularly the very costly moon program. That program should be judged on whether any gains from it are proportionate to the totality of its costs. Some scientists and laymen believe the costs are disproportionate. At any rate, that is the proper criterion for weighing the future of space exploration, not the admitted and awesome danger of death in eternal orbit." (*WSJ*, 4/20/70)

- ComSatCorp Chairman James McCormack submitted annual *Report to the President and the Congress*, including information reported to ComSatCorp shareholders March 9. (Text)
- NASA announced appointment of Assistant Administrator for Legislative Affairs Robert F. Allnutt as Assistant to Dr. Thomas O. Paine, NASA Administrator, and as member of *Apollo 13* Accident Review Board. Allnutt would be succeeded by H. Dale Grubb, President's Special Assistant for Legislative Affairs. (NASA Release 70-60)
- Firemen in Evendale, Ohio, had worked two years to perfect new firefighting uniform of aluminum-beta glass, material used in astronauts space suits, UPI said. (*W News*, 4/20/70, 4)
- Flow-visualization technique using laser as light source was enabling AFSC engineers to obtain cross-sectional view of entire flow field around scale model aircraft in wind tunnels at Arnold Engineering Development Center at Tullahoma, Tenn., AFSC announced. Technique, using saturated air and continuous-wave ruby laser, had been used to photograph shock waves around "hypersonic cruise vehicle"—theoretical aircraft that would fly twice as fast as fastest jet-powered research aircraft. (ASFC Release 113.70)
- Secretary of Defense Melvin R. Laird discussed "basic asymmetry" between U.S. and U.S.S.R. efforts in nuclear weapons field at AP annual luncheon in New York: From 1965 to 1967 U.S.S.R. had more than tripled its inventory of strategic offensive nuclear weapon launchers—from 500 to 1700, including 200 heavy bombers—and quadrupled total megatonnage in its strategic offensive force. U.S. in same period had not increased established level of 1710 strategic nuclear missile launchers, had reduced heavy bomber strength of 780 by over 200, and had reduced megatonnage by more than 40%. "Except for the minimum 'hedge' that SAFEGUARD will provide, we have not responded to the Soviet strategic offensive buildup with new deployment programs. We did not respond in past years because the United States deliberately chose to assume that the Soviet buildup was aimed at achieving a deterrent posture comparable to that of the United States. We have not responded this year because...we fervently hope that SALT can render such a response unnecessary." (Text)

April 20-24: American Geophysical Union held annual meeting in Washington, D.C.

Cal Tech physicist Dr. Gerald J. Wasserburg said detailed studies of *Apollo 12* moon rocks had shown preliminary estimates of their age to be off by some 800 million yrs. Evidence from *Apollo 11* and *12* lunar sample studies indicated moon—and possibly earth and some major planets—had at least one period of sustained turbulence, lasting about 300 million yrs and beginning some billion years after moon and rest of solar system originated. Solar system was estimated to be 4.6 billion yrs old. At news conference following meeting, Dr. Wasserburg said he and collaborator, Cal Tech physicist Dr. D. A. Papanastassiou, had found age of lava-born *Apollo 12* rocks to be 3.4 billion yrs, rather than preliminary estimate of 2.6 billion yrs. *Apollo 11* rocks were found to be 3.65 billion yrs old. Evidence indicated both areas of moon had been formed from same sustained process lasting some 300 million yrs. There was "widespread outpouring" of lava over sea regions of moon about 3.5 billion yrs ago. Evidence suggested Sea of Storms, where *Apollo 12* rocks were found, had almost same age as Sea of Tranquility, *Apollo 11* lunar site. (AP, *LA Times*, 4/23/70)

Dr. Rhodes W. Fairbridge, Columbia Univ. geologist, reported finding territory which was earth's south polar region in Upper Ordovician period, about 450 million yrs ago, in middle of Sahara desert. Finding supported belief that phenomenon of sliding land masses known as continental drift had occurred. (Blakeslee, *NYT*, 4/21/70, 1)

At closing session scientists, with exception of ESSA's Dr. S. Fred Singer, maintained earth and moon had simultaneous and similar origin despite differences between lunar rocks and those of earth. Dr. Singer agreed moon had most likely been captured in earth's gravitational field "very shortly" after earth's formation about 4.6 billion yrs. ago. Dr. John A. O'Keefe of GSFC thought earth and moon were once single planet that broke apart. Dr. A. G. W. Cameron of Yeshiva Univ. felt earth and moon had formed in same gaseous cloud. (UPI, *NYT*, 4/27/70)

April 21: *Apollo 13* Astronauts James A. Lovell, Jr., John L. Swigert, Jr., and Fred W. Haise, Jr., described their initial reaction to *Apollo 13* accident to press at MSC. Lovell explained: "Fred was still in the lunar module, Jack was back in the command module in the left-hand seat, and I was halfway in between the lower equipment bay...when all three of us heard a rather large bang, just one bang. Now, before that, Fred...had actuated a valve which nominally gives us the same sound...so when this bang came we really didn't get concerned right away, but I looked up at Fred and Fred had that expression like it wasn't his fault, and we suddenly realized that something else had occurred, but exactly what we didn't know."

Swigert said he had felt vibration accompanying bang and then noticed master alarm and main bus B undervolt light. "I transmitted to Houston that we had a problem and proceeded to...look at the voltage...[which] at that time was completely normal...." Swigert thought bang had occurred in LM and Haise, in LM, thought it had occurred in CM.

Lovell said his first impression was "that we had had an electrical problem that caused this bang.... That quickly went away and then I looked out the window and saw this venting and my concern was increasing all the time. It went from I wonder what this is going to do to the landing to I wonder if we can get back home again. It was apparent...that the only way to survive the situation was to transfer to the LM...."

Lovell said decision on date, crew, and destination of Apollo 14 would be made after debriefings and investigation by *Apollo 13* Review Board. If NASA "wants this crew to go back to Fra Mauro, we'll be glad to go. If they decide to send another crew or not to go to Fra Mauro, that's their decision." (Transcript)

- NASA announced appointment of members of *Apollo 13* Review Board, headed by LaRC Director Edgar M. Cortright: Astronaut Neil A. Armstrong; Robert F. Allnutt, Assistant to NASA Administrator; Dr. John F. Clark, GSFC Director; B/G Walter R. Hedrick (USAF), Director of Space, DCS/R&D; Vincent L. Johnson, OSSA Deputy Associate Administrator (Engineering); Milton Klein, Manager of AEC-NASA Space Nuclear Propulsion Office; and Dr. Hans Mark, ARC Director. (Off NASA Administrator; Wilford, *NYT*, 4/22/70, 27)
- Rep. Lou Frey, Jr. (R-Fla.), discussed space program spinoff on House floor: "Between 1965 and 1971 NASA awarded grants and research contracts for 1640 programs to 223 colleges and universities in all 50 states and the District of Columbia. The total university expenditures by NASA from 1959 through 1969 totaled $2.7 billion and contributed to this Nation's educational goals by the development of new scientific disciplines, technologies and educational facilities." NASA's educational efforts in secondary and elementary schools had been "many and varied, but one of the unpublicized and most important has been its programs to assist underprivileged youth in the inner city. The problem has been tackled in a number of locations by a wide variety of techniques, including special materials, instructors, and so forth." (*CR*, 4/21/70, H3317-25)
- New standards for certification of transport aircraft over 5670 kg (12 500 lbs), effective May 8, were announced by Secretary of Transportation John A. Volpe. Standards included warning system to alert pilots to failing power-operated control system; operating procedures for flight in turbulence; design of airframe for dynamic loads from continuous turbulence, with flight load measurements required for new aircraft with unreliable methods of predicting load intensities; system for preventing condensation on inside of windshield during flight and taxiing; greater redundancy in control system designs; flight tests to demonstrate freedom from flutter through transonic speed range; protection of empennage from bird strikes; more comprehensive airframe lightning protection criteria; additional fire protection for powerplant; dual locking for all critical removable fasteners; protection of fuel system components in event of gear-up landing; design precautions to minimize hazards of turbine rotor failure; fail-safe criteria for aircraft systems; changes in design standards to accommodate modern instrumentation and to require basic flight and navigation instrument at each pilot station; and improvements in reliability of hydraulic systems. (FAA Release 70-35)

April 22: *Intelsat-III F-7* was launched by NASA for ComSatCorp on behalf of International Telecommunications Satellite Consortium (INTELSAT). Satellite, launched from ETR by three-stage, long-tank Thor-Delta booster, did not reach predicted transfer orbit because booster underperformed. Engineers fired onboard hydrazine thrusters to push satellite into desired synchronous orbit with 35 772.9-km (22 233.0-mi) apogee and 35 737.7-km (22 211.1.-mi) perigee at 19° west longitude over Atlantic. Remaining fuel was expected to maintain satellite in orbit over Atlantic for five years. Full-time commercial service was scheduled to begin May 8.

Intelsat-III F-7 was fifth successful launch in Intelsat III series. Last successful mission, *Intelsat-III F-6*, had been launched Jan. 14. (ComSatCorp Release 70-26; NASA Proj Off)
- USAF's X-24A lifting-body vehicle, piloted by Maj. Jerauld R. Gentry (USAF), successfully completed third powered flight, reaching 18 300-m (60 000-ft) altitude and mach 0.9 after air launch from B-52 aircraft at 12 200-m (40 000-ft) altitude from FRC. Objectives of flight, 12th in X-24A series, were to expand envelope to mach 0.9 and to obtain lateral-directional stability derivatives at mach 0.85. (NASA Proj Off)
- Senate Committee on Aeronautical and Space Sciences ordered favorably reported H.R. 16516, FY 1971 NASA authorization bill, with amendments which cut $15 million from $692.3-million Administration budget request for research and program management and $2.05 million from $34.6-million Administration request for construction of facilities. Total $3.315 billion NASA authorization recommended by Senate committee was $17.05 million below $3.333 billion Nixon budget request. (*CR*, 4/23/70, D388)
- Fairchild Hiller Corp. Vice President Edward G. Uhl told stockholders at annual meeting that company had filed protest with NASA, U.S. Comptroller General, and President Nixon over April 8 NASA award of $50 000 000 contract to General Electric Co. for design and manufacture of experimental commercial spacecraft, according to Washington *Evening Star*. Fairchild Hiller General Counsel John F. Dealy said company was protesting on four points: GE submitted proposal six days after deadline and made technical changes contrary to NASA regulations; GE's cost savings were not genuine savings to Government; NASA's selection procedure on contracts was not adequate; and Fairchild Hiller was not given fair and proper evaluation of its technical innovations for spacecraft. (Harris, *W Star*, 4/23/70, D9)
- Bruce Biossat discussed *Apollo 13* in *Washington Daily News:* "Our greatest hangup is the dreams we cling to on earth. And no one is more guilty of fostering and perpetuating these than the men who today are assailing our space undertaking as costly distraction.... What we have long needed, as historian Dr. Daniel Boorstin suggests, is to forget about fixing and fulfilling some great national purpose. We will do better in small human clusters, working with dedication and fuller knowledge, in assaults on an ever-changing roster of important but quite specialized goals. Far from adding to crippling illusion, our men in space have shown us—in a very cruel laboratory—that such hard goals can be attained." (*W News*, 4/22/70, 19)

April 23: U.S.S.R. launched *Cosmos CCCXXXIV* into orbit with 482-km (299.5-mi) apogee, 271-km (168.4-mi) perigee, 92-min period, and 70.9° inclination. Satellite reentered Aug. 9. (GSFC *SSR*, 4/30/70; 8/31/70)
- House passed by 229 votes to 105 H.R. 16516, FY 1971 NASA authorization, after agreeing to recommit bill to House Committee on Science and Astronautics with instructions to reduce Apollo program funding by $14.5 million and space flight operations funding by $15.5 million. Action brought authorization total to $3.601 billion, $30 million below $3.631 billion reported by House Committee on Science

April 23

and Astronautics on March 12, but $268 million more than $3.333 billion requested by Administration. (CR, 4/23/70, H3377-23)

- Dr. Wernher von Braun, NASA Deputy Associate Administrator for Planning, addressed American Newspaper Publishers Assn. in New York: "Our mental concepts and life habits are Earth-oriented. The ability to go the Moon or Mars is regarded as something less than important, and this is only natural. But I am firmly convinced that the space program will prove to be one of the most important, creative, and beneficial to mankind ever undertaken by the United States.... I think we must clearly establish in this new decade our goals and objectives as an agency first, and then define the hardware needed only after we know exactly where we are headed." Hardware "on hand or that would be fine to develop" should not "tell us where we're going." (Text)

  U.S. detection in 1950s of Soviet radar three times size of football field was described before Association by Dr. John S. Foster, Jr., DOD Director of Defense Research and Engineering: "It was so big that when we first found out about it we were hesitant even to identify it as a radar. Much later we learned that it was far too powerful for most of the applications we could imagine. Assuming it was a radar, many argued that it made sense only as part of a detection, tracking and control network for satellites. The non-radar proponents even tried to make it into such things as a space communications system. The location of the first such radar argued that...this equipment probably would form part of an ABM system." (Text)

- Earth resources data on selected areas of Texas, gathered for MSC data bank, were presented to representatives of Texas Committee for the Study of Land Use and Environmental Control during MSC briefing on NASA earth resources aircraft program. Committee also received pertinent earth resources data collected by NASA while developing remote sensing techniques for spacecraft applications. (MSC Release 70-42a)

- Rep. J. Glenn Beall, Jr. (R-Md.), on House floor called for Government investigation of a selection by NASA of General Electric Co. for negotiation of ATS-F and ATS-G contract announced April 8. He claimed that technical innovations developed by Fairchild Hiller Corp. "eventually became a part of the competitor's proposal." (CR, 4/23/70, H3435-6)

- USAF resumed testing at WSMR after seven-month pause for remodeling with launch of 15-m (50-ft) Athena test missile from Utah Launch Complex at Green River. Remodeling permitted facility to accept Athena H—longer, heavier, improved Athena—scheduled for launch in late 1970. Athena was used in SAMSO's Advanced Ballistic Reentry Systems (ABRES) program to test reentry concepts and phenomena with subscale models of ballistic reentry vehicles. (AFSC Release 105.70)

- Scientists at Smithsonian Institution's Astrophysical Laboratory said studies of 10-kg (22-lb) meteorite that fell near Lost City, Okla., Jan. 9 had provided firmest evidence to date of intensity of cosmic rays beyond sun. Data obtained had raised possibility that manned spacecraft venturing to planets beyond Mars might have to be armored against swarms of high-energy "space bullets." Meteorite was estimated to have weighed 227 kg (500 lbs) originally and was

assumed to be fragment of explosive collision between two huge objects 6 million yrs ago. (AP, B *Sun,* 4/24/70, A3; Smithsonian PAO)
- MSC announced award of $3 000 000 NASA contract to Garrett Corp. AiResearch Manufacturing Co. Div. for portable astronaut life support assembly (ALSA) to support extravehicular and intravehicular activity in Skylab program. (MSC Release 70-43)
- Former Presidential science advisers testified on NAS funding before Senate Committee on Labor and Public Welfare's NSF Subcommittee. Dr. James R. Killian, Jr., MIT Board Chairman, said he could recall "no time when financial outlook was so bleak" in 30 yrs of college administration work. Dr. Jerome B. Wiesner, MIT Provost, said cutbacks in scientific research might cost U.S. world leadership in science and technology and economic well-being in future. Dr. George B. Kistiakowsky, Harvard Univ. chemist, said budget stringency and "selective and uncoordinated pruning" by Government agencies had damaged U.S. scientific effort. Dr. Donald F. Hornig, President-elect of Brown Univ., concurred with colleagues. (Schmeck, *NYT,* 4/24/70, 9)

April 24: Communist China launched her first earth satellite (designated *Chicom I* by NORAD) from Shuang-ch'eng-tsu, east of Lop Nor nuclear test site. The 173-kg (380.6-lb) satellite entered orbit with 2387-km (1483.2-mi) apogee, 439-km (272.8-mi) perigee, 114-min period, and 68.4° inclination, broadcasting telemetry data and revolutionary song "Dong Fang Hong" ("The East is Red") on frequency of 20.009 megacycles. Launch made Communist China fifth nation to orbit satellite with own booster. Radio Peking, announcing launch on evening newscast, credited feat to Communist Party Chairman Mao Tse-tung's leadership. (GSFC *SSR,* 4/30/70; *SBD,* 4/28/70, 271; UPI, W *Star,* 4/25/70, A1; *Newsweek,* 5/4/70)
- U.S.S.R. launched *Cosmos CCCXXXV* from Kapustin Yar into orbit with 398-km (247.3-mi) apogee, 249-km (154.7-mi) perigee, 91-min period, and 48.4° inclination. Satellite reentered June 22. (GSFC *SSR,* 4/30/70; 6/30/70; *SBD,* 4/28/70, 272)
- Testimony on *Apollo 13* mission was given before Senate Committee on Aeronautical and Space Sciences by Dr. Thomas O. Paine, NASA Administrator; *Apollo 13* Astronauts James A. Lovell, Jr., and John L. Swigert, Jr.; Dr. Rocco A. Petrone, Apollo Program Director; and Apollo Flight Director Glynn S. Lunney. Astronaut Fred W. Haise, Jr., was unable to attend.

Dr. Paine said: "The Apollo 13 mission was a failure. We did not succeed in America's third lunar landing attempt and we were therefore unable to explore the moon's Fra Mauro formation. This hilly region remains a high priority objective in our lunar program.... We cannot yet say which of the remaining Apollo missions will be sent there. Although the Apollo 13 mission failed, we regard the recovery actions that followed...as a gratifying success which will contribute greatly to the nation's space flight competence." There was no reason "why this setback should be—or should be made to be—the occasion for a major change in the course of the nation's space program. We have clearly demonstrated the basic soundness of the Apollo system and of our operational procedures.... From a technical standpoint, there is no question but that we are prepared to move forward in the 1970's with the space program we have outlined to you. . . ." Dr. Paine said President Nixon "fully shares this view." (Testimony)

Dr. Petrone testified: "This long arduous voyage continuously presented the challenge to balance the spacecraft systems required to perform necessary functions against the availability of consumables, of water, electrical power, oxygen, and the lithium hydroxide to remove the carbon dioxide. The options available to flight and ground crews in every case permitted acceptable courses of action to be selected. During this emergency situation, the remaining systems performed in an outstanding manner meeting the unusual demands placed upon them. At the same time, the flight and ground crews demonstrated exceptional confidence in meeting these sets of unusual circumstances."

Mission ended with "set of consumables which were within about two percent of what we had predicted back at 63 hours or about seven hours after the problem occurred," Lunney told Committee. "If the consumable posture. . .had worsened or if we had absorbed another failure. . .we would have had to resort to scheduling communication times with the crew." Although probably within 12 hours to 24 of running out of most critical consumable, water, "had we proceeded into. . .duty cycling the communications gear onboard, we probably could have stretched that period."

Astronaut Lovell described crew's feelings during crisis: "We soon realized that our only hope of survival was to go into Aquarius, power up its systems and use the systems and the consumables on board. . .for a return home. And it was here that we realized that it was now a case of survival." Spacecraft was on trajectory that "would have entered us in a sort of permanent orbit, some 230. . .thousand miles [370 000 km] in apogee, and maybe a thousand [1600 km] or so in perigee. . . . Again, the ground came through, and gave us. . .the free trajectory burn, which was required to get us back to earth. My main concern at this point was to get this spacecraft back within the earth's atmosphere." (Transcript)

• LeRC Special Projects Div. had begun systematic evaluation of lift fan-in-wing configurations for VTOL aircraft in $750 000 V/STOL wind-tunnel facility, *Lewis News* said. Concept was one of several being studied for use in VTOL aircraft. Current studies were to determine variation in fan thrust and efficiency as horizontal airflow over fan increased. Subsequent tests would use more efficient version of existing rotor to study effect of devices to prevent boundary-layer separation during change to horizontal flight and louvers to direct exhaust rearward. October tests of three small fans in longitudinal array would simulate installation in actual aircraft. Noise-measuring capability was being added to facility. (*Lewis News*, 4/24/70, 1)

• Sen. Charles McC. Mathias, Jr. (R-Md.), told press he had forwarded file on NASA ATS-F and ATS-G contract with General Electric to Senate Committee on Government Operations for referral to its permanent Investigating Subcommittee. (Rovner, B *Sun*, 4/25/70, A14)

• Tests to determine full capabilities of satellite communications were being conducted by AFSC Rome Air Development Center engineers with orbiting *Les VI* (launched Sept. 26, 1968) and *Tacsat I* (Feb. 9, 1969), AFSC announced. Using two parabolic antennas and new equipment at Verona Test Site near Griffiss AFB, N.Y., engineers simulated traffic conditions of a comsat system being used simultaneously by hundreds of separate stations. Simulation permitted measurement of number of possible communications circuits through satellite before actual construction of stations. Tests were being conducted with terminals on ground, aircraft, surface vessels, submarines, and land vehicles, with participation by U.K., Canada, and Netherlands. (AFSC Release 72.70)

- MSC announced selection of Wackenhut Services, Inc., for negotiation of $1.2-million, one-year, cost-plus-award-fee contract for security, safety, fire protection, and emergency ambulance services at MSC. Contract would cover July 1, 1970, through June 30, 1971, with two one-year options. (MSC Info Sheet)

- Gen. James Ferguson, AFSC Commander, discussed role of military in current controversies over cost and efficiency of F-111 and C-5A aircraft during Seventh Space Congress at Cocoa Beach, Fla.: "If we in military development have been remiss, I would say it is because we not only failed to fully communicate the difficulties involved in pushing the state of the art, but we also neglected to stress the uncertainties between known design parameters on paper and the hardware that ultimately results. To the intense competition for finite resources, and to a flood of paper studies, we reacted with overoptimism and neglected to properly qualify the normal uncertainties of development or the technical risks involved." Advancing state of the art was "synonymous with working beyond the boundaries of the certain and the known. As long as we operate at these outer limits of science and technology, performance, schedule and cost are inevitably going to be, at best, elastic parameters. The alternative is to stand pat with the F-4, the B-52, and the C-141. Such an alternative would be hardly supportable in even a static world." (Text)

- AIAA Board of Directors meeting in Denver Colo., created two new committees: Technical Committee on Application of Aerospace Technology to Society would report through Technical Activities Committee (TAC); Committee on International Cooperation in Space Flight would "foster international cooperation in space activities among the engineers and scientists of the world." (A&A, 6/70, 76)

April 25: U.S.S.R. launched eight Cosmos satellites with single booster for continued space research. Orbital parameters: *Cosmos CCCXXXVI*, 1489-km (925.2-mi) apogee, 1464-km (909.7-mi) perigee, 115.4-min period, and 73.9° inclination; *Cosmos CCCXXXVII*, 1554-km (965.6-mi) apogee, 1469-km (912.8-mi) perigee, 116.2-min period, and 74° inclination; *Cosmos CCCXXXVIII*, 1517-km (942.6-mi) apogee, 1472-km (914.7-mi) perigee, 115.8-min period, and 74° inclination; *Cosmos CCCXXXIX*, 1468-km (912.2-mi) apogee, 1450-km (901.0-mi) perigee, 115-min period, and 74° inclination; *Cosmos CCCXL*, 1468-km (912.2-mi) apogee, 1412-km (877.4-mi) perigee, 114.6-min period, and 74° inclination; *Cosmos CCCXLI*, 1470-km (913.4-mi) apogee, 1345-km (835.7-mi) perigee, 113.9-min period, and 74° inclination; *Cosmos CCCXLII*, 1470-km (913.4-mi) apogee, 1312-km (815.2-mi) perigee, 113.5-min period, and 74° inclination; and *Cosmos CCCXLIII*, 1468-km (912.2-min) apogee, 1379-km (856.9-mi) perigee, 114.2-min period, and 74° inclination. All systems were functioning satisfactorily. (GSFC *SSR*, 4/30/70; AP, W *Star*, 4/27/70, Al)

- Dr. Thomas O. Paine, NASA Administrator, released statement: "Red China has apparently become the fifth nation to launch an Earth satellite with its own booster following the Soviet Union, the United States, France and Japan. Working in close cooperation with the U.S., Canada, Britain, France, Italy, Australia, and the European Space Research Organization (ESRO) have flown Earth satellites. This new development has been anticipated for some time.... It is a dramatic demonstration of the determination of the Chinese leaders, despite many problems, to put science and technology very high on their list of national priorities." (Text)
- DOD and USAF scientists showed "marked lack of surprise at launch of first Communist Chinese satellite," AP said. It quoted unidentified DOD scientist as saying launch did not indicate, by itself, that Communist Chinese had achieved capability to fire ICBMs at U.S. Secretary of Defense Melvin R. Laird had told Congress Feb. 20 that he believed Chinese would attempt launch, AP said. (B *Sun*, 4/26/70, A1)

  U.S. was engaged in close electronic monitoring of Communist Chinese satellite and booster launch to determine its characteristics and capabilities, UPI reported. Intelligence analysts estimated Communist China might be able to develop and deploy 10 to 25 intercontinental ballistic missiles by mid-1975, which could threaten U.S. (Daniloff, UPI, *P Inq*, 4/26/70)
- Announcement of launch of first Communist Chinese satellite was greeted in Peking with crashing of drums and cymbals. Thousands of Chinese converged on Tien An Men Square in demonstration which *Washington Post* said had been "well organized" since launch had been expected for "some time." Groups of teenagers, children, and even grandmothers marched behind red banners and large portraits of Chairman Mao Tse-tung. People's Liberation Army group carried six-foot-high model of Chinese character for double happiness. Downtown office buldings were outlined in white lights when news was relayed by TV at 9:30 pm Peking time; colored bulbs lit trees along Boulevard of Eternal Peace. Firecrackers exploded everywhere, and crowds, said *Post*, "seemed very relaxed and grinned happily at foreigners." (Webster, *W Post*, 4/28/70, A12)
- Vice Premier Chiang Ching-kuo of Nationalist China watched technicians track Communist China's first satellite at NORAD Hq. in Colorado Springs, Colo. (*W Post*, 4/26/70, A14)
- NASA was "bracing itself for the inevitable backlash of opinion," now that "public jubilation" over *Apollo 13*'s safe return had subsided, *Economist* said. "But so far the reaction has been surprisingly mild. The newspapers have generally argued that people had become too over-confident about the dangers of space travel and that the exploration of the moon must continue. President Nixon has made it clear where he stands; he announced to Dr. Thomas Paine...that he had 'a great organization' and that he himself was a firm believer in manned space flight." (*Economist*, 4/25/70, 43-4)

April 26: International reaction to first Chinese satellite launch:

In U.S.S.R. Pravda devoted 10 words on inside page to report of launch. *Washington Post* said it was "continuation of a war of words"; Chinese had barely mentioned Soviet space activities since nations became enemies. Romania congratulated Chinese for "fresh successes in the field of scientific investigation of outer space." Albania also hailed achievement.

U.K. Minister of Technology Anthony Wedgwood Benn sent message to Chinese government congratulating engineers and space workers on launch. West Germany's Bochum Observatory called event greater technological feat than 1957 launch of Soviet *Sputnik I*.

North Vietnam news agency said launching "brings out the supremacy of socialism and contributes to the development of science and technology in the interests of progressive mankind." Pakistan President Yahya Khan in message to Chinese Premier Chou En-lai called launch "great step forward," demonstrating progress of China's technology. He said feat "by our friendly neighbor has been received by us with a feeling of admiration."

Japanese officials said if China had developed guidance system for satellite launching, it soon would be able to fire intercontinental missiles with high precision. (*W Post*, 4/27/70, A9; *WSJ*, 4/27/70)

- Vice President Spiro T. Agnew and Astronaut Thomas P. Stafford attended dedication ceremony of Stafford Field, small airfield in Stafford's hometown, Weatherford, Okla. Airfield was Weatherford's memorial to Stafford who had piloted two Gemini missions. At dedication Vice President said: "While Stafford Field may not be equivalent in size to the giant airports of this country and the world, it...symbolizes the best of American traditions...harnessing that pioneering spirit of old." (*W Post*, 4/27/70, A4; Text)

- ComSatCorp had been "bouncing from one controversy to another" since its 1962 inception, Robert J. Samuelson said in *Washington Post*. "Last week it was the International Telephone and Telegraph World Communications, Inc., that asked the FCC to scrub the satellite launch. ITT argued that the Commission was favoring satellite communications over undersea cable systems.... Others share ITT's irritation. The remaining communications carriers with cable interests...are similarly unhappy with their new satellite competitor." ComSatCorp stock holders were displeased that company had yet to pay dividend. "Even foreign countries, which have benefited by Comsat's experience in establishing an international satellite network now resent Comsat's dominant position in that system." But international satellite traffic had been growing at annual rate of 20%. With initial investment nearly completed, ComSatCorp's operating revenues were rising rapidly. (*W Post*, 4/26/70, L3)

- F-14, "first completely new American fighter plane...since the XF-4H1—prototype of the F4 Phantom," would "skim down the runway" at Grumman Aircraft Engineering Corp. before end of January 1971, Washington *Sunday Star* said. USN would attempt to buy 710 F-14s over next decade at $8 billion including R&D costs. Aircraft would have titanium wings. It was first time any manufacturer had used as much as 2700 kg (6000 lbs) of titanium in aircraft production. (Kelly, *W Star*, 4/26/70, A7)

- Anthropologist Dr. Margaret Mead in *New York Times Magazine* discussed prefigurative society in which young became mentors of elders. "A scientist friend of mine was having a conversation with his son and wasn't getting anywhere until he realized that he was standing on the earth looking at the moon, and the boy was standing on the moon looking at the earth." (*NYT Magazine,* 4/26/70, 102)

April 26-30: During American Physical Society's spring meeting in Washington, D.C., Cornell Univ. physicist Dr. James R. Houck told news conference series of rocket and balloon observations by MIT in autumn of 1969 had confirmed that unexplained infrared radiation impinged on earth's upper atmosphere, apparently from all parts of sky. In meeting he expressed hope that 5400-kg (six-ton) earth satellite, High Energy Astronomy Observatory, could be funded for 1974 launch to observe extremely high phenomena that could not be detected from earth because of intervening atmosphere. (Sullivan, *NYT,* 4/30/70, 21)

Dr. Reuven Ramaty of GSFC reported measurements of low-energy cosmic-ray positrons by *Ogo II* satellite (launched June 6, 1966) had indicated that relatively young and close cosmic-ray source—a supernova explosion—might have occured 100 000 yrs ago and 300 light years from earth. (Crab Nebula, closest known remnant of any supernova previously discovered, was 5000 light years from earth.) (Press Conf Summary; Text; AP, *CSM,* 5/2/70)

During meeting, 32 APS members marched on DOD to oppose Vietnam War, ABM and MIRV missile systems, and DOD ties with American universities. (*W Post,* 4/30/70, A35)

April 27: Senate passed S.J.R. 193 appointing former NASA Administrator James E. Webb as Citizen Regent of Board of Regents of Smithsonian Institution. (*CR,* 4/27/70, S6234-5)

- Aerospace Medical Assn. held 41st annual scientific meeting in Houston, Tex. Dr. Charles A. Berry, MSC Director of Medical Research and Operations, presided.

  NASC Executive Secretary William A. Anders, former astronaut, delivered Louis H. Bauer Lecture: Space program had contributed "in innumerable ways to new advances in biotechnology and human factors research." Developments included miniature radiation probes size of clinical thermometer, EKG transmitting systems, computer enhancement of x-ray films using techniques developed to bring out details from Mars photographs, and "supersensitive devices to detect the tiny tremors of early Parkinson's disease, which were developed to register micrometeoroid hits on spacecraft." (MSC Release 70-42; Text)

- Assistant Under Secretary of State for Public Affairs and former astronaut Michael Collins addressed meeting of Foreign Service Wives Assn. in Washington, D.C.: "By the year 2000 or 2100, people will look back on the 1970s and find it incredible that cancellation of the space program was proposed. . . . It is inevitable there will be colonization of other planets, despite the risks." Collins told group he hadn't been sure of success of first manned lunar landing until he and colleagues were safely enroute home; "that was the first time I felt any certainty about it." (*W Star,* 4/28/70, B7)

- Dr. John S. Foster, Jr., DOD Director of Defense Research and Engineering, testified before House Committee on Appropriations' Subcommittee on DOD Research, Development, Test, and Evaluation concerning Communist Chinese satellite: April 24 launch of *Chicom*

I by Communist China indicated "Chinese commitment to a large space program" rather than "attempt to make a show by launching a single object into space." Chinese had "benefited from two findings by the Soviets and ourselves...enormous political impact that such a launch has around the world...[and] military applications. I believe the Chinese statement made it rather clear that they associate this space effort with their future needs from a military point of view." (Transcript)

- F-5D "Skylancer" aircraft used by *Apollo 11* Astronaut Neil A. Armstrong in training was being retired by FRC and eventually would be turned over by NASA to Ohio Historical Society for display in future Armstrong Museum at Wapakoneta, FRC announced. Until completion of museum in Armstrong's hometown, aircraft, which he flew in 1960s to practice spacecraft emergency maneuvers, would be displayed at USAF Museum at Wright-Patterson AFB, Ohio. (FRC Release 7/70)
- MSFC announced award of $4 435 275 contract modification to North American Rockwell Corp. for changes in seven Saturn V 2nd stages, S-II-9 through S-II-15. Work, which included testing and replacing components that caused corrosion, would be completed by March 31, 1973. (MSFC Release 70-79)
- Appointment of John P. Olsen, Assistant to Senior Vice President of U.S. Smelting, Refining and Mining Co., as Deputy Under Secretary of Transportation was announced by Secretary of Transportation John A. Volpe. He would succeed Charles D. Baker, who had been appointed Assistant Secretary of Transportation for Policy and International Affairs, effective April 28. (DOT Release 8970)
- At NAS annual meeting in Washington, D.C., Dr. Philip H. Abelson, Director of Carnegie Institute geophysical laboratory and editor of *Science*, said cutbacks in Government research funds during last five years had dealt devasting blow to U.S. colleges and reduced Nation's total scientific research effort by at least 30%. If cutbacks continued, U.S. would become "second-class nation in science."

  Harvard Univ. scientist Dr. Harvey Brooks said surge of science funding in Japan, Western Europe, and U.S.S.R. had cost U.S. its research lead in astronomy, radioastronomy, biology, and physical and organic chemistry. It was doubtful U.S. would ever regain supremacy it once enjoyed in virtually every scientific field. (Kirkman, *W News*, 4/28/70, 20)
- Stewart Alsop said in *Newsweek*: "The story of the failure of the Russian moon-exploration project...is interesting and rather puzzling. The first test of the Soviet space vehicle comparable to Apollo was operationally successful in March 1967. The most recent SL-12 mission—an unmanned, nonrecoverable lunar probe—aborted in February, the eleventh to do so.

  "The Soviet Government and the Soviet press had not a word to say about this string of failures.... But in the American intelligence community there is no doubt at all about the failures, and no doubt either that the Soviet space-exploration program has been a thoroughly miserable performance." (*Newsweek*, 4/27/70)
- Communist China's first satellite was believed to have been planned by U.S.-trained scientist Tsien Hsueh-shen, *Wall Street Journal* said. Former MIT and Cal Tech scientist had been high-ranking Pen-

tagon adviser during World War II and had left U.S. in 1955 after having been named in deportation order charging him with being Communist Party member. (*WSJ*, 4/27/70)

- Philadelphia *Evening Bulletin* editorial on Communist Chinese satellite launch: "To identify Peking's satellite as proof of a perfected weapon is to orient debate on the ABM to expansion of that system—and the proliferation of weapons. This at a time when the world's two prime producers of nuclear weapons are trying to agree on curtailing them. For these talks [SALT] to fail is to deliver the world to witless expansion of arsenals, with the always accompanying threat to use them. The ultimate promise is catastrophe far beyond anything the Chinese could execute." (*P Bull*, 4/27/70)

April 27-30: American Nuclear Society held annual meeting in Huntsville, Ala. Dr. Thomas O. Paine, NASA Administrator, told members U.S. technological edge over rest of world depended on joint nuclear-space developments. Presence of highly skilled personnel at MSFC made it logical that program be centered there. (*Huntsville News*, 4/29/70)

April 28: U.S.S.R. launched *Meteor IV* from Plesetsk to photograph snow and cloud cover and study "atmospheric thermal energy radiated by the earth." Orbital parameters: apogee, 743 km (461.7 mi); perigee, 636 km (395.2 mi); period, 98.2 min; and inclination, 81.2°. Equipment was functioning satisfactorily. (GSFC *SSR*, 4/20/70; AP, *W Post*, 4/30/70, A34; *SBD*, 5/6/70, 25)

- *Apollo 13* manned lunar landing mission (April 11-17), aborted after 56 hrs of flight, was officially adjudged unsuccessful by NASA. (NASA Proj Off)

- MSC announced award of $34 065 390 cost-plus-fixed-fee contract to Boeing Co. Southeast Div. to define letter contract for systems engineering and conduct flight readiness assessment for Apollo program. Contract would cover Jan. 1, 1970, through Dec. 31, 1970, with options for 1971 and 1972. (MSC Release 70-45)

- Appointment of *Apollo 8* commander Frank Borman as Vice President of Eastern Airlines was announced by EA Vice President and Executive Officer Floyd D. Hall. Borman would assume new duties after completing course for executives at Harvard Univ. (EA Release)

- Aerobee 150 sounding rocket launched by NASA from Churchill Research Range carried ARC experiment to study dust particles in atmosphere. (SR list)

- U.S.S.R. was using long- and medium-range bombers and, possibly, communications satellites to coordinate global naval exercises, DOD sources were reported as saying. No violations of territorial air space of any nation had been reported. (UPI, *C Trib*, 4/26/70, A4; AP, *NYT*, 4/29/70, 8)

- James McCormack announced he would not continue as ComSatCorp Chairman and Chief Executive Officer beyond annual shareholders meeting scheduled for May 12. He would continue as a director. (ComSatCorp Release 70-5)

- Baltimore *Sun* editorial commented on Communist Chinese satellite launch: "Of course Peking still runs far behind America and Russia in nuclear and space development.... But not nearly so far behind as only a few years ago. And apart from conjectures about progress in missilery, 'Following Star' [satellite] is a reminder of the potential military dimensions of space which may acquire a new importance now that China has entered the extraterrestrial realm." (B *Sun*, 4/28/70, A10)

April 29: Members of *Apollo 13* Review Board entered Apollo simulator at MSC and, using recorded data, reconstructed tense moments after explosion disabled spacecraft April 13. Board member Astronaut Neil A. Armstrong said purpose of reconstruction was to give investigators better appreciation of accident from crewmen's point of view. (*W Post*, 4/30/70, A27)

- Dr. Robert C. Seamans, Jr., Secretary of the Air Force, presented Gen. Thomas D. White Space Trophy to *Apollo 11* crew—astronauts Neil A. Armstrong, Edwin E. Aldrin, Jr., and Michael Collins, since named Assistant Secretary of State for Public Affairs—in Washington, D.C., ceremony. Award, established by National Geographic Society in 1961 for USAF members and units, honored crew for making most outstanding contribution to U.S. aerospace progress during 1969 by accomplishing first manned lunar landing. (DOD Release 359-70)

- ESSA announced award of $1.6-million contract to Gulton Industries Data Systems Div. to construct infrared temperature profile radiometer (ITPR), advanced satellite instrument for obtaining temperature sounding through atmosphere to earth's surface. ESSA was de-

April 29: Apollo 11 *Astronauts Michael Collins (left) and Edwin E. Aldrin, Jr. (right), were presented the Gen. Thomas D. White Space Trophy by Dr. Robert C. Seamans, Jr. (center left), Secretary of the Air Force. Dr. Thomas O. Paine, NASA Administrator, represented Astronaut Neil A. Armstrong, who was unable to be present.*

veloping ITPR for use by NASA on Nimbus-E, scheduled for launch in 1972. (ESSA Release 70-28)
- Vice President of Soviet Academy of Sciences Mikhail D. Millionshchikov said at second national Convocation on Challenge of Building Peace in New York that exchange of experience in space between U.S. and U.S.S.R. would be "very fruitful." It was "very dangerous to make a race . . . because the lives of human beings are at stake." Astronaut Frank Borman, chairman of panel discussion, suggested "Soviet-American committee to chart the exploration of the solar system, a group to develop and formulate policy and interchange of ideas, both in the manned and unmanned fields." (Wilford, *NYT*, 4/30/70)
- Development of world's most powerful continuous-beam laser, with energy output capable of driving sports car, was described by Dr. Edward T. Gerry of Avco Research Lab. during annual meeting of American Physical Society in New York. Laser could produce 30 kw of power in narrow beam and 60 kw in divergent beam. Values were triple those of most powerful continuous-beam lasers in operation. Avco laser resembled rocket engine. Nitrogen-carbon dioxide gas was heated to 1900 K (3000°F) in one chamber and shot through nozzle into second chamber at supersonic speed. Carbon dioxide molecules emitted photons which mirrors focused into beam of light emitted from chamber.

    At news conference following meeting, scientists denied lasers were being built for use as "death rays." Wayne State Univ. scientist Dr. Alexander J. Glass said scientists in U.S., U.S.S.R., and France were developing increasingly larger lasers that produced 100 kw of light energy in few trillionths of second. (Lyons, *NYT*, 4/30/70, 20)
- *Christian Science Monitor* editorial: "China's successful orbiting of her first satellite is further proof of her continuing effort in space and rocketry—even if her entering of the space age comes nearly 13 years after Sputnik I. It is proof, too, that even if China still lacks industrial brawn and remains basically an underdeveloped country, she is not lacking in scientific and technical brilliance. What she has achieved is all the more remarkable because—as the Chinese themselves report it—the Soviet Union reneged in 1959 on its 1957 agreement to help China with 'new technology for nuclear defense.' Thus what China has done has been done on her own—and by her scientists and technologists in spite of the self-inflicted torment of Chairman Mao's cultural revolution." (*CSM*, 4/29/70)

April 30: Arnold W. Frutkin, NASA Assistant Administrator for International Affairs, described Brazilian proposal for educational broadcast experiment using NASA ATS-F or ATS-G satellite to House Committee on Foreign Affairs' Subcommittee on National Security Policy and Scientific Development: "The project is intended . . . as the development prototype of a system that will broadcast television and radio instructional material to the entire country through a government-owned geostationary satellite. The experiment itself contemplates TV and radio broadcasting into some 500 schools in the Rio Grande de Norte region. It would test the use of satellites for central broadcasting using a balance of TV and radio, to upgrade teacher quality, permit the use of local monitors with less training,

- and so facilitate rapid expansion of the educational system." Proposal was undergoing revision before final technical evaluation by NASA. (Testimony)
- In experiment to test validity of Einstein theory of relativity, JPL astronomer Dr. John D. Anderson and Cal Tech astronomer Dr. Duane O. Muhleman measured radio signal from *Mariner VI* just before it swung behind sun. If Einstein theory was correct, precise distance-measuring system built by JPL engineer and coinvestigator Warren L. Martin and 64-m (210-ft) antenna at Goldstone tracking station in Mohave Desert would register slowup of 200 millionths of second in roundtrip signal from satellite. Einstein said velocity of light was slower in gravitational field near sun than in interplanetary space where gravitational fields were weaker. (NASA Release 70-62)
- Saturn V 2nd stage (S-II-13) was successfully captive-fired for 367 secs at MTF. Stage would be used to launch Skylab space station in late 1972. (MSFC Release 70-85)
- NASA announced selection of Aerojet Liquid Rocket Co., NR Rocketdyne Div., and United Aircraft Corp. Pratt & Whitney Div. for final negotiations of $6-million, fixed-price contracts for 11-mo parallel studies for design definition of space shuttle main propulsion system. Contracts would be managed by MSFC. (NASA Release 70-64)
- *Apollo 13* Astronaut James A. Lovell, Jr., at National Press Club Luncheon in Washington, D.C., denied interest in running for office of Senator from Wisconsin: "I'm a public servant and I intend to be that way." He hoped to devote "next two years at least" to helping develop strong, well-balanced space program. (Schmeck, *NYT*, 4/30/70)
- NAE announced election of retired General Electric Co. Vice President Clarence H. Linder as first full-time NAE President and Dr. Chauncey Starr of UCLA as Vice President. Dr. Thomas C. Kavanagh of Praeger-Kavanagh-Waterbury and of Madigan-Hyland was reelected Treasurer and J. H. Mulligan, Jr., Secretary. Among those elected to Council were Dr. T. Keith Glennan, first NASA Administrator, and Dr. Eugene G. Fubini, NASA consultant. (NAE Release; NAE *Bridge*, 6/70, 1)
- President Nixon's address to Nation on situation in Southeast Asia was televised live by satellite to Australia, Japan, Taiwan, Thailand, the Philippines, and Hong Kong. (*PD*, 5/4/70, 596-601; UPI, W *Star*, 5/1/70, A10)
- USAF announced F-111 wing structure undergoing ground fatigue testing at General Dynamics Corp.'s San Diego, Calif., facility cracked at point equivalent to 7½ yrs of service life. (AP, *W Post*, 5/1/70)
- Australian cabinet decided to lease U.S. F-4E Phantom aircraft as stand-in for 24 F-111s it had ordered. F-111s continued to be grounded for further tests. (UPI, NY *News*, 5/1/70, 8)
- Rep. Jerry L. Pettis (R-Calif.) announced resignation from House Committee on Science and Astronautics to fill seat on House Committee on Ways and Means. (*CR*, 4/30/70, H3719-20)
- *New York Times* editorial deplored decline of Federal support for research: "The search for peace and knowledge, the rescue of man's environment and the rebirth of the decaying cities are certain to become illusory goals unless they are pursued with as much hard-hitting scholarship and scientific purpose as the perfection of mis-

siles or exploration in space. The Pentagon has never failed to impress the political leaders with its dependence on the creativity of the ablest minds on campus; it would be a devastating commentary on the nation's values if history were to record that the military establishment was the only dependable patron of American research and scholarship, and that when military support was cut off, scientific research withered and died." (NYT, 4/30/70, 34)

- *During April: Science and Technology: Tools for Progress*, report to President Nixon by President's Task Force on Science Policy, was released. Report, submitted Dec. 10, 1969, recommended that President enunciate as national policy, need for vigorous, high-quality science and technology and continuing leadership in science and technology "relevant to our other national goals and purposes"; that President direct Federal departments and agencies to strengthen capability to use science and technology in attack on social, urban, and environmental problems; and that U.S. achieve effective and consistent commitment to long-range research, uninterrupted support of graduate education, improved use of Federal laboratories, and improved process for establishing priorities in Federal support of science.

  Report further recommended President enunciate policy of increasing long-term participation by private institutions—particularly business—in social, urban, and environmental programs and direct appropriate departments and agencies to identify deterrents to private investment of capital and technology in these programs; suggest incentives for action and remedies for each deterrent; enunciate policy of increased emphasis on R&D for national security, if necessary; and continue encouragement of new science-based foreign policy initiatives and opportunities for international cooperation. Administration should make clear policy of technical assistance to underdeveloped nations to help them establish their own scientific research, education, and technical training institutions. Task Force recommended President direct his Science Adviser to develop, for his approval, broadly-based program for continuing development of national science policy. (Text)

- NASA published *The Terrestrial Environment: Solid-Earth and Ocean Physics* (NASA CR-1579), report of study sponsored by NASA and MIT at Williams College. Report concluded that greater understanding of earth from measurements by satellites would have significant applications to "entire effect of the atmospheric and oceanic development, including protection against (storm) hazard and pollution."

  Recommended program goals (which were supported by technology recommendations) were: identification of forces and mechanisms that accounted for motions and earthquakes and variations of gravitational field and similar solid-earth phenomena; discovery of general circulation of oceans at all depths; increased knowledge of how earthquakes occur; improved understanding of global heat balance by study of ocean currents and heat transport of air-sea interaction; identification of internal driving forces and complex mechanisms of interaction of earth's core, mantle, and crust; definition of locations and mechanisms of energy dissipation in oceans; and explanation of mechanisms associated with variations in rotation rate

and wobble of earth's rotation axis and their possible association with major earthquake events. (Text; NASA Release 70-63)
- NAE President Eric A. Walker announced election of 51 engineers to NAE membership. They included Dr. George M. Low, NASA Deputy Administrator, for "contributions to the proposal, formulation, and management of Project Apollo"; Gerhard Neumann, Vice President and Group Executive of General Electric Co. Aircraft Engine Group, for "contributions to the development of variable stator compressors and high by-pass turbofan engines for jet aircraft"; Kendall Perkins, Corporate Vice President, Engineering and Research, of McDonnell Douglas Corp., for "contributions to aerospace technology, and engineering management in the design of aircraft and spacecraft"; and Dr. Paul Rosenberg, President of Paul Rosenberg Associates, for "pioneering contributions to space photogrammetric systems, radar prediction techniques, information storage and retrieval, and the kinematics of human spine motion." (NAS-NRC-NAE News Rpt, 4/70, 1)
- In *Astronautics and Aeronautics*, Herbert Friedman discussed pulsars and their implications in new astronomy: "Even with the modest payloads of small rockets it is possible to instrument for pulsar detection at periods as short as 1 millisecond." X-ray astronomers had been eager to have NASA provide program of high-energy astronomy payloads in 4500- to 9000-kg (10 000- to 20 000-lb) class that would be launched by vehicles of Titan-Centaur capability. Large x-ray detector could "very likely detect thousands of weak sources and discover high-frequency pulsar characteristics, if present. Such a program," HEAO (High Energy Astronomy Observatory), "was to have begun in 1970 but has been a casualty of the present budgetary squeeze." (A&A, 4/70, 22-5)
- Society that could afford to spend as much as U.S. did on cokes, cosmetics, tobacco, and liquor could afford its space expenditure, William H. Bayley, NASA Assistant Director for Tracking and Data Acquisition, said in JPL *Lab-Oratory*. Between 1959 and 1969 U.S. had devoted $30 billion to space program, $40 billion each for cosmetics and soft drinks, almost $80 billion each for cosmetics and soft drinks, almost $80 billion for tobacco, and some $110 billion for alcoholic beverages—about four times cost of space program. Annual space expenditure had been about one half of 1% of GNP, and about 1/20th of defense expenditure. Most expensive space program had been Apollo at $24 billion, 93% of which was spent for goods and services provided by private industry. To date, "no money at all has been spent on the Moon or planets; it has all been spent right here on earth." (JPL *Lab-Oratory*, 4/70, 4-19)
- *Astronautical Multilingual Dictionary* of the International Academy of Astronautics, conceived by the late Dr. Theodore von Kármán, was published by Academia Publishing House of Czechoslovak Academy of Sciences in Prague. Executive editor was Ing. J. Vlachý and scientific editor Prof. R. Pešek. U.S. coeditors were Dr. Woodford A. Heflin of USAF's Air University Aerospace Studies Institute and William H. Allen of NASA. Other coeditors were in U.S.S.R., Germany, France, Italy, Argentina, and Czechoslovakia. (Text; *Air University Dispatch*, 4/10/70, 18)

During April

- *Teknika i Vooruzheiniye* (Moscow) quoted Soviet Cosmonaut Georgy T. Beregovoy on future U.S.S.R. space program: "The Soviet Union is successfully carrying out a space research program designed for many years in the future. Its aims are determined by the needs of science and the national economy and the requirements of scientific and technical progress." Prospects for development were "unfolding with the establishing of orbiting stations. . . . Cosmodrome stations, extra atmospheric bases for geophysicists and astronomical observatories, and space laboratories for chemists will appear." (JPRS 50505, 5/13/70)
- Article on birth pangs of first satellite, written by Dr. Clifford C. Furnas shortly before his death in April 1969, was published in *Research Trends*. Dr. Furnas had been member in 1955 of DOD's Ad Hoc Advisory Group on Special Capabilities, a panel appointed to investigate feasibility of U.S. satellite program. (*Research Trends*, Spring 1970, 15-8)
- Failure of Government to define priorities in science and technology and stay with them was one cause of unemployment among scientists and engineers in U.S., Richard S. Lewis said in *Bulletin of the Atomic Scientists*. Cutback in civilian space program was prime example: "The consequences of this retrenchment in the dissipation of human resources could hardly be more pronounced than if the Nixon Administration was phasing out space exploration—which it is not." (*Bull of Atomic Scientists*, 4/70, 28-9)

May 1970

May 1: Dr. George M. Low, NASA Deputy Administrator, held press briefing at MSC on *Apollo 13* Review Board's progress in investigation of accident that aborted April 11-17 mission: Combination "of all the data from telemetry, from all the testing of all the analytical work and perhaps information from photographs" would be required "to determine the most probable cause or causes" of April 13 accident. Board believed most probable sequence of events was: Short circuit in oxygen tank No. 2 caused combustion within tank. "This in turn caused the pressure and a temperature within the tank to increase. The tank then ruptured. This rupture... caused the pressure in the compartment in which the tank is located to increase which then caused... the big covering panel in the Service Module to blow off." Panel hit high-gain antenna and knocked it out, causing brief loss of data.

Preflight events of possible significance included change of motors of fans inside tank at Beech Aircraft Co. plant early in manufacture. "Later on the tank, itself, was...moved from one spacecraft and installed in spacecraft 109 and during the removal from spacecraft...106, it was jarred or dropped an inch or two [5-10 cm].... Finally, during the loading and unloading of the tank during the countdown demonstration tests at the Cape there was an anomaly which made it very difficult to get the oxygen out of the tank. This was several weeks before the flight and a new procedure not previously tried was used in the detanking"—gaseous oxygen "was pumped into the tank and released again, the heaters were turned off and on...."

Board was investigating connection between events and accident and would submit final report to Dr. Low and Dr. Thomas O. Paine, NASA Administrator, about June 1. (Transcript)

• Senate Committee on Aeronautical and Space Sciences favorably reported H.R. 16516, NASA FY 1971 authorization bill, with amendment reducing total authorization to $3.316 billion—$285 million less than $3.333 billion requested by Administration. Committee's recommended total—providing $2.606 billion for R&D, $32.5 million for construction of facilities, and $677.3 million for research and program management—was lowest recommended by Senate Space Committee since 1961 and $399.6 million less than total recommended by Committee in FY 1970.

Committee rejected House increase of $130.5 million over Administration request for Apollo program. It also rejected increases of $139.5 million for space flight operations, $5.6 million for ATS project, $2.3 million for selected aeronautics research efforts, $500 000 for technology utilization program, and $2.05 million for construction of Earth Resources Technology Laboratory "at this time." It restored House cuts of $1.5 million to advanced manned missions

program, $5.6 million to Explorer satellite project, and $4.2 million to tracking and data acquisition. Committee strongly urged Executive Branch to develop formal interagency agreement defining responsibilities for experimental and operational ERTS systems. (Senate Rpt 91-833)
- NASA established NASA Planning Board under direction of Dr. Wernher von Braun, Deputy Associate Administrator for Planning, to provide focal point for coordinating and integrating NASA planning and related activities. Board and its activities would be supported by central staff under Dr. von Braun and Dr. DeMarquis D. Wyatt, new Assistant Administrator for Planning. Six members, one executive secretary, and five observers had been selected from top NASA Hq. administrators. (NASA Release 70-65)
- NASA communications network (NASCOM) service was transferred from *Intelsat-II F-2* (launched Jan. 11, 1964) to *Intelsat-III F-7* (launched April 22, 1970). (ComSatCorp Release 70-26)
- Aerobee 150 sounding rocket was launched by NASA from Churchill Research Range carrying ARC experiment to collect dust particles. Rocket and instruments functioned satisfactorily. (SR list)
- H. Dale Grubb, former Special Assistant to President, was sworn in as NASA Assistant Administrator for Legislative Affairs. (NASA Off Leg Aff)
- *Apollo 13* Astronauts James A. Lovell, Jr., and John L. Swigert, Jr., were given ticker-tape parade welcome in Chicago, where they addressed 800 at luncheon and later spoke to 2500 high school students. Astronaut Fred W. Haise, Jr., third *Apollo 13* astronaut, was indisposed. (UPI, *H Post*, 5/2/70)
- FAA reported operations at 300 airports in U.S. with FAA-staffed control towers had increased two percent in 1969—smallest annual increase since 1961. In 1968 increase was 11%. Decline was attributed primarily to hourly flight quotas established June 1, 1969, by FAA at O'Hare International Airport in Chicago and four other major airports to reduce congestion and delays. O'Hare, with 676 473 operations in 1969, remained Nation's busiest airport despite two percent drop in total 1969 operations. Second and third busiest were Los Angeles International with 613 938 operations and Long Beach (Calif.) with 550 867, with Santa Ana (Calif.) following with 545 299. (FAA Release 70-39)
- Dr. Glenn T. Seaborg, AEC Chairman, addressed International Joint Conference of American Geographical Society and American Div. of World Academy of Art and Science in New York: "It is obvious that for some people science and technology are among the best scapegoats of the time. They are said to be the cause of most of our ills today. By conquering disease and extending life they have been responsible for an explosion of population. By increasing productivity and raising living standards they have been responsible for depleting resources and polluting nature. By expanding knowledge and emphasizing efficiency, they have been responsible for deflating myths and diminishing man. And by placing enormous power in the hands of man they have brought him to the brink of his own destruction."

Science had been "victim of its own single-minded success" in going "from the broad and general philosophy from which it origi-

May 1: *The city of Chicago welcomed the Apollo 13 astronauts and flight controllers with ticker tape, confetti, and cheers during a parade around the Loop. Honorary citizenship awards were given John L. Swigert, Jr., and Fred W. Haise, Jr. (absent), and the city's Medal of Merit was given to flight commander James A. Lovell, Jr.*

nated into a growing number of more precise disciplines—each becoming more productive the narrower its focus became." But "in recent years this process has been reversed and we are now seeing the growth of interdisciplinary sciences and a striving for a more all-encompassing grasp of the physical world and even broader relationships. . . . This type of growth is essential if science is to be the guiding force—as it must be—behind our evolving mankind." (Text)

- Samuel K. Hoffman, President of NR Rocketdyne Div., retired after leading company's work in development of rocket engine technology for more than 20 yrs, including 10 yrs as Rocketdyne President. Hoffman received American Rocket Society's Robert H. Goddard Memorial Award in 1959 for outstanding work in liquid-propellant rockets. (*NR News*, 5/1/70, 1; *Aeronautics and Astronautics, 1915-60*, 176)
- Dr. Philip H. Abelson, editor of *Science* magazine and codiscoverer of element neptunium, was named President of Carnegie Institution of Washington. He would succeed Dr. Caryl P. Haskins, who would retire at end of June 1971. (*NYT*, 5/6/70; Carnegie Inst Release)
- Holt, Rinehart and Winston published *Operation Overflight*, in which author Francis Gary Powers said information supplied by Lee Harvey Oswald, identified by Warren Commission as assassin of President John F. Kennedy, might have enabled Russians to shoot down his U-2 espionage aircraft in 1960. Powers believed Oswald had supplied U.S.S.R. with data on MPS-16 height-finding radar equipment. (AP, B *Sun*, 4/20/70, 6)

May 2: Dr. Thomas O. Paine, NASA Administrator, spoke during 10-day "Emphasis Space" program at Nebraska Wesleyan Univ. on "very basic question" of why U.S. should be spending "all of this money out in space when we have so many problems here on earth." He said answer was "indeed that human values have the highest priority. I would submit that the American space program is indeed a program that has tremendous human values. . . . we in the space agency who are effecting technical change feel that history shows that the interaction between technical change and social change is extremely close. We believe that the activities that we are carrying out will have a tremendous impact on the way our children and our children's children will live. We are in the business of creating new wealth and new power that can be used by all men."

Apollo 12 Astronaut Charles Conrad, Jr., said that lunar flights were "enhancing man's knowledge of the environment in which he lives. . . .space has proven to date that it is something that man can use that is of benefit to man. . . . The question is do you want it used." Space program was at point "where it is up to the people to decide through the government what uses we will put space to." (Neb Wes Univ's *New Dimensions*, 9/70; AP, *San Diego Union*, 5/3/70)

- Impact of space requirements on U.S. computer industry was discussed by Dr. George M. Low, NASA Deputy Administrator, at ARCS (Achievement Rewards for College Scientists) Foundation Annual Scholarship Awards Banquet in Houston, Tex.: "In Project Mercury, ground-based computers were only required to determine quickly and accurately booster cut-off conditions. In Apollo, however, computers are used throughout the mission in real time, to calculate the trajectory to the moon and back, to compare three separate solutions for the lunar descent, to record and analyze thousands of bits of telemetered spacecraft information, to compare these to predicted values to detect trouble, and . . . to monitor the well-being of the crew. For Mercury, the computer program contained 40 000 computer words; for Apollo, a 1 500 000-word program was needed, while at the same time, the speed of the computers has increased sevenfold."

U.S. computer industry was doing $8 billion worth of business annually, paid highest average wages of any U.S. industry, and was "one of the most rapidly growing." In 1960, U.S. exported $48 million worth of computers; in 1969, $728 million dollars worth. "U.S. computer exports have increased by over 1400 percent in the first decade of the space age. This impressive record was built on excellence of performance through continuing technological superiority. In a large measure it was the stimulus of NASA's requirements that brought about these technological advances in the computer industry." (Text)

May 3: With exception of Mission Control, MSC buildings normally open to public had remained open during *Apollo 13* crisis, *New York Times* commented. Visitors could watch—from overhead catwalk in Building 5—scientists and astronauts devising techniques for *Apollo 13*'s return to earth. (Gordon, *NYT*, 5/3/70, 7)

- Although Communist China officially had entered space age with April 24 launch of satellite, much of the country's industrial machinery was "product of an earlier age," *New York Times* said. Workers at Shanghai factories had said they "used the equipment of the twenties to turn out products of the seventies." While Communist China's space success reflected "application of a more advanced technology," even this was said to have been achieved "because the Chinese people adhered to the 'principle of independence and self-reliance.'" (*NYT*, 5/3/70, 18)

- Col. Gen. Nikita V. Yegerov, ideological chief of U.S.S.R.'s strategic rocket forces, died at age 62. (*W Post*, 5/7/70, B14)

May 4: MSC successfully conducted initial system verification tests of 1/10 size, dynamically scaled, experimental model of proposed 12.5K Space Orbiter Shuttlecraft in 2400-m (8000-ft) drop from helicopter at Fort Hood, Tex. Initial test and continuing drop tests at WSMR were to demonstrate transition from high angle of attack reentry to level cruise attitude, to demonstrate stability in stalled conditions, and to obtain free-flight data. (MSC Release 70-53; NASA Photo 70-H 813)

- House passed S.R. 193, to provide for appointment of former NASA Administrator James E. Webb as Citizen Regent of Board of Regents of Smithsonian Institution. (*CR*, 5/4/70, H3828-9)

May 5: Ninth anniversary of first U.S. manned space flight, by Astronaut Alan B. Shepard, Jr., in *Freedom 7* Mercury spacecraft.

Apollo 13 astronauts visited Grumman Aerospace Corp. plant at Bethpage, N.Y., where LM *Aquarius* was built. Astronauts received rousing cheer from 15 000 workers. *Aquarius* received rousing cheer from astronauts. (UPI, *W Star*, 5/6/70, A2; MSC Hist Off)

- ESSA announced that photos from NASA's orbiting *Ats I* over Pacific and *Ats III* over Atlantic were being transmitted to National Severe Storms Forecast Center in Kansas City, Mo., providing new source of information on tornado and severe storm development during daylight hours. Beginning June 1, ATS photos would also be sent to

National Hurricane Center in Miami, Fla., for tropical storm prediction. (ESSA Release 70-31)
- Propeller-driven XC-142 V/STOL aircraft on loan to NASA by USAF ended experiments in support of research into V/STOL techniques at LaRC and was flown by NASA pilots to Air Force Museum at Dayton, Ohio. Built by Ling-Temco-Vought, Inc., four-engine tilt-wing aircraft had evolved from wind-tunnel models tested at LaRC and other NASA centers starting in 1956. (NASA Special Release)
- John L. Steele, Senior Correspondent of Time-Life News Service, praised James E. Webb's book *Space Age Management* in speech at Time-Life and Kaiser Corp. luncheon to direct attention to need for pollution control. Former NASA Administrator Webb had written that space accomplishments could "usher in new era of great advances for us and for mankind...because our accomplishments indicate a fundamental improvement in man's ability to find and use new knowledge." These included new ways in which large-scale organized efforts were managed; encouragement of multidisciplinary efforts using science, engineering, and management; new techniques and tools for research in social and physical sciences; and "creative manner in which these are applied to the solution of age-old problems." With these accomplishments U.S. had shown, Webb had written, "that we can perform the most difficult and challenging tasks without departing from or damaging our fundamental values or our democratic institutions." (*CR*, 5/28/70, S8008-9)

May 6: Senate passed, by vote of 69 to 15, H.R. 16516, $3.316-billion NASA FY 1971 authorization, after rejecting amendment by Sen. Walter F. Mondale (D-Minn.) to delete $110 million for definition and design of space shuttle and space station. Senate requested conference with House. (*CR*, 5/6/70, S6768-817)
- *Apollo 11* Astronauts Neil A. Armstrong, Edwin E. Aldrin, Jr., and Michael Collins received Robert J. Collier Trophy for "their high courage and stunning success in accomplishing man's highest adventure in recorded history—the first moon landing." Award was presented at Washington, D.C., luncheon given by National Aeronautic Assn. and National Aviation Club. (NAA Release; *NR News*, 5/15/70, 1)
- ARC research scientist Dr. Dale R. Lumb received National Telemetry Conference's National Telemetry Man of the Year Award at NTC luncheon in Los Angeles. Award was for development of advanced coding system for spacecraft telemetry signals that had enabled *Pioneer IX* (launched Nov. 8, 1968) to double amount of data returned with no increase in power and little added equipment. (ARC *Astrogram*, 5/7/70, 1)
- NASA announced issuance of *Analysis of Apollo 8 Photography and Visual Observations* (NASA SP-201). Book contained black-and-white photos taken on 1968 mission to furnish data on approach topography and landmarks for early Apollo landings, on scientific merit and roughness of areas for possible follow-on Apollo landings and on broad structure and characteristics of lunar surface. (NASA Special Release)
- Sen. Howard W. Cannon (D-Nev.) said on Senate floor that in FY 1971 "Federal Government will spend more for human resources programs than anything else. This amounts to $81.9 billion, or 41

percent of the total budget. This is $8.3 billion more than for national defense. In sharp contrast is the outlay for space research and technology which is only $3.4 billion, or a scant 1.7 percent of the total." National defense outlays had declined by 4.8% in 1969-1971; outlays for human resources had increased 13.5%; and space funding had decreased 10.5%—"greatest decrease of any item in the entire Federal budget." (CR, 5/6/70, S6805)
- FAA released aviation forecast covering FYs 1970-1981: Passengers carried in scheduled service by U.S. carriers would increase after 1971 at average rate of 12% annually. Figure more than doubled that forecast for remainder of national economy, but was below 18% annual rate 1965-1969. Passengers carried by U.S. scheduled carriers would reach 522 million in 1981—more than triple 1969 total of 168 million. Domestic travel would account for 460 million of 1981 total; international travel, 62 million. Revenue passenger-miles flown by U.S. scheduled carriers would grow from 120 billion in 1969 to 450 billion in 1981. (FAA Release 70-40)

May 6-7: NASA held Conference on Materials for Improved Fire Safety at MSC to demonstrate to industry NASA-developed technology available to consumers. Rep. Jerry L. Pettis (R-Calif.) said in keynote address: "The successful R&D program conducted by NASA in the field of nonmetallic fireproof materials has produced a 'quantum jump' that can greatly affect many major industries and has profound life-preserving implications." (CR, 5/13/70, E4249-51; MSC *Roundup,* 5/22/70,1)

May 7: Dr. Thomas O. Paine, NASA Administrator, announced that Apollo 14 would be launched no earlier than Dec. 3 toward landing on moon's Fra Mauro, intended landing site of aborted *Apollo 13* mission. "Our present assessment is that the modifications to the oxygen tanks in the Service Module that have already been identified will require several months.... We will take whatever time is necessary and not commit to a specific launch date until the Apollo 13 Review Board completes its work and makes its findings and recommendations."

Flight crew for Apollo 14 would be same as originally announced: Alan B. Shepard, Jr. (commander), Stuart A. Roosa (CM pilot), and Edgar D. Mitchell (LM pilot). (NASA Release 70-67)

- House Committee on Appropriations approved H.R. 17548, FY 1971 Independent Offices and HUD appropriations bill containing $3.197-billion NASA appropriation—$136 million less than $3.333-billion NASA budget estimate. Appropriation provided $2.500 billion for R&D, $18.3 million for construction of facilities, and $678.7 million for research and program management. Report recommended postponement of Apollo 14 mission until 1971 "to give ample time for study, report, and correction of the Apollo 13 problems."

  Bill also allocated $497 million to NSF—$16 million less than NSF budget estimate of $513 million. (House Rpt 91-1060; Subcom on Independ Off)

- George C. Wilson said in *Washington Post* that "top secret" DOD report based on photos taken by U.S. Samos satellites indicated U.S.S.R. had built fewer SS-9 missile sites in 1969 than in 1965. (*W Post,* 5/7/70, A7)

May 6-7: *Conference on Materials for Improved Fire Safety*, held at MSC. demonstrated NASA-developed technology available to industry. Among materials developed to reduce spacecraft fire hazards, Fluorel could be sprayed on materials to fireproof them against temperatures to 1500 K (2200°F). Fluorel-coated Durette fabric could provide lightweight, durable suits (suits to the right) that would permit firemen to work close to fires. Added aluminum layer would reflect heat outward and provide a vapor barrier to protect firemen from steam. Fluorel sprayed on chair (lower left) charred in the direct flame of a torch but when the material was scraped off the wood was revealed to be undamaged.

May 8: Aerobee 170 sounding rocket, launched by NASA from WSMR, carried MIT payload to 146.2-km (90.9-mi) altitude to obtain precise position of two or more x-ray sources. Instruments functioned satisfactorily, but timer on rocket closed shut-off valves at 46 secs. Good data were collected from first target, but data from second target were not expected to be useful. (NASA Rpt SRL)
- *Intelsat-III F-7*, launched April 22, began full-time regular commercial service. (ComSatCorp PIO; ComSatCorp Release 70-26)
- Four ground test models of Apollo Telescope Mount manned solar observatory were under construction at MSFC Manufacturing Engineering Laboratory, MSFC announced. Flight versions—prime spacecraft and backup—would follow. MSFC engineers were modifying laboratory, previously used for Saturn V assembly and early Skylab mockup work, for ATM fabrication role. Changes included installation of clean room where air would wash downward over unit under construction, installation of airlock 14 m (45 ft) high for moving hardware to and from building without contaminating air, and installation of second clean room and automatic checkout equipment in Quality and Reliability Assurance Laboratory area. (MSFC Release 70-87)
- DOT announced plans to staff new Transportation Systems Center at Cambridge, Mass., with 425 employees beginning July 1, when agency would acquire former NASA ERC facility. DOT would retain as many ERC staff members as possible. Of 826 persons employed by NASA at ERC on Dec. 29, 1969, when ERC closing was announced, 159 had left and additional 56 had indicated they planned to leave. (DOT Release 10070)
- In *Science*, Project Tektite I scientific team—Dr. H. Edward Clifton, Conrad V. W. Mahnken, John G. Vanderwalker, and Richard A. Waller—discussed results of February-April 1969 experiment sponsored by NASA, USN, Dept. of Interior, and General Electric Co. to provide data for behavioral, biomedical, and engineering studies and marine sciences program. Between 10% and 20% of waking time had been devoted to psychological tests for biomedical studies. Minor external ear infections contracted by all divers was only health problem to affect scientific program. Experiments had demonstrated advantages of underwater habitation and saturation diving for biological and geological research: opportunity for continuous monitoring of organisms or processes and for more research time in water. (*Science*, 5/8/70, 659-63)
- NASA and DOT announced award of $221 000 contract to George Washington Univ. to analyze potential social impacts of future aviation technology and new air transportation systems. Analysis would support part of Civil Aviation R&D Policy Study being conducted jointly by NASA and DOT. (DOT Release 9870)
- NASA and DOD budget cuts had contributed to dismissal of 60 persons —chiefly chemists, physicists, and physical science technicians—at National Bureau of Standards, *Science* reported. Bureau received about one third of its funding from NASA, DOD, and other agencies. (*Science*, 5/8/70, 682)

May 9: *Washington Daily News* said SEC was completing study to determine if Lockheed Aircraft Corp. officials had sold $1.2 million worth of company stock after having confidential information that Lockheed contract with USAF to build C-5A cargo aircraft was in financial trouble. (Dietsch, *W News*, 5/9/70)

May 10: In second attempt to test validity of Einstein theory of relativity [see April 30], JPL astronomer Dr. John D. Anderson and Cal Tech astronomer Dr. Duane O. Muhleman measured radio signal from *Mariner VII* just before it passed almost directly behind sun. During April 30 attempt, astronomers measured *Mariner VI* signal. (Sullivan, *NYT*, 5/10/70; NASA Release 70-62)

• Dr. Robert Jastrow, Director of NASA's Institute for Space Studies, discussed eventual colonization of outer planets in *New York Times Magazine:* While most Homo sapiens would remain on earth in world of future, small percentage—"restless, inquisitive, innovative, continually seeking out challenges and testing the limits of the environment"—would "move out to become a small, hardy population on Mars. Within a few decades of the year 2000, a pioneering band of men—and women—will be living on Mars. . . . Children will be born on Mars. Later, the spacefarers will go beyond Mars and beyond the solar system. Some day they will find the earth and its debilitating gravitational pull as difficult a place in which to survive as today's land-adapted men find the water out of which the ancestral fishes emerged 300 million years ago. Eventually they will constitute a new species, evolved out of Homo sapiens, but linked to the ancestral planet only by sentiment." (*NYT Magazine*, 5/10/70, 30)

May 11: NASA launched Aerobee 150 sounding rocket from Churchill Research Range carrying Univ. of Minnesota payload to study atmospheric composition. Rocket and instruments functioned satisfactorily. (SR list)

• SST program had experienced $76-million cost overrun during past six months, Rep. Sidney R. Yates (D-Ill.) said in testimony before Joint Economic Committee's Subcommittee on Economy in Government. Government would have to absorb some $57 million of SST cost growth. (Testimony)

• Kenneth M. Smith, former President and Chief Executive of Windecker Research Inc., was sworn in as FAA Deputy Administrator. (FAA PIO)

May 12: U.S.S.R. launched *Cosmos CCCXLIV* into orbit with 326-km (202.6-mi) apogee, 204-km (126.8-mi) perigee, 87.3-min period, and 72.0° inclination. Satellite reentered May 20. (GSFC *SSR*, 5/31/70)

• House passed H.R. 17548, FY 1971 Independent Offices and HUD appropriations bill containing $3.197-billion NASA appropriation. (*CR*, 5/12/70, H4219-70)

• NASA selected McDonnell Douglas Corp. and NR Space Division to negotiate 11-mo, $8-million, fixed-price contracts for parallel definition and preliminary design studies of two-stage reusable space shuttle vehicle to transport crew, passengers, and cargo between earth and near space. MSFC would manage McDonnell Douglas work and MSC would manage NR work. (NASA Release 70-68)

• Russell E. Train, Chairman of President's Council on Environmental Quality, testified before congressional Joint Economic Committee's Subcommittee on Economy in Government during hearings on SST: "Current design of the U.S. supersonic transport and of the [Anglo-French] Concorde leads to a noise field radiated perpendicular to the runway, called 'sideline noise,' that is substantially greater than that of the conventional subsonic jets." In terms of FAA measures to assess annoyance, "SST would be three to four times louder than current FAA sideline noise standards and four to five times louder than the [Boeing] 747. In terms of noise pressure, the sideline noise level would also be substantially higher than that of subsonic jets meeting the FAA requirements." (Testimony)

- ComSatCorp had made formal proposal to FAA for satellite system to enable FAA and foreign air controllers to maintain radar surveillance of commercial aircraft over oceans, *Washington Post* reported. System would be used exclusively to improve airline communications. (Samuelson, *W Post*, 5/12/70)
- Secretary of Defense Melvin R. Laird testified before Senate Armed Services Committee on strategic balance and arms limitation: "Today, we believe that 64 Moscow ABM launchers are operational. In addition, testing for new and/or improved ABM systems continues, while several of the large surveillance radars, that have an important early warning and tracking function in the Soviet ABM weapons system, are already deployed. The United States has no operational ABM components in place. We have reoriented and slowed down the deployment of the ABM system authorized by Congress in 1967....
  "We are concerned about the future because of the momentum in this Soviet buildup.... Advances in Soviet deployments and technology could threaten the survivability of our ICBMs and bombers." (Testimony)

May 13: MSC announced selection of Lockheed Missiles & Space Co. to receive 18-mo, $1 075 000, cost-plus-fixed-fee contract for large space station solar-array technology-evaluation program. Program objectives were to conduct technical evaluation of current state of the art of solar systems and design analysis and test evaluation of components. Work would be conducted under MSC direction. (MSC Release 70-48)

- *Sovetsky Voin* in Moscow published article on conquering universe, by Soviet Academician Boris N. Petrov: "During the next few years' studies of circumterrestrial space, the upper atmosphere and earth from space will undoubtedly develop further with the use of satellites with automatic instrumentation, periodically launched manned spaceships and large orbiting scientific stations operating for long periods and having a replaceable crew, as well as by sounding the atmosphere with geophysical rockets." (JPRS, 50505, 5/13/70, 53-4)
- AFSC announced participation with MIT's Lincoln Laboratory in Project Seek Storm to provide basis for design and development of improved weather radar system. USAF WC-130 Air Weather Service aircraft, using specialized weather instrumentation, were to gather flight data for system to provide information on where storms would strike and to provide maps showing where heaviest rainfalls would occur. Project was under way at AFSC's Electronic Systems Div. at L. G. Hanscom Field, Mass. (AFSC Release 122.70)

May 14: NASA's X-24A lifting-body vehicle, piloted by NASA test pilot John A. Manke, completed 13th flight from FRC. Two of four rocket chambers failed to fire after air launch from B-52 aircraft at 12 800-m (42 000-ft) altitude and desired speed (mach 0.95) was not

achieved. Some data were obtained at about mach 0.7 and below. (NASA Proj Off)
- Dr. Thomas O. Paine, NASA Administrator, delivered commencement address at Univ. of New Brunswick in Fredericton, New Brunswick, Canada: "...the Space Age, being only a decade old, has not yet had time to develop a professional university curricula. Thus, we have not been able to hire graduates specifically trained in space and astronautical engineering. . . . This parallels the experience we had when solid-state electronics replaced vacuum tubes, and we had to move ahead by retraining people. . . . It was also true at the start of the nuclear age, when we had to develop new nuclear power plants without the benefit of professionally trained graduates. The message for you here is that all of these programs were successful because the primary discipline that universities teach is intellectual discipline itself." (Text)
- Arcas sounding rocket was launched by NASA from Pacific Missile Range carrying GSFC experiment to study ozone in atmosphere. Mission did not meet minimum scientific requirements. (SR list)
- *Chicago Daily News* said student antiwar demonstrations at Univ. of Illinois in Urbana might have caused DOD to reevaluate decision to locate world's largest computer—Illiac IV—on University campus. Sources in Washington, D.C., had said DOD was reevaluating whether to locate computer center in Urbana. Computer, being funded by DOD at cost of more than $24 million, would be used two thirds time for military work, such as plotting operations of U.S. missile defense systems under enemy attack. During remaining time, computer would be available for academic work by university scientists. Computer had been target of student and faculty protests at Urbana since Nov. 15, 1969, when university newspaper revealed it would be built on campus. (*C Daily News*, 5/14/70, 5)

May 15: Saturn V and Saturn IB launches scheduled for 1972-1973 Skylab program would be launched from KSC Launch Complex 39 instead of Launch Complex 34, NASA announced. Change was made to take advantage of Launch Complex 39's modern facilities, including Vehicle Assembly Building; to save money by consolidating manpower and spacecraft support and checkout equipment; and to reduce transportation costs. Launch Complex 34, part of AFETR shared by NASA and DOD, had been operating since 1961 and would require extensive updating and repair for use in Skylab program. (NASA Release 70-70)
- Analysis of photo of Phobos, largest of two Martian moons, taken 138 000 km (86 000 mi) from Mars by *Mariner VII* was discussed in *Science* by New Mexico State Univ. astronomer and Mariner project scientist Bradford A. Smith. JPL photo-enhancement techniques had indicated Phobos was nonspherical and was larger and had darker surface than previously thought. Satellite's limb profile measured 18 by 22 km (11.2 by 13.7 mi) and was elongated along orbital plane. Phobos had average visual geometric albedo of 0.065, lower than that known for any other body in solar system. It seemed probable Phobos had not formed by accretion around primordial Mars, but was captured at some later time. (*Science*, 5/15/70, 828-30; NASA Release 70-66)
- FAA was increasing required separation between jumbo jets and small-

er aircraft on landing and takeoff from 4.8 km (3 mi) to 8 km (5 mi) as result of February tests with USAF F-104 and C-5A. Tests had shown severe effects of wake turbulence caused by larger aircraft on trailing smaller aircraft. However, earlier requirement that aircraft following Boeing 747 stay 16 km (10 mi) behind and 600 m (2000 ft) below 747 had been found unnecessary. FAA had since reverted to normal enroute requirement of 8-km (5-mi) and 300-m (1000-ft) separation for all aircraft, including 747. (*WSJ*, 5/15/70)

- AFSC announced M/G James T. Stewart, AFSC Deputy Chief of Staff for Systems, had been nominated by President Nixon for promotion to grade of lieutenant general. He would assume new duties as Commander of Aeronautical Systems Div. at Wright-Patterson AFB, Ohio, replacing M/G Lee V. Gossick, who would assume Gen. Stewart's current post. (AFSC Release 147.70)
- Charles Scribner's Sons published first two volumes of 13-volume *Dictionary of Scientific Biography*, first comprehensive biographical reference work covering science exclusively. Project was under auspices of American Council of Learned Societies and had NSF financial support. (Whitman, *NYT*, 7/11/70; Charles Scribner's Sons)

May 16: Shrinking profits and widespread attacks on military programs were making 1970s "age of uncertainty" for U.S. defense industry, Max Lehrer, Vice President of RCA Defense Electronics Products Div., said in speech before Philadelphia Chapter of Federal Government Accountants Assn. Logistic Management Institute study had shown tax profits for defense firms had dropped from 5.1% average in 1959 to 3.9% in 1968, while profits for comparable nondefense companies over same period rose from 8.9% to 9.4%. (Text)

May 17: Sert II (Space Electric Rocket Test), launched Feb. 3, had reached halfway point of six-month endurance test in space with ion engine operating satisfactorily for more than 2000 hrs, NASA announced. Experiment to determine whether electric rocket operation would interfere with radio communications between spacecraft, electric propulsion system, and nearby space indicated there would be no problem. Experiment to measure effects of ion thruster efflux on silicon-cell solar array showed solar cells were being adversely affected. As ions were emitted through molybdenum grid, small particles of grid broke off (sputtered) and were deposited on nearby surfaces, reducing ability of test patches to convert solar energy into electric power.

Controllers had turned off electric bombardment engine for several hours March 7 to avoid damage during solar eclipse when *Sert II* intersected eclipse path on two orbits with sharp loss in electric power from solar cells. After engine was restarted it continued to function satisfactorily. Only problem had been Feb. 7 failure of miniature electrostatic accelerometer, one of four methods of measuring engine's thrust. (NASA Release 70-69)

May 18: NASA announced appointment of Astronaut Neil A. Armstrong, *Apollo 11* commander, as NASA Deputy Associate Administrator for Aeronautics, Office of Advanced Research and Technology, effective July 1. He would succeed Charles W. Harper. Harper had been appointed Special Assistant to the Director for Interagency Affairs at ARC, NASA announced June 25.

May 18

Armstrong had served as Naval aviator 1949-1952 and after 1955 as aeronautical research pilot for NACA and NASA. Flight testing included piloting X-15 and X-1 rocket aircraft, F-100, F-101, F-102, F-104, F-5D, and B-47. He became NASA astronaut in 1962, was command pilot for *Gemini VIII* mission's first space docking of two vehicles March 16, 1966, and was first man to step onto moon July 20, 1969.

Harper had left ARC in 1964 to become NASA Director of Aeronautical Research and had been appointed Deputy Associate Administrator for Aeronautics in 1967. He had first joined ARC staff in 1941. (NASA Releases 70-72, 70-105)

- Mutual assistance agreement between Army Missile Command and MSFC was signed in Huntsville by M/G Edwin I. Donley, Commanding General of Missile Command, and Dr. Eberhard F. M. Rees, MSFC Director. Agreement on operational and support relationships included provision of facilities, services, materials, and equipment to carry out respective missions of the two organizations. (*Marshall Star*, 5/20/70, 1)

- Rep. Henry S. Reuss (D-Wis.) told House that 1969 DOT economic analysis of SST had predicted market for 420 aircraft rather than 500 required for Government to recoup its investment plus four percent interest. Reuss quoted from summary of DOT report: "If the government has as its primary objective recovery of past SST program expenditures ($633.4 million by the end of FY 1969) as well as future investment, the principal would be recovered plus a small return on investment. The profits to industry in excess of the normal industry return are not sufficient to cover the federal sunk costs plus future planned federal expenditures at either the interest rate specified in the present contract or recommended by the Bureau of the Budget." (*CR*, 5/18/70, H4480-2)

- L/G John W. O'Neill, AFSC Vice Commander, delivered keynote address at AFSC Air Force Materials Symposium in Miami Beach, Fla.: USAF was finding it "essential to push harder against the material and structural limits" of knowledge than at any time in its history. "We are forced to look over and beyond the horizon for new materials; new combinations; new structural concepts; and new methods of fabricating, testing, inspecting—and predicting." New aircraft had to "fly higher and faster than ever before—in some cases at the fringes of space—and still be large enough to carry an effective combat payload over extended ranges." Missiles had to "do incredible things under even worse conditions, with the added proviso that no humans will be aboard to make in-flight corrections. . . . With our spacecraft. . .we simply combine the harshest requirements imposed on both aircraft and missiles. And the difficulties will be compounded as the Air Force and NASA jointly develop reusable boosters, shuttles and tugs for the space transportation system." (Text)

- "Ailing aerospace industry" was described by Leroy F. Aarons in *Washington Post:* Since early 1968, aerospace employment in California had dropped from high of 616 000 to 525 000 in March 1969. "Roughly 30 percent of these were professional and technical personnel." North American Rockwell Corp. had accounted for "nearly a third of the layoffs." Since December 1969, NR had cut Los

Angeles Div. to "skeleton force of 4660, half of whom took pay cuts." Spokesman for State Unemployment Office in Van Nuys, Calif., had said desperation had driven out-of-work professionals into variety of positions at pay cuts of 50% to 60%. Aerospace engineering association official had explained that base technology had been developed. "We don't need propulsion groups of thousands of people to figure out how to make a missile fly any longer. Maybe we need 100 to come up with a modification." (*W Post*, 5/18/70, A1)

May 18-20: At annual convention of Aviation/Space Writers Assn. in Las Vegas, Nev., L/G James H. Doolittle (USAF, Ret.)—pioneer of instrumented flight, holder of aircraft flight records, and winner of four major aviation trophies—received FAA's Extraordinary Service Award. Gen. Doolittle was cited for "distinguished contributions to virtually every segment of American aviation."

Daniel Z. Henkin, Assistant Secretary of Defense (Public Affairs), disclosed in speech the existence of second "Chinese object" orbiting earth about 3200 km (2000 mi) ahead of active satellite launched by Communist Chinese April 24. He said DOD believed "that this is part of the final stage of the missile which orbited the satellite." Significance of launch was "that the Communist Chinese have the capability to put in orbit a considerable payload," which confirmed technological progress; "prudence requires that we continue to credit them with a near term capability to fire their first ICBM." (FAA Release 70-48; Text)

May 19: Dr. Thomas O. Paine, NASA Administrator, commented on May 7 action of House in reducing by $136 million NASA FY 1971 budget request, during appropriations hearing before Senate Committee on Appropriations' Subcommittee on Independent Offices: "I cannot state specifically now what actions NASA would have to take to meet a reduction of this amount, but coming on top of the actions already taken to accommodate to the $3.333 billion budget request...the effect would surely be a major blow to this nation's efforts to carry out a balanced space program, hold our teams together, and move out, however slowly, toward the goals of the future. We would have to reexamine again all aspects of the program and decide what further delays, terminations, and retrenchments have to be made."

In R&D, NASA would have to forego all but two of 11 construction projects needed to satisfy requirements of approved projects, reduce hazards, or improve basic operational efficiency in key activities. In research and program management, cut would require reduction-in-force of at least 1300 NASA civil service employees in addition to reduction planned in FY 1971 budget. "Based on past experience, a reduction-in-force...has an impact on three to five employees for every employee reduced. This means that 20 percent of the NASA work force would be disrupted, and our already hard-hit plant communities further depressed." (Testimony)

May 20: USAF launched two unidentified satellites from Vandenberg AFB by Thor-Agena booster. One entered orbit with 236.6-km (147-mi) apogee, 178.6-km (111-mi) perigee, 88.6-min period, and 83° inclination and reentered June 17. Second entered orbit with 503.7-km (313-mi) apogee, 489.2-km (304-mi) perigee, 94.5-min period, and 83.1° inclination. (Pres Rpt 71; GSFC *SSR*, 5/31/70; 6/30/70)

May 20

- U.S.S.R. launched *Cosmos CCCXLV* into orbit with 257-km (159.7-mi) apogee, 187-km (116.2-mi) perigee, 88.9-min period, and 51.7° inclination. Satellite reentered May 28. (GSFC *SSR*, 5/31/70)
- MIT announced it was divesting itself of Charles Stark Draper Laboratory. Institute would retain Lincoln Laboratory in Lexington, Mass., which performed $65-million, annual, long-range DOD research. Decision culminated year-long debate over appropriateness of academic institution's sponsoring weapons research. Laboratory would convert to private corporation. (*W Post*, 5/21/70)
- Univ. of Chicago scientist Dr. Albert V. Crewe announced his team of scientists had obtained images of individual atoms within organic compounds in verifiable manner for first time. Images, made with microscope having 35 000-volt scanning electron beam, showed geometric structure of compounds. (Sullivan, *NYT*, 5/21/70, 16)
- *Pravda* editorial had charged Communist China's first satellite was part of warlike anti-Soviet program, UPI reported. *Pravda* said Communist China was building "military psychosis" and splitting world Communist movement. (*Knoxville News-Sentinel*, 5/20/70)
- AP quoted Soviet geophysicist Sergey Timofeyev as saying Europe had entered period of colder summers and winters because moon was closer to earth than usual. Situation created "atmosphere tide" that disturbed normal movements of warm and cool air. (*NYT*, 5/20/70)

May 20-29: Thirteenth annual meeting of Committee on Space Research (COSPAR) of International Council of Scientific Unions (ICSU) was held in Leningrad. Soviet Premier Aleksey N. Kosygin sent message: "International cooperation in space exploration and in the use of outer space for peaceful purposes must be based on the development of mutual understanding and trust among the peoples." There was "growing cooperation on an international scale in space research" and "further progress in this field can open up still greater prospects for mankind."

Among 32 NASA officials and scientists attending meeting were Dr. George M. Low, Deputy Administrator; Gerald M. Truszynski, Associate Administrator for Tracking and Data Acquisition; and *Apollo 11* Astronaut Neil A. Armstrong.

Armstrong received tumultuous welcome from predominantly Russian audience, AP reported. During question-and-answer period before his report, Armstrong said that when he viewed earth from moon, "I thought it was impossible that a little island like that in the sky could have so many problems on it." He had found it "more pleasant on the moon than on earth as far as gravity is concerned." In report on experiences on moon, Armstrong said, "Since the sky is black the impression is somewhat like being on a sandy athletic field at night that is very well illuminated with flood lights." Further improvement in mobility of pressure garments would improve ability of astronauts to carry out lunar experiments. At later session when Armstrong attended as observer, guards had difficulty restraining cheers of audience, which included 800 Soviet delegates and 420 representatives of 30 countries. AP said it was "dramatic and unusual" tribute to foreign visitors in U.S.S.R.

Dr. Leonard D. Jaffe, JPL scientist, reported to COSPAR meeting on study and comparison of 60 Surveyor photos taken in April 1967 and 20 *Apollo 12* photos that showed moon's extreme stillness. Photos indicated only one small pebble had moved during 31 mos at *Surveyor III* site. Rate of meteoroid impact or surface movement was low in area of Ocean of Storms where *Apollo 12* and *Surveyor III* had landed. Camera and several portions of Surveyor spacecraft had coating of fine dust stirred from lunar surface as *Apollo 12* LM descended nearby. Only other change observed in Surveyor was that its white paint had turned to tan, "partly due to deposition of material on the spacecraft."

NAS-NRC submitted *United States Space Science Program*, comprehensive summary of scientific research in space science in U.S. during 1969: "Popular interest in the space program, which reached a high point with the first successful lunar landing, has diminished somewhat. Furthermore, a reassessment of national priorities has resulted in reduced financial support for space research and technology during the past year, and the support will probably be even more limited this year and next. It is our intention, despite these limitations, to try to maintain a reasonably balanced program, with continuing effort in all areas of space research in which we are currently engaged and to support mutually advantageous international cooperation in space research wherever possible." (Text)

Soviet academician Anatoly A. Blagonravov told meeting U.S.S.R. would launch two French SRET earth resources satellites as part of 1966 French-Soviet agreement on scientific cooperation. French ROSEAU (Radio Observation par Satellite Excentrique á Automatisme Unique) program, with multisatellite payload to be launched in 1971-1972, had been postponed because of budgetary restrictions and SRET (Satellite de Recherches et d'Environment Technique) had been retained. Joint French-Soviet space program also included plans for emplacement of French laser reflector on lunar surface by Soviet Luna spacecraft; STERO, launching of French experiments for measurement of solar radio emission; and ARCADE, study of solar energy particles with French and Soviet equipment on board Soviet spacecraft. (NASA Release 70-73; UPI, *NYT*, 5/24/70; Reuters, *NY News*, 5/24/70; AP, *NYT*, 5/26/70, 31; Reuters, *B Sun*, 5/26/70; AP, *W Star*, 5/27/70, A11; *SBD*, 5/27/70, 125; JPL, Release 554)

May 21: Operation of electron-bombardment thruster on board NASA's *Sert II* satellite (launched Feb. 3) was interrupted when excessive arcing triggered automatic shutdown. LeRC controllers returned engine to full thrust within 11 hrs. (NASA Release 70-79)

- *Apollo 12* Astronaut Charles Conrad, Jr., said in Memphis, Tenn., that U.S. could probably explore Mars or other planets in 13 to 15 yrs but "it will be terribly expensive." Following speech at annual Navy League convention, he told press: "I don't believe we will ever see one nation do it. I look for a multiple-nation group.... Some have as good or better technology and such a combination would not impose abnormal costs on any one." (AP, *Knoxville News-Sentinel*, 5/22/70)

- NASA announced award of $1-million contract to TRW Inc. for initial design of subsatellite to be placed in lunar orbit on Apollo 16 and 18 missions. Satellite would operate for year, providing information on

charged particles and earth's magnetic field in lunar vicinity. (MSC Release 70-56)
- President Nixon signed into law H.R. 14465, to provide for expansion and improvement of U.S. airport and airways system and for imposition of airport and airway user charges.(*PD*, 5/25/70, 680)
- DOT issued statement defending SST: There was "little or no substance to the claims of SST critics that harm will result from the plane's development, [and] there are some very real threats to the national interest if it is not put into production on schedule." Planned SST flight rules would confine aircraft's noise to airfield. There was "no scientific support" for allegation that supersonic flights at 18 300 m (60 000 ft) would pollute upper atmosphere and change world's weather. Overpass of SST at supersonic speeds would not disrupt earth's natural environment.

    Without SST program, "United States would surely lose its world leadership as an aircraft supplier." Loss of SST sales, "combined with the increased purchase by U.S. airlines of foreign-built supersonics, is estimated to result in an unfavorable swing of some $16 billion in balance of payments through 1990." Halt or slowdown in SST program "would be equally serious to the domestic economy." (Text)

May 22: NASA's M2-F3 lifting-body vehicle, piloted by NASA test pilot William H. Dana, successfully completed captive flight from FRC to verify compatibility of B-52 aircraft, adapter, and vehicle; check out systems; and verify check list. (NASA Proj Off)
- LRL scientist Frederick J. Mitchell reported microorganism—Streptococcus mitis—thought to have been accidentally deposited in *Surveyor III* camera before April 17, 1967, lunar launch had been recovered from inside camera after it was returned to earth by *Apollo 12* crew in November 1969. Streptococcus mitis was common respiratory organism and had produced no disease on animal inoculation. Mitchell said microorganism was probably deposited on camera when camera's shroud was removed for repairs and then replaced. Procedures and high vacuum of space were thought to have been responsible for preservation of organism in lyophilized (freeze-dried) state during launch, 3-day moon journey, and 950-day stay in hostile lunar environment. Organism had been found after laboratory testing of camera parts at MSC. Investigators had predicted possibility of organisms surviving under such circumstances. (NASA Release 70-76)
- NASA announced it had installed remote terminal for high-speed computer service in offices of House Committee on Science and Astronautics. RECON (remote console) system—linked to computer at NASA Scientific and Technical Information Facility in College Park, Md.—would permit Committee members to make literature search of NASA's worldwide collection of scientific and technical documents. Pilot project initiated by Dr. Thomas O. Paine, NASA Administrator, and Rep. George P. Miller (D-Calif.), Committee Chairman, would evaluate equipment over several months. Miller said it was first time agency of Executive Branch had been tied directly by computer to a committee of Congress. (NASA Releases 70-75, 68-118)

- MSFC announced selection of Grumman Aerospace Corp. and TRW Systems Group for negotiation of $250 000 fixed-price contracts for one-month definition and preliminary design studies of 10 000-kg (22 000-lb) High Energy Astronomy Observatory (HEAO), scheduled for launch in 1972.

  MSFC also announced award of three contracts for continuation of nuclear shuttle definition studies: $343 000 to McDonnell Douglas Astronautics Co. for completely modular concept with earth-to-orbit shuttle to transport modules for in-space assembly; $245 000 to NR Space Div. for nuclear stage with 10-m (33-ft) diameter to be orbited by Saturn V and refueled by earth-to-orbit shuttle; and $282 000 to Lockheed Missiles & Space Co. for both types. (MSFC Releases 70-97, 70-98)
- Selection of Boeing Co. for final negotiation of $4.5-million contract to modify de Havilland CV-7A Buffalo aircraft so that it could demonstrate augmentor wing, jet-flap concept, was announced by NASA. (NASA Release 70-77)
- Lightning and new-generation aircraft were discussed in *Science* by Rice Univ. space scientist A. A. Few: "It has been estimated that there are approximately 500 lightning strikes per year to commercial jet airplanes operating in the United States alone. Most...are triggered by the aircraft, as was very true of the lightning flashes that occurred during the launch of Apollo 12. Because of the larger size of the new-generation aircraft...this lightning hazard will increase. If the new aircraft are permitted to fly under the same meteorological conditions that are considered allowable for present aircraft, the probability of the aircraft's being hit by lightning will be considerably increased." New aircraft would not be as safe as conventional aircraft in lightning strikes "because of their extensive use of more sophisticated hardware, which is more susceptible to damage from lightning." (*Science*, 5/22/70, 1011-2)
- William M. Magruder, Director of DOT's Supersonic Transport Program, told Washington, D.C., press conference that he was satisfied with SST design, which would produce aircraft profitable for airline operators and Nation. With sale of 500 SSTs, U.S. Government would recover entire cost of its share of development program. Magruder produced letter from NASC Executive Secretary William A. Anders saying SST production was sound. (AP, W *Star*, 5/23/70, A2)

May 23: President Nixon issued statement on U.S. oceans policy: "I am today proposing that all nations adopt as soon as possible a treaty under which they would renounce all national claims over the natural resources of the seabed beyond the point where the high seas reach a depth of 200 meters [650 ft] and would agree to regard these resources as the common heritage of mankind. The treaty should establish an international regime for the exploitation of seabed resources beyond this limit. The regime should include the collection of substantial mineral royalties to be used for international community purposes, particularly economic assistance to the developing countries. The regime should also establish general rules to prevent unreasonable interference with other uses of the ocean, to protect the ocean from pollution, to assure the integrity of the

May 24: Radio interference experiment on board NASA's *Sert II* satellite (launched Feb. 3) stopped operating. Experiment had supplied more than three months of useful data and indicated operation of electric engine in space was unlikely to interfere with radio communications between ground and spacecraft at signal strengths measured. Halt would not affect mission's primary objective, operation of ion engine in space for six months. (NASA Release 70-79)

- FRC announced NASA had assumed operational control of one of two YF-12 aircraft being flown in joint NASA-USAF program to obtain information from sustained cruise flight at mach 3 and altitudes in 23 000-m (75 000-ft) range for use in development and operation of commercial and military supersonic aircraft and proposed space shuttle. Two NASA flight crews, each consisting of pilot and flight test engineer, had been assigned to program. Pilots, Fitzhugh L. Fulton, Jr., and Donald L. Mallick, were former XB-70 pilots for NASA. (FRC Release 9/10)

- JPL engineers had completed preliminary design for model of long-life unmanned space vehicle for exploring outer planets in late 1970s, *New York Times* reported after interviewing Dr. Frank E. Goddard, JPL Assistant Director for Research and Advanced Development. Spacecraft would have dish antenna 4 m (14 ft) wide and 9-m (30-ft) handle bar with instrument packages and nuclear power generators at either end. Antenna would be mounted on area housing computer designed to operate spacecraft, detect its own failures automatically, and repair itself. Dr. Goddard had said that flight model might differ in minor ways, "but what we're doing should point the ways it should be different." Called TOPS for Thermoelectric Outer Planet Spacecraft, craft was designed to draw electricity from four radioisotope thermoelectric generators. (Wilford, *NYT*, 5/24/70)

- Technical progress made through space program was described by Dale D. Myers, NASA Associate Administrator for Manned Space Flight, in commencement address at Whitworth College in Spokane, Washington: Chrysler Corp. was using computer-controlled equipment derived from Apollo checkout equipment to test distributors in its automobile ignition systems. U.S. Bureau of Public Roads had found automotive safety device that originated in Apollo spacecraft couches was trimming 97-km-per-hr (60-mph) impact to equivalent of 8 km per hr (5 mph) when used in conjunction with special highway guardrails. Nationwide insurance company had announced collision premium reductions of 20% for cars equipped with device. (Text)

May 25: Dedication ceremonies were held at Goddard Park in Auburn, Mass., memorial built by Auburn Rotary Club to honor late rocket pioneer Dr. Robert H. Goddard. Park contained replica of first liquid-fuel rocket invented by Dr. Goddard and Polaris rocket donated by USN. *(CR, 6/3/70, E5170)*

- House adopted H.J.R. 1117, to establish Joint Committee on the Environment and Technology. (*CR,* 5/25/70, H4774-95)

May 26: Discovery among *Apollo 12* lunar samples of 4.6-billion-yr-old radioactive rock from moon's Ocean of Storms was announced simultaneously by Dr. Paul W. Gast, Chief of MSC Lunar and Earth Sciences Div., at MSC press conference and by Cal Tech scientist Dr. Gerald J. Wasserburg at COSPAR meeting in Leningrad. Rock—oldest lunar material discovered to date—was first solid evidence that solar system was 4.6 billion yrs old. Rock was chemically unique and had highest concentration of radioactive elements yet observed in lunar samples. MSC analysts said 85-g (3-oz) rock, size of lemon, had 20 times as much uranium, thorium, and potassium as any other *Apollo 11* or *Apollo 12* rock. (NASA Release 70-80; Wilford, *NYT*, 5/27/70, 1)

- Astronaut Donn F. Eisele would leave MSC in mid-June to become Technical Assistant (Manned Flight) in LaRC's Space Systems Research Div., NASA announced. Eisele had been astronaut since October 1963 and was CM pilot on *Apollo 7* mission Oct. 16-22, 1968. (NASA Release 70-81)
- MSFC Advanced Systems Analysis Office was investigating possible uses of space tug, multipurpose vehicle to be developed simultaneously with larger space shuttle, MSFC announced. Only space vehicle that would work with and connect all existing and future vehicles and systems, tug would first be used as link between space shuttle and space station to taxi cargo and passengers in earth orbit. Both MSFC and MSC were working to develop space tug plans. (MSFC Release 70-102)
- MSFC announced selection of McDonnell Douglas Astronautics Co. and TRW Inc. for final negotiations leading to four contracts (two per company) totaling $1.2 million for space shuttle auxiliary propulsion system definition. (MSFC Release 70-103)
- Aerobee 150 sounding rocket launched by NASA from Woomera Rocket Range, Australia, carried MSC experiment to study UV spectra. Mission did not meet minimum scientific requirements. (SR list)
- DOD Deputy Secretary for Public Affairs Jerry W. Friedheim said in Washington, D.C., that deployment of Minuteman III missiles had begun in North Dakota on April 18 but their MIRVs would not be installed until June. (Roberts, *W Post*, 5/27/70, 3)
- In speech before Young Communist League in Moscow, Cosmonaut Aleksey Yeliseyev said: "Every country follows its own path in space exploration. The main trend in Soviet space research is the construction of orbital stations that can make long flights. The United States in recent years concentrated its efforts on putting a man on the moon. In the future these paths will merge into a wide road of mankind far into space." (UPI, *NY News*, 5/27/70, 34; UPI *NYT*, 5/31/70)

May 26-27: Critical design review of lunar roving vehicle's mobility system was held at AC Electronics Defense Research Laboratory in Santa Barbara, Calif. Meeting and second LRV meeting scheduled for June 8 at MSFC were preliminaries to final LRV critical design review scheduled for June 16-17 at MSFC. *(Marshall Star, 5/27/70, 1)*

May 27: Special detanking procedure applied to *Apollo 13* SM No. 2 oxygen tank before launch "probably resulted in major damage to the wiring insulation in the tank," Edgar M. Cortright, *Apollo 13* Review Board Chairman, said in Status Report. It appeared that two thermal switches designed to protect tank's heaters from overheating might have failed. Tests had shown heater tube could have reached 800 K (1000°F) and that temperature would seriously damage insulation around heater wires. Cortright said damage could have resulted in arcing short circuits that were believed to have initiated combustion of insulation inside tank during flight. Burning, in turn, raised pressure of supercritical oxygen and caused tank to rupture. (Status Rpt No 14)

- One-tenth-size model of MSC 12.5K Space Orbiter Shuttlecraft was air-dropped from CH-54 helicopter at low speed from 2400-m (8000-ft) altitude at WSMR in series of tests to demonstrate vehicle's transition from high angle of attack reentry to level cruise attitude and

May 27: *Space Orbiter Shuttlecraft model was air-dropped from helicopter at White Sands Missile Range to demonstrate transition from a high angle of attack reentry to cruise attitude and stability in stall conditions. For the tests, the nose was made of crushable material to attenuate landing shock; the wings were replaceable.*

stability of vehicle in stall conditions and to obtain free-flight data to assist in aerodynamic analytical transition prediction techniques. Test vehicle was 4 m (13 ft) long with fuselage 1/2-m (2 ft) in diameter, weighed 270 kg (600 lbs), and had 2 1/2-m (8-ft) wing span. Test was adjudged "limited success" because vehicle's entry angle was too steep. (MSC Release 70-53; *C Trib*, 5/28/70; NASA Proj Off)

- House, by voice vote, passed H.R. 17755, $7.052-billion FY 1971 DOT appropriations bill, after rejecting by vote of 176 to 162 amendment which would have deleted $290 million for SST. (*CR*, 5/27/70, H4870-6; 5/28/70, H4877-902)
- Four-engine USAF RB-57F jet aircraft assigned to NASA Science and Applications Directorate at MSC had conducted surveys at 16-km (10-mi) altitude for U.S. Geological Survey over 26 cities, NASA announced. Purpose was to gather standardized data on urban areas and to correlate land usage with statistical data gathered during 1970 U.S. Census. (NASA Release 70-74)
- Financial predicament of Lockheed Aircraft Corp. was described by Deputy Secretary of Defense David M. Packard in testimony before Senate Armed Services Committee: Cash flow analysis had determined Lockheed could pay cost of work it would be doing through 1970. "One uncertainty during the short term is whether the banks will honor their $400 million commitment. . . ." Lockheed was so badly overextended that it had "essentially no reserve financial resources available." Most desirable course was for Lockheed "to seek additional financial strength through commercial channels. The alternatives available would include a merger with a strong partner, substantial additional underwriting of their financial requirements by commercial banks or investment bankers." Packard said Lockheed was exploring alternatives. (Testimony)
- MSC announced award to North American Rockwell Corp. Space Div. of $250 000 contract for lunar orbit station study to define configuration, refine functions of station, detail feasible scientific investigations, and develop rescue capabilities. Station would operate in zero gravity but would be modifiable to artificial-gravity environment. (MSC Release 70-60)

May 28: In letter to Rep. Olin E. Teague (D-Tex.), Dr. Thomas O. Paine, NASA Administrator, replied to inquiries on design, cost, and operation of space shuttle: "NASA and DOD have been working for more than three years on the preliminary analysis of alternate approaches and concepts, and on the research and technology effort needed to determine whether it is appropriate to develop reusable vehicles that will substantially reduce the cost of operating in space. We have concluded that this is an achievable objective. We are convinced that availablility of these vehicles will lead to significant changes in our concepts of operation in space environment as well as reductions in costs. Accordingly, we selected contractors on May 12 to proceed into the second or definitive phase of detailed study, comparative analysis, and preliminary design directed toward facilitating the choice of a single program approach . . . . We will decide at the conclusion of this phase whether it is appropriate to settle on a single design or continue competitive approaches." (*CR*, 6/23/70, E5878-80)

May 28

- Dr. Eli L. Whitely, President of U.S. Congressional Medal of Honor Society, presented NASA with plaque honoring agency for "making America first in space." Dr. Thomas O. Paine, NASA Administrator, accepted plaque on behalf of NASA and other organizations that had contributed to space program. (NASA Release 70-87; Citation)
- Apollo launch operations contractors at KSC were reducing work forces from 17 500 to between 16 500 and 17 000 by July 1, NASA announced. NASA and contractors would continue to review future manpower requirements and final determination on civilian work force would be made by Sept. 1. Factors influencing final determination included consolidation of manned launch operations at Launch Complex 39; changes in 1970, 1971, and 1972 Apollo and Skylab schedules; and internal cost-reduction effort of Government and contractors. *Apollo 13* Review Board results might also affect planned work schedule. (NASA Release 70-84)
- African nation of Senegal became 76th member of INTELSAT. (ComSatCorp PIO; *SBD*, 6/1/70, 147)
- DOD issued broad policy directive that placed greater emphasis on conceptual and development stages of new weapon systems and suggested greater flexibility in negotiating ultimate contract prices and weapon requirements. (Text)
- *New York Times* editorial commented on 4.6-billion-yr-old rock brought from moon by *Apollo 12:* "Rock 13's appearance at this moment in history is more than a major scientific coup for the Apollo program. The recovery of this staggeringly ancient rock strengthens the argument of those who have pressed for extensive exploration of the lunar surface on the ground that the moon will provide the clues needed to understand the origin and development of the solar system. Thus rock 13 fortifies the scientific case for going ahead with additional Apollo flights despite the dangers which were made so vivid by the hazardous journey of Apollo 13." (*NYT*, 5/28/70)
- Rep. Gillespie V. Montgomery (D-Miss.) entered in *Congressional Record* article from Oct. 23, 1936, *Meridian* (Miss.) *Star* which described unsuccessful efforts of R. G. V. Mytton of Meridian to fly with aid of wing-like machine in 1900. (*CR*, 5/28/70, E4748)

May 29: NAS released *Infectious Disease in Manned Spaceflight: Probabilities and Countermeasures.* Report of NAS-NRC Space Science Board recommended biological precautions taken after lunar flights be taken before flights as well. Preflight quarantine period should be instituted to permit acute disease to express itself, prevent contact and infection of astronauts by general population, and permit cross-contact of flora and exchange of microorganisms among spacecraft crew. During last two weeks before launch, astronauts' contacts should be limited to essential technical crew held under same isolation regulations.

Report called for extensive and periodic screening of astronauts by immunological tests and "microbiological surveillance" of astronauts' families and close associates before and after missions. Astronauts with even minor respiratory infections should be barred from mission because they might "result in greatly accelerated clinical responses under space-cabin conditions if transmitted to other astronauts." (Text)

- Aerobee 150 sounding rocket was launched by NASA from Woomera Rocket Range, Australia, carrying Univ. of Wisconsin experiment to conduct x-ray studies. Rocket and instruments functioned satisfactorily. (SR list)
- India had successfully completed five flight tests in Rohini multistage-rocket development program at TERLS, including one test of booster to be used for Dart payload, *Space Business Daily* reported. India had reportedly completed feasibility study of design and development of carrier rocket that could launch 30-kg (66-lb) satellite into near polar orbit and planned to launch first satellite by 1973. (*SBD*, 5/29/70, 143)
- President Nixon in San Clemente, Calif., announced intention to nominate Dr. T. Keith Glennan, former NASA Administrator, to be U.S. Representative to International Atomic Energy Agency with rank of Ambassador. Dr. Glennan would replace Henry DeWolf Smyth, who had resigned. Dr. Glennan had served as member of Atomic Energy Commission from 1950 to 1952. Nomination was submitted to Senate June 1. (*PD*, 6/1/70, 703; 6/8/70, 734)
- Bell Telephone Laboratories had indicated it would give up missile contracting and shift some of 16 000 employees working on ABM into civilian areas, *Science* said. Company spokesman had said rising protests against defense contractors had played minor role in decision. (*Science*, 5/29/70, 1076)

May 30: AAAS Board of Directors selected William Bevan, Jr., Vice President and Provost of Johns Hopkins Univ., to become AAAS executive officer Oct. 1. He would succeed Dael Wolfle, who would retire June 30 to become professor of public affairs at Univ. of Washington in Seattle. During interim, *Science* editor Dr. Philip H. Abelson (named to become President of Carnegie Institution of Washington at end of June) would serve as acting AAAS executive officer. (*Science*, 6/26/70, 1555)

May 30-31: Apollo 11 Astronaut Neil A. Armstrong visited U.S.S.R. space center near Novosibirsk, Siberia, with Cosmonauts Georgy T. Beregovoy and Konstantin P. Feoktistov following May 20-29 COSPAR Conference in Leningrad. He flew from Siberia to Moscow. (UPI, *W Star*, 6/1/70, A1)

During May: In *Astronautics and Aeronautics*, JPL scientists William G. Breckenridge and Thomas C. Duxbury discussed applications of spacecraft-based navigation data for outer-planet missions and described processing of spacecraft-based science and engineering data from 1969 Mariner Mars missions. They concluded: "Stringent navigation requirements for proposed outer-planet missions make even the projected ability of a solely Earthbased system marginal, and so have emphasized the need for an onboard system." (*A&A*, 5/70, 44-9)

- JPL *Lab-Oratory* reproduced crystallotype print from daguerrotype of moon made by John A. Whipple in 1851. Accompanying article included description by Whipple of difficulties in obtaining photo. Whipple account and daguerrotype had first appeared in July 1853 issue of *Photographic Art-Journal*. Daguerrotype was "one of the oldest close-ups of the Moon on record," JPL said. (*Lab-Oratory*, 5/70, cover, 3)

- Avco Systems Div. engineer M. Mitrovich reviewed in *Journal of Aircraft* efforts at man-powered flight culminating in Nov. 9, 1961, 64-m (70-yd) "flight at maximum 2-m (6-ft) altitude by Southampton Univ. Man Powered Aircraft Group—"first flight with manpower." Since successful efforts relied "on wings with extraordinary aspect ratios," point might be reached "where the corresponding wing spans will become intolerable." Mitrovich diagrammed and described application of particular wing configuration to alleviate this dilemma. (*JA*, 5-6/70, 246-51)
- *Space/Aeronautics* editorial commented on *Apollo 13* accident: "It has shown us how vulnerable are the men who go into space in their tiny, fragile capsules. It has stunned us into the realization that accidents can and will happen in space, that not every mission will run off like clockwork, that manned space flight is, and will always be, dangerous. Public awareness, finally, that this is so could immeasurably strengthen the space program. If so, Apollo 13 may prove to be one of the most important missions we have flown to date." (*S/A*, 5/70, 21)
- *Armed Forces Management* editorial by Michael Getler said nuclear test ban treaty "came only after the Soviets unilaterally broke a testing moratorium to gather what they needed to know in the vital area of weapons effects, and the ink was barely dry on the Peaceful Uses of Outer Space Treaty when the Pentagon announced the Russians were testing a new Fractional Orbital Bombardment System that was certainly a spiritual, if not technical, violation of that pact." (*AFM*, 5/70, 13)
- Medical experience in *Apollo 7* through *11* was summarized by Dr. Charles A. Berry, MSC Director of Medical Research and Operations, in *Aerospace Medicine:* "The 3105 hours of exposing man to spaceflight during the Apollo program have added greatly to knowledge of man's response to space travel." Spacecraft cabin environment had been "suitably maintained"; radiation environment had been "benign," with no solar flares occurring; crews had adapted well to weightlessness; and improvements had been made in inflight food. Body-weight losses which continued to occur were not entirely due to body fluid loss. Supplying potable water had been effective and "great strides have been made in removing gas bubbles in the water." Work-sleep cycles had been improved and bioinstrumentation had continued to function well.

  "Although a preflight preventative medicine program has been difficult to conduct, it has been effective in the later Apollo missions in reducing pre-, in-, and postflight illnesses, which had occurred in all flight phases and were usually viral upper respiratory and gastrointestinal infections." Although crews had adapted to motion environment, "this area will require continuing attention."

  Cardiovascular deconditioning had been identified after flight, along with "significant decrement in work capacity in immediate postflight period." Postflight neutrophilia had been observed after crew recovery. Loss of red cell mass observed in Gemini program had reoccurred only during *Apollo 9*, indicating that hyperoxia was responsible. Even small amount of nitrogen in atmosphere might protect red cells from lytic action of oxygen. Microbiological studies had shown that organisms transferred between crewmembers and that organism's growth was favored by shifts. EVA on lunar surface during *Apollo 11* was conducted within expected energy costs, at average of 1200 BTU per hour. It appeared lunar surface time could be extended safely. (*Aero Med*, 5/70, 500-18)

- *Atlantic* magazine article said U.S.S.R. had "used aid and trade, backed by its newly acquired sea power, to build a vast arc of influence around the northern rim of the Indian Ocean and up as far as the home islands of Japan." Included were "two large mooring buoys laid near the Seychelles Islands in connection with the Soviet space program." They provided "refueling base for the Soviet flotilla." (*Atlantic*, 5/70)

- NSF published *Resources for Scientific Activities at Universities and Colleges, 1969* (NSF 70-16), report of survey of 1969 employment and 1968 expenditures: In 1969, 253 500 full- and part-time scientists and engineers were employed in U.S. universities and colleges, increase of 8.1% over 217 200 employed in 1967. Yearly increase rate 1958-1967 had been 6.9%. Growth rate in teaching had averaged 10.5% per year 1967-1969 and 6.7% per year 1958-1967. Growth rate for number in R&D was 5.7% per year 1958-1967, but averaged only 2.9% per year 1967-1969, reflecting leveling off of Federal R&D support to universities and colleges in late 1960s. Life scientists made up 41% of professional science and engineering staff, social scientists 21%, physical scientists 14%, and engineers 10%. (Text)

- NSF issued *Scientific and Technical Personnel in the Federal Government, 1968* (NSF 70-24). Federal professional scientific and technical personnel numbered 207 000 in October 1968, about 1% more than in previous year. Engineers, largest of three major groups, numbered 82 800, scientists 78 100, and health professionals 46 200. Physical scientists accounted for 88% of scientist category. DOD personnel total of 79 800 was largest number at any agency. Most—93%—were scientists or engineers. About 29% of all Federal scientists and engineers were engaged in R&D. (Text)

- NSF issued *Research and Development in State Government Agencies, Fiscal Years 1967 & 1968* (NSF 70-22). State government agencies spent $131 million in FY 1967 and $155 million in FY 1968 for R&D and additional $5.1 million in 1967 and $4.5 million in 1968 for supporting R&D plant. Expenditures had grown at 20% annual rate since 1964, when expenditures were $77 million. Federal agencies provided funds for nearly one half of state agency R&D expenditures in both years. Applied research activity represented 50% of state agency R&D expenditures; development accounted for 20% of total in 1968 and basic research 23%. (Text)

June 1970

June 1: NSF said high-altitude photos of planet Uranus taken by 91-cm (36-in) telescope on Stratoscope II balloon March 26-27 might reveal whether Uranus had cloud cover. Photos, five times as distinct as ones taken from earth's surface, showed Uranus with none of previously reported surface details, but with darkening at edges that might give clue to whether planet had clear or cloudy atmosphere. (W *Star,* 6/2/70, A10)

- NASA was relocating several Saturn IB and Saturn V stages and instrument units to avoid possible damage or destruction by hurricanes in area of Michoud Assembly Facility in New Orleans, La., MSFC announced. Three Saturn IB boosters would be moved from Michoud to MSFC in mid-July; two Saturn V vehicles were stored at MTF; and three Saturn instrument units had been moved from IBM plant in Huntsville, Ala., to MSFC for storage. (MSFC Release 70-106)

- Dr. Louis Morton, professor of history at Dartmouth College and Chairman of American History Assn. Committee on the Historian in Federal Government, had been named Chairman of NASA Historical Advisory Committee for 1970-71, NASA announced. Dr. Morton was formerly with U.S. Army Office of Military History. Other members of NASA Historical Advisory Committee were Dr. Daniel J. Boorstin, Director, Museum of History and Technology, Smithsonian Institution; Prof. David Bushnell, Univ. of Florida; Prof. A. Hunter Dupree, Brown Univ.; Prof. Melvin Kranzberg, Case Western Reserve Univ.; Prof. Rodman W. Paul, Cal Tech; Robert L. Perry, RAND Corp. Economics Div.; and Prof. John B. Rae, Harvey Mudd College. Executive Secretary of Committee was Dr. Eugene M. Emme, NASA Historian. (NASA Release 70-78)

- John D. Hodge, Manager of MSC Advanced Missions Program Office, left NASA to become Director of Transportation Systems Concepts for DOT's Transportation Systems Center at Cambridge, Mass. Hodge had received NASA Medal for Exceptional Service twice during 10-yr NASA service. (MSC Release 70-60)

- Victor C. Clarke, Jr., JPL engineer, received 1970 award from Gravity Research Foundation of New Boston, N.H., for essay "The Application and Principle of Gravity-Assist Trajectories for Space Flight." Clarke, mission analysis and engineering manager for 1973 Mariner Venus-Mercury project, had first demonstrated possibility of bouncing spacecraft from planet to planet in 1961. He had received 1969 NASA Exceptional Service Medal for contribution to Mariner Mars 1969 mission design. (NASA Release 70-112; JPL PIO)

June 2-19: U.S.S.R.'s *Soyuz IX*, carrying Cosmonauts Andrian G. Nikolayev and Vitaly I. Sevastyanov, was launched from Baykonur at 12:09 am Baykonur time into orbit with 249-km (154.7-mi) apogee, 236-km (146.6-mi) perigee, 89.3-min period, and 51.6° inclination. Tass said cosmonauts would conduct extensive program of scientific and technical research in "solitary orbital flight," including medical and biological research on effects of space flight on man, studies of earth's geography and atmosphere, research into "physical processes in near-earth space," and checkout of spacecraft systems. Launch and pictures of crew in spacecraft were televised in Moscow about one hour after liftoff.

Five hours after liftoff Tass reported cosmonauts were "coping well with conditions of weightlessness and carrying out the flight program." Only anomaly had been dirtying of windows from engine fired for midcourse maneuver after fifth orbit, Tass said. Systems were functioning normally and Nikolayev had begun "experiments linked with the investigation of human peculiarities as an element of the control system in different dynamic operations."

Tass issued daily bulletins that described routine of space flight chores and experiments: "While one cosmonaut runs a vacuum cleaner, the other brews coffee or shaves." Cosmonauts performed strenuous physical exercises each day in special "load suits," practiced manual maneuvers of spacecraft, and photographed earth and sun. By June 8 *Soyuz IX* had completed 100 orbits and 147 hrs of flight—record for man-hours in space in Soviet space program—with all systems functioning satisfactorily and with crew demonstrating it could "withstand well the complex impact of the factors of space flight and retain high efficiency." Insomnia, only problem reported by crew, had been overcome.

On June 19 Tass announced *Soyuz IX* had softlanded 76 km (47 mi) west of Karaganda, Kazakhstan, after orbiting in space for 17 days 16 hrs 59 min and setting new world record for manned space flight endurance. Tass said that crew had "fully fulfilled its programs of research" and that preliminary medical examination showed cosmonauts "had withstood the prolonged space flight well." Most important part of program, Tass said, was medical-biological experiments on effects of prolonged flight on man. (GSFC SSR, 6/30/70; SBD, 6/2/70, 152; Mills, B *Sun*, 6/2/70, A1; 6/20/70, A1; AP, *NYT*, 6/3/70, 18; UPI, W *Star*, 6/7-8/70; 6/12/70)

June 2: *Apollo 11* Astronaut Neil A. Armstrong, on goodwill trip to U.S.S.R., watched TV broadcast of *Soyuz IX* launch during party in his honor in Moscow and was told by Cosmonaut Georgy Beregovoy that mission was "especially in honor of your trip here."

Armstrong later presented moon fragment and small Soviet flag carried on *Apollo 11* mission to Soviet Premier Aleksey N. Kosygin. Kosygin said he would "always cherish this gift as a symbol of a great achievement." He told Armstrong, "The Soviet peoples are second to none in admiring your courage and knowledge." (UPI, W *Star*, 6/2/70, A1; AP W *Post*, 6/3/70, A19)

- NASA's M2-F3 lifting-body vehicle, piloted by NASA test pilot William H. Dana, successfully reached mach 0.7 after air launch from B-52 aircraft at 13 700-m (45 000-ft) altitude. Objectives of glide flight—first in M2-F3 series—were to check out pilot, evaluate center-fin modification, determine aileron characteristics, evaluate stability augmentation system modifications, and evaluate speed brakes. (NASA Proj Off)

- NASA launched three Aerobee 150 sounding rockets. Two from WSMR carried MIT experiment to study stellar spectra and Johns Hopkins

Univ. experiment to study UV spectra. One launched from Woomera Rocket Range, Australia, carried Univ. of Wisconsin experiment to conduct x-ray studies. (SR list)

- Review of *Apollo 13* accident was nearing completion and understanding of accident was good, LaRC Director Edgar M. Cortright, Chairman of *Apollo 13* Review Board, told press at MSC. Final report had been delayed pending completion of "critical tests being carried out which will help pin down some of the details of what took place." One of the key tests at MSC was related to special detanking and checkout procedures before launch. Tests had found faulty thermal switches and "demonstrated that if these thermal switches had failed . . . the temperatures . . . in the heater tube assembly could have exceeded 1000 degrees F. [800 K] in some spots," hot enough to bake teflon-coated wires and destroy insulation.

    Actual flight tank at Beech Aircraft Corp. would be "cycled back through the same series of detanking operations that took place on the 02 tank . . . from Apollo 13." Other tests were being conducted at MSC, ARC, and LeRC on "the ignition and combustion processes in the tank." Preliminary data suggested explosion resulted from failure of switches before launch and overheating of heater tube during detanking which destroyed insulation and sparked fire in tank. (Transcript)

- MSFC announced negotiation of $1.8-million supplemental agreement with IBM for Saturn-launch-vehicle programming computer requirements and development of computer programs through December 1972 and $2-million modification to contract with Martin Marietta Corp. for Skylab payload integration. Modification was for systems integration for multiple docking adapter (MDA). (MSFC Releases 70-109, 70-110)

- Dr. Alfred J. Eggers, Jr., NASA Assistant Administrator for Policy, delivered Minta Martin Lecture "Interactions of Technology and Society" at Stanford Univ. Lecture had also been delivered by Dr. Eggers at MIT on May 1 and at Johns Hopkins Univ. on May 21. Wendell Willkie's "One world" was rapidly becoming Buckminster Fuller's Spaceship Earth, socially, economically, technically and environmentally, Dr. Eggers said. "It is essential, therefore, that we begin to view the earth in all its finite reality and recognize that 'exponential projections' of the society of man and his works on earth are not only undesirable (if not impractical), but indeed they become unreal carried into the indefinite future. Accordingly, we do better to think in terms of the ultimate equilibrium we would like to see man achieve with and on spaceship earth, and from this vantage point we are drawn naturally to considerations of how best to approach and find the desired equilibrium." Population projections thus "would be expected to evolve . . . in accordance with satisfying the requirements of dynamic stability of the society of man in relation to his environment both natural and artificial. Technology, from earth-based to space-based, must play a massive role in this evolution because it will be crucial to effective earth resources management and distribution, and yet it must be compatible with the earth's environment." (Text; *A&A*, 10/70, 38-50)

- Discovery that blazing stars collided at rate of one every four months in heart of at least one kind of galaxy was reported from evidence obtained by Stratoscope II balloon-borne optical telescope March 26-27. Princeton Univ. physicist Dr. Martin Schwarzchild said collisions might be source of brightness fluctuations in Seyfert galaxies —rare heavenly masses with small, dense nuclei that radiated light and radio energy and bright lines produced by hot gases in motion. Stratoscope II had observed Seyfert galaxy at range of 30 light years from earth. (Lannan, W *Star,* 6/2/70)

- NASA announced publication of *The Moon as Viewed by Lunar Orbiter* (NASA SP-200). Book by Leon K. Kosofsky of NASA Apollo Lunar Exploration Office and Farouk El-Baz of Bellcomm, Inc., contained stereoscopic and other photos of Aristarchus, Schroter's Valley, Rimae Parry, and Tobias Mayer Dome—lunar craters and valleys. Photos were taken on five Orbiter missions preparatory to Apollo flights. (NASA Release 70-82)

June 3: Apollo 11 Astronaut Neil A. Armstrong, during U.S.S.R. visit, told Soviet Academy of Sciences that he favored increased space cooperation and that development of space stations and shuttles was "most important" method of practical space usage. "I have found in my discussions with my Soviet cosmonaut colleagues that their objectives in space are very much the same as ours." Armstrong said he would be glad to be member of U.S.-U.S.S.R. space crew. He felt *Soyuz IX* experiments in earth measurements would be useful to U.S. space program. In answer to question, Armstrong said he would ask "if I could take my family along" in volunteering for three-year Martian mission. (Clarity, *NYT,* , 6/4/70, 27)

- Edward H. White II Memorial Museum was dedicated at Hangar 9 of Brooks AFB, Tex., in honor of Astronaut White, who died in Jan. 27, 1967, Apollo spacecraft fire. Museum, near White's home town of San Antonio, Tex., would be devoted largely to aerospace medicine.

  In dedication address, L/G John W. O'Neill, AFSC Vice Commander, said: "Space is our new world, a final frontier. . . . it is also a medium through which attack is possible, which means that we must learn everything we can about it for the defense and security of the nation." Space science and technology had spawned "new generations of materials, systems and processes, for both military and consumer goods." Of 12 000 new products and techniques developed during last decade, "very large percentage derives directly from space and missile development."

  Most "satisfying and gratifying" space spinoffs were in health and medical field—"new type of wheel chair adapted from the moon walker. . .or chemicals used in the treatment of tuberculosis and mental illness that were originally derived from missile fuels. Or spray-on instrumentation for remotely monitoring the condition of heart patients. Or the numerous new materials and techniques being applied to. . .health care." (Text)

- Dr. Rafael Caldera, President of Venezuela, addressed joint session of U.S. Congress: "I am convinced that the future of the hemisphere depends on the extent to which this great nation reaches a decision to become a pioneer in social international justice. The measure to which your people. . .become conscious of the fact that with the cost of one of its Apollo moon shots it could contribute to the prosperity and happiness of nations like ours on whose security its own security depends." Venezuelans hoped Apollos would continue exploring space, "but the result of these explorations make the need for a better life for men on earth more urgent." (CR. 6/3/70, H5025-7)

- Baltimore *Sun* editorial on Soviet space platform program: "How far along towards realization it is and what Soyuz 9 will contribute to it can only be matters of conjecture. What is impressive is the methodical way the Russian space directors are concentrating their efforts and resources. While there has been nothing eyecatchingly novel or adventurous about the progress so far, the program seems concerned not with the public response, either in Russia or the rest of the world, but with a gradual, systematic achievement of a single, selected goal." (B *Sun*, 6/3/70)

June 3-5: Dr. Thomas O. Paine, NASA Administrator, headed group of NASA and industry officials who met in Paris at request of European Space Research Organization (ESRO) to brief ESRO representatives and European scientific, industrial, and government representatives on space station and related systems. Dr. Paine told meeting U.S. wished to expand cooperation with European nations. He said space station should grow over next decade into very important international research facility. (NASA Release 70-83; *Av Wk*, 6/8/70, 16)

Arnold W. Frutkin, NASA Assistant Administrator for International Affairs, said: "Where men are concerned in utilizing the space facilities of the future...we are prepared to accommodate astronauts of other countries in these programs." NASA assumed there would be "an interest in training pilot-astronauts in conjunction with some expectation to use them for your own experimental or

June 3-5: *NASA and industry officials briefed European Space Research Organization representatives in Paris on the space station and related systems. Left to right at the speakers' table were Charles W. Mathews, NASA Deputy Associate Administrator for Manned Space Flight; Prof. Hermann Bondi, Director General of ESRO; Dr. Thomas O. Paine, NASA Administrator; Prof. Giampietro Puppi, Chairman of the ESRO Committee of Senior Officials; Jean-Albert Dinkespiler, ESRO Director of Planning and Programs; and Dr. Johannes Ortner, ESRO Director of Space Missions. (ESRO Photo)*

applications objectives. We would be very happy to cooperate in that kind of training, and we would be prepared to see pilot astronauts fly the space shuttle after a test period establishes the operational status of the vehicle." (Transcript)

June 4: Apollo 13 Astronauts James A. Lovell, Jr., Fred W. Haise, Jr., and John L. Swigert, Jr., were honored by City of New York at luncheon in Philharmonic Hall. Astronauts were greeted by Mayor John V. Lindsay and serenaded by N.Y. Police Dept. band. (AP, B *Sun,* 6/4/70, A3)

- MSC announced contract awards: NR Space Div. would receive $305 700 000 cost-plus-fixed-fee/award-fee supplemental agreement to Apollo spacecraft contract for four CSM modules for Skylab program. Agreement definitized March 1969 letter contract and brought estimated cost of contract for both Apollo and Skylab to $3 618 006 813. NR Space Div. also would receive $250 000 for preliminary planning study of reusable space tug with multipurpose applications.

  ILC Industries would receive $13-million extension to contract for Apollo space suits and support through 1971, bringing total contract value to $46 543 299. (MSC Releases 70-61, 62, 63; NASA Release 70-88)

- AFSC announced senior staff changes: Gen. James Ferguson, AFSC Commander, would retire Sept. 1. Gen. George S. Brown, Deputy Commander, Seventh Air Force (PACAF), would succeed Gen. Ferguson. L/G James T. Stewart, AFSC Deputy Chief of Staff for Systems, had been reassigned as Commander, Aeronautical Systems Div. (ASD), Wright-Patterson AFB, Ohio, effective June 1, replacing M/G Lee V. Gossick who assumed Gen. Stewart's previous duties. M/G Clifford J. Kronauer, Jr., Commander, Space and Missile Test Center, would become AFSC Chief of Staff on Aug. 1. M/G Louis L. Wilson, Jr., Vice Commander, Armament Development and Test Center, Eglin AFB, Fla., would become Vice Commander, Ninth Air Force on Aug. 1. B/G Jack Bollerud, Deputy Chief of Staff, Bioastronautics and Medicine, had announced retirement effective July 31. (AFSC Release 156.70)

- Robot developed by UCLA engineer Dr. Amos Freedy mimicked human motions, *Huntsville Times* reported. Autonomous control subsystem (ACS) learned, thought, and made decisions. Consisting of manipulator arm and claw in rigging operated by small analog computer connected to IBM 1800 process-control computer several floors away, machine might be adapted to use on unmanned space flights, in rehabilitation of handicapped persons, and in housework. (Latty, *Huntsville Times,* 6/4/70)

June 5: NASA and Astronomical Netherlands Satellite Program Authority agreed on cooperative project to launch first Netherlands scientific satellite, ANS, on four-stage solid-fueled Scout booster from WTR into near sun-synchronous orbit in August 1974. Satellite would carry U.S. and Netherlands experiments to study stellar UV and x-ray sources. Spacecraft would be designed and developed in Holland. NASA would provide one experiment, the booster, launching, tracking, and emergency data acquisition. (NASA Release 70-91)

- Team of NASA, Agency for International Development (AID), and university contract personnel left U.S. for 10-day visit to Korea in project to test feasibility of transferring NASA-sponsored industrial technology into developing country. Team—which included Ronald J. Philips, Director of NASA Technology Utilization Div.—would work out guidelines for Korean team to study way U.S. agencies transferred technology to industrial, educational, and medical communities. (NASA Special Release)

- *Apollo 11* Astronaut Neil A. Armstrong's May 24-June 5 visit to U.S.S.R. was reviewed in *Christian Science Monitor:* "To those few Russians who saw him, the first man to walk on the moon was unquestionably a big hit. But overall, his 12-day goodwill visit was an unobtrusive affair that stayed very much in the realm of Soviet officialdom."

    Armstrong had arrived in Leningrad May 24 to attend May 20-29 COSPAR Conference. "Although he was in Leningrad for five days, only one Leningrad newspaper made any mention of him and Russians for a long time were not aware he was even in the country." It was only after his arrival in Leningrad that Armstrong was invited by Soviet government to stay longer in U.S.S.R. He had "no exposure to Soviet public" during Novosibirsk visit. In Moscow, press had given more coverage. Armstrong had told press he had been "most emotionally moved" by meeting with Mrs. Yuri Gagarin and Mrs. Vladimir Komarov, widows of Soviet cosmonauts. (Saikowski, *CSM*, 6/5/70)

- Dr. George M. Low, NASA Deputy Administrator, spoke at dedication ceremony for $2.8-million Life Support Technology Laboratory at LaRC. He said decade of space exploration had been characterized by progress in scientific knowledge and practical benefits to man. New laboratory would contribute to future U.S. efforts to learn more about universe. Laboratory was dedicated "to the welfare of man in space and to the solution of the problems that must be overcome before he can stay there." (*Langley Researcher*, 6/12/70, 1)

- Authorization to proceed with B-1 Advanced Strategic Bomber development programs was issued by Deputy Secretary of Defense David M. Packard in memo to Secretary of the Air Force, Dr. Robert C. Seamans, Jr. It would be "several years" before production decision would be made. "Factors to be considered . . . will include the progress and success of the engineering development program, the progress of the SALT talks, and the relationship of the program to these talks."

    Dr. Seamans announced selection of North American Rockwell Corp. and General Electric Co. as airframe and propulsion contractors for engineering development of bomber. Total estimated cost of airframe contract was $1 350 814 739 exclusive of avionics. Total estimated propulsion cost was $406 654 000. (Text; DOD Release 479-70)

- Discovery that glassy portion of 4.6-billion-yr-old lunar sample 12013 [see May 26] was chemically more like some tektites from Java than like any terrestrial igneous rock was reported in *Science* by GSFC scientist Dr. John A. O'Keefe. Glassy portion of *Apollo 12* moon rock satisfied all chemical criteria for tektite. "Tektites are relatively recent and acid rocks, whereas the moon is chiefly ancient and basaltic; hence, tektites are probably ejected volcanically, rather than by impact, from the moon." (*Science*, 6/5/70, 1209-10)

June 5
- Selection of General Electric Co. Space Div. and Link Div. of Singer-General Precision, Inc., for competitive negotiations leading to $3.5-million contract for development of crew training simulator for Skylab program at MSC was announced by NASA. Cost-plus-award-fee contract would include design, fabrication, installation, checkout, simulation programs, on-site systems engineering, and supporting documentation. (NASA Release 70-93)
- MSC announced award of $210 000 fixed-fee contract to Lockheed Aircraft Corp. for study of safety methods and rescue techniques for lunar operations. (MSC Release 70-65)
- AP said Tass had reported construction of meteorological rocket launch complex at Molodezhnaya, Soviet station in Antarctica. Powerful radio was being constructed to speed information to Moscow and other world weather centers. (*C Trib*, 6/6/70)
- Tass said Cosmonaut Valentina Nikolayeva-Tereshkova, wife of *Soyuz IX* Commander Andrian G. Nikolayev, was candidate for U.S.S.R.'s Supreme Soviet. Mrs. Nikolayeva-Tereshkova, who orbited earth 48 times aboard *Vostok VI* mission June 16, 1963, had been meeting with electorate in preparation for June 14 election. (Reuters, *B Sun*, 6/6/70)
- Reuters reported *Red Star* interview during which *Soyuz IX*'s unidentified chief designer said, "The Soyuz series are already working spacecraft. They are not being tested but are a means of exploration...." One of Soyuz's tasks was to work out systems for future orbiting stations. Designer described Soyuz interior: "Convenient sofa stands at one end of the cabin. Opposite is an original 'working office' with a table and sideboard. Finished in mahogany, they are handsome and harmonious. A shelf with books and microfilms, control desks, portholes and scientific instruments complete the interior." (*B Sun*, 6/6/70)
- Exhibit of 23 paintings of space subjects by Chesley Bonestell opened at American Museum-Hayden Planetarium in New York. (Am Museum-Hayden Planetarium Release)

June 6: Apollo 11 Astronaut Neil A. Armstrong and wife, Janet, participated in ceremonies in Rome, Italy, during which Mrs. Armstrong christened Alitalia Boeing 747 named for her husband. Later Armstrong and wife flew in 747 *Neil A. Armstrong* from Rome to New York. (*C Trib*, 6/6/70)
- NASA released *Ecological Surveys from Space* (NASA SP-230). Monograph listed feasible uses of spacecraft in 1970s and described scientific and technical studies of new images and concepts of earth acquired by venture into space. Publication contained photos taken from aircraft and from Gemini and Apollo spacecraft. (Text)
- Alaska Airlines inaugurated "Gold Samovar" service between Anchorage, Alaska, and central U.S.S.R. Series of 10 summer flights marked first commercial route and charter authority negotiated by private airline company with Soviets, according to Alaska Airlines Board Chairman and Chief Executive Officer Charles F. Willis, Jr. (Valentine, *W Post*, 6/16/70, C2)
- Aegir, 180-metric-ton (200-short-ton) undersea habitat with six-man crew, reached surface after more than 120 hrs on ocean floor off coast of Oahu, Hawaii. Aegir was to have ascended from 157-m (516-ft) depth on June 5 but pressure valve in ballast tank malfunc-

tioned. Habitat was refloated by rigging bypass into ballast tanks and feeding compressed air from surface support ship into tank. (*W Post*, 6/8/70, A4)

June 7: Astronaut Charles Conrad, Jr., on behalf of *Apollo 12* crew, received special Emmy Award from National Academy of Television Arts and Sciences in New York for TV photography during *Apollo 12* lunar landing. (AP, W *Star*, 6/8/70, B9)

June 7-10: American Astronautical Society held 16th annual meeting in Anaheim, Calif. Francis M. Stewart, MSFC Shuttle Engine Manager, and Ryndal L. Wetherington, NASA Hq. Space Shuttle Propulsion Manager, presented paper on main propulsion requirements for fully reusable space shuttle engine concept and new approaches to engine development and testing. Throttling to only 50% thrust satisfied system's needs and was baseline requirement. Value of one second of specific impulse had been calculated at $25 000 000. Payload sensitivity of space shuttle concepts under study was 6900-8900 newtons (1500-2000 lbs per sec) of impulse. Minimum impulse per second in basic requirements represented 96% efficiency, "which has not been achieved in any rocket engine to date."

Reusability requirement might be most difficult of all. "One can speculate that if we can do it in aircraft engines, why not rocket engines? But we are talking about heat fluxes and turbomachinery characteristics in the shuttle engine that extend well beyond those experienced in the engines of the [Boeing] 707, 747, or even the military high performance aircraft. It is yet to be shown if we can fly 100 missions (or even 10 or 20) with this high performance engine, and achieve the confidence and maintainability which the airline operations have in hand. . . .reusability adds a new dimension to rocket engine development." (Text)

Prof. Yu. K. Khodarev, Deputy Director of Institute for Space Research of Soviet Academy of Sciences, and Dr. Valery Yevdokimov, head of Academy's data processing dept., called for studies of what U.S. and U.S.S.R. could do to establish and operate scientific base on moon. Prof. Khodarev said: "We would like to see concrete lists from American scientists of instruments that we could place on the moon for research. Then we can explore further avenues for cooperation in space. We can't decide how to transport things there until we first decide what we want to take." There was general acceptance of U.S.S.R. space efforts among Soviet people "but people are looking for a direct payoff—they want something for their money." Khodarev said U.S. and U.S.S.R. could "cooperate in a Martian program as well as in intense investigations on the moon." He suggested area for primary investigation from lunar station would have to be decided. "Do we want to look at the moon, or do we want to look into the cosmos from the surface?" During panel discussion Robert L. Lohman of NASA's Space Station Task Force invited Prof. Khodarev to attend NASA meeting on advanced space programs at ARC later in year.

Dr. Takao Ueda of Japan said Japan was working on comsats, TV satellites, and geosynchronous weather spacecraft. (AAS Paper No. 70-044; Bane, LA *Her-Exam*, 6/11/70; NASA Sci and Tech Info Div)

June 8: First radar receiver developed by Naval Research Laboratory scientists was presented to Smithsonian Institution's Hall of Electricity by Capt. James C. Matheson, Director of NRL. Device had been successfully tested first on April 28, 1936, when it detected and ranged aircraft in flight at 40-km (25-mi) distance. (*Naval Research Reviews*, 7/70, 31)

- Milky Way galaxy was losing gravitational energy at rate equivalent to mass of 1000 suns annually, Maryland Univ. physicist Dr. Joseph Weber said during third Cambridge Conference on Relativity at Institute for Space Studies in New York. Dr. Weber had measured flow of gravitational energy and computed energy being lost from source of waves in galaxy's center. Results indicated that eventually a star like the sun, at galaxy's edges, would break free of weakening gravity and fly off into space. One estimate had placed event at hundreds of millions of years in future. (AP, B *Sun*, 6/10/70, A3)

June 9: NASA successfully conducted second drop test of 12.5K Space Orbiter Shuttlecraft at WSMR. [See May 27]. (NASA Proj Off)

- USMC Sikorsky S-65 helicopter set record for intercity travel in 1-hr 17-min 11-sec flight between Wall Street heliport in New York and Washington Hospital Center in Washington, D. C. Flight was organized by Pan American World Airways and Sikorsky Aircraft Corp. to demonstrate feasibility of inter-inner-city helicopter transportation. (Cushing, *W Post*, 6/10/70, D10)

- Nomination by President Nixon of *Apollo 10* astronauts Cdr. John W. Young and Cdr. Eugene A. Cernan, both USN, for permanent promotion to captain was announced by White House.

  President and Mrs. Nixon honored *Apollo 13* astronauts and wives at White House dinner. (*PD*, 6/15/70, 756)

- USAF launched experimental reentry vehicle down AFWTR from Vandenberg AFB. Launch was part of Advanced Ballistic Reentry System (ABRES) program. (UPI, *NYT*, 6/10/70, 6)

- MSC announced selection of United Aircraft Corp's Pratt & Whitney Div. and General Electric Co.'s Direct Energy Conversion Business Section to negotiate parallel cost-plus-fixed-fee contracts for development of fuel-cell technology leading to design, fabrication, and testing of engineering-model fuel-cell system. Contracts, which would run for 13 mos beginning July 1, were valued at $825 000 each. MSC program to provide electrical power for space shuttle was aimed at development of hydrogen-oxygen fuel cell with 5000-hr lifetime, 5000-w power output, and weight of 333 kg (735 lbs) or less. (MSC Release 70-66)

June 10: U.S.S.R. launched *Cosmos CCCXLVI* from Baykonur into orbit with 351-km (218.1-mi) apogee, 206-km (128.0-mi) perigee, 90.0-min period, and 51.8° inclination. Satellite reentered June 17. (GSFC *SSR*, 6/30/70; *SBD*, 6/11/70, 196)

- NATO's *Nato I*, launched by NASA March 20, arrived on station over Atlantic with all systems operating satisfactorily. (NASA Proj Off)

- U.S. position in V/STOL development was described by Calvin B. Hargis, Jr., Deputy for Development, Office of Assistant Secretary of the Air Force (R&D), before American Helicopter Society at Wright-Patterson AFB, Ohio: "Over the past 20 years, the United States has spent between 500 and 750 million dollars on V/STOL programs. . . ." USAF had spent $250 million since 1962. "U.S.-FRG [Federal Republic of Germany] fighter program was terminated largely due to high cost and direct competition for funds with the F-

15. The LIT received a severe setback this year due to, again, lack of priority in the funding arena and lack of support from the using command; and the XC-142, while partially successful, was not approved for production. Only the British Harrier, of all the recent V/STOL programs, is in a production status in the free world." U.S. had "built a large number of test vehicles, evaluated many different concepts, established a technology base, and spent hundreds of millions of dollars," but had "no operational systems in being yet." (Text)

- New Soviet agency, Institute of the United States of America, created by Kremlin mandate as arm of Soviet Academy of Sciences, was examining U.S. economic, political, and social matters for edification of Soviet political scientists, economists, and government leaders, *Wall Street Journal* said. U.S. diplomat in Moscow had said institute "seems to be trying to present a more accurate picture of America than people here have been getting. And it doesn't seem to be a spy outfit." (Vicker, *WSJ*, 6/10/70)

June 11: NASA's HL-10 lifting-body vehicle, piloted by Maj. Peter A. Hoag (USAF), successfully completed 36th flight at FRC, demonstrating and evaluating powered approach to landing for first time. After air launch from B-52 aircraft at 13 700-m (45 000-ft) altitude, HL-10 reached mach 0.07 and fired three hydrogen-peroxide rocket engines to approach landing on 6° glide slope instead of usual 18° slope used on unpowered flights. (NASA Proj Off)

- Arrangement for mutual satellite tracking support had been made by NASA and French Centre National D'Études Spatiales (CNES), NASA announced. Each agency would provide free network support in tracking, data acquisition, and command activities of space vehicles when workload permitted. For first time, French equipment was compatible with U.S. Arrangment was applicable to French stations in Canary Islands, Upper Volta, Congo, South Africa, French Guiana, and France. NASA stations would be in STADAN network in Australia, Madagascar, South Africa, Ecuador, Chile, England, and U.S. (NASA Release 70-94)

- SAMSO, Aerospace Corp., and NR Space Div. had completed study of solution to in-space emergencies involving near-orbiting spacecraft and space shuttle vehicles, AFSC announced. Study was based on requirements that space escape vehicle (SEV) must perform all functions required for crew safety from decision to abandon spacecraft until successful recovery. In delayed return, ground station would supply data to aid selection of recovery site; sufficient oxygen and water to last 24 hrs after departure from spacecraft would be available. In "quick return," life support would be provided for only six hours. SEV would become autonomous following issue of "May Day" signal and would splash down without ground-based advice. Crew would plot recovery course using wristwatches and information indicating orbital paths for 24 hrs. Study recommended three kinds of escape vehicle: spherical heat shield, rigid vehicle thermally protected by ablative heat shield; rib-stiffened expandable vehicle which would function like umbrella; and two-man conical reentry vehicle like rubber raft, inflatable and made of double-wall fabric connected by woven drop threads. (AFSC Release 137.70)

- Feasibility of high-speed, efficient, digital message transmission through troposphere had been confirmed by AFSC scientists, AFSC announced. Findings were result of research program conducted at AFSC Rome Air Development Center near Niagara Falls, N.Y. (AFSC Release 151.70)
- AFSC announced successsful flight testing of night strip photography, which produced continuous strip picture rather than individual frames. System, designed and fabricated by Linde Div. of Union Carbide Corp., used reflector that projected narrow beam to ground to illuminate moving strip of terrain and lamp with spiraled-gas-flow-through-arc technique, providing intense "line of light." System had been tested in B-47E aircraft from May through August 1969. (AFSC Release 13.70)
- Wallops Station announced award of $450 000, cost-plus-fixed-fee contract to Computer Sciences Corp. for engineering support services at Wallops. Contract covered one year with provisions for two one-year extensions. (WS Release 70-8)
- House passed by vote of 308 to 57, H.R. 17970, FY 1970 DOD military construction appropriations bill totaling $637.9 million, after rejecting amendment to strike out $353.8 million for Safeguard ABM deployment. (CR, 6/11/70, H5437-57)
- U.S.S.R. had established underwater launching area for submarine missiles targeted on U.S. in Atlantic Ocean off Greenland, *Washington Post* said. Soviet Y class submarines—equivalent of U.S. Polaris missile submarines—had begun to patrol area in last few weeks. (Wilson, *W Post*, 6/11/70, A1).
- NAS and Bulgarian Academy of Sciences announced beginning of two-year exchange program providing for visits of up to one year by individual scientists to lecture and conduct research. (NAS Release)

June 12: U.S.S.R. launched *Cosmos CCCXLVII* into orbit with 2005-km (1245.9-mi) apogee, 217-km (134.8-mi) perigee, 107.5-min period, and 48.4° inclination. (GSFC *SSR*, 6/30/70)
- Three-stage ELDO booster built by U.K., France, and West Germany successfully lifted off launch pad at Adelaide, Australia, but Italian-built test satellite failed to enter orbit. ELDO spokesman said launch would still be considered successful. (Reuters, *W Post*, 6/13/70, A2)
- FAA announced proposal of new rule requiring all U.S. powered aircraft to be equipped with approved anticollision lights for night operation and higher intensity standards for anticollision lights on newly certified aircraft. (FAA Release T 70-26)
- Activation of new high-speed telecommunications message switching center—key element in North Atlantic and Caribbean Aeronautical Fixed Telecommunications Network (AFTN)—was announced by FAA. Center eliminated intermediate relay points, stored messages for delivery to users under priority system, and retransmitted urgent messages almost instantaneously. (FAA Release T 70-27)
- USAF K-30, one of largest aerial reconnaissance cameras, had been retired to Air Force Museum at Wright-Patterson AFB, Ohio, AFSC announced. Camera, which had more than 20 yrs service, measured 0.9 by 1.4 by 1.5 m (3 by 4½ by 5 ft), weighed 300 kg (665 lbs), and used 254-cm (100-in) focal-length optical system. (AFSC Release 153.70)

June 13: Cosmos *CCCXLVIII* was launched by U.S.S.R. into orbit with 540-km (335.5-mi) apogee, 196-km (121.8-mi) perigee, 91.8-min period, and 70.9° inclination and reentered July 25. (GSFC *SSR*, 6/30/70; 7/31/70)
- In LaRC test to develop regenerative life support systems for advanced manned spacecraft, four crewmen were sealed into space station simulator at McDonnell Douglas Astronautics Co. Div. of McDonnell Douglas Corp. in Huntington Beach, Calif. During 90-day experiment, part of NASA OART program, crew would drink reclaimed water and breathe regenerated oxygen. Information obtained would be applied to operation of three-man Skylab workshop scheduled for 1971 launch and to design of 12-man space station for later in 1970s. (*Langley Researcher,* 6/26/70, 1)
- UPI reported in *Washington Post* that NASA report not yet made public acknowledged General Electric Co. was permitted to submit bid for controversial $50-million ATS-F and ATS-G contract one week later than the deadline its competitor Fairchild Hiller Corp. was required to meet. NASA denied knowledge of any leak to GE of Fairchild Hiller's design and price. Report said Dr. Thomas O. Paine, NASA Administrator, had been told before he announced award of contract to GE on April 8 that both companies had submitted bids on same date —March 6. Actually, UPI said, Fairchild Hiller proposal had been submitted Feb. 27—NASA deadline—and copies were circulated among NASA personnel before GE proposal was submitted March 6. (Cassels, UPI, *W Post,* 6/13/70, E8)

June 14: Success of Oct. 8, 1969, experiments to test validity of Einstein theory of relativity with radiotelescopes for first time was reported to press by team of Cal Tech astronomers. Einstein equations inferred that light, radio, and other electromagnetic waves were deflected by sun's gravitational field at rate of 1.7 secs of arc. Cal Tech team, using 64-m-dia (210-ft-dia) dish-shaped radio antennas at two observatories near Los Angeles, had obtained figures of 1.82 and 1.77 secs, which astronomers credited with accuracy within 10%. Previous optical measurements had contained only 20% accuracy. (AP, *W Post,* 6/15/70, A8)
- Nationwide TV viewers in U.S.S.R. saw *Soyuz IX* Cosmonauts Andrian G. Nikolayev and Vitaly I. Sevastyanov in spacecraft cast ballots in Soviet parliamentary election. (*W Post,* 6/15/70, A24)
- Robert H. Gray, Director, Unmanned Launch Operations at KSC, became Deputy Director of KSC Launch Operations. John J. Neilon, Deputy Director, Unmanned Launch Operations, assumed post of Acting Director, Unmanned Launch Operations. (KSC Ann)

June 15: *Apollo 13* Review Board accident report was released by NASA at Hq. news conference. LaRC Director Edgar M. Cortright, Chairman of Review Board, said accident "was not the result of a chance malfunction in a statistical sense but, rather, it was the result of an unusual combination of mistakes coupled with a somewhat deficient and unforgiving design."

After assembly and acceptance testing, oxygen tank No. 2 that flew on *Apollo 13* had been shipped from Beech Aircraft Corp. to NR in apparently satisfactory condition. "It is now known, however, that the tank contained two protective thermostatic switches on the heater assembly which were inadequate and would subsequently fail during ground test operations. . . . In addition, it is probable that the tank contained a loosely fitting fill tube assembly. . .which was loose when it left the plant. . .[and] was subsequently displaced during handling. . .[and] led to the use of improvised detanking procedures. . .which almost certainly set the stage for the accident."

Special detanking procedures at KSC subjected tank to extended heater operation. "These procedures had not been used before and the tank had not been qualified by test for the conditions experienced. However the procedures did not violate the specifications which governed the operation of the heaters. . . . In reviewing these procedures before the flight, officials. . .did not recognize the possibility of damage due to overheating.

"The thermostatic switch discrepancy was not detected. . . . Nor did tests identify the incompatibility of the switches with the ground support equipment at KSC. . . . As shown by subsequent tests, failure of the thermostatic switches probably permitted the temperature of the heater tube assembly to reach as much as 1000 degrees Fahrenheit [810 K] in spots during the continuous eight-hour period of heater operation. From that time on the oxygen tank. . .was in a hazardous condition when filled with oxygen and electrically powered.

"However it was not until nearly 56 hours into the mission that the fan motor wiring possibly moved by the fan stirring of the fluid contents of the tank short circuited and ignited the wiring insulation by means of an electric arc. The resulting combustion in the oxygen tank probably overheated and failed the wiring conduit where it enters the tank and possibly a portion of the tank itself. The rapid expulsion of high pressure oxygen which followed possibly augmented by combustion of insulation in the space surrounding the tank blew off the outer panel. . .of the service module, caused a leak in the high pressure system of Oxygen Tank Number 1, damaged the high gain antenna, caused other miscellaneous damage and aborted the mission."

Board recommended that cryogenic oxygen storage system be modified to minimize use of combustible materials and remove materials that might short-circuit from contact with potential ignition sources and that system be rigorously requalified; onboard warning systems be reviewed and modified; consumables and emergency equipment be reviewed to determine whether potential for use in "lifeboat" mode should be enhanced; MSC continue special tests and analyses under way to understand better the accident details; standard procedures require "presentation of all prior anomalies" whenever significant anomalies occurred in critical subsystems during final preparations for launch; NASA thoroughly reexamine all spacecraft, launch vehicles, and ground systems to identify and evaluate potential combustion hazards and conduct additional research on materials compatibility, ignition, and combustion in strong oxidizers at various g levels; and MSC reassess all Apollo spacecraft subsystems and engineering organizations responsible for them to ensure adequate understanding of subsystems at subcontractor and vendor level.

Cortright said thermostatic switch discrepancy and incompatibility of switches with KSC ground support equipment was "a serious oversight in which all parties [NASA, NR, and Beech] shared. . . . The accident is judged to have been nearly catastrophic. Only outstanding performance on the part of the crew, Mission Control and other members of the team which supported the operations successfully returned the crew to earth." (Transcript; text)

- U.S.S.R.'s *Soyuz IX* (launched June 2) set new manned flight endurance record at 6:35 pm Baykonur time, when it surpassed 13-day 8-hr 35-min record set by NASA's *Gemini VII* (Dec. 4-18, 1965).

    Apollo Astronauts Frank Borman and James A. Lovell, Jr., sent congratulatory telegram to *Soyuz IX* crew: "Your accomplishments are adding new evidence supporting the capability of men to live and work in extended periods in space. We wish you continued success in this important mission and a safe return to earth."

    Lovell told press in Houston that mission suggested U.S.S.R. would develop permanent earth-orbiting space station: "Soyuz now seems to be purely an aeromedical flight, pushing men to their limits to see how weightlessness and the pure oxygen environment affects their work in space. It all fits very naturally with the space station concept." (Astrachan, *W Post*, 6/16/70, A1; *SBD*, 6/16/70, 211)

- *Itos I* (Tiros M), Improved Tiros Operational Satellite launched by NASA from WTR Jan. 23, 1970, was turned over to ESSA for operational use after five-month engineering checkout by NASA. (NASA Release 70-96)

- House and Senate conferees on H.R. 16516, FY 1971 NASA authorization, reached agreement and filed conference report H. Rpt. 91-1189. Report recommended $3.411-billion NASA authorization containing $2.693 billion for R&D, $34.5 million for construction of facilities, and $683.3 million for research and program management, of which not more than $506.1 million would be for personnel costs.

    R&D total included $994.5 million for Apollo program (increase of $38 million over NASA request), $565.2 million for space-flight operations (increase of $50 million), $1.5 million for advanced missions program (decrease of $1 million), and $116 million for physics and astronomy program (as requested by NASA), restoring $5.6 million which House had deleted for Explorer satellites.

    Conferees approved NASA request of $167 million for Space Applications and agreed to authorize revised plan for construction of experimental Earth Resources Technology Laboratory at GSFC at $1.9 million cost. (*CR*, 6/15/70, H5490-3; Text)

- NASA announced it would negotiate with Grumman Aerospace Corp., Lockheed Aircraft Corp., and Chrysler Corp. for 11-mo Phase A (feasibility) contracts to study space shuttle concepts. Boeing Co. would be major subcontractor to Grumman on $4-million contract for three shuttle concepts: stage-and-a-half shuttle of reusable manned spacecraft with onboard propulsion system and droppable tanks for supplementary propellants; reusable orbiter with expendable booster; and reusable 1st stage using existing J-2S engine technology, solid-propellant auxiliary boosters, and reusable 2nd-stage orbital shuttle with J-2 engine.

Lockheed's $1-million fixed-price contract would define alternate stage-and-a-half shuttle system including high and low cross-range designs.

Chrysler Corp. would study reusable vehicle capable of placing payload into earth orbit with single stage, under $750 000 contract. (NASA Release 70-97)

- Dr. Eberhard F. M. Rees, MSFC Director, welcomed 59 science and engineering professors from universities and colleges in 26 states who would participate in Summer Faculty Fellowship programs sponsored by NASA and American Society for Engineering Education. Research program from June 15 to Aug. 21 would be followed by 11-wk design program, both conducted by Auburn Univ. and Univ. of Alabama. Participants would perform individual research in MSC laboratories. Work would be documented for NASA use. (MSFC Release 70-115)

- Staran IV system to reduce midair collisions—by computer performing more than 40 million mathematical operations per second to predict collision courses and determine evasive maneuvers—was introduced at Washington, D.C., press conference by developer, Goodyear Aerospace Corp. System could single out aircraft on collision course and show them to air traffic controllers on viewing screen as though they were only aircraft in air. FAA was studying proposal for adoption of system. (Goodyear Aerospace Corp Release)

- Intention to create separate FAA office to deal with total air transportation security problems, including aircraft hijacking, was announced by Secretary of Transportation John A. Volpe. (FAA Release 70-57)

June 16: Edgar M. Cortright, LaRC Director and Chairman of *Apollo 13* Review Board, summarized report of Review Board before House Committee on Science and Astronautics and commented: "Total Apollo system of ground complexes, launch vehicle, and spacecraft constitutes the most ambitious and demanding engineering development ever undertaken by man. For these missions to succeed, both men and equipment must perform to near perfection. That this system has already resulted in two successful lunar surface explorations is a tribute to those men and women who conceived, designed, built, and flew it. Perfection is not only difficult to achieve, but difficult to maintain. The imperfections in Apollo 13 constituted a near disaster, averted only by outstanding performance on the part of the crew and the ground control team which supported them. The Board feels that the Apollo accident holds important lessons which, when applied to future missions, will contribute to the safety and effectiveness of manned space flight." (Testimony)

- Sensor developed primarily for V/STOL aircraft by ERC's Aircraft Hazard Avoidance Programs Office was being considered by Dept. of Interior's Bureau of Mines for use as mine safety instrument, NASA announced. Intended to measure airspeed during hovering, sensor would provide accurate reading of slow air flow within mine. (NASA Release 70-90)

- Discovery of hydrogen cyanide in radio emissions from space was described by Dr. David Buhl of National Radio Astronomy Observatory at Kitt Peak, Ariz., during interview with *New York Times*. Dr. Buhl, with Dr. Lewis E. Snyder of Univ. of Virginia, had used 11-m-dia (36-ft-dia) dish antenna at Kitt Peak to locate second or-

June 16: *Rep. George P. Miller, Chairman of the House Committee on Science and Astronautics, examined an Apollo spacecraft oxygen tank with Dr. Thomas O. Paine (right), NASA Administrator, and Director Edgar M. Cortright (left) of Langley Research Center before the Committee convened to hear a report of the Apollo 13 Review Board. Cortright, Chairman of the Review Board, summarized investigations into the oxygen tank explosion that had aborted the Apollo 13 mission on its way to a lunar landing.*

ganic substance identified in radio emissions. They had discovered existence of formaldehyde in same well-defined clouds of dust and gas in 1969 and believed both substances might have been key ingredients in early evolution of life. (Sullivan, *NYT*, 6/17/70, 30)
- British geophysicist Dr. Sydney Chapman died in Boulder, Colo., at age 82, after heart attack and stroke. Author of classic theory to explain how eruptions on sun caused magnetic storms and auroral displays on earth, Dr. Chapman had been President of Comité Spéciale de l'Année Géophysique Internationale (CSAGI) which managed International Geophysical Year of 1957-58. (Sullivan, *NYT*, 6/20/70, 29)
- Astronauts Alan B. Shepard, Jr., and Edgar D. Mitchell tested Apollo LM in vacuum chamber at KSC in preparation for Apollo 14 lunar landing mission. Mission was scheduled for Dec. 3 pending evaluation of *Apollo 13* Review Board Report. (UPI, *NYT*, 6/18/70, 9)

June 16-17: Final critical-design review of manned lunar roving vehicle was held at MSFC to decide on final design configuration leading to hardware manufacture and testing. First of four flight-model lunar rovers was scheduled to fly aboard Apollo 16 LM, set for 1971 launch. (MSFC Release 70-112; MSFC PAO)

June 16-17: *Final critical design review of the manned lunar roving vehicle was held at MSFC to decide on a configuration leading to hardware manufacture and testing. Photographed June 11, Astronauts Charles M. Duke, Jr. (left), and John W. Young ran through practice exercises with a full-scale model in crew station review.*

June 17: U.S.S.R. launched *Cosmos CCCXLIX* from Plesetsk into orbit with 350-km (217.5-mi) apogee, 203-km (126.1-mi) perigee, 89.8-min period, and 64.4° inclination. Satellite reentered June 25. (GSFC *SSR*, 6/30/70; *SBD*, 6/18/70, 224)

- NASA's X-24A lifting-body vehicle, piloted by NASA test pilot John A. Manke, successfully completed 14th flight from FRC. Vehicle was launched from B-52 aircraft at 12 800-m (42 000-ft) altitude and reached 19 800 m (65 000 ft) and mach 0.95. Data on lateral directional derivatives at mach 0.9 and longitudinal trim and lift-to-drag ratio were obtained. (NASA Proj Off)

- USN successfully launched 19th Poseidon missile from ETR toward Atlantic Ocean target in preparation for first launch from submarine in late July. (UPI, *NY News*, 6/18/70, 8)

- James H. Wyld Propulsion Award was presented to Hans G. Paul, Chief of MSFC Propulsion Div. and Joseph G. Thibodaux, Jr., Chief of MSC Power and Propulsion Div. during AIAA 6th Propulsion Specialist Conference in San Diego, Calif. Paul was cited for "outstanding leadership in research and development of Saturn launch vehicle propulsion"; Thibodaux, for "outstanding leadership in research and development of Apollo spacecraft propulsion." (MSFC Release 70-114)

- NASA announced it would initiate studies and analyses of satellite and other data on Gulf Coast area in FY 1971 at MTF. In Washington, D.C., Dr. Thomas O. Paine, NASA Administrator, said activities would use data from Earth Resources Technology Satellites A and B, Apollo manned flights, and aircraft operations. Study would be supported by MTF laboratory and Slidell, La., computer facility in line with earlier NASA invitation to other Government agencies to collocate study and analysis activities at MTF. (NASA Release 70-98)
- Maintenance and operation of NASA Deep Space Facility 61, near Madrid, had been assumed by Spanish engineers and technicians, NASA announced. Similar facility nearby had been consigned to Spanish operation in 1969. Both tracking facilities, major units of NASA's Deep Space Network, were operated for NASA in cooperation with Spain's Instituto Nacional de Técnica Aeroespacial (INTA) under Jan. 29, 1964, agreement. (NASA Release 70-95)
- USAF released summary of findings of C-5 Review Board headed by Dr. Raymond L. Bisplinghoff, Dean of MIT School of Engineering. Board concluded: C-5A flight performance met Lockheed Corp. contract guarantees at current weight of aircraft; change actions applying to contract had not reduced required flight performance; landing gear design system was "based on good engineering principles backed by adequate tests and is basically sound"; landing gear problems that had occurred could be related to large number of components required to meet aircraft's landing and ground handling requirements. "It is essential that correct procedures be established, documented, and made available to train crews in proper procedures."

  Major recommendations on C-5A subsystems were: reconfigure triple-redundant pitch, roll, and heading sensor assemblies from two attitude heading reference units (AHRUs) and one inertial measurement unit (IMU) to three AHRUs, retaining IMU for navigation only; provide triple redundancy for fail-operational pitch, roll, and heading information fed to cockpit displays; substitute state-of-the-art weather and mapping radar for multimode radar (MMR) for transport mission until reliability for other modes had been demonstrated by MMR. Board found static tests had demonstrated 80% of design ultimate strength, while full-scale test article had accumulated only 1500 simulated hours of life. It recommended additional modifications to wing structure and program for tracking use of each individual C-5A aircraft by load recorders, to adjust inspection intervals and monitor remaining life. It also recommended acceleration of all tests, to identify problems in laboratory rather than in service; second wing-fatigue test; and choice by USAF of one of several proposals to enable wing to meet static strength requirements. (DOD Release 507-70)
- Physicist Dr. Ralph E. Lapp testified before Subcommittee on Economy in Government of Joint Economic Committee on changing national priorities: "If we eliminate the space shuttle and the space station, then the NASA budget can be trimmed to a level below $3 billion per year throughout the rest of this decade. Such a decision will shift NASA's priorities from sensational manned space ventures to space science and its applications. If the U.S. space program is restructured to this new pattern and the Apollo manned lunar program

is allowed to run out, it should be possible to effect further economies in the NASA budget...." When NASA shifted to science base "it would be desirable that the scientific community give some expression of its sense of priorities for space science with respect to the rest of science. The quest for new knowledge of the solar system must be reckoned as pure or basic or fundamental research. As such, it should be related to the federal support of all other basic research." (Testimony)
- MSC announced award of $100 000 NASA contract to Grumman Aircraft Engineering Corp. for study of station operating techniques during space station mission to define crew command structure necessary to ensure successful operations. (MSC Release 70-68)
- AFSC announced it was testing T-38 aircraft escape system on High Speed Test Track at Holloman AFB, N. Mex., in effort to reduce time from starting ejection sequence to parachute inflation. Tests changed sequence of events during ejection process and used explosive devices to deploy new seat stabilization-retardation parachute and speed main parachute inflation. (AFSC Release 161.70)
- Senate confirmed nomination of Adm. Thomas H. Moorer (USN) to be Chairman of Joint Chiefs of Staff for two-year term. (*CR*, 6/17/70, S9235)
- U.S. Conference of Mayors, meeting in Denver, Colo., called for immediate cutbacks in military and space budgets to provide more money for beleaguered cities. Conference supported resolution by Milwaukee, Wis., Mayor Henry W. Maier and Cleveland, Ohio, Mayor Carl Stokes that money should be taken from "waste" in military, space, agricultural, and highway budgets and channeled to cities. (Moritz, *NY News*, 6/18/70, 28)

June 18: MSFC announced award of $320 000 contract to NR Space Div. for lunar base synthesis study to define and analyze lunar exploration missions, establish requirements, and develop conceptual descriptions of semipermanent lunar surface bases. Study would appraise mobile systems for long traverses of lunar surface, examine different surface drills and other tools and equipment, and relate capabilities of flying and surface roving vehicles to mission needs. Study also would examine operation of semipermanent lunar base without lunar orbit space station, operation of base while space base was in lunar orbit, missions that could be performed from base, and surface missions that could be performed from either lunar orbit or surface base. (MSFC Release 70-119)
- Arcas sounding rocket launched by NASA from PMR carried GSFC experiment to study ozone in atmosphere. Rocket and instruments functioned satisfactorily. (SR list)
- NASA announced selection of Honeywell, Inc., for $3-million cost-plus-incentive-fee contract to design, develop, and deliver 10-band multispectral scanner for Skylab Workshop experiment to develop techniques and interpretive methods for earth survey from orbital vehicles. Scanner would detect and measure radiated and reflected solar energy from materials on earth. (NASA Release 70-99)
- Continued problems with Boeing 747 engines were causing postponements and interruptions in service on all U.S. airlines operating jumbo jet 747 aircraft, *Wall Street Journal* said. Main problem had been malfunction in blade-retaining mechanism which had damaged

- some turbine blades. Spokesman for Pratt & Whitney Div. of United Aircraft Corp., manufacturer of 747 engines, had said modifications were under way. (*WSJ*, 6/18/70, 23)
- U.S. exhibit including *Apollo 10* spacecraft was most popular attraction at Poznan International Fair in Poland, AP quoted U.S. pavilion director Kendall Niglis as saying. He had estimated 6000 visitors an hour were filing past capsule. (*W Post*, 6/19/70, A8)
- AEC announced selection of Dr. Norris E. Bradbury, Director of AEC's Los Alamos Scientific Laboratory, to receive $25 000 Enrico Fermi Award for 1970 for outstanding achievement in atomic energy. Dr. Bradbury was cited for postwar research leadership which "completely revolutionized nuclear weapon technology" and made U.S. nuclear capability "cornerstone of free world's security." (AEC Release N-103)

June 19: USAF launched by Atlas-Agena booster from AFETR first in series of satellites equipped with TV cameras and x-ray sensors to provide early warning of enemy attack from land or sea bases. Satellite entered orbit with 35 840.1-km (22 270-mi) apogee, 35 791.8-km (22 240-mi) perigee, 1426.5-min period, and 0.1° inclination. USAF did not announce launch, but press reported informed sources said satellite would be stationed over Southeast Asia at about 32 000-km (20 000-mi) altitude. (Pres Rpt 71; GSFC *SSR*, 6/30/70; AP, *NYT*, 6/20/70,8)

- Largest solar-cell-array system ever devised to power spacecraft had been designed by MSFC engineers, NASA announced. Each of arrays, to turn sunlight into electric power for Saturn Workshop and Apollo Telescope Mount, would provide 10 500 w at 328 K (55°C) during 58- to 69-min sunlight portion of each orbit. Skylab power-generation systems would produce 21 000 w at peak operation. Number of solar cells in workshop array had not been decided; 164 160 cells would be used in ATM array. (MSFC Release 70-121)
- NASA announced selection of General Electric Co. for $5-million cost-plus-incentive-fee contract to design, develop, and deliver microwave radiometer-scatterometer/altimeter for Skylab Workshop. Experiment was to evaluate usefulness of passive and active microwave systems in obtaining data on sea and land conditions, wave heights, oceanic wind fields, snow cover, and ground and soil moisture. Data would be correlated at MSC with measurements in visible and infrared bands to be made aboard Workshop. (NASA Release, 6/19/70)
- LeRC scientists were studying metal shielding to protect reusable space vehicle from intense heat as it reentered earth's atmosphere, NASA announced. Shielding would have to withstand temperatures to 1900 K (3000°F) as orbiter traveled more than 27 000 km per hr (17 000 mph). Best candidate for heat shield material was dispersion-strengthened nickel-chromium alloy, according to Charles Blankenship, head of LeRC Materials Processing Section. LeRC also was conducting research into coatings for refractory metal alloys to shield leading edges of orbiters at temperatures above 1500 K (2200°F) and act as backup material at 1300-1500 K (1800°-2200°F). (NASA Release 70-100)
- First measurements of spectral reflectivity (0.30 to 1.10 μ) of several asteroids using double-beam photometer on 152-cm (60-in) tele-

scope of Cerro Tololo Inter-American Observatory, Chile, and 152-cm and 254-cm (100-in) telescopes of Mount Wilson Observatory, were described in *Science* by MIT astronomers Thomas B. McCord and Torrence V. Johnson and astronomer John B. Adams of Caribbean Research Institute College of the Virgin Islands. Asteroid Vesta's reflection spectrum contained strong absorption band centered near 0.9μ and weaker absorption feature between 0.5 and 0.6μ. Reflectivity decreased strongly in UV. Reflection spectrum for Pallas and probably for Ceres asteroids did not contain 0.9μ band. Vesta showed "strongest and best defined" absorption bands yet seen in reflection spectrum for solid surface of object in solar system. Comparisons with measurements of meteorites and *Apollo 11* samples indicated that Vesta surface had composition similar to that of certain basaltic achondrites. (*Science*, 6/19/70, 1445-7)

June 20: *Soyuz IX* Cosmonauts Andrian Nikolayev and Vitaly Sevastyanov returned to Moscow and complained of having "heavy bodies" in adjusting to earth gravity after 17-day space flight and 1 day on ground. Doctors escorted cosmonauts to hotel for extensive physical examinations and start of 10-day debriefing. Later Tass said cosmonauts were "quickly adapting. What was heavy an hour ago is becoming light." Initial results of debriefing under precautionary quarantine indicated "condition of health, working efficiency and the general tone of the cosmonauts are much higher than medical men expected." (UPI, *C Trib*, 6/21/70; *W Post*, 6/21/70, A22)

June 22: NASA launched two sounding rockets. Nike-Cajun launched from Wallops Station carried GSFC payload to 132-km (82-mi) altitude to obtain data on temperature, pressure, and wind between 35 and 95 km (22 and 59 mi) by detonating grenades and recording their sound arrivals on ground. Rocket and instruments functioned satisfactorily; 18 of 19 grenades exploded as planned. Launch was in conjunction with ESSA program to study temperature-sensing methods using balloon sonde, Arcas and Loki-Dart rockets, and *Nimbus III* and *IV* satellites.

Aerobee 150 launched from WSMR carried Hawaiian experiment to conduct solar studies. (NASA Rpt SRL; SR list)

• Laser ranging retroreflector (LRRR) similar to one placed on lunar surface by *Apollo 11* crew had been added to ALSEP package for Apollo 14, NASA announced. LRRR would be placed in Fra Mauro region of moon to make laser distance measurements from earth to moon with 15-cm (6-in) accuracy. (MSC Release 70-71)

• House and Senate agreed to conference report on H.R. 16516, $3.411-billion FY 1971 NASA authorization bill, thereby clearing bill for President. Conference Report total was $190 million less than total passed by House and $94.9 million more than total passed by Senate. (*CR*, 6/22/70, H5856-9, S9497-8)

• Aug. 15 closing of Antigua Tracking Station in Caribbean would reduce Manned Space Flight Network to 13 stations, 4 aircraft, and 1 ship, NASA announced. At peak complement—during July 17-24, 1969, *Apollo 11* mission—network had had 15 stations, 8 instrumented aircraft, and 4 tracking ships. (NASA Release 70-101)

• President Nixon announced approval of International Air Policy Statement by interagency committee appointed to review U.S. international air transportation policies. Statement would supersede 1963 statement.

White House released conclusions, including recommendations that system of exchanging air transport rights through bilateral agreements should be retained but excessive price should not be paid for rights for which there was little near-term need; attempts to restrict U.S. carrier operations abroad should be opposed; both scheduled and supplemental carriers should be permitted to compete in bulk transportation market; intergovernmental agreements should be sought on charter services; and U.S. should continue to accept IATA as machinery for pricing scheduled services. (*PD*, 6/29/70, 804-5)

- American Society of Mechanical Engineers announced selection of Dr. Robert R. Gilruth, MSC Director, to receive ASME Medal for "his distinguished service in aeronautics and space research and for his outstanding engineering leadership by which he inspired and directed...manned space flights and successful landings on the moon." Award would be presented during ASME annual meeting Nov. 29-Dec. 3. (MSC *Roundup*, 7/17/70, 1; ASME)
- *Aviation Week & Space Technology* editorial by Robert Hotz commented: "Aerospace technology stands on the brink of another multi-faceted explosion that is already producing revolutionary new developments.... The decade ahead will pose another of those challenges that alter the course of history. The opportunity for military technological surprise has never been greater. The need to build a revolutionary new economy based on the technology developed in aerospace was never more acute. The U.S. policy makers have a clear choice. They can seize the opportunity offered by the technical revolution of the 1970's and exploit it fully to provide U.S. security and economic and social leadership in this troubled world. Or they can economize to the point where the Soviet Union—as it did in the 1950's—can gain sufficient technological momentum to open another technological gap that will have dire consequences for the western world." (*Av Wk*, 6/22/70, 21)

June 23: U.S.S.R. launched *Meteor V* weather satellite into orbit with 888-km (551.8-mi) apogee, 830-km (515.7-km) perigee, 102.0-min period, and 81.2° inclination. Spacecraft would photograph cloud and snow cover and collect weather data, Tass announced. (GSFC *SSR*, 6/30/70; AP, *C Trib*, 6/25/70)

- Astronaut L. Gordon Cooper, Jr., one of seven original Mercury astronauts, was resigning effective July 31 to become President of National Exhibits, Inc., and member of Board and engineering consultant to Intersales, Inc., in Washington, D.C., NASA announced. Cooper had participated in *Faith 7* mission launched May 15, 1963, and eight-day *Gemini V* mission launched Aug. 21, 1965. (NASA Release 70-103)
- Tass said doctor attending *Soyuz IX* Cosmonauts Andrian G. Nikolayev and Vitaly I. Sevastyanov had found "certain instability" in their cardiovascular systems, indicating cosmonauts had not yet adapted to earth gravity. Soviet medical authorities were speculating that simulated gravity might be necessary in spacecraft. (AP, *W Post*, 6/24/70, A3)

- USNS *Vanguard*, seagoing tracking station of NASA Manned Space Flight Network, joined OSSA-GSFC-Wallops Station mapping expedition to measure cavity in ocean surface over deepest spot in Atlantic —five-mile-deep Puerto Rico Trench. *Vanguard* would sail across cavity making precise navigation and gravity measurements while its radar tracked NASA's *Geos II* satellite orbiting overhead. Underwater transponders would provide local geodetic control point for referencing ship's position in support of onboard navigation system. Purpose of mission was to provide calibration standard for altimeter planned for use aboard GEOS satellite scheduled for 1972 launch. (NASA Release 70-108)
- At Detroit press conference, *Apollo 13* Astronaut Fred W. Haise, Jr., commenting on *Soyuz IX* mission, said he was "very happy to see the Russians back in space." He predicted U.S.-U.S.S.R. cooperation in space stations. "In the hostile environment of space it will be like the stations in the Antarctic where all nations help each other." Haise was in Detroit to address 55th annual Kiwanis International Convention. (AP, *B News*, 6/25/70)

June 24: Tass reported comments by chief designer of *Soyuz IX* spacecraft at press conference for Soviet journalists at Zvezdny Gorodok (Star City), U.S.S.R. Designer, whose name was withheld, said: "The first results of the [June 2-19 *Soyuz IX*] flight show that man can work in space for a long time, for at least a month. The new space experiment was a complete success. The information that was obtained will make possible another step towards the construction of orbital stations." (Gwertzman, *NYT*, 6/25/70)
- MSC announced it had awarded Bendix Corp. $8-million, cost-plus-fixed-fee contract modification for fabrication of fifth Apollo lunar surface experiments package (ALSEP) for deployment on moon during Apollo 16 mission. Modification brought total estimated value of contract to $76 million. (MSC Release 70-73)
- Senate Committee on Appropriations approved, with amendments, H.R. 17548, Independent Offices and HUD appropriations bill containing $3.319-billion NASA appropriation. (CR, 6/24/70, D675)
- U.S. civil aircraft fleet had increased 90% during 1960s, though 5% growth rate in 1969 was lowest recorded in recent years, FAA reported. At end of 1969, 133 814 civil aircraft were eligible to fly; at end of 1968, 127 164; and at end of 1959, 70 747. General-aviation aircraft accounted for more than 97% of total eligible aircraft, with 130 806 at end of 1969, 124 237 at end of 1968, and 68 727 at end of 1959. Air carrier fleet included 3008 eligible aircraft at end of 1969, 2927 at end of 1968, and 2020 at end of 1959. (FAA Release 70-62)
- Col. Roscoe Turner—winner of Harmon Trophy for aviator in 1932 and of Thompson Trophy Race in 1934, 1938, and 1939—died in Indianapolis, Ind., after long illness. He was 74. Col. Turner had held cross-country speed record as winner of Bendix Race, 11 hrs 30 min New York to Los Angeles, 1933-1938. He later had been Chairman of Board of Roscoe Turner Aeronautical Corp. flight school and sales service facility at Indianapolis Weir Cook Airport. (AP, *W Star*, 6/24/70, B5; *Who's Who*)

June 25: USAF launched unidentified satellite from Vandenberg AFB by Titan IIIB-Agena D booster. Satellite entered orbit with 410.4-km (255-mi) apogee, 118.9-km (73.9-mi) perigee, 89.8-min peri-

od, and 108.8° inclination and reentered July 6. (Pres Rpt 71; GSFC SSR, 6/30/70; 7/31/70)
- Aerospace Safety Advisory Panel submitted report to NASA in form of letter to Dr. Thomas O. Paine, NASA Administrator, from panel chairman, Dr. Charles D. Harrington. Panel found that *Apollo 13* Review Board procedures and scope of inquiry had "proved effective in their task." Board had "performed a thorough and technically competent analysis in the reconstruction of the factors contributing to the Apollo 13 abort." Panel "found no evidence and no reason to doubt the technical validity of their determination and findings." (Hearing, Sen Com on Aeronautical and Space Sciences, Apollo 13 Mission Review, 6/30/70, 51, 56)
- MSC announced NR had been awarded $4 785 130 supplemental agreement for changes in Apollo CM and SM contract. Agreement formally incorporated 26 changes previously authorized by NASA for modification to ground support equipment, for test and checkout of CSM, modification of flight and ground test hardware, and additional test-and-effect-analysis changes and brought total value of contract to $3.5 billion. (MSC Release 70-73)
- Nike-Cajun sounding rocket launched by NASA from Wallops Station carried GSFC experiment to obtain data on atmosphere by detonating grenades and recording their sound arrivals on ground. Rocket and instruments functioned satisfactorily. (SR list)
- Construction began on 64-m (210-ft) tracking antenna near Madrid, Spain, final link in worldwide tracking system to triple useful distance of spacecraft in deep space missions, NASA announced. Antenna, to be operative in 1973, would duplicate largest fully steerable antenna in U.S. at Goldstone, Calif. Another 64-m antenna was under construction at Tidbinbilla, Australia. Combined facilities would provide continuous tracking of spacecraft carrying experiments several hundred million kilometers into space and might reach to edge of solar system. Madrid antenna was being built by NASA under agreement with Spain's Instituto Nacional de Téchnica Aeroespacial (INTA). (NASA Release 70-104)
- ERS data-handling problem was discussed by Dr. John M. DeNoyer, Director of Earth Observations Programs, NASA OSSA, in testimony before House Committee on Merchant Marine and Fisheries' Subcommittee on Fisheries and Wildlife Conservation: "Fact that over one million photographic frames are currently on file indicates the scope of the developing ERS data handling problem, particularly in view of the forthcoming ERTS A&B data." NASA had established Data Management Working Group. At request of OST, NASA had been conducting, with user agencies, study of basic approach to be used. "The primary initial demand will be for imagery, as few users will be equipped for automatic data processing. The basic approach agreed upon is one designed to store, retrieve, and disseminate imagery. The approach is a multi-agency network, all participants using standard methods, and a common retrieval system based upon a central computer linked to many input/output remote consoles by telephone lines. The retrieval system will be similar to the NASA RECON." (Testimony)
- Post-*Apollo 13* poll showed U.S. public was opposed by 64% to 30% to major space funding over next decade, Louis Harris reported in

June 25

Chicago Tribune. Of 1520 persons polled, 55% said they were very worried about fate of *Apollo 13* astronauts following mission abort, 24% were somewhat worried, 20% were not very worried, and 1% were not sure. Total of 71% expected fatal accident would occur on future mission, 18% thought it would not occur, and 11% were unsure. (*C Trib,* 6/25/70, F12)
- NASA announced award of $9 406 000 cost-plus-incentive-award-fee contract to Lockheed Missiles & Space Co. for development and construction of improved shroud system for Centaur vehicle. Contract provided for delivery of six shroud systems, one to be tested, and others to be flown on Titan IIID-Centaur rockets. (LeRC Release 70-28)
- Senate confirmed nomination of Dr. T. Keith Glennan, first NASA Administrator, to be U.S. representative to International Atomic Energy Agency with rank of Ambassador. (*CR,* 6/25/70, S9950)
- Lockheed Aircraft Corp. announced 13% cutback, amounting to 13 000 jobs, by end of 1970. Officials blamed USAF reduction in order for C-5 Galaxy transport aircraft from 115 to 81, completion of tooling and development phases, and delay in production go-ahead for short-range attack missile (SRAM) propulsion system. (*W Post,* 6/25/70, G14)
- Dr. Kenneth S. Pitzer resigned as President of Stanford Univ. In letter of resignation released by university, he said situation at Stanford "represents another manifestation of the destructive nature of the current conflict" in U.S. society. "Both on campus and off campus, support for reasoned discourse and nonviolent change has steadily diminished." (Turner, *NYT,* 6/26/70, 15)
- NAS announced formation of Committee for International Environmental Programs (IEPC) to facilitate participation of U.S. scientists in international environmental activities, advise Federal Government, act as adhering group in U.S for international ecological programs on nongovernmental level, and develop and maintain clearinghouse for information on nongovernmental ecological activities. (NAS Release)

June 26: U.S.S.R. launched two satellites: *Molniya I-14* comsat entered orbit with 39 233-km (24 378.2-mi) apogee, 468-km (290.8-mi) perigee, 704.5-min period, and 65.4° inclination. *Cosmos CCCL* entered orbit with 249-km (154.7-mi) apogee, 200-km (124.3-mi) perigee, 89.0-min period, and 51.7° inclination and reentered July 8. (GSFC *SSR,* 6/30/70; 7/31/70)
- *Aurorae (Esro I)* satellite, launched by NASA for ESRO Oct. 3, 1968, reentered atmosphere after investigating auroras and related phenomena for 21 mos. Satellite—designed, developed, and constructed by ESRO—was second ESRO spacecraft launched under NASA-ESRO agreement. (NASA Proj Off; GSFC *SSR,* 6/30/70)
- NASA-developed fire-proofing materials were "stirring great hopes among professionals in the field of fire prevention and soon may be in widespread use," AP reported. Airline industry was "excited" about materials to reduce fire hazards in aircraft cabins. USAF had asked NASA for advice in fire-proofing T-39 jet trainers and light transports. International Assn. of Firefighters had asked NASA help in developing new firefighting apparel. Post Office Dept. was seeking fireproof mailbags. National Assn. of Home Builders, with HUD,

was preparing for advance use of new fire-resistant materials in home construction. (*P Inq*, 6/27/70)
- Harry H. Gorman had accepted, in permanent capacity, position of NASA Deputy Associate Administrator for Manned Space Flight (Management) to which he acceded in October 1969, MSFC announced. Richard W. Cook had been named to fill Gorman's former post as MSFC Deputy Director, Management, pending Hq. approval. (MSFC Release 70-129)
- President Nixon in San Clemente, Calif., announced intention to nominate Clay T. Whitehead, Special Assistant to the President and White House contact for INTELSAT, as first Director of Office of Telecommunications. Nomination was submitted to Senate June 29. (*PD*, 6/29/70, 822; 7/6/70, 871)
- NASA announced award by LaRC of $1 077 000 contract to General Electric Co. Space Div. for flight experiment to make year-long global measurements of carbon monoxide concentrations using advanced spaceborne sensor. By mapping portions of earth's atmosphere with high, low, and average concentrations of poisonous gas, scientists hoped to identify "removal sinks" in which carbon monoxide was changed into another compound. By defining removal mechanisms, scientists hoped to be able to predict whether carbon monoxide concentration would increase in future and to what degree. (NASA Release 70-102)
- NASA announced it had issued $8 965 000 supplemental agreement to $57-million Delta procurement contract with McDonnell Douglas Corp. to incorporate Delta inertial guidance system (DIGS) to replace radio inertial system in Delta rocket. DIGS system would use Apollo LM abort sensor package and computer developed for Centaur vehicle and would increase overall vehicle performance by eliminating need to select Delta trajectories within sight of ground command stations while radio guidance system was steering rocket. (NASA Release 70-106)
- MSFC announced issuance of RFPs for study of inflight venting of space shuttle vehicles to identify areas of vehicle requiring venting, establish venting schemes, outline data requirements for venting analyses, and develop test program to analyze shuttle vehicles during launch and reentry. Proposals were due July 17. (MSFC Release 70-126)
- Discovery that velocity of sound waves through *Apollo 11* and *12* moon rocks was roughly one third that through earth's rocks was reported in *Science* by Edward Schreiber of Queens College, City Univ. of New York, and Orson L. Anderson of Lamont-Doherty Observatory. Scientists compared sound velocity in various earth materials with that of lunar samples. Materials which exhibited "compressional velocities that are in consonance with those measured for lunar rocks" included provolone, emmenthal, muenster, and cheddar cheese—"which leads us to suspect that perhaps old hypotheses are best, after all, and should not be lightly discarded." (*Science*, 6/26/70, 1579-80)
- Nixon Administration policy toward Federal graduate aid was discussed in *Science* by John Walsh: There were signs "the Administration is veering away from a policy of changing the primary form of direct federal support to graduate education from fellowships and

training grants to guaranteed loans. For those who feared that the prevailing system of federal grants was to be supplanted there is some consolation that the grants are headed down but not out." While NASA had liquidated sustaining university programs, agency would still provide indirect support of graduate students through research assistantships financed under $75 million in research grants to universities approved in past budgets. (Science, 6/26/70, 1559-61)
- Possible pollution of stratosphere by SST was discussed in Science: "An environment issue that has been intensifying in recent weeks is the question of whether exhaust products added to the stratosphere by the SST could produce significant environmental changes." Most meteorologists agreed gaseous exhaust products other than water would present no special problems. Water was "more likely to be a problem because stratospheric water concentrations are generally low as the result of condensation in the troposphere." Water accumulated in troposphere could change radiation balance of earth; initiate cloud formation, thus increasing amount of solar energy reflected into space; or lead to decrease in ozone concentration of stratosphere, "thereby allowing more ultraviolet radiation to reach the earth's surface." However, there was not enough information available to predict whether persistent clouds would form, since no research flights had been made in cold areas of tropical tropopause and above Arctic and Antarctic, which were believed to be saturated. (Nuessle, Holcomb, Science, 6/26/70, 1562)
- Molecular oxygen supply in atmosphere and open ocean were not threatened by man's activities in foreseeable future, Columbia Univ. geologist Dr. Wallace S. Broecker said in Science. "Molecular oxygen is one resource that is virtually unlimited.... We are faced with so many real environmental crises that there is no need to increase the public concern by bringing out bogeymen." (Science, 6/26/70, 1537-8)
- Pravda said new 3-m (10-ft) wide telescopic mirror being installed at Zelenchuk Observatory in northern Caucasus could pick up glimmer of candle 24 900 km (15 500 mi) from earth and enable astronomers to see parts of universe never before viewed by man. Largest of its kind, mirror was cut from 64-metric-ton (70-short-ton) mass of glass and was cooled in furnace from 518 K (472°F) to room temperature during 18-mo casting period. (Reuters, B Sun, 6/27/70)
- MIT oceanographer Dr. Henry Stommel commented in Science, on prospects for physical oceanography: "Is the dynamics of the ocean similar to or fundamentally different from that of the atmosphere? Only measurement will tell, and it will have to be measurement at a level of technological sophistication quite beyond the present level and beyond that proposed for routine monitoring. Access to somebody's monitoring system, or to a few years of ship time, will not suffice to obtain the kind of information the scientist needs about the oceanic velocity field. Some carefully designed measurement programs are going to be needed—on a scale larger than an oceanographic institution can manage but smaller than the space program. To be useful scientifically, these programs will have to give first priority to questions of hydrodynamics. To date there is little indication they will do so." (Science, 6/26/70, 1531-6)

June 27: Cosmos CCCLI was launched by U.S.S.R. into orbit with 464-km (288.3-mi) apogee, 270-km (167.8-mi) perigee, 91.8-min period, and 70.9° inclination. Satellite reentered Oct. 13. (GSFC *SSR*, 6/30/70; 10/31/70)
- Tenth anniversary of MSFC, NASA's largest installation. MSFC facilities consisted of 700 hectares (1800 acres) with 370 structures and buildings totaling about 370 000 sq m (4 million sq ft). During decade, MSFC scientists and engineers had designed Saturn I, Saturn IB, and Saturn V boosters. (*Marshall Star*, 6/24/70; NASA PAO)
- Actions in response to *Apollo 13* Review Board Report [see June 15] were recommended by Dr. Rocco A. Petrone, Apollo Program Director, in memo to Dr. Thomas O. Paine, NASA Administrator. They included oxygen tank modifications, addition of third oxygen tank on Apollo 14 and 15, modification of fuel-cell oxygen reactant valve to separate Teflon-coated wires from oxygen environment, caution and warning system modifications, addition of second-level limits sensing in Mission Control Center, comprehensive review of consumables and emergency equipment in LM and CM, and tests and analysis recommended by Review Board. (Hearing, Sen Com on Aeronautical and Space Sciences, Apollo 13 Mission Review, 6/30/70, 52, 57)
- NASA launched two sounding rockets from WSMR. Aerobee 170 carried American Science and Engineering, Inc., payload to 175.2-km (108.9-mi) altitude to collect data from cosmic x-ray sources to determine location of sources and flux levels in range 0.2-1 kev. Rocket and instruments functioned satisfactorily.

    Nike-Apache carried Cal Tech experiment to conduct x-ray studies. Rocket and instruments functioned satisfactorily. (NASA Rpt SRL; SR list)
June 29: Sen. Walter F. Mondale (D-Minn.), for himself, Sen. Clifford P. Case (R-N.J.), Sen. William Proxmire (D-Wis.), and Sen. Jacob K. Javits (R-N.Y.), submitted amendment to H.R. 17548, FY 1971 Independent Offices and HUD appropriations bill which would reduce NASA R&D appropriation by $110 million—amount requested by NASA for design and definition of space shuttle and station. Sen. Mondale said: "This project represents NASA's next major effort in manned space flight. The $110 million...is only the beginning of the story. NASA's preliminary cost estimates for development of the space shuttle/station total almost $14 billion, and the ultimate cost may run much higher. Furthermore, the shuttle and station are the first essential steps toward a manned Mars landing...which could cost anywhere between $50 to $100 billion. I have seen no persuasive justification for embarking upon a project of such staggering costs at a time when many of our citizens are malnourished, when our rivers and lakes are polluted, and when our cities and rural areas are decaying." (*CR*, 6/29/70, S10057-8)
- Ninth anniversary of launch of SNAP-3A, first atomic battery in space. Generator launched by USN on *Transit IV-A* June 29, 1961, had already operated four years beyond its five-year design life. (AEC Release N-108; *Aeronautical and Astronautical Events of 1961*)
- MSFC announced award of $143 987 contract to McDonnell Douglas Astronautics Co. to establish requirements and procedures for field

repair of thermal protection panels for reusable space transportation system. (MSFC Release 70-130)

- USN F-14A jet fighter, in developmental stage, had encountered problem with movable wing, same design feature that had plagued USAF F-111, *Wall Street Journal* said. F-14, being built by Grumman Aircraft Corp., would substitute for USN's canceled version of F-111. (*WSJ*, 6/29/70, 1)
- FAA announced publication of *R&D Plan To Increase Airport and Airway System Capacity*, to meet future demands of air transportation. Plan assumed new approach and landing system was necessary to achieve accurate guidance and minimize site sensitivity problem at high-density airports; reduction in collision potential must be provided as capacity increased; many manual air traffic control functions must be automated; design and configuration of system must reduce vulnerability to subsystem or component failures; increase in IFR capacity required higher precision and greater reliability of navaids; expanded airport accommodations required investigation of new concepts like satellite terminal facilities and all cargo airports; use of more flexible navigation and ATC systems was necessary to reduce noise pollution effects around airport; and, to increase aircraft handling capacity, close-spaced, dependent and independent parallel runways were necessary. (FAA Release T 70-29)
- President Nixon in San Clemente, Calif., issued E.O. 11538 delegating to Secretary of Transportation authority to establish and conduct international aeronautical exposition. (*PD*, 7/6/70, 840)

June 29-30: NASA held "Skylab and Beyond" press briefing and tour of production facilities at MSFC. William C. Schneider, Skylab Program Director, said project was "in the very critical phase of firming up our designs" and predicted 1972 launch date would be met. Three missions were planned for eight-month lifetime of 14.6-m (48-ft) Workshop. Primary task of first mission would be to study physiological and psychological aspects of space flight for 28 days. Second mission, for 56 days, would operate telescopes. Third mission, for 56 days, would survey earth resources with highly sensitive cameras. (NASA Note to Editors, 6/8/70; Wilford, *NYT*, 7/1/70, 1)

June 30: Dr. Thomas O. Paine, NASA Administrator, reviewed results of *Apollo 13* accident investigations and announced postponement of Apollo 14 mission to Jan. 31, in testimony before Senate Committee on Aeronautical and Space Sciences. LaRC Director Edgar M. Cortright, Chairman of *Apollo 13* Review Board, summarized prepared statement before Committee. Recommendations of *Apollo 13* Review Board [see June 15] would be implemented before Apollo 14 was approved for launch, Dr. Paine said. CSM systems would be modified to eliminate potential combustion hazards in high-pressure oxygen. Unsealed fan motors would be removed from oxygen tanks and additional oxygen tank would be added to Apollo 14 CSM. Electrical wiring within high-pressure oxygen systems would be limited to stainless-steel-sheathed wires, and Teflon, aluminum, and other potentially reactive materials would be used as little as possible and kept away from possible ignition sources. Spacecraft and Mission Control warning systems would be modified to provide more immediate and visible warnings of system anomalies. MSC was reviewing spacecraft emergency equipment and procedures and use of CSMs and LMs in "lifeboat" modes.

OSSA, OMSF, and OART had been directed to apply Review Board recommendations throughout NASA and submit report on assessment and actions taken or proposed by Aug. 25.

Lessons of *Apollo 13* would be disseminated throughout industry and technical community. Dr. Paine had forwarded copy of Review Board report to Soviet Academician Mstislav V. Keldysh "so that lessons which might be learned from our accident can be applied to prevent a similar hazard to Soviet Cosmonauts."

Aerospace Safety Research and Data Institute at LeRC would conduct additional research on materials compatibility, ignition, and combustion at various g levels and on characteristics of supercritical fluids, expanding its review of oxygen handling in aerospace programs. Research would "be of direct long-term benefit to NASA in...future programs, and will help other sectors of the economy."

Aerospace Safety Advisory Panel would review NASA management processes in implementing Review Board recommendations and *Apollo 13* Review Board would reconvene later in year.

Postponement of Apollo 14 mission would move Apollo 15 launch date to July or August 1971, Dr. Paine testified. It was too early to detail costs and budgetary impact of modifications and program changes necessitated by *Apollo 13* accident but best current estimate would be "in the range of $10 to $15 million of increased costs, which we plan to handle within our total Apollo budget."

NASA Associate Administrator for Manned Space Flight Dale D. Myers presented specific plans for reassessing all Apollo spacecraft subsystems. Dr. Rocco A. Petrone, Apollo Program Director, outlined actions planned to carry out Review Board recommendations. (Transcript)

- Electronics Research Center, Cambridge, Mass., was officially closed as NASA installation. Transfer to DOT as Transportation Systems Center, announced by President Nixon March 25, would be completed July 1. Dr. Thomas O. Paine, NASA Administrator, had announced decision Dec. 29, 1969, to close ERC because "NASA could not afford to continue to invest broadly in electronics research" in face of budget reductions. Center had opened Sept. 1, 1964. In December 1969 it had 850 employees engaged in advanced research in electronics for aeronautics and space, under Director James C. Elms. Six buildings were in final phases of construction. (NASA Steering Group, *Rpt on Closing of NASA ERC*, 10/1/70; PD, 3/30/70, 446; NASA Release 69-171)

- Dr. Wernher von Braun, NASA Deputy Associate Administrator for Planning, received 1970 "World Citizen" Award of Civitan International at organization's 50th annual convention in Atlanta, Ga. Citation was for "accomplishments as leading pioneer scientist and engineer" in space program "which has already resulted in successful flights to the moon and untold scientific advances for the benefit of mankind." Only previous recipients of award were Sir Winston Churchill and President Dwight D. Eisenhower. (Thompson, *Atlanta Constitution*, 7/1/70)

- During FRC Space Shuttlecraft Symposium, Milton O. Thompson, FRC Director of Research Projects and NASA test pilot, proposed construction and flight-testing of subscale version of selected space shuttlecraft configuration before construction of actual spacecraft. Test could provide confidence in configuration to fly at low supersonic and subsonic speeds and to land successfully and could discover and solve unpredicted problems. Proposed manned tests would be similar to X-15 and lifting-body flights. They would be launched from B-52 aircraft, use propulsion system to climb to higher speeds and altitudes, and glide to landing pattern. (FRC Release 13-70)

- NASA announced appointment, effective Oct. 1, of Richard C. McCurdy, consultant and member of NASA Management Advisory Panel, to post of Associate Administrator for Organization and Management. He would have responsibility for evaluation and strengthening of organization and management policies and practices. McCurdy retired as President and Chief Executive Officer of Shell Oil Co. in December 1969. (NASA Release 70-110)
- NASA announced award of $856 000 contract to Cutler-Hammer, Inc., for design of L-band microwave radiometer for use in Skylab program. Experiment would measure brightness temperature of terrestrial surface of Skylab ground track in L-band portion of spectrum. (MSC Release 70-86)
- USAF contract awards: Aerojet-General Corp. was being issued $12 100 000 initial increment to $33 409 607 fixed-price-incentive contract for procurement of Titan IIIB, C, and D stage 1 and 2 liquid-rocket engine systems. TRW Inc. was being issued $351 000 initial increment to $4 669 800 contract for advanced development of electronic propulsion system for space satellites. General Dynamics Corp. was receiving $250 000 initial increment to $2 585 000 contract for development, assembly, test, and launch support of integrated unmanned spacecraft for space flight-testing of DOD experiments. (DOD Release 544-70)

During June: Hypersonic aircraft technology and applications were discussed in *Astronautics & Aeronautics* by Dr. Alfred J. Eggers, Jr., NASA Assistant Administrator for Policy; Richard H. Petersen, Senior Research Scientist in ARC Mission Analysis Div.; and Nathaniel B. Cohen, special assistant in NASA policy research program. Because of long-life requirements of its structure, it was unlikely that hypersonic transport would be developed until operational experience was available on some other hypersonic aircraft. Therefore, hypersonic transport would benefit not only from technology developed for space-shuttle vehicle, but very probably from experience gained during shuttle operation. Operating experience with high-temperature structures would help immensely in developing longer life structures for hypersonic transport and gain considerable new knowledge of hypersonic aerodynamics. In contrast to situation in aerodynamics and structures, technology program being planned by NASA to support development of rocket propulsion for shuttle would not contribute significantly to hypersonic airbreathing-propulsion-system technology. Current generalized scramjet research program had amply demonstrated feasibility of such engines. Developing airbreathing propulsion for either hypersonic transport or second-generation airbreathing 1st stage for space shuttle would require applied program focused specifically on scramjet. (*A&A*, 6/70, 30-41)

- Aerospace said USAF's Vela nuclear test satellite program had been credited with longest continually operating spacecraft, saving Government $26 million and achieving 100% of its objectives. April 8 launch of two Vela satellites had been final launch in program. (*Aerospace*, 6/7/70)
- North American Rockwell Corp. received applications from more than 1000 unemployed aerospace workers during week following June 5 announcement that NR had been awarded $1.35-billion USAF B-1 bomber contract. (Wright, *NYT*, 6/10/70, 63)
- Two NAE reports to NSF had proposed that NSF's authority to support applied research not be used to assume burdens of Federal agencies, NAS-NRC-NAE *News Report* said. *Federal Support of Applied Research* by task force headed by Cal Tech President Harold Brown and *Priorities in Applied Research: An Initial Appraisal* by NAE Committee on Public Engineering Policy had urged that NSF support stop short of developmental work. Support should be directed toward projects for which there was widespread need but little or no other support. Highest priorities were recommended for research "intended to deepen our understanding of the structure and dynamics of the biosphere," with emphasis on atmospheric and hydrological processes and perturbed ecosystems. (NAS-NRC-NAE *News Rpt*, 6-7/70; NAE Release, 7/1/70)
- Potential for spotting pollution from space was discussed in *Space/Aeronautics* by Kurt R. Stehling, author and former NASA Office of Program Planning and Evaluation scientist: Full-scale program would require development of new and improved sensors and establishment of highly efficient data-management system. "The answer lies either in larger aircraft, capable of flying at higher altitudes, or...in spacecraft complemented by aircraft." (*S/A*, 6/70; 47-9)
- George Washington Univ. published *International Aspects of Earth Resources Survey Satellite Programs* by John Hanessian, Jr., reprint of article published in *Journal of the British Interplanetary Society*: International participation in experimental earth resources survey satellite program was possible in several forms, each both advantageous and disadvantageous for U.S. and participating country. "The technical constraints and realistic expectations need to be carefully considered in any feasibility evaluation that may be carried out...." Technical assistance from industrialized countries or regional or international multilateral agencies would be necessary for developing countries. "It is also quite possible that private industry could play a leading role in such efforts." Potential participation of U.S.S.R. "must be viewed not only with the optimism engendered by recent statements, but also by reflection on the experience of a decade of space history." Survey satellite concept "could provide an ideal opportunity for the technologically advanced nations of the world to converge their interests with the aspirations of the many developing countries in their effort to build a just, peaceful, and economically progressive world community." (Text)

July 1970

July 1: Astronaut Neil A. Armstrong was sworn in as Deputy Associate Administrator for Aeronautics, NASA OART, at Hq. ceremony attended by Dr. Thomas O. Paine, NASA Administrator, and NASA officials. (Daily Appointments Calendar)
- Loss of 0.56-g (0.02-oz) fragment of 0.45-kg (1-lb) *Apollo 11* lunar rock from GSFC was reported by NASA. Sample, one of several used by GSFC scientists to study effect of shock waves in lunar material, had been reported missing by GSFC scientist Dr. Nicholas M. Short. NASA was investigating. (NASA Release 70-113)
- NASA announced award of $43-million cost-plus-award-fee contract to General Electric Co. for development of Nimbus E and F spacecraft, including responsibility as prime contractor for associated systems, integration, and testing. Nimbus E and F would be similar in structural design to previous Nimbus spacecraft and would each carry 10 experiments, with launch scheduled for 1972 and 1973. (NASA Release 70-111)
- MSC announced NASA contract awards: Grumman Aerospace Corp. received $3 248 000 supplemental agreement for changes in Apollo LM contract. Agreement formally incorporated 18 changes previously authorized by NASA for modification of vehicle readiness review procedures, vibrational testing, rendezvous radar, quality assurance and reliability procedures, and descent engine updating. Agreement brought total estimated value of contract to $1 673 119 200. Federal Electric Corp. received 10-mo, $3.29-million, cost-plus-award-fee contract extension for support services at MSC. Extension brought total value of contract to $7.394 million since June 1, 1969. (MSC Releases 70-75, 70-76)
- MSFC announced NR Rocketdyne Div. had received $22 841 941 contract for Saturn engine support work between July 1, 1970, and June 30, 1971. (MSFC Release 70-134)
- Senate passed and returned to House H.R. 16595, $550-million FY 1971 NSF authorization bill. (*CR*, 7/1/70, S10389-93)
- ComSatCorp announced Bartlett earth station for comsats at Talkeetna, Alaska, had been placed in commercial operation, with 80 circuits for telephone and record service between Alaska and lower 48 states. Service would also be available between Alaska and Hawaii. Circuits to Japan were expected to be established later in year. (ComSatCorp Release 70-38)
- Blue Ribbon Defense Panel, with Gilbert W. Fitzhugh as Chairman, submitted to President Nixon *Report to the President and the Secretary of Defense on the Department of Defense.* Report offered 113 recommendations, including top-to-bottom DOD reorganization, removing Joint Chiefs of Staff from involvement with day-to-day problems of war, streamlining military staffs, reorganizing all U.S.

combat forces under three main operating commands, and abandoning contracting policies that led to cost overruns and weapons that didn't work. Panel had been appointed by President and Secretary of Defense in 1969 to study DOD problems. (Text)
- Nomination of Dr. Glenn T. Seaborg, AEC Chairman, to serve additional five-year term as member of AEC was submitted to Senate. President Nixon asked that Dr. Seaborg continue as AEC Chairman. (*PD*, 6/6/70, 871; AEC PIO)
- Merger of USAF Office of Aerospace Research and AFSC placed responsibility for all USAF R&D in AFSC. OAR field installations were kept intact but OAR Hq. in Arlington, Va., was absorbed into AFSC Hq. at Andrews AFB, Md. (AFSC Release 209.70)

July 2: In letter to Dr. Thomas O. Paine, NASA Administrator, Elmer B. Staats, U.S. Comptroller General, recommended that award of ATS-F and ATS-G contract to General Electric Co. be reconsidered because of ambiguity in NASA's instruction to competitors and one-week extension granted GE to submit its revised proposal. "It is our [GAO's] opinion that the established award selection procedures were not followed and that the procedures which were followed were defective."

Within two hours of receipt of GAO recommendation, Dr. Paine issued statement: "On April 9 I requested the GAO to conduct a review of the events leading to the selection of the General Electric Co. rather than the Fairchild Hiller Corp. to build two Applications Technology Satellites.

"The General Accounting Office has reported to me that the Fairchild Hiller Corp. may not have been accorded an equal opportunity to submit a winning bid because of a one-week extension granted to General Electric who turned in the lower bid.

"At the time the award was made, this time disparity was not known to the selecting officials.

"NASA will therefore reopen the bidding to Fairchild Hiller and General Electric in strict accordance with our procurement regulations." (*CR*, 7/9/70, S10975-7; NASA Release 70-119; Text)

Daniel J. Fink, Vice President and General Manager of General Electric Co. Space Div., issued statement in Washington, D.C., and Philadelphia, Pa. Conclusion of GAO report "seems to rest on the single detail that Fairchild's proposal was submitted one week earlier than ours. . . .both parties had the opportunity to inform NASA when they could submit their proposal and both parties responded accordingly. The Fairchild implication that this difference in time could have been used unfairly by General Electric by our learning our competitor's price and changing ours is a shocking allegation. No such 'leak' occurred, and the GAO report specifically states that there is no evidence to support that allegation." Fink said facts were "that both parties competed hard for this important space program which has the potential of contributing greatly to our well being in this country and in several foreign countries where this communications satellite will be used." (Text)
- President Nixon approved H. R. 16516, $3.411-billion FY 1971 NASA Authorization Act. Bill became P.L. 91-303. (*PD*, 7/13/70, 931)
- Aerobee 150MII sounding rocket, launched by NASA from WSMR by VAM-20 booster, carried Univ. of Wisconsin payload to 168.3-km

(104.6-mi) altitude to measure UV flux from three stars in constellations Lyra, Cygnus, and Ursa Major to provide check on prelaunch calibration of Wisconsin OAO payload. Rocket functioned satisfactorily, but program stars were not acquired and experiment collected no data. (NASA Rpt SRL)
- MSC announced award of $600 000, fixed-price contract to Sperry Rand Corp. Univac Div. for lease and maintenance of four Univac 1108 computer systems used in management, administration, analysis of test data, mission support, and trajectory analysis. Contract covered April 1 through June 30, with two one-year options. (MSC Release 70-78)
- UCLA nuclear physicist Dr. Darrell J. Dickey said the four members of his team for scientific exchange program with U.S.S.R. to study peaceful uses of atom had been tentatively selected. Under U.S.-U.S.S.R. agreement, Dr. Dickey and team would be based at Serpukhov, south of Moscow, site of world's largest nuclear particle accelerator with 78-bev power rating. Two Soviet physicists already were studying at Batavia, Ill., where U.S. was building even larger nuclear accelerator. (AP, *W Post*, 7/3/70, A6)

July 3: *Soyuz IX* Cosmonauts Andrian G. Nikolayev and Vitaly I. Sevastyanov emerged from 13-day postflight quarantine and were honored at special reception in Moscow. (AP, *Minneapolis Trib*, 7/4/70)
- USAF C-5 Galaxy, world's largest aircraft, flew from Charleston AFB, S.C., in first of series of flights to airlift cargo between U.S. and U.S. overseas military bases to demonstrate aircraft's operational capability. Aircraft was to stop at U.S. and Pacific bases to provide orientation and training for enroute command and support personnel and route familiarization for Military Airlift Command crews before beginning regular cargo flights. (DOD Release 552-70; USAF PIO)
- *New York Times* editorial hailed decision by U.S. and U.S.S.R. to permit exchange of scientists between world's largest nuclear accelerators at Serpukhov in U.S.S.R. and at Batavia, Ill. [see July 2]: "This increased cooperation in particle physics contrasts sharply with the continued—and even rising—Soviet-American competition in space." NASA's Skylab project was case in point. "The semipermanent manned space station...is precisely the same goal toward which the Soviet Soyuz experiments are directed. The expensive and wasteful rivalry for priority in reaching the moon that preceded the Apollo 11 flight is being repeated in the effort to create manned orbiting vehicles for earth reconnaissance and scientific experiments in space." (*NYT*, 7/3/70, 24)

July 4: LeRC announced F-102 supersonic all-weather fighter-interceptor obtained from Bradley Air National Guard Base at Windsor Locks, Conn., would be used as chase aircraft to observe research flights by LeRC's F-106 modified USAF jet. Flights would test advanced exhaust nozzles in under-wing nacelles. (LeRC Release 70-30)
- New York State Attorney-General Louis J. Lefkowitz had filed suit against 58 domestic and foreign airlines using John F. Kennedy International and LaGuardia Airports to require improved noise abatement devices on jet aircraft, *Washington Post* said. (*W Post*, 7/4/70, A4)

- Tass reported 15-kg (33-lb) meteorite had crashed into street in Ukranian city of Yagotin. Kiev Univ. meteorologists had determined object was comet. (UPI, W *Star*, 7/5/70)
- Otto Hirschler, last U.S. Army employee among 118 German V-2 rocket experts brought to U.S. by Dr. Wernher von Braun, had lost his $25 000-a-year job as electronics expert with Army Missile Command in USA personnel cutback, Reuters reported. Most of other German rocket experts had transferred to NASA. (*NYT*, 7/5/70, 2)

July 5: *Chicago Tribune* published interview with Dr. Thomas O. Paine, NASA Administrator, in which he said average age of space employees had risen rapidly as result of cutbacks in aerospace employment. "We're going from a bunch of rather swinging young engineers doing things in the forefront of science and technology back to an older, more conservative group, because young people get laid off first." Continued cutbacks in technology and research could cost U.S. leadership in these fields. "Russia is coming on strong and is spending a greater percentage of her GNP on space defense and technology than the U.S. is." (*C Trib*, 7/5/70)

July 6: Senate, debating H.R. 17548, FY 1971 Independent Offices and HUD appropriations bill, rejected by vote of 32 to 28 June 29 Mondale amendment to eliminate $110 million for space shuttle and station from $3.319-billion NASA appropriation. (*CR*, S10604-24)
- U.K. Minister of Technology Geoffrey Rippon told Parliament that British Concorde supersonic transport prototype was expected to fly at month's end, with flight tests at cruising speeds scheduled for later in year. French prototype would resume test flights in September. (UPI, *W Post*, 7/7/70)
- General Dynamics Corp. test pilots had successfully flown three USAF F-111 aircraft from Fort Worth, Tex., USAF said. Aircraft had been grounded since fatal Dec. 23, 1969, crash and had undergone wing-structure tests. (W *Star*, 6/6/70, A7)
- World's first team of female aquanauts—Peggy Lucas, Alina Szmant, Dr. Ranate True, Ann Hartline, and Dr. Sylvia E. Mead—descended to project Tektite II habitat 15 m (50 ft) beneath surface of Great Lameshur Bay, V.I., for two-week study of fish and plant life. (AP, *W Post*, 7/7/70, A9)

July 7: U.S.S.R. launched *Cosmos CCCLII* from Baykonur into orbit with 292-km (181.4-mi) apogee, 205-km (127.4-mi) perigee, 89.4-min period, and 51.7° inclination. Satellite reentered July 15. (GSFC *SSR*, 7/31/70; *SBD*, 7/10/70, 41)
- Senate, by vote of 58 to 4, passed H.R. 17548, FY 1971 Independent Offices and HUD appropriations bill containing $3.319-billion NASA appropriation. Bill was approved by Senate following Senate rejection of three amendments: one by Sen. Barry M. Goldwater (R-Ariz.) to increase research and program management funds by $4.575 million, one by Sen. William Proxmire (D-Wis.) to reduce NASA R&D funds from $2.606 billion to $2.5 billion and construction of facilities funds from $34.4 million to $18.2 million, and one by Sen. J. W. Fulbright (D-Ark.) to reduce NASA funds by $300 million. (*CR*, 7/7/70, S10681-713)
- Dr. Wernher von Braun, NASA Deputy Associate Administrator for Planning, and eight other NASA Hq. officials began attending series

of meetings at MSFC to discuss Skylab, HEAO, and future scientific space projects. (MSFC Release 70-135)

- MSFC announced award of $369 478 NASA contract to General Electric Co. Space Div. for 12-mo space-base nuclear-systems safety study to identify potential and inherent radiological hazards of space-base program and recommend approaches for eliminating hazards or reducing them to acceptable risk level. Space base would be exposed to radiation from radioactive materials and systems on board plus natural radiation from outer space. Major source of electrical power would be output from nuclear-electric-reactor power systems with up to 100-kw capacity. Several small isotope power sources might be carried for use in experiments and equipment. (MSFC Release 70-136)
- DOD announced General Dynamics Corp. was being awarded $1 098 000, cost-plus-fixed-fee contract for operation, maintenance, and improvement of radar-target-scatter facility to obtain characteristic radar data of signal return from various aerospace vehicles. Work would be managed by Air Force Missile Development Center and would be conducted at WSMR. (DOD Release 561-70)
- AFSC announced reassignment of M/G Edmund F. O'Connor, Vice Commander of Aeronautical Systems Div., Wright-Patterson AFB, Ohio, to become Deputy Chief of Staff for procurement and production at AFSC Hq. Aug. 1. He would be succeeded at Wright-Patterson by M/G John B. Hudson. (ASFC Release 180.70)
- Peter Ross Murray, AFSC Deputy Director of Laboratories, received DOD Distinguished Civilian Service Award from Secretary of Defense Melvin R. Laird at DOD. Citation was for "exceptionally meritorious devotion to duty and his significant accomplishments during a career of over 34 years with the Air Force." His "outstanding career has been characterized...by major technical contributions in the field of avionics as well as by superior management achievements as Director of the Air Force Avionics Laboratory and Deputy Director of Laboratories for the Air Force." (DOD Release 559-70)

July 7-8: NASA delegation attended European Space Conference in Bonn, West Germany, to brief members on propulsion for reusable space shuttle. Delegation headed by Dr. Homer E. Newell, NASA Associate Administrator, included Dale D. Myers, Associate Administrator for Manned Space Flight, and Arnold W. Frutkin, Assistant Administrator for International Affairs. NASC Executive Secretary William A. Anders also attended. At press conference following meeting, Frutkin said time for European decision to participate in $6-billion program was growing short. (NASA Release 70-115; AP, *NYT*, 7/12/70, 4; *Marshall Star*, 7/15/70, 1)

July 8: Apollo 11 commander Neil A. Armstrong held press conference on his reactions to first anniversary of manned lunar landing July 20, 1969: "I really had hoped, I think, that the impact [of lunar landing] would be more far-reaching than it has [been]. I had hoped that it might take our minds away from some of the more mundane and temporal problems that we as a society face.... I think if we could...direct our attention a little farther into the future, to try to foresee problems that are going to occur in the next years and decades, then perhaps we could be attacking these solutions from the front side when it's...more easy to provide a viable solution and a meaningful one."

While scientific discoveries resulting from *Apollo 11* had been numerous and were still continuing, Armstrong felt "actual discovery, or revelation, was that man could in fact live on the moon in a rather normal fashion." This was "necessary key to all future plans for the use of our natural satellite, whatever they may be and however they might develop."

Armstrong described his feelings in looking at moon from earth: "I see the moon from the back yard as a friendly place. And I see a lot of landmarks. I can no longer look at it just as a spot in the sky; but, rather, I actually pick out, of course, the Sea of Tranquility, Clavius, Aristarchus, and places I know and have flown by."

Asked if he expected to fly in space again, Armstrong said, "I'd be surprised if I don't have the option of buying a ticket." (Transcript)

- Dr. Lee A. DuBridge, Director of OST and Presidential Science Adviser, testified before House Committee on Science and Astronautics' Subcommittee on Science, Research, and Development on need for adequate Federal science policy: "Science and technology policy...must be formulated on an evolutionary basis, continuously under review and subject to change with changing conditions. It must be a policy that can be implemented both by the Aministration and by Congress and one which will be supported by the public." Policy should provide "strong base of science" and "mechanisms for using science and technology to meet changing problems of national concern." It should "recognize that policies and events wholly outside the field of science and technology may strongly influence our R&D expenditure level."

  Dr. DuBridge told Subcommittee some of nation's "most valuable and productive research projects" had been lost because of FY 1970 amendment to DOD appropriation bill which banned defense-financed studies unless they had direct relation to military missions. (Testimony; Cohn, *W Post*, 7/9/70, A7)

- Dr. Robert C. Seamans, Jr., Secretary of the Air Force, announced selection of Boeing Co. as prime contractor for Airborne Warning and Control System (AWACS). Boeing was issued $16 500 000 initial increment of $169 982 522 cost-plus-incentive-fee contract for Phase I, which required modification of Boeing 707 aircraft to accommodate 30-ft-wide, mushroom-shaped radome on top of fuselage aft of wings. (DOD Release 564-70)

- Secretary of Transportation John A. Volpe announced award of $97 880 cost-sharing contract to Northern Research and Engineering Corp. to establish design criteria for control and reduction of oxides of nitrogen emitted from jet aircraft engines. (FAA Release 70-65)

- Undersea nuclear probe to detect valuable minerals on ocean floor had been successfully tested at AEC's Pacific Northwest Laboratory in Richland, Wash., AEC announced. Device, designed to operate from surface ship or submersible vehicle, could identify 20 to 30 elements in mineral deposits in three to five minutes using ultra-low-level and rapid neutron-activation analysis as opposed to laboratory analysis. (AEC Release N-119)

July 9: U.S.S.R. launched *Cosmos CCCLIII* from Plesetsk into orbit with 304-km (188.9-mi) apogee, 205-km (127.4-mi) perigee, 89.6-min period, and 65.4° inclination. Satellite reentered July 21. (GSFC *SSR*, 7/31/70; *SBD*, 7/10/70, 41)

- Agreement under which U.S. Coast Guard would conduct National Data Buoy Development Project at MTF had been signed by Dr. Thomas O. Paine, NASA Administrator, and Adm. T. R. Sargent, Acting Commandant of USCG, NASA announced. Agreement established basic operational support and reimbursement relations between NASA and USCG. Slidell, La., NASA computer facility also would be available to project. (NASA Release 70-114)

- Dr. Thomas O. Paine, NASA Administrator, received President's Safety Award on behalf of NASA in White House ceremony. Award recognized Federal Executive departments and agencies which best safeguarded lives and health of employees. Other 1970 winners were USAF and Civil Service Commission. (NASA Hq *WB*, 7/13/70, 2)

- Thomas O'Toole reported in *Washington Post* that NASA was considering canceling three of six remaining Apollo lunar landing missions so it could use Saturn V boosters to orbit Skylab space station. Although firm decision had not yet been made, NASA had "undertaken a top priority study to see if it might be feasible to cancel . . .the landings to divert funds into. . .an 'intermediate' space station that could be launched in 1976." (*W Post*, 7/9/70, A12)

- Sen. John L. McClellan (D-Ark.) on Senate floor accused NASA of secretly altering low bid of Fairchild Hiller in competition for $50-million contract for ATS-F and G spacecraft and of permitting General Electric Co. to reduce overhead costs in its final offer so that total was below that of Fairchild Hiller Corp. He compared procurement project with DOD TFX competition and contract awards. "I believe that a thorough review of the phased procurement source selection system should be undertaken. . .in order to determine whether the system should be modified, changed, or eliminated." (*CR*, 7/9/70, S10974-5)

- Mstislav V. Keldysh, President of Soviet Academy of Sciences, said at Moscow press conference that U.S.S.R. had not received proposal for general cooperation in space research and in docking systems from NASA Administrator, Dr. Thomas O. Paine. "But in the event that we do, we will treat them with maximum attention."

    Keldysh said *Soyuz IX* had provided foundation for new missions of more than 30 days, "quite long enough for a long-term orbital mission, because such stations presuppose crew changes." Earlier Soyuz missions devoted to rendezvous, maneuvering, docking, and transfer had been programmed to provide essential crew change knowledge. U.S.S.R.'s main task now was "orbital stations and. . .the study of the solar system and the earth by unmanned vehicles." (Clarity, *NYT*, 7/10/70; *SBD*, 7/10/70, 40)

- President Nixon sent to Congress Reorganization Plan No. 3 to establish Environmental Protection Agency (EPA) and Reorganization Plan No. 4 to establish National Oceanic and Atmospheric Administration (NOAA) within Department of Commerce. Plans would be effective in 60 days unless Senate or House disapproved.

NOAA would bring together functions of ESSA and its major elements—Weather Bureau, Coast and Geodetic Survey, Environmental Data Service, National Environmental Satellite Center, and Research Laboratories—with Bureau of Commercial Fisheries, Marine Game Fish Research Program, and Marine Minerals Technology Center (formerly of Dept. of Interior); National Oceanographic Data Center and National Oceanographic Instrumentation Center (formerly under USN); National Data Buoy Development Project (formerly of Coast Guard and DOT); National Sea Grant Program (formerly of NSF); and elements of U.S. Lake Survey (formerly of Army Corp of Engineers). New agency was to provide unified approach to problems of oceans and atmosphere; better understanding, development, and conservation of marine resources; consolidated efforts towards greater knowledge of oceanic and atmospheric phenomena and those of solid earth; and balanced Federal program toward more effective environmental monitoring control.

Environmentally related functions of other Federal agencies would be moved to new EPA to establish and enforce protection standards, study pollution effects and control, assist others in arresting pollution, and assist Council on Environmental Quality in recommending policies to President. (*PD*, 7/13/70, 908-21; Dept. of Commerce Release G 70-122; Text)

- Melvin R. Laird, Secretary of Defense, said at DOD press conference that U.S.S.R. had "gone forward with new starts" in build-up of ICBM force after opening of SALT in Helsinki in November 1969 and after transfer of talks to Vienna in April. Increase had included deployment of more SS-11 missiles and smaller SS-11 and SS-13 missiles. (Sheehan, *NYT*, 7/10/70,1)

July 9-11: NASA and Univ. of California at Berkeley scientists successfully launched 300 000-cu-m (10.6-million-cu-ft) balloon carrying 1400-kg (3000-lb) scientific payload from National Center for Atmospheric Research at Palestine, Tex. Balloon, part of High Altitude Particle Physics Experiment (HAPPE) project to study cosmic rays in upper atmosphere, reached 341 000-m (104 000-ft) altitude and drifted westward to Odessa, Tex., where it was brought down and recovered July 11. (MSC *Roundup*, 7/17/70,1)

July 10: Appointment of ATS Procurement Review Committee headed by LeRC Director Bruce T. Lundin to reconsider April 7 selection [announced April 8] of General Electric Co. for ATS-F and G contract was announced by Dr. Thomas O. Paine, NASA Administrator. Committee would make recommendation to new Selection Panel of senior NASA officials. Panel would reaffirm original selection or reverse it. Work done by GE or competitor, Fairchild Hiller Corp., since April 7 would not be considered relevant by Committee. NASA had discontinued funding of Fairchild work April 16 and would stop GE funding on July 15. Dr. Paine said NASA, Fairchild Hiller, and GE had agreed that reconsideration of selection was preferable to recompetition. (NASA Release 70-119)

- NASA announced publication of *Research and Technology: Objective and Plan* (N-7029204). It described new NASA method of summarizing research projects within agency to facilitate idea exchange among researchers and project managers and included research in progress to support NASA planning for next several years. RTOP

Summary replaced NASA Flash Index published in previous years and was designed to improve communications and coordination among technical personnel in government, industry, and universities. (NASA Release 70-116)

- AFSC announced award by SAMSO of $9 228 319 contract to Lockheed Aircraft Corp. for space experiments support program (SESP) launch scheduled for late 1971 from Vandenberg AFB, Calif. Contract called for integration of several satellites onto spacecraft that would be orbited by Thor-Agena booster. Largest experiment was 113-kg (250-lb) flexible solar array designed to test deployment of flexible panels in zero-g environment and verify long-term power-generation characteristics of new solar cell. (AFSC Release 132.70)
- Postage stamp commemorating record-breaking flight of *Soyuz IX* (June 2-19) had been issued by U.S.S.R., AP reported. (*CSM*, 7/10/70)
- Apollo 14 Commander Alan B. Shepard, Jr., said in Houston that he was concerned that cutbacks in space program, with ensuing layoffs, might lead to poor workmanship at KSC. "I think we would be naive if we didn't assume that people are unhappy." (*W Post*, 7/12/70, A5)
- In *Science* editorial, Brown Univ. scientist Dr. A. Hunter Dupree outlined changes that would shape new U.S. science policy: DOD had lost ability to justify support for basic research and to attract many scientists. "If the scientists knowledgeable in military research, who provide one of the groups with the best chance to change the course of events with competent criticism, lose touch with the Department of Defense completely, an unparalleled disaster could ensue. Yet, a reordering of the relation of the scientific community to the Department of Defense cannot be postponed." Scientific community needed to pay more attention to environmental problems. Space program "must find a role for itself with predominantly scientific objectives and a steady state of funding." Social sciences must receive greater emphasis. Justification of Federal support for research in universities must emphasize goal of building healthy institutions in national interest. And support for education "must contemplate a national research program with a radically different mix of disciplines from that recently prevailing." (*Science*, 7/10/70, 131)

July 11: USAF Athena missile launched from Green River, Utah, toward WSMR target, veered off course and apparently landed 290 km (180 mi) southeast of Chihuahua City, Mexico. Objective of test was to evaluate atmospheric reentry capabilities. (*W Post*, 7/11/70, A5)

July 12: NASA Lockheed Electra remote-sensing aircraft equipped to acquire black-and-white, color, and color-infrared photos and thermal infrared imagery, arrived in Lima, Peru, to help assess damage from severe earthquake. Mission data would be processed at MSC and transmitted to Peruvian government for analysis. Data also would contribute to application of remote-sensing science to natural disasters. (NASA Release 70-122)

- Four crewmen sealed into space station simulator at McDonnell Douglas Astronautics Co. [see June 12] successfully completed first 30 days of scheduled 90-day confinement. (*Langley Researcher*, 7/24/70, 1)

- William Hines in Washington *Sunday Star* praised NASA handling of *Apollo 13* investigation: Handling "contrasted sharply" with that following Jan. 27, 1967, Apollo spacecraft fire. "Administrator Thomas O. Paine is undoubtedly directly and personally responsible for the change. Paine seems to be one of the few high officials in Washington with a decent respect for the opinions—and the intelligence—of the taxpaying public."(*W Star*, 7/12/70, B4)
- Egyptian 13.7-m (45-ft) papyrus boat, *Ra II*, piloted by Norwegian anthropologist Thor Heyerdahl and seven-man crew arrived at Bridgetown, Barbados, after successfully completing 5100-km (3200-mi), 57-day journey from Safi, Morocco, to prove Egyptians could have crossed Atlantic 3000 yrs before Columbus. (*W Post*, 7/13/70, A1)
- Installation of Soviet antiaircraft missiles—SAM-2s and 3s—near Suez was threatening Israel's command of air, *New York Times* reported. While SAM-2s did not become operative until reaching 900-m (3000-ft) altitude, SAM-3s could knock down aircraft flying as low as 90 m (300 ft) and reached top speed of mach 3.5 much more quickly. SAM-3s also had better radar and guidance systems, which allowed faster reactions necessary to hit low-flying aircraft. (*NYT*, 7/12/70)

July 13: Survey of U.S. news coverage in five most important Soviet newspapers Jan. 1 to June 30, 1970, was published by *Washington Post*. U.S. space program, including *Apollo 13*, had been mentioned 30 times. U.S. economic problems had received 89 mentions and reports of U.S. "sick society," including crime, had appeared 82 times. (Astrachan, *W Post*, 7/13/70, A1)
- L/G Leslie R. Groves (USA, Ret.), former member of Army Corps of Engineers and head of Manhattan Project that developed atomic bomb, died of heart attack at age 73. Gen. Groves, who had described first atomic test in his 1962 book *Now It Can Be Told*, had been awarded USA's Distinguished Service medal for his contributions to the project and to shortening World War II. He had also supervised construction of numerous DOD facilities—including Pentagon and laboratories at Oak Ridge, Tenn.; Hanford, Wash.; and Los Alamos, N. Mex.—before his retirement from USA in 1948. (Ward, *W Post*, 7/15/70, C8)

July 14: NASA announced it had issued RFPs to potential experimenters for scientific investigations based on data acquired during ERTS missions A and B, scheduled for 1972 and 1973 launch. Missions would be first in U.S. space program devoted exclusively to study of earth's environmental resources from space. They were designed to improve methods for managing resources. ERTS A and B sensors would obtain image data in visible and infrared spectrum using spectral ranges previously studied by instrumented aircraft and verified by *Apollo 9* experiment in March 1969. Proposals were due April 15, 1971. (NASA Release 70-117)
- MSFC engineer Hans F. Wuenscher was awarded U.S. patent 3 520 496 for Serpentuator—snake-like instrument to move men and tools under weightless conditions around space station. Series of metal tubes with motors at joints to control angles could be regulated from control panel inside or outside space station. (Jones, *NYT*, 7/18/70, 29; Patent Off)

- U.K. launched Skylark SL-971 sounding rocket from Woomera Rocket Range, Australia, carrying rotation collimator x-ray telescope to detect x-ray stars in region of Scorpius and Sagittarius. (*Spaceflight*, 12/12/70, 484)
- Selection of Harrison A. Storms, NR Aerospace Div. Vice President, as recipient of 1970 Aircraft Design Award was announced by AIAA. He would receive $500 honorarium sponsored by Northrop Corp. for "direction of the engineering effort that led to the design and development of the XB-70 aircraft" at July 21 AIAA meeting in Los Angeles. (AIAA Release)
- Siegfried H. Reiger, ComSatCorp Vice President-Technical, died at age 50. Reiger had joined ComSatCorp in 1963 and had been directly responsible for technical activities, including launching of comsats for global comsat system. (ComSatCorp Release 70-41)

July 14-28: First meetings of joint working group established under Sept. 18, 1969, U.S.-India agreement on instructional television (ITV) satellite experiment project were held. Indian Space Research Organization (ISRO) and NASA personnel visited Bombay, New Delhi, and facilities in project. One-year experiment would provide first community broadcasting of TV by NASA ATS-F satellite to small village receivers without use of large ground relay stations. (NASA Release 70-121)

July 15: NASA announced it would reduce civil service employment by about 900 persons by Oct. 1. Reduction included 200 planned in FY 1971 Budget and 700 required by FY 1971 Authorization Act limitations [see July 2]. Reduction would bring total NASA civil service reductions in last three years to above 5200 and give NASA 29 850-member staff—lowest since 1963. (NASA Release 70-124)
- Dr. Thomas O. Paine, NASA Administrator, said at Univ. of Colorado press conference in Denver that Apollo 14 could have been pushed through for 1970 launch. "But we're still doing the undone, the untried, every time we venture into space—and we're not prisoners of a time program." NASA had decided to take extra month with less strain on all concerned. "There is much to be changed because of the lessons learned from Apollo 13." (Lindberg, *D Post*, 7/16/70)
- NASA announced selection of General Electric Co. for negotiation of $50-million, cost-plus-award-fee contract for hardware development of Earth Resources Technology Satellite (ERTS) system. Work —to be directed by GSFC—would include development of two flight spacecraft, equipment and services for ground data-handling system, spacecraft receiver, and six ground platforms for remote-site data-collection-system experiment. (NASA Release 70-123)
- Meteorological rockets were being launched from EC-121 "Super Constellation" aircraft at PMR for first time to test new sounding rocket system, PMR announced. EC-121 carried line of six rockets that were launched at 45° angle to rear of aircraft, curved to 91 000 m (300 000 ft) above flight level, and transmitted atmospheric data to aircraft while parachuting to ground. (PMR Release 1084-70)
- Joint DOD-DOT test program to determine value of helicopters in providing medical assistance to automobile accident victims and others requiring emergency medical care was announced by Secretary of Defense Melvin R. Laird and Secretary of Transportation John A. Volpe. Program would use UH-1 helicopters of Army's 507th Air

Ambulance Corps at Fort Sam Houston, Tex., in San Antonio area July 15 through Dec. 31. (DOD Release 577-70)
- Sen. Gordon L. Allott (R-Colo.) introduced, for himself and cosponsors, S. 4085, Technology Assessment Act of 1970, to establish Office of Technology Assessment for Congress as aid in identification and consideration of existing and probable impact of technological application. (CR, 7/15/70, S11336-40)
- House Subcommittee on Science, Research, and Development submitted to Committee on Science and Astronautics *Technology Assessment: Annotated Bibliography and Inventory of Congressional Organization for Science and Technology.* Bibliography was designed to identify selected items on technology assessment as part of public policy. Citations dealt with concept of technology assessment, proposals made for organization of technology assessment mechanisms in Government, and examples of technology assessments completed or in progress. (Text)

July 15-17: Conference on Space Shuttle Technology at LeRC was attended by more than 600 representatives of industry, universities, and foreign countries. (NASA Release 70-107; NASA PAO)

July 16: Nike-Cajun sounding rocket, launched by NASA from Wallops Station, carried GCA Corp. and Univ. of Illinois payload to 142.8-km (88.7-mi) altitude to measure electron concentration, electron collision frequency, electron temperature, and electric field intensity during daytime intense-blanketing sporadic E at altitudes of 100-110 km (62-67 mi). Rocket and instruments functioned satisfactorily and usable data were obtained. (NASA Rpt SRL)
- Dr. George M. Low, NASA Deputy Administrator, testified before House Ad Hoc Subcommittee on Aerospace Museum Study on H.R. 10771, introduced April 30, 1969, to authorize NASA study of advisability of establishing permanent NASA Aerospace Museum in western U.S.: "It is our view that a study for a Western Aerospace Museum should be the responsibility of the Smithsonian Institution, and not of NASA." NASA would support study. Dr. Low submitted copy of 1967 NASA-Smithsonian agreement on custody and management of NASA historical artifacts. NASA felt arrangement was "working well" and believed construction of National Air and Space Museum in Washington, D.C., "should proceed as expeditiously as possible when funding can be made available. NASA recommended existing museums, such as Goddard Rocket and Space Museum in New Mexico, be considered in study and that existing legislation for museum be taken into account so that legislation establishing second museum would provide for administration of both by Smithsonian. Study was expected to cost $50 000 to $75 000 and take slightly more than one year. (Testimony)
- New mineral found in *Apollo 11* samples had been named "armalcolite," incorporating letters of *Apollo 11* astronauts' names, Neil A. Armstrong, Edwin E. Aldrin, Jr., and Michael Collins, MSC announced. Name had been included in *Proceedings of Apollo 11 Lunar Science Conference of Geochimica et Cosmochimica Acta (cq.),* dedicated to *Apollo 11* crew, and had been approved by Nomenclature Committee of International Mineralogical Assn. (MSC Release 70-82)

- Sen. William B. Spong, Jr. (D-Va.), introduced S.J.R. 221, providing for construction of monument in Washington, D.C., commemorating *Apollo 11* lunar landing. Resolution was referred to Senate Committee on Interior and Insular Affairs. (*CR*, 7/16/70, S11535)
- Twenty-fifth anniversary of first atomic bomb explosion near Alamogordo, N. Mex. *New York Times* commented: "The destructive power that thundered at Alamogordo, was almost trivial in comparison with that which exists in any one of thousands of atomic and hydrogen bombs now in the arsenals of five nuclear powers. But the ingenuity and industry which have added so much to the effectiveness of the military atom have not yet been matched by corresponding political accomplishment aimed at protecting all mankind, at guaranteeing there will be no successors to Hiroshima and Nagasaki. Today's reality among the major powers is a peace based on mutual terror. . . ." (*NYT*, 7/16/70, 32)
- For first time in 25 yrs of nuclear age, "powerful and once-unquestioned interests that have grown up around this nation's atomic technology are being forced onto the defensive by a rising public clamor for reform," Anthony Ripley said in *New York Times*. "Bitter, often emotional accusations of carelessness, duplicity and indifference are being sounded over a broad range of subjects by scientists, politicians, peace groups, environmentalists and laymen concerned about health and safety." Government and industry had established more than 100 atomic-electric plants in 31 states during last few years. AEC had been "gearing up its weapons plants to increase Defense Department's supply of strategic nuclear warheads" by almost 300%.

  Public knowledge about atomic energy was focused on weapons, power plants, and Project Plowshare to find peaceful uses for atomic explosions but much of AEC's funding went for projects like atom smashers, including 200-bev accelerator under construction at Batavia, Ill., for $250 million. U.S. had spent about $49 billion on atomic energy since 1940, with about two thirds for weapon development. While five-man AEC watched over "massive operation," real power was wielded by Congressional Joint Committee on Atomic Energy. President Nixon had proposed that AEC Div. of Radiation Protection Standards be shifted to new Environmental Protection Agency. Plan was expected to meet strong opposition in Joint Committee as was Administration plan for remaking AEC into agency dealing with all forms of energy. (*NYT*, 7/16/70, 1)
- In letter to President Nixon, Walter Binaghi, President of ICAO Council, conveyed declaration made by ICAO Assembly during June 16-30 extraordinary session to consider unlawful interference with international civil aviation and its facilities. Declaration condemned all acts of violence directed against aircraft, crews, passengers, and civil aviation personnel; recognized need for international cooperation for civil aviation safety; requested action toward suppressing acts that jeopardized safety and development of international civil air transport; and requested application of Assembly recommendations to prevent and deter such acts. (*PD*, 8/17/70, 1063-3)
- *Washington Post* editorial commented on July 13 death of Manhattan Project head, L/G Leslie R. Groves: "The ultimate appraisal of Les-

lie Groves' contribution to his country and to mankind must await a future judgment—perhaps a Judgment Day. He was a superb soldier. The talents for organization and administration, the daring and imagination he brought to his mammoth assignment were peculiarly American, reflective in a sense of the genius of America. Their imprint on the world is, in any event, indelible." *(W Post, 7/16/70, A22)*

July 17: NASA's HL-10 lifting-body vehicle, piloted by Maj. Peter C. Hoag (USAF), successfully completed 37th flight from FRC. Objectives—evaluation and documentation of powered approach, evaluation of straight-in GCA approach, and acquisition of longitudinal stability and control data in subsonic configuration—were met. (NASA Proj Off)

- NASA announced it was seeking proposals for use of six orbiting satellites which had fulfilled their original scientific objectives but were still operational. Satellites—*Ogo I* (launched Sept. 4, 1964), *Ogo III* (launched June 6, 1966), *Explorer XXXI* (launched Nov. 28, 1965), *Explorer XXXIII* (launched July 1, 1966), *Oso III* (launched March 8, 1967), and *Oso IV* (launched Oct. 18, 1967)—had exceeded one-year design lifetimes with most of onboard experiments still functional and were available for new or expanded experimentation. (NASA Release 70-120)

- House Committee on Science and Astronautics had tentatively approved funding of $6.5-million project to convert antenna 300-m (1000-ft) wide at Arecibo Observatory in Puerto Rico to radar 2000 times more sensitive than any in existence, *New York Times* reported. Modification of world's largest radiotelescope would enable telescope to map features on Venus down to mile in width. (Sullivan, *NYT*, 7/17/70)

- Situation at KSC and in area on first anniversary of *Apollo 11* launch was described by John N. Wilford in *New York Times*. Since July 1969, employment at KSC had dropped from 23 500 to 16 500 with major Apollo contractors hardest hit. At least 750 workers would be laid off as result of $12.2-million cut in Patrick AFB contract with Pan American World Airways, Inc., for operation of tracking stations. "For the entire Cape Kennedy area, it has been a year of economic woe, frustrated growth plans, for-sale signs and broken personal dreams. This is no ghost town, but the space-rush boom—which in a decade doubled the Brevard County population to 230 000—is definitely over." Dr. Kurt H. Debus, KSC Director, had said, "We are on a downward trend but this should reach a level as of September." (*NYT*, 7/17/70, M27)

- Senate passed with amendment S. 209, Smithsonian Institution facilities construction bill, which authorized sum not to exceed $500 000 for preliminary planning and design of museum support and depository facilities to be constructed on Federally owned land within District of Columbia. (*CR*, 7/17/70, S11613)

- ComSatCorp reported increase in 1970 second-quarter net income to $3 974 000 (40 cents per share), up from $1 976 000 (20 cents per share) for second quarter of 1969. Earnings for first six months of 1970 totaled $7 319 000 (73 cents per share), increase from $3 501 000 (35 cents per share) for first six months of 1969. (ComSatCorp Release 70-42)

July 18: President Nixon issued statement on Congressional action and Government spending: There was "persistent and growing tendency" in Congress to approve increase in expenditures "without providing the revenue to pay the costs." It had become "almost a cliché to say that all we need do to resolve this dilemma with regard to our Federal budget is to cut space and defense outlays and 'change our national priorities.' Let's set the record straight. We *have* changed our national priorities."

In proposed FY 1971 budget, spending for human resources (45%) exceeded defense spending (32%) for "the first time in 20 years." To accomplish "this massive change in emphasis, military and space expenditures were cut by some $6 billion." But continuation of pattern of Congressional action would substantially increase expenditures. President asked Congress to establish "*firm* ceiling on total expenditures," within which he could determine priorities. (*PD*, 7/20/70, 940-1)

- White House released summary of *Toward Balanced Growth: Quantity with Quality*. Report to President by National Goals Research Staff said basic scientists were being asked to shift their work "from the development of knowledge for its own sake to working on basic problems which have relevance for today's issues." U.S. needed "forum in which the partially conflicting needs for maintaining the integrity of the core of basic research and the practical needs of the society are resolved." In technology assessment, "we may have to accelerate our efforts to detect new benign technological opportunities and facilitate their rapid introduction to offset the impact of inhibiting the introduction and use of harmful technology." (*PD*, 7/20/70, 941-7)

- Two-year battle for multibillion-dollar "air bus" market would enter new phase with July 23 roll-out of McDonnell Douglas Aircraft Co.'s three-engine DC-10, Robert A. Wright said in *New York Times*. First of wide-bodied trijets was appearing "at a time when orders from the depressed, money-short airlines have almost dried up. But the competition between Douglas and the Lockheed Aircraft Corporation, which is producing the rival L-1011 TriStar is heating up." Douglas had "decided lead" over Lockheed in production schedules and orders for "generation of aircraft that is expected to be a workhorse of the airlines in the 1970's." Lockheed, which would roll out L-1011 in September, was sensitive about "growing public speculation" that its financial distress stemming from four defense contracts might impair L-1011 program. Bankruptcy or merger had been suggested as solutions to Lockheed's financial problems. "But if Lockheed is bloodied its L-1011 team. . .is talking and acting aggressively." (*NYT*, 7/18/70, 29)

July 19: Alabama Space and Rocket Center in Huntsville had attracted 60 000 visitors since March 17 opening, *New Haven Register* reported. Averaging 770 visitors per day, Center had registered visitors from 49 states and 35 foreign countries. (*New Haven Register*, 7/19/70)

- Tass announced U.S.S.R. would conduct carrier rocket tests in Pacific from July 25 to Aug. 25 and warned ships and aircraft to avoid area 966 km (520 nm) northwest of Midway Island with 74-km (40-

nm) radius between noon and midnight local time. (UPI, W *Star*, 7/20/70, A1)
- Widespread hostility to science was evident in broad sectors of public, particularly the young, and "reflected in the attitudes of legislators who control funds for university and laboratory research," Walter Sullivan said in *New York Times.* While public disaffection was partly to blame, "this drop in research support is also attributable to the Vietnam war and other budget demands." There was "growing feeling in Washington and on the campuses that efforts to explain science to the young, as well as to the adults, have failed—that new ways must be found to do so and that the old clichés are not enough." (*NYT*, 7/19/70, 5, 7)

July 20: President Nixon issued statement on anniversary of first manned lunar landing: "This triumph of unique achievement...brought...a moment of greatness in which we all shared, a priceless moment when the people of this Earth became truly one in the joy and wonder of a dream realized."

Later, at White House press conference, President said he had not figured out any travel plans that would "top" July 20, 1969, trip to greet *Apollo 11* astronauts in mid-Pacific. (*PD*, 7/27/70, 962, 969)

- On first anniversary of *Apollo 11's* first manned lunar landing, *Apollo 11* Astronauts Neil A. Armstrong, Michael Collins, and Edwin E. Aldrin, Jr., with NASA Administrator, Dr. Thomas O. Paine, visited *Apollo 11* spacecraft exhibit at state capitol in Jefferson City, Mo., and received plaque from Missouri Gov. Warren E. Hearnes. Later they flew to New York, where they presented 113 g (4 oz) lunar sample to U.N. Secretary General U Thant. Sample would remain on exhibit at U.N.

Presenting sample, Dr. Paine said: "What the full significance of Apollo will be to the world of the future, no one can yet predict. We must await the verdict of history. But as we continue to learn and to explore, I am inclined to think that our first moon landing may some day be thought of as the turning point when man first demonstrated that he can open new worlds, where eventually exciting new extraterrestrial societies will be founded. Mankind will move onward to settle and colonize other worlds, and the United Nations will encompass new territories, if not new dimensions."

NASA centers and tracking stations around world celebrated anniversary throughout week. (NASA Release 70-125; UPI, *NYT*, 7/21/70; Text)

- Newspaper editorials marked lunar landing anniversary:

New York Times: "...the Eagle's landing was one of mankind's finest hours. The scientific dividends from the new era of lunar exploration have already been very rich, and they will be even more abundant in the future when numerous astronauts from many nations will work on the lunar surface. Men now inhabit the earth-moon system, not earth alone, and much in tomorrow's history will be shaped by the developments that will follow the Eagle's triumph." (*NYT*, 7/20/70, 26)

Washington Post: "The basic issue involved in NASA's appropriations this year, as it will be in the next year and for many years to come, is whether this country should give up something it has done and which is going well, something that has brought it great international prestige and internal pride, and something that we believe may well hold the key to man's future." (*W Post*, 7/20/70)

- India announced plans to launch first satellite within 3½ yrs. Vikram Sarabhai, Secretary of Indian Atomic Energy Dept., said satellite would weigh 29.9 kg (66 lbs). be football size, and be launched by four-stage booster. (Reuters. *W Post*, 7/21/70, A9)
- Formation of Environmental Advisory Council to ensure environmental acceptability of SST operations was announced by William M. Magruder, DOT Director of SST Development, in speech before National Press Club in Washington, D.C. Council—chaired by Dr. Myron Tribus, Assistant Secretary for Science and Technology, Dept. of Commerce—would "suggest or plan research in any areas where doubts or uncertainties exist. Where further research is necessary it will be conducted, so that the definitive data and unequivocal answers on which the environmental acceptability of the SST can be determined will be in hand before a decision is made on the production program." (Text)
- Coalition Against the SST—23 groups including Sierra Club, Friends of the Earth, and United Auto Workers Union—urged Federal law banning SST sonic booms over U.S. Coalition spokesman, Dr. Karl M. Ruppenthal of Stanford Univ., told Washington, D.C., news conference Government officials "state that no annoying overland flights are contemplated, but they refuse to cooperate when meaningful noise legislation is proposed." (Cohn, *W Post*, 7/21/70, A2)
- World's first team of female aquanauts surfaced after two weeks of research in Project Tektite habitat and entered decompression chamber for 21-hr stay [see June 6]. (AP, *W Post*, 7/21/70, A6)

July 21: Nimbus IV meterorological satellite, launched April 8, was adjudged successful by NASA. Spacecraft had acquired sufficient number of global samples of atmospheric radiation measurements to compare vertical temperature, water vapor, and ozone profiles, and to compare merits of several instrument approaches. (NASA Proj Off)

- Examination of Federal science policy was essential, Dr. Philip Handler, NAS President, said in testimony before House Committee on Science and Astronautics' Subcommittee on Science, Research, and Development. "Substantial changes are being made in the organization of some arms of the executive branch of government that deal with science; the Mansfield Amendment which limits the manner in which research-supporting funds may be utilized by the Department of Defense has begun to influence. . .behavior of other agencies of the government as well; federal funding for fundamental studies has remained essentially plateaued in absolute dollars for four consecutive fiscal years while, in constant dollars, such funding has declined by perhaps 25%.". NASA appropriation "has been reduced significantly and one frequently hears that federal military and space expenditures for R&D are to be reduced for some years to come; the nation seems determined to mitigate the damage which has been done to our natural environment, but flounders in the attempt; our nation is engaged in the painful exercise of assessing and reassembling our priorities; and our country now seems uncertain in what light, and with what resolution, it should view the pace of future scientific progress. Our national apparatus for the conduct of research and scholarship is not yet dismantled, but it is falling into shambles. Morale of the scientific community is lower than at any time since World War II. Yet, new fields of scientific exploration and application clamor for attention and funding." U.S. lead in science "is in jeopardy." Loss of lead "bodes ill for our future national security and for the vigor of our economy." (Testimony)

- Louisville, Ky., architect Carl D. Russell received patent for emergency warning system that would permit President of U.S. to reach 95% of U.S. public within 60 secs. System would automatically turn on radios at full volume and would activate broadcasting stations that were off the air. System could be activated in individual sections of country in local emergencies and could be installed within year for $1 billion. (AP, B Sun, 8/17/70, A4)

- President Nixon submitted to Senate nomination of Rudolph A. Peterson, Chairman of Executive Committee of Bank of America, to succeed William A. Hagerty as member of ComSatCorp Board of Directors until 1973. (PD, 7/27/70, 969, 986)

- Legacy of *Apollo 11* was described by Walter Sullivan in *New York Times:* Collection and analysis of lunar samples had "not only reinforced the earlier deductions. . .but also made possible basic new discoveries." Many "fruits of Project Apollo" were "yet to be harvested." Records of earth-moon distances furnished by laser reflector left on moon by *Apollo 11* might eventually reveal whether or not gravity was weakening and moon was increasing its distance from earth. "Examination of lunar material that was subjected to bombardment by gas from the sun during various periods of the past should also help outline the history of the earth's parent star." Apollo findings "obtained at high cost and high risk" had not thus far led to "revolutionary discoveries." They had produced one suggestion that would be revolutionary if confirmed—proposal by Cornell Univ. astronomer Dr. Thomas Gold that tektites on lunar samples had been formed when sun flared to 100 times its normal brilliancy every few thousand years. (NYT, 7/21/70, 18)

- Lockheed Aircraft Corp. Vice President F. A. Cleveland received $1000 honorarium and $1500 travel allowance for 1970 AIAA Wright Brothers Lecture at AIAA Aircraft Design and Operations Meeting in Los Angeles. He delivered lecture, "Size Effects in Conventional Aircraft Design," during meeting. (AIAA Release; AIAA PIO)

July 22: Development of light source 20 times more powerful than largest light source commercially produced was announced by LeRC. Operating at 400 kw, 76-mm (3-in) electric arc radiated through 125 lenses and lit 47-sq-m (500-sq-ft) area. It had been installed in Space Power Facility vacuum chamber at Plum Brook Station, Ohio, to simulate deep space environment for testing power-generation systems on communications satellites or manned orbiting laboratories. (LeRC Release 70-36)

- National Air and Space Museum was victim of "shocking lack of attention" by Smithsonian Institution management, Sen. Barry M. Goldwater (R-Ariz.) said on Senate floor. He said that museum received "extremely meager share," about 1.7% of $40-million-plus annual Smithsonian appropriation, and was undermanned, with "only 31 people on board." Museum attracted at least one third of visitors to Smithsonian Park. It had no director "even though it has been 24 months since the last Director gave notice of his planned retirement"; museum had been put into arts and humanities wing of Smithsonian for years rather than being considered science and technology component; in 1969, Smithsonian management had threatened to expel museum from Mall. Most museum exhibitions were badly housed and deteriorating rapidly, and program for construction of permanent Air and Space building was "at dead center." In words of "Paul Johnston, the last Director of the museum, the project 'may never get off the ground.'" (CR, 7/22/70, S11911-6)

- Outline of financing plan to keep Lockheed Aircraft Corp. in business in "substantially its present form" had been agreed on by 24 of major U.S. banks, *New York Times* reported. But bulk of funds Lockheed was thought to require—$300 million out of total $430 million—would have to be provided either by Government directly or as Government-guaranteed loan. (Heinemann, *NYT*, 7/22/70, 1)
- Anglo-West German agreement to develop combat aircraft for service in late 1970s was announced in House of Lords in London by U.K. Minister of Defence, Lord Carrington. Aircraft would replace U.S. F-104s of West German Air Force and U.S. Phantoms in service with RAF. (Middleton, *NYT*, 7/23/70, 13)

July 22-24: European Space Conference held fourth meeting in Brussels. Conference decided to continue ELDO programs with Europa I and II boosters and preliminary work on Europa IIIB and endorsed ESRO scientific satellite programs and preliminary study of large meteorological satellite. It approved new ESRO programs on applications satellites toward development by 1975 of 200-kg (440-lb) Europa II comsat and by 1980 of advanced 500-kg (1100-lb) satellite to be launched by Europa IIIB and agreed to develop, with NASA, satellite for air-traffic control and navaid. Conference also approved merger of ELDO, ESRO, and CETS into new European Space Organization (ESO) and decided to continue negotiations with NASA on European participation in post-Apollo programs. Conference endorsed ELDO's space tug studies as form of participation in post-Apollo program and approved measures for consolidating European position in negotiations for permanent INTELSAT constitution. (Cleaver, *A&A*, 10/70, 70-2)

July 23-24: Intelsat-III F-8 was successfully launched from ETR by NASA for ComSatCorp on behalf of International Telecommunications Satellite Consortium (INTELSAT). Three-stage, long-tank, thrust-augmented Thor-Delta launch vehicle boosted satellite into planned elliptical transfer orbit with 36 245.1-km (22 526.5-mi) apogee, 260.3-km (161.8-mi) perigee, 641-min period, and 27.8° inclination. Apogee motor began scheduled 27-sec burn at 27:00 GET to place *Intelsat-III F-8* into planned synchronous orbit over Pacific, but motor cut off after burning for 14.5 secs and contact with satellite was lost abruptly. Repeated efforts to command or locate satellite were unsuccessful and ComSatCorp appointed committee to investigate failure.

Intelsat-III F-8, planned as spare for Pacific and Indian Ocean areas, was final launch in Intelsat III series. Of seven Intelsat III satellites launched from Sept. 18, 1968, to April 22, 1970, four were operational with 1200 voice circuits each. (ComSatCorp PIO; GSFC *SSR*, 7/31/70; NASA Proj Off)

July 23: USAF launched unidentified satellite on Titan IIIB-Agena booster from Vandenberg AFB into orbit with 404-km (251-mi) apogee, 118.9-km (73.9-mi) perigee, 90.1-min period, and 59.9° inclination. Satellite reentered Aug. 19. (GSFC *SSR*, 7/31/70; 8/31/70; Pres Rpt 71)

- ESSA announced two-year program to complete U.S.-Canadian satellite triangulation program—highly accurate geodetic-survey network with stations 1000 to 1300 km (600 to 800 mi) apart linking Canada with Alaska and lower 48 states. Program had been initiated in 1964 with *Echo I* and *II* balloon satellites and suspended when June 23, 1966, launch of *Pageos I* signaled start of worldwide program in which simultaneous observations from 41 portable camera stations would be used to construct three-dimensional geodetic reference system. System would measure distance between two surface points on earth 4800 km (3000 mi) apart with 9.8-m (32-ft) accuracy. *Pageos I* would be photographed from U.S. and Canadian sites in new program to enable land surveyors and engineers to start their surveys from geodetic control points with greater accuracy. Completion of North American network could give measurements with accuracy of 1:1 000 000, comparable to error of 305 mm (12 in) in approximately 322 km (200 mi). (ESSA Release 70-46)

- McDonnell Douglas DC-10, first of wide-bodied trijet transport aircraft, was rolled out in ceremonies at Long Beach, Calif. Aircraft was 55 m (180 ft) long, with 5.2-m-dia (17-ft-dia) fuselage. Tail was 17.7 m (58 ft) from ground. (McDonnell Douglas Release 70-111)

- Apollo 15 prime and backup crews were among 14 astronauts who observed explosion by Canadian Defence Research Establishment of 453 500 kg (500 tons) of TNT in international test near Suffield, Alberta. Blast produced 70-m-dia (230-ft-dia) crater 4.6 m (15 ft) deep, with central uplift similar to that observed in many lunar craters. Astronauts had been invited to view processes in crater formation. Explosion was second 453 500-kg detonation in series to determine effects on military equipment and civilian shelters of blast forces and ground shock equivalent to those from small tactical nuclear weapon. (MSC *Roundup*, 7/31/70, 1)

- USAF science policy was outlined by Dr. Robert C. Seamans, Jr., Secretary of the Air Force, in testimony before House Committee on Science and Astronautics' Subcommittee on Science, Research, and Development: "Air Force will still support long-range, fundamental research, but in fewer areas than previously." Research "in aerodynamics, aeropropulsion, and structures will always be directly related to the Air Force missions." Research "in solid state physics and chemistry will be directly related to the Air Force mission as long as we need new electronic systems and lighter, higher temperature materials. But research in fields such as astrophysics (other than solar phenomena that do affect our operations) is probably permanently removed from the Air Force program." (Testimony)

- Lawrence Radiation Laboratory scientists Dr. John W. Gofman and Dr. Arthur Tamplin were engaged in "bitter, name-calling battle" with AEC while nuclear power industry "watched uneasily," Donald Rothberg said in *Washington Post*. Dr. Gofman and Dr. Tamplin claimed Federal standards allowed too much exposure to radiation and that if every American had maximum permissible dose of radiation, cancer deaths would increase by 32 000 per year. AEC said claim was unfounded and incorrect. Watching debate was power industry, "with its 17 nuclear plants now generating electricity, 47 under construction and 48 more proposed." (*W Post*, 7/23/70, A1)

July 23-24: No. 1 thruster on board NASA's orbiting *Sert II* (Space Electric Rocket Test) satellite failed after 3785 hrs operation—about one month short of design goal—causing early shutdown of one ion engine. Seven attempts to restart it were unsuccessful. No. 2 thruster, turned on July 24, would be turned off in late August during solar eclipse. If attempts to restart No. 1 were unsuccessful after eclipse, No. 2 would be turned on again to complete six-month test. (NASA Special Release)

July 24: President Nixon spoke at Pioneer Day ceremonies in Salt Lake City, Utah: "Just a year ago. . .I welcomed back from the moon three men who had been pioneers in landing on the moon. . . . I can only say that the spirit that took those three men to the moon. . .has built the greatest country on earth." (*PD*, 8/3/70, 995)

- USAF C-5A Galaxy, world's largest aircraft, flew first cargo flight between U.S. and Europe—from Charleston AFB, S. C., to Dover AFB, Del.; Rhein Main Air Base, Germany; Lakenheath RAF Station, England; and Torrejon Air Base, Spain. Regular U.S.-Europe C-5 cargo flights would begin in August. (DOD Release 594-70; USAF PIO)

- Western Europe wanted full partnership in U.S. post-Apollo space program and access to any inventions that emerged from it, Theo Lefevre, President of European Space Conference, said at Brussels meeting. Conference of 13 nations asked Lefevre to visit Washington, D.C., in September to examine financial and political conditions of European participation. Conference agreed to establish new space agency to replace ELDO, ESRO, and CETS and develop program to give Western Europe operational systems of telecommunication satellites, 1978-1980; begin joint program with U.S. for air traffic control satellites; finance studies for weather satellites; and proceed with construction of Europa 1 and 2 rockets and development of Europa 3. (AP, *NYT*, 7/25/70)

- International cooperation in space was described in *Science* by Arnold W. Frutkin, NASA Assistant Administrator for International Affairs: NASA had developed extensive program which "opens the entire range of its space activities to foreign participation and benefit." Limitations in program lay in areas of manned space flight, large-booster development, and cooperation with U.S.S.R. Low level of European space budgeting accounted for gaps in manned space flight and booster development. Lack of Soviet cooperation "seems to rest with Soviet political views rather than with technical problems or any lack of interest by the United States." Despite restricting factors abroad, there was "every reason to persevere with existing and improved programs for international cooperation." (*Science*, 7/24/70, 333-8)

- NASA announced selection of Philco-Ford Corp. for negotiation of $12 500 000, cost-plus-award-fee contract to develop Synchronous Meteorological Satellite (SMS) system that would become part of

National Operational Meteorological Satellite System (NOMSS), managed and operated by ESSA. (NASA Release 70-127)
- Public exhibit of lunar samples by NASA was described by George M. Low, NASA Deputy Administrator, in letter to Rep. Olin E. Teague (D-Tex.), Chairman of Special Ad Hoc Subcommittee of House Committee on Science and Astronautics: "Even with a reasonable allowance for overlap in our attendance figures for individuals who may have seen exhibits more than once, these figures indicate that over six and a half million people in the United States have seen a lunar sample." Public displays had been accompanied by exhibit and supporting material so that viewers "come away...with some appreciation of its broader meaning." NASA believed "exhibits at 15 major science museums to date, with 41 more planned for the coming year, in addition to the exhibit accompanying the Apollo 11 capsule to all 50 state capitals and the Smithsonian exhibit are especially effective." (CR, 8/7/70, E7445-6)
- President Nixon approved H.R. 16595, $550-million NSF Authorization Act of 1971, which became P.L. 91-356. (PD, 8/3/70, 1007)
- Dr. Lee A. DuBridge, Presidential Science Adviser, said in *Science* editorial: "Granted that many of our current problems must be cured more by social, political, and economic instruments than by science and technology, yet science and technology must still be the tools to make further advances in such things as clean air, clean water, better transportation, better housing, better medical care, more adequate welfare programs, purer foods, conservation of resources, and many other areas. The discovery and use of knowledge has always been relevant to a humane future. They are equally relevant today." (*Science*, 7/24/70, 331)

July 25: Delivery of FB-111 bomber to Carswell AFB, Tex., marked return of first aircraft of F-111 designation to operational flying status since Dec. 22, 1969, grounding of all F-111 aircraft—more than 230—following fatal crash. (AP, *W Post*, 7/28/70, A6)
- Japan's Institute of Aeronautical Science had announced that Japan on Aug. 19 would launch 61.7-kg (136-lb) satellite equipped with four telemetric antennas, two solar-electric-wave antennas, and solar batteries, UPI reported. (*NYT*, 7/26/70)
- *New York Times* said U.S. had formally presented package proposal for strategic arms agreement to U.S.S.R. Terms included overall numerical limitation on strategic launching systems, limitation within quota on numbers of giant missiles developed by either side, and limitation of antimissile defense systems to fewer than 100 launchers. (Smith, *NYT*, 7/25/70, 1)
- USN announced first launch of Poseidon missile from nuclear submarine, scheduled for July 27, would be postponed pending additional preflight tests. (AP, *B Sun*, 7/28/70, A1)

July 26: Apollo seismic experiment scientists, after comparing thousands of signals received by radio from sensitive seismometer left on moon by *Apollo 12* astronauts, determined that one or more moon quakes were occurring every 28.4 days, each time moon came closest to earth. Quakes were centered in Fra Mauro highland crater 80 km (50 mi) south of scheduled Apollo 14 landing site. Chief lunar seismologist Dr. Gary V. Latham later said quakes might be source of escaping gases that produced red or orange flashes often seen by astronomers. He intended to ask astronomers to concentrate on Fra Mauro area in future search for flashes. Dr. Latham reported in interview with *Washington Post* that "seismic record since November now produces strong evidence that the moon has at least some internal heat and seismic activity. (ALSEP Status Rpt; Cohn, *W Post,* 7/28/70, A1)

- "Potentially fatal financial crisis" facing Lockheed Aircraft Corp., "nation's top defense contractor" with total sales in 1969 of $2.07 billion, was described in *Washington Post* by Michael Getler. "As Lockheed's problem unfolds, it sends new shudders through an aerospace industry already shaken by a $3-billion dip in sales last year and the loss of more than 200,000 jobs." DOD officials believed Lockheed's fate depended upon Senate authorization of $200 million in contingency funds for C-5A in military spending bill and Lockheed's raising estimated $250 million from private sources without Government guarantees, to see it through first deliveries of L-1011 airbus in autumn 1971. (*W Post,* 7/26/70, A1)

July 27: NASA-Indian Space Research Organization project to broadcast TV from satellite to 5000 small Indian villages had been postponed from 1972 to 1974 because of NASA budget cuts, *New York Times* reported. Postponement, which also meant year's delay in launch of India's first comsat, had been announced in New Delhi by five-man NASA team headed by Dr. Richard B. Marsten, NASA Director of Communications Programs. Team was visiting India for consultations. Launch of Indian comsat depended on data gathered from educational TV experiment. (Schanberg, *NYT,* 7/27/70, 7)

- NASA published *Apollo 12 Preliminary Science Report* (NASA SP-235): Nov. 14-24, 1969, mission had accomplished "wide variety of preplanned tasks and paved the way for planning future missions to smaller, more selected landing areas with the possibility of significant scientific returns." By changing CSM orbital plane twice, crew had demonstrated capability to explore new lunar areas during orbital operations. "Future flights will take advantage of this capability to photograph additional potential landing sites and to make scientific observations of the surface, both visually and photographically." *Apollo 12* success in lunar orbit had allowed increase in planned orbital activities for *Apollo 13.* Large quantity of lunar soil and rocks from *Apollo 12* mission would add to scientific information obtained from *Apollo 11* samples. (Text)
- Aerobee 150 sounding rocket was launched by NASA from WSMR carrying GSFC payload to conduct stellar studies. Mission did not meet minimum scientific requirements. (SR list)
- World's first female aquanaut team received special awards from Walter J. Hickel, Secretary of the Interior, during Washington, D.C., ceremony. Later Richard H. Sprince of NASA, who was project manager for team, said test of women's ability to function in hostile environment was one reason NASA had joined Tektite program. Asked if NASA would consider bringing women into astronaut program, Sprince said, "We've got to start looking in that direction." (AP, B *Sun,* 7/28/70, A3)
- First launch of Poseidon missile from submarine had been postponed by USN [see July 25] because of proximity of heavily instrumented Soviet trawler to U.S. nuclear submarine *James Madison,* AP re-

ported. Informed source had said there was nothing wrong with Poseidon or support systems and that DOD officials "became a little apprehensive after the Russian trawler...moved in so close [within 183 m (200 yds)]" during test July 24. (B *Sun*, 7/28/70, A1)
- Secretary of Defense Melvin R. Laird and Deputy Secretary of Defense David Packard met with President Nixon in San Clemente, Calif., to discuss report of Blue Ribbon Defense Panel on DOD reorganization [see July 1]. Later, Secretary Laird told press U.S. was instituting new practice of "fly before we buy" to prevent costly defense overruns. Single package contract used to purchase C-5A and TFX was bad, Laird said, because it provided no protection to taxpayer against huge overcosts. Under new plan, DOD would let contracts for individual components of new weapon systems, which would permit cancellation of any one if component proved not feasible. (*PD*, 8/3/70, 1005; Beckman, *C Trib*, 7/28/70, 2)
- William J. Normyle commented on July 15-17 Conference on Space Shuttle Technology at LeRC in *Aviation Week & Space Technology*: "Problems facing the development of the system were far more evident than solutions." Lack of resolution had led to decision to hold another conference in nine months, by which time NASA "will have produced some key solutions." It was clear at meeting "that experts in a number of fields differ with their colleagues on selection of such basics as thermal protection, design configuration, materials, integrated avionics, operations and crew systems." (*Av Wk*, 7/27/70, 17)

July 27-31: Crew station review of Skylab airlock and multiple docking adapter was held at McDonnell Douglas Astronautics Co. in St. Louis, Mo. McDonnell Douglas was developing airlock; MSFC was building multiple-docking-adapter structure; and Martin Marietta Denver Div. was integrating equipment and experiments. (MSFC Release 70-146)

July 28: Dr. Thomas O. Paine, NASA Administrator, submitted his resignation, effective Sept. 15. In letter to President Nixon at Western White House, Dr. Paine said: "Now is an appropriate time for a change of command at NASA, and this coincides with my wish to return to private life. During my direction Americans orbited the moon and walked on its surface, achieving our boldest national goal on time and within budget. We have made the transition to the post-Apollo internationally oriented space program of the 1970's, and the Congress has approved the new direction and pace in the 1971 budget. We will shortly publish a prospectus for man's conquest of space through the year 2000 which charts a long-range plan for future progress. . . . It has been a privilege and honor to have led the nation's space program through critical times. . . . I am most grateful to you for having given me this unique opportunity to serve my country during mankind's first journey to another world."

Dr. Paine later told press in San Clemente, Calif., he had accepted challenging position with former employer, General Electric Co., involving "important national problems and technical opportunities outside aerospace and defense" fields and assured them that resignation had "absolutely nothing" to do with budget cuts.

President accepted resignation with deep regret and expressed gratitude for Dr. Paine's "outstanding leadership" of U.S. space program: "Your contribution to man's knowledge of the Earth as well as the heavens has been major, and the course you have done so much to set will help guide our efforts for years to come. . . . You have earned a unique and permanent place of honor in the history of man's exploration." (*PD*, 8/3/70, 997; Lyons, *NYT*, 7/29/70)

- X-24A lifting-body vehicle, piloted by Maj. Jerauld R. Gentry (USAF), successfully completed 15th flight from FRC. Objectives were to obtain lateral directional derivatives at mach 0.85 to 0.9 and longitudinal trim and lift-to-drag data with 40° upper flap. (NASA Proj Off)
- Rep. Joe L. Evins (D-Tenn.) submitted to House conference report on H.R. 17548, FY 1971 Independent Offices and HUD appropriations bill including $3.269-billion NASA appropriation. Report appropriated $2.565 billion to NASA for R&D instead of $2.500 billion proposed by House and $2.606 billion proposed by Senate. NASA appropriation for construction of facilities was $24.9 million instead of $18.2 million as proposed by House and $34.4 million as proposed by Senate. Research and program management remained at $678.7 million.

  Report also authorized $20.5 million for NSF program development and management as proposed by Senate instead of $19.5 million as proposed by House, in total $513 million NSF appropriation. (*CR*, 7/28/70, H7326-8; House Rpt 91-1345)
- NASA announced selection of seven scientific investigations for Mariner-Venus-Mercury 1973 (MVM-73) mission to photograph the planets, measure particles and fields surrounding them, and study their atmospheres and ionospheres. Experiment package would weigh 51 kg (113 lbs) and would include imaging science, radio science, plasma science, magnetometer, UV spectrometer, infrared radiometer, and energetic particles experiments. The 400-kg (900-lb) spacecraft would be launched in fall 1973 and would make gravity-assist swing within 5300 km (3300 mi) of Venus in February 1974 and fly within 1000 km (625 mi) of Mercury in March 1974. (NASA Release 70-126)
- Canada's plans to launch first comsat by end of 1972 had been set back because of controversy over who should build it, Peter Dempson reported in *Christian Science Monitor.* Hughes Aircraft Co. of California had bid $30 million for 12-channel comsat with 12% Canadian components. RCA Canada Ltd. of Montreal had bid $35 million for 8-channel comsat with 65% Canadian components and insisted on cost-plus contract that could raise cost to $100 million.

  Telesat, government agency formed in 1969 to develop Canada's comsat program, preferred Hughes because of lower bid and because Hughes was pioneer in spacecraft construction. RCA supporters argued that selection of RCA would enable Canada to build up competent scientific team, thus adding substantially to space knowhow that constructed *Alouette I* and *Alouette II*, with extra cost well worth it. If kept in Canada, they argued, project would provide employment for up to 1200 skilled and semiskilled workers in Quebec, where unemployment was high. (*CSM*, 7/28/70)
- Point-to-point second-generation space shuttle was possibility and could be "transportation system of the future as opposed to the hypersonic aircraft," Dale D. Myers, NASA Associate Administrator for Manned Space Flight, said in speech before Aero Club in Washington, D.C. (Text)

- USAF-USN feasibility studies had proved that land-based ICBMs could easily be deployed on surface ships, Chicago *Sun Times* quoted DOD sources as saying. Sources said studies were basis of decision to propose missile and bomber freeze at Vienna SALT meetings. (Ross, C *Sun Times*, 7/28/70)
- MSFC announced it had awarded $38 979 000 contract modification to McDonnell Douglas Astronautics Co. for additional work on Skylab airlock. (MSFC Release 70-147)

July 29: Cosmos *CCCLIV* was launched by U.S.S.R. from Baykonur into orbit with 208-km (129.3-mi) apogee, 144-km (89.5-mi) perigee, and 50° inclination. Satellite reentered same day. (GSFC *SSR*, 7/31/70; *SBD*, 7/30/70, 128; *Spacewarn*, 8/11/70, 1)

- Dr. Thomas O. Paine, NASA Administrator, held press conference at NASA Hq. to comment on his resignation, submitted July 28: "When I faced up to the job in the Space Agency. . .there were two major responsibilities. . . . One. . .was the meeting of the commitment. . .[to] land on the Moon within the decade. . .and the second was to lay out the post-Apollo program which would follow. . . . The action that was taken yesterday by the Congress in conference, which essentially gave us more than ninety-eight percent of the monies that the President had approved for the 1971 budget, but more importantly followed a very intensive review by Congress of the proposals we had made for the new space program of the seventies, in many ways represents a major step forward in the second concern which I had, which was to lay out the post-Apollo program and secure initial approval for it in the Administration and the Congress."

  NASA "had reached the point where the person who heads up the space program should be prepared now to devote very many months to the prosecution of this new program. . .and it was my conclusion that this was an excellent time for a change of command at NASA, since, for personal reasons, I could not give it the many months of continuing activity which would be required."

  Questioned about future space projects, Dr. Paine said he had found "great depth of support" for space program in U.S. and in foreign countries. Space program was valuable to Nation and world, was on solid foundation, and would probably be funded by Government on its present level or a little higher. He identified key issues facing NASA in FY 1972 budget as future of Apollo program, particularly with suspension of Saturn V production and continuing need to reassess best use of each booster, and question of relative priority within limited resources of space shuttle, space station, space tug, scientific missions, application of satellite programs, and increased aeronautical R&D. "It's very important. . .that we take a very long-range view because the direction that the space program takes in the 1970's is indeed going to cast a shadow through to the end of the century." (Transcript)

- House adopted conference report on H.R. 17548, FY 1971 Independent Offices and HUD appropriations bill, which contained $3.269-billion NASA appropriation and $513-million NSF appropriation after defeating motion to reconsider by vote of 227 to 156. (CR, 7/29/70, H7368-82)

- Astronaut James A. Lovell, Jr., *Apollo 13* commander, received insignia of Chevalier of French Legion of Honor from French Ambassador Charles Lucet during ceremonies at French Embassy in Washington, D.C. (*CR*, 7/31/70, S12536-7; Dixon, W *Star*, 7/30/70, C1)
- Western Union Telegraph Co. announced it had filed proposal with FCC to build domestic satellite system to serve all 50 states. Application called for launching of three comsats, initial construction of six earth stations, and construction of 31 terrestrial microwave stations. First comsat would be launched two years after FCC approval. (Western Union Release)
- MSC announced award of contract supplements:

    North American Rockwell Corp. received $92 449 970 for changes in Apollo CSM contract. Agreement formally incorporated changes to increase mission duration, add scientific instrument module (SIM), and provide experiment integration and brought total value of cost-plus-fixed-fee/award-fee contract since August 1963 to $3.724 billion.

    Grumman Aerospace Corp. received $7 974 400 for changes in Apollo LM contract. Agreement formally incorporated changes previously authorized by NASA for modification to contractor's maintenance and repair program. Modifications brought total value of cost-plus-incentive-fee contract since January 1963 to $1.681 billion. (MSC Releases 70-83, 70-84)
- Dr. W. D. McElroy, NSF Director, called for reevaluation of U.S. science and technology policies in testimony before House Committee on Science and Astronautics' Subcommittee on Science, Research, and Development. He proposed smooth transition between science and its applications by ensuring continued support of basic science by mission-oriented agencies; raising level of support for academic science from 18% to 35% or 40% of total Federal expenditure; elimination of conflict between "big science" (high-energy physics, etc.) and "little science" (projects of equal importance but smaller scale); long-term, level funding for science to avoid periodic ups and downs; Government sponsoring and funding of research on national problems to avoid competition with "normal" flow of research; and system of priorities with built-in "safeguards to hedge against imprudent judgments." (Testimony)
- AP quoted anti-Communist Hong Kong newspaper *Ming Pao* as reporting Communist China would test her first ICBM before Oct. 1. Chinese army officer in Peking had said missile had sufficient range to reach North America. (AP, W *Star*, 7/29/70, A11)
- USN had identified Soviet trawler cruising off ETR as electronic spy ship *Laptev*, UPI reported. Vessel's presence reportedly would not affect launch of Poseidon missile from U.S. nuclear submarine *James Madison*, which had been rescheduled for as early as Aug. 3, (*W Post*, 7/30/70, A4)
- Mexican Foreign Relations Secretary Antonio Carrillo Flores said he had received assurances from U.S. State Dept. that test-firing of Athena missiles from Green River, Utah, would be suspended until correction of mechanical failure that had caused Athena to veer off course and crashland in Mexico July 11. Search for missile was continuing. (UPI, *W Post*, 7/30/70, A14)

July 30: Finance was one of main reasons for resignation of Dr. Thomas O. Paine as NASA Administrator, Thomas O'Toole said in *Washington Post*. At 48, Paine had four children in college and private school. His NASA salary was $42 500 per year "which after taxes barely pays for his children's schooling, his housing and his everyday expenses." While "nobody at GE will discuss it," Dr. Paine apparently would move into "executive suite" at General Electric Co. "whose chairman (Fred Borch) makes $275,000 a year and whose four top executives under Borch all make more than $200,000 a year." (W *Post*, 7/30/70, A2)

- Nike-Tomahawk sounding rocket was launched by NASA from Wallops Station carrying Univ. of New Hampshire payload to study energetic particles. Mission did not meet minimum scientific requirements. (SR list)
- DOD Deputy Secretary for Public Affairs Jerry W. Friedheim confirmed that U.S.S.R. had resumed flight-testing of Fractional Orbital Bombardment System (FOBS) after nine-month hiatus. Most recent mission had been *Cosmos CCCLIV* which reentered 90 min after launch July 29. Friedheim told press mission was "evidence of the continuing momentum of the Soviet development and test program for strategic weapons." (W *Post*, 7/31/70, A22; DOD PIO)
- Completion of "Compass Link" reconnaissance photo-relay system and its turnover to Air Force Communications Service was announced by AFSC. System, developed by AFSC Electronic System Div., used three ground stations and two satellites. Laser beam scanned picture for light variations and converted image to electronic signals for transmission through satellite to ground station. Laser beam also was used for reconstruction of picture at destination. (AFSC Release 98.70)
- General Dynamics Corp. announced Convair Div. had received $2 585 000 USAF contract for OV I-20 and OV I-21 Orbiting Vehicle research satellites, scheduled for launch in 1971. Contract was managed by SAMSO. (General Dynamics Release 1529)
- ESSA scientist Elmer Schuman was using computer to determine exact limits of safe flight through electrically charged clouds, in study to help avert lightning strikes such as occurred during Nov. 14, 1969, launch of *Apollo 12*, ESSA reported. (ESSA Release)
- President Nixon in San Clemente, Calif., announced appointment of Rolf Eliassen, and reappointment of Howard G. Vesper and William Webster, as members of General Advisory Committee to AEC for terms expiring Aug. 1, 1976. (PD, 8/3/70, 998)

July 31: DOT announced reassignment of management responsibilities of joint DOT-NASA civil aviation R&D study to ensure its maximum impact on Federal aviation planning. Study was started in 1969 to develop single national policy for aeronautical R&D. Lawrence P. Greene, organizer and original Executive Director of study, would work fulltime with aviation planners in DOT, NASA, other Federal agencies, and aviation community. Clarence A. Syvertson, ARC Deputy Director, had joined DOT to direct study's final phase. (DOT Release 16370)

- No harmful effects had been found in lower animals exposed to lunar material, LRL Lower Animal Test Team of 14 investigators reported in *Science*. Selected species of fish and invertebrates had been exposed to *Apollo 11* lunar samples for 38 days in tests to detect extraterrestrial replicating agents possibly harmful to life on earth. No pathological effects or evidence of replicating organisms were detected. (*Science*, 7/31/70, 470-2)

- Next to last Saturn V 2nd stage (S-II-14) was ground-tested by NR engineers for 374 secs at MTF and developed thrust equivalent to 4.9 million newtons (1.1 million lbs) in space. All test objectives were met. (*Marshall Star*, 8/5/70, 1)
- Discovery (through infrared radiometric observations) of local temperature in Jupiter's North Equatorial Belt far in excess of those at level of solid ammonia clouds and (through visual observations) of orange-brown coloration within belt were described in *Science* by MIT astronomers John S. Lewis and Ronald G. Prinn. Astronomers concluded that high-resolution spectroscopic examination of spots might result in detection of water vapor in Jupiter's atmosphere. "In the design of Jupiter flyby and orbiter missions for the 1970's, the possibility of conducting related experiments in the immediate vicinity of the planet should be considered." Data could reduce uncertainties in design parameters of entry probes. Astronomers also suggested that, in multilayer cloud model, solar UV photolysis of hydrogen sulfide in regions where ammonia clouds were sparse or absent should lead to production of inorganic chromophores—molecular coloring agents. (*Science*, 7/31/70, 472-3)
- *Atlanta Constitution* editorial commented on resignation of Dr. Thomas O. Paine as NASA Administrator: "Dr. Thomas O. Paine didn't ride the Apollo spaceships. But he will be remembered in the history of space exploration as the man who guided America's space agency through its great adventure that ended in a moon landing." Dr. Paine had "served his country well and deserves its thanks." (*Atlanta Constitution*, 7/31/70)
- New York Airways' helicopter passengers between Kennedy International, La Guardia, and Newark Airports during 1970 to date totaled 134 600 for monthly average of more than 19 228. (*VertiFlite*, 9/70, 2)

During July: Use of infrared telescopes lifted by balloons and NASA's high-flying jet aircraft to penetrate "last portion of the electromagnetic spectrum to remain unexplored by astronomers" was described by Rice Univ. scientists in *Astronautics & Aeronautics*. In last 10 yrs, astronomers had built "super-cooled scanning telescopes that peered through atmospheric windows letting in infrared wavelengths between 1μ and 25μ." But water vapor had blocked wavelengths from 25-100 μ from reaching ground. Now jet aircraft and balloons lifted telescopes above vapor to "bridge gap between ground-based infrared and radio observations" and find "most powerful radiators in the universe." (Low, Aumann, Gillespie, *A&A*, 7/70, 26-30)

- U.S. space program was "in a kind of limbo, waiting, really, to be reborn and reconstituted," William Leavitt said in *Air Force and Space Digest*. As Apollo stretched out, "NASA space planners are already working toward the first launch in 1972 of Skylab, which is really the unheralded beginning of the long-awaited US manned space-station program." (*AF/SD*, 7/70, 60-4)

- Experimental InterAmerican Meteorological Rocket Network (EXAMETNET)—joint project of U.S., Argentina, and Brazil—released *Annual Report, 1968* (NASA SP-231), describing 1968 activities with summary of meteorological rockets and data and list of publications, reports, and related handbooks. EXAMETNET provided scientific measurements of upper atmosphere to facilitate research into structure and circulation of atmosphere in Northern and Southern hemispheres. (Text)
- "Development of Airplane Stability and Control Technology," 1970 von Kármán Lecture by Courtland D. Perkins, Associate Dean of Princeton Univ. School of Engineering and Applied Sciences, was published by *Journal of Aircraft:* "As the airplane becomes larger, faster, and as a result, more complex, the balance between inherent aerodynamic stability and stability provided through automatic systems must be made on an overall aircraft performance judgment. There is much yet to be done with the technology...." (*JA,* 7-8/70, 290-301)
- Sen. Mike Gravel (D-Alaska) writing in *Aerospace* dismissed ecological arguments against SST program: Greatest single consideration "in America's Supersonic Transport program is technological: Whether the United States will continue to maintain its dominant position in the development and production of commercial transport aircraft for the world—with all that this means to the health of the domestic economy and U.S. position in international markets." In next two decades, level of the world's commercial airline traffic was expected to rise to six times as high. SST was "logical and inevitable step forward to meet the demands of this new volume of air travel—regardless of who builds it." (*Aerospace,* 7/70, 3-5)
- Interview with Dr. John S. Foster, Jr., DOD Director of Defense Research and Engineering, was reported by *Air Force and Space Digest.* Dr. Foster believed U.S. held considerable lead over U.S.S.R. in basic space technology and application of that technology to military requirements. U.S. programs in communications, mapping, warning, surveillance, weather, and other activities related to military had been extremely rewarding. In manned military space operations, "we will come to realize that there are military needs in space that cannot be accomplished without placing much larger payloads in orbit." (*AF/SD,* 7/70, 31-6)
- European space accomplishments were reviewed in *Science Journal:* ESRO was evolving as center for scientific and applications satellites and was interested in developing satellites for meteorology, navigation, and air traffic control. Less successful had been European efforts to obtain independence in satellite launching through ELDO. U.S. had introduced new element into European deliberations by inviting collaboration in post-Apollo program. "So far the U.S. State Department has ruled against the purchase of American launch vehicles for European commercial ventures, such as the regional communications satellite. This could be changed by cooperation in the post-Apollo programme. Europe could reserve the right to buy boosters for any non-military application." Cost might be problem since "second league nations together spend approximately $300 million a year on space—an order of magnitude less than the

United States and probably the Soviet Union." (Gatland, *Science Journal*, 7/70, 77-80)

- NSF Released *Research and Development in Industry, 1968: Funds, 1968; Scientists & Engineers, January 1969* (NSF 70-29). of 36% of industrial scientists and engineers in R&D activities in January 1969, about two fifths worked on Federally financed R&D projects. For first time since 1956, companies financed over half of their R&D work with their own funds, spending $8.9 billion in 1968. Company funds increased 23% between 1966 and 1968, while Federal spending remained nearly level. Pattern was expected to continue from 1968 to 1970, with company R&D financing expected to rise about 20% and Federal support possibly declining slightly.

  In 1968, 90% of $8.6 billion in R&D funds furnished to industries by Federal agencies came from DOD and NASA. DOD funded 104 600 R&D scientists and engineers in 1968, at an average cost of $51 800 per R&D professional. NASA projects averaged $66 000 per R&D professional and included 35 000 persons. Industrial firms spent $3.7 billion on research, both basic and applied. Federal Government furnished 32% of these funds. In 1968, 72% of applied R&D funds was devoted to five product fields—guided missiles and spacecraft, electrical equipment and communication, aircraft and parts, machinery, and chemicals (not including drugs and medicine). (Text)

August 1970

August 1: *The Economist* commented on resignation of Dr. Thomas O. Paine, NASA Administrator, "just at a time when the governments of western Europe are considering whether to throw what passes for their space programmes in with Nasa's." European governments "scare easily and do not co-operate easily with each other, let alone with the Americans, whom they subconsciously suspect of always having the technical emasculation of Europe in mind. Dr. Paine, placid, patient and unusually sensitive—for a Washington man—to other people's susceptibilities, handled them as gently as nervous horses. A successor who is less imaginative, or more inclined to wear his emotions on his sleeve, will undo much of this work. This will not do much harm to the American space programme, but for Europe to miss this chance would be tragic." (*Economist*, 8/1/70, 43-4)

- MIT-sponsored group of U.S. and European scientists concluded month-long environmental study at Williams College to advise 1972 U.N. Conference on Human Environment and other national and international conferences. In preliminary report, *Study of Critical Environmental Problems,* group recommended that large-scale operation of supersonic transport aircraft be delayed until serious questions about potential for environmental contamination could be answered. Study had indicated increase of water vapor induced into stratosphere by supersonic aircraft favored formation of clouds and that fine particles from jet engine exhaust would double global averages of particles, with unknown effects. Particles reflected sunlight back into stratosphere and tended to warm it. Report recommended program to measure lower stratosphere for water vapor quantities and sulfur dioxide, nitrogen oxide, and hydrocarbons—fine particle matter.

    Study also found increasing carbon dioxide in atmosphere had caused little climate change. Earth's oxygen supply remained fairly constant. (Webster, *NYT*, 8/2/70, 1)
- Operation of NASA Motion Picture Film Depository and Film Loan Library was transferred by formal agreement to National Audiovisual Center of National Archives and Records Service of General Services Administration. (NASA Hq WB, 8/24/70, 1)

August 2: An 85-g (3-oz) palladium generator-separator developed for use in space could be used for air-pollution detection on earth, JPL announced. Unit was heart of gas chromatograph system developed by NASA and JPL for use by Viking Molecular Analysis Science Team in analyzing Martian soil and atmosphere in mid-1970s. It was particularly sensitive to major U.S. air pollutants like sulfur dioxide, carbon monoxide, and oxides of nitrogen. (JPL Release 558)

- European role in future NASA plans was discussed by London *Sunday Telegraph:* "White House officials appear to share the belief that

Congress will react more favorably to post-Apollo if it can be launched internationally, on the grounds that it would increase stability between nations, and in the long term cut the cost of major space programmes. For N.A.S.A. it would have the added advantage that such a programme, once begun, would be difficult for a new American administration to halt or cut back dramatically." Europe held "high card" in negotiations toward international space program. "N.A.S.A. needs a decision early next year so that the post-Apollo programme can be placed before Congress as an international venture." Europe would insist on conventional U.S. launch vehicles for commercial and scientific satellites in decade it took to make space shuttle operational. "Then the European Space Launcher Development can be scrapped and the money invested in post-Apollo." If Europe and U.S. "start the ball rolling in space cooperation, there are still more exciting projects." (London *Sunday Telegraph*, 8/2/70)

- *Apollo 11* Astronaut Edwin E. Aldrin, Jr., was named "Swedish-American of the Year" and presented Vasa Medal during Swedish-American Day ceremony in Stockholm. Aldrin and his family were beginning 10-day visit to Sweden. (UPI, *P Inq*, 8/3/70; AP, *W Star*, 8/3/70)
- *Papers of Robert H. Goddard*, edited by Esther C. Goddard and G. Edward Pendray, were reviewed in *New York Times Book Review* by I. Bernard Cohen: "Goddard was not universally appreciated during his lifetime; few dreamers are. He was often ridiculed as 'the moon man.' Today his name is commemorated in laboratories and professorships, but his greatest lasting memorial does not carry his name. It is the background of theoretical principles, engineering devices and practical experience, in the decades between World War I and II, that are embodied in every rocket that flies today.... These splendid volumes, handsomely printed and well illustrated, should go a long way to making known the personality of this shaper of our times." (*NYT Book Review*, 8/2/70)
- Officials of Environmental Defense Fund—group of scientists, lawyers, and conservationists incorporated in New York State—had decided to ask Federal court to order FAA to set noise and environmental standards for SST before certifying its airworthiness, *New York Times* reported. Fund had already obtained court orders blocking immediate construction of haul road for Alaska pipe line and requiring Dept. of Agriculture to give reasons why use of pesticide DDT should not be suspended. (Kenworthy, *NYT*, 8/2/70, 25)

August 3: Nike-Apache sounding rocket was launched by NASA from Wallops Station carrying Univ. of Michigan payload to study atmospheric structure. Rocket and instruments functioned satisfactorily. (SR list)

- USN successfully conducted first underwater launch of two-stage, solid-propellant Poseidon C-3 missile from nuclear submarine submerged off Florida coast. New, 30 000-kg (65 000-lb) missile was expected to become operational January 1971. Soviet ship *Laptev* watched launch closely and tried to pick up debris. (DOD Press Release; *W Post*, 8/4/70, 1)
- U.S. Athena missile that veered off course after launch from Green River, Utah, July 11 had been found 320 km (200 mi) from U.S. bor-

der near Carrillo, Mexico, Mexican Foreign Ministry said. Missile had gouged out trench 15 m (50 ft) long, 5 m (17 ft) wide, and 3 m (10 ft) deep. Mexican troops cordoned off site while scientists waited for equipment to recover radioactive pieces of missile. (AP, *NYT*, 8/4/70, 12)

- Pan Am Boeing 747 en route from New York to San Juan, Puerto Rico, was hijacked to Havana, Cuba, with 378 persons aboard. Aircraft was met in Havana by Cuban Premier Fidel Castro who inspected it before helping Capt. Augustine Watkins take off within 53 min. Aircraft arrived safely in San Juan. (Suarez, *NYT*, 8/3/70, 1)
- MSFC announced award of $13 460 726 contract modification to Martin-Marietta Corp. for continuing work on Skylab multiple docking adapter. (MSFC Release 70-152)
- *Aviation Week & Space Technology* editorial on recommendations of Blue Ribbon Defense Panel [see July 1]: "We think the Fitzhugh panel deserves an accolade from its fellow citizens for providing a shrewd and modern insight into problems of operating the national defense system and a series of specific points that require improvement. This is a process that should be repeated at regular intervals to insure that the national defense keeps pace with the constantly changing challenges of technology and the international chess board." (*Av Wk*, 8/3/70)

August 3-21: Joint NASA and Univ. of Virginia Bio-Space Technology Training Program was held at Wallops Station to give 46 life scientists information on engineering and operational procedures in planning and conducting biological experiments in space. Scientists were selected by NASA from more than 150 applicants representing universities, Government, and industry. Laboratory phase of program included launching five modified Arcas sounding rockets, carrying live animal and plant payloads, on 20-min flights. All payloads were recovered. (WS Release 70-10)

August 4: Senate adopted, by vote of 70 to 8, conference report on H.R. 17548, FY 1971 Independent Offices and HUD appropriations bill, containing $3.269-billion NASA appropriation and $513-million NSF appropriation. Bill would go to President for approval. (*CR*, 8/4/70, S12741-8)

- Edwin C. Kilgore, Deputy Chief of LaRC Office of Engineering and Technical Services, was appointed to new position of Deputy Associate Administrator (Management) in OART at NASA Hq. (NASA Release 70-130)
- NASA announced award of $1 366 400 contract to General Electric Co. for construction of fuel systems for research into operation of turbojet engines with liquid-methane and liquid-oxygen fuels. Contract would be managed by LeRC. (NASA Release 70-129)
- *Christian Science Monitor* editorial discussed future TV projection from foreign countries by comsat: "Do people want such invasions? Suppose the Russians or the Chinese loft communications satellites overhead which could beam propagandist television programs directly into American receivers?" Direct foreign-initiated TV broadcasts "could allow the viewer to see many more of the world's events directly as they are happening. The United States could participate in a worldwide agreement on the interchange of import pro-

gramming via satellite. But this electronic proliferation has its complications, of which we should be early aware." (*CSM*, 8/4/70)
- U.S. and U.S.S.R. were nearing agreement to curb quantitative part of nuclear arms race, but Aug. 3 successful test of Poseidon missile "is a vivid reminder that the qualitative aspect of that race goes on unimpeded," Chalmers M. Roberts said in *Washington Post*. Unless SALT in later phase could curb such qualitative improvements as Poseidon, "a host of new ideas now under research and engineering or simply in men's minds will come into being in future decades." (*W Post*, 8/4/70, A12)
- USAF Museum at Wright-Patterson AFB, Ohio, was described in *Christian Science Monitor* article: "Oldest and largest military aviation museum in the world, the display...portrays the history of the United States Air Force in such a way as to interest all ages. Nearly half of the aircraft and missiles have to be exhibited outdoors, so construction is under way on a new $6 million facility...to contain the collection of 8,000 planes and other items." (*CSM*, 8/4/70)

August 5: President Nixon submitted to Congress NASA's *Twenty-first Semiannual Report to Congress, January 1-June 30, 1969.* (PD, 8/10/70, 1032)
- Discovery by ARC biochemist Dr. J. Ken McDonald that enzyme DAP I (dipetidyl aminopeptidase) could be used to determine primary structure of many proteins was announced by NASA. Discovery could be used to determine amino acid sequence of immunoglobulins, body's main defense against disease. DAP I was enzyme with unique ability to take protein molecules apart one piece at a time. Order in which amino acid pairs, called dipeptides, were released revealed their sequence in original protein molecules. (NASA Release 70-128)
- FAA announced formation of interagency committee including representatives of NASA, DOD, and DOT to prepare five-year national plan for development and implementation of new microwave-scanning-beam instrument landing system for civil-military use. System was major recommendation of DOT's Air Traffic Control Advisory Committee in March 5 report to Secretary of Transportation John A. Volpe. (FAA Release T 70-32)
- NASA announced selection of RCA Corp. Astro-Electronics Div. to receive $1.62-million, cost-plus-fixed-fee contract for ground-command TV system for last four Apollo lunar exploration missions. System, which would permit color cameras to be operated either manually by astronauts on moon or remotely from earth, would be used with portable lunar communications relay unit (LCRU) being developed under separate RCA contract. (NASA Release 70-131)
- MSFC Technology Utilization Office had studied 1200 items of new technology from private contractors and MSFC R&D activities during past year, MSFC announced. NASA Hq. had published 165 MSFC items as NASA Tech Briefs, mailed to more than 8000 companies and individuals throughout U.S. Almost 5500 requests for technical information or advice had been received from U.S. and abroad. (MSFC Release 70-155)
- *New York Times* quoted French military sources as saying France's nuclear production team would be able to provide hydrogen bombs for French forces by 1972. Paper said move would place France

ahead of U.K., with only U.S. and U.S.S.R. deploying greater nuclear strength. (Middleton, *NYT*, 8/5/70, 2)
- Air Line Pilots Assn. had asked U.S. District Court of Appeals for District of Columbia to stop simultaneous landings on intersecting runways at National Airport in Washington, D.C., Friendship Airport in Baltimore, and other airports around U.S., Washington *Evening Star* reported. ALPA said procedures authorized by FAA in April lowered safety standards. (Kadis, W *Star*, 8/5/70, F9)
- Secretary of Transportation John A. Volpe announced that Dulles International Airport in Chantilly, Va., would be site of June 1972 U.S. International Aeronautical Exposition. One million visitors were expected to view more than 500 exhibits. (DOT Release 16670)

August 6: Photographs of area of Peru damaged by May 31 earthquake, taken by instrumented NASA Electra aircraft, were turned over to Peruvian Government in Lima. Data had been processed at MSC and were given to Peruvian authorities for possible use in restoration of area. (NASA Release 70-134)
- Expansion of Project MAST (military assistance for safety in traffic), joint test program to determine value of helicopters in providing medical assistance in emergencies [see July 15], was announced by Secretary of Defense Melvin R. Laird and Secretary of Transportation John A. Volpe. Program would be expanded to area surrounding Colorado Springs, Colo., and Seattle, Wash. (DOD Release 634-70)
- Twenty-fifth anniversary of world's first atomic bomb explosion by U.S. in Hiroshima, Japan. Hiroshima marked event by tolling solitary bell and stopping all activities for moment of silence and prayer. Prime Minister Eisaku Sato said in message that Japan, as only nation to suffer nuclear attack, now aspired to establish permanent world peace to protect mankind from nuclear war. (AP, W *Post*, 7/6/70, A3)

August 7: U.S.S.R. launched two satellites. *Intercosmos III,* launched from Kapustin Yar, carried Soviet and Czechoslovak experiments into orbit with 1192-km (740.7-mi) apogee, 200-km (124.3-min) perigee, 98.7-min period, and 48.4° inclination. Satellite reentered Dec. 6. *Cosmos CCCLV* entered orbit with 321-km (199.5-mi) apogee, 198-km (123-km) perigee, 89.6-min period, and 65.3° inclination and reentered Aug. 15. (GSFC *SSR*, 8/31/70, 12/31/70; *SBD*, 8/10/70, 173)
- U.S. Geological Survey announced it was photographing 26 cities in U.S. and Puerto Rico from aircraft at altitudes exceeding 15 200 m (50 000 ft) to test feasibility of detecting urban changes from earth-orbiting satellites in 1972. Study to develop ERTS-A techniques to provide data for urban planning organizations and other potential clients was being conducted with NASA cooperation. RB-57F aircraft on loan from USAF Weather Service were based at MSC and operated by NASA personnel. Project, part of Dept. of Interior's Earth Resources Observation Systems program, had been under way since April and was scheduled for completion in fall. (Interior Dept Release 27013-70)
- FAA announced it would propose rule to encourage public participation in establishing engine noise standards for SST. (FAA Release 70-75)

- President Nixon appointed *Apollo 8* Astronaut Frank Borman his Special Representative on Prisoners of War. Borman would immediately undertake special mission to seek help of third parties in countries around world in "securing the humane treatment and earliest possible release of all Americans held by the enemy forces in Southeast Asia." Borman would be on leave of absence from his executive position with Eastern Air Lines. (*PD*, 8/10/70, 1031)
- President Nixon submitted to Congress nomination of Adm. Willard J. Smith (USCG, Ret.) to be Assistant Secretary of Transportation, replacing Walter L. Mazan, who had resigned. Secretary of Transportation John A. Volpe established Office of Assistant Secretary of Transportation for Safety and Consumer Affairs and announced nomination of Adm. Smith to new post. (*PD*, 8/10/70, 1033; DOT Release 17070)
- More foreign-born and U.S. scientists were going abroad from U.S. to work, while changes in immigration laws had stemmed flow of foreign scientists to U.S., Thomas P. Southwick said in *Science*. Result was "drastic slackening of the brain drain" from Europe to U.S. NSF had reported 12 523 scientists and engineers were granted immigrant status in U.S. in 1967—increase of 74% over 1966 and 134% over 1965. In 1968 number of immigrants rose only 4% over 1967 figure and in 1969 number dropped by 21%, to 10 300. (*Science*, 8/7/70, 565-6)

August 8: Four Nike-Apache sounding rockets were launched by NASA from Kiruna, Sweden, carrying AFCRL and Dudley Observatory payloads to study noctilucent clouds. Rockets and instruments functioned satisfactorily. (SR list)
- Washington *Evening Star* editorial on 25th anniversaries of U.S. dropping of atomic bomb on Hiroshima and Nagasaki: "We can never know what might have happened if the terrible reality of the death of two Japanese cities had not branded itself on the collective consciousness of man. We know only that since August 9, 1945, the two greatest powers of the world have repeatedly confronted one another with mutual suspicion and distrust. And we know that, regardless of conflicting ideologies and interests, the superpowers have both been guided by one overriding principle: Nuclear war is not acceptable." (W *Star*, 8/8/70, A4)

August 9: LeRC was studying substitution of thermal reactor for standard exhaust manifold to reduce pollution of air from automobile exhaust, NASA announced. National Air Pollution Control Administration had asked LeRC to study reactor use to correct incomplete fuel combustion that led to emission of one quarter to one half ton of carbon-monoxide and hydrocarbons annually by each car. Industry tests had shown that thermal reactor, which also acted as afterburner, could reduce pollutants to within predicted 1980 Federal requirements. LeRC had been asked to develop technology for long-life, inexpensive reactor. (NASA Release 70-133)
- Next three months would be critical for Anglo-French Concorde supersonic transport, Don Cook said in *Washington Post*. Two prototypes would begin advance tests to determine whether aircraft could carry 11 300 kg (25 000 lb)—125 passengers and baggage—from Paris to New York at 2250 km per hr (1400 mph). If tests were successful, airlines with options on 76 aircraft would be asked to supply

August 9: NASA announced that Lewis Research Center was studying substitution of a thermal reactor for the standard manifold on an automobile engine, to reduce pollution from exhaust. In the photo technician Warren A. Moore connected a temperature sensor to an experimental reactor during testing on a V-8 engine at LeRC. The reactor burned up the carbon monoxide and hydrocarbons formed during combustion.

funds for firm orders. British and French governments had put $1.5 billion into project in seven years and were "increasingly reluctant" to see Concorde spending spiral. Result might be "that everybody concerned (except the French and British manufacturers) will be content to see the program slow down."

British prototype 002 had been fitted with more powerful Olympus 593 engines and was undergoing runway testing. French prototype 001 was undergoing engine refitting and was scheduled to fly again in September. Together, two prototypes had logged 225 hrs in 120 test flights. French model had logged 170 hrs in air, 30 at supersonic speeds. Neither had flown at mach 2 but this would become possible with engines now being fitted. Testing would be over 1290-km (800-mi) "boom corridor" from Northwest Scotland to tip of Cornwall over Irish Sea. Flight plan called for 25 runs at mach 2 within 50 days. Measurements would be taken of sonic boom, noise level at takeoff, and smoke emission. (*W Post*, 8/9/70, E1)

August 10: *Cosmos CCCLVI* was launched by U.S.S.R. into orbit with 502-km (311.9-mi) apogee, 224-km (139.2-mi) perigee, 91.7-min period, and 81.9° inclination. Satellite reentered Oct. 2. (GSFC *SSR*, 8/31/70, 10/31/70)

- Dr. Gary V. Latham, principal seismic investigator for Apollo program, asked astronomers worldwide to study moon each month at its closest approach to earth to determine whether surface cracks were opening and closing under gravitational attraction. Request was made through Smithsonian Astrophysical Observatory's Center for Study of Short-Lived Phenomena, which mailed "alert" cards to subscribers. Dr. Latham and NASA team at MSC had determined from lunar seismograph left on moon's surface by *Apollo 12* crew that moon's surface might show 7.6-cm (30-in) bulge earthward during these periods. It was also during these periods that astronomers studying moon with ground-based telescopes had reported sudden flares of color. If optical observations from telescopes of opening and closing of lunar surface cracks could be correlated with detection of moonquakes [see July 26], Dr. Latham said, "it would be strong evidence that the glows are produced by the escape of gases from the lunar interior through cracks which open and close each month." (SAO-PAO; Alert card; W *Star*, 8/9/70, A19)
- Space researchers at Bad Godesberg Institute for Aeronautical Medicine in West Germany had reported tests showed athletic achievement was no guarantee against physical hazards in space, Baltimore *Sun* said. Two groups of 12 persons each had been subjected to simulated space conditions including oxygen shortage, low air pressure, centrifugal force, and constantly shifting positions. Group of highly trained athletes had done no better than group of healthy but untrained students. Tests contradicted Soviet findings in same field and additional experiments would be necessary to verify German results. Institute would conduct similar tests to determine body reaction to weightlessness and to determine whether body adaptation at 4000-m (13 000-ft) altitude would increase physical resistance to space conditions. (B *Sun*, 8/10/70)
- Use of radiotelescopes to identify pollutants in atmosphere had been advocated by astronomers Dr. David Buhl of National Radio Astronomy Observatory and Dr. Lewis E. Snyder of Univ. of Virginia, Washington *Evening Star* reported. Astronomers believed pollutants from oxides of nitrogen to sulfur dioxide and small particles could be measured day and night in all weather at low cost, but radioastronomers did not have background "to do the chemistry." They recommended modification of equipment to survey regional atmosphere or known polluting sources unattended, by use of receivers feeding into computerized data banks via microwave links. Identification and quantification of pollutants could be instantaneous and single-dish antenna could be electronically controlled to "sweep" wide range of pollutants at regular intervals. (Lannan, W *Star*, 8/10/70, A4)
- President Nixon transmitted to Congress *Environmental Quality: The First Annual Report of the Council on Environmental Quality*. In message published in report, he said: "Our environmental problems are very serious, indeed urgent, but they do not justify either panic or hysteria. The problems are highly complex, and their resolution will require rational, systematic approaches, hard work and patience." (PD, 8/10/70, A7)
- Australia had sent RAAF contingent to MacDill AFB, Fla., to learn to fly F-4E Phantom aircraft, George C. Wilson reported in *Washington*

Post. It was "fresh evidence," he said, "that Australia is backing away from buying the controversial F-111 fighter-bomber from the United States." (*W Post*, 8/10/70, A7)
- U.S. and U.S.S.R. had reached tentative agreement to expand commercial airline service between New York and Moscow to permit Pan Am and Aeroflot to land at Leningrad and Washington, D.C., AP reported. Agreement also would increase flights from two to six weekly. Diplomatic sources had reported U.S. rejection of Soviet bid to permit transcontinental flights over both nations. (*NYT*, 8/11/70, 55)

August 10-14: International Atomic Energy Agency and AEC held Symposium on Environmental Aspects of Nuclear Power Stations at U.N. in New York. In opening address Dr. Glenn T. Seaborg, AEC Chairman, said: "Tremendous worldwide concerns about environmental matters are certain to cause significant changes in our technical approaches to energy problems. Looking well into the future one can imagine world-wide power transmission networks which will take full advantage of time zone differences and seasonal diversities to equalize the overall global daily demand for electricity. Transmission of large power over such distances may be brought about by cryogenic superconducting cables which are presently under development. Also, power may in the future be transmitted by microwaves or by light." It might be possible "to convert large amounts of electrical power into lights with a laser beam, transmit this light between continents by satellite and then reconvert the energy to electricity." (Text)

August 11: NASA's X-24A lifting-body vehicle, piloted by NASA test pilot John A. Manke, reached mach 0.98 at 20 000-m (66 000-ft) altitude after air launch from B-52 aircraft from FRC. Objectives of powered flight, 16th mission in X-24A series, were to evaluate handling qualities at 5° angle of attack at mach 0.95, determine lateral directional derivatives at mach 0.94 with 40° upper-flap deflection, and determine longitudinal trim and lift-to-drag data with 40° upper flap. (NASA Proj Off)
- President Nixon vetoed and returned to Congress H.R. 17548, FY 1971 Independent Offices and HUD appropriations bill, which contained $3.269-billion NASA appropriation. President said in message to House that he was vetoing bill because HUD portion would "help drive up the cost of living, harming the people it is most designed to help. This kind of excessive spending would also help cause the kind of huge deficits that drive up interest rates, which would make it impossible to speed the recovery of the housing industry." (*PD*, 8/17/70, 1056-7; House Doc 91-377)
- USAF Aerospace Defense Command reported that 141 satellites were orbited and 142 decayed during first six months of 1970. ADC had cataloged 4428 objects between Oct. 4, 1957, and June 29, 1970. (AP, *W Star*, 8/11/70)
- Two alternative plans for future Apollo lunar exploration were being developed by OMSF for NASA Lunar and Planetary Missions Board and NAS-NRC Space Science Board evaluation, NASA announced. First plan was to fly remaining six Apollo missions as planned; second was to delete two missions, thus freeing two Saturn Vs and

spacecraft for possible future use such as in space station mission. (NASA Release 70-135)
- Apollo 14 Astronauts Alan B. Shepard, Jr., Stuart A. Roosa, and Edgar D. Mitchell visited Noerdlingen Crater near Tuebingen, West Germany, during field trip to familiarize themselves with kinds of rock they expected to find on moon. (AP, *W Post,* 8/12/70, A7)

August 12: Development by MSFC scientists John W. Kaufman, Dennis W. Camp, and Robert E. Turner of "maxometer" that could record peak hurricane velocities of more than 320 km per hr (200 mph), was described in *Marshall Star.* One version of maxometer was antispark severe-environment sensor that could withstand extreme temperatures of Saturn rocket engine exhausts and acquire peak exhaust-gas-flow speeds during launch. Other version was ambient environment sensor for recording peak wind speeds in thunderstorms, squall line activity, tornadoes, and hurricanes. Lockheed Missiles & Space Co. had turned concept into dynamic pressure disc-spring device to measure 555-km-per-hr (345-mph) speeds to study air flow from launch of Saturn vehicles. ESSA had installed maxometer at National Hurricane Center at Miami and at Boothville, La., to gather hurricane data. (*Marshall Star,* 8/12/70, 1)
- Rep. Charles H. Wilson (D-Calif.) introduced in House H.C.R. 710, "to utilize more effectively the expertise and abilities" of NASA scientists and engineers in "fight against environmental pollution." (*CR,* 8/12/70, H8191)
- NASA announced award of $1.5-million contract extensions to McDonnell Douglas Corp. and NR Space Div. to continue parallel space station program-definition studies through Feb. 1, 1971. (NASA Release 70-136)
- USAF had dropped plans for advanced aerial tanker capable of high speeds, for accompanying bomber formations, in favor of modified large aircraft of Lockheed C-5 or Boeing 747 class, *Aerospace Daily* said. In meantime, USAF would attempt to modernize equipment aboard Boeing KC-135A. (*Aero Daily,* 8/12/70)

August 13: House, by vote of 203 to 196, sustained President Nixon's veto of H.R. 17548, FY 1971 Independent Offices and HUD appropriations bill containing $3.269-billion NASA appropriation. (*CR,* 8/13/70, H8238-55)
- NASA launched two Aerobee sounding rockets. Aerobee 350 launched from Wallops Station carried Univ. of Michigan experiment to conduct radiation belt studies. Rocket and instrument functioned satisfactorily.

  Aerobee 150 launched from WSMR carried NRL experiment to 185.6-km (115.3-mi) altitude to study extreme UV spectra. Experiment, model of solar experiment scheduled to fly on NASA's Apollo Telescope Mount, successfully photographed sun for 300 secs and recorded excellent scientific data. (SR list; MSFC Release 70-213)
- In letter to NASA, TRW Inc. asked review of July 15 award of $50-million contract for hardware development of Earth Resources Technology Satellite (ERTS) system to General Electric Co. TRW, losing bidder, said there had been "significant information available to our competitor and not available to TRW during the period of competition." TRW said it had not received Bendix Corp. study of computer

and ground components of system until 10 days before preliminary bid was due. (Nossiter, *W Post*, 8/29/70, A1)

- President Nixon submitted to Congress nomination of Dr. Raymond L. Bisplinghoff, Dean of MIT School of Engineering, as NSF Deputy Director. Dr. Bisplinghoff had been NASA Associate Administrator for Advanced Research and Technology, 1963-1965, and Special Assistant to NASA Administrator, 1965-1966. (*PD*, 8/17/70, 1068)
- AEC announced selection of Gulf General Atomic to receive $2 000 000, one-year contract to begin development of technology for thermionic-reactor power system for use in space. Reactor could be used in manned orbiting space station, lunar base, manned or unmanned electric-propulsion probe to explore solar system, and advanced communications satellites. Program was expected to take 8 to 10 yrs. (AEC Release N-155)
- President Nixon replied to July 16 letter of ICAO Council President Walter Binaghi that had conveyed ICAO Assembly Declaration on unlawful interference with international civil aviation: "You may be sure that my Government, in its effort to stop hijacking of aircraft and other acts of violence against international air transport, will give serious consideration to both technical and legal measures recommended by the Assembly." (*PD*, 8/17/70, 1061)

August 13-23: Apollo 11 CM *Columbia*, lunar samples, and spacesuits worn by *Apollo 11* crew were exhibited at Illinois State Fair in Springfield. (*C Trib*, 8/9/70; NASA PAO)

August 14: Election of Dr. Thomas O. Paine, NASA Administrator, as a Vice President and his appointment as Group Executive, Power Generation Group, were announced by General Electric Co. Both positions were effective Sept. 15, date on which Dr. Paine's resignation from NASA would become effective. (GE Release)

- Sen. Edward M. Kennedy (D-Mass.) introduced S. 4241 to authorize NSF "to conduct research and educational programs to prepare the country for conversion from defense to civilian, socially-oriented research and development activities." NASA was named, with AEC and DOD, to participate in conversion efforts. (*CR*, 8/14/70, S13457-66)
- Dr. Eberhard F. M. Rees, MSFC Director, told MSFC employees that number of center staff members to be released in NASA-wide manpower reduction had been cut from 190 to 121 by voluntary retirement of 69 employees. MSFC had been directed to reduce employment by 190 positions, to leave 5804 permanent positions. (MSFC Release 70-160)
- NASA announced selection of Philco-Ford Corp. to receive $26-million, cost-plus-award-fee contract for operation and maintenance of Deep Space Network (DSN). Contract would cover 33 mos with two one-year options. (NASA Release 70-138)
- Proportion of national R&D total supported with Federal funds had dropped from almost 66% in 1964 to 55% in 1970, NSF said. In 1970 Federal Government was scheduled to finance 62% of basic research effort, 51% of applied research, and 55% of development.

  Federal obligations for R&D (plant excluded) totaled $15.6 billion in FY 1969 and were expected to total $15.7 billion in FY 1970 and $15.6 billion in FY 1971. (NSF Science Resources Studies *Highlights*, 8/14/70, 1)

- U.S. and U.S.S.R. completed second round of Strategic Arms Limitation Talks and agreed to begin third round in Helsinki, Finland, Nov. 2. (AP, W *Star*, 8/14/70, A5)
- U.S. had offered to give up its ABM defense system entirely if U.S.S.R. would agree to limit production of its SS-9 offensive missiles and abandon its own missile defenses, Nixon Administration spokesman said at background briefing by White House for news editors in New Orleans, La. (Smith, *NYT*, 8/17/70, 1)

August 15: North American Rockwell Corp. had begun testing world's largest known rotational machine under NASA contract to evaluate human performance in rotating environment, *New York Times* reported. Four-man crew would live in machine for one week and perform mechanical chores and decision-reaction tests in LaRC study, to approximate conditions in space stations of late 1970s. Bullet-shaped crew module 12 m (40 ft) long rotated at 4 rpm, which was best approximation of space station obtainable, according to NR Program Manager, Dr. James A. Green. (Wilford, *NYT*, 8/15/70)

August 16: Full-scale mockup of General Electric Co. GE4 engine for SST had taken position beside full-scale mockup of aircraft at Boeing plant near Seattle, Wash., *New York Times* said. Largest and most powerful jet aircraft engine in world, according to Boeing, GE4 measured 7.9 m (26 ft) overall with 2.3-m (7.5-ft) diameter. It weighed 5 metric tons (5 1/2 short tons) and would develop estimated 30 400-kg (67 000-lb) takeoff thrust. GE had already logged more than 1000 hrs of testing on GE4 prototypes. SST would be equipped with four GE4s.

Timetable for SST called for shipment of first flight engines to Rohr Corp. in Chula Vista, Calif., for pod buildup early in 1972. Final assembly of SST would begin about same time. Roll-out would follow in summer 1972 and first flight test would be in late 1972 or early 1973. (*NYT*, 8/16/70, 78)

- *Washington Post* quoted German police as saying Astronaut Joseph H. Engle had escaped uninjured from head-on automobile collision in Southern Bavaria. Engle was in Germany to visit Noerdlingen Crater with Apollo 14 crew on mission to familiarize crew with rocks similar to moon rocks expected to be encountered. (*W Post*, 8/16/70, A30)

August 17: U.S.S.R. launched 1178-kg (2596-lb) *Venus VII* probe into heliocentric orbit from Baykonur at 10:38 am local time. Tass said 244-sec burn 81 min after launch placed spacecraft on trajectory to Venus where it would "continue exploration of the planet...which was earlier carried out by Soviet automatic stations." Spacecraft was on trajectory "close to the preset one" with all systems functioning normally and would reach vicinity of Venus around mid-December [see Dec. 15].

Venus VII was seventh spacecraft in Venus series. *Venus IV* (launched Oct. 18, 1967), *Venus V* (launched Jan. 5, 1969), and *Venus VI* (launched Jan. 10, 1969) had transmitted data about planet and landed on surface. (AP, *W Post*, 8/18/70, A12; *SBD*, 8/18/70, 208; GSFC *SSR*, 8/31/70)

- Manufacture of 15th and final Saturn V 1st stage (S-IC-15) had been completed by Boeing Co., prime contractor, and stage was being shipped to MTF, MSFC announced. Stage, scheduled to boost Apollo

19 toward moon in 1974, would be static-fired in late September. S-IC program had begun in December 1961. Eight of 15 S-IC stages had flown on Apollo missions, 6 manned and 2 unmanned. (MSFC Release 70-162)
- Dr. Lee A. DuBridge tendered resignation as Presidential Science Adviser and OST Director. In letter to President Nixon he said, "I have always been convinced I should retire well in advance of my 70th birthday in 1971." (PD, 8/24/70, 1084)
- Senate rejected, by vote of 43 to 22, amendment to H.R. 17123, $19.3-billion FY 1971 military procurement authorization, to ensure that weapon systems and subsystems were thoroughly tested before production was begun. Amendment, proposed by Sen. William Proxmire (D-Wis.), had been called "fly-before-you-buy." (CR, 8/17/70, S13529-44; SBD, 8/17/70, 202)

August 17-20: Space Flight Orientation Course at MSFC was attended by 60 Civil Air Patrol cadets from 45 states, District of Columbia, and Puerto Rico. (MSFC Release 70-158; NASA PAO)

August 18: USAF launched unidentified satellite from Vandenberg AFB by Titan IIIB-Agena booster into orbit with 395.9-km (246-mi) apogee, 151.3-km (94-mi) perigee, 89.9-min period, and 110.9° inclination. Satellite reentered Sept. 3. (Pres Rpt 71; GSFC SSR, 8/31/70; 9/30/70)
- NASA announced it was inviting 450 scientific and technical leaders to Space Station Utilization Conference to be held at ARC Sept. 9-11. Invitees would participate in planning of manned earth orbital space station scheduled for 1976-1978 launch. Laboratory was to be operated in space for 10 yrs. At conference, panels would be formed to screen potential space station uses and panel recommendations would provide material for related study in summer 1971. (NASA Release 70-137)

August 18-27: International astronomers attending triennial assembly of International Astronomical Union in Brighton, England, named lunar craters after 513 persons—including crews of *Apollo 8* and *11;* astronauts who died in Jan. 27, 1967, Apollo fire; six living Soviet cosmonauts; and late Cosmonaut Yuri A. Gagarin, who became first man in space on April 12, 1961. One crater was named Apollo to commemorate U.S. space program. All craters named were on moon's far side except three near Sea of Tranquility—*Apollo 11* landing site—named for mission's crew Neil A. Armstrong, Edwin E. Aldrin, Jr., and Michael Collins.

Among those honored were *Apollo 8* Astronauts Frank Borman, James A. Lovell, Jr., and William A. Anders; Astronauts Virgil I. Grissom, Roger B. Chafee, and Edward H. White II, killed in Apollo fire; Cosmonauts Andrian G. Nikolayev and wife Valentina V. Nikolayeva-Tereshkova, Gherman S. Titov, Aleksey A. Leonov, Vladimir A. Shatalov, and Konstantin P. Feoktistov; and late Cosmonauts Vladimir M. Komarov and Pavel I. Belyayev.

Others honored included Dr. Hugh L. Dryden, former NASA Deputy Administrator; N. E. Golovin, White Sands rocket scientist; Willy Ley, rocket scientist and author; Dr. W. R. Lovelace, former NASA Director of Space Medicine for Manned Space Flight; Theodore T. von Karman, aerodynamics expert; Joseph A. Walker,

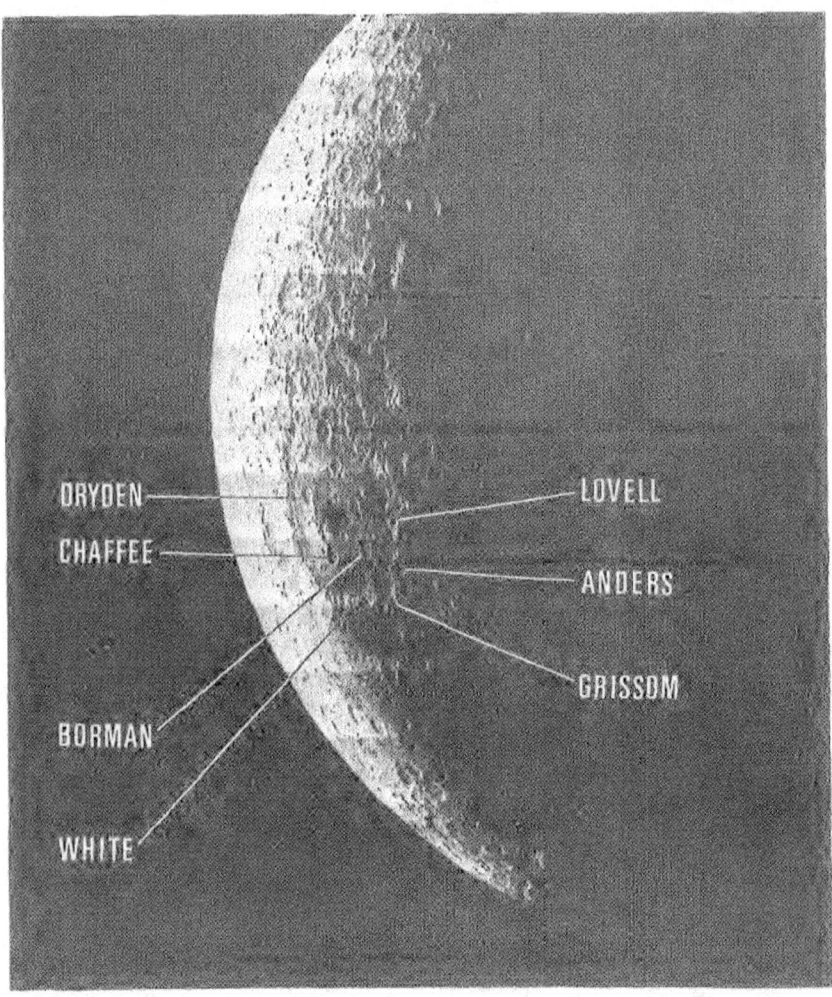

August 18-27: *The International Astronomical Union's triennial assembly named lunar craters after 513 persons, including crews of Apollo 8 and 11; astronauts who died in the Jan. 27, 1967, Apollo fire; Soviet cosmonauts; Dr. Hugh L. Dryden, NASA's first Deputy Administrator; and other pioneers. This photograph of the moon's far side was taken by NASA's Lunar Orbiter V, which was launched Aug. 1, 1967.*

NASA X-15 test pilot; and Alan T. Waterman, NASA consultant and first NSF Director.

Names had been selected by Working Group on Lunar Nomenclature and were approved by IAU officials. Other names approved included those of medieval Persian poet-astronomer Omar Khayyam, U.S. physicist Dr. J. Robert Oppenheimer, U.S. scientist George Washington Carver, mythological figures Icarus and Daedalus, and apocryphal Chinese rocket inventor Wan Hu. (McElheny, *W Post*,

8/15/70, A1; AP, W *Star*, 8/22/70, A2; SAO PIO; *Sky and Telescope*, 11/70, 262-6)

August 19-21: U.K.'s *Skynet B* military comsat was successfully launched by NASA from ETR for USAF and U.K. by long-tank, thrust-augmented Thor-Delta (DSV-3M) booster. Spacecraft entered transfer orbit with 37 426.5-km (23 260.7-mi) apogee, 264.9-km (164.6-mi) perigee, and 26.02° inclination. Primary NASA objective was to place spacecraft in transfer orbit accurate enough for apogee motor to place it in satisfactory synchronous equatorial orbit over Kenya.

Communications with *Skynet B* were lost suddenly during apogee motor firing Aug. 22. Search for spacecraft was being conducted by NORAD.

Skynet B was second of two U.K. military comsats launched in IDCSP-A program under DOD-U.K. agreement. USAF managed project for DOD and would reimburse NASA for launch services. U.K. would reimburse USAF. *Skynet A*, launched Nov. 21, 1969, was operating satisfactorily over Indian Ocean. (NASA Proj Off; B *Sun*, 8/20/70, A3)

August 19: U.S.S.R.. launched *Cosmos CCCLVII* into orbit with 466-km (289.6-mi) apogee, 271-km (168.4-mi) perigee, 91.9-min period, and 70.9° inclination. Satellite reentered Nov. 24. (GSFC *SSR*, 8/31/70; 11/30/70)

- Dr. Thomas O. Paine, NASA Administrator, sent memorandum to NASA officials on NASA's receipt of President's Safety Award for reduction of injuries to NASA employees for past three years: "NASA has earned this award in the face of the ever present and peculiar hazards that are associated with the pioneer research operations of our aeronautical and space programs.... Safety is reflected in the successes that we...have realized in our ventures in space. It is gratifying to realize that this award was based on achievements during the period of preparation for and completion of our first moon landings." (NASA Hq *WB*, 8/31/70)
- President Nixon accepted resignation of Dr. Lee A. DuBridge as OST Director and Science Adviser with "deep regret": "The skill, the wisdom and the seasoned judgment you have brought to your responsibilities here deserve a special accolade...and I trust that you will take pride in the outstanding contribution you have made to the welfare of all of your fellow citizens."

  President announced intention to nominate Dr. Edward E. David, Jr., Executive Director of Communications Systems Research at Bell Laboratories, as successor. Dr. David had been with Bell Laboratories since 1950. Previously he had been on MIT research staff. Nomination was submitted to Senate Aug. 20. (*PD*, 8/24/70, 1083-4, 1097)
- Dr. Wernher von Braun, NASA Deputy Associate Administrator for Planning, urged strong role for a centralized science and technology agency but also continued strong mission agency roles in support of basic research—in statement to Subcommittee on Science, Research, and Development of House Committee on Science and Astronautics: "...in the formulation of a national science policy—or perhaps more appropriately...of guidelines in this area—I would

recommend a strengthened role for a centralized agency, with larger responsibility for the development of academic science, and, concurrently, a continuation of the role of mission agency support of basic research deemed important to their missions. Increased responsibility must also be borne by the central agency to help coordinate the efforts of the mission agencies in their dealings with the academic community, a coordination which opens up new opportunities for academic science, and the various agencies, rather than inhibits new approaches. The coordination must not become simply another means of attempting to eliminate duplication, but a mechanism to spotlight areas of opportunity and need where new research, new technology, and new applications can be initiated." (*NASA Activities*, 9/15/70)

- U.S.S.R. was spending two percent of GNP on space and U.S. one half of one percent, Dr. Charles S. Sheldon II, Chief of Science Policy Div. of Library of Congress, said in speech before National Space Club in Washington, D.C. U.S.S.R. GNP in 1969 was estimated at $420 billion; U.S. total was $931 billion. Successful Soviet launches had totaled 44 in 1966, 74 in 1968, 70 in 1969, and 40 thus far in 1970. Largest element in Soviet space program had been photo-reconnaissance. U.S.S.R.. had led U.S. in payload launch weight every year since launch of 83.5-kg (184-lb) Sputnik in 1957 except 1969. Total payload weight for U.S.S.R. was 2 million kg (4.5 million lbs); for U.S. total was 1.9 million kg (4.25 million lbs). (B *Sun*, 8/20/70. A3)

August 20: *Cosmos CCCLVIII* was launched by U.S.S.R. into orbit with 538-km (334.3-mi) apogee, 515-km (320.0-mi) perigee, 95.1-min period, and 74.0° inclination. (GSFC *SSR*, 8/31/70)

- Smithsonian Institution's Museum of Natural History opened exhibition of photographic enlargements of moon particles, including *Apollo 11* nickel-iron pellet with 4-mm (0.15-in) diameter discovered by Smithsonian scientists while sifting through moon rock samples. Pellet, called "mini-moon" because of its resemblance to miniature moon in enlarged photo, was magnetic and was believed by Smithsonian scientist Dr. Brian H. Mason to have been formed eons earlier when meteoric fragment crashed on lunar surface. (Schaden, W *Star*, 8/20/70, B4)

- FAA study had shown that if rush-hour commuters traveled from suburban areas to downtown business sections by aircraft, they would generate one eighth of air pollution emitted by private automobiles, Secretary of Transportation John A. Volpe said. Study completed by Center of Transportation at Rutgers Univ. had found from research in Connecticut-New York-New Jersey area that 36 metric tons (40 short tons) of pollutants emitted in morning and evening by commuter automobiles could be reduced to 4.5 metric tons (5 short tons) if aircraft were used instead. Number of commuters who would voluntarily change to aircraft, if available, had been determined at approximately 30%. (FAA Release 70-78)

- Documents had shown USAF had amended 1965 contract with Lockheed Aircraft Corp. on Jan. 17, 1969, to give Lockheed additional $200 million for C-5A, *Washington Post* said. Lockheed had denied revision had increased payment. Later USAF said contract wording change did not alter amount to which Lockheed would be entitled under repricing formula. It merely "changed the timing of repricing and it has allowed the payment of interim repricing adjustments." (Nossiter, *W Post*, 8/20/70, A1; USAF statement for correspondents)

- Secretary of Transportation John A. Volpe appointed Herbert W. Richardson, professor of mechanical engineering at MIT, as DOT Chief Scientist. (DOT Release 17670)

August 21: Results of *Apollo 12* experiment to measure permanent and induced magnetic fields on lunar surface with magnetometer deployed on eastern edge of Ocean of Storms during EVA were reported in *Science* by ARC scientists Palmer Dyal, Curtis W. Parkin, and Charles P. Sonett. Magnetometer had measured steady magnetic field of 36 ± 5 gammas on lunar surface. Surface gradient measurements and data from lunar orbiting satellite *Explorer XXXV* (launched July 19, 1967) indicated field was localized rather than global. Data suggested source of field was large magnetized body that had acquired field during epoch in which inducing field was much stronger than any now on moon. (*Science*, 8/21/70, 762-4)

- International Academy of Astronautics announced selection of Soviet Cosmonauts Andrian G. Nikolayev and Vitaly I. Sevastyanov to receive Daniel and Florence Guggenheim International Astronautics Award for 1970. Award and $1000 stipend, presented annually for outstanding contribution to space research and exploration, would be presented to cosmonauts at Oct. 4-10 IAA Congress, for 18-day *Soyuz IX* mission (June 2-19). (IAA Release 38)

- Scripps Howard newspaper survey had shown NASA's rocket building program "all but grounded" because of budget cuts, *Washington Daily News* reported. With last rocket of $9.3-billion Saturn project nearing completion, layoffs of scientific and technical personnel had reached 80% in some areas. At peak of 11-yr Saturn project, NASA had employed more than 125 000 engineers, scientists, and technicians to build 37 Saturns including 15 Saturn Vs. "Today only 10 000...are still with the space program. And the boom towns that built, tested and launched the rockets...are starting to deflate." At Seal Beach, Calif., NR had said only 1743 out of 10 800 engineers and technicians who built Saturn V 2nd stage were still employed, an 84% decline. At New Orleans, Boeing Co. and NASA had laid off all but 1800 of 11 994 Saturn V 1st-stage builders. NASA was constructing only 21, smaller and less expensive, rockets for planetary exploration. Two of five kinds of rockets used to explore space and launch satellites were out of production and no new rockets were being developed. (Kirkman, *W News*, 8/21/70)

- *New York Times* editorial commented on appointment of Dr. Edward E. David, Jr., to succeed Dr. Lee A. DuBridge as President's Science Adviser: "Not surprisingly for one who has been engaged in industrial research for many years, Dr. David has indicated he is most concerned with applied areas such as health-care delivery, transportation, housing and defense. But he is certainly aware that the long-term scientific and technological health of the nation depends on the vitality of its basic research enterprise. Even the best medical organization in the world will be able to do little to conquer cancer...until molecular biologists and other basic scientists understand why cells grow out of control and how to prevent them from doing so. As science adviser, Dr. David will have to help assure that basic research is not sacrificed to the urgent demands for solving society's immediate problems." (*NYT*, 8/21/70)

- AT&T had urged FCC to adopt overseas communications policy that would permit building of 10 additional undersea cables by end of 1979, *Wall Street Journal* reported. Company had told FCC it did not object to FCC approval of pending applications for four additional satellites beyond four new ones already approved for launching, but asked that FCC also approve new cable projects "to avoid a serious imbalance between cable and satellite facilities in the future." (*WSJ*, 8/21/70)
- Washington *Evening Star* editorial on SST: "Backers of the SST program say it is necessary for maintenance of technological primacy and a favorable world trade position, and that, besides, this bird is another of those inevitabilities. But certainly the nation that placed the first men on the moon isn't hurting for technological status. As to the other arguments, the United States is capable of putting the kibosh on everybody's SST ambitions, through denial of U.S. landing rights, if the environmental sacrifices appear too forbidding."(*W Star*, 8/21/70)
- U.S. District Court Judge John H. Pratt, in Washington, D.C., refused to order release of top-secret report prepared by team headed by physicist Richard L. Garwin at request of President Nixon and said to be critical of SST. Judge Pratt dismissed suit brought by Sierra Club, Friends of Earth, and cities of New York and Boston to compel release of report. (Ungar, *W Post*, 8/22/70, A3)
- Dr. Theodore P. Wright, Civil Aeronautics Administrator from 1944 to 1948 and former Vice President and Director of Engineering with Curtiss-Wright Corp., died in Ithaca, N.Y., at age 75. (Bachinski, *W Post*, 8/24/70, C4)

August 22: U.S.S.R. launched *Cosmos CCCLIX* into orbit with 853-km (530.0-mi) apogee, 205-km (127.4-mi) perigee, 95.2-min period, and 51.1.° inclination. Satellite reentered Nov. 6. (GSFC *SSR*, 8/31/70; 11/30/70)

- Aerobee 170 sounding rocket was launched by NASA from WSMR carrying NRL experiment to study stellar UV. Mission did not meet minimum scientific requirements. (SR list)
- Concorde 002, British prototype of Anglo-French supersonic transport aircraft, reached its fastest speed—1600 km per hr (1000 mph) during test flight. French Concorde 001 had already reached 1600 km per hr. (Reuters, *NYT*, 8/24/70, 6)
- U.S.S.R. announced completion of Pacific carrier rocket tests begun July 25. Tass said area was "free for sea navigation and air traffic as of August 23." (AP, *NYT*, 8/24/70, 5)

August 23: Canadian Ministry of Transport had reported almost flawless operation during 1970 of radar system linking balloons 30 500 m (100 000 ft) above Pacific with spherical tracker on weather ship *Vancouver, New York Times* said. Since 1967 *Vancouver* and sister ship *Quadra* had alternated seven-week patrols at Ocean Station Papa, 1545 (960 mi) west of Vancouver, B.C., to collect data on upper-air conditions and provide meteorological, navigational, and oceanographic information for jet aircraft and surface vessels. System, developed by Sperry Rand Corp.'s Sperry Gyroscope Div., was shipborne version of that used by USN at PMR. (*NYT*, 8/23/70, 74)

- Harry Schwartz commented in *New York Times* on Aug. 17 launch of *Venus VII* by U.S.S.R.: "Some of the space age's richest scientific dividends have been the surprising findings of Soviet and American instrument-carrying rockets sent to Venus in the past decade. The excitement generated by this new information explains the widespread approval from the world's scientific community of the news last week that Moscow had launched still another unmanned spaceship toward Earth's nearest planetary neighbor." (*NYT*, 8/23/70, 4:5)

August 24: Two USAF HH-53 rescue helicopters successfully completed pioneer 14 500-km (9000-mi) ferry flight from Eglin AFB, Fla., to Da Nang, South Vietnam. Flight took nine days and included six stops. Mission fell two days behind when aircraft were grounded by bad weather in Alaska for 48 hrs. Longest nonstop leg was flown between Sheyma Island in Aleutians and Misawa Air Base in northern Japan, distance of 2700 km (1700 mi). USAF announcement had said crossing demonstrated capability of Military Airlift Command's aerospace rescue service to deploy rescue aircraft with inflight refueling "to meet any contingency worldwide." (USAF News Service Release 517; AP, *P Inq*, 8/25/70, 2)

- South Korea inaugurated world's 45th earth station for comsats at Kumsan, south of Seoul. Station, costing $6 million, was partly financed by U.S. Export-Import Bank loan. (AP, *C Trib*, 8/25/70)

August 24-28: Critical design review of Skylab's multiple docking adapter was held at Martin-Marietta facility in Denver, Colo. (MSFC Release 70-168; MSFC PIO)

August 25: DOD announced General Dynamics Corp. was receiving $191 297 000 as final payment for cost overruns on production models of F-111 aircraft. Original contract, managed by Aeronautical Systems Div., had been for $3 251 947 744. (DOD Release 693-70; *WSJ*, 8/26/70, 4)

- Nomination of Thomas W. Moore as member of Board of Directors of Corp. for Public Broadcasting for term expiring March 26, 1976, was submitted to Senate. Term of John D. Rockefeller III had expired. (*PD*, 8/31/70, 1116)

August 26: USAF launched unidentified satellite from Vandenberg AFB by Thor-Agena booster into orbit with 503.7-km (313-mi) apogee, 484.4-km (301-mi) perigee, 94.4-min period, and 74.9° inclination. (Pres Rpt 71; GSFC *SSR*, 8/31/70)

- NASA's X-24A lifting-body vehicle, piloted by Maj. Jerauld R. Gentry (USAF), completed 17th flight, reaching 13 700-m (45 000-ft) altitude and mach 0.75 after air launch from B-52 aircraft. Objectives of flight from FRC—to expand flight envelope to mach 1.1 and obtain additional stability and control data—were not realized because of igniter failure. Two unsuccessful postlaunch attempts were made to start four igniters; on third attempt, two igniters were started and pilot followed alternate flight plan based on two-chamber operation. Postlanding inspection revealed fire had caused minor damage to base of vehicle. Investigation would be conducted to determine source. (NASA Proj Off)

- ATS Procurement Review Committee released report on selection of contractor for ATS-F and ATS-G, unanimously recommending that April 7 award to General Electric Co. be reversed in favor of Fairchild Hiller Corp. PRC did not believe that the difference in time that contracts had been submitted was significant or that a leak on cost figures had occurred from GSFC. It found "that Fairchild was superior in nearly all aspects of project organization and overall management. Their more highly projectized organization, the existence of a top-level steering committee, the experience of their key personnel, the proposed method of liaison and control of subcontractors, and their lower composite overhead rates were notable features...." PRC said Fairchild was "superior in the important areas of systems engineering and in several features of their subsystem designs" and demonstrated "a skillfull integration of the several subsystems into an effective spacecraft design. . . . While GE was found to be superior in the areas of the electric power system and in antenna feed designs, several features of the Fairchild design. . .were considered to provide a greater assurance of mission success. . . .Taken altogether, the margin of superiority in both management and technical areas was in Fairchild's favor." (Text)

- NASA announced it has reached agreement with Dept. of Interior's Bureau of Commercial Fisheries under which Bureau would conduct Marine Resources Assessment and Harvesting Program at MTF. NASA would provide technical and support services. (NASA Release 70-141)
- Moon once had magnetic field six percent as strong as earth's, according to studies of Apollo moon rocks by Colorado Univ. geologist Dr. Edwin E. Larson and Univ. of Toronto scientists William Pierce and David Strangway, *Denver Post* reported. Studies had revealed moon rocks had stable magnetization, *Apollo 11* dust samples were 4½ billion yrs old, and solidified crystalline rocks were 3½ billion yrs. Presence of nickel-iron in *Apollo 11* rocks indicated that frequent meteoroid activity had fragmented them. Little or no free water or oxygen existed on moon. (*D Post*, 8/26/70)
- Local 9 of American Federation of Technical Engineers, AFL-CIO affiliate, had filed suit in District Court in Washington, D.C., against NASA to compel space agency to fire contract employees before it fired civil service employees, *Washington Post* reported. Union attorney William Peer had said NASA fired civil service employees and then hired other workers through private contractors. NASA had denied that it had added private contract employees to payroll following Aug. 17 notice of dismissal issued to 51 civil service employees. (Latham, *W Post*, 8/26/70)
- MSFC announced award of $7 932 440 contract modification to IBM for changes in ATM digital computers. (MSFC Release 70-166)
- *Apollo 11* Astronauts Neil A. Armstrong and Michael Collins attended unveiling of painting "Crew of Apollo 11" by Ronald Anderson of Dallas, Tex., at National Portrait Galley in Washington, D.C. Oil painting would be presented to Smithsonian Institution. (AP, *W Star*, 7/27/70, B4)
- U.S.S.R. had made first tests of warheads on SS-11 ICBM, Secretary of Defense Melvin R. Laird reported at Washington, D.C., press conference. Laird also announced decision to establish all-service Blue Ribbon Action Commission to assist in implementing DOD decisions in response to recommendations of Blue Ribbon Defense Panel [see July 1]. (Transcript)

- NASA announced selection of General Dynamics Corp. Convair Div. for negotiations leading to eight-month, $250 000, fixed-price contract for design study for "versatile upper stage" (VUS), high-energy stage that would increase orbital capacity of current launch vehicles and could be used as expendable upper stage with space shuttle. (NASA Release 70-140)

August 27: USAF launched USN's *Oscar XIX* navigation satellite from WTR at 6:23 am PDT by Scout booster. Orbital parameters: apogee, 1218.3 km (757 mi); perigee, 957.6 km (595 mi); period, 106.9 min; and inclination, 90°. Satellite, which could locate objects in ocean to 91-m (300 ft) accuracy, was part of Navy Navigation Satellite System for positioning fleet, ballistic-missile submarines, attack carriers, and oceanographic and miscellaneous vessels. *Oscar XVIII* had been launched March 1, 1968. (Pres Rpt 71; GSFC *SSR*, 8/31/70; USN CNO PAO)

- Paul A. Volcker, Acting Secretary of the Treasury, clarified stand of Treasury Dept. on SST development in letter to Sen. John C. Stennis (D-Miss.), Chairman of Senate Appropriations Committee's Subcommittee on Transportation: "When members of the [Treasury] Department first commented on this issue better than a year ago, the prevailing opinion was that the overall balance-of-payment effect would probably be negative, in the absence of a viable foreign competitive aircraft. Now...Concorde flight tests reportedly have been quite successful, and it appears likely that the British-French SST will be in commercial service by 1974. On this basis...balance-of-payments argument...loses force." Potential balance-of-payments impact "supports the advisability of going forward with the U.S. SST." (Text)

- ATA President Stuart G. Tipton testified on attitude of U.S. airlines toward SST development before Senate Appropriations Committee's Subcommittee on Transportation: "Airline support of the SST research program is bottomed on the belief that in the course of the prototype program, solutions to [SST's] problems will be found. The airlines have supported this belief in American ingenuity with over $50 million in risk capital. . . ." Airlines "will insist that these problems be solved before they will be ready to operate SST's." (Testimony)

August 28: NASA announced it had signed agreement with Italy under which Italy would launch three NASA satellites—one Small Scientific Satellite (SSS) and two Small Astronomy Satellites (SAS)—from San Marco platform in Indian Ocean off coast of Kenya. Use of Italy's equatorial launch facility would enable NASA to orbit satellites with Scout booster rather than with larger boosters required for launches from KSC.

Contract, signed by NASA and University of Rome's Centro Ricerche Aerospaziali (CRA), implemented Memorandum of Understanding signed by U.S. and Italy in February 1969. NASA would supply satellite and Scout booster. NASA-trained CRA team would assemble, check out, and launch satellite and provide some tracking and data acquisition and would be reimbursed by NASA. (NASA Release 70-139)

- Group of MSFC engineers successfully completed week-long testing of Skylab program hardware in simulated weightlessness aboard USAF

August 28

KC-135 four-engine-jet research aircraft. Tests included operation of flight-configuration doors for film cassette compartments, retrieval and replacement of film cassettes, and evaluation of handrails and foot restraints. Lunar soil penetration experiment aboard aircraft studied load-bearing characteristics of lunar soil and how LRV wheels would perform on moon. KC-135 was flown in parabolas, with 30 secs of weightlessness achieved on each parabola in technique which most clearly duplicated zero *g*. (MSFC Release 70-169; MSFC PIO)

- Nike-Apache sounding rocket launched by NASA from WSMR carried Cal Tech experiment to conduct x-ray studies. Rocket and instruments functioned satisfactorily. (SR list)
- First inflight interception of ICBM by ABM was successfully conducted by DOD, demonstrating use of Safeguard ABM system to defend U.S. offensive missile sites. Minuteman I test vehicle, launched from Vandenberg AFB, was detected by prototype missile site radar at Kwajalein Atoll in Pacific, which controlled launch of Spartan missile and flight of Spartan to intercept point outside atmosphere. Range instrumentation, which determined successful intercept, indicated Spartan's final stage (which in operational situation would carry nuclear warhead) had flown close enough to nosecone to have caused its destruction. (DOD PIO; *NYT*, 9/1/70, 18)
- White-hot metal slabs, each weighing approximately 45 kg (100 lbs), fell from sky and landed on Adrian, Tex., farm; near farmer at Beaver, Okla.; and near oil crew at Pratt, Kansas. AP later said Government team would investigate.

    On Aug. 31 USAF announced that chunks were probably part of U.S.S.R.'s reentering *Cosmos CCCXVI* (launched Dec. 23, 1969). Decay of satellite along path where pieces were found had been predicted by NORAD. Parts were being analyzed by USAF and would be returned to U.S.S.R. if positively identified. (*W Post*, 8/31/70, A8; UPI, *NYT*, 9/1/70, 59; AP, *CSM*, 9/2/70, 3)
- Experiment to interpret seismic record produced by impact of *Apollo 12* LM on lunar surface was described in *Science* by Univ. of Dayton, MIT, and USAF Institute of Technology scientists. In laboratory simulation, plastic pellets were fired into sand targets at 4° launch angle and 1.68-km-per-sec (1.04-mps) velocity. Shallow elliptical craters were formed, similar to certain lunar craters. Analysis of ejecta suggested LM debris skipped and, with some crater ejecta, reimpacted far downrange but ballistic rain did not account for unusual seismic signal. Experimenters had concluded that deliberate low-angle impact of spacecraft on lunar surface might create serious hazard to men and instruments on and near lunar surface in downrange direction. (Swift, Preonas, *et al.*, *Science*, 8/28/70, 851-4)
- Observation of nighttime lightning activity from *Oso II* orbiting solar observatory (launched Feb. 3, 1965) by monitoring light in 10° field of view of four telescopes was described in *Science* by Univ. of California at Berkeley physicists. Positions of nighttime thunderstorms determined from detection of optical radiations by *Oso II* had revealed that 10 times as many lightning storms occurred over land areas as over sea. (Vorpahl, Sparrow, Ney, *Science*, 8/28/70, 860-2)
- Accomplishments of Dr. Lee A. DuBridge, retiring Presidential Science Adviser and OST Director, were described in *Science*. Dr.

DuBridge was credited with influencing Nixon Administration to moderate space program, initiating steps to meet electric-power generating-capacity crisis, contributing to President's renunciation of biological warfare, and assisting "modest recovery in the fortunes" of NSF. (Walsh, *Science*, 8/28/70, 843-3)

- *Apollo 11* Astronaut Edwin E. Aldrin, Jr., and Dr. Thomas D. Barrow, President of Humble Oil & Refining Co., had been named co-chairmen of Goals for Texas in the Coastal Zone Conference to be held in Houston, Sept. 10-11, *Houston Chroicle* said. (*H Chron*, 8/28/70)
- One hundred twenty-fifth anniversary of *Scientific American*. Magazine had been founded in New York by Rufus Porter, New England electroplater, who had invested $100 in venture. It now had circulation of 450 000—90 000 of it overseas, including Italian language edition. Soviet Academy of Sciences distributed facsimile. Magazine had published works of 45 Nobel Prize winners, 25 of them before they became laureates. (Lipman, *W Post*, 8/30/70, C5)
- U.N. Committee on the Peaceful Uses of the Seabed adjourned after four weeks of meetings in Geneva with no concrete results. (Hamilton, *NYT*, 8/30/70)

August 29: U.S.S.R. launched *Cosmos CCCLX* into orbit with 286-km (177.7-mi) apogee, 207-km (128.6-mi) perigee, 89.4-min period, and 64.9° inclination. Satellite reentered Sept. 8 . (GSFC *SSR*, 8/31/70; 9/30/70)

- Concorde 002, British prototype of Anglo-French supersonic transport, reached mach 1.54 speed in test flight over North Sea. Concorde 002 flew for two hours at 12 500-m (41 000-ft) altitude in seventh flight of series to prove aircraft could fly at up to 2250 km per hr (1400 mph). (AP, *B Sun*, 8/30/70)
- McDonnell Douglas Corp.'s DC-10 trijet airbus made first flight, from Long Beach, Calif., Municipal Airport to Edwards AFB for further testing. DC-10, 55.5 m (182 ft) long and 6 m (20 ft) wide, could seat 345 passengers. It was designed for shorter flights than Boeing 747. (*NYT*, 8/31/70, 54; McDonnell Douglas Release, 8/29/70)
- Dr. Glenn T. Seaborg, AEC Chairman, presented $25 000 Enrico Fermi Award to Dr. Norris E. Bradbury, who was retiring after 25 yrs as director of Los Alamos Scientific Laboratory. Citation, presented in ceremony in Los Alamos, N. Mex., was for "great contribution to the national security and peacetime applications of atomic energy." Dr. Seaborg said laboratory had made "and will continue to make, significant contributions to what will be the next great step in nuclear power—controlled thermonuclear fusion." He said laboratory also was developing nuclear rocket engine for service within 10 yrs. (UPI, *P Inq*, 8/30/70; 4)

August 30: Rep. James G. Fulton (R-Pa.), ranking minority member of House Committee on Science and Astronautics, confirmed to press in Washington, D.C., that he had proposed George M. Low, NASA Deputy Administrator, for post of NASA Administrator being vacated by Dr. Thomas O. Paine on Sept. 15. Rep. Fulton said he had made recommendation in telegram to President Nixon July 31. (Lannan, *W Star*, 8/31/70, A1)

August 31: USAF launched unidentified satellite from ETR by Atlas-Agena booster into orbit with 39 831.3-km (24 750-mi) apogee, 31 913.3-km (19 830-mi) perigee, 1441.9-min period, and 10.3° inclination. (Pres Rpt 71; AP, *W Post,* 9/2/70; GSFC Proj Ops Support Div)

- USAF satellite launched July 23 "was placed in an unusual orbit which is especially suited to taking photos which could be used to police the ceasefire in the Middle East," *Aviation Week and Space Technology* editorial noted. "The orbit was an inclination of only 60 deg., in contrast to most reconnaissance satellites.... Since the satellite went into its unusual orbit two weeks prior to the [Middle East] cease-fire, it should have been possible to obtain photos of Soviet missile deployment prior to the truce, if that is its intended function. If the U.S. has obtained such photos by satellite, this fact as well as the pictures themselves would be closely guarded.... Reconnaisance satellite pictures are among this nation's most closely guarded secrets." (*Av Wk,* 8/31/70)

- FAA had ordered airlines operating Boeing 747s to inspect some of their newer engines, AP reported. Order followed request by National Transportation Safety Board prompted by explosion of high-pressure turbine section of Air France 747 engine after takeoff from Montreal Aug. 17. Aircraft had landed safely after being diverted to New York. AP quoted spokesman for Pratt & Whitney Div. of United Aircraft Corp. as saying inspections had been ordered for all JT9D3 and JT9D3A engines that had operated for less than 100 complete trips. Air France engine failure had been caused by improper installation of part of turbine wheel. (B *Sun,* 9/1/70, A5)

- Apollo CSMs would be modified to enhance potential use in emergency mode, NASA announced. Modifications included addition of 400-amp-hour battery in SM as alternative power source, provision of storage space for additional 9 kg (20 lbs) of potable water, and addition of third oxygen tank for SM. (NASA Release 70-144)

- MSFC announced NR had been awarded two modifications to contract for Saturn V 2nd stage (S-II): $1 768 228 for changes to original contract awarded in 1962 for manufacture and test of 15 S-II flight versions and $2 429 005 to identify S-II's capabilities for launching space stations and determine what production impacts would result from MSFC space station studies. (MSFC Release 70-172)

- Development of inexpensive, pocket-size laser for possible use in communications systems was announced at Bell Telephone Laboratories news conference at its Murray Hill, N.J., facility. Laser was semiconductor device that operated continuously at room temperature and could be powered by flashlight batteries. Single high-frequency light beam could carry thousands of TV signals, telephone calls, or other messages and might find greatest application when picture phones, high-speed computer conversations, and general communications needs expanded beyond existing carrying capacity. Bell scientists predicted laser ultimately would be pen-light size, cost a few dollars, and last a human lifetime. Laser included crystal semiconductor smaller than grain of sand, which operated on small currrent, eliminating need for extreme cooling. (Brody, *NYT,* 9/1/70)

- NSF transmitted to President Nixon *Federal Support to Universities, Colleges, and Selected Nonprofit Institutions, Fiscal Year 1969* (NSF

70-27). Federal obligations to universities and colleges totaled $3453 million in FY 1969, second consecutive year in which Federal support increased by 2%. Federal obligations for academic science programs including R&D increased by 0.05%, from 2350 million in 1968 to $2361 million in 1969. R&D amounted to $1495 million in Federal obligations during FY 1969, 5% increase over 1968. Increase was accompanied by $47-million drop (45%) in Federal R&D plant investment, to $57 million, lowest level in 1963-1969 reporting period. (Text)

- *Time* commented on resignation of Dr. Lee A. DuBridge as OST Director and President's Science Adviser: "Although he had initially warned the Administration that [budget] economies could turn the U.S. into a second-rate scientific power in some areas, he later seemed too willing to accept the cuts. Still, DuBridge could be an effective behind-the-scenes advocate." He had helped persuade President Nixon to curtail use of defoliants in Vietnam, played key role in decision to ban germ warfare, and helped focus attention on environmental problems. "But in the face of the Administration's tightfisted mood, it is doubtful whether he could have staved off the research cutbacks even if he had protested more vigorously." (*Time*, 8/31/70, 48)

During August: Boeing Co. announced it had built operational prototype of noise-suppressing SST engine inlet for testing with GE4 turbojet engine. Tests, to be completed by end of November, were to demonstrate compatibility and installed performance characteristics of propulsion system components. Inlet included movable centerbody that, when moved to aft position, created sonic wave in throat region, preventing emanation of most high-frequency sound from engine compressor components during landing approach. Boeing said movable centerbody feature, combined with SST's rapid takeoff climb, would result in better takeoff and landing characteristics than those of Boeing 707 subsonic aircraft. (Boeing Release S-0683)

- Mariner-Mars missions in 1971 were described by team of JPL scientists headed by Dan Schneiderman in *Astronautics & Aeronautics:* "The first of these voyages of exploration begins...when, between May 6 and June 3, two Mariner-type spacecraft will rise from launch pads at Cape Kennedy on the way to Mars to become the first spacecraft to orbit that planet." To obtain maximum scientific information, two identical spacecraft would perform separate missions, A and B. Each spacecraft would orbit Mars for 90 days in differing inclinations and periods. "Together they will cover 70% of the surface, permitting the study of seasonal effects. The information gained will help in selecting suitable landing zones for the unmanned Viking project which will land an instrument package on Mars."

  Missions would carry six science experiments including TV experiment. TV pictures would enlarge understanding of dynamic characteristics, history, environment, and surface physiography of Mars and improve maps of surface. They were not expected to provide direct evidence of possibility of life, but to furnish indirect evidence of Mars' suitability as habitat for life. (*A&A*, 8/70, 64-77)

- Lawrence Lessing said in *Fortune* that satellites were "only hope in sight" for comprehensive air traffic control: "Growing congestion and delays are seriously impairing U.S. and international airline operations, and air-traffic control is at the nerve center of the congestive problem. For nearly a decade leading scientists have urged the development of an advanced space-satellite system for navigation and traffic control. Indeed, such a system is the only hope in sight if air traffic is to grow to its full potential through the rest of this century. Only now, under the pressure of events, is this proposal reaching a critical point of decision." (*Fortune*, 8/70, 115-86)

- International participation in ERTS program was discussed in *Astronautics & Aeronautics* by John Hanessian, Jr., and John M. Logsdon, of George Washington Univ.: "Broad participation in ERTS...makes sense only if a global operational Earth-resources-survey system follows. The design and institutionalization of such a system will test the ingenuity of national and international policy makers, and any steps which may make their task less complex are desirable. Widespread international involvement in the ERTS program would be such a step. And it would...serve [both] the needs of many nations of the world and the national interest of the U.S. Many constraints on broad participation in ERTS stem from the level of funding and the pace now planned for the program. Thus it falls within the power of the U.S. to eliminate them. This we should do." (*A&A*, 8/70, 56-63)

- Significance of *Soyuz IX* in Soviet efforts toward permanent space station was discussed in *Science Journal:* Although engineering tests and scientific observations had been made during June 2-19 mission, "major emphasis was on bio-medical problems of weightlessness and post-flight adaptation to normal terrestrial conditions." Indications that cosmonauts' systems were slow in readapting to normal gravity had prompted Soviet doctors to suggest it might be necessary to create artificial gravity in permanent space stations and in spacecraft on protracted journeys to planets. Soyuz had been described as "prototype space bus able to accommodate three cosmonauts in comfort and to commute with ease to orbital stations disembarking and taking on passengers through a forward hatch." How rapidly U.S.S.R. advanced to this goal depended on many factors, "not the least the perfection of the new launch vehicle." (Gatland, *Science Journal*, 8/70, 5, 7)

- Use of aerospace management to solve social problems was advocated by Albert J. Kelley, Dean of Boston College School of Management, in *Astronautics & Aeronautics*. "In the quest for application of technology to domestic and other internal problems, aerospace management with its ability to marshal resources to solve complex problems may well be the greatest [space program] spinoff. In addressing current and emerging problems, the most urgent tasks will be to identify the parts of the problem that can be solved by existing technology, to recognize the parts of the problem that need more R&D, to identify the resources needed to solve all elements of the problem, and to organize the resources and efforts and get the job done." (*A&A*, 8/70, 46-52)

- *Air Force and Space Digest* editorial by John L. Frisbee advocated "show and tell" method of alerting U.S. public to Soviet nuclear threat: "Today's public skepticism about the realities of Soviet nuclear power, and its intended uses, is somewhat reminiscent of the skepticism that existed in the fall of 1962, concerning allegations of Soviet missile-site construction in Cuba. That earlier skepticism disappeared when photos of the Cuban missile sites, taken from high-flying U-2 aircraft, were made public." One way to awaken U.S. public to current Soviet threat "might be to release some of the satellite photography on which US analysis of the threat is based." (*AF/SD*, 8/70, 8)

September 1970

September 1: Concorde 002, British prototype of Anglo-French supersonic transport, reached mach 1.68 (1770 km per hr; 1100 mph) and created double sonic boom in test flight over Oban, Scotland. It was first time British Concorde had broken sound barrier in current test series. No sonic boom damage was reported. (AP, *W Post,* 9/2/70, A19)

- Apollo 14 flight crew completed mock flight to 60 900-m (200 000-ft) simulated altitude in altitude chamber at MSC, in preparation for Jan. 31, 1971, launch to moon. (AP, *NYT,* 9/2/70, 9)
- Moon watchers in California and Oregon had reported splotches on moon's surface during its closest approach to earth in August, Dr. Gary V. Latham, Apollo program chief seismic investigator, said in telephone interview with Washington *Evening Star.* Reports were in response to Dr. Latham's Aug. 10 alert to astronomers. Observations, coupled with recent series of simultaneous moonquakes, "may indicate the moon is not as cold and dead as we thought." Dr. Latham said current sightings had not yet been verified, but he and colleagues at MSC and at Lamont Doherty Geological Observatory at Columbia Univ. had detected 160 moonquakes, 27 at time of moon's closest approach to earth, in analysis of seven months' data. Quakes had been mild, "about 1.4 on the Richter Scale." (Lannan, *W Star,* 9/1/70, A4)
- Javelin sounding rocket launched by NASA from Wallops Island carried experiment to study helium geocorona. Mission did not meet minimum scientific requirements. (SR list)
- NASA announced increase from $312.2 million to $359.8 million in contract with Martin-Marietta Corp. to build Viking lander spacecraft and integrate entire system. Revision reflected delay of scheduled launch date from 1973 to 1975. (NASA Release 70-145)
- AIAA announced selection of 29 new fellows who would be honored at annual meeting in Houston Oct. 19-22. Fellows included Astronaut Neil A. Armstrong, cited for *Apollo 11* mission; M/G Lee V. Gossick, USAF Aeronautical Systems Div. Commander, for military-aeronautical leadership; and I. Irving Pinkel, LeRC, for design and implementation of aircraft safety systems. (AIAA Release)
- USAF helicopters of Military Airlift Command at Mountain Home AFB, Idaho, and at Luke AFB, Ariz., began flying MAST missions in joint DOT-DOD project announced July 15. (DOD Release 708-70)
- Gen. George S. Brown—former Commander of Seventh Air Force and Deputy Commander for Air Operations, U.S. Military Assistance Command, Vietnam (MACV)—assumed command of AFSC. He succeeded Gen. James Ferguson, who retired Aug. 31. (AFSC Release 246.70)
- U.S. and U.S.S.R., preparing draft treaty to prohibit emplacement of nuclear weapons on seabed, had agreed in Geneva that signature

nations should be allowed to ask U.N. help in policing violations, *New York Times* said. (Hamilton, *NYT*, 9/1/70, 9)

September 2: NASA FY 1971 interim operating plan news conference was held in Washington, D.C. Dr. Thomas O. Paine, NASA Administrator, said principal decision was "how best to carry out the Apollo and other existing programs to realize maximum benefits from them while preserving adequate resources for the future.... In our discussions, it became clear that the vitality of our national space program depends on a determined and vigorous continuation of plans for a reusable space shuttle followed by a space station in the manned flight program"; early development of ERTS, ATS, and HEAO; Grand Tour unmanned flights to distant planets; unmanned Viking Mars landers; Pioneer flights to Venus, Mercury, and Jupiter; and "a healthy aeronautical research program."

"With all of these considerations in mind...actions have been taken that will effectively reduce NASA's FY 1971 budget request of $3.333 billion to an interim operating plan of $3.2687 billion...and to minimize NASA's budgetary requirements for future years." Budget request for Apollo was being reduced by $42.1 million to $914.4 million by canceling Apollo 15 and 19 lunar missions and by phasing down manpower levels at major Apollo facilities more rapidly. Although Lunar and Planetary Missions Board and NAS-NRC Space Science Board had strongly recommended, after special review, that NASA carry out all six remaining lunar landing missions, NASA had "most reluctantly concluded that a reduction...should be made." Remaining Apollo missions would be redesignated 14 through 17 and would be flown before start of Skylab operations in November 1972. Schedule would allow additional flexibility by providing Apollo hardware for possible use in Skylab, space station, or other programs where manned operations or heavy boost capability would be required. Advanced mission studies would be reduced by $1 million.

Tracking and data acquisition network operations would be reduced $8 million by reducing number of manned flights and by more rapid phasedown in operating levels. Construction of facilities would be reduced to $25 million by deferring Polymer Research Laboratory at ARC, Calibration Laboratory at MSC, and Multispectral Photo Laboratory at MSFC and by rehabilitation and modification effort and facility planning and design. Research and program management minimum operating budget would be established at $678.7 million, amount included in previous appropriations bill and $13.6 million below NASA's FY 1971 request.

Dr. Paine said NASA had concluded "that it is prudent and realistic...to make these decisions now...to avoid more drastic and disruptive actions which would result if the decisions were postponed until later in the fiscal year. We also no longer have the capability to prepare in parallel two Apollo spacecraft, as would be required to maintain an option to reach a later decision on the number of Apollo missions."

Associate Administrator for Manned Space Flight Dale D. Myers explained Apollo schedule changes: Apollo 14 would be launched as planned Jan. 31, 1971, followed at six-month intervals by Apollo 16 (renamed Apollo 15) with advanced capability for greater exploitation of lunar exploration, Apollo 17 (renamed Apollo 16), and Apollo 18 (renamed 17), last launch in series. Beginning with Apollo 15 launch in July, EVA would be increased from two periods to three and man-hours per mission, from 18 hrs to 40 hrs. Weight of landed scientific payload would be doubled, from 227 kg to 454 kg (500 lbs to 1000 lbs); SM experiment package would be added; and range and efficiency of surface operations would be increased by improved suit mobility, improved life support system, and addition of lunar roving vehicle. Crew for new Apollo 15 would be same as previously planned—Astronauts David R. Scott, Alfred M. Worden, and James B. Irwin.

Skylab core module would be launched by Saturn V in November 1972, as scheduled, and would be revisited by Saturn IBs and CSMs up through June 1973. Beyond that, activities would be "at a standstill as far as orbital operations are concerned until the shuttle comes on line, hopefully in the 1976-77 time period." (Text; Transcript)

- U.K.'s first attempt to launch satellite with own booster failed when pressurization system in three-stage Black Arrow booster malfunctioned. Booster was to have orbited 82-kg (180-lb) satellite to study upper atmosphere. Previous U.K. satellites had been orbited by NASA boosters. (Berger, B Sun, 9/3/70)
- Recommendations on conduct of "comprehensive review and analysis of the complete NASA acquisition process from determination of requirements through final contract award" were requested by Dr. Thomas O. Paine, NASA Administrator, in memo to Daniel J. Harnett, Assistant Administrator for Industry Affairs. NASA procurement policies were "sound ones which have in general fairly treated our contractors while giving NASA excellent value for its money," Dr. Paine said. But NASA was now planning programs in "changed economic environment with more limited resources." (NASA Release 70-146)
- Ion engine on board orbiting *Sert II* (launched Feb. 3) was returned to full thrust after two-day interruption and was operating satisfactorily. Engine had been turned off Aug. 31 about one hour before spacecraft passed through solar eclipse shadow over South Pacific. Four attempts during shutdown to restart ion engine No. 1, which had failed July 23, were unsuccessful. (NASA Release 70-149)
- MSFC announced McDonnell Douglas Astronautics Co. had received $60 918 000 modification to $97 340 000 NASA contract for conversion of original orbital workshop to be launched by Saturn IB booster to completely outfitted workshop for launch by Saturn V. (MSFC Release 70-178)
- MSC awarded LTV Aerospace Corp. Missiles and Space Div. and McDonnell Douglas Astronautics Co. $215 000, fixed-price contracts to conduct six-month parallel development studies of radiative nonmetallics for high-temperature heat protection on space shuttle. (MSC Release 70-95)
- NASA said it had confirmed Aug. 12 observation by astronomer V. E. Bell at Lodi (Calif.) Observatory of bright blue-white flare on lunar surface near Fra Mauro highland area, when moon was closest to earth. International astronomical group Astronet had reported similar sighting in crater Lassell K Sept. 1. Observation reports were in

response to Aug. 10 alert by Dr. Gary V. Latham, Apollo program's principal seismic investigator, made through Smithsonian Astrophysical Observatory. (W *Star*, 9/2/70, A6)

September 3: USAF launched unidentified satellite from Vandenberg AFB by Thor-Burner II booster into orbit with 872.3-km (542-mi) apogee, 764.4-km (475-mi) perigee, 101.2-min period, and 98.7° inclination. (Pres Rpt 71; GSFC *SSR*, 9/30/70)

- AIAA announced selection of MSC Deputy Director Christopher C. Kraft, Jr., to receive Louis W. Hill Space Transportation Award for 1970. Certificate and $5000 honorarium for "outstanding management and leadership in directing the planning and operational control of all United States manned space flight missions from the first Mercury suborbital mission through the first Apollo lunar landing" would be presented Oct. 22.

  AIAA also named three honorary fellows who would receive certificates Oct. 21: Dr. Grover Loening, retired aircraft inventor; Dr. Wernher von Braun, NASA Deputy Associate Administrator for Planning; and Dr. Stanley G. Hooker, Technical Director of Rolls-Royce Bristol Engine Div. in England. (AIAA Releases)

- Dozen veterans of 118-member German V-2 rocket team of Dr. Wernher von Braun had received notice of demotion or layoff from NASA at MSFC, *New York Times* reported. At least seven would lose their jobs altogether before Oct. 1 because of manpower reductions. Notices had come as team was planning October reunion to mark 25th anniversary of its arrival in U.S. (Wilford, *NYT*, 9/3/70, 1)

- Federal financing of SST would be "complete misplacement of national priorities," Mayor John V. Lindsay of New York said in letter to Sen. John C. Stennis (D-Miss.), Chairman of Senate Appropriations Committee's Subcommittee on Transportation. "I am prepared to do all in my power to prevent any SST from landing at New York's airports until it is proven safe both to our environment and to the health of our citizens." (UPI, W *Star*, 9/4/70, A5)

September 4: MSC's cosmic ray emulsion plastic experiment—balloon 182.9 m (600 ft) long with 544.3-kg (1200-lb) instrument package—was launched from Minneapolis, Minn., by Winzen Research Corp. in cooperation with National Center for Atmospheric Research. Payload of 6.1- by 3.7-m (20- by 12-ft) package housing 22.3 sq m (240 sq ft) of detectors would record intensity and direction of transiron [of particles heavier than iron] primary cosmic rays in upper atmosphere during 48-hr flight. (NASA Release 70-154)

- Map of portion of Venus larger than Asiatic continent made by using 64.0-m (210-ft) dish antenna at Goldstone Tracking Station was described in *Science* by JPL astronomers Dr. Richard M. Goldstein and H. Rumsey, Jr.: Map covered planet from 90° west longitude to 30° east longitude and from 45° south latitude to 35° north latitude. Area extended 13 700 km (8500 mi) at Venus equator and was up to 7400 km (4600 mi) wide. Resolution was "about two times as good as the moon can be seen with the naked eye" and was "best definition of Venus yet obtained." (*Science*, 9/4/70, 975-7; JPL Release 561)

- Venus radius controversy was discussed in *Science* by MIT scientist W. B. Smith: Scientists had finally arrived at a consensus in interpreting certain data from *Mariner V* and *Venus IV* after "ebb and flood of recent opinion about Venus's lower atmosphere as seen from a moderately invariant (and biased) point of view—that of the 'radar radius.' " Smith and associates had reported accurate determination of Venus's radius in mid-1966. Value, derived from earth-based radar observations, was 6055.8 km (3762.9 mi). "In early 1967, two Mariner 5 experimenters visited us to ascertain our very latest value and our confidence in it, stressing the importance of the radius for the forthcoming Mariner 5 flyby." Earliest *Mariner V* reports had shown *Venus IV* data inconsistent by about 24 km (14.9 mi) and asymmetry between night and day refractivity profiles of Venus. "The central question was: are the surface pressure Ps and Temperature Ts 19 atm and 553 K [536°F] as derived from Venera 4 data, or about 100 atm and 700 K [800°F] as derived from Mariner 5 data plus the radar radius?" Soviet scientist V. R. Eshleman had suggested that 2:1 ambiguity had occurred in altimeter. After reexamination of time systems used by *Mariner V* experimenters, Venus-centered distance of ray path had been decreased by 8.85 km (5.49 mi), "confirming our independent estimate of 10 km [6.21 mi] for this displacement. As a result, the refractivity profiles became nearly symmetric." (*Science*, 9/4/70, 1001-2)

- U.S.S.R. announced it had found Apollo experimental capsule (used for training and experiment) and would turn it over to USCG cutter *Southwind* at port of Murmansk. Cutter was on oceanographic expedition in area. U.S. Embassy sources in Moscow had said Soviet government had informed Embassy three weeks before that capsule had been found in Bay of Biscay, off France, by Murmansk fishermen. (*NYT*, 9/5/70, 5; NASA OMSF PIO)

- *New York Times* published comments of three lunar scientists on NASA's cancellation of Apollo 15 and 19: Dr. Thomas Gold, Cornell Univ. astronomer and member of President's Science Advisory Committee, had said, "It's like buying a Rolls-Royce and then not using it because you claim you can't afford the gas."

   Dr. Harold C. Urey, lunar expert and Nobel Prize winner, had said, "I think the American people and Congress should realize that the moon is an extremely old object....this gives scientists a way of studying an object that goes back to the very beginning of the solar system." Saving of $40 million by cancellation was "chicken feed" in view of $25 billion already spent on Apollo program.

   Former scientist-astronaut Dr. Brian T. O'Leary said, "The scientific community has become disenchanted with NASA. The present decision seems ridiculous." (*NYT*, 9/4/70, 9)

- Editorials on Apollo program cutback:
   New York Times: "Throughout the last decade this newspaper opposed the top priority then accorded Project Apollo on the ground that too much money was being diverted from urgent social needs. But now that these huge sums have been spent the need is to obtain the maximum yield, scientifically and otherwise, from that investment. Surely, NASA, which has been able to reach the moon, can find a better solution than the one now offered for adjusting to austerity in space research. One desirable alternative would be to enlist foreign resources in the exploitation of Apollo technology, perhaps by offering to send teams of British, French or Soviet astronauts on the journey pioneered by Apollo 11." (*NYT*, 9/4/70)

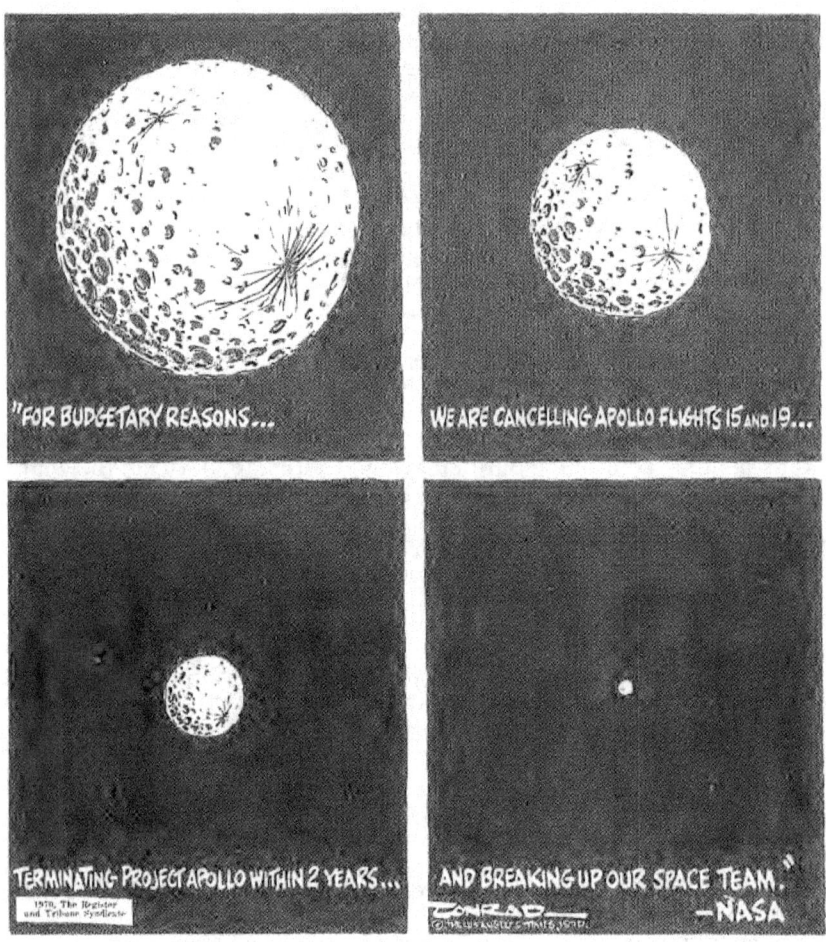

September 4, 6, to 9: *Scientists and the press reacted to* NASA*'s cancellation of two lunar landing missions, announced Sept. 2. The* Los Angeles Times *published the cartoon by Paul Conrad Sept. 8. (Courtesy of Register and Tribune Syndicate)*

Baltimore *Sun:* "...the thrust into space is the most open-ended of all human enterprises and there is no cause for desperate hurry. So far as science goes, time is not a major factor. Money is, and when it comes to spending many billions consideration has to be given to other needs, many of them much more immediately urgent than any extraterrestrial venture. We have gone far and shall certainly go farther—more surely, in the long run, if we keep a sense of proportion and a wise balance among our choices." (B *Sun*, 9/4/70, A10)

• President Nixon at San Clemente, Calif., signed E.O. 11556 establishing responsibilities and authority of new Office of Telecommunications Policy and announced intention to nominate Dr. George F. Mansur, Jr., as Deputy Director. Nomination was submitted to Senate Sept. 10. Dr. Clay T. Whitehead, whose nomination had

been confirmed by Senate, would be sworn in after President returned from San Clemente. (PD, 9/7/70, 1147; 9/14/70, 1197)
- Institute for Strategic Studies in London had released annual report, *The Military Balance, 1970-71*, Reuters said. July survey had shown U.S.S.R. could deploy 1300 ICBMs—246 more than U.S. But U.S. "promises to increase its warhead lead" with three-warhead system on Minuteman III rocket with 12 900-km (8000-mi) range. U.S. also was converting submarines to take 10-warhead Poseidon missile. U.S.S.R. had been testing three-warhead system. Report estimated that U.S. had more than 6000 nuclear warheads fitted to missiles capable in theory of reaching Soviet territory, while some 2000 Soviet warheads were fitted to missiles with range to hit continental U.S. U.S. had 656 submarine-launched ballistic missiles; U.S.S.R., 280. (Maynes, *W Post*, 9/4/70, A18)

September 5: NASA announced it had reversed April 8 decision to award $50-million ATS-F and ATS-G contract to General Electric Co. and would award contract to Fairchild Hiller Corp. instead. Decision was based on Aug. 26 report of ATS Procurement Review Committee (PRC) and recommendations of Selection Panel appointed by Dr. Thomas O. Paine, NASA Administrator, July 16. Panel concurred with PRC finding that procedural discrepancies had not affected outcome of competition and that cost difference between final proposals was not significant as basis for selection, but that important technical differences existed in which Fairchild Hiller was superior. It did not concur with PRC conclusion that Fairchild was superior in nearly all aspects of organization and management. (NASA Special Release; Memo of Decision)
- Name "tranquilite" had been given to new mineral discovered in *Apollo 11* moon rock by scientists Paul Ramdohr and Josef Zaehringer of Max Planck Institute for Nuclear Physics at Heidelberg, West Germany, Baltimore *Sun* said. Tranquilite—named for moon's Sea of Tranquility, *Apollo 11* landing site—was compound of titanium, iron, and magnesium related to earth mineral titaniferous magnetite. International Nomenclature Commission would review designation of new mineral before name became permanent. German team also had confirmed theory that lunar rocks had lain on moon's surface for 500 million yrs and had never penetrated deeper than 0.9 m (3 ft). (B *Sun*, 9/5/70, A10)
- *Venus VII*, unmanned U.S.S.R. interplanetary probe launched Aug. 17, was on course 4 990 000 km (3 100 000 mi) from earth with all equipment functioning normally, official Soviet progress report announced. (AP, *P Inq*, 9/6/70)

September 6: Arab extremists of Popular Front for Liberation of Palestine (PFLP) hijacked Swissair DC-8, TWA Boeing 707, and Pan Am Boeing 747 with passengers and crews. Attempted seizure of fourth aircraft, El Al Israeli Airlines plane, was prevented; security guards on plane shot and killed one hijacker and wounded other. Swissair and TWA aircraft were flown to desert airport at Zerqua, Jordan. PanAm 747 was flown to Beirut, refueled, and then flown to Cairo where it was blown up on runway after evacuation of passengers and crew. Passengers were later flown to Rome on another aircraft. PFLP held occupants of the two aircraft in Jordan as hostages for imprisoned commandos in Israel, U.K., Switzerland, and West

September 6

Germany. PFLP spokesman in Beirut said TWA and Pan Am aircraft had been seized "to give the Americans a lesson" for supporting Israel. (*FonF*, 9/3-9/70, 637-8)

- First data from three-week NASA-Purdue Univ. study to detect corn blight infestation using sensing devices aboard aircraft had shown blight could be identified from aircraft in both severe and earlier stages, NASA announced. Color photos and multispectral scanner pictures and data had shown signature keys to crop disease, southern leaf blight. MSC's RB-57F aircraft was providing coverage, from 18 300-m (60 000-ft) altitude, of area from Michigan City, Ind., to Evansville, Ind. Purdue, Univ. of Michigan, and USAF aircraft covered area at lower altitudes. Successful identification by remote sensing could warn farmers of approaching blight or of presence of blight in crops before they would be otherwise aware of it. (NASA Release 70-148)
- Work of Grover Moreland, "Neil Armstrong of the rockcutters, the man entrusted with...stretching the first precious lumps of moon matter into as many research laboratories as...possible...without losing a particle," was described in Washington *Sunday Star* magazine Sunday. Moreland had prepared 60 or 70 thin sections from *Apollo 11* samples and 15 or 20 from *Apollo 12*. They were studied under petrographic microscope for mineral content and origin and under electron probe for detailed chemical analysis. He shaved sections as thin as 0.15 mm (0.006 in) and had perfected method of impregnating porous matter with plastic to prevent crumbling. (Schaden, *Sunday*, 9/6/70, 14-5)
- Battle over choice of undersea cables and comsats had begun at FCC, Washington *Evening Star* said. AT&T was urging quick decision authorizing sixth Atlantic cable. ComSatCorp's satellite facilities in Atlantic area were operating at 50% of capacity and company could benefit from more volume if existing five undersea cables became overcrowded. (Aug, W *Star*, 9/6/70, C5)
- Press comment on Apollo program cutback:

    Washington Post editorial: "There has rarely been a major program that has gone up the hill and down again as fast as has the manned space program. Just a year ago NASA was riding high on the success of Apollo 11. Now it is scratching operations and scrambling for money to keep even a portion of the original program and its progeny in business. This, in our view, is bad because if it is carried too far it threatens to turn the entire manned space effort into the most spectacular (and most expensive) stunt in history instead of the first step in a genuine effort to learn a little more about some of the secrets of the universe. NASA's decision to focus on the future rather than on a full study of the moon will help at least to keep the option of a worthwhile space program open." (*W Post*, 9/6/70, B6)

    Richard D. Lyons in *New York Times:* When U.S. politicians first proclaimed national effort to land men on moon influential scientific leaders had scoffed. "A decade later the scythe of discontent with Apollo has swung full circle. Now it is the politicians who are attacking the program while the scientists wring their hands over lack of support." Nixon Administration had "not been kind to the space program, although the President repeatedly stated that he supports

it enthusiastically. The main reason is money, or the lack of it, which was almost no object when Apollo was born." (*NYT*, 9/6/70, 4:7)
- End of "galloping science" was predicted by Kenneth W. Boulding of Univ. of Colorado in *Washington Post:* "The scientific revolution and science-based technology represent a kind of takeoff from the old world of classical civilization. The 'flight' of development cannot go on forever. At some point there must be a re-entry into Spaceship Earth. This re-entry will present acute difficulties. If, however, we have a clear view of the nature of the problem, a certain optimism about our power to solve it is entirely reasonable. The one great cause for optimism indeed is the clear fact that the evolutionary potential of the human nervous system is very far from having been exhausted and that there is no nonexistence theorem at present about continued human learning." (*W Post*, 9/6/70, B1)

September 7: Second volume of third edition of new Soviet encyclopedia *Bolshaya Sovetskaya Entsklopediya*—to go on sale in October—displayed "factual approach" to U.S. space program, AP said. *Apollo 11* Commander Neil A. Armstrong received half-column biography and small photo as first man to walk on moon. Apollo missions were given nearly two pages, including two-column photo of *Apollo 11* Astronaut Edwin E. Aldrin, Jr., on moon and diagram of lunar landing. (AP, *LA Times*, 9/7/70)
- If Anglo-French supersonic transport Concorde emerged from test series with reputation intact, it would "in its builders' eyes confirm its image as the spearhead of European technology, and the last defense against American domination of the aviation industry," Stephen V. Armstrong said in *Washington Post*. Public attention was focused on aircraft's sonic boom. West Germany, Sweden, Switzerland, Norway, and Holland had said they were likely to ban supersonic flight over their territories because of boom. Instruments in Oban Cathedral in Scotland during recent Concorde 002 tests had shown boom produced between 10 300 and 20 700 newtons of pressure per sq m (1.5 and 3 psf)—equivalent of thunder. BAC officials had predicted Concorde would produce no more noise than this in commercial service. Concorde opponents were concerned about U.S. reports that supersonic airliners could change weather patterns by condensing water vapor in atmosphere. Airlines had options on only 74 Concordes but BAC was saying it could sell minimum of 200 by mid-1970s for $24 million each and was considering construction of larger model, which would hold 200 rather than 150 passengers. Concorde had already cost more than $1 billion and was expected to cost $650 million more by 1973. No one expected it to make enough to cover this expenditure. (*W Post*, 9/7/70, A13)
- U.S. Secretary of State William P. Rogers discussed Sept. 6 aircraft hijackings with representatives of Israel, U.K., West Germany, and Switzerland in Washington, D.C., and met with President Nixon. (*FonF*, 9/3-9/70, 638)
- R/A Donald B. MacMillan (USN, Ret.), last survivor of 1909 Peary expedition that discovered North Pole, died at age 95. (*NYT*, 9/9/70, 48)

September 8: U.S.S.R. launched *Cosmos CCCLXI* from Plesetsk into orbit with 338-km (210.0-mi) apogee, 195-km (121.2-mi) perigee,

89.8-min period, and 72.8° inclination. Satellite reentered Sept. 21. (GSFC *SSR*, 9/30/70; *SBD*, 9/9/70, 38)

- MSC and NR Space Div. announced selection of Messerschmitt-Boelkow-Blohm of Munich, West Germany, and BAC of Bristol, England, to conduct subsystems studies for NASA's space shuttle. In major step toward international cooperation in space. Messerschmitt would study attitude-control system and BAC would study structures, aerodynamics and flight-test instrumentation, and data handling under contracts financed by their governments. Work would be performed in Downey, Calif. (MSC Release 70-98; NR Release NS-59)

- NASA study of advanced exhaust nozzles for jet aircraft had shown "surprising potential ability" of plug nozzle to reduce noise without affecting aircraft performance, NASA reported. Experimental inlets and nozzles had been ground-tested and then flight-tested on modified supersonic jet F-106 aircraft operated by LeRC at Selfridge AFB, Mich. Aircraft was flown as low as about 90 m (300 ft) to obtain good noise signals at microphone-tape-recorder ground station. (NASA Special Release)

- Series of reviews to select flight design of Saturn Workshop, part of Skylab program, was announced by MSFC. Government engineers, astronauts, and industry representatives would determine any changes necessary before final approval of Workshop scheduled for 1972 launch into earth orbit. In Sept. 9-10 review at MSFC, astronauts would study Workshop procedures in mockup. Sept. 14-18 critical design review at McDonnell Douglas Astronautics Co. Huntington Beach, Calif., facility would be attended by more than 200. At Workshop crew station review Sept. 21-24 at MSFC, astronaut crewmen would "walk through" Skylab tests. Results of reviews would be considered during preliminary review at Huntington Beach Sept. 28-29 and at Workshop Critical Review Board meeting at Huntington Beach Oct. 2. (MSFC Release 70-181)

- U.S., U.K., Switzerland, and West Germany began efforts through International Red Cross toward release of passengers held hostage in Jordan after Sept. 6 multiple aircraft hijacking. U.S. Secretary of State William P. Rogers met in Washington, D.C., with representatives of 10 Arab countries. International Federation of Airline Pilots met in London on Israeli proposal of aviation boycott of Arab countries in reprisal. U.N. Secretary General U Thant urged international community "to adopt prompt and effective measures to put a stop to this return to the law of the jungle." (*FonF*, 9/3-9/70, 638)

- MSFC announced McDonnell Douglas Astronautics Co. had received $97 057 455 supplement to $1 026 393 830 NASA contract for S-IVB program realignment caused by schedule stretchouts. Contract covered storage and maintenance through July 31, 1972. (MSFC Release 70-182)

- *Apollo 13* Astronaut John L. Swigert, Jr., told news conference at Coast Guard Academy in New London, Conn.: "In the past it has taken a war to force technology to advance at an abnormal rate. The space program did it in peacetime, and to me this makes much more sense." Swigert defended proposals for Mars exploration: "If we can learn more about our solar system and have those benefits ac-

crue to the economy, I would say it is worthwhile." (AP, *W Post*, 9/10/70)
- Electrical engineer J. Dominguez Rego received U.S. patent No. 3527 for Electrojet engine that produced electricity by magnetoplasmadynamics—generating electricity directly from fuel burning at high temperature. Some particles of combustion became ionized, or electrically charged, and current could be drawn off. Engine eliminated gas turbine, thus removing limitations on heat and quality of exhaust gases and increasing efficiency while permitting use of cheaper fuels. (Jones, *NYT*, 9/12/70, 35; Pat Off PIO)
- Establishment of FAA Office of Air Transportation Security to cope with criminal acts against aircraft and airports was announced by Secretary of Transportation John A. Volpe. Office would replace FAA Office of Investigations and Security and would emphasize deterrence of aircraft hijacking. Carl F. Maisch, Director of Office of Investigations and Security and former FBI agent, had been named director of new office. (FAA Release 70-81)

September 9: Arab PFLP guerrillas hijacked BOAC VC-10 jet with 105 passengers and crew of 10. Aircraft was ordered to land at Beirut to pick up commando and then was flown to desert airstrip where two other hijacked aircraft were being held [see Sept. 6]. U.N. Security Council asked immediate release of all hostages held.

U.S. sent six USAF transports with medical teams to base at Incirlik, Turkey, for possible use in evacuation of hostages. (*FonF*, 9/10-16/70, 653-5)

- NASA announced it was establishing Earth Resources program regional activity at MTF to complement and supplement automated and manned Earth Resources programs at GSFC and MSC. Robert O. Piland, Deputy Director for Science and Applications at MSC, would direct new activity stressing research in applications of remote-sensing techniques and using data from earth resources experiments on aircraft flying out of Houston, ERTS planned for 1972 launch, and manned orbiting Skylab spacecraft scheduled for 1972 launch. Estimated 75 Government and contract personnel would be employed at MTF center by July 1, 1971, with planned growth to 185 over next two years. (NASA Release 70-147)
- Cancellation of Apollo 15 and 19 had eliminated six more astronauts from chance at moon trip, NASA Director of Flight Crew Operations Donald K. Slayton said in MSC interview. Of 49 astronauts on flight status, 39 had not yet flown in space. Many would not have opportunity for at least five years because of NASA budget cutbacks. Five astronauts had resigned since *Apollo 11* lunar landing and further resignations were "very possible." Slayton said he would not deter astronauts from resigning. "The facts of life are, we don't need them." (UPI, *NYT*, 9/10/70, 54)
- FAA and Eastern Air Lines demonstrated "smokeless jet" at National Airport in Washington, D.C. Boeing 727 with JT8d engines modified to reduce carbon particulates by 75% took off with thin trail of smoke. Pratt & Whitney engineer Gerald R. Daly, who designed pollution-fighting equipment, explained that modified engines reduced smoke but did not eliminate it. (Latham, *W Post*, 9/10/70, A27)
- USN had "taken the wraps off" highly sophisticated $12-million control center being constructed for NATO comsats at Naval Radio Re-

ceiving Station near Chesapeake, Va., New York *News* reported. "The site joins a widespread network controlled from Supreme Headquarters, Allied Powers, Europe, in Casteau, Belgium." (*NY News*, 9/9/70)
- UPI said it had learned U.S. had agreed to sell 16-18 new F-4 Phantom jet fighter-bombers to Israel "in apparent response to the Soviet-supported buildup of Egyptian anti-aircraft missile defenses along the Suez Canal." Sale, with 6 Phantoms already purchased by Israel earlier in year, would come close to filling Israel's request for 25 supersonic jets. (*P Inq*, 9/10/70, 3)
- *Wall Street Journal* editorial on Apollo program cutback: "Although the Soviet Union has never said so, it appears that its officials decided some time ago that their [space] program was too much of a drain on resources and had to be slowed. At a time when the U.S. Government's budget is sinking deeper into the red, it's certainly reasonable that this nation too should start getting a grip on its priorities." (*WSJ*, 9/9/70)
- Aircraft hijacking was discussed by David Hoffman in *Washington Post:* "That hijacking entails much more today than the Havana detour, a sidetrip costing passengers frayed nerves and the airline $3000, has begun to sink home in official Washington. It has now become an instrument of international policy, a play within a play about the cold war. Quite apart from whether a treaty might help, there exists no treaty. That testifies to the concern of governments, this one included." (*W Post*, 9/9/70, A15)

September 9-11: Lunar Science Institute in Houston, Tex., sponsored seminar on moon findings.

At meeting on future of Apollo, scientists proposed crater Tycho near moon's south pole and Descartes, hilly region near Ocean of Storms, as sites for two of remaining four Apollo program lunar landings. Cal Tech geologist Dr. Leon T. Silver said, "If you had only two missions left to make, you'd send one of them to Tycho." Tycho offered samples from area of one of last collisions of large meteor with lunar surface, samples from moon's original southern hills, and samples from deep within moon that had come to surface after impact that had created Tycho. Flight path to and from Tycho would enable astronauts to photograph and examine almost half visible face of moon from lunar orbit. Astronauts could leave scientific instruments less than 1609 km (1000 mi) from moon's south pole, to give earth-bound scientists listening post on lunar region that would otherwise go unexplored. *Surveyor VII* photos had shown Tycho landing to be strewn with many varieties of rocks. Difficulty with Tycho landing was that Apollo crew might have to use so much fuel getting to it that they would not have enough to return if they encountered difficulty on approach, Astronaut Charles M. Duke, Jr., said. Donald K. Slayton, MSC Chief of Flight Crew Operations, said NASA was studying possibility of Tycho landing: "Our position so far is that Tycho is a bad place. . . . But, if we can figure the reasonable way to do it. . .and decide it's a reasonable place to go, then Tycho becomes a candidate for landing."

Descartes, second site favored by scientists, contained relatively young volcanic material and also crust on surface away from volcanic highlands that might be as old as moon itself. It was thought to be one of two Apollo sites where moonquakes might be occurring; other was hills of Fra Mauro, Apollo 14 landing site.

Dr. Alfred E. Ringwood of Australian National Univ. told seminar he had concluded from study of *Apollo 11* and *12* lunar samples that moon once had heat source 160-480 km (100-300 mi) beneath surface that was emitting 1500 K (2200°F) temperatures. Heat had emanated from abundance of radioactive minerals in moon's depths, he proposed, and partially melted moon's inner rock, which then flowed to lunar surface through pores or forced its way upward as it expanded where pores did not exist. Upswelling rock, Dr. Ringwood said, caused lunar features visible on moon's face. He speculated that moon had lost most of its radioactivity and heat 2 billion yrs earlier.

Dr. Charles P. Sonnett, ARC scientist, said he had found "little bumps" of magnetism along lunar surface when studying signals from instruments left on moon. Bumps suggested that radioactive heat was coming to surface and causing parts of moon to behave as if they had magnetic field. Moon was "strong and rigid to depth of 200 mi [322 km]," Dr. Sonnett said, "but I believe that all you need is a porosity of about 10 per cent in that crust to have a warmer and more plastic material leak to the top."

Dr. Rolf Meisner of Univ. of Hawaii said analysis of data from *Apollo 12* lunar seismometer led him to believe that moonquakes could not be caused by breaks in moon's crust but were caused by warm gases escaping from beneath lunar surface in response to earth's increased gravitational pull. He had recorded 14 moonquakes, all of which seemed to come from about 240 km (150 mi) east-southeast of seismometer in Fra Mauro hills, where Apollo 14 was to land in February 1971.

Dr. Thomas Gold, Cornell Univ. astronomer, argued that lunar maria contained dust and not lava. Lack of electrical activity of surface material and the way Apollo moonquake detectors had reacted to tremors and impacts could mean there was deep deposit of powder. Acoustical signal from *Apollo 12* lunar seismometer was "compatible with a deep layer of a very homogenous nature in which the speed of sound is uniform." (O'Toole, *W Post*, 9/12/70, A3; 9/13/70, A27; Hill, *H Chron*, 9/13/70)

September 9-10: Space Station Utilization Conference was held by NASA at ARC. Some 450 scientists invited to plan potential functions of manned orbital research station had included 60 representatives of foreign universities and industrial firms. (ARC *Astrogram*, 8/27/70, 1; ARC PAO; *KC Star*, 8/19/70)

September 10: Construction and launch of orbiting sun reflector with diameter of 914 m (3000 ft), to light entire metropolitan areas of large cities with five times the brightness of full moon was technologically feasible, NR Space Div. scientific adviser Dr. Krafft A. Ericke said during NASA conference on space application technology at Mountain View, Calif. Reflector could be important crime deterrent. (AP, *Glendale News-Press*, 9/11/70)

• Misgivings about NASA's cancellation of Apollo 15 and 19 missions were expressed by 39 leading lunar scientists in letter to Rep. George P. Miller (D-Calif.), Chairman of House Committee on Science and Astronautics: "The NASA policy leading to the cancellation appears to be one favoring the early construction of large manned earth orbital systems following after Skylab A, and the effort and funds saved by the curtailment will probably go towards these. The merit of these programs for science or applications should be investigated, and the very important decision regarding their funding should in our view be made a separate step. At present, it appears that the approved and scientifically most fruitful lunar program will suffer in favor of an as yet unapproved program for whose scientific value there is no consensus, and whose purpose is unclear." (Text)

- MSFC announced award of six-month, $195 000 contract to Ball Brothers Research Corp. to define requirements for 66-cm (26-in) solar telescope to be used in new ATM experiment on Skylab missions. (MSFC Release 70-183)
- City of Boston, Mass., had filed $10.2-million noise pollution suit in U.S. District Court against 19 airlines and Massachusetts Port Authority, *Washington Post* reported. Boston Mayor Kevin H. White had said suit was to pay city $4.5-million cost of soundproofing 15 East Boston schools near Logan International Airport. Remainder would reimburse city for usurping air rights over city property without compensation, depreciating land values, and making land unfit for educational purposes. (*W Post*, 9/10/70, A24)
- First charting of exact locations of six continental land masses as they drifted across globe for millions of years at rate of about 10 cm (4 in) annually was reported by ESSA marine geologists Dr. Robert S. Dietz and John C. Holden in *Journal of Geophysical Research*. Calculations, made by using geological landmarks to plot maps, showed that original "supercontinent" of Pangaea had split midway through Triassic geological period, about 200 million yrs ago, into land masses of Laurasia (now North America, Europe, and Asia) and Gondwana (now Africa, South America, Australia, Antarctica, and India) through sea floor's spreading, welling up of molten material called magma from earth's mantle through rift in ocean bed. Subsequent spreading had separated continents. (*Journ of Geophysical Research*, 9/10/70, 4939-55)
- London *Times* report that British historian Charles H. Gibbs-Smith had discovered drawing of toy helicopter that predated design of helicopter by Leonardo da Vinci by 150 yrs was published by *New York Times*. Drawing had been discovered in Flemish manuscript dated about 1325 in Royal Library in Copenhagen. Gibbs-Smith had said it was "earliest known illustration in history of a powered aircraft. The fact that a fairly sophisticated drawing of it appeared in a manuscript of this date suggests that this toy was in use before 1300." (*NYT*, 9/10/70, 30)
- Invention of hovercraft in 1950s by British boat builder Sir Christopher Cockerell had been described by *In Britain*, publication of British Tourist Authority, *New York Times* said. Sir Christopher had eliminated problem of "drag" on boat hull from friction of water by interposing air cushion as lubricant between hull and water. In 1956 he had demonstrated invention at British Ministry of Supply in London. First public demonstration in 1958 had led to purchase by British government of experimental manned model from Saunders-Roe, Ltd. Launched on July 25, 1959, the 3175-kg (3.5 ton), 19-m (30-ft) SRN 1 became first hovercraft to cross English Channel. (*NYT*, 9/10/70, 29)

September 11: Seismic signals obtained from seismograph carried by *Apollo 12* crew at epicentral distance of 76 km (47.2 mi) from LM impact differed from any seen on earth, Cornell Univ. radiophysicists Dr. Thomas Gold and Dr. Steven Soter reported in *Science*. Signal caused by *Apollo 12* LM was interpreted in terms of propagation between source and receiver through layer of powder in which sound velocity increased with depth. Increase, due to compaction, extended over several kilometers and led to concentration of seismic waves toward surface. Computer simulations using ray acoustics and based on assumption of randomly undulating lunar surface approximated the observed signal. Seismic amplitudes were enhanced in this medium over solid rock, so that observed signal required less power to be transmitted than previously estimated. (*Science*, 9/11/70, 1071-5)

- Four-man LaRC team emerged from space station simulator at McDonnell Douglas Astronautics Co., Huntington Beach, Calif., after successful completion of 90-day test to develop regenerative life support systems for advanced manned spacecraft [see June 13]. *Apollo 11* Astronaut Neil A. Armstrong greeted crewmen John H. Hall, Terry Donlon, Stephen G. Dennis, and Wilson Wong. Project chief engineer Karl Houghton said test could have continued beyond scheduled 90 days. (AP, *NYT*, 9/12/70, 8; *Langley Researcher*, 9/18/70, 1)
- President Nixon announced program to deal with air piracy by placing armed Government personnel on U.S. commercial airliners, extending use of electronic surveillance equipment to all gateway airports in U.S. and in other countries where possible, developing new security measures for detecting weapons and explosive devices, studying antihijacking techniques of foreign airlines, and supporting multilateral convention to extradite or punish hijackers (to be considered at International Conference of ICAO) and Tokyo convention for prompt return of hijacked aircraft, passengers, and crew. President called for suspension of airline services to countries refusing to extradite or punish international hijackers. (*PD*, 9/14/71, 1193-4)
- Twenty-five USAF Phantom jet fighters and four additional transport planes [see Sept. 9] arrived at Incirlik, Turkey. White House Press Secretary Ronald L. Ziegler said aircraft were "routine precaution" in connection with multiple hijacking by PFLP. (*FonF*, 9/10-16/70, 655)
- Number of unidentified space objects had reentered atmosphere and landed in U.S., U.S. Ambassador to U.N. Charles W. Yost said in letter to U.N. Secretary General U Thant. Three objects—including one 90.7 kg (200 lbs) in weight, 76 cm (30 in) long, 25 cm (10 in) wide, and 13 cm (5 in) thick—had been recovered and reports of other sightings were being investigated. It had not yet been established whether fragments had caused any damage. (AP, *NYT*, 9/12/70, 13)
- DOD announced appointment of Dr. James W. Mar, professor of aeronautics and astronautics at MIT, as USAF Chief Scientist, effective Sept. 14. Dr. Mar would succeed John J. Welch, Jr., who would return to LTV, Inc. (DOD Release 740-70)
- President Nixon issued "call for cooperation" in message to Congress: "Our present problems in large degree arise from the failure

to anticipate the consequences of our past successes. It is the fundamental thrust of technological change to change society as well. . . . The task of government in the future will be to anticipate change; to prevent it where clearly nothing is to be gained; to prepare for it when on balance the effects are to be desired; and above all to build into the technology an increasing degree of understanding of its impact on human society. With this in mind, the *National Commission on Productivity* [appointed by Nixon July 10] will be evaluating the impact of technology and other factors related to achieving higher levels of productivity vital to the healthy growth of our economy." (*PD*, 9/14/70, 1172-90)

- President Nixon submitted nominations to Senate: Dr. Glenn T. Seaborg, AEC Chairman, as U.S. Representative to 14th Session of General Conference of International Atomic Energy Agency; Dwight J. Porter as Deputy U.S. Representative to IAEA; and Dr. T. Keith Glennan, first NASA Administrator, as alternate representative. President announced appointment of George M. Wolfe as an alternate U.S. representative on ICAO Council and accepted resignation of J.D. Braman as Assistant Secretary of Transportation for Environment and Urban Systems, effective Oct. 1 (*PD*, 9/14/70, 1196, 1197)

- Management of U.S. and Soviet space programs were compared in *Science* by Dr. Foy D. Kohler, former U.S. Ambassador to U.S.S.R., and Dr. Dodd L. Harvey, both of Univ. of Miami: Programs had been roughly equal in complexity and in input of resources. "If, as has been asserted, the U.S. moon undertaking represented a task equal in technological complexity to the total of all the great tasks performed by man from the building of the pyramids through explosion of the atom bomb, hardly less can be said of Soviet space enterprise." Soviet investment in space program had been comparable to that of U.S., "if not substantially greater." Both programs had depended for success on effective organization and use of existing organizations and expertise and both had required extensive organizational and administrative innovations and new management techniques. In both programs most of basic scientific knowledge and technology already existed. Organization and management had been complicated by necessity of serving both military and nonmilitary objectives and by pressures of diverging national interests. Both programs represented "organizational triumphs of a high order." Aspect of U.S. moon landing now generally forgotten was "that a number of knowledgeable people were convinced it could not be done."

  Programs differed in that "to the Soviet regime. . .the name of the game in space has been 'the system,' not scientific and technological excellence as such." In U.S., National Aeronautics and Space Act "reflected the operation of the representative process at what could be called its best." Its provisions "subjected space activities to continuing congressional control in accord with long-established processes." U.S. space effort represented "new sort of partnership between government, industry, and universities. . . a partnership that has benefited and strengthened each of the participants without impinging on their interests and integrity."

Soviet space effort had had less impact on national economy than had U.S. effort. Soviet sources had revealed continuing shortcomings in computer technology and hardware upon which space program was dependent. Soviet scientists and technologists not directly engaged in space program did not receive direct knowledge of space program's technical discoveries and innovations and were denied use of research tools and techniques employed. "The Soviets themselves are evidencing increasing agitation over their inability to match the United States in transferring 'great prestige' and 'great accomplishments' in science and technology to meet the general needs of society." Authoritative statement indicated U.S.S.R. intended to triple space expenditures in next few years, while in U.S. "prevailing national mood seems to be moving strongly against investments for the future in favor of maximum utilization of resources for the immediate betterment of conditions of life."

This raised question, "Has the whole operation represented but another highly successful one-shot exercise in crisis management, or has it represented incorporation into American society of a new way to organize, systematically and purposefully, the development and use of scientific and technological resources to the furtherance of national goals?" (*Science*, 9/11/70, 1049-56)

September 12-24: U.S.S.R. launched *Luna XVI* unmanned lunar probe from Baykonur at 6:26 pm local time (9:26 am EDT). Tass said purpose of mission was to explore moon and near-moon space and reported all systems were functioning normally.

On Sept. 17 *Luna XVI* entered near-circular lunar orbit with 110.1-km (68.4-mi) altitude. Tass said ground controllers put spacecraft into elliptical orbit with 104.6-km (65-mi) apolune and 14.5-km (9-mi) perilune. Rocket firing thrust spacecraft toward moon and braking rockets were fired 295 m (968 ft) from lunar surface. Main braking rockets were shut down 19.8 m (65 ft) above surface and *Luna XVI* softlanded on moon's Sea of Fertility at 10:18 am Baykonur time (1:18 am EDT) Sept. 20. Scientists at Bochum Observatory in West Germany monitored spacecraft's signals and reported TV pictures transmitted during and after landing were of excellent quality.

After collecting lunar samples and performing other undisclosed experiments, *Luna XVI* lifted off from moon. It had spent 26 hrs 25 min on lunar surface. Tass announced that spacecraft's mission was the "solution of a radically new task—automatic unmanned delivery of lunar ground to earth." Samples, Tass said, were obtained with earth-operated electric drill that penetrated to 350 mm (13.7 in) and were placed in container by a "manipulator" and hermetically sealed in. Spacecraft's landing stage was used as launch pad for liftoff from moon on command from earth. After liftoff spacecraft entered "ballistic trajectory of flight to the earth close to the predetermined one."

Luna XVI entered earth atmosphere at 10:00 am Baykonur time (1:00 am EDT) Sept. 24 and parachuted onto steppes of Kazakhstan 26 min later, after 11-day 16-hr mission. Capsule containing lunar samples was recovered by helicopter for delivery to Soviet Academy of Sciences.

Luna XVI was first unmanned spacecraft to land on moon and return to earth with lunar samples. In last mission in series *Luna XV* had landed on moon July 21, 1969, but had not returned to earth. (*NYT*, 9/13-25/70; B *Sun*, 9/22/70, A1; *SBD*, 9/18/70, 78; 9/25/70, 102)

September 12: Arab PFLP commandos blew up TWA Boeing 707 and Swissair DC-8 hijacked to desert airport in Jordan Sept. 6 and BOAC VC-10 aircraft hijacked to same airport Sept. 9. Majority of more than 300 passengers and crewmen was permitted to leave Jordan soon after destruction of aircraft but at least 54 were kept as hostages to demands for release of imprisoned commandos. (*FonF*, 9/10-16/70, 653-4)

- Dr. Thomas O. Paine, NASA Administrator, discussed further opportunities for U.S.-U.S.S.R. cooperation in space with Soviet academician Dr. M. D. Millionshchikov during luncheon given by Dr. Edward E. David, Jr., nominee for posts of OST Director and Science Adviser to President Nixon, at Cosmos Club in Washington, D.C. (NASA Off of Admin; Paine speech text, 9/14/70)
- Two USAF F-111E tactical jet fighters completed nonstop transatlantic flight from Langley AFB, Va., to join NATO aircraft team at RAF base at Upper Heyford, England. Aircraft were part of conversion program during which 20th Tactical Fighter Wing would be equipped with 72 new F-111E aircraft. (DOD Release 738-70)
- Despite "seeming disenchantment among the young with the so-called 'military-industrial complex,'" DOD was attracting younger and better qualified civilian scientists and engineers than ever before, *New York Times* said. Federal surveys had indicated "competence of new technical talent is steadily improving." Among reasons cited were reduction in demand by private industry because of economic slowdown, recent pay raises for Federal technical personnel, Federally supported programs that allowed graduate study by DOD civilian scientists and engineers, faster promotions for younger men, and intellectual attraction of working for Government. (Lyons, *NYT*, 9/13/70, 74)
- *Washington Post* editorial on President's Sept. 11 statement on air piracy: "It is particularly important that Mr. Nixon recognizes the cardinal need to deprive hijackers of a safe haven. One of the scandals of international diplomacy is that the proposed treaty to require all countries either to extradite or punish hijackers did not go into effect long ago. It would be unwise, however, to await ratification of this treaty by all the countries concerned: airline security is too important to wait on national politics. What is required is suspension of airline services to countries that cannot or will not crack down on hijackers. Mr. Nixon calls upon the international community to take 'joint action' to this end. This...does not go far enough. The United States should unilaterally suspend air services to offending countries. By its example and influence, it should constrain other governments to follow suit." (*W Post*, 9/12/70, A18)

September 13: Concorde 002, British prototype of Anglo-French supersonic transport, piloted by BAC chief test pilot Brian Trubshaw, made unscheduled landing at Heathrow Airport in London during flight from Bristol, England, to demonstrate aircraft over Farnborough Air Display grounds. Bad weather on return sector of flight forced aircraft to make low approach with high power in first appearance at major international airport. (AP, B *Sun*, 9/14/70, A2)

September 14: Dr. Thomas O. Paine addressed Economic Club of Detroit in last public appearance as NASA Administrator. He summarized NASA projections of potential space advances to year 2000—advances U.S., U.S.S.R., Western Europe, and Japan, alone or in concert, would have industrial and economic resources to realize. For both America and other countries, response to challenge would "determine their futures."

New space transportation system—with reusable space shuttle to make return trips to orbit with passengers and cargo, permanent orbiting space station, and automated and manned space tugs for emplacing and recovering satellites and for transportation between orbits—was "bridge to our future in space." Space shuttle would be "NASA's major single effort during the next ten years."

New approach to large payloads based on shuttle booster stage would offset termination of Saturn V after completion of Apollo lunar landings by 1972. Skylab program scheduled to start during 1973 would provide basis for final design of space station. Integrated lunar exploration in early 1980s could use station in lunar orbit as platform for descent to any portion of moon's surface for up to 14 days. Broad, vigorous, unmanned program to explore planets in 1970s would be start of orderly program to visit every planet in solar system, blazing trail for manned trips to Mars and Venus before end of century.

"The space in orbit two hundred miles [320 km] above the earth can be thought of as a vast new continent, equal in area to the surface of the entire globe." Discovered only a dozen years ago, new continent would "become increasingly important as the leading world source of global information and communication services and then, inevitably, as a new home of men from earth." Beyond solar system, automated and manned astronomical probes would "extend our understanding of the universe through more intensive investigations in optical and radio astronomy and space physics." Before century's end, man might find answers to "the overwhelming questions of man's origin, his destiny, and of his uniqueness as the bearer of intelligence in the universe." Man "may indeed be unique—but the odds against it approach a thousand billion billion to one.... Every experiment so far has been negative, but the search will continue.... We do not know what may exist, but we are upon the verge of knowing."

Of Apollo program Dr. Paine said, "History may record that the substitution of major international technical competition, in place of war, was the great and unprecedented peacekeeping breakthrough of our era." Benefit on international social and technical change was "evident to all." (Text)

• Three of nine USAF volunteers began test at MSC to determine man's physiological tolerance to reentry loads expected on space shuttle missions. Subjects, whose reactions to shuttle reentry profile had been measured after 24 hrs rest Sept. 9-11 would rest in bed seven days before riding centrifuge in "eye balls down" reentry configurations at 2.5-g to 4.5-g levels for up to 6 min 10 secs to determine what effect prolonged "eye balls down" reentry acceleration would have on space shuttle crew members and passengers after long periods of weightlessness. Normal space shuttle reentry mode of MSC straightwing shuttle produced downward through-the-head acceleration as opposed to "eye balls in" acceleration through the chest. (MSC Release 70-94)

- Dr. Edward E. David, Jr., was sworn in as OST Director and Science Adviser to President Nixon in Washington, D.C., ceremony. President Nixon said, "This nation needs to strengthen its support of basic science so that the practical applications which will benefit us all can be forthcoming." (PD, 7/21/70, 1213-4)

- President Nixon sent formal request to Congress for $28 million to recruit and train 2500 security guards to ride commercial aircraft. Funds also would go to other measures for ensuring "aircraft security on United States flag carriers." President said DOT was transmitting bid for required legislative authority. (AP, NYT, 9/15/70, 28)

- International tribune to prosecute aircraft hijackers was advocated by U.N. Secretary General U Thant in New York speech on eve of opening of 25th General Assembly: "The world has no other choice but to surrender some national sovereignty. Nations and people must have the courage to resort to adequate new methods of international law and order." (Sherman, W Star, 9/15/70, A1)

- House Committee on Science and Astronautics published *For the Benefit of All Mankind: A Survey of the Practical Returns from Space Investment* (H. Rpt. 91-1446): Space Research "already has produced an extremely broad range of concrete benefits, not only to the American citizenry but to the people of many nations. The flow of 'hard benefits' has grown from a trickle to a stream, and it is widening to a river, as expanding technology uncovers more and more ways of improving man's mode of existence." Report cited tangible benefits in business, medicine, transportation, construction, new fabrics, and new materials. Future benefits "of even greater impact" included use of satellites for more efficient management of earth's natural resources, use of satellites in crime detection, and possible development of hospitals in space "which would offer gravity-free treatment to patients with heart ailments." Illustrations, based on existing or impending technology, were the "*extra* dividends which are a fallout of ingenious application of space experience by business, industry, commerce, Government, the medical profession, and the academic community. Those dividends already paid, coupled with those in sight for the near-term future, affect practically every facet of human convenience and concern. They promise continuing and increasing return on the space investment for the benefit of mankind on earth today." (Text)

- House concurred in S.C.R. 49, providing for congressional recognition of Goddard Rocket and Space Museum. Action cleared measure for President's approval. (CR, 9/14/70, H8620)

- NASA issued policy announcement directing that "measurement values" used in NASA scientific and technical publications be expressed in International System of Units. (NASA Release 70-157)

- Lockheed Aircraft Corp. had announced private financing plan that would provide immediate $30-million loan and $250 million in additional funds, *Wall Street Journal* reported. Additional financing was contingent on settlement of firm's contractual disputes with DOD. (WSJ, 9/14/70, 5)

September 15: Dr. Thomas O. Paine's resignation as NASA Administrator became effective. He was honored at Bolling AFB reception and testimonial dinner given by Dr. George M. Low, who would be Acting Administrator until appointment of successor.

Dr. Paine told guests he believed requirements of U.S. space program "are not understood at the highest level of our official life" and said he was confident NASA budget would be approved despite President's Aug. 11 veto of Independent Offices and HUD appropriations bill. U.S., he said, "must be the first nation to put in the first skylab and space stations. We must look to the post-dog-sled era of lunar exploration and develop a more permanent type of lunar exploration." Dr. Paine warned that NASA must "face the reality that in the 70s the Soviet Union will attempt to overtake us. . . . We will be tested against a strong and vigorous Soviet program." U.S. had "held out the olive branch. . .but if the cold war is extended into space, then NASA will be ready to meet any threat that may develop in space." According to Washington *Evening Star*, it was first time Dr. Paine had publicly cited threat of Soviet military supremacy in space. Later, in interview following dinner, Dr. Paine said he meant NASA's charge was to keep U.S. in forefront of space technology and to be ready to hand that technology to DOD if required. He believed defense against ICBM might yet be space-based. If so, NASA should be in technologically superior position because of space station, shuttle, and other concepts. (W *Star*, 9/16/70, B8, C7)

- In profile of Dr. Edward E. David, Jr., new OST Director and Science Adviser to President, *New York Times* said Dr. David's goal was "to combine science and technology with the human factor in society." Associate had said Dr. David was skeptical of value of manned space program and felt space program should be pushed in "very studied fashion." (*NYT*, 9/15/70, 26)
- Boeing Co. would begin flight testing of Boeing 747B—upgraded version of 747—on Oct. 1, AP reported. Aircraft, already rolled out, had more powerful engines and enlarged center-wing tank. Changes had been made to improve strength of wing, fuselage, and landing gear. Alterations had increased aircraft's weight from 322 000 kg to 351 500 kg (from 710 000 lbs to 775 000 lbs) and range from 9000 km to 10 600 km (from 6000 mi to 6600 mi). New version would cost $1 million more than standard 747, which cost $23 million. (*NYT*, 9/15/70, 39)
- MSC announced selection of Singer-General Precision, Inc., Link Div. to receive $4-million, cost-plus-award-fee contract to design, develop, install, and support Skylab simulator for astronaut and ground crew training at MSC. (MSC Release 70-101)
- Rep. Robert N. Giaimo (D-Conn.) introduced for himself and cosponsors bill to authorize NSF to conduct research and educational programs to prepare U.S. for conversion from defense to civilian, socially oriented, R&D activities. NASA, DOD, and AEC were cited as agencies that should participate. (*CR*, 9/15/70, H8735)
- Sen. J. W. Fulbright (D-Ark.) on Senate floor summarized views of 17 economists polled earlier in year by members of his staff in cooperation with conservationist Sierra Club on subject of SST. There was

"near-unanimous belief that proceeding with the SST would be unwise and is not justified on economic grounds." (CR, 9/15/70, S15397-405)

- New York Times editorial on air piracy: "Is it conceivable that the aerial hijackings by the Palestinian terrorists could have one enduring positive result? Could they provoke or inspire the 126 member governments of the United Nations to concur on a modest advance of the rule of law in international affairs to deal effectively with this crime?" Secretary General U Thant had warned that U.N. had perhaps only another decade in which to become effective or perish. U.N. pioneer Lester Pearson had predicted it would not survive 50 yrs unless it could strengthen machinery to maintain peace and security. "An effective move to restore peace and tranquility to international air travel could be a modest first response to those warnings —if member governments, including the United States, are ready." (NYT, 9/15/70, 44)
- George A. Derbyshire, Executive Secretary of NAS-NRC Space Science Board, died in Bethesda, Md., at age 52. He had been responsible for more than 20 scientific reports on lunar programs, space goals, and physics goals, and had directed scientific conferences on space program and planetary exploration. (W Post, 9/18/70, C10)

September 16: U.S.S.R. launched *Cosmos CCCLXII* from Plesetsk into orbit with 817-km (507.7-mi) apogee, 269-km (167.2-mi) perigee, 95.4-min period, and 70.9° inclination. (GSFC SSR, 9/30/70; SBD, 9/17/70, 77)

- Summary of confidential DOT report on SST to President's Council on Environmental Quality was released to press by "qualified sources." Report recommended that SST development proceed pending study of possible minor harmful effects on environment. More research was needed on temperature increases that would be caused by water vapor and exhaust from SST; airport noise from SST would be higher than desired; and occupants of SST would receive radiation doses higher than occupants of subsonic jets flying at lower altitudes. However, SST's steep angle of climb should make overall community noise no higher than noise from existing jets. Exposure of SST passengers and crews to high-altitude radiation would be lessened by SST's greater speed, which would shorten transit time. SST might increase UV radiation reaching earth's surface and decrease ozone in upper atmosphere, but these effects were "insignificant" and within present-day variations. Possible effect on temperature from increased water vapor should be of no practical significance in temperate and tropical zones, but could be of significance in some isolated polar areas. (Thelen, W Post, 9/17/70, A3)
- President Nixon in Alfred M. Landon Lecture on Public Issues at Kansas State Univ. noted: "When Palestinian guerrillas hijacked four airliners in flight, they brought to 250 the number of aircraft seized since the skyjacking era began in 1961. And as they held their hundreds of passengers hostage under threat of murder, they sent shock waves of alarm around the world to the spreading disease of violence and terror and its use as a political tactic." (PD, 9/21/70, 1227-33)
- Jordanian King Hussein announced formation of military cabinet to meet mounting civil strife. Government of Premier Abdel Moneim

al-Rifai had been dismissed in midst of unrest over Arab commandos' hijacking and destruction of four Western airliners and their challenge to Hussein's authority. Civil War followed. (FonF, 9/17-23/70, 669)

- USAF launched Minuteman III missile from Cape Kennedy to Atlantic target 8047 km (5000 mi) away. (UPI, *NYT*, 9/18/70, 2)
- LeRC was honored with IR-100 award in Chicago for "development of one of the 100 most significant technical products of the year." Sponsored by Industrial Research Inc. of Beverly Shores, Ind., award cited LeRC development of electron-bombardment ion thruster. Award was fourth won by LeRC since it entered IR-100 competition in 1966. (LeRC Release 70-47)
- NAA President Frederick B. Lee announced selection of C. R. Smith, former Secretary of Commerce and President of American Airlines, to receive Wright Brothers Memorial Trophy. Smith would be honored at annual memorial dinner Dec. 17 for outstanding contributions to aviation over four decades. (W *Star*, 9/16/70, A2)
- NASA announced publication of *Apollo 12: Preliminary Science Report* (NASA SP-235), 227-page book containing extensive black-and-white photos, charts, and initial scientific findings of *Apollo 12* mission. (NASA Special Release)

September 16-17: European and U.S. representatives met on cooperation in post-Apollo space program, at Dept. of State in Washington, D.C. Participating were Theo Lefevre, President of European Space Conference; Lord Bessborough, U.K. Minister of State; J. F. Denisse, representing French Minister for Scientific Research; U. Alexis Johnson, Under Secretary of State for Political Affairs; Dr. Edward E. David, Jr., Presidential Science Adviser; Dr. George M. Low, Acting NASA Administrator; and William A. Anders, NASC Executive Secretary. Meetings explored development and use of advanced space transportation system and space station and continued use by European Space Conference members of existing U.S. launch vehicles for their own space programs. (State Dept. Release 272)

September 17: U.S.S.R. launched *Cosmos CCCLXIII* from Baykonur into orbit with 288-km (179.0-mi) apogee, 201-km (124.9-mi) perigee, 89.3-min period, and 65.0° inclination. Satellite reentered Sept. 29. (GSFC *SSR*, 9/30/70; *SBD*, 9/21/70, 88)

- NASA launched series of six sounding rockets from Wallops Station—two Arcas, Nike-Cajun, Nike-Apache, Viper Dart, and Loki Dart—to obtain data on atmosphere. Data would be compared with that from orbiting *Nimbus III* and *IV* satellites to improve understanding of relations between various data. (NASA Rpts SRL)
- Univ. of California physicist and lunar expert Dr. Harold C. Urey pleaded for reinstatement of Apollo missions canceled by NASA because of lack of funds in letter to *Washington Post:* "We scientific people in great numbers are working on many aspects of this problem. We wish that we could talk more about our conclusions. A bag of rocks requires much study and much other evidence must be carefully considered. The remaining Apollo missions are needed to secure an understanding of the moon which may be the only planetary object with the very ancient history recorded in its surface rocks." (*W Post*, 9/17/70, A23)

- *Christian Science Monitor* editorial: "Mankind must add one more category to its pollution roster: outer space pollution. Which means all those left-over, burned-out, still-extant bits and pieces and whole units of spacecraft now circling the earth. There's more of this space junk than we might imagine. Something should be done about it, preferrably by a treaty at the United Nations." (*CSM,* 9/17/70)
- Preliminary results of uncompleted proof-test program had shown "F-111 fleet will be structurally sound, and that it will indeed perform its intended mission," Dr. Robert C. Seamans, Jr., Secretary of the Air Force, said in letter to Sen. John C. Stennis (D-Miss.), Chairman of Senate Armed Services Committee. (Text)

September 18: Astronauts Alan B. Shepard, Jr., and Edgar D. Mitchell completed testing of Apollo 14 LM in vacuum chamber at KSC, in preparation for launch to moon Jan. 31, 1971. (*W Post,* 9/19/70, A7; *SBD,* 9/21/70, 85)

- NASA announced formation of steering group, headed by NASA Associate Administrator for Organization and Management Richard C. McCurdy, for comprehensive study of "the complete NASA acquisition process." Study would include the "definition of requirements; the review and approval process; the source evaluation and selection process; contracting methods and governing policies; and management of the acquisition process by project, program, and functional organizations." (NASA Release 70-152)
- *Science* editorial by Dr. Preston E. Cloud, Jr., geologist of Univ. of California at Santa Barbara, scored cancellation of Apollo 15 and 19: "The moon is the only other planet we can hope to study in sufficient detail for close comparison with our own. We have just begun that study. It is as if we were trying to understand North America by examining Plymouth Rock." Billions had been spent to get within reach of primary scientific goals on moon. "Only a small fraction of the investment already made would see the job to a fruitful conclusion. To stop short for reasons within our control would, in retrospect, be seen as one of history's most irresponsible follies. Nothing less than the early institution of a comprehensive automated program to get similar information and sample return could begin to ameliorate such a failure." (*Science,* 9/18/70, 1159)
- After Sept. 17 and 18 press reports of hints from President Nixon that U.S. would be prepared to intervene in Jordan if Syria and Iraq entered civil war, *Washington Post* reported from Beirut warning by PFLP that U.S. hijack hostages could be killed if U.S. entered conflict. (*FonF,* 9/17-23/70, 670)
- *Introduction to the Solar Wind* by John C. Brandt was reviewed in *Science* by Roger A. Kopp of National Center for Atmospheric Research. In 1958 it had been argued that sun's 995 800 K (1 000 000°F) corona must expand continuously into space with supersonic velocities near earth's orbit. Subject had remained in controversy until 1962 observations from *Mariner II* indicated plasma outflow from sun with 400-km-per-sec (249-mps) velocities and mean particle density of about 10 per cc (0.06 cu in) at orbit of earth. "The obvious place to begin a discussion of the solar wind is at the sun, and the author sets the stage by summarizing current knowledge of classical solar physics." (*Science,* 9/18/70, 1195-6)

- Twenty-third anniversary of USAF. Official anniversary was established Sept. 18, 1947, with swearing-in of first secretary of the Air Force W. Stuart Symington. Gen. Carl Spaatz was sworn in as first USAF Chief of Staff on Sept. 26, 1947). (*Airman*, 9/70, 15)
- Communist China was producing medium-range, twin-engine jet bomber that could carry nuclear weapons to targets 2400 km (1500 mi) beyond its borders, *Washington Post* reported U.S. sources had said. Aircraft was copy of Soviet Tu-16 Badger in service with Soviet air force since 1955. Production versions of first jet bomber manufactured inside mainland China had been spotted by high-flying U.S. reconnaissance craft, "either satellites or U2s." (Getler, *W Post*, 9/18/70, A1)

September 19: Instrument package from MSC's cosmic ray emulsion plastic experiment, launched Sept. 4, crashlanded in flax field 32 km (20 mi) west of Regina, Saskatchewan, after its 183-m (600-ft-long) balloon crashed into power lines. Balloon, still containing portion of 588 990 cu m (20.8 million cu ft) of helium, drifted eastward and descended near Fork River, Ontario. Payload was recovered "miraculously. . .in excellent condition" by NASA, National Center for Atmospheric Research, and Winzen Research Corp. scientists and taken to Univ. of California at Berkeley for examination and preliminary analysis.

Balloon and payload had drifted west for several days after launch from Minneapolis, crossed over coastline, drifted westward 1207 km (750 mi), returned over U.S., and then drifted over Canada. (NASA Release 70-154)

- Complex space tug NASA was planning as part of integrated space transportation system was described in Washington *Evening Star* by John Lannan. Tug would be assembled like "astronaut's erector set." It would consist of propulsion, living, cargo, and electronic control modules and could be "put together or broken down to meet any immediate need." Tug would be used to land men on moon for long-term stays, to push and pull large pieces of space hardware to assemble them into space stations or interplanetary vehicles, and for astronaut rescue. Tug, launched into orbit aboard shuttle, would never return to earth. It was designed for 3-yr, 10-mission lifetime and would be chemically re-refueled with oxygen and hydrogen replenished by tanker shuttle. NASA expected tug to deliver 27 200-31 750 kg (60 000-70 000 lbs) of gear to lunar surface during 28-day lunar surface mission. NASA wanted vehicle to be capable of 90° orbit change and of servicing geosynchronous satellites at 35 900-km (22 300-mi) altitude. (*W Star*, 9/19/870, A1)
- Dr. W. Ross Adey, head of UCLA Brain Research Laboratory and principal investigator for NASA *Biosatellite III* primate experiment launched June 28, 1969, had filed $2-million defamation of character suit against United Action for Animals, Inc. (antivivisectionist group), *Washington Post* reported. Group had accused Dr. Adey of "savagery" in death of Bonny, pigtailed monkey used in experiment. Suit had been prompted by feeling among researchers that attacks on scientists by antivivisection movement should be discouraged. (Auerbach, *W Post*, 9/19/70)
- Five U.S. nuclear scientists were enroute to Moscow for six months research in peaceful uses of atomic energy, AEC announced. Sixth

scientist would begin assignment in October. Reciprocal opportunities would be provided for Soviet scientists in U.S. laboratories at later date in exchange program between AEC and U.S.S.R. State Committee on Atomic Energy. (AEC Release N-169)
- Dr. James G. Allen, professor of history at Univ. of Colorado, authority on history of space, and NASA consultant, died of heart ailment at age 69. He had authored several books and articles on space history and had received special citation from American Astronautical Society for work on historical and philosophical background of space age. (*Boulder Daily Camera*, 9/19/70)

September 20: Apollo 14 lunar landing crew would conduct active seismic experiment during first EVA period, NASA announced. Crew would set off 21 small explosions while on lunar surface and would arm mortar to rocket-launch 4 grenades by ground command after crew left moon. Explosions would create seismic waves for detection by geophones emplanted in surface to provide data on structure of moon. (NASA Release 70-150)
- Attempt at first free balloon crossing of Atlantic Ocean was initiated by New York broker Rodney Anderson, his wife Pamela Brown, and British aeronautical engineer Malcolm Brighton with takeoff from East Hampton, N.Y., in balloon *Free Life*, 24 m (80 ft) high, 15 m (50 ft) wide. If planned 4800-km (3000-mi) journey was successful, it would break existing balloon distance record of 3050.8 km (1896 mi) set by eight-man German team in 1914. (Horseley, *NYT*, 9/21/70, 39; AP, B *Sun*, 9/21/70, A3)
- Visit to Project Tektite II undersea habitat was described in Washington *Sunday Star* magazine *Sunday*. NASA had found "perfect place to see if man can put up with his fellow men on the long trip to Mars—at the bottom of the Caribbean." Aquanauts' home, built by General Electric Co., consisted of two steel cylinders, each 5.5 m (18 ft) high and 3.8 m (12½ ft) wide, connected by tunnel of 1.4 m (4½ ft) dia. Cables and tubes provided communications with surface, electricity, and nitrogen-oxygen breathing gases. (Berry, *Sunday*, W *Star*, 9/20/70, 11-6)
- At time when nation's big cities were often choked by too much air traffic, scores of small U.S. towns had lost their only airline service, *New York Times* reported. Over past five years, airlines had stopped serving 66 communities. In last year, 22 small towns had lost air service. CAB was considering airline applications to suspend flights at 8 more towns and industry experts estimated that the nine subsidized regional airlines would drop 100 additional communities if CAB would allow them to. Airlines contended $36-million subsidy in 1969 was insufficient to offset rising operational costs and passenger demand was too weak in many areas. (Lindsey, *NYT*, 9/20/70, 1)
- *The Wartime Journals of Charles A. Lindbergh* were reviewed by Eric F. Goldman in *New York Times Book Review*: Journals covered 1937-1945, "when the celebrated flyer was in Europe surveying military aviation; the battle between President Franklin Roosevelt and the isolationists, during which Lindbergh stumped the United States as the star speaker of the anti-interventionists. . .; war years when, denied an Air Corps commission, he served as a civilian aeronautical expert in private industry and in the Pacific, also managing to

work in 50 combat missions; and the weeks just after the Nazi surrender, which found him again in Europe, attached to a Naval Mission studying wartime developments in plane design and missiles."

Journals showed Lindbergh as "a first-rate mind who was widely informed yet retained certain key areas of naïveté, an instinctive tinkerer with a passion for science as an idea and a gifted scientist who worried over the 'narrow-mindedness' of the pursuit, a man who read Plato the night before inspecting aviation factories and Dostoevsky in between speeches before roaring crowds, a hard-hitting bombadier screeching down over Japanese installations while he contemplated the beauty of God's nature and the dignity of man." (*NYT Book Review*, 9/20/70, 1, 42)

September 20-25: Fifth Intersociety Energy Conversion Engineering Conference was held in Las Vegas, Nev. JPL engineers Dr. David G. Elliott and Jack F. Mondt reported that spacecraft of 1980s and 1990s could employ multikilowatt power plants linking wastebasket-sized nuclear reactor and cluster of ion-thruster engines to reach beyond solar system. Nuclear electric rocket represented "most advanced form of space propulsion likely to be available in this century," Dr. Elliott said. Two systems under development and study at JPL were liquid-metal magnetohydrodynamic (MHD) system for nuclear power conversion and thermionic reactors system.

Owen S. Merrill of JPL Nuclear Power Sources Group outlined progress of radioisotopic thermoelectric generator (RTG) being developed by AEC for NASA and JPL. RTG was only known source considered capable of producing 500 w of power over 9- to 12-yr period required for NASA Grand Tour missions to outer planets.

Melvin Swerdling, JPL spacecraft power systems specialist, described improved nickel-cadmium batteries with increased discharge-charge cycle capability, designed by JPL for NASA Mariner Mars 1971 spacecraft. (JPL Release 563)

September 21: NASA launched series of four sounding rockets from Wallops Station—one Loki Dart, one Nike-Cajun, one Viper Dart, and one Arcas—to obtain atmospheric data for comparison with data from orbiting *Nimbus III* and *IV* satellites. Rocket and instruments functioned satisfactorily. (NASA Rpts SRL)

- Achievements of U.S.S.R. *Luna XVI* mission were praised in Washington, D.C., by Dr. George M. Low, Acting NASA Administrator: "The Soviet Union's reported success in obtaining lunar material, depositing it in a sealed container and launching it from the moon's surface on a return trip to earth with the Luna 16 unmanned spacecraft is a major engineering and scientific achievement. We wish the Soviet Union and its scientists success in completing this exciting mission." NASA would "look forward to sharing in the information which will be developed by an analysis of the samples." (Schmeck, *NYT*, 9/22/70, 11)

- Rep. George P. Miller (D-Calif.), Chairman of House Committee on Science and Astronautics, replied to Sept. 10 letter by 39 lunar scientists protesting NASA's cancellation of Apollo 15 and 19: "Forced into the development of an interim operating plan occasioned by the Presidential veto of the Independent Offices Appropriations Bill and reduced funding levels, the NASA management had to determine how best to carry out the Apollo and other ongoing programs to realize maximum benefits from them while preserving adequate resources for the future programs. Had your views on the Apollo program been as forcefully expressed to NASA and the Congress a year or more ago, this situation might have been prevented." (Text)

- Balloon 24 m (80 ft) high, carrying crew attempting first transatlantic balloon crossing [see Sept. 20], fell into Atlantic 965 km (600 mi) southeast of Newfoundland in rain and rough weather after cold air had made helium in balloon contract. USCG instituted search. (*NYT*, 9/22/70, 15)
- Nine astronauts, headed by Richard H. Truly, began walk-through of Skylab Workshop tasks in week-long Saturn Workshop crew-station review at MSFC. (MSFC Release 70-189)
- DOD announced approval of development of heavy-lift helicopter (HLH), first new U.S. military helicopter development approved in last five years. HLH would have 22 680-kg (25-ton) maximum lift capacity—more than twice capacity of existing U.S. helicopters—and would be used by USN and USA. (DOD Release 767-70)
- Potential hazards from funnel-shaped winds spinning off wings of jumbo jet aircraft had been revealed by series of FAA flight tests, Washington *Evening Star* said. Tests had shown that swirling winds trailed behind huge aircraft for many miles and did not readily dissipate at high altitudes. Invisible turbulence could endanger passenger transports as well as small aircraft. Small aircraft which penetrated turbulence within 4.8 km (3 mi) of giant aircraft could be forced into sudden 75° roll. (Lang, W *Star*, 9/21/70, A9)
- Presidential Management Improvement Award was presented to Van A. Wente of NASA Information Systems and Development Branch for conception, planning, design, and implementation of NASA RECON computer system for on-line retrieval of scientific, technical, and management information. Presentation was made at Presidential Awards Dinner in Washington, D.C. (NASA Org & Mgmt Off)
- Secretary of Transportation John A. Volpe testified on DOT action in President Nixon's program against air piracy, before House Committee on Ways and Means. As first step, DOT had urged ICAO to establish basic principles for effective sanctions: State in which hijacked aircraft landed was obligated to permit passengers and crew to continue journey and to return aircraft and cargo to owners. State to which hijacker fled was obligated to extradite or prosecute hijacker. Multilateral sanctions should be taken against states that failed to fulfill obligations. U.S. had urged ICAO to draft international convention providing legal basis for effective sanctions.

  To finance program of placing armed Government employees on U.S. carriers and extension of surveillance equipment at airports, DOT had submitted $28-million FY 1971 budget amendment to Senate Appropriations Committee. Agency proposed costs of program be borne by 0.5% increase in 8% ticket tax on domestic airfares and increase of $2 in $3 head tax paid by passengers on international flights. (Text)
- Appointment of L/G Benjamin O. Davis, Jr. (USAF, Ret.), as Director of Civil Aviation Security for DOT was announced by Secretary of Transportation John A. Volpe. (DOT Release 20370)

- Since July 15, 106 patients had been airlifted by helicopter to civilian hospitals in 68 missions flown under DOD-DOT project MAST (military assistance for safety in traffic). Project in Texas, Colorado, Washington, Idaho, and Arizona was testing use of helicopters to aid accident victims. (DOD Release 776-70)
- Library of Congress Legislative Reference Service released *The SST: The Issues of Environmental Compatibility*. Report, prepared at request of Sen. Mike Gravel (D-Alaska), said SST, with acoustical correction equipment system, "will emit about one half the noise of the subsonic Boeing 707." Without acoustical correction aircraft would produce "excessive noise during the run-up and take-off phase." SST afterburners would require further work to diminish noise. Weather would not be significantly affected by SST (Text)
- Discovery of 2500-yr-old solar observatory at ancient ceremonial center of Monte Alto near Escuintla, Guatemala, was reported in London *Times* dispatch printed in *New York Times*. Harvard Univ. anthropologist Edward M. Shook had dated observatory provisionally to 500 B.C. (*NYT*, 9/21/70)

September 21-23: Air Force Assn. held 1970 National Convention in Washington, D.C.

During seminar on threat to U.S. security, Dr. John S. Foster, Jr., DOD Director of Defense Research and Engineering, said U.S.S.R. had "more than 300 large SS-9 ICBMs operational or under construction."

Dr. Robert C. Seamans, Jr., Secretary of the Air Force, said in speech that USAF must modernize its strategic offensive and defensive forces to remain creditable deterrent to aggression. He was concerned over increased numbers and total payload of Soviet ICBMs and Soviet deployment of initial ABM systems and extensive ABM research.

Apollo 11 Astronauts Edwin E. Aldrin, Jr., and Michael Collins received H. H. Arnold Trophy, Air Force Assn.'s highest award, on behalf of Apollo team "responsible for achieving man's first landing on the moon—the flight crew, industry and government." L/G Samuel C. Phillips, SAMSO Commander and former Apollo Program Director, accepted award on behalf of Federal Government and J. L. Atwood, former NR president and chief executive officer, accepted on behalf of industry. (*SBD*, 9/11/70, 52; *W Star*, 9/9/70, A2; Getler, *W Post*, 9/24/70, A27; *P Inq*, 9/24/70, 2)

September 22: U.S.S.R. launched *Cosmos CCCLXIV* from Plesetsk into orbit with 294-km (182.7-mi) apogee, 201-km (124.9-mi) perigee, 89.4-min period, and 65.4° inclination. Satellite reentered Oct. 2. (GSFC *SSR*, 9/30/70; *SBD*, 9/28/70, 118)
- NASA launched two sounding rockets from WSMR. Nike-Apache carried Univ. of Colorado experiment to study airglow, but mission did not meet minimum scientific requirements. Aerobee 170 carried GSFC experiment to study x-ray spectra. Rocket and instruments functioned satisfactorily. (SR list)
- Soviet press and radio underlined importance of *Luna XVI* mission. Tass carried interview with unidentified deputy chief designer of spacecraft, who demonstrated prototype of drill used to obtain lunar

samples and said similar device "will bring one day rock samples from Venus, Mars, and other planets of our solar system to earth. Man will go to other planets sometime in the future, but before this is possible, these planets will have to be explored for a long time by automatic stations. At present, an apparatus of a radically new type has been designed in principle. In the long run it will be modified to suit specific tasks....A flight to Mars will require somewhat different technical solutions than a flight to the moon. But the main complex elements of Luna 16—its ability to take off in the directon of a desired celestial body, make a soft-landing on the surface of the planet, operate on it and then start off in the direction of the earth or the planet from which it started—will remain unshakable."

In *Izvestia* Georgy I. Petrov, head of Soviet Institute for Space Research, said man would still play a role in lunar research, but only in limited way. He said short visits by man appeared useful but he proposed manned orbiting moon stations equipped with automatic spacecraft capable of descending to moon, moving over surface, conducting scientific research, and returning to orbiting station. (Gwertzman, *NYT*, 9/23/70, 11)

- Search by USCG aircraft and cutters of 12.5-million-hectare area (50 000 sq mi) of Newfoundland failed to produce trace of missing balloon or crew attempting first transatlantic balloon crossing in 24-m (80-ft) balloon. (*W Post*, 9/25/70, A5)
- House Ways and Means Committee in Executive session agreed basically with revenue aspects of President Nixon's antihijacking proposal and agreed to creation of special force of 2500 air marshals through June 30, 1972. (*CR*, 9/22/70, D1040)
- Washington Airways, Inc., had told CAB it was abandoning its application for helicopter route approval between downtown Washington, D.C., and Washington's three major airports, *Washington Post* reported. District zoning commission had refused to act on request by consortium of nine airlines to amend zoning law to permit rooftop helipads. (*W Post*, 9/22/70, C1)
- Formation of Convair Aerospace Div. and Electro Dynamic Div. was announced by Roger Lewis, General Dynamics Corp. President. Both new divisions would report to Dr. George E. Mueller, former NASA Associate Administrator for Manned Space Flight who had joined General Dynamics as Senior Vice President in December 1969. (General Dynamics Release 1537)
- Dr. Glenn T. Seaborg, AEC Chairman, addressed Fourteenth General IAEA Conference in Vienna: "The industrial and technological advances being made in the world are creating increasing demands for electric power that is cheap, abundant, safe, reliable and clean. In my country, the estimated total electrical generating capacity for 1970 is about 300 000 megawatts. I believe that a realistic estimate of the projected growth of our electrical generating capacity by the year 2000 is about 2 100 000 megawatts, which will require adding the equivalent of a total of 1 800 power plants of 1000 megawatts capacity each." On global scale, "some 5000 new electric generating plants of 1000 megawatts capacity each would be needed over the next 30 years." (Text)

September 22-23: Meetings were held at MSFC to select requirements for main engine design of proposed space shuttle. Dr. Karl Reinhold of ELDO attended. (MSFC Release 70-193)

September 22-25: Eurospace, organization of European industries concerned with space, sponsored conference on cooperation in space in Venice, Italy. Theo Lefevre, Belgian Minister for Scientific Policy and Planning and head of European Space Conference, reported on his Sept. 16-17 negotiations with U.S. Government. He confirmed that, in return for European commitment to participate in development of space shuttle and related hardware, U.S. would launch European spacecraft until shuttle was available. Charge for each launch would not exceed cost of rocket and its firing.

Predominant theme of conference, was that reusable shuttles would substantially lessen cost of practical uses of space. There was little talk of manned space stations or exploration of other planets. Discussions centered on use of satellites for communications, air traffic control, and weather forecasting. European public had become "cool to manned space flight," *New York Times* said. ESRO head, Dr. Herman Bondi, said at conference, "We have grown up with the impression that everything in space is devilishly expensive." Belgian aerospace executive Maurice Desirant said European efforts at joint space programs had been handicapped by pressures to perform national feats. Now, however, European countries had become "more and more aware of the dangers." They had seen money wasted on duplicated efforts which had achieved limited results.

Lefevre said that, to be effective, European contribution to major space effort must be under leadership strong enough to override any differences arising from "divergent technical points of view, reasons of industrial competition, economic whims for national politics or finances."

NASA Associate Administrator for Manned Space Flight Dale D. Myers outlined shuttle plans. Payload of 22 680 kg (50 000 lb) had been halved. (Sullivan, *NYT*, 9/23-24/70; *NYT*, 9/26/70, 11)

September 23: Roy P. Jackson, Vice President and Assistant General Manager of Northrop Corp. Aircraft Div., was named NASA Associate Administrator for Advanced Research and Technology, effective Nov. 2.

Dr. George M. Low, Acting NASA Administrator, also announced appointment of Oran W. Nicks, who had been Acting Associate Administrator of OART, as Deputy Director of Langley Research Center, effective Nov. 2.

Vincent L. Johnson was named Deputy Associate Administrator for Space Science and Applications, effective immediately. (NASA Release 70-155)

• Dr. George M. Low, Acting NASA Administrator, accepted Franklin Institute's William M. Vermiyle Medal on behalf of NASA at Philadelphia ceremony. Award cited NASA for "a decade of unprecedented managerial, scientific and technological achievement culminating in the incredible triumph of landing men on the Moon and returning them safely to earth." NASA was first organization to receive medal; previous awards for outstanding contributions in industrial management had been made only to individuals. (Franklin Institute Release)

- Convair engineers presented results of "Phase A" experiment module concepts study during meeting at MSFC of NASA representatives and Convair Aerospace Div. of General Dynamics Corp. Meeting was also attended by French representatives of ESRO. Module would be extension of space station that could be outfitted on ground and transported into space by shuttle vehicle. Program technical director Max E. Nein said module was attractive to European scientific commmunity because country desiring to participate could either build module complete with experiment package or build experiment package for integration into U.S. module. (MSFC Release 70-196)
- President Nixon approved H.R. 16539, amending National Aeronautics and Space Act of 1958 to provide that Secretary of Transportation shall be member of NASC. (PD, 9/28/70, 1273)
- White House announced that President Nixon had asked *Apollo 13* astronauts James A. Lovell, Jr., John L. Swigert, Jr., and Fred W. Haise, Jr., to visit Iceland, Switzerland, Greece, Malta, and Ireland as his personal representatives. Mission would include visit to 21st International Astronautical Congress in Constance, Germany, where astronauts would address session in October. They would be accompanied on tour by Mrs. Lovell and Mrs. Haise; Swigert was bachelor. Group would depart Oct. 1 and return Oct. 15. (PD, 9/28/70, 1258)
- Agreement under which 50 astronauts and wives would share $100 000 for each program in series of TV documentaries on U.S. space program had been announced by David L. Wolper, President of Wolper Productions Inc., and Louis Nizer, New York attorney for astronauts, AP reported. (W Post, 9/23/70, B4)
- Baltimore *Sun* editorial on *Luna XVI:* "No machine yet within our conceiving, no matter how many tasks it can perform with high precision, can replace direct human observation. Nor can it reproduce the actual human experience, in all its variety, of conditions on remote objects and the effects of those conditions on body and mind. It cannot equal man's power to see for himself, let alone end the instinctive desire to do so. Luna 16, however, promises to provide an immensely useful tool for the ever deeper study of the planetary system, and one which may ultimately reduce the great dangers of man's own exploratory travels." (B Sun, 9/23/70)

September 23-25: "Fort Bliss Oldtimers"—group of rocket pioneers which included many of 118 former Germans who came to U.S. with Dr. Wernher von Braun after World War II to continue rocketry development—held reunion in Huntsville, Ala., and visited MSFC. (MSFC Release 70-191; MSFC PAO)

September 24: NASA's congratulations on *Luna XVI* achievement were telegraphed by Dr. George M. Low, Acting Administrator, to Mstislav V. Keldysh, President of Soviet Academy of Sciences: "This impressive technical achievement promises to add significantly to man's knowledge of the Moon. Along with scientists throughout the world, we look forward to the first reports on analyses of these samples." (NASA Release 70-156)

- DOD and AEC conducted nonnuclear, high-altitude test with Thor booster at Johnston Island in Pacific. Test vehicle carried test simulator and stellar x-ray package to locate and identify stellar x-ray sources and transmit data to earth. Purpose of launch was to ensure that delivery system could transport test device for detonation at predetermined time and point in space. Test was part of DOD-AEC readiness tests conducted since limited test-ban treaty became effective in 1963. (AEC-DOD Release N-66; DASA Release)
- MSC announced RCA had received $3-million, cost-plus-fixed-fee contract modification for final development and production of lunar communications relay units (LCRU), bringing total contract to $10 million. LCRU would be carried on Apollo missions beginning with Apollo 15 to transmit astronaut voice, TV, and telemetry communications directly from moon to earth. On earlier Apollo lunar landing missions astronauts communicated with earth via LM on moon. (MSC Release 70-106)
- John Leyton, President of Professional Air Traffic Controllers Organization, said in Washington, D.C., that addition of SST to existing mix of aircraft would necessitate creation of entire new traffic control system in U.S. To enable SST to respond to traffic control instructions would require 320 km (200 mi) of maneuver space. (W Star, 9/25/70, A4)

September 25: U.S.S.R. launched *Cosmos CCCLXV* into orbit from Baykonur with 210-km (130.5-mi) apogee, 144-km (89.5-mi) perigee, and 49.5° inclination. Satellite reentered same day. (Spacewarn, 10/20/70, 2; GSFC SSR, 9/30/70; Spaceflight, 3/71, 94)
- Black Brant IV sounding rocket was launched by Brazilian space team for NASA from Natal, Brazil, carrying MSC experiment to study energetic particles in South Atlantic Anomaly region. Rocket and instruments—including positive ion telescope, integral flux spectrometer, and two magnetometers—functioned satisfactorily. Launch was first of two in series; second would be launched Sept. 29. (SR list; MSC Release 70-105)
- Cause of July 23 failure of ion engine on board orbiting *Sert II* had been identified and could be prevented in future, LeRC announced. Project Manager Raymond J. Rulis said beam from neutralizer had eroded and broken off small piece of metal from one of two grids at back of engine. Metal, attracted by forces of electric field between screen grid and accelerator grid, had been welded between grids by high voltages and had caused short circuit that shut down grid power lines. Although damaged thruster on *Sert II* could not be repaired, erosion problem could be prevented in future by relocating neutralizer and redirecting its beam to eliminate localized wear on grid.

  Second ion engine on board *Sert II* was operating satisfactorily after more than 1300 hrs. Engine would be shut down in late November for three months while spacecraft was in earth's shadow and then restarted to complete six-month test. (LeRC Release 70-49)
- Japan's attempt to launch second satellite, 61.8-kg (136.4-lb) MS-F1 solar monitor, failed when Mu 4S1 4th stage malfunctioned after launch from Uchinora Space Center. Instrument package went dead 16 min 10 secs after launch. Mission had been postponed seven times since original Aug. 19 launch date because of inclement

weather and technical problems. First satellite, *Ohsumi*, had been launched Feb. 11. (SBD, 9/28/70, 119; AP, W *Star*, 9/25/70)

- President Nixon announced reorganization of Council on Physical Fitness and appointment of members. *Apollo 13* Astronaut James A. Lovell, Jr., President's Consultant on Physical Fitness, would serve as Council Chairman. (*PD*, 9/28/70, 1264)
- *Science* editorial by Philip H. Abelson discussed U.S. energy shortage: "The longer-term solutions to our energy problems involve becoming more prudent in the use of energy. The solutions also demand the skillful employment of coal and atomic energy. In principle, all our energy needs could be met for a long time with coal. This raw material could be processed to yield sulfur-free fuel, liquid hydrocarbons, and methane. In practice, however, the development of the use of coal is limping along and is underfinanced. A few hundred million dollars a year devoted to research, development, and demonstration plants could be the most valuable expenditure the government could make." (*Science*, 9/25/70, 1267)

September 25-29: Remaining airline hijack hostages in Jordan [see Sept. 12] were freed by Arab PFLP terrorists as part of deal for release of Arab terrorists held in Europe. (*FonF*, 9/24-30/70, 691)

September 26: First published eyewitness account of lunar samples brought from Sea of Fertility by Soviet *Luna XVI* was given in *Izvestia* by B. Konovalov: Lunar dust turned greenish and sometimes reddish under direct light but in general looked like "dry mud." Later *New York Times* said *Izvestia* correspondent Konovalov had been permitted to observe lunar material at special Soviet Academy of Sciences laboratory in Moscow, where initial examination was taking place. Exact location of laboratory and names of top officals in *Luna XVI* program were being kept secret.

Konovalov had reported container of samples had been placed in cylindrical stainless steel chamber with portholes. Pumps had sucked oxygen from chamber and sterilizing gas had been injected. Chamber had been filled with helium gas to prevent reaction with moon samples. Then container, which included moon rock and its electrical drill, had been opened for observation. Rock had been placed on steel trough containing scale divisions. Some material would be taken in special bags to other laboratories for examination. Konovalov's reference to lunar "dust" had been first use of term by Soviet journalist, *New York Times* said. Previous references had been to "lunar surface," "lunar ground," or "lunar rock."

Soviet scientist Aleksey Turgarinov said on Moscow TV that lunar samples were blue and had been removed from hermetically sealed container in low vacuum and sterile conditions. Tass said rocks would be kept in quarantine and would be given to Soviet Academy of Sciences for special research program. Research results would be published. (*W Post*, 9/26/70, A7; Gwertzman, *NYT*, 9/27/70, 12)

September 27: U.S. space program problems were described in *Washington Post* by Thomas O'Toole: U.S. needed to begin development of space shuttle "now, which in a way is the worst possible time for it." Country was "weary of big new space projects, particularly while the spectacular Apollo project is still unfolding. . . . Also, the space agency is without an appointed administrator, which means that NASA cannot exercise its full role with Congress or the

White House." NASA managers "go on sailing a skipperless ship. But what they find more disturbing is the feeling that nobody at headquarters cares whether the ship gets a captain or not." It was no secret that "men who run the space program are deeply disappointed in the Nixon Administration." They felt "that the White House neither understands nor cares to understand what the space program means or what it can do for the country." There was feeling Congress had ceased to care, too. House Committee on Science and Astronautics' Subcommittee on NASA Oversight had canceled hearings on Apollo future because, "outside of Rep. Olin Teague (D-Tex.), chairman of the Subcommittee on Manned Space Flight, none of the subcommittee members was concerned enough. . . .to agree to the urgency of the hearings." (*W Post*, 9/27/70, B1)

September 28: Aerobee 170 sounding rocket was launched by NASA from WSMR carrying American Science & Engineering, Inc., experiment to study x-ray spectra. Rocket and instruments functioned satisfactorily. (SR list)

- President Gamal Abdel Nasser of United Arab Republic died after heart attack in Cairo, following Sept. 27 Arab summit meeting that ended civil war in Jordan. Fighting that had followed Arab commandos' hijacking and destruction of four Western airliners appeared about to be ended and last hostages from aircraft were being freed. (*FonF*, 9/24-30/70, 689)

- House rejected H.R. 1210, resolution against accepting Reorganization Plan No. 4 of 1970. Plan, transmitted to Congress by President Nixon July 9, called for establishment of National Oceanic and Atmospheric Administration. (*CR*, 9/28/70, H9281-92)

- LeRC Director Bruce T. Lundin announced appointment of G. Mervin Ault as Assistant Director of Power and Materials Directorate, Dr. Bernard Lubarsky as Assistant Director for Aeronautics, and Edward F. Baehr as Technical Consultant to the Director. (LeRC Release 70-51)

- U.S.S.R. and West Germany signed agreement in Moscow calling for exchange of scientists and direct contacts between institutes and laboratories. At news conference after signing, Chairman Vladimir A. Kirillin of Soviet State Committee for Science and Technology said countries planned to cooperate in physics, physical chemistry, astronomy, biology, medicine, oceanography, computing techniques, problems of information, and education. (*NYT*, 9/29/70, 3)

- F-111 horizontal tail had broken when load applied to it had reached 191% of critical-design load limit during tests at General Dynamics Corp's Convair Aerospace Div., General Dynamics announced. USAF required that F-111 structure withstand loads of up to 150% of critical design load. (General Dynamics Release 1538)

- House and Senate conferees submitted conference report on H.R. 17123, FY 1971 DOD military procurement authorization bill. Report set authorization total at $19.929 billion, $642.4 million lower than House total, $686.2 million more than Senate total, and $676.4 million less than Administration request of $20.605 billion. Report limited Safeguard ABM expansion to one additional site at Whiteman AFB, Mo., and preparation of additional site at Warren AFB, Wyo. (*CR*, 9/28/70, H9320-8)

September 29: U.S.S.R. launched *Molniya I-15* comsat into orbit with 39 300-km (24 420-mi) apogee, 480-km (298-mi) perigee, 11-hr 46-min period, and 65.5° inclination. Tass said objective of mission was continued operation of long-range telephone and telegraph radiocommunications and transmission of Soviet Central Television programs to stations in Orbita network. (*Spacewarn*, 10/20/70, 2; GSFC *SSR*, 9/30/70; UN Gen Assembly Release 70-29113)

- Black Brant IV sounding rocket launched by Brazilian space team for NASA from Natal, Brazil, carried MSC experiment to study energetic particles in South Atlantic Anomaly region. Rocket and instruments functioned satisfactorily. Launch was second in two-rocket series; first had been launched Sept. 25. (SR list; MSC Release 70-105)
- General Electric Co. said it would not challenge NASA decision to reverse award of $50-million contract for ATS-F and ATS-G [see Sept. 5]. In statement to press, Leon L. Farnham, General Manager of Spacecraft Programs at GE Valley Forge Space Technology Center, said GE still thought NASA decision was wrong but that the "further delay [of a court challenge] would be injurious to NASA and the total space program." (Jones, *W Post*, 9/30/70, D9)
- *Washington Post* editorial on *Luna XVI*: "...the flight of Luna 16 is a major achievement for the Soviet Union. Its scientists have concentrated in recent years on a system of landing and retrieving unmanned spacecraft, an activity that the United States had not emphasized, and Luna 16 shows that the system works." In long-term space exploitation "it seems to us that the manned approach will be most useful. Men will be out there some day, following the trail marked by the Apollo craft, and in the meantime the American effort can properly be devoted to finding a way to get them out of the earth's atmosphere more cheaply and to making the best possible uses of earth-orbiting space stations and laboratories." (*W Post*, 9/29/70, A18)

September 30: NASA successfully launched 134.7-kg (296-lb) RAM C-III (RAM C-C) spacecraft from Wallops Station at 4:06 pm EDT by four-stage Scout booster. Spacecraft reached 209.2-km (130-mi) altitude, reentered atmosphere at 7.6 km per sec (25 000 fps), and splashed down in Atlantic 277.8 km (172.6 mi) northeast of Bermuda after eight-minute flight. Primary objective was to determine relative effectiveness of water and freon in restoring communications during reentry blackout period. Preliminary data, obtained by Langmuir probes on board, indicated all downrange telemetry stations received signals before and after blackout and intermittent signals during blackout.

RAM C-III was third and last mission in NASA's RAM (Radio Attenuation Measurement) program to study problems of communicating through plasma sheath around reentering spacecraft. Results could be applied to advanced lifting-body vehicles, space shuttle, and other reentering spacecraft. RAM C-I had been launched Oct. 19, 1967, and RAM C-II Aug. 22, 1968. (NASA Proj Off; WS Release 70-13)

- Last in series of 15 Saturn V 1st stages (S-IC-15) was successfully captive-fired for 2 min 15 secs at MTF. Captive testing of S-IC stages had begun at MTF May 16, 1967, to check out and flight-certify stages for manned lunar missions. (MTF Release 70-35)

September 30: *The 15th and last Saturn V 1st stage (S-IC-15) was successfully captive-fired for 2 minutes 55 seconds at NASA's Mississippi Test Facility. At right, the 14th S-IC stage was photographed on the test stand earlier. The captive tests began May 16, 1967, to flight-certify stages for lunar missions.*

- Flight Crew Health Stabilization program to minimize possible exposure of Apollo flight crews to disease or illness would be introduced before Jan. 31, 1971, Apollo 14 mission, NASA announced. Program provided that crew health be stabilized through epidemiology, clincial medicine, and immunology programs and limitation of outside contacts with flight crews; prime and backup crew members reside solely in KSC crew quarters for 21 days before launch; access to primary training areas used by crew members be controlled; access to areas during crew occupancy be limited; crew activities be limited to primary areas of Manned Space Operations Building and Flight Crew Training Building, flight line at Patrick AFB, Fla., and launch pad white room during 21-day prelaunch period; and crew members use personal vehicles when traveling between primary areas. If crew members were required to go to MSC for training, they would live in crew reception area of LRL or in their own residences, where they would be excluded from contact with everyone but their wives. During 60 days before launch, all illnesses in family members of prime, backup, and support crews and close contacts of these families would be reported to medical officials. (NASA Release 70-159)

September 30

- ESSA announced it had assumed control of Solar Particle Alert Network (SPAN) from NASA. Network, which originated in 1966 in support of NASA space flight programs, included observatories in Canary Islands and Carnarvon, Australia, and at ESSA's Space Disturbance Laboratory in Boulder, Colo. Under ESSA management, principal observatories would have identical optical and radio solar telescopes to provide continuous coverage of sun. Three stations were connected by voice-quality communications links to allow real-time transmission of data and observations, and all stations were equipped to photograph sun in hydrogen (H-alpha) light. Stations could be supplemented by standby observatory at MSC. ESSA also had accepted responsibility for processing, storing, and distributing photographic data and radio data from SPAN observations. (ESSA Release 70-74-238)
- *Space Business Daily* reported DOD had confirmed that U.S.S.R.'s *Cosmos CCCLXV* (launched Sept. 25) had been FOBS test launched by modified SS-9 booster. (*SBD*, 9/30/70, 129)
- NASA announced selection of Bendix Field Engineering Corp. to receive $85-million, cost-plus-award-fee contract for operation of 10 stations of NASA's Manned Space Flight Network. Contract covered three years with two one-year options. (NASA Release 70-161)
- Soviet and West German talks on future cooperation in science and technology would begin in Moscow in November, Hans Leussink, Minister of Science and Education in Bonn, West Germany, said at news conference. (*NYT*, 10/1/70, 6)

During September: Scientists at FRC, LaRC, ARC, and MSFC were conducting flight-test programs and studies into ways to alleviate trailing vortex hazards from large transport aircraft, NASA reported. Research was concentrating on understanding trailing-vortex-system movement and attenuation for different aircraft, operation modes, and meteorological conditions; developing remote monitoring technique for trailing-vortex position and wind intensity; and discouraging formation of high-density vortices by aircraft design or artificially induced impedances.

Theoretical descriptions of trailing-vortex-system behavior would be developed from flight-test programs at FRC and LaRC. ARC was experimenting with wing design to reduce trailing-vortex intensity and with effects of induced impedances near aircraft path. MSFC was attempting to adapt laser Doppler technology to flight research programs and to development of remote monitor for trailing vortices in airport areas. (NASA Aviation Fact Sheet, 9/70)

- Twenty-fifth anniversary of AFCRL was noted in AFSC *Newsreview* with historical summary. Formed after World War II to retain research teams employed at MIT Radiation Laboratory, AFCRL had amassed "a record of contributions that few research institutions. . .can equal" and had become leading USAF center for research in environmental and physical sciences. "With a budget of $55 million in FY70 AFCRL continued to conduct experiments and research through. . .flying laboratories, rocket launches, and balloon programs," with 140 balloons launched annually from permanent launch sites at Holloman AFB, N. Mex., and Chico, Calif. Basic components of space environmental forecasting system had been put into operation during June to enable USAF to predict more accurately the anticipated degradations in performance of electronic systems and to provide knowledge about effects of sun on earth. Research was also being conducted on air pollution in connection with SST program to discover specific dangers, effect on solar radiation reaching earth, and general make-up of "atmospheric aerosols" that would also provide valuable information on ecology of earth. (AFSC Newsreview, 9/70, 6)

- Transition of World War II aircraft industry into 1970 aerospace industry was traced in *Air Force* magazine by Karl G. Harr, Jr., AIA President. Increased performance had increased complexity. Average 1944 combat aircraft had carried 454 kg (½ ton) of avionic gear; 1948-1950 counterpart needed three to five times as much.

  Weapon revolution did not signify end of manned aircraft era; it brought forth "new family of aircraft of substantially improved capability." First operational "barely supersonic" 1954 fighter had been followed by succession of more advanced fighters. "The complexity curve took a sharp upward turn, taking the cost curve with it," with corollary decline in numbers of aircraft produced. Fewer aircraft were needed because of individual aircraft's superior performance.

  With advent of supersonic aircraft production, guided missile output became significant portion of industry workload. Big push had come in 1954 when missile procurement topped billion-dollar level for first time.

  "Undoubtedly, the space program was the dominant influence in history's most explosive decade of technological advance. There were two primary contributing factors: breadth and acceleration." Breadth of program required continual probing of new research frontiers across spectrum of almost every scientific and technological discipline. "The acceleration of effort imposed by the lunar-landing timetable necessitated a *forcing* of technology, compressing into one decade the normal advance of several." (*AF*, 9/70, 84-8)

- VTOL aircraft linking new Tokyo International Airport at Narita, Japan, with downtown Tokyo were scheduled for 1975 service and would make 59.6 km (37-mi) flight in 20 min, *Air Force and Space Digest* said. Aircraft, seating 150 passengers, would also link Tokyo's old and new airports. Aeronautical and Space Technological Research Institute of Japanese Science and Technology Agency was testing to develop domestic VTOL. Experimental aircraft was powered by two JR 100F lift-jet engines with 1243-kg (1.4-ton) thrust each. (Golden, *AF/SD*, 9/70, 31)

- *Boeing Magazine* suspended publication with September issue after 40 yrs of covering Boeing Co. products, people, skills, and history. (*Boeing*, 9/70)

- Naval Air Development Center (NADC) had refined laser beam so that contents of large city library could be inscribed on salt crystal, *Ordnance* reported. Center had also developed system by which pilot could eject from crippled aircraft in seat capable of carrying him 80 km (50 mi) and had developed system analysis for complicated weapons that could be adapted to provide complete traffic control system for highway. Center, near Philadelphia, was developing more than 800 projects to keep "Navy's air arm abreast of the latest technological innovations." (Clark, *Ordnance*, 9-10/70, 181-3)

During September

- Sales of military and civilian aircraft in 1970 would drop to $14 billion, down $3.4 billion from 1968 peak, Dan Cordtz wrote in *Fortune*. Employment was off 15% from 1969, "with further substantial layoffs already announced." Evidence was mounting "that future volume simply will not support the present number of competitors." Outlook was "ruinous scramble for a permanantly shrunken market, with some companies squeezed out of aircraft production and even the survivors forced to subsist on meager rations. A decade ago, a dozen airframe manufacturers were able to thrive; a few years hence, its is difficult to see profitable business for more than three or four." (*Fortune*, 9/70, 114-7, 199-201)
- NSF published *Federal Funds for Research, Development, and Other Scientific Activities, Fiscal Years 1969, 1970, and 1971* (NSF 70-36). Federal funds for R&D reached peak $16.5 billion in 1967, $15.6 billion in 1969, and were expected to remain same in 1970 and 1971. In terms of constant dollars, 1970 estimate was 17% decrease from 1967. Of Federal total, $1.4 billion was provided to universities and colleges in 1969 for support of academic research, with $1.5 billion expected in 1971. Basic research obligations had leveled off since 1967, remaining at $2.1 billion to $2.2 billion. NASA was leading basic research supporter, accounting for 35% of 1970 total.

    Applied research obligations remained at about $3.3-billion funding level from 1967 through 1970, with rise to $3.7 billion projected for 1971 subject to congressional actions. In constant dollars, funding decreased 11% from 1967 to 1970. DOD, HEW, and NASA accounted for 79% of Federal agency total in 1970. From 1964 to 1970 only HEW among these agencies showed growth. (Text)
- NSF released *Graduate Student Support and Manpower Resources in Graduate Science Education, Fall 1969* (NSF 70-40): Graduate enrollment in 2894 science doctorate departments of 224 reporting universities and colleges in fall 1969 totaled 184 845, with 28% in engineering. Enrollment was 2.8% higher than in fall 1968. (Text)

October 1970

October 1: U.S.S.R. launched *Cosmos CCCLXVI* from Baykonur into orbit with 288-km (179-mi) apogee, 202-km (125.5-mi) perigee, 89.4-min period, and 64.9° inclination. Satellite reentered Oct. 13. (GSFC SSR, 10/31/70; SF, 4/71, 138)

- At NASA Hq. press briefing, NASA Associate Administrator for Manned Space Flight Dale D. Myers and program officers gave status of programs, plans for 1970s, and studies leading into 1980s.

  Dr. Rocco A. Petrone, Apollo Program Director, announced selection of Hadley-Apennine site for lunar landing of Apollo 15, scheduled for launch July 25, 1971. Site, cut by large gorge along base of mountains, was 3° east of moon's center and 25° north of lunar equator on edge of Sea of Rains. Apollo 15 crew would collect material from base of Apennine Mountains, examine Hadley Rille area, sample fresh-looking mare and volcanic-like features, and deploy ALSEP. Apollo 16 would be launched in January 1972 and Apollo 17, in June 1972. High-priority landing sites were Copernicus, Davy, Descartes, and Marius Hills.

  To improve CSM return capability in emergencies after landing on moon, an LM descent battery and third tank would be added to Apollo 14 SM. Additional potable water would be generated and stored on way to moon.

  Douglas R. Lord, Deputy Director of Space Station Task Force, described plans for 12-man space station weighing 49 985-54 400 kg (110 000-120 000 lbs) that could remain in earth orbit for up to 10 yrs.

  Richard J. Allen of Space Shuttle Task Force said NASA was developing plans for shuttle with 100-mission reusable capability. Myers said DOD was "very much involved" with shuttle activities. "We have a joint committee operating with the Air Force. . .for continuous communication and review of requirements." He saw "possibility of international cooperation on certain flights and [U.S.] military use of the vehicle on others." Myers said he had discussed with European officials possibility of participation in space shuttle program on three levels: orbit-to-orbit space tug, "cleanly separable and easily interfaced piece of equipment" that would probably be managed by European consortium and financed with European dollars; subcontracts from American companies to European contractors without transfer of American money; and use of technological contributions from European national space activities and national aircraft establishments. (Transcript; NASA Release 70-162)

- Twelfth anniversary of NASA, established by the National Aeronautics and Space Act of 1958. (Space Act)
- *Report on the Closing of the NASA Electronics Research Center* was released by Planning Steering Group for ERC closing. Report was prepared by Deputy Administrator for Administration Boyd C.

Myers II, Steering Group Chairman, and members of group; ERC Task force; and DOT. Following ERC closing on June 30, 1970, new DOT center functioned "reasonably well." All but 85 former ERC employees had jobs; NASA had programs and equipment it needed. DOT had ample equipment for new center and 27 universities were pursuing new work relevant to NASA interest. "Most important, the government and the nation were in a position to fully utilize a national capability of nearly $60 million in facilities and equipment and over 740 highly capable people on new programs. NASA had retained the highest priority work and related equipment." Move was almost completed in six months but personal equipment packing and shipping and disposition of leased space would take about 60 days.

Dr. Thomas O. Paine, NASA Administrator at time of ERC closing, had directed disposition priorities: people first, program second, and physical facilities third, with every employee placed in appropriate job at higher salary. "While these objectives could not be entirely achieved sequentially, there was no question about the prime importance Dr. Paine placed on the actions affecting each individual employee." (Text)

- Fire protection system for large passenger aircraft had been tested by ARC scientists, ARC reported. System, using ARC-developed fire-retardant paints and foams, was believed capable of protecting aircraft during 10 min of maximum-intensity fuel fire on ground, permitting firemen to subdue fire and remove passengers and crew. (ARC Release 70-11)
- MSC announced it was negotiating with Lockheed Missiles & Space Co. on $699 000 cost-plus-fixed-fee contract for 18-mo study of space shuttle cryogenic systems. (MSC Release 70-110)
- ICAO Council approved U.S. proposal for multilateral agreement on sanctions against countries that failed to extradite or prosecute aircraft hijackers or to permit passengers and crew of hijacked aircraft to continue their journey or to return aircraft and cargo to their owners [see Sept. 21]. Vote was 14 to 3, with 10 countries abstaining. Minimum support of 14 countries was needed for approval. (AP, *W Post*, 10/2/70, A2)
- Senate voted 47 to 5 against S.R. 433, which would disapprove Reorganization Plan No. 4 of 1970, calling for establishment of National Oceanic and Atmospheric Administration. (*CR*, 10/1/70, S16943-64)
- National Transportation Safety Board released contents of its letter to FAA recommending safety measures to protect Boeing 747 passengers. And Reuben B. Robertson III of Ralph Nader's Center for Study of Responsive Law, in letter to Secretary of Transportation John A. Volpe, charged FAA with suppressing news of "near tragedies" aboard 747s. Later *Washington Post* said both actions stemmed from Sept. 17 and 18 incidents in which 747 engines had disintegrated in flight. There had been no injuries. (Hoffman, *W Post*, 10/2/70, A3)
- USCG called off search for three balloonists downed Sept. 21 in storm south of Newfoundland after attempting to make first transatlantic balloon crossing. (AP, *W Post*, 10/2/70, A3)

October 2: Launch of Orbiting Frog Otolith (OFO) satellite, originally scheduled for Aug. 19 and postponed because of booster malfunction and countdown discrepancies, had been postponed until at least

Oct. 19, NASA announced. Bullfrogs being trained for mission had been killed or disabled by redleg, virus infection common to water life, and new set of frogs was being trained. OFO, with Radiation Meteoroid (RM) spacecraft as secondary payload, would carry two bullfrogs into orbit for five-day mission to study adaptability of vestibule in inner ear to sustained weightlessness and to acceleration. (NASA Release 70-160)
- Julian M. West, former MSC Special Assistant for Long Range Planning, was sworn in as consultant to Acting NASA Administrator, Dr. George M. Low. West would advise on potential space and aeronautics development programs. (NASA Release 70-164)
- Sen. George S. McGovern (D-S.D.) introduced S. 4430, "Economic Conversion Act" to create Economic Conversion Commission composed of government and industry representatives including NASA Administrator. Measure would require that 12½% of each contractor's profits from space and defense work be held in trust by commission to finance implementation of conversion plans developed by contractor and to pay benefits to workers who might suffer during transfer to civilian production. (CR, 10/2/70, S16995-7003)
- NASA announced award of $380 000 contract to Ralph M. Parsons Co. for engineering services in development of overall plan for space shuttle ground facilities. (NASA Release 70-165)

October 3: U.S.S.R. launched Cosmos CCCLXVII from Baykonur into orbit with 1026-km (637.5-mi) apogee, 919-km (571-mi) perigee, 104.5-min period, and 65.2° inclination. (GSFC SSR, 10/31/70; SF; 4/71, 138)
- U.S.S.R. published report on initial examination of lunar samples brought to earth by Luna XVI Sept. 25. Material consisted primarily of gray, loosely structured, "fine-granular mineral particles" with "appreciable forces of cohesion between the particles." Intensity of gamma radiation was "not considerably greater than that of earth rock, with a small content of natural radioactive elements." (AP, NYT, 10/5/70, 6)
- Discovery of cyano-acetylene molecules in gas cloud Sagittarius B2 near center of Milky Way was announced by Dr. Barry E. Turner of National Radioastronomy Observatory in Green Bank, W. Va. Discovery, made through radiotelescopes on earth, suggested that planets and life were being created in Milky Way galaxy and brought to seven the number of chemical radicals (incomplete molecules) and chemical molecules found in deep space. Dr. Turner said radio emissions from cyanoacetylenes, most complex chemicals yet found in space, were stronger than background radiation filling all of space. He thought they were remnants of radiation from fireball that created the universe billions of years ago. (O'Toole, W Post, 10/4/70, A5)
- National Oceanic and Atmospheric Administration came into being as result of Reorganization Plan No. 4 of 1970 proposed by President Nixon July 9. [See also Sept. 28 and Oct. 1.] (Dept of Commerce Release G 70-122)

October 4: Four major research directorates were established at LaRC: Aeronautics, directed by Laurence K. Loftin, Jr.; Space, directed by Clifford H. Nelson; Electronics, directed by George B. Groves, Jr.; and Structures, directed by Dr. George W. Brooks. Supporting

Centerwide directorates were established in Systems Engineering and Operation, directed by Percy J. Crain; Administration, directed by T. Melvin Butler; and Center Development, directed by Eugene C. Draley. Reorganization was to implement goals outlined in staff meeting by LaRC Director Edgar M. Cortright: expanded aeronautics program with increased emphasis on civil aircraft, new era of manned space flight with permanent space stations supplied by reusable space shuttles, planetary exploration, and maintenance of NASA's basic research skills. (*Langley Researcher*, 10/2/70, 1)

- U.S. and France had agreed on program to launch series of 18 Nike-Cajun sounding rockets from France's space center in French Guiana, French National Space Study Center announced. Launches would begin in February or March 1971. (UPI, *W Post*, 10/5/70)
- Washington *Sunday Star* editorial commented on SST: "What the SST adds up to is a large question mark, a gamble with public funds and the environment that comes at time of crisis in both areas. We should improve the odds for survival by pulling out of the SST race until the facts—not the suppositions—are in. We should gamble not on a questionable investment, but on the ability of our economy and our prestige to withstand the impact of losing one part of one industry's market if the Concorde lives up to its sponsors' expectations. The SST program should be shelved." (*W Star*, 10/4/70, B1)

October 4-10: Twenty-first International Astronautical Federation Congress was held at Constance, West Germany. During opening session Klaus von Dohnanyi, Parliamentary Secretary to West German Minister of Education and Science, appealed to nearly 1000 participants to preserve peaceful nature of space research and prevent space technology from producing "fresh weapons escalation." *Apollo 13* Astronauts James A. Lovell, Jr., John L. Swigert, Jr., and Fred W. Haise, Jr., and *Apollo 9* Astronaut (now MSC Apollo Spacecraft Program Manager) James A. McDivitt attended Congress.

Soviet scientist Dr. K. Y. Kondratyev described October 1969 experiment in which *Soyuz VI* and *VII* spacecraft had been observed simultaneously from one aircraft at 7900-m (26 000-ft) altitude, from one at much lower altitude, and from ground, to determine atmospheric distortion of observations of earth surface from space. Additional observations from *Soyuz IX* in June 1970 and automated measurements from unmanned *Cosmos CCXLIII* launched Sept. 23, 1968, had combined to show fully equipped space station could clearly identify forms and densities of vegetation; chart geologic formations in inaccessible regions; identify soil suitable for agriculture; trace origin of fungus blights; sight outbreaks of blight and pinpoint forest fires; chart ground temperatures, identifying trends in volcanic activity and mountain building; detect erosion patterns invisible from ground; record seasonal flow patterns in watershed; monitor spread of pollutants; monitor seas for fish movement and evidence of pollution; and transmit photos indicating safe shipping lanes in Arctic ice pack areas. Dr. Kondratyev said orbital space station could be developed within next 5 to 10 yrs and be equipped with sensor system to scan earth at wide range of wavelengths for remote sensing of environment. (*Congress Bulletin*, 10/6/70, 1; *NYT*, 10/6/70, 21; NASA PAO)

Dr. Charles A. Berry, MSC Director of Medical Operations, and Prof. Oleg G. Gazenko, Director of Soviet Institute of Biomedical Problems, said selection of crews for longer space missions would require special personality testing to identify persons of more dominant type for assignment of roles compatible with that trait. Dr. Berry said he thought use of multinational flight crews was good idea but chances for Orbiting International Laboratory (OIL) in near future seemed slim because of problems of language, attitudes, and prejudices such as food preferences. "We can't have a restaurant with a very large menu." In interview, Dr. Berry and Prof. Gazenko discussed flashes of light experienced by astronauts in space with eyes closed or in darkened spacecraft. Dr. Berry believed flashes might be produced by heavy cosmic ray particles—atomic nuclei of intermediate weight—that produced ionization in eye retina, affected nervous system, or caused light flashes within eye fluid. Dr. Gazenko said heavy cosmic rays were insufficiently frequent to account for flashes. Soviet cosmonauts had not experienced phenomenon. (Sullivan, *NYT*, 10/12/70, 19)

Cosmonauts Andrian G. Nikolayev and Vitaly I. Sevastyanov said muscles of *Soyuz IX* crew were so atrophied after June 2-19 mission that they stood up with difficulty and walked with peculiar stamping steps. On landing they were so weak that ground crew had to assist them from hatch. Sensation of heaviness lasted 10 to 11 days. Crew had performed antiatrophy exercises during mission but more intensive exercise program might be needed for flights to other planets, with simulation of gravity included. Cosmonauts felt effects of weightlessness were not dangerous unless experienced for more than one month. *Soyuz IX* cosmonauts displayed photos of June 2 launch and colored films taken inside spacecraft during 18-day orbit. Nikolayev said cosmonauts had slept with head down before launch, to become accustomed to rush of blood to head that had bothered previous cosmonauts and astronauts on early part of flights. Soyuz spacecraft had three sections: orbital module in which cosmonauts slept, exercised, and played chess; central module with three windows and couches for use in liftoff and reentry; and third module containing service equipment. Both orbital and service modules were jettisoned before reentry. (Sullivan, *NYT*, 10/10/70)

Indian government scientists H. G. S. Murthy, E. V. Chirnis, and K. S. Karnik urged U.S. and U.S.S.R. to share economic benefits of space exploration with developing nations. "International cooperation in global planning of assistance for the operation of meteorological communications and navigation satellites will go a long way in helping developing countries. National space programs in developing countries are not a luxury, but a necessity." (Benedict, AP, *Huntsville Times*, 10/7/70)

Third International Symposium on Space Rescue was held as part of IAF congress. Cochairman P. A. Campbell said letter from Soviet Academy of Sciences had nominated for first time four Soviet engineers to serve on IAF's 11-nation committee studying space rescue. Later AP quoted IAF Congress sources as saying Soviet engineers might meet with U.S. experts at MSC in November.

Herbert Schaefer and Dr. Jack W. Wild of NASA Advanced Manned Missions program said space rescue vehicles would have to be versatile, with several payloads kept at hand to deal with variety of emergencies. Standardized docking system would be needed on all future manned spacecraft to permit rescue by any nation. NASA Assistant Safety Director Philip H. Bolger said, "The rescue of stranded astronauts is a concern recognized by all nations and therefore provides a common base for establishing international cooperation." It was too late to develop rescue system for four remaining Apollo flights, but reliability of Apollo spacecraft was being improved. Bolger said Skylab three-man space station scheduled for 1972 launch might have double protection—spacecraft that could be separated and returned to earth and standby rescue rocket at KSC. (AP, W *Star*, 10/8/70, A4; Sullivan, *NYT*, 10/8/70, 4)

LeRoy E. Day, Manager of NASA Space Shuttle Task Group, described system characteristics of space shuttle: "It should be a fully reusable two-stage vertical take-off and horizontal landing space vehicle" that could transport 11 340 kg (25 000 lbs) to the design reference orbit of 500 km (310 mi) at 55° orbital inclination. Shuttle "must have a large internal cargo bay which will give it the capability of carrying a variety of manned and unmanned payloads to low earth orbit. The large internal payload compartment will allow the shuttle to deliver to low earth orbit both a satellite and a high energy stage for a synchronus orbit or a planetary mission. We expect the gross lift-off weight. . .to be approximately 1.6 million kilograms [3.5 million lbs] fully fueled and with the payload on-board." Rocket engines would be one of most critical subsystems "For both the booster and the orbiter we will use high pressure hydrogen/oxygen rocket engines which can be throttled to keep the acceleration during ascent to less than 3 g's." (Text)

Dr. George E. Mueller, General Dynamics Corp. Vice President and former NASA Associate Administrator for Manned Space Flight, said U.S. space shuttle could be "golden transport" to carry passengers between any two cities on earth in an hour in 1980s. Shuttle could be converted easily to transport flying to 130-km (80-mi) altitude and plunging back to target city at approximately 18 500 km per hr (11 500 mph). Rocket-boosted aircraft would take off vertically, eliminating noise problem. Other nations might purchase U.S. shuttle for $50 million. (AP, B *Sun*, 10/10/70, A3)

Fourth International History of Astronautics Symposium was held, with Dr. Eugene M. Emme, NASA Historian, as Chairman. Papers were presented by I. Saenger-Bredt of Stuttgart, Oleg G. Gazenko of Soviet Academy of Sciences, and R. Cargill Hall, JPL Historian. Papers were submitted by J. Kaplan of UCLA, G. H. Osborn and R. Gordon of Aerojet Liquid Rocket Co. (Program; NASA Hist Off)

IAF officials elected during Congress were: President, Prof. A. Jaumotte of Belgium; Vice Presidents, Dr. George E. Mueller of U.S., H. G. S. Murthy of India, Prof. L. G. Napolitano of Italy, and Prof. L. I. Sedov of U.S.S.R. (IAF Release 20)

October 5: Javelin sounding rocket, launched by NASA from Wallops Station, ejected barium cloud at 900-km (560-mi) altitude in cooperative experiment with Max Planck Institute of Munich, Germany, to

study earth's electric and magnetic fields. Red-tinged green cloud was visible for hundreds of kilometers along East Coast. Similar launches would be conduced Oct. 7 and in spring 1971. (WS Release 70-14)
- Impending consolidation of launch support functions was announced by AFETR and KSC. Consolidations, resulting from joint studies in cost reduction, were expected to save more than $1 million annually. Included would be KSC medical services, life support services, nondestruct testing, frequency control, and electromagnetic compatibility and precision cleaning; also ETR timing, meteorological support, and ordnance storage. (NASA Release 70-167)
- Space applications of plutonium 238 were discussed by Dr. Glenn T. Seaborg, AEC Chairman, at Fourth International Conference on Plutonium and Other Actinides in Santa Fe, N. Mex. Most noteworthy use of plutonium-238-powered, radioisotope thermoelectric generator (RTG) had been as SNAP-27, power source for *Apollo 12* experimental package left on moon. RTG's were "of particular interest as space electric power sources for missions requiring a long life and high reliability, where there are long periods of darkness or great distance from the sun, or where orientation or other problems tend to reduce the effectiveness of solar cells as an energy source." Significant number of future space missions might require plutonium 238 power source.

  Element "finds another extremely important use in space as the power source for waste disposal." Heat from plutonium 238 decay would both produce bacteria-free drinking water from waste products and incinerate waste itself. After Apollo, exploration of moon might continue with robot vehicles. "The maintenance and servicing of space stations and NERVA engines in space may be done more easily and safely by robots than by man. . . . Nuclear power would be an obvious candidate power system." (Text)
- American Institute of Physics held meeting at Rockefeller Univ. in New York. Dr. Frederick Seitz, President of Rockefeller Univ., received Karl Taylor Compton Award for distinguished statesmanship in science.

  During interview following meeting Dr. William D. McElroy, NSF Director, said NSF had begun testing use of large teams of scientists from different disciplines under Federal funding as new approach to basic scientific research. NSF had allocated $70 million in 1970 for studies of pressing national and social matters.

  Dr. Jerome B. Wiesner, MIT Provost and science adviser to President Kennedy, said prioritites had been misplaced. He advocated a national science policy. "If I invented a new military system to stop tanks, I could get $100 million at the snap of my fingers. But if I say we want to develop a new method of medical care, I don't know if I could raise $10 million."

  Dr. Harvey Brooks, Harvard Univ. Dean of Engineering, said: "The central problem is how science can be reasonably responsive to social problems without destroying itself as science. The health of science depends on its capacity to develop its own intellectual laws. If it attempts to become too responsive, it loses its integrity, and I see a danger of that happening now." (Reinhold, *NYT*, 10/6/70, 1)

October 6: Termination of manned space flight within three years was urged at Washington, D.C., press conference by four Senators and space scientists Dr. James A. Van Allen of Univ. of Iowa and Dr. Thomas Gold of Cornell Univ. Sen. William Proxmire (D-Wis.) said, "I want to shift spending from space extravaganzas to needy programs." Sen. Clifford P. Case (R-N.J.) said, "The space shuttle...should not be allowed to go forward until the proper role of manned versus unmanned exploration has received a fuller examination than it has to date." Both Senators, with Sen. Jacob K. Javits (R-N.Y.) and Sen. Walter F. Mondale (D-Minn.), favored continuation of four remaining Apollo flights and three-mission Skylab program but called for ending manned flight on their completion. Dr. Van Allen advocated unmanned space program after Skylab that would devote two thirds of its funds to direct applications of space with remainder devoted to "scientific experiments to examine the solar system." Dr. Gold said, "If we fritter away our limited resources on unwanted further demonstrations of manned flights [U.S.S.R.] may, with much smaller means, get way ahead in all the areas that really count." He predicted that, if space station was built, it would be "focal point of anti-science and anti-intellectualism which is very much in predominance today." (Lyons, *NYT*, 10/7/70, 21)

- Mark IV eddy current proximity gauge developed by MSFC engineers for space use was being used by Pennsylvania Highway Dept. in experiment to measure new highway thicknesses, MSFC reported. Instrument, designed to measure paint thickness on launch vehicles and foam insulation thickness on Saturn V 2nd stage, was placed atop wet concrete and over embedded metal target strips to measure distance of strips from top surface. Other potential application for device was to detect firearms or grenades in airline passenger luggage to prevent hijacking of aircraft. (MSFC Release 70-206)

- President Nixon signed E.O. 11564, implementing Reorganization Plan No. 4 of 1970 that established National Oceanic and Atmospheric Administration in Dept. of Commerce [see Oct. 3]. Order transferred to Secretary of Commerce functions of DOD's Oceanographic Instrumentation Center, Ocean Station Vessel Meteorological Program, Trust Territories Upper Air Observation Program, and DOT's National Data Buoy Development Project. (*PD*, 10/12/70, 1344-5)

- Afghanistan was allowing New Mexico State Univ. team to track satellites from mobile tracking station in area near Kabul that bordered on U.S.S.R. and Communist China, AP said. Project, supported by AID, had begun after U.S. closed base in Peshawar on orders from Pakistan government in November 1969. Embassy spokesman had said project was directed by Afghan Cartographic Institute and data collected was "probably available to the Russians." (AP, *C Trib*, 10/7/70)

- *Pravda* said Soviet test pilot Lt. Valentin I. Danilovich had been killed after ejection from supersonic aircraft at high altitude while testing new supersonic catapault system. Danilovich had helped develop techniques for world's first space walk, by Soviet Cosmonaut Aleksey A. Leonov during *Voskhod II* mission, March 18-19, 1965. (UPI, *W Post*, 10/7/70)

October 7: Nike-Tomahawk sounding rocket, launched by NASA from Wallops Station, ejected barium cloud at 257.5-km (160-mi) altitude to provide data on electric and magnetic fields in geomagnetosphere. Barium-release payload was first to use liquid fluorine. Fuel tank contained hydrazine mixed with barium salts. Fluorine oxidizer, carried in separate tank under cryogenic conditions, reacted hypergolically with fuel via valve, manifold, and burner system. Resulting barium cloud was photographed and tracked from several sites along East Coast. Similar launch had been conducted Oct. 5. (WS Release 70-15)

- NASA released preliminary time line for Apollo 14 manned lunar landing mission. Spacecraft would lift off from KSC at 3:23 pm EST Jan. 31, 1971; land LM north of moon's Fra Mauro crater at 4:14 am EST Feb. 5 for 34-hr stay; and splash down in Pacific at 4:01 pm EST Feb. 9. (NASA Release 70-166)

- *Apollo 13* Astronauts James A. Lovell, Jr., Fred W. Haise, Jr., and John L. Swigert, Jr.—in Lucerne, Switzerland, on European goodwill tour—visited Swiss Transport Museum and Longines Planetarium. They were shown representation of view of earth from Mars. (NASA Hist Off)

- USAF F-111 crashed at Carswell AFB, Tex., killing pilot and copilot. (*WSJ*, 10/28-70, 25)

October 8: U.S.S.R. launched two Cosmos satellites. *Cosmos CCCLXVIII*, launched from Baykonur, entered orbit with 383-km (238-mi) apogee, 202-km (125.5-mi) perigee, 90.4-min period, and 65° inclination. Satellite, whose objective was to conduct biological investigations and study physical characteristics of outer space, reentered Oct. 14. *Cosmos CCCLXIX*, launched from Plesetsk to investigate upper atmosphere and outer space, entered orbit with 467-km (290.2-mi) apogee, 261-km (162.2-mi) perigee, 91.8-min period, and 70.9° inclination. It reentered Jan. 22, 1971. (GSFC *SSR*, 10/31/70; 1/31/71; *SBD*, 10/9/70, 169; UN Gen Assembly Release 70-29113)

- House, by vote of 274 to 31, passed H.R. 19590, $66.8-billion FY 1971 DOD appropriations bill, after rejecting amendments to cut $651 million for ABM deployment, delete $389.4 million in F-14 aircraft procurement funds, cut $200 million from C-5A aircraft program, and cut $548 million from F-111A program. Bill was approximately $2 million less than requested and $6 million less than FY 1970 appropriation. (*CR*, 10/8/70, H9823-87)

- USAF was modifying more than 8000 in-service aircraft in retrofit program to ensure that all aircraft were equipped with AIMS (air traffic control radar system, identification friend or foe, military identification system) by Jan. 1, 1973, AFSC announced. System would meet FAA requirement for automatic altitude reporting on all aircraft by that date. (AFSC Release 252.70)

- Employee Assistance Center at NASA Hq. had mailed resumés of available former NASA employees to 600 prospective employers in effort to place about 200 personnel affected by budget cutbacks, NASA reported. Engineering was largest single category of jobs eliminated by reduction in force. Administrative positions made up second largest category. (NASA Release 70-168)

October 9: *Supersonic Planetary Entry Decelerator (SPED) experiment, launched from Wallops Station, simulated Mars reentry spacecraft speeding through thin atmosphere. A June photograph shows the parasol for SPED deployed in systems check-out before mating to a single-stage Castor rocket motor. The flight experiment aided study of deployment characteristics required for a parachute to land instruments on Mars.*

October 9: Cosmos CCCLXX was launched by U.S.S.R. from Baykonur into orbit with 264-km (164.0-mi) apogee, 185-km (115.0-mi) perigee, 88.9-min period, and 64.9° inclination. Satellite reentered Oct. 22. (GSFC *SSR,* 10/31/70; *SF,* 4/71, 138)

- NASA launched 1384.4-kg (3052-lb) SPED (Supersonic Planetary Entry Decelerator) experiment from Wallops Station at 1:57 pm EDT, by one-stage Castor booster with two Recruit rockets strapped on. Payload separated from booster at 72.4-km (45-mi) altitude and coasted to 91.7-km (57-mi) altitude before descending. At 70.8-km (44-mi) altitude the conical aeroshell, 4.6-m (15-ft) in diameter, popped open like a parasol to simulate Mars entry spacecraft speeding through thin atmosphere. Aeroshell splashed down in Atlantic and was recovered by USNS *Range Recoverer*. Onboard cameras and test equipment parachuted toward earth 112.7 km (70 mi) from Wallops and were recovered in midair by helicopter. (NASA Release 70-171)

- Discovery of tektite glass in *Apollo 12* sample 12013, reported June 5th by GSFC scientist Dr. John A. O'Keefe, was debated in *Science* by Dr. O'Keefe, Univ. of Houston geologist Elbert A. King, Jr., and LRL scientists R. Martin and Weldon B. Nance.

  Analyses of sample by King, Martin, and Nance had led them to conclude that "there is *no* existing chemical or mineralogical observation or data that uniquely support the idea that tektites originate from the moon. There *are* abundant chemical data that support a close genetic relationship between tektites and terrestrial rock materials. These data have been presented or summarized by numerous authors. Tektite glass has not been found in samples from Apollo 12 (or Apollo 11)."

  Dr. O'Keefe responded: "They find that the material is crystalline rather than glassy. The crystals are, however, evidently rather fine; from statistical considerations it can be judged that the agreement of the two major-element analyses given for samples of only a few tens of milligrams is likely only if the crystals are of the order of 100μ m [0.04 in] or smaller in diameter. This raises the question whether they are the products of devitrefication. In any case, the former existence of a liquid of this composition seems to be implied. Evidence of somewhat similar liquids was found in the Apollo 11 sample." (*Science*, 10/9/70, 199-200)

- Analyses of argon in *Apollo 11* crystalline rocks at LRL were reported in *Science* by MSC scientist Dr. Donald D. Bogard, State Univ. of New York scientists Dr. Oliver A. Schaeffer and Dr. John G. Funkhouser, and Max Planck scientist Dr. Josef Zaehringer. Crystalline rocks from Sea of Tranquility had yielded potassium-argon dates as old as 3.8 billion yrs. Crystalline rocks from Ocean of Storms gave potassium-argon ages as old as 2.8 billion yrs. Scientists concluded that maria were ancient lunar features. Potassium-argon ages of rocks from Ocean of Storms showed more argon loss than rocks from Sea of Tranquility, indicating Ocean of Storms rocks had experienced more severe shock effects or longer cooling rates. (*Science*, 10/9/70, 161-2)

- Thermal properties of lunar magnetic rocks at high temperatures were evaluated in *Science* by Univ. of Oregon volcanologists Tsutomu Murase and Alexander R. McBirney: "Thermal conductivity of a synthetic lunar rock in its melting range is about half that of a terrestrial basalt. The low conductivity and increased efficiency of insulating crusts on lunar lavas will enable flows to cover great distances without being quenched by high radiant heat losses from the surface. For a given rate of heat production, the thermal gradient of the moon would be significantly steeper than that of earth." (*Science*, 10/9/70, 165-7)

- Implications of failure of *Mariner VI* and *VII* UV experiments to detect emissions associated with nitrogen in Martian atmosphere were discussed in *Science* by Harvard College Observatory and SAO astronomer A. Dalgarno and M. B. McElroy of Kitt Peak National

Observatory. Examination of physical and chemical processes causing nitrogen emission in Martian airglow had shown that if atmosphere was uniformly mixed, mixing ratio of nitrogen to carbon dioxide of 5% was consistent with observational data on UV dayglow of Mars. If magnitude of the eddy coefficient in atmosphere was similar to that for earth, this limit was reduced to less than 0.5%. (*Science*, 10/9/70, 167-8)

October 10: Arcas sounding rocket, launched by NASA from Cold Lake Range, Primrose Lake, Canada, carried GSFC payload to 67.5-km (41.9-mi) altitude to obtain ozone measurements in conjunction with *Nimbus IV* satellite overpass. Rocket performed satisfactorily. Payload transmitter stopped transmitting 40 secs after ejection, but prefailure data were of good quality. Launch was first in two-rocket series; second would be conducted Oct. 17. (NASA Rpt SRL)

October 12: U.S.S.R. launched Cosmos *CCCLXXI* from Plesetsk into orbit with 757-km (470.4-mi) apogee, 750-km (466.0-mi) perigee, 99.8-min period, and 74.0° inclination. (GSFC *SSR*, 10/31/70; *SF*, 4/71, 138)

• NASA officials would meet with Soviet counterparts to discuss possible compatible space docking arrangements in Moscow Oct. 26-27, NASA announced. Discussions were outgrowth of correspondence during year between NASA and Soviet Academy of Sciences. Academy President Mstislav V. Keldysh had proposed Moscow meeting in Sept. 11 letter to NASA and had invited NASA to select dates. Dr. George M. Low, Acting NASA Administrator, had accepted invitation and proposed dates in Sept. 25 letter. Keldysh had confirmed acceptability of dates. NASA officials attending meeting would be: Dr. Robert R. Gilruth, MSC Director; Arnold W. Frutkin, Assistant Administrator for International Affairs; George B. Hardy, Chief of MSFC Program Engineering and Integration Project; Caldwell C. Johnson, Chief of MSC Spacecraft Design Office; and Glynn S. Lunney, Chief of MSC Flight Director's Office. (NASA Release 70-173)

• FRC announced tests had demonstrated that unpowered approaches and landing maneuvers of space shuttle could be made safely and readily by qualified professional pilots and did not require highly trained test pilots. United Air Lines captains Donald C. McBain and James V. Mitchell had flown landing approaches of four-engine jet transport configured to simulate space shuttle during NASA study of energy-management techniques for proposed shuttle orbiter. (FRC Release 18-70)

• NASA announced selection of Brayton power conversion system for development of long-life space power source independent of sunlight and more compact than large solar-cell arrays. Two power-conversion systems had been under study by OART, Brayton cycle and SNAP-8 mercury Rankine system, both of which provided closed-loop system without bringing in new fuel or emitting exhaust. Working fluid was circulated through heat exchanger (in Brayton cycle) or boiler (in Rankine cycle), turbo-alternator, condenser, pumps, back to heat, and on around. Although Rankine cycle could be developed for space use, Brayton equipment had inherent flexibility and potentially longer life capability for high-powered applications. (NASA Release 70-172)

- Sen. Joseph M. Montoya (D-N. Mex.) introduced S. 4453, "to establish a Department of Science and Technology, and to transfer certain agencies and functions to such Department." Bill provided that NASA, NSF, and AEC be transferred to new department. (CR, 10/12/70, S17689-95)
- FAA announced it had adopted three-bar VASI (visual approach-slope glide indicator) to keep pilots of long-body jet aircraft on proper guide slope during approach for landing. Bicolor VASI system alongside touchdown or aiming point of runway would indicate by bars of light whether pilot was on proper slope. System would be installed primarily for runways not equipped with instrument landing system. (FAA Release T 70-36)
- President Nixon in Stamford, Conn., announced decision to continue Turbotrain demonstration experiment between New York and Boston beyond scheduled Oct. 22 termination date: "This administration has been trying to apply some of the space age technology and expertise to practical problems such as rail transportation." (PD, 10/19/70, 1366-7)
- With business getting more scarce, Government contractors were "fighting harder for the business, squeezing harder for more bucks, and contesting bidding procedures," *Purchasing Week* magazine said. In 1969 claims before Armed Services Board of Contract Appeals had jumped from 900 to 1100. Protests filed through GAO had increased from 554 to 583 despite downward trend in contract letting. (*Purchasing Week*, 10/12/70)

October 13: NASA Hq. press briefing cited background events leading to scheduled U.S.-U.S.S.R. Moscow meeting on compatible space docking procedures and hardware [see Oct. 12]. Arnold W. Frutkin, Assistant Administrator for International Affairs, said government-to-government discussions with U.S.S.R. had begun in 1962 "when some substantive possibilities were explored and...first agreements were reached." These had included coordinated meteorological satellite projects, joint mapping of earth's magnetic field in space for joint use of Echo satellites, and joint approach to publication of U.S. and U.S.S.R. achievements in space medicine and biology. There had been "continuing overtures" since 1962, which "moved to a more accelerated pace beginning in...April 1970, when Dr. Paine took a personal interest in the problem and addressed a number of letters...to academician Keldysh...President of the Academy of Sciences of the Soviet Union."

Compatible docking procedures had been discussed for first time by Dr. Paine and Soviet academician Dr. Anatoly A. Blagonravov in New York in April 1970. Subject had been discussed on subsequent Moscow visits by Dr. Philip Handler, NAS President; Dr. George M. Low, NASA Deputy Administrator; and *Apollo 8* Astronaut Frank Borman. First formal proposal for exploration of subject had been made by Dr. Paine in letter to Keldysh July 31, 1970.

Frutkin said U.S.S.R. had chosen compatible docking for discussion out of "long list of initiatives from the U.S. side." Space rescue and joint experiments had not been discussed. Release of news of impending U.S.-U.S.S.R. discussions [at future IAF space rescue committee meeting] from IAF conference in Constance, Germany,

Oct. 8 had been coincidence; NASA overtures had been direct, not through IAF. (Transcript)

- NASA announced it was conducting experiments with NCAR to locate high-flying balloons and measure atmospheric pressure and temperature via IRLS (interrogation, recording, and location system) on board orbiting *Nimbus IV* satellite. Thirteen ballons, carrying 4.5-kg (10-lb) balloon interrogation packages (BIP), had been launched May 27 through July 8 from Ascension Island and had been floating around earth at altitudes of 20 700-24 000 m (68 000-79 000 ft). Seven were still flying and four were still transmitting to *Nimbus IV*. Fifteen more balloons were being readied for mid-October launch. Experiment was expected to provide extensive data on global wind circulation for research to improve worldwide weather forecasting. (NASA Release 70-169)
- President Nixon praised ICAO's action calling on member states to take strong measures against international aircraft hijacking, in letter to ICAO Council President Walter Binaghi: "I have instructed my representatives to put before the Organization's Legal Committee a draft convention which would implement these principles." (*PD*, 10/19/70, 1372-3)
- *Apollo 13* astronauts, on European goodwill tour, arrived in Dublin for three-day visit that would include meeting with clergyman who had held all-night vigil for them during *Apollo 13* mission abort. (Reuters, *NYT*, 10/14/70)
- NASA sources quoted by *New York Times* said *Soyuz IX* Cosmonauts Andrian G. Nikolayev and Vitaly I. Sevastyanov—due in Washington, D.C., Oct. 18 for 10-day NASA tour—had not yet accepted NASA invitation to visit KSC. Acceptance would signify thaw in U.S.-U.S.S.R. space relations, since there was tacit understanding that U.S.S.R. would have to reciprocate with invitation to tour Soviet launch facilities. (*NYT*, 10/14/70)

October 13-15: Technical symposium on blackout of radio communications with spacecraft and missiles during atmosphere entry was held at LaRC. (*Langley Researcher*, 10/16/70, 1)

October 14: U.S.S.R. launched *Intercosmos IV* from Kapustin Yar into orbit with 602-km (374.1-mi) apogee, 255-km (158.5-mi) perigee, 93.1-min period, and 48.4° inclination. Purpose of launch was investigation of UV and x-radiation of sun and its effect on structure of earth's upper atmosphere, U.S.S.R. said. Satellite reentered Jan. 17, 1971. (GSFC *SSR*, 10/31/70; 1/31/71; UN Gen Assembly Release 70-29113; *SF*, 4/71, 138)

- USAF X-24A lifting-body vehicle, piloted by NASA test pilot John A. Manke in joint program, successfully completed 18th flight, reaching mach 1.15 after air launch from B-52 aircraft at 13 700-m (45 000-ft) altitude from FRC. Objectives of powered flight—first at supersonic speed—were to expand flight envelope to mach 1.1, obtain lateral directional derivatives at mach numbers greater than 1.0, obtain longitudinal-trim and lift-to-drag-ratio data with 40° upper flap, obtain longitudinal derivatives, and make 270 knots-indicated-airspeed approach. (NASA Proj Off; UPI, *W Post*, 10/15/70, A4)
- Nike-Cajun sounding rocket, launched by NASA from Wallops Station, carried NOAA and Univ. of Michigan payload to 115.7-km (71.9-mi) altitude to measure ambient atomic oxygen between altitudes of 80

and 110 km (50 and 68 mi). Payload—including four silver probes calibrated to measure atomic oxygen, lateral magnetometer, solar-aspect sensor, and photocell to monitor nosecone ejection—was being flown for first time. Apogee and impact were less than predicted; instrumentation performance was satisfactory. (NASA Rpt SRL)

October 15: U.S.S.R. launched *Meteor VI* meteorological satellite from Plesetsk into orbit with 647-km (402.0-mi) apogee, 627-km (389.6-mi) perigee, 97.4-min period, and 81.2° inclination. Satellite would study clouds and weather changes. (GSFC *SSR*, 10/31/70; *SBD*, 10/19/70, 209)

- House Committee on Science and Astronautics' Subcommittee on Science, Research, and Development published *Toward a Science Policy for the United States*. Report summarized hearings held during July, August, and September and recommended that Federal Government establish national policy for support and furtherance of science and technology. Report further recommended that OST be separated from direct administrative connections with President's Science Adviser or President's Science Advisory Committee; OST submit annual report to President and Congress reviewing status of R&D in U.S. and recommending R&D program for following year; National Institutes of Research and Advanced Studies (NIRAS) be inaugurated; and Congress seek centralized Senate jurisdiction over science and technology and establish Office of Technology Assessment. Report also recommended, pending establishment of NIRAS, responsibility for basic research be centered in NSF, which should provide one third of all Federal support; OST develop criteria for support of basic research by mission-oriented agencies; Office of Management and Budget develop "stable funding" procedure for basic research; and NIRAS, when established, emphasize science education. Smithsonian Institution should have responsibility for science-information system, with backup from OST's Committee on Scientific and Technical Information. NAS and NAE should assist in explaining to public science and technology's role in solving major problems (Text)

- NSF released *Impact of Changes in Federal Funding Patterns on Academic Institutions* (NSF 70-39). Total expenditures for academic science increased 7% in FY 1969 and 8.5% in FY 1970 but effective support had declined 5% to 10% since FY 1968 because of inflation and undergraduate enrollment increases. Public institutions were faring better than private ones because of expansion of Federal support. Federal funds increased annually about 2.5%, with public institutions' receipt increasing 5%, while those of private institutions declined 1.5% in each year. (Greenberg, *Science*, 11/6/70, 609-12)

- Dr. Charles H. Townes of Univ. of California was awarded $5000 Michelson-Morley science award by Case Western Reserve Univ. in Cleveland, Ohio, for his work in creating laser and maser. (AP, W *Star*, 10/16/70, A2; JPL Hist Off)

October 16: U.S.S.R. launched *Cosmos CCCLXXII* from Plesetsk into orbit with 806-km (500.8-mi) apogee, 785-km (487.8-mi) perigee, 100.7-min period, and 74.0° inclination. (GSFC *SSR*, 10/31/70; *SF*, 4/71, 138)

October 16

- Astronaut Edgar D. Mitchell described planned EVA and equipment for Apollo 14 mission at Washington, D.C., press briefing. After landing LM on moon's Fra Mauro, Mitchell and Astronaut Alan B. Shepard, Jr., would leave LM and explore lunar surface with equipment stored on 9-kg (20-lb) mobile equipment transporter (MET). On previous lunar landing missions astronauts had ended up with more equipment than they could carry. MET would carry cameras, special sample containers, work table, lunar portable magnetometer, trenching tools, hand tool carrier, extension handles, tether, and other tools, thus freeing astronauts' hands. MET would be pulled over lunar surface by special handle that could be held with pressurized glove without being gripped. (Transcript; Sehlstedt, B *Sun*, 10/17/70, A5)
- Decline to 2.3%—lowest point in several years—of U.S. aerospace industry profits in first quarter of 1970 was reported by *Space Business Daily*. Net income of entire manufacturing industry had declined, but profits remained at 4.0%. Aerospace backlog at first quarter's end was $27.1 billion, off $1.2 billion from end of preceding quarter, and down $4.3 billion from end of first quarter of 1969. (*SBD*, 10/16/70, 199)

October 17: No. 2 electron bombardment ion engine on board *Sert II* (launched Feb. 3) shut down unexpectedly at 7:30 pm EDT after operating successfully for 2011 hrs since it was turned on July 24, with one brief interruption for eclipse of sun. LaRC engineers were attempting to restart engine. Primary goal of *Sert II* mission was to accumulate six months of engine operation. No. 1 engine had operated for a little more than five months before shutting down July 23. (NASA Release 70-177)

- Arcas sounding rocket, launched by NASA from Cold Lake Range, Primrose Lake, Canada, carried GSFC payload to 62-km (38.5-mi) altitude to measure ozone in conjunction with overpass of *Nimbus IV* satellite. Rocket and instruments functioned satisfactorily and good data were obtained. Launch was second in two-rocket series; first had been lanched Oct. 10. (NASA Rpt SRL)

October 17-18: Apollo lunar samples were exhibited at PMR Space Fair in Point Mugu, Calif. Fair attracted more than 200 000 visitors. (PMR *Missile*, 10/9/70, 1; 10/23/70, 1)

October 18: C-5A Galaxy, first of world's largest aircraft built by Lockheed-Georgia Co., was destroyed on ground at Marietta, Ga., by series of explosions followed by fire. Maintenance man was killed and another seriously injured while preparing to empty fuel tanks in right wing to search for suspected fuel leak. Lockheed spokesman said probable cause of fire was spark which ignited fuel tank fumes. Aircraft, valued at $20 million, had been rolled out March 2, 1968. (Reuters, *W Post*, 10/18/70, A1)

- *Soyuz IX* Cosmonauts Andrian G. Nikolayev and Vitaly I. Sevastyanov arrived in Washington, D.C., to begin 10-day goodwill tour in U.S. as NASA guests. (*W Post*, 10/19/70; NASA Release 70-170)

October 19: Dr. George M. Low, Acting NASA Administrator, addressed Inland Daily Press Assn. in Chicago on U.S. achievement and leadership in science and technology and "danger that exists today that we will soon lose the capability that gave us this achievement and leadership." Capability built by NASA in 1960s was rapidly diminishing. Recent study had shown 35 000 aerospace scientists and en-

October 18: *Soviet Cosmonauts Andrian G. Nikolayev (right) and Vitaly I. Sevastyanov (center right), who established a manned-flight endurance record in their 18-day Soyuz IX mission June 2-19, were welcomed at Washington National Airport by Astronauts Edwin E. Aldrin, Jr. (left), and Neil A. Armstrong. Cosmonauts would tour U.S.*

gineers unemployed "and this number could easily double next year." Major concern was "that in the future a new shortage of scientists and engineers will arise, based on the growing disenchantment with the unstable labor market for such skills." Implications were clear: "We are losing the capability to react to a national crisis, should the need arise; we are losing the capability to undertake an Apollo-like adventure, should we in the future desire to do so; and we are losing a capability that today provides an inflow of dollars that exceeds the net positive balance of payments."

Cost of U.S. space program in 1970 was $3.4 billion—decrease of $2.5 billion since 1966. In terms of Federal budget, $3.4 billion represented 1.7% or $17 for every person in U.S. "At the same time we are spending $400 per person for our domestic programs.... In fact, this year's increase in our expenditures for domestic programs was more than twice the NASA budget." Cancellation of entire space program would not significantly affect Federal spending on domestic programs. "But without the space program, without the capabilities it represents, what would become of the 'scientific and technological lead' this generation of Americans gave so much of themselves and their resources to achieve?" (Text)

- AT&T and ComSatCorp formally announced details of proposed satellite system under which ComSatCorp would furnish two 10 800-circuit satellites that would be leased entirely to AT&T for $205 million over seven years.

ComSatCorp filed with FCC application to provide advanced high-capacity satellites for domestic communications that would double capacity by doubling use of limited frequency spectrum. Com-

October 19

- SatCorp said it would file in near future for separate multiservice system to provide services to other customers. (AT&T Release; ComSatCorp Release 70-54)
- *Soyuz IX* Cosmonauts Andrian G. Nikolayev and Vitaly I. Sevastyanov, on 10-day U.S. visit, toured Washington, D.C., and placed wreaths on graves of Astronauts Virgil I. Grissom and Roger B. Chaffee in Arlington National Cemetry. At NASA Hq. press conference they said U.S.S.R. did not plan to put man on moon and had disbanded training program for female cosmonauts. (Valentine, *W Post*, 10/20/70, A2)
- MSFC engineer Hans F. Wuenscher was awarded U.S. patent 3 534 926 for machine to manufacture materials and articles in space. Cylindrical tank would be used in space station to produce new metal alloys and glasses, metal membranes, metal foam, high-precision ball bearings of hollow steel, and optical components in zero-g condition, where heavy materials did not settle out because of lack of buoyancy. Astronaut outside tank would control admission of liquid metal, gases, and other materials, and regulate heat, ventilation pressure, spinning, and agitation. (Jones, *NYT*, 10/23/70, 43)
- Secretary of Transportation John A. Volpe announced appointment of William M. Huey, special consultant to FAA Administrator and Deputy Administrator, to fill new position of FAA Deputy Assistant Administrator for General Aviation Affairs. (FAA Release 70-89)

October 19-22: AIAA Seventh Annual Meeting and Technical Display was held in Houston, Tex., under theme "Aerospace for Man's Needs." Display included space shuttle and Skylab mockup. *Soyuz IX* Cosmonauts Andrian G. Nikolayev and Vitaly I. Sevastyanov toured exhibits as AIAA guests and delivered same paper they had presented at Oct. 4-10 IAF Congress in Constance, Germany. Cosmonauts said in Houston that U.S.S.R. would attempt longer missions in space before beginning to build orbiting space station. *Soyuz IX* mission had ended after completion of scheduled experiments but spacecraft was capable of flight longer than 18 days. Both cosmonauts said they had been ill for days after return to earth. U.S. scientists said later that Astronauts Frank Borman and James A. Lovell, Jr., had suffered similar disorientation after 14-day *Gemini VII* mission Dec. 4-18, 1965. (AIAA Release; AP, *W Star*, 10/23/70, A7; Reuters, B *Sun*, 10/23/70, A5)

Dr. George M. Low, Acting NASA Administrator, said in speech at meeting: "NASA and the aerospace industry are not uniquely qualified to solve our domestic problems. The domestic problems, including those of the physical environment, are largely social, economical, and political. The need for new technology, for research and development, does exist in some areas, but certainly is not the driving force in obtaining solutions." Solutions "must come from sociologists, economists, and politicians—not engineers and scientists. We in aerospace should. . .assist wherever and whenever we can: We have developed techniques of management that may apply; there are technological problems that we are qualified to tackle; and there are areas where space technology, and the applications of space technology, apply directly. We should assist, but we should not think or encourage others to believe that we can solve all of our domestic problems."

Space program had "one major deficiency—a gap in manned space flight after Skylab, a hiatus of four years or more when no American will be in earth orbit, or anywhere in space." Gap had to be accepted because of lack of funding. Decision to cancel Viking or delay space shuttle would be wrong. "Viking is *the* major scientific experiment of the decade, and a delay in starting the shuttle would only delay the hiatus, not eliminate it." If shuttle development was advancing when gap occurred, Dr. Low believed gap would not be "unacceptable." NASA was building "the next generation of space vehicles, a generation that will far surpass existing capabilities to explore space, to use space, to live and work in space." (Text)

Sen. Mike Gravel (D-Alaska) told meeting: "Our space program is being viewed with increasing skepticism. Future historians will smile at the irony of our situation. Immediately following the fantastic feat of sending men to the surface of the Moon and back in safety, our national resolve in expanding this effort faced a mounting wave of domestic criticism. The irony of homo sapiens for the first time standing upright on his planet and not availing himself of his full ability to explore and experiment in the universe. The irony of not moving forward as aggressively as possible from a new beginning almost as fundamental as the beginning of life itself." (*SBD*, 10/23/70, 23)

Work of DOT Transportation Systems Center, formerly NASA ERC, was described by Secretary of Transportation John A. Volpe: Center was "already working on harbor advisory radars for the Coast Guard, phased array radars for the FAA, crash sensors for the Federal Highway Administration, and alcohol detectors for the National Highway Safety Bureau. In communications we are studying wayside systems for tracked air-cushion vehicles and data links from ground controllers to cockpit displays. Computer systems are being adapted to automated scanning of highway traffic films and simulation of ground and air traffic patterns controllers must learn. The architecture of large, real-time computers for air traffic control systems is also under study." (Text)

JPL engineers Ronald F. Draper and Thomas R. Gavin said technology for long-lived, unmanned spacecraft for Grand Tour flybys and separate orbiting missions to Jupiter and outer planets was well advanced for 1977 and 1979 missions. Thermoelectric outer planet spacecraft (TOPS) under development at JPL would use radioisotope thermoelectric generators, with nuclear energy supplying power beyond range where solar panels and batteries were effective. Scientific experiments and spacecraft control problems would be handled by self-test and repair (STAR) computer with sufficient backup units for 10 to 15 yrs operation. (JPL Release 564)

Richard S. Johnston, MSC Deputy Director for Biomedical Engineering, announced that final two Apollo missions would land at Descartes, upland valley surrounded by craters, and Marius Hills, near Imbrium Basin. Sites had been selected because material there would furnish new data on age and origin of moon. (AP, W *Star*, 10/20/70, A4)

Christopher C. Kraft, Jr., MSC Deputy Director, received $5000 AIAA Louis W. Hill Space Transportation Award for "outstanding management and leadership in directing the planning and operational control of all United States manned space flight missions from the first Mercury suborbital mission through the first Apollo lunar landing." Dr. Maxime A. Faget, MSC Director of Engineering and Development, received $500 Spacecraft Design award for "outstanding leadership and technical ability which resulted in the conception and design of this nation's first manned spacecraft, the Mercury capsule system." Glynn S. Lunney, Chief of MSC Flight Director's Office, received $500 Lawrence Sperry Award for "outstanding performance as a flight director during the Gemini and Apollo manned space flight missions; for exceptional leadership, professional skill and personal dedication which contributed to the success of this Nation's first manned lunar mission, and particularly during the flight of Apollo 13." Dr. Walton L. Jones, Director of Biotechnology and Human Research Div. of NASA OART, received John Jefferies Award of $500 and certificate for "outstanding contributions to the advancement of aerospace medical research." Von Kármán Lecture award of $1000 and travel allowance of $1500 was awarded to Dr. Erik L. Mollo-Christensen, MIT meteorologist. (AIAA Release; MSC *Roundup*, 9/25/70, 1)

NASA and DOT sponsored session on national aeronautical programs, with Charles J. Donlan of NASA OMSF as theme chairman. Exhibit was prepared by LaRC, LeRC, and ARC. Dr. Alfred J. Eggers, Jr., NASA Assistant Administrator for Policy, chaired panel discussion on problems facing aeronautics in U.S., including vehicles, airport systems, and airway systems.

Historical session, "American Advancement into Space," was chaired by Royal D. Frey of USAF Museum at Wright-Patterson AFB and Dr. Loyd Swenson of Univ. of Houston. Organized by Dr. Eugene M. Emme, NASA Historian, session heard description by Col. John A. MacReady (USAF, Ret.) and Sally MacReady Liston of early experiments to combat known and unknown hazards of first attempts to climb into space, narrative by L/G William E. Kempner (USAF, Ret.) of flight of *Explorer I* into stratosphere, description by Cdr. George W. Hoover (USN, Ret.) of USN developments leading to space, and story by Col. Joseph F. Cotton (USAF, Ret.) of XB-70. (Program; NASA Hist Off)

October 20-27: *Zond VIII* automatic space station was launched by U.S.S.R. from Baykonur and placed on lunar trajectory. Tass, breaking traditional secrecy surrounding Soviet space missions, announced launch and said purpose was "to carry out physical research along the flight path and the near-moon space, take pictures of the lunar surface, of the earth and the moon at different distances, check on improved on-board systems, units and the construction of the spacecraft. Under the flight program Zond 8 will round the moon on Oct. 24, then change to an eastward path and return to earth on Oct. 27."

On Oct. 24 Tass announced that *Zond VIII* had circled within 1120 km (696 mi) of moon, studied space, photographed moon, and was returning to earth. Spacecraft splashed down in Indian Ocean Oct. 27 and was recovered by Soviet vessel *Tama*. Tass said spacecraft brought back "extremely important photographs needed for studies of the geological and morphological qualities of the moon."

Zond VIII was eighth spacecraft in Zond series and second spacecraft to land in water; first had been *Zond V* (Sept. 15-22, 1968). *Zond VI* (Nov. 10-17, 1968) and *Zond VII* (Aug. 8-14, 1969) had circled moon and returned to earth with data and photos. (GSFC *SSR*, 10/31/70; Gwertzman, *NYT*, 10/22/70, 1; *SBD*, 10/28/70, 249; 10/29/70, 256; B *Sun*, 10/28/70, A5; *SF*, 4/71, 138-9)

October 20: U.S.S.R.. launched *Cosmos CCCLXXIII* from Baykonur into orbit with 562-km (349.2-mi) apogee, 459-km (285.2-mi) perigee, 94.8-min period, and 62.9° inclination. *Spaceflight* magazine later reported that satellite made close approach to two other Cosmos satellites on days they were launched: *Cosmos CCCLXXIV* Oct. 23 and *Cosmos CCCLXXV* Oct. 30. Fragments of the other satellites were cataloged. (GSFC *SSR*, 10/31/70; *SF*, 4/71, 138-9)

- Orbiting *Nimbus IV* satellite (launched April 8) was relaying temperature of steam emitted from 4392.2-m (14 410-ft) Mt. Ranier mountain on command eight times weekly, NASA and Dept. of Interior announced. Satellite received radio signals from special temperature-monitoring station on mountain's peak and relayed them to ground station at Fairbanks, Alaska, for transmission to GSFC. Temperature had varied only few degrees from 343.2 K (158°F) recorded at time of installation. Experiment marked first time volcanic temperatures had been relayed through satellite. (DOI Release)

- Full-scale hypersonic research engine (HRE) under development by NASA was tested for first time in 8-Foot High-Temperature Structures Tunnel at LaRC. Mach-7.4 speed was reached at temperature of 1800 K (2000°F) and pressure of 6.2 meganewtons per sq m (900 psi). NASA project would demonstrate technology needed for engines to propel future aircraft at hypersonic speeds—more than five times speed of sound. (Pres Rpt 71; NASA Photo Release 70-H-1362)

- NASA announced it had issued "Invitation for Participation in Mission Definition for Grand Tour Missions to the Outer Solar System" in late 1970s to U.S. and foreign scientists. Purpose was to obtain intensive participation of experimental and theoretical scientists with mission engineering team in missions' definition stages, not proposals for flight missions. Missions included series of three-planet swingbys, which would result in flybys of all five outer planets. Scientific objectives for planets were to study physical properties and atmospheric composition; geological features; thermal regimes and energy balances; charged particles and electromagnetic environments; periods of rotation, radii, and other body properties; and gravitational fields. Scientific objectives in interplanetary space included studies of solar wind plasma and magnetic field, solar energetic particles, galactic cosmic rays, interplanetary dust, and theorized boundary between solar wind and interstellar medium. Proposals from scientists were due Jan. 18, 1971. NASA would select participating scientists by March 15, 1971. (NASA Release 70-175; Text, 10/15/70)

- *New Haven Register* editorial commented on U.S.-U.S.S.R. space talks: "Those who despair about relations between the two powers have at least a straw to grasp at. While they may be miles apart in some areas, which, because of their nature, make the headlines, Washington and Moscow are quietly working together in less spec-

October 20: *Full-scale hypersonic research engine (HRE) was tested for the first time in Langley Research Center's 8-Foot High-Temperature Structures Tunnel, which produced a stream of hot gases moving faster than seven times the speed of sound.* NASA *project would demonstrate technology needed for future hypersonic aircraft. The test engine incorporated advanced technology in structure and materials and a cooling system that circulated liquid hydrogen through engine parts to protect them from the intense heat of hypersonic flight. Coolant was then burned as fuel.*

tacular areas. And while politicians may glare at each other, the technologists are finding common interests." (*N Hav Reg*, 10/20/70)

October 21: Two solid fuel rockets, carrying cloud-measuring payload triggering device, were launched from KSC in lightning study, but electrical field was too low for triggering payload to ignite lightning discharge. Launches were first in NASA Mighty Mouse program being conducted by National Oceanic and Atmospheric Agency to study feasibility of discharging lightning potential from clouds before lightning could damage vulnerable areas, such as launch pads. (KSC Releases 354-70, 337-70)

October 21-22: Soyuz IX Cosmonauts Andrian G. Nikolayev and Vitaly I. Sevastyanov toured MSFC and Alabama Space and Rocket Center during two-day visit to Huntsville, Ala. (*Huntsville Times*, 10/22/70)

October 22: New comet—13th observed in 1970—had been sighted by three amateur Japanese astronomers in western Japan and north of Tokyo on Oct. 19 and in western Japan on Oct. 20, Tokyo Observatory said. (AP, *W Post*, 10/23/70, A5)

- Environment Defense Fund, Inc., New York-based organization with environment lawsuits pending throughout U.S., petitioned airport authorities at seven major U.S. airports to bar SST if it could not meet noise standards applicable to subsonic aircraft. (Hornig, W *Star*, 10/22/70, A1)

October 23: USAF launched unidentified satellite by Titan IIIB-Agena booster into orbit with 399.1-km (248-mi) apogee, 130.4-km (81-mi) perigee, 89.8-min period, and 111° inclination. Satellite reentered Nov. 11. (Pres Rpt 71; GSFC *SSR*, 10/31/70; 11/30/70)

- U.S.S.R. launched *Cosmos CCCLXXIV* from Baykonur into orbit with 2140-km (1329.7-mi) apogee, 521-km (323.7-mi) perigee, 112.2-min period, and 62.9° inclination. Close approach was made same day to *Cosmos CCCLXXIII* (launched Oct. 20), *Spaceflight* magazine reported, and 15 fragments were cataloged to Dec. 11. (GSFC *SSR*, 10/31/70; 12/31/70; *SF*, 4/71; 138-9)

- Twenty-fifth Anniversary Session of U.N. General Assembly. President Nixon paid tribute to U.S. founders and called on all nations to strive for peace: "Across this planet let us attack the ills that threaten peace. In the untapped oceans of water and space, let us harvest in peace." (*PD*, 10/26/70, 1434-40)

- Moon had been formed as separate planet outside earth orbit and was later captured by earth according to calculations published by Interior Dept. scientist S. Fred Singer and Univ. of Hawaii astronomer L. W. Bandermann in *Science*. Depletion of volatile substances, which were of lower abundance in lunar surface rocks than in terrestrial rocks, had been explained by late accretion of volatile materials from solar nebula with falling temperature. Scientists had concluded moon had accumulated not in earth orbit but as separate planet. (*Science*, 10/23/70, 438-9)

- FAA extended until Oct. 25, 1971, flight quota rule at four Washington, Chicago, and New York high-density airports and suspended quota for Newark Airport, N.J. Helicopters, previously included in quotas, were exempted, since their operation had had insignificant impact on operation of fixed-wing aircraft. Rule, in effect since June 1, 1969, would have expired Oct. 25. (FAA Release 70-92)

- U.S. had already trained 1000 sky marshals to apprehend would-be hijackers on domestic and international flights and expected to have 300 more by Oct. 31, Baltimore *Sun* said. DOT spokesman had said about 54 marshals a day were being turned out after training at Fort Dix, N.J. Eventually, Government hoped to establish 2500-man force hired from outside Government agencies and armed forces. (Sehlstedt, B *Sun*, 10/23/70, A7)

- Lockheed Aircraft Corp. had settled dispute with DOD over SRAM (short-range attack missile) motor subcontract for $20 million, *Wall Street Journal* reported. In one of four disputes with DOD, Lockheed had originally asked $54.3 million. Settlement resolved legal issues on SRAM subcontract held by Lockheed Propulsion Co. with Boeing Co. Still to be settled were Lockheed disputes with DOD over C-5A transport, AH-56A Cheyenne helicopter, and USN ship-building contracts. (*WSJ*, 10/23/70, 5)

- Appointment of Dr. Edward E. David, Jr., as OST Director and presidential science adviser had evoked surprise, D. S. Greenberg said in *Science*, because "unlike his five predecessors" he was not "member or protege of the fairly small group that had dominated the upper levels of government science since Sputnik." Dr. David had been "picked by members of President Nixon's personal staff without consultation with the then-incumbent, Lee A. DuBridge." There had been "no inclination to ask the White House science advisory apparatus to assist in finding a successor when it was time for Dr. DuBridge to go." (*Science*, 10/23/70, 417-19)

October 24: NASA delegation to U.S.-U.S.S.R. space talks—headed by Dr. Robert R. Gilruth, MSC Director—arrived in Moscow for Oct. 26-27 meetings to discuss cooperation in developing manned space stations and linking spacecraft in orbit. (Clarity, *NYT*, 10/25/70, 24)

- *Soyuz IX* Cosmonauts Andrian G. Nikolayev and Vitaly I. Sevastyanov arrived in Los Angeles with Astronaut Edwin E. Aldrin, Jr., to tour Cal Tech and visit Disneyland. They would leave for San Francisco Oct. 26, after touring ARC. (*LA Times*, 10/25/70; ARC *Astrogram*, 10/22/70, 1)

October 25: Record distance for photographing a spacecraft was claimed by U.S.S.R. Tass announced that scientists at Sternberg Astronomical Institute in Alma, U.S.S.R., had photographed *Zond VIII* (launched Oct. 20) in space 267 000 km (166 000 mi) away. "This is a record distance for optical instruments in photographing artificial space bodies." (UPI, *W Star*, 10/26/70, A1)

- Communications system on board NATO's *Nato I* (launched by NASA March 20) failed. Satellite was switched to redundant communications repeater system and was operating satisfactorily. (NASA Proj Off)

- NASA team in Moscow for Oct. 26-27 U.S.-U.S.S.R. space talks visited Zvezdny Gorodok (Star City), home of Soviet cosmonauts, as guests of Soviet team including Cosmonaut Konstantin P. Feoktistov, Deputy Director of Soviet manned flight program. NASA team was taken into Soviet spacecraft, where systems were explained to establish information base for space talks. (Transcript, NASA background press briefing, 10/29/70)

- Data from millimeter-wave experiment on board *Ats V* (launched Aug. 12, 1969) had indicated scientists might be able to "open up" extremely overcrowded microwave band to frequencies above 10 000 mhz for communications, NASA announced. Millimeter-wave frequencies, which operated in range above 10 ghz, offered better gain-to-antenna characteristics and extremely wide bandwidths, allowed reduction of component size and weight, and could offer more than 10 times the frequency spectrum currently available. Experiment was providing amplitude and phase measurements on two independent test links at 15.3 ghz (satellite-to-earth) and at 31.65 ghz (earth-to-satellite) during measured and defined meteorological conditions. Results showed that only very heavy rainfall would affect reliable communications and that frequency band "windows" existed where water vapor and oxygen absorption were low. (NASA Release 70/178)

- NASA officials attended opening of art exhibit by Robert Rauschenberg at Fendrick Gallery in Washington, D.C. Exhibit included 25 lithographs from "Stoned Moon" series. Rauschenberg had been one of artists invited by NASA to witness *Apollo 11* launch and convey impressions of first lunar landing mission. (Richard, *W Post*, 11/30/70, D1; Fendrick Gallery)

- Fortieth anniversary of all-air transcontinental service in U.S. TWA flew first all-air coast-to-coast flight Oct. 25, 1930, from New York with 13-passenger Ford trimotor aircraft, which stopped at Philadelphia, Harrisburg, Pittsburgh, Columbus, Indianapolis, St. Louis, Kansas City, Wichita, Amarillo, Albuquerque, and Winslow, Ariz. Total elapsed time for east-west service was 36 hrs including 10½-hr overnight in Kansas City. (*TWA Today*, 11/2/70, 1)

- AP quoted USAF as saying 89 officers and airmen conducting upper-atmosphere research would be withdrawn from Chile by year's end because "requirement no longer exists." (*C Trib*, 10/27/70, 14)

October 26: White House announced appointment of R/A Don A. Jones as Acting Director of National Ocean Survey and of Dr. John W. Townsend as Acting Associate Administrator of National Oceanic and Atmospheric Administration (NOAA) until offices were filled under provisions of NOAA reorganization plan. Adm. Jones was member of ESSA Commissioned Officers Corps and had been Director of Coast and Geodetic Survey since 1965. Dr. Townsend had been ESSA Deputy Administrator since 1968, Chief of NASA Space Sciences Div. 1958-1959, and GSFC Assistant Director and Deputy Director 1959-1968. (*PD*, 11/2/70, 1457)

- Federation of American Scientists held Washington, D.C., news conference to protest extensive advertising campaign by American Security Council asserting U.S.S.R. had taken lead in strategic nuclear weaponry. Dr. Herbert Scoville, Jr., head of FAS Strategic Weapons Committee presented chart showing U.S. had 1710 long-range missiles and 550 strategic bombers, for total of 2260 delivery vehicles. Soviet force had 1518 missiles and 150 bombers, or 1668 delivery vehicle total, and was building 390 additional vehicles. (Beecher, *NYT*, 10/27/70, 13)

October 26-27: U.S.-U.S.S.R. talks on possibilties for compatible rendezvous and docking arrangements in space were held in Moscow between NASA delegation headed by MSC Director, Dr. Robert R. Gilruth, and Soviet Academy of Sciences team headed by Academician Georgy I. Petrov.

In exchange of basic information on docking systems NASA officials described Gemini and Apollo techniques, procedures, and docking adapters and Skylab project. Soviet team described plans for future system similar to Apollo's, with tunnel between spacecraft to accommodate docking apparatus. Agreement was reached that 12 specific technical elements required further joint study, including guidance systems for rendezvous, docking hardware, coordinate systems, and reference markings. Teams agreed on November exchange by mail of supplementary technical information and subsequent meeting in March or April to organize three joint working groups, to meet alternately in U.S. and U.S.S.R. to develop compatible hardware and procedures. (Transcript, NASA background press briefing, 10/29/70)

October 27: USAF X-24A lifting-body vehicle, piloted by NASA test pilot John A. Manke, successfully completed 19th flight in joint NASA-USAF research program from FRC. Purposes of flight were to expand flight envelope to mach 1.25, obtain lateral directional derivatives at mach numbers greater than 1.15, obtain longitudinal trim and lift-to-drag ratio data with 30° upper flap, and determine effect of upper flap on approach and landing. (NASA Proj Off)

- NASA announced reorganization of OART to improve aeronautical research and give more support of space activities. Effective immediately, reorganization created new aeronautical research units: Short Landing and Take-Off (STOL) Program Office, Advanced Technology Experimental Transport (ATET) Program Office, Lifting Body Program Office, Aeronautical Operating System Div., Aeronautical Research Div., and Aeronautical Propulsion Div. New units to support technology in space or atmosphere were: Shuttle Technologies Office, Technology Applications Office, Nuclear Systems Office, Space Propulsion and Power Div., Environmental Systems and Effects Div., Guidance, Control and Information Systems Div., and Materials and Structures Div. Remaining units were: Resources and Institutional Management Div., Advanced Concepts and Mission Div., Safety and Operating Systems Office, and Research Council for planning future research efforts. Council members would include directors of LaRC, LeRC, ARC, and FRC. OART would retain its name and additional staffing would not be required. (NASA Release 70-183)
- Brain sensor and radio transmitter system developed for space medical research was being used by scientists at ARC and Agnews State Hospital in San Jose, Calif., in diagnosis and treatment of schizophrenics, NASA announced. System, developed by ARC, employed wire clip headset with two small electrodes that sensed brain waves through hair with no scalp preparation. Tiny battery-powered radio transmitter broadcast brain signals to computer for analysis. Light weight and simplicity of system eliminated disquieting procedures required by past diagnostic methods. (NASA Release 70-179)
- *Apollo 12* seismometer had detected moonquakes at least twice when moon was at apogee, 405 555 km (252 000 mi) from earth, *Washington Post* reported. Univ. of Hawaii seismologist Dr. Rolf Meissner had said quakes, which had occurred during seven months ending in August, might be caused by sudden venting of trapped gas when tidal stresses from earth's pull were suddenly relaxed. Finding followed Aug. 10 finding by Dr. Gary V. Latham and NASA team at MSC that moon quaked at perigee. Stress pattern was reversed during apogee, Dr. Meissner had said. "What might be happening is the lunar surface layer is pushing down, pulling the interior apart and unblocking gas chambers at some depths in the moon." While apogee quakes did not appear to emanate from same place producing perigee quakes, they did emanate from same general location—in Fra Mauro hills, where Apollo 14 astronauts were scheduled to land. (O'Toole, *W Post*, 10/27/70, A3)
- Nobel Prize for physics for 1970 was awarded jointly to Hannes O. Alfven of Univ. of California at San Diego and Louis E. Neel of Grenoble Univ. in France. They would share $78 000 cash award. Alfven was cited for "fundamental work and discoveries in magnetohydrodynamics, with fruitful applications in different parts of plasma physics." Neel was cited for "fundamental work and discoveries concerning ferromagnetism, which have led to important applications in solid-state physics."

    Dr. Luis F. Leloir of Argentine Institute of Biochemical Research received Nobel Prize in chemistry for study of how sugars were metabolized and stored in the body and for work that had provided "real knowledge in wide fields of biochemistry, where earlier we had to resort to vague hypothesis." (O'Toole, *W Post,* 10/28/70, A13)

- NASA announced award of $4 660 000, cost-plus-fixed-fee contract to RCA for flight-model videotape recorders and associated equipment for ERTS program. Recorder system would be used to record output of return-beam videcon high-resolution TV system and multispectral scanner system when ERTS was not in contact with a ground station and to relay data to earth on command. (NASA Release 70-184)
- Oral history of space flight on 647 reels of tape assembled over six years had been presented to Smithsonian Institution's Air and Space Museum, Washington *Daily News* reported. Assembled by Michael Kapp, Vice President of Capitol Records Co., tapes began with reminiscences of Mrs. Esther Goddard, widow of rocket pioneer Dr. Robert H. Goddard, and went through *Apollo 11* mission. Six-record LP album had been excerpted from tapes. (*W News,* 10/27/70)
- Visiting Soviet *Soyuz IX* Cosmonauts Andrian G. Nikolayev and Vitaly I. Sevastyanov took turns with Astronaut Edwin E. Aldrin, Jr., flying Boeing 747. They were accompanied by Boeing test pilot during flight from Boeing facility near Seattle, Wash. (AP, *B Sun,* 10/28/70)
- Pan Am Boeing 747 engine caught fire minutes after landing at London after flight from Los Angeles. Several of 89 passengers aboard were injured slightly in evacuating aircraft via escape chutes. FAA spokesman later said fire was not related to conditions cited by National Transportation Safety Board on Oct. 1 (*W Post,* 10/28/70, A18)
- U.S.S.R. announced it would conduct month-long series of missile firing tests in Pacific Oct. 28 to Nov. 30. Target area would be circle 257.5 km (160 mi) in diameter about 482.8 km (300 mi) from Midway Islands. (AP, *W Post,* 10/28/70, A14)
- "Our World in Space—Yesterday, Today and Tomorrow," illustrated lecture by artist Robert McCall, was presented by Smithsonian Associates in Washington, D.C. (Program)

October 28: NASA delegation to Oct. 26-27 U.S.-U.S.S.R. space talks was taken by Soviet Academy of Sciences hosts to see *Luna XVI* lunar samples at Moscow geochemical institute before delegation departed for U.S. Academy President Mstislav V. Keldysh later said it was first time foreigners had seen samples of 100 grams (3.5 oz) of lunar soil picked up by unmanned *Luna XVI* during Sept. 12-24 mission.

At Moscow press conference on *Luna XVI* Academician Keldysh said Soviet scientists would continue to emphasize unmanned missions "in the near future." He said fact that U.S.-U.S.S.R. space talks took place "means that both sides think it advisable to make the docking units of manned spacecraft compatible," but problem was "complex." It would require more research before countries could decide even to what degree they could cooperate.

U.S. Embassy in Moscow issued statement describing space talks as "preliminary exchange of views and information." Both sides had agreed to continue discussions. (Mills, *B Sun,* 10/29/70, A4)

- Edgar M. Cortright, LaRC Director, addressed National Space Club in Washington, D.C., on NASA's time for decision: "We in NASA recognize that our program no longer enjoys the number one priority but is now but one of a number of priority areas competing for scarce funds. But we also recognize that our program, in addition to its intrinsic value, sets the pace for much of the technological development of this country and promises to do so for the foreseeable future. ... We do not believe that we could long survive and prosper without this preeminent technological position. Therefore, we have to stand and fight off the stampede by other groups to acquire funds now being channeled into our national technology through NASA and the Department of Defense. It is a veritable 'run on the bank.' The withdrawal slips carry such catch terms as relevancy, poverty, pollution, and housing, and are designed to shame us into sacrificing our space program and national technology in the name of 'humanity.' I would reverse the logic and say—in the name of humanity we should maintain our national technology."

Cortright also described NASA-DOT-DOD experimental program to introduce V/STOL aircraft into commercial use: "We have chosen the externally blown flap and augmentor wing jet STOL airplanes for initial attention. Jet VTOL aircraft will come later. ... We envision a combination of V/STOL and high-speed ground transportation as offering the best hope for continued mobility as the megalopolis continues to spread. We are also preparing a program to develop an Advanced Technology Experimental Transport using Whitcomb's [Dr. Richard T. Whitcomb, Jr., Head of LaRC's 8-Foot Tunnels Branch] supercritical aerodynamic concepts to explore the problems of efficient cruise at Mach numbers approaching one." (Text)

- Aerobee 150 sounding rocket was launched by NASA from WSMR carrying NCAR experiment to collect air samples. Mission did not meet minimum scientific requirements. (SR list)
- Motorized wheelchair controlled by sight-switch perfected by space engineers was delivered to Texas Medical Center in Houston by MSFC for evaluation. Second wheelchair would be delivered to Roncho Los Amigos Hospital in Downey, Calif., in November. Switch, worn on head, enabled user to open and close circuit controlling chair's wheel movement with his eyes and move easily without assistance. Switch had been developed to help astronauts in training operate essential equipment under extreme flight conditions. Manufacturer, Hayes International Corp., was extending application of switch as control device for handicapped. (MSFC Release 70-218)
- Agreement under which NASA would provide FAA with bibliographies in aeronautical engineering and make searches for scientific technical reports required by FAA was announced by NASA. FAA-developed scientific and technical information would be processed into NASA's scientific and technical facility at College Park, Md., and be available to other users of NASA's retrieval services. Joint use of NASA information facility would save FAA expense of developing and operating retrieval system. (NASA Release 70-180)
- U.K. Minister of Aviation Supply Frederick V. Corfield announced in House of Commons that Anglo-French supersonic transport Concorde would cost $1.98 billion to develop. Latest estimate increased cost by $280 million over previous $1.7 billion. Original estimate in 1962 had been $420 million. (AP, W Post, 10/29/70)

- USAF investigating board had found that F-111 crash Oct. 7 at Carswell AFB, Tex., had not been caused by inflight structural failure, *Wall Street Journal,* reported. No malfunction was found that would require further inspection of F-111 fleet. USAF declined to disclose cause. (*WSJ,* 10/28/70, 25)
- Ecuador became 77th member of INTELSAT, formed in 1964 with 11 member countries. (INTELSAT PIO; *SBD,* 11/5/70, 22)

October 29: NASA and Soviet Academy of Sciences reached agreement in Moscow for joint efforts to design compatible space docking and rendezvous arrangements. Agreement called for exchange of technical materials by correspondence during November 1970, exchange of technical requirements for systems to ensure compatibility in January and February 1971 for familiarization, and meetings of working groups in March and April 1971 to refine lists of systems to provide compatibility and discuss common requirements. Working group meetings would be held alternately in U.S. and U.S.S.R. After agreements on technical requirements, each side would independently prepare preliminary system designs to be considered by NASA and Soviet Academy. (NASA Release 70-210)

- At NASA Hq. press briefing Arnold W. Frutkin, Assistant Administrator for International Affairs, reported on Oct. 26-27 U.S.-U.S.S.R. talks on compatible rendezvous and docking: "We would hope that the next year would show very substantial progress toward the definition of possible compatible systems; the implementation of those systems is another matter which rests entirely on the pace of our respective programs." Space shuttle would be more plausible craft for compatible docking than Skylab because of timing. (Transcript)
- Dr. Donald J. Williams, Director of National Oceanic and Atmospheric Administration's Space Disturbances Laboratory in Boulder, Colo., revealed plans to study space weather by squirting small blobs of electrically charged gas out of earth-orbiting satellite. He told 1970 Meeting of Interagency Conference on Weather Modification at Virginia Beach, Va., NOAA would fly simple experiment on small geophysical monitoring satellite. "When activated 22 300 miles [35 900 km] above the nighttime earth, the device will trigger a self-quenching flow of electrons into the upper atmosphere. Our proposed experiment in modifying space weather requires interacting with our environment in a scientifically significant yet totally non-destructive way." If initial release of plasma from satellite caused artificial aurora scientists would "be able to deduce the exact shape of the magnetic field line that passes through the satellite's position. In addition, the...experiment should provide information on diffusion of plasma along field lines, diffusion across field lines, and electric fields deep in the magnetosphere."

Naturally occurring electron-precipitation events were accompanied by many terrestrial effects, including auroral displays, radio-propagation disturbances, and intense, short-lived fluctuations of earth's magnetic field. If NOAA experiment enhanced precipitation from magnetosphere into ionosphere scientists "would then be more certain of the natural processes that cause storms and should be able to improve our predictions of radar clutter and the quality of polar radio communications in future, naturally occurring storms. Once we know more about the magnetosphere, the triggering mech-

anism of electron precipitation events and their enhancement into geomagnetic storms, we can then consider ways to modify or control such storms when they occur naturally. Thus, the small NOAA experiment riding piggyback on a geophysical satellite may be our first step toward ultimate control of our space environment." (NOAA Release 70-85-285)
- Heart-rate tachometer invented by MSFC engineers James R. Currie and Ralph R. Kissel to monitor astronauts' heart beat in space might have medical application on earth, MSFC said. Device, which could measure heart rates between 40 and 200 beats per minute and furnish almost instantaneous rate after measuring only two heart beats, might be used to monitor heart rates of cardiac patients. (MSFC Release 70-226)
- MSC had accepted delivery of three of eight new supersonic Northrop T-38 trainers for astronauts, purchased at $800 000 each, *Washington Post* reported. Two-engine aircraft would replace 10 Lockheed subsonic, single-engine T-33s in NASA service for past 10 yrs. (O'Toole, *W Post*, 10/29/70)
- Early Birds of Aviation held annual convention in Washington, D.C. Group, organized in 1928, consisted of aviators who flew solo before Dec. 17, 1916—13 yrs after Wright brothers first flight, but before U.S. entry into World War I. Convention participants included Georgia (Tiny) Broadwick, 77, who parachuted from aircraft in 1913; Walter Waterman, 76, who made his first solo flight July 1, 1909, in glider he built from plans published in *Popular Mechanics*; Roy Waite, 86, who made one of world's first aerial bomb runs in 1912, when he dropped 20 bags of flour onto two battleships in Boston Harbor; and William Diehl, 79, who claimed to have carried first regular passengers. (Weil, *W Post*, 10/30/70, A22)

October 30: U.S.S.R. launched two Cosmos satellites. *Cosmos CCCLXXV*, launched from Baykonur, entered orbit with 2098-km (1303.6-mi) apogee, 525-km (326.2-mi) perigee, 111.8-min period, and 62.8° inclination. It made close approach same day to *Cosmos CCCLXXIII* (launched Oct. 20), *Spaceflight* magazine reported, and 25 fragments were cataloged to Dec. 11.

Cosmos CCCLXXVI, launched from Plesetsk, entered orbit with 286-km (177.7-mi) apogee, 206-km (128-mi) perigee, 89.4-min period, and 65.3° inclination. It reentered Nov. 12. (GSFC *SSR*, 10/31/70; 11/30/70; 12/31/70; *SF*, 4/71, 138-9)
- NASA combined Office of Industry Affairs with Office of Technology Utilization to form new Office of Industry Affairs and Technology Utilization. Daniel J. Harnett, former Assistant Administrator for Industry Affairs, was named Assistant Administrator for Industry Affairs and Technology Utilization. George J. Vecchietti continued as Deputy Assistant Administrator for Industry Affairs and Melvin S. Day, former Acting Assistant Administrator for Technology Utilization, became Deputy Assistant Administrator for Technology Utilization. (NASA Ann)
- LeRC scientists had perfected production of isotope iodine 123, diagnostic aid, in form 100 times less radioactive than commonly used iodine 131, NASA announced. In cooperative program with HEW's U.S. Public Health Service in Cincinnati, Ohio, technique for producing very pure iodine 123 with LeRC cyclotron had been pioneered

by LeRC scientist James W. Blue. AEC was building two linear accelerators that could produce enough iodine 123 to supply entire U.S. (NASA Release 70-185)
- Study of *Apollo 12* lunar samples had confirmed earlier conclusion from *Apollo 11* sample study that lunar soil contained 1.9% meteoric material of similar composition to carbonaeceous chondrites and that moon had formed in earth's neighborhood, Enrico Fermi Institute and Univ. of Chicago chemists reported in *Science*. Four *Apollo 12* core and soil samples were "enriched in a number of trace elements of meteoric origin to virtually the same degree as Apollo 11 soil." Meteoritic average influx rate of 4 billionths g per sq cm (1.4 billionths oz per sq in) seemed valid for entire moon. (Ganapathy, Keays, Anders, *Science*, 10/30/70, 533-5)
- Dr. George N. Constan, former manager of Michoud Assembly Facility, had returned from private industry to MSFC as Special Assistant to the Director of Program Management, MSFC announced. (MSFC Release 70-227)
- NASA announced selection of Computer Sciences Corp. for $5 million, cost-plus-award-fee contract to provide programming and computer support services for Manned Space Flight Network (MSFN) Central Computing System at GSFC. Contract covered two-year period, preceded by six-month phase-in, with option for one-year extension. (NASA Release 70-187)

October 31: Aerobee 170 sounding rocket was launched by NASA from WSMR carrying GSFC experiment to study stellar UV. Rocket and instruments functioned satisfactorily. (SR list)
- FAA air traffic controllers would delay final descent for landing of turbojet aircraft until close to their destination airport and climb aircraft out as rapidly as possible after takeoff, in program announced by Secretary of Transportation John A. Volpe to increase safety and reduce noise at U.S. airports. (FAA Release 70-94)
- *Christian Science Monitor* editorial on U.S.-U.S.S.R. space discussions: "While not wishing to minimize the long road which still lies ahead before America and Russia have peacefully resolved all their major disagreements, we see the new space pact as an extraordinarily heartening development. It has created a new bond, a new mutual interest, a new feeling of joint participation. It is from the multiplication of such ties that international peace and friendship grow. And from such a vision of space togetherness there can come a stronger desire for greater brotherliness on earth." (*CSM*, 10/31/70)

During October: Scientists at Boeing Space Center in Kent, Wash., were studying piece of *Surveyor III*, 5 cm (2 in) long, returned from moon by *Apollo 12* astronauts, to determine why original white paint had discolored, Boeing Co. said. White paint reflected intense solar rays in space and helped prevent heat buildup inside spacecraft. Boeing researchers thought paint degraded in space because of contaminating film from rocket exhaust, constant proton bombardment, or other radiation from sun. If discoloring came from contamination, paint could be restored to whiteness through oxygen plasma process. If paint had changed through radiation damage, Boeing could replenish depleted atoms. Results of study would enable

Boeing to renew reflective paint of future spacecraft which might need to function for 10 to 20 yrs. (Boeing Release S-0788)

- Time was ripe for U.S.-European space partnership, William Leavitt said in *Air Force and Space Digest.* "Europe...seems to be working, at long last, toward the creation of a European 'NASA' that would bring together the expertise, and create a focus for energies and resources, of the principal European countries." In U.S., "prestige-fired fiscal support of multibillion-dollar space projects is rapidly shrinking." There was opportunity for real partnership. "More cooks in today's kitchen would enrich the broth. To such a combination, the U.S. could bring its vast experience of the past decade and certainly some significant funding, even in a tightened money situation. The Europeans could bring fresh enthusiasm, a good deal of skill, financial support, and developmental philosophies that might be a good deal more frugal, since they have been accustomed to small budgets. Money and time could be saved all round." (*AF/SD*, 10/70, 66-7)

- NAS announced appointment of special committee for celebration of Copernicus Quinquecentennial in 1973 to "facilitate participation of U.S. science in national and international observations...and to propose and implement plans...to explain...the significance of the work of Copernicus in the history of science." (NAS-NRC-NAE *News Rpt*, 10/70, 1)

- Progress of Global Atmospheric Research Program (GARP) was described in *Astronautics & Aeronautics* by ESSA scientist David S. Johnson. NASA's *Itos I*, launched Jan. 23, was in sun-synchronous orbit. Operational sounding capability would be added to system in 1972. Third-generation operational satellite was planned for 1975 launch. Synchronous Meteorological Satellite scheduled for 1972 launch would be operational prototype of Geostationary Operational Environmental Satellite that would complement *Itos I*. Initially, U.S. would keep one geostationary spacecraft operating continuously. It was hoped second would be launched before first GARP global experiment scheduled for one-year period about 1976.

  France and Japan were considering synchronous satellites for first GARP global experiment. ESRO and U.K. might build polar-orbit meteorological satellites for experiment. U.S.S.R. was operating meteorological satellites in quasi-polar orbit and was developing sounding instruments. Space-based observing system apparently would furnish major portion of data for GARP global experiment, by which time "U.S. operational meteorological satellite system will approach the coverage needed, except for the four synchronous satellites. Promising signs of significant contributions towards these have come from other countries." (*A&A*, 10/70, 28-32)

- Work of NASA's six regional Dissemination Centers was described in *Mechanical Engineering*. Part of NASA's specialized information services, centers had been established at universities and nonprofit research institutes "to help potential users of new technology obtain it in packages appropriate to their needs." Cost of individual retrospective searches of full NASA file averaged $200, charged against minimum annual client fee commitment. Client could be kept current in particular subject for about $80 per month. Computer tapes bearing 6000 or more new monthly citations of scientific and

technical reports were searched each month by machine matching "interest profile" of client's objectives against indexed descriptions of aerospace researchers' findings. (Kotel, *Mechanical Engineering*, 10/70, 16-23)
- First Doctor of Civil Law degree in field of air and space law was awarded by McGill Univ. in Montreal to George S. Robinson, Jr., of Alexandria, Va. (*Vienna* [Va.] *Globe*, 1/7/71,1)
- Stages in development of Soviet space stations were described by Boris Petrov in Soviet Academy of Sciences journal *Vestnik Akademii Nauk SSSR:* "First...small stations...with three to twelve men, with a period of existence of from one month to a year or slightly longer, will be put into circumterrestrial orbit.... Well elaborated and tested compartments of space vehicles and individual stages of carrier rockets will be used as the main units of those stations." Orbital stations could be orbited in assembled state by carrier rockets or in parts, with one or two dockings. Station crew could be delivered by transport space vehicle and crews exchanged. One of main tasks of stations would be medical and biological experiments. Orbital stations would be created "of block design, assembled in circumterrestrial orbit in parts," with maximum 10-yr life and 12- to 20-member crew. Petrov saw advisability "of very large multipurpose orbital base stations" for crews of 50 to 70 and, ultimately 100 to 120. U.S.S.R. had in view not simply space station but system of space stations. (*AF Mag*, 6/71, 54-9)

November 1970

November 1: Nixon Administration was "moving to make the hydra-headed Federal science and technology apparatus more responsive to the nation's needs," *New York Times* reported. New effort, to be tested in early 1971, would identify priorities in science and technology and weigh them to determine "where they will rank in the struggle for Federal dollars." OST and NSF officials had said in interviews that they were determined to bring better order into system of Federal research agencies. Reorganization efforts would focus on $14-billion-a-year R&D programs to translate scientific knowledge into better housing, mass transportation, cleaner air, and "healthier people." Dr. Edward E. David, Jr., Presidential Science Adviser, had emphasized that pure and basic research, currently allotted $2 billion per year, would be specifically exempted from reorganization. Current R&D budget was lowest in five fiscal years and smallest in decade in terms of constant dollars. (Lyons, *NYT*, 11/1/70, 1)

November 2: NASA's M2-F3 lifting-body vehicle, piloted by NASA test pilot William H. Dana, successfully completed third flight from FRC. Captive portion—usually conducted on separate flight—was completed at 9100-km (30 000-ft) altitude to verify B-52/adapter/heavyweight M2-F3 compatibility, check out propellant system, and check out jettison system. Flight portion followed air launch from B-52 aircraft at 13 700-m (45 000-ft) altitude to expand launch envelope, study transonic configuration, investigate lateral phugoid present on M2-F2, check out reaction control system, and evaluate nose window requirements. (NASA Proj Off)

• NASA and U.K. Science Research Council (SRC) signed agreement for NASA to launch fifth U.K. satellite in 1973. UK-5, carrying six scientific experiments, would be launched by NASA Scout booster into orbit with 555-km (435-mi) altitude, where it would study cosmic x-ray sources. Satellite would be U.K.'s first employing magnetic-core data storage, pulse-code-modulation telemetry, and gas-jet attitude-control system that could be programmed by ground command. U.K.'s *Ariel I* (launched April 26, 1962), *Ariel II* (launched March 27, 1964), and *Ariel III* (launched May 5, 1967), had been launched by NASA under cooperative agreements. Fourth satellite, UK-4, would be launched into near-polar orbit in 1971. (NASA Release 70-195)

• Kenneth W. Gatland, Vice President of British Interplanetary Society, told press in London that U.S.S.R. had been testing satellite designed to destroy U.S. military satellites. Claim, he said, was substantiated by studies from Royal Aircraft Establishment at Farnborough. *Cosmos CCCLXXIII* (launched Oct. 20) had been a target; *Cosmos CCCLXXIV* (launched Oct. 23) had been orbited to same altitude as *Cosmos CCCLXXIII* and then exploded, apparently on ground command. Six large fragments had been identified. Gatland said objective could have been to destroy *Cosmos CCCLXXIII* with the fragments or to use radiation from a nuclear explosion to make it inoperable. *Cosmos CCCLXXV* (launched Oct. 30) had been test of same system. (AP, B *Sun*, 11/3/70, A2; W *Star*, 11/3/70)

- Battelle-Northwest Laboratories team under Dr. Franklin E. Roach was investigating "diffuse galactic light" to measure material between stars, UPI reported. Dr. Roach hoped to determine where dust was clumping and coalescing in early stages of star formation. (*NYT*, 11/2/70)
- Report to NATO by Secretary of Defense Melvin R. Laird placed number of Soviet land-based ICBMs ready for use or under construction at about 1400, AP said. Figure—increase of about 100 ICBMs over Laird's estimate given to Congress earlier in year—put U.S.S.R. about 350 ICBMs ahead of U.S. Laird had also reported that U.S.S.R. had about 30 missile-launching nuclear submarines of Polaris class operational or under construction and warned that U.S.S.R. could overtake U.S. in production by 1974. (*W Post*, 11/2/70, A11)
- Editorials commented on U.S.-U.S.S.R. space talks.

    New York Times: "The first concrete steps toward arranging direct Soviet-American cooperation in space would be important under any circumstances. But the timing of last week's [Oct. 26-27] agreement aimed at standardizing both nations' space ships...makes that development doubly welcome. The talks succeeded when most aspects of Washington-Moscow relations are roiled by tensions ranging from the Middle East to the continued imprisonment in Russia of two American generals whose plane strayed across the Turkish border." (*NYT*, 11/2/70)

    Christian Science Monitor: "While not wishing to minimize the long road which still lies ahead before America and Russia have peacefully resolved all their major disagreements, we see the new space pact as an extraordinary heartening development. It has created a new bond, a new mutual interest, a new feeling of joint participation. It is from the muliplication of such ties that international peace and friendship grow. And from such a vision of space togetherness there can come a stronger desire for greater brotherliness on earth." (*CSM*, 11/2/70)

    Milwaukee Journal: "The economic benefits and savings of international co-operation are obvious. The intangible benefits, however, are just as great, particularly defusing space of the rivalries and antagonisms that have been so destructive on earth. NASA must be commended for its efforts so far, and urged to press on in this important business." (*Milwaukee Journal*, 11/2/70)

November 3: Lunar samples returned by U.S.S.R.'s *Luna XVI* (launched Sept. 12) contained less titanium and zirconium oxide and more ferric oxide than *Apollo 11* samples, Soviet Academy of Sciences Vice President Aleksander Vinogradov was quoted as saying. Luna and Apollo samples "have an equally high content of space originated inert gases" but in overall composition the *Luna XVI* samples more closely resembed those from *Apollo 12*. *Luna XVI* material was fragmented basaltic type containing 70 chemical elements; granularity of soil increased with depth. Spherical, glazed formations of congealed droplets similar to those found by Apollo

crews were present. Specific heat of lunar soil did not depend on density, Vinogradov said, and, "on the average, corresponds to terrestrial rocks, while heat conductivity is characterized by extremely low values, considerably below those of the best heat-insulating materials on Earth.... It may be assumed that the general course of differentiation of the substance of the Earth and of the Moon and probably of other planets of the terrestrial type proceeded by similar routes, although reached different stages of development." (SBD, 11/3/70, 8-9)
- Supercritical wing developed by Dr. Richard T. Whitcomb, Jr., head of 8-Foot Tunnels Branch at LaRC, to increase speeds of U.S. transport aircraft by more than 160 km per hr (100 mph) without causing sonic booms was delivered to FRC for testing. Wing, which would be flown on modified F-8 jet fighter aircraft in spring 1971, was built by NR Los Angeles Div. at $1.8-million cost. (Miles, *LA Times*, 11/2/70)
- Chemist Dr. Gustaf O. S. Arrhenius of Univ. of California at San Diego said in interview that he was working with NASA on plans to expand U.S. space program by sending unmanned spacecraft on "mission to an asteroid." He said idea had been presented to NASA in 1969 by Dr. Hannes O. Alfven, 1970 Nobel Prize winner in physics. (UPI, *NYT*, 11/4/70)
- U.S. and U.S.S.R. resumed SALT in Helsinki after adjournment of nearly three months. (Gwertzman, *NYT*, 11/4/70, 11)

November 4: Sud-Aviation test pilot André Turcat flew Concorde 001, French prototype of Anglo-French supersonic transport to mach 2 (2125 km per hr; 1320 mph) for first time during 90-min, 102nd test flight from Toulouse-Blagnac Aerodrome. Soviet supersonic transport Tu-144 had flown at 2150 km per hr (1336 mph) May 26. (Farnsworth, *NYT*, 11/5/70, 94)
- U.S.S.R.'s *Zond VIII* spacecraft was flown to Moscow from Bombay, India, where it had been taken by ship after splashdown in Indian Ocean Oct. 27. (Reuters, *W Post*, 11/5/70, A6)
- AP interview of Dr. Charles H. Townes, Univ. of California at Berkeley physicist and Nobel laureate, on possibility of life in outer space was published in Washington *Evening Star:* "It is clear that we already have found most of the molecules that biologists have postulated are necessary for the beginning of life." Scientists were also finding particles of dust in "interstellar smog" that were similar to earth rocks. "...we can see in space many of the materials from which planets are made and which life needs to begin." (W *Star*, 11/4/70, A2)

November 4-5: European Space Conference in Brussels discussed European participation in $13-billion, 10-yr U.S. post-Apollo program. At Nov. 4 session, U.K. delegation rejected U.S. proposal to join program, because program's costs and timing were too imprecise and program had not yet been approved by U.S. Congress. French, West German, and Belgian delegations declared they would accept U.S. invitation. At close of Nov. 5 session, Conference President Theo Lefevre, Belgian Minister for Scientific Policy and Planning, announced 12 countries participating had set up committee to examine U.S. proposals with view to accepting it on suitable financial and political terms. Delegation would visit Washington, D.C., at undecided date for further talks. (Reuters, *NYT*, 11/6/70, C9)

November 5: Success of NASA-Purdue Univ. study to detect corn blight infestation with sensing devices [see Sept. 6] was announced by NASA. . Most precise data had come from multispectral scanner flown at 900 m (3000 ft) by Univ. of Michigan's C-47 aircraft. Combined with ground computer readouts, scanner data classified corn as healthy or as suffering very mild blight, mild blight, moderately severe blight, or severe blight. While infrared photos showed little difference, scanner data could make distinction. (NASA Release 70-188)

- *Nato I*, satellite, launched by NASA March 20 for NATO, had met mission objectives and was operating satisfactorily, NASA announced. (NASA Proj Off)
- Four-man crew successfully completed 159-day, 11 000-km (7000-mi) journey from Guayaquil, Ecuador, to Mooloolaba, Australia, in 7.9-m (26-ft) raft to prove South American Indians could have sailed to Australia centuries ago. Expedition was headed by Prof. Vital Alzar of Spain, who was making second attempt to cross Pacific on raft. Others were Marco Modena of France, Norman Terenault of Canada, and Gabriel Salas of Chile. (UPI, *W Post*, 11/6/70, A18)
- Dr. James A. Van Allen, Univ. of Iowa physicist and discoverer of radiation belts encircling earth, said in Durham, N.H., that U.S. should phase out manned space programs. Maintenance of men in spacecraft "increases the cost of a given mission enormously and risks human life unnecessarily, and in a conspicuous and dramatic way." Dr. Van Allen was in Durham to deliver first Spaulding Distinguished Series lecture at Univ. of New Hampshire. (AP, *W Post*, 11/6/70, A7)
- N.Y. State Supreme Court Appellate Div. ruled unanimously that Cornell Univ. had right to sell Cornell Aeronautical Laboratory to private, profit-making company, EDP Technology, Inc. Decision upset 1969 State Supreme Court ruling that laboratory had been created as public trust and therefore could not be sold to private company. Appellate Div. held that laboratory had been transferred to Cornell as gift without any restrictions. (AP, *NYT*, 11/6/70)

November 6: USAF launched 500th successful U.S. satellite into orbit from ETR by Titan IIIC booster. Orbital parameters: apogee, 35 886.8 km (22 299 mi); perigee, 26 048.8 km (16 186 mi); period, 1197.1 min; and inclination, 7.8°.

U.S. press reported that satellite, carrying several new sensors, was part of early warning system to provide 30-min warning of ICBMs fired from U.S.S.R. or Communist China. (Pres Rpt 71; GSFC *SSR*, 11/30/70; Beecher, *NYT*, 11/7/70, 1)

- Arcas sounding rocket, launched by NASA from Ft. Sherman Canal Zone, Panama, carried GSFC payload to 59.5-km (37-mi) altitude to obtain ozone measurements in conjunction with *Nimbus IV* satellite overpass. Rocket was first of two-rocket series; second would be launched Nov. 13. Rocket and instruments functioned satisfactorily and good quality data were obtained. (NASA Rpt SRL)
- Seismic data from impacts of *Apollo 12* LM and 3rd stage of *Apollo 13* booster (S-IVB), relayed by seismic station installed on lunar surface by *Apollo 12* astronauts, were described in *Science*. Team of

lunar scientists headed by Dr. Gary V. Latham, principal seismic investigator for Apollo program, made report. Unusually long reverberations suggested "that the lunar mare in the region of the Apollo 12 landing site consists of material with very low seismic velocities near the surface with velocity increasing with depth to 5 to 6 kilometers per second [3.1-3.7 mps] (for compression waves) at a depth of 20 kilometers [12.4 mi]. Absorption of seismic waves in this structure is extremely low relative to typical continental crustal materials on earth. It is unlikely that a major boundary similar to the crust-mantle interface on earth exists in the outer 20 kilometers [12.4 mi] of the moon. A combination of dispersion and scattering of surface waves implies the presence of heterogeneity within the outer zone of the mare on a scale of from several hundred meters (or less) to several kilometers. Seismic signals from 160 events of natural origin have been recorded during the first 7 months of operation of the Apollo 12 seismic station. At least 26...are small moonquakes. Many...are thought to be meteoroid impacts." (*Science*, 11/6/70, 620-6)

- MSC *Roundup* described Skylab food system being developed "to compensate partially for astronauts' long absence from the usual fare of Earthlings and the warmth and delight of home-cooked meals...." Skylab food system would maintain caloric level between 2000 and 2800 calories per day and provide at least minimum dietary allowances recommended by NAS. Meals would be selected from a variety of foods and cooked by each crewman on special food tray. Four kinds of food would be prepared: dehydrated, including scrambled eggs, beverages, and desserts; intermediate moisture, including peanuts and cookies; wetpack, including turkey with gravy and meatballs with sauce; and frozen, including filet mignon and lobster Newburg. (MSC *Roundup*, 11/6/70, 1)
- President Nixon announced intention to nominate William D. Ruckelshaus as Administrator of Environmental Protection Agency. Agency would bring together major Federal pollution control programs, effective Dec. 2. Nomination was submitted to Senate Nov. 16. (*PD*, 11/9/70, 1545-6; 11/23/70, 1583)

November 7: U.S.S.R. celebrated 53rd anniversary of Bolshevik Revolution with display of military equipment including four SAM-3 missiles. Western observers said they noticed no new equipment in parade. (Clarity, *NYT*, 11/8/70, 1)
- New European satellite consortium was pooling talents to strengthen chances for large share of European space business, including possible work on U.S. shuttle program, *Nature* reported. Seven-member group with D. W. Malim of Marconi Space and Defence Systems of U.K. as Chairman of Managing Board, included French Société Nationale Industrielle Aérospatiale, Études Techniques et Constructions Aerospiatales of Belgium, Société Anonyme de Telecommunications of France, Messeschmitt-Boelkow-Blohm and Siemens of West Germany, and Selenia of Italy. (*Nature*, 11/7/70, 494)

November 8: NASA announced selection of two 5400-kg (12 000-lb) prime candidate payloads for HEAO-A and HEAO-B High Energy Astronomy Observatories. HEAO-A payload to observe x-rays over wide variety of wavelengths at same time would include x-ray study, low-energy x-ray study with modulation collimators, high-energy x-ray study with modulation collimators, long-wavelength cosmic x-ray study, low-energy gamma-ray sky survey, extremely heavy nuclei in cosmic rays study, and composition and spectra of high-energy cosmic ray study.

HEAO-B would carry new instruments of better spectral resolution and would observe higher energy gamma rays up to 20 mev and the diffuse background of radiation at x-ray and gamma-ray wavelengths. Experiments included high-resolution x-ray Bragg crystal spectroscopy, spatial and spectral structure of x-ray sky study, high-energy gamma-ray astronomy study, superconducting magnetic spectrometer for primary cosmic rays, and flux and spectrum of primary cosmic ray electrons study. (NASA Release 70-191)

- Cuts in Federal R&D programs had "pole-axed the employment rolls of scientists and engineers and brought the nationwide demand for their professional services to the lowest point in at least a decade," Richard Lyons said in *New York Times*. "Grist coming from Washington rumor mills" indicated R&D "crunch" would worsen, with further layoffs expected in projects of DOD, NASA, AEC, and other agencies. Latest statistics of Deutch, Shea & Evans' Engineer Scientist Demand Index, "barometer of the need for professional technical talent," showed demand during September was lowest in 10-yr history of index. (*NYT*, 11/8/70)

November 9: Ofo Orbiting Frog Otolith and secondary payload, *RM* Radiation/Meteoroid satellite, were successfully launched by NASA from Wallops Station at 1:00 am EST by four-stage Scout booster. Two spacecraft separated 9 min after launch and entered orbit with 528-km (328-mi) apogee, 299.7-km (186.4-mi) perigee, 92.8-min period, and 37.4° inclination. Countdown had been aborted five times since original August launch date because of bad weather, instrument malfunctions, and death of frogs.

Primary objective of 132.5-kg (292-lb) *Ofo* was to obtain information on functioning and adaptability in weightlessness of vestibule, portion of inner ear which controlled balance, by studying two male bullfrogs (*Rana castebianca*) on board. Through microelectrodes surgically implanted in vestibular nerves leading from sensor cells in frogs' otoliths, scientists would be able to study for the first time electrical response of otolith sensors under weightlessness and simulated gravity provided by centrifuge for at least three days. Frogs had been chosen for flight because their inner ear mechanism functioned similarly to man's and because they were small and amphibious. Onboard spacecraft they were submerged in water with oxygen composition controlled by instruments. After several hours in orbit frogs were in good condition and spacecraft was transmitting data satisfactorily.

Primary objective of 21-kg (46-lb) *RM* satellite was to demonstrate feasibility and accuracy of advanced radiation dosimeter concept in space and to verify, in flight, operation of instruments to be used on long-duration flights to study meteoroids for at least three months. Satellite reentered Feb. 7, 1971.

Ofo experiment was managed by ARC and *RM* by MSC. OFO project was managed by WS. (NASA Proj Off; NASA Release 70-132; WS Release 70-18)

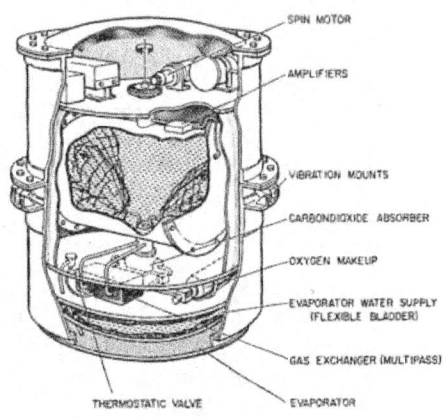

November 9: NASA Orbiting Frog Otolith was launched into orbit. Above, principal investigator Dr. Torquato Gualtierotti of the University of Milan stands by part of the spacecraft. Experiment package is at right.

- NASA's Apollo 14 spacecraft was moved from assembly building to launch pad at KSC in preparation for launch toward moon Jan. 31.

    Astronaut Stuart A. Roosa told press at briefing that launch preparations had "gone quite smoothly" so far. Apollo 14 crew, he said, was "the only crew that's been six months from launch four times. I personally will have close to 1000 hours in the command module simulator by the 31st of January." In response to press query Roosa said he would attempt to photograph evidence of moonquakes detected by seismometers: "...the current consensus is that these quakes are coming from around the Rima Parry Five Rille and the southern edge of crater Fra Mauro. So this will be a specific photo target. Both with the Hy-con and with the Hasselblad to see if we can see anything new." (Transcript; UPI, *NYT*, 11/10/70, 33)
- Awards ceremony to observe successful completion of more than four years of Saturn V launch vehicle test firings was held at MTF. MTF employees, contractors, and groups of Government workers received special plaques and awards. Guests included local, state government, and NASA officials. *Apollo 13* Astronaut Fred W. Haise, Jr., presented flag flown around moon on board *Apollo 10* to MTF employees to express Apollo astronauts' gratitude, as facility marked last day of space operations. (MSFC Release 70-229)
- During individual trips to Europe in past two weeks, Pan Am President Najeeb E. Halaby and TWA President F. C. Wiser had not departed from Aviation industry's support for a supersonic transport, *Washington Post* reported. "Nevertheless, their reservations on Concorde could have implications for...SST program." Halaby had made "straight-forward appeal" for delay of Concorde development and proposed mass production of aircraft be postponed until consortium of international airlines had subjected Concorde to extensive checks by flying it on commercial routes. Wiser, in Geneva, had said Concorde would be money loser unless travelers were willing to pay ticket surcharge of 30%-40% above current fares. (Samuelson, *W Post*, 11/9/70, A1)
- L/C Joe Klass and Maj. Joseph Gervais, both former USAF officers, said in New York that 10-yr investigation indicated Aviatrix Amelia Earhart, who had disappeared over Pacific in 1937, was alive and living in U.S. They believed woman called Mrs. Guy Bolam, who claimed to have flown with Earhart, was Earhart herself and that Earhart and navigator Fred Noonan had been intercepted in 1937 mission for U.S. Government by aircraft from Japanese carrier, held captive throughout World War II in Imperial Palace in Tokyo, and released when Emperor Hirohito bartered her life for his exemption from trial as war criminal. Col. Klass had written book *Amelia Earhart Lives: A Trip Through Intrigue To Find America's First Lady of Mystery*, published by McGraw-Hill. Mrs. Bolam later held press conference to issue denial, calling book "poorly documented hoax" and "utter nonsense." (AP, *W Post*, 11/10/70, D2: AP, *W Star*, 11/11/70, A2)

November 10-22: U.S.S.R. launched *Luna XVII* unmanned lunar probe from Baykonur at 7:44 pm local time (9:44 am EST). Tass said purpose of mission was to explore moon and near-moon space and check new systems and reported all systems were functioning normally. During flight to moon 36 communications sessions were held with spacecraft and data on parameters and systems were transmitted. Midcourse maneuvers were conducted Nov. 12 and 14 and on Nov. 15 *Luna XVII* entered lunar orbit with 85-km (53-mi) altitude, 1-hr 56-min period, and 141° inclination.

Luna XVII softlanded in moon's Sea of Rains Nov. 17 at 8:47 am Baykonur time (10:47 pm EST Nov. 16). Spacecraft released *Lunokhod I*, self-propelled vehicle placed on moon to conduct scientific investigations. According to drawing and models shown on Moscow TV, *Lunokhod I* looked like large silver pot-bellied tub about the size of small automobile with eight spoked wheels apparently powered by solar energy and batteries. Tass said *Lunokhod I* carried flags and pennants with Soviet coat of arms and Lenin's bas relief and was equipped with scientific apparatus, instruments and control systems, and radio and TV devices. *Lunokhod I* left *Luna XVII* descent stage at 11:28 am Baykonur time (1:28 am EST) Nov. 17 by special gangway and traveled 20 m (66 ft) from spacecraft to begin experiments. Tass said vehicle's movement could be controlled automatically or by ground controllers following televised information on position and character of surrounding lunar surface.

Between Nov. 17 and Nov. 22, 10 radio and TV sessions were held and large amount of data was transmitted. Chemical composition of surface layer of lunar soil was analyzed with x-ray spectrometer and mechanical properties of lunar soil and transportability of vehicle were investigated. In addition, Tass said, "an instrument was used which has a punch for soil intrusion, and a system of sensors for measuring forces acting on the wheels." Panoramic pictures of lunar landscape were taken, in which tracks made by *Lunokhod I*, details of lunar surface, and parts of vehicle were clearly visible. Tass said TV cameras, "designed for surveying and evaluating the route traveled, provided the necessary information for travel in the conditions of uneven, broken terrain. At some points...the incline reached 14 deg. Along the path, Lunokhod encountered...deep craters the detours to which were ensured by the operators (on Earth)."

On Nov. 22, after five days on lunar surface, *Lunokhod I* was commanded to terminate operations in preparation for lunar night. Tass said vehicle had competed "a complex of tests...and a wide program of tests of its mobility" and had traveled 197 m (646 ft). Vehicle was moved to an "assigned position" where it would remain stationary during lunar night and solar panels were closed.

Luna XVII was first unmanned spacecraft to land on moon and deploy automatic lunar explorer. Last mission in series, unmanned *Luna XVI* (Sept. 12-24), had landed on moon and returned to earth with lunar samples. (*Sov Rpt*, 12/4/70; GSFC *SSR*, 11/30/70; *SBD*, 11/20/70, 81-2; Clarity, *NYT*, 11/18/70, 1)

November 10: *Nimbus IV* meteorological satellite, launched April 8 to acquire samples of atmospheric radiation measurements, was adjudged successful by NASA. Primary and secondary objectives had been achieved. Except for filter-wedge spectrometer, which had failed June 7, instruments were functioning satisfactorily and providing good data. (NASA Proj Off)

- Nike-Tomahawk sounding rocket was launched by NASA from Fairbanks, Alaska, carrying Univ. of Minnesota experiment to conduct

auroral studies. Rocket and instruments functioned satisfactorily. (SR list)
- Automated instrument that detected bacteria in urine hundreds of times faster than standard culture method had been developed by GSFC scientists in cooperation with Johns Hopkins Hospital in Baltimore, NASA announced. Technique was derived from GSFC study to detect life on other planets with chemicals from firefly. Photoelectric devices would measure adenosine triphosphate (ATP), high-energy biochemical found in all living organisms, using luciferase and luciferin, chemicals in firefly tails that flashed on contact with ATP. It would then differentiate between bacterial ATP and that from other sources.(NASA Release 70-190)
- *Apollo 11* Astronaut Neil A. Armstrong, NASA Director of Aeronautical Research, described Soviet Tu-144 as "striking" supersonic aircraft "every bit as attractive" as Anglo-French Concorde, during luncheon meeting of Aero Club of Washington and Aviation-Space Writers Assn. in Washington, D.C. Armstrong said visit to Tu-144 hangar had been highlight of Moscow visit. He said Tu-144's four engines were similar to those on U.S. B-70 bomber and interior resembled McDonnell Douglas DC-9. Tu-144 pilots had told him aircraft had flown at twice speed of sound and handled well at all speeds. (Sehlstedt, B *Sun*, 11/11/70, A3)
- Pros and cons of SST debate were cited by David Hoffman in *Washington Post:* "SST advocates blend airline economics with international balances of trade and payments. They mix Atlantic alliance politics with trans-Siberian flight routes, and they link the SST to the health of the American economy." Environmentalists agreed aircraft might generate 50 000 jobs, improve balance of trade, and guarantee U.S. leadership in commercial aviation, but said nothing would justify "additional noise irritation and risk to man's surroundings." SST files contained ammunition to support any conclusion. "If indeed an American SST goes into airline service in 1978, the files started at the plane's birth will be at least 20 years old." (*W Post*, 11/10/70, A3)
- Secretary of Transportation John A. Volpe announced appointment of Robert H. Binder, Director of Office of International Transportation Policy and Programs, to succeed Hugh J. Gownley as Deputy Assistant Secretary of Transportation for International Affairs. Gownley had retired July 31. (DOT Release 23870)

November 11: U.S.S.R. launched *Cosmos CCCLXXVII* from Baykonur into orbit with 286-km (177.7-mi) apogee, 204-km (126.8-mi) perigee, 89.3-min period, and 65.0° inclination. Satellite reentered Nov. 23. (GSFC *SSR*, 11/30/70; *SBD*, 11/13/70, 53)
- Semiannual AIA employment analysis said aerospace industry payrolls would drop by 374 000 persons to 1 044 000 by March 1971. Industry would remain largest manufacturing employer in U.S. despite drop from 1968 peak of 1 418 000 employees. (Corddry, B *Sun*, 11/12/70, A7)

November 12: Delays caused by sun's gravity in return of radio signals by Mariner spacecraft had verified Einstein theory of relativity, JPL astronomer Dr. John D. Anderson said in report to Conference on Experimental Tests of Gravitational Theories at Cal Tech. In tests begun April 30, Anderson and team including Dr. Duane O. Muhleman of Cal Tech and Dr. Pasquale B. Esposito and Warren L. Martin of JPL had taken several hundred radio measurements of *Mariner VI* and *VII* in wide-range solar orbit using precise distance-measuring system and 64-m (210-ft) Goldstone antenna. Experiments tested validity of theory that velocity of light was slower in gravitational field near sun than in interplanetary space where gravitational fields were weaker. "Our experiments to date show that Einstein's predictions are right, to within 2 to 4 per cent. The best measurements we have thus far indicate a maximum delay of about 204 microseconds for Mariner VI, as compared with an expected 200 microseconds using Einstein's theory." Findings cast doubt on more recent theories of Dr. Charles Brans and Dr. Robert H. Dicke that Einstein's formulations could be incorrect by 7% "If the Brans-Dicke theory were correct, the time delay would be much less...perhaps about 186 microseconds....if the much older Newtonian theory were correct, there would have been no time delay at all." (NASA Release 70-200)

- Advanced version of Boeing 747 transport set world heavyweight record at Edwards AFB, Calif., with takeoff at 372 400-kg (820 700-lb) gross weight. Weight exceeded unofficial record of 362 000 kg (798 000 lbs) set by Lockheed C-5A Galaxy Oct. 14, 1969. (AP, B *Sun*, 11/13/70, A6)
- *Ofo* Orbiting Frog Otolith experiment, launched by NASA Nov. 9, was reported successful by NASA. Frogs had completed 72 hrs of flight, during which primary and secondary objectives had been met. Since mission was so successful experiments would be continued few days past originally planned three to five days. (NASA Release 70-199)
- Soviet supersonic transport Tu-144 was tested for first time near its 2400-km-per-hr (1500-mph) maximum rated speed. Aircraft had reached 2330 km per hr (1448 mph) at 15 500-m (51 000-ft) altitude. (UPI, *NYT*, 11/17/70)
- NASA announced selection of experiments for ATS-F Applications Technology Satellite, scheduled for launch in 1973: very-high-resolution radiometric experiment; position-location and aircraft communications experiment (PLACE); television relay using small terminals (TRUST); radio-frequency-interference measurement experiment; 10.6μ laser experiment; millimeter-wave-propagation/communication experiment; data-relay experiment (Nimbus); radio beacon experiment; low-energy electron-proton experiment; measurement of low-energy protons; solar cosmic rays and geomagnetically trapped radiation; particle-acceleration mechanisms and dynamics of outer trapping region; magnetometer experiment; auroral-particle experiment; solar-cell-radiation flight experiment; flight demonstration of self-adaptive precision-pointing spacecraft attitude-control systems; and cesium bombardment ion engine experiment. (NASA Release 70-194)
- Proposal for space station to be assembled in orbit one module at a time was being studied at MSFC and MSC, NASA announced. Station, to provide centralized facility for research, applications, and operations for 6 to 12 men over 10 yrs, would be composed of cylindrical modules 4.3 m (14 ft) in diameter and 17.7 m (58 ft) long, joined to form variety of shapes. Individual modules would be carried into orbit on reusable shuttle vehicle. Station could be assembled within

months in circular orbit 320-480 km (200-300 mi) high with 55° inclination. (NASA Release 70-192)

- Achievement described as "first artificial synthesis of a living cell" was reported by State Univ. of New York at Buffalo in letter to press signed by Dr. Raymond H. Ewell, Vice President for Research, and Dr. James F. Danielli, Director of Center for Theoretical Biology, who had led experimenters. Amoebas had been partly dismembered and then reassembled, using components from other amoebas. Reconstituted amoebas had reproduced themselves and were indistinguishable from normal amoebas. Work opened "new era for artificial life synthesis, now being explored" and helped clear way for "synthesis of new micro-organisms, new egg cells and an organism capable of living on Mars." *New York Times* said officials at NASA, which was supporting the work, had described experiment as "exciting" and "a big step." (Sullivan, *NYT*, 11/13/70, 1)
- Space Development Committee of Japan had requested equivalent of $67 918 900 in space appropriations for FY 1971, *Space Business Daily* reported. Increase was 63% over FY 1970 budget. (*SBD*, 11/12/70, 43)

November 13: NASA held 12th Annual Honor Awards Ceremony in Washington, D.C., with Dr. George M. Low, Acting NASA Administrator, as keynote speaker. Vice President Spiro T. Agnew presented NASA Distinguished Service Medal to Dr. Thomas O. Paine, former NASA Administrator; *Apollo 11* Astronauts Neil A. Armstrong, Michael Collins, and Edwin E. Aldrin, Jr.; *Apollo 12* Astronauts Charles Conrad, Jr., Richard F. Gordon, and Alan L. Bean; and *Apollo 13* Astronauts James A. Lovell, Jr., John L. Swigert, Jr., and Fred W. Haise, Jr.

Vice President said award ceremony was "perhaps the most significant since NASA was founded. Those individuals we honor have accomplished goals that required over a decade of vision, hard work, determination and courage by thousands in the NASA, industry, and university research and development team.... NASA has had an outstanding beginning, successfully meeting its challenging mission and goals.... Though some reordering of national priorities and some fiscal constraints will not permit us to do all the things we would like to do in the immediate future, the future still holds many challenges and wonderful opportunities.... I am proud of you, the President is proud of you. We salute you."

Dr. Low said NASA could look forward to new achievements in program designed "to maintain that hard-earned lead, but one that also recognizes the many other problems that we as a nation must face. Today, we are moving forward on all...fronts—we have accepted the challenge of the future—a challenge in aeronautics, a challenge in science, a challenge in applications, and above all, a challenge in the exploration of the unknown."

Presidential Medal of Freedom was presented to *Apollo 11* and *Apollo 13* astronauts and to *Apollo 13* Mission Operations Team.

Herbert W. Grandy, Stationary Engineer at GSFC, received Exceptional Bravery Medal for responding quickly and "without regard for personal safety" to smoke alarm in GSFC building housing satellites, computers, and clean rooms—preventing loss of thousands of dollars and sustaining serious injuries.

Group Achievement Award was presented to LeRC Dynamic Power Systems Group; FRC, ARC, and LaRC Lifting-Body Team; Wallops Station Solar Eclipse Sounding Rocket Project; and USN Solar Eclipse Underwater Salvage Team Associated with Recovery of Payload.

Outstanding Leadership Medal was presented to Robert L. Krieger of Wallops Station and to former ERC Director James C. Elms.

Other awards included Exceptional Service Medal to 18 persons and Exceptional Scientific Achievement Medal to 17. (Program; NASA Release 70-196; *NASA Activities*, 12/15/70)

- NASA launched two sounding rockets. Arcas launched from Ft. Sherman, Canal Zone, Panama, carried GSFC payload to 60-km (37.3-mi) altitude to obtain ozone measurements in conjunction with *Nimbus IV* satellite overpass. Rocket and instruments functioned satisfactorily. Launch was second in two-rocket series; first had been launched Nov. 6.

  Nike-Tomahawk was launched from Andoeya, Norway, carrying Norwegian experiment to conduct auroral studies. Rocket and instruments functioned satisfactorily. (NASA Rpt SRL; SR list)

- Influence of high-altitude aircraft on stratospheric ozone by adding water vapor from exhausts was predicted in *Science* by Boeing Co. scientist Halstead Harrison: "Simple, steady-state models for ozone photochemistry, radiative heat balance, and eddy-diffusive mass transport can be combined to estimate water-induced changes in the stratospheric ozone concentrations and temperatures, the integrated ozone column, the solar power transmitted to the earth's surface, and the surface temperature." He found: "With added water from the exhausts of projected fleets of stratospheric aircraft, the ozone column may diminish by 3.8 per cent, the transmitted solar power increase by 0.07 per cent, and the surface temperature rise by 0.04 K [0.072°F] in the Northern Hemisphere." Temperatures in lower stratosphere would remain essentially unchanged. (*Science*, 11/13/70, 734-6)

November 15: Comment of Dr. Paul W. Gast, Chief of MSC Lunar and Earth Sciences Div., on learning of chemical composition of lunar samples brought from Sea of Fertility by Soviet *Luna XVI* spacecraft was quoted by *New York Times*. Dr. Gast had said Soviet analysis of material, reported at Oct. 28 Moscow news conference, indicated same relative abundances of some 11 major elements as found by U.S. scientists in *Apollo 12* lunar samples from Ocean of Storms. Both differed significantly from relative abundances found in Sea of Tranquility rocks and soil returned by *Apollo 11*. Most striking difference in Sea of Fertility and Ocean of Storms material was titanium content. Sea of Tranquility material contained 10% titanium dioxide; all soil and rocks from Ocean of Storms and Sea of Fertility contained under 5%. Dr. Gast said Sea of Fertility sample provided 50% additional data to that known previously on chemical composition of lunar surface. It was important and remarkable that analyses of Ocean of Storms and Sea of Tranquility data were so similar though sites were more than 1600 km (1000 mi) apart. (Schwartz, *NYT*, 11/15/70)

- Document circulated in Moscow announced formation of Committee for Human Rights by Soviet scientists Andrey D. Sakharov, V. N.

Chalidze, and A. N. Tverdokhlebov. Purposes were "to help the organs of state power create and apply guarantees of human rights"; assist persons who "want to conduct constructive research into the theoretical aspects of the problem of human rights and the study of the specifics of the problem in a socialist society"; and provide legal instruction, "specifically propaganda" for documents of international and Soviet law on problem of human rights. Committee would offer "constructive criticism of... system of legal guarantees for the freedom of the personality in Soviet law." (Astrachan, *W Post* 11/16/70, A1)

- Nation's "real brain drain" was described in *New York Times* editorial: "Large numbers of scientists, engineers and educators are being shaken out of their jobs in an upheaval that must be viewed as something far more serious than a temporary dislocation in employment, born of the business slowdonwn. . . .The deeply troubling fundamental question is whether a nation can allow itself to be the pawn rather than the master of its destiny. . . . The inevitable conclusion, if the downgrading and disuse of highly educated manpower gathers force, will be that advanced skills are relevant only to competition for hegemony in arms or space. Such a concept is absurd when there is such apparent need for the application of scientific, technological and philosophical brain power in the pursuit of peace and dignified living conditions." (*NYT*, 11/15/70, 4:10)

November 15-17: ARC meeting on planet habitability and cell synthesis was attended by 20 scientists, including Dr. James F. Danielli, whose team at State Univ. of New York at Buffalo had reported first artificial synthesis of living cell [see Nov. 12]. In statement to press NASA said agency was considering "planetary engineering," creation of new life forms to be placed on Mars and other planets. (UPI, *LA Times*, 11/15/70; ARC PIO)

November 16: Lockheed L-1011 trijet airbus made maiden flight from Lockheed-California Co. facility at Palmdale, Calif. Large-capacity jet designed for medium- and long-haul operations attained planned speed of 478 km per hr (297 mph) at 6100-m (20 000-ft) altitude during 2-hr 25-min flight. (*Av Wk*, 11/23/70, 16-8; AP, *NYT,* 11/17/70, 77; Pres Rpt 71)

- FAA announced publication of new taxiway system design standards to increase airport capacity by speeding ground movements of large aircraft, including Boeing 747, McDonnell Douglas DC-10, and Lockheed L-1011. (FAA Release 70-100)

November 16-19: AIAA and AAS held related conferences on solar physics in Huntsville, Ala., and participants toured MSFC. (MSFC Release 70-233)

November 16-20; EVA critical design review was held at Skylab mockup area and Neutral Buoyancy Simulator (NBS) at MSFC. Astronaut Russell L. Schweickart headed 10-astronaut team performing under normal earth gravity in Saturn Workshop mockup and NBS. (MSFC Release 70-235)

November 17: U.S.S.R. launched *Cosmos CCCLXXVIII* from Plesetsk into orbit with 1723-km (1070.6-mi) apogee, 234-km (145.4-mi) perigee, 104.6-min period, and 74.0° inclination. (GSFC *SSR*, 11/30/70; *SBD,* 11/20/70, 86)

- Lunar roving vehicle training model, 1-G trainer, was delivered by Boeing Co. to MSFC representatives at Santa Barbara, Calif., marking major milestone in manned LRV program. Two-man trainer was 3.2 m (10 ft 7 in) long and 1.8 m (6 ft) wide with 2.3-m (7.5-ft) wheelbase. It looked like flight version scheduled for use on Apollo 15, 16, and 17 missions, but weighed twice as much and was equipped with both automobile-type and wire mesh tires. Extra weight came from additional strength in frame, wheels, drive motors, and suspension system, which permitted it to carry 362.9 kg (800 lbs), including two men, equipment, and samples. Battery-operated trainer was driven with an aircraft-like stick hand controller and could stop, turn, move forward, and reverse at variable speeds. Top speed was 14.5 km per hr (9 mph) on smooth surface. Trainer would be shipped to MSC in about one week. (NASA Release 70-201)
- Fire in KSC rocket gantry extensively damaged "clean room" under reconstruction at Launch Complex 17, Pad B, one of two pads used to launch unmanned spacecraft by Delta boosters. There were no injuries. (*W Post*, 11/18/70, A8)
- U.N. General Assembly approved, by vote of 91 to 2, treaty to prohibit placement of nuclear weapons on seabed outside any nation's 19.3-km (12-mi) limit. Treaty, which required ratification by 22 governments, was outcome of U.S.-U.S.S.R. agreement. *New York Times* later reported diplomats had likened treaty to 1967 pact barring nuclear weapons from space or celestial bodies. (Teltsch, *NYT*, 11/18/70, 1)
- U.S. and Soviet delegates to SALT observed first anniversary of meetings in Helsinki, Finland. (*W Post*, 11/18/70, A27)
- Newly elected Sen. Adlai E. Stevenson III (D-Ill.) was assigned to serve on Senate Committee on Aeronautical and Space Sciences in seat left vacant by retiring Sen. Ralph T. Smith (R-Ill.). (NASA *LAR*, IX/121; Stevenson Off)

November 17-19: Skylab Subsystems Review Team meeting was held at MSFC. Team—chaired by Charles W. Mathews, NASA Deputy Associate Administrator for Manned Space Flight and Acting Space Station Task Force Director—inspected Skylab mockup area, toured simulation facilities, and viewed assemblage of ATM hardware. (MSFC Release 70-235)

November 18: USAF launched two unidentified satellites from Vandenberg AFB on one Thor-Agena D booster. First entered orbit with 225.9-km (140.4-mi) apogee, 177-km (110-mi) perigee, 88.5-min period, and 82.9° inclination and reentered Dec. 11. Second entered orbit with 511-km (317.5-mi) apogee, 486-km (302-mi) perigee, 94.5-min period, and 83.1° inclination. (Pres Rpt 71; GSFC *SSR*, 11/30/70; 12/31/70; Baltimore *News-American*, 11/19/70)
- News conference on results of *Apollo 12* lunar surface experiments was held at NASA Hq. Dr. Rocco A. Petrone, Apollo Program Director, said first anniversary of ALSEP was "scientific and technological triumph. The design specification was for one lunar day and night. Our design goal was for one calendar year." Year after deployment ALSEP had received and processed over 6500 commands, had undergone 2600 533-K (500°F) swings of temperature, and was still being operated remotely 24 hrs a day, 7 days a week.

Astronaut Alan L. Bean said unmanned vehicle, such as U.S.S.R.'s *Lunokhod I,* could not explore surface as well as men could because men could deploy more equipment more effectively, perform more experiments, and collect more samples. He noted that *Apollo 11* had collected 22.7 kg (50 lbs) of samples; *Apollo 12,* 34 kg (75 lbs); and *Luna XVI,* 90 g (0.2 lb).

Dr. Gary V. Latham of Columbia Univ.'s Lamont-Doherty Geological Laboratory said seismic events were being recorded at rate of about one per day; some were impacts from grapefruit-size meteoroids and some appeared to be small moonquakes. In region of *Apollo 12* station there were at least nine places at which moonquakes occurred "every month at or near the time when the moon comes closest to earth in its monthly orbital cycle." Most active zone for moonquakes was about 193 km (120 mi) southeast of ALSEP, near rilles in Fra Mauro crater, Apollo 14 landing site.

Dr. Charles P. Sonett of ARC said data indicated moon's temperature increased very slowly with depth below surface, probably rising no higher than 1273 K (1332°F)—too cool for molten core characteristic of volcanic body.

Dr. Conway M. Snyder of JPL described behavior of thin cloud of ionized gas detected after ascent stage was crashed into moon. Solar wind of ionized particles should have blown the gas downwind; it traveled upwind and was detected by ALSEP instruments. Cause of acceleration of the particles was unknown. (Transcript; Schmeck, *NYT,* 11/19/70, 32; O'Toole, *W Post,* 11/19/70, A14)

- Nike-Tomahawk sounding rocket was launched by NASA from Fairbanks, Alaska, carrying Univ. of Minnesota experiment to conduct auroral studies. Rocket and instruments functioned satisfactorily. (SR list)
- Flight model of Skylab multiple docking adapter was flown from MSFC to Martin Marietta Corp. Space Center in Denver, Colo., aboard Super Guppy aircraft. It would be outfitted with controls and display panels for solar astronomy and earth resource experiments, storage vaults for experiment film, and thrust-attitude control system. When completely equipped, adapter would be mated with Skylab airlock flight version at McDonnell Douglas Astronautics Co. in St. Louis, Mo., and unit would perform simulated mission in altitude chamber. (MSFC Release 70-237; MSFC PIO)
- *Apollo 12* Astronaut Alan L. Bean received special citation from AEC, commemorating first anniversary of deployment of nuclear-powered generator on moon Nov. 19, 1969, during *Apollo 12* mission. Bean accepted similar citations for fellow *Apollo 12* crewmen Charles Conrad, Jr., and Richard F. Gordon, Jr., during ceremonies at AEC Hq. in Germantown, Md. (AEC Release N-203)
- *Apollo 11* Astronaut Neil A. Armstrong received award for first lunar landing at second international congress of World Wildlife Fund in London. He said earth was oasis of life that must be protected "from its own population." Wielders of modern technology were not deliberately destructive of earthly environment, but they were as apathetic as most humans about consequences of their actions. Time had come for them to be concerned. (Lewis, *NYT,* 11/19/70, 4)

- NASA announced that Charles J. Donlan, Deputy Associate Administrator for Manned Space Flight (Technical), would also become Acting Director of Space Shuttle Task Force, replacing Dale D. Myers, Associate Administrator for Manned Space Flight. Appointment was effective immediately. (NASA Ann, 11/18/70)
- *Wall Street Journal* quoted USAF as saying Lockheed Aircraft Corp.'s C-5A cargo aircraft was undergoing additional tests and modifications costing approximately $28 million over three years, "to comply with the intent" of recommendations of C-5 Review Board released June 17. Work included $15.9-million fatigue-test program on wing and engineering study of alternate measurement device in navigation system. (*WSJ*, 11/18/70)
- Editorials commented on U.S.S.R.'s *Luna XVII.*

    New York Times: "In a rational world, this task of lunar exploration would be taken on by an international agency using the resources and capabilities of all nations. But so far, unfortunately, the grandeur of the challenge has not elicited any far-sighted statesmanlike response from the community of nations. Man is still a petty, earthbound creature with narrow, nationalistic horizons." (*NYT*, 11/18/70)

    Baltimore *Sun:* "Such information as Luna 17 gathers will, in the long run, be less important to the future than its demonstration of the feasibility of sending out a space ship that can release a component which moves, works and returns, all on man's orders but without a crew." (B *Sun*, 11/18/70)

November 18-19: Advances in propulsion were discussed at LeRC conference by 420 representatives of U.S. and foreign aircraft engine industry: NASA fan and compressor technology program was providing machines and tests to demonstrate high efficiency, light weight, appropriately wide operating limits, low noise, and high tolerance to flow distortions in fans and compressors for wide range of airbreathing propulsion applications. LeRC program encompassed studies with experimental rotors and their applications to fans and multistage compressors. Program objective was high pressure ratio per stage to produce lighter, more compact engines. (*Lewis News*, 12/4/70, 1)

November 19: House Appropriations Committee favorably reported H.R. 19830, FY 1971 Independent Offices and HUD appropriations bill containing $3.269-billion NASA appropriation. Bill provided $2.565 billion for R&D, $25 million for construction of facilities, and $679 million for research and program management. It replaced H.R. 17548, vetoed by President Nixon Aug. 11. NASA appropriation was unchanged. (*CR*, 11/19/70, D1194; H Rpt 91-1616)

- George M. Low, Acting NASA Administrator, addressed Eighth Annual Briefing on New Horizons in Science of Council for Advancement of Science Writing in Raleigh, N.C.: "While many of today's ills can be attributed to our use of modern technology, the long cure —and more important, the prevention of future ills—can only be based upon facts not now well established, and in many cases even unknown. That requires a search for knowledge. That is science, and the application of science, on earth and in space. That was NASA's mission before men began to state their ecological concerns in social, political, and economic terms. And it will be NASA's mission long after the air is clear again, and the lakes and rivers pure again, and the balance of man and nature restored more nearly to the conditions man's sense of values deems necessary." (Text)

- SNAP-27 radioisotope thermoelectric generator carried to moon on board *Apollo 12* (Nov. 14-24, 1969) completed one year of operation on lunar surface. SNAP-27 weighed 19.5 kg (43 lbs) and was designed to provide 63.5 w electrical power for one year. Scientists now expected lifetime to be five or six years. (*SBD*, 11/18/70, 68)

- U.S. laser reflector placed in moon's Sea of Tranquility by *Apollo 11* crew was measuring variations in distance between site and McDonald Observatory in Texas with accuracy within less than 0.3 m (1 ft), *New York Times* quoted Univ. of Maryland scientist Dr. Carroll O. Alley as reporting. Accuracy was determined by shortness of pulses and accuracy with which travel time of pulses to moon and back could be measured. New lasers being developed might cut accuracy margin to less than 25 mm (1 in) and could produce reflections bright enough to be observed in telescopes of small observatories. Dr. Alley was chairman of committee of international specialists seeking to encourage widespread ranging of lunar reflectors such as French laser reflector landed on moon by Soviet *Luna XVII*. (Sullivan, *NYT*, 11/19/70)

- MSFC announced that Dr. Leslie W. Ball, formerly Director of Reliability and Director of Product Assurance for Boeing Co.'s Aerospace Group, had been appointed Assistant to MSFC Deputy Director, Technical. (MSFC Release 70-238)

- MSFC awarded McDonnell Douglas Astronautics Co. $2-million, one-year modification to space shuttle study contract. Under new agreement McDonnell Douglas would recommend program to test all structural components of proposed space shuttle's booster and orbiter craft, including verification of design and advanced state-of-the-art testing of materials, wing leading edges, and propellant tanks.

    MSC, which had companion space shuttle study under way, was negotiating similar amendment with NR. (MSFC Release 70-239)

- Some 750 scientists—including six Nobel Prize winners—from 138 universities, medical schools, hospitals, and research institutes in 35 states and District of Columbia had signed petition against threat to their laboratory work in House-passed defense and industry security bill, *Washington Post* reported. Bill, which cleared House in January by 274-to-65 vote, would empower President to institute personnel security screening program to determine individual's eligibility for access to sensitive information. (*W Post*, 11/19/70, A27)

- *Philadelphia Inquirer* editorial commented on *Luna XVII:* "It is a time for congratulations to the Soviet effort this time. . .and perhaps time to take a lesson. For future probable explorations to lethal Mars, Venus and other planets almost certainly cannot employ manned landings. All the same, there is no reason to abandon plans NASA has made to land men and a wheeled Rover vehicle they can drive on the moon. There is always the possibility that the man on the scene can pick up more than back-seat drivers 240,000 miles [386 000 km] away." (*P Inq*, 11/19/70)

- Marshal Nikolay Krylov, Commander of U.S.S.R.'s strategic rocket forces, said in *Pravda* that forces had carried out "exceptionally important" military construction "in the last few years."

Equipment in all branches of service was being continuously improved for "final victory in contemporary war." Strategic rocket troops would become "main shock force" in nuclear war. (Starr, *C Trib*, 11/20/70, 12)

- *November 20:* USAF's X-24A lifting-body vehicle, piloted by Maj. Jerauld R. Gentry (USAF) in joint NASA-USAF program, successfully completed 20th flight at FRC. X-24A reached 21 300-m (70 000-ft) altitude and mach 1.33 in powered flight after launch from B-52 aircraft at 13 700-m (45 000-ft) altitude. Objectives were to obtain lateral directional derivatives, obtain longitudinal trim and lift-to-drag data with 40° upper flap, and determine effect of upper flap on approach and landing. (NASA Proj Off)
- NASA launched two Nike-Apache sounding rockets. Rocket launched from WSMR carried Univ. of Pittsburgh experiment to study atmospheric composition. Mission did not meet minimum scientific requirements. Rocket launched from Eglin AFB, Fla., carried Univ. of Michigan experiment to study atmospheric structure. Rocket and instruments functioned satisfactorily. (SR list)
- Ten MSC employees had applied for joint patent on device and process for recovering water and its constitutent elements from lunar soil, NASA announced. Chemical process used hydrogen and solar energy to reduce iron oxides and produce water vapor, which could be electrolyzed to yield oxygen and hydrogen. Calculations showed process could produce 0.5 kg (1 lb) of water from 45.4 kg (100 lbs) of lunar soil; if iron oxides were first concentrated magnetically, yield increased to nearly 6.4 kg (14 lbs) of water from 45.4 kg (100 lbs) of lunar material. (NASA Release 70-202)
- *Washington Post* editorial commented on Soviet *Lunokhod I:* "The appearance of this vehicle, as well as the landing on the moon and return to earth of Luna 16, makes it clear that the Russians hope to regain some of the prestige they lost in the race to land men on the moon.... In terms of the American space program, this new success of the Russians should have little impact other than to persuade the administration and Congress not to cut back its funds any further. There is no logical reason for NASA to go like gang-busters in an effort to match the Russians in unmanned space exploration since the decision has already been made to proceed with a scaled down and more rational space program. The first big trophy in space activities went to the Russians, the most recent big one to the Americans, and it is reasonable to suspect that future ones will be divided between the two nations." (*W Post*, 11/20/70, A26)
- *Science* editorial commented on comsats: The communications satellites have a variety of applications. . . . One application that will probably be implemented fairly soon is aeronautical and ship communication. Radio works well on line of sight but is undependable at distances greater than 300 miles [483 km]. Many ships have been lost while radio operators were trying in vain to communicate their peril. Midocean aeronautical navigation represents another growing need which could also be filled. Satellite enthusiasts point to other potential uses. One of these is Picturephone, which now requires about 100 telephone circuits for a single two-way conversation. High-capacity, low-cost satellite circuits would help to make feasible large-scale use of this device. Perhaps more distant but of great

consequence would be exploitation of future inexpensive communications for use in educational television. Constructive applications of satellites are now a reality, and extension of technological potentials will have profound and global consequences." (*Science*, 11/20/70)

November 21: Initial plans for international docking system that would eliminate current "key-in-lock" approach of U.S. and Soviet space programs had been presented by NASA representatives at Oct. 26-27 U.S.-U.S.S.R. space cooperation meetings in Moscow, Walter Sullivan said in *New York Times*. U.S.S.R. had reciprocated with detailed plans of docking system similar to Apollo's. "While the proposed American scheme is a long-term project," NASA engineers believed "the Apollo system and the new Soviet one are enough alike to be modified for compatibility." (*NYT*, 11/21/70)

November 22: Apollo 11 Astronaut Neil A. Armstrong and *Apollo 12* Astronaut Charles Conrad, Jr., received gold medal from Fédération Aéronautique Internationale in New Delhi for promoting peaceful space exploration. (AP, *H Post*, 11/22/70)

• Pioneering efforts by Univ. of Rochester and other U.S., Soviet, and French laboratories to use lasers to control thermonuclear fusion were described in *New York Times* by Harry Schwartz. Experiments to tame hydrogen bomb's thermonuclear reaction to provide limitless, cheap, and pollution-free electricity had begun about 1950 before laser was invented. "But two decades of hard work have shown that had been anticipated. Now the laser is providing an alternative." (*NYT*, 11/22/70, 4:8)

November 23: USAF satellite launched from ETR Nov. 6 "did not achieve the desired orbit and for all accounts and purposes is of no practical use," *Aerospace Daily* reported. "Reliable sources said the Transtage. . .did not achieve the second burn which would circularize the transfer elipse at synchronous altitude—21,000 miles [33 800 km]. . .where it would trace a figure eight pattern over a specific portion of the earth. Instead. . .the spacecraft is in a highly elliptical orbit" in which ground stations "could do little more than check out a few on-board housekeeping systems." (*Aero Daily*, 11/23/70)

• NSF released figures on slackening growth rate of R&D activities in independent nonprofit institutions during 1966-1969. Intramural R&D expenditures totaled $845 million in 1969, increase of 6% per year over 1966 figure; annual rate of increase during 1953-1966 had been 16.3%. Major factor accounting for R&D growth during 1953-1966 and slackening of growth since 1966 was Federal R&D financing. (NSF *Highlights*, 11/23/70, 1)

November 24: U.S.S.R. launched two Cosmos satellites. *Cosmos CCCLXXIX,* from Baykonur, entered orbit with 14 026-km (8715.4-mi) apogee, 176-km (109.4-mi) perigee, 259.5-min period, and 51.6° inclination.

Cosmos CCCLXXX, launched from Plesetsk, entered orbit with 1491-km (926.5-mi) apogee, 198-km (123.0-mi) perigee, 101.7-min period, and 81.9° inclination. (GSFC *SSR*, 11/30/70; *SBD*, 11/25/70, 106; 11/30/70, 114)

• NASA launched two rockets from Wallops Station. Four-stage Trailblazer II launched at 6:56 pm EST carried 34-kg (75-lb) AFCRL payload to 321.9-km (200-mi) altitude to measure plasma effects upon microwave systems during reentry. Launch supported USAF program to study techniques for improving transmission of radio signals from aerospace vehicles during reentry. Rocket and instruments functioned satisfactorily and payload reentered at 5.5 km per sec (18 000 fps).

Three-stage, solid-propellant rocket, launched at 10:17 pm EST, carried 103.4-kg (228-lb) NRL payload to 1600-km (1000-mi) altitude before it impacted in Atlantic 25 min after launch. Primary objective was to measure, over wide range of wavelengths, the intensity, spatial distribution, and altitude variation of the diffuse far-UV radiation of the night sky and thereby determine its source. Secondary objective was to measure, over a wide range of wavelengths, the far-UV radiation from stars and other discrete sources. (WS Releases 70-21, 70-22)

- Aerobee 150 sounding rocket launched by NASA from WSMR carried American Science and Engineering, Inc., experiment to conduct solar x-ray studies. Rocket and instruments functioned satisfactorily. (SR list)
- NASA's thick supercritical test wing, mounted on USN T2-C Buckeye aircraft, was flight-tested for first time on flight from NR plant at Columbus, Ohio. New airfoil was test version of model that might permit structural weight savings on moderate-speed aircraft. Thin version of airfoil, aimed at maximum-speed performance objectives, was being fitted to modified F-8 jet for spring 1971 tests at FRC [see Nov. 3]. (NASA Release 70-204)
- House passed, by vote of 375 to 10, H.R. 19830, $17.7-billion Independent Offices and HUD FY 1971 appropriations bill that contained $3.269 billion NASA appropriation. NASA appropriation was $64.3 million below Administration's request of $3.333 billion. (CR, 11/24/70, H10711-24)
- Senate Committee on Appropriations approved H.R. 17755, $2.7-billion DOT FY 1971 appropriation bill after incorporating into bill $290 million for further construction of two SST prototypes. (CR, 12/24/70, D1189; NYT, 11/24/70, 65; Text)
- SAMSO had proposed development of USAF System 621B, navigation system employing several clusters of satellites and system of ground stations, ASFC announced. System would provide three-dimensional position and velocity information for users, reduce equipment weight problem, and solve military logistics problem through uniformity. SAMSO had awarded eight contracts totaling more than $1.5 million for additional verification of system design concepts and for mathematical model of ionosphere. (AFSC Release 263.70)
- LeRC and Northern Ohio section of American Nuclear Society held combined meeting at LeRC. Dr. Abe Silverstein, Director of Environmental Planning for Republic Steel Corp. and of Lake Erie International Jetport Project, discussed technology and society. Dr. Silverstein had retired as LeRC Director in 1969. (LeRC Release 70-69)
- Washington *Evening Star* editorial commented on Nov. 12 confirmation by JPL radioastronomers of Einstein theory of relativity: "NASA scientists who have confirmed the general relativity theory as well as the critical scientists who have thought it to be in error, have alike based their calculations on the most sophisticated and expensive

November 24: NASA's thick supercritical test wing was flight-tested on a USN T2-C aircraft for the first time, in flight from the North American plant in Columbus, Ohio. Joint NASA-USN program was testing the new airfoil that might save structural weight on moderate-speed aircraft. Thin version of the airfoil for maximum-speed performance, fitted to a modified F-8 jet fighter (below), was delivered to the Flight Research Center Nov. 3 for test flights beginning in the spring of 1971.

pieces of modern equipment, from computers to the Mariner vehicles themselves. And they were operating in the light of the theory itself. Einstein, drastically revising received thought, created the theory with a blackboard and a piece of chalk and a mind able to talk to Isaac Newton and correct him." (W Star, 11/24/70)

November 25: NASA's M2-F3 lifting-body vehicle, piloted by NASA test pilot William H. Dana, completed fourth flight—and first powered flight—from FRC. Objectives were to expand flight envelope to mach 0.8, check out engine operation, obtain stability and control data at mach 0.8, investigate lateral phugoid present on second flight but not on third, check reaction control operation, and investigate landing visibility. Engine shut down prematurely, after launch from B-52 aircraft, precluding completion of all planned maneuvers. Investigation was under way. (NASA Proj Off)

- Significance of Soviet flights to moon was discussed by Dr. George M. Low, Acting NASA Administrator, in letter to Sen. Clinton P. Anderson (D-N. Mex.), Chairman of Senate Committee on Aeronautical and Space Sciences. *Luna XVI* and *XVII*, launched Sept. 12 and Nov. 10, had been "technically impressive; but as isolated events, their importance to science and technology is relatively minor." U.S. still held technological leadership with *Apollo 11*. U.S. had had four men on moon and U.S.S.R. none. U.S. had returned 56 kg (123 lb) of lunar material and U.S.S.R. had returned 85-113 g (3-4 oz). U.S. had placed automated station on moon that had functioned for more than year. U.S. had gained "enormous amount" of lunar information; U.S.S.R. had added very little.

  In overall trends, however, U.S. launch rate was decreasing while Soviet rate was increasing. In 1970 to date, U.S.S.R. had placed 74 payloads in space, to 31 by U.S. "They have seldom missed a launch window in the exploration of Venus and Mars; our planetary program has many gaps." U.S.S.R. had continuing manned space program and capability, with Zond spacecraft, for manned circumlunar flight. Soviets had stated intention to orbit space station before U.S. and reportedly were developing booster of Saturn V class while U.S. had suspended Saturn V production. U.S.S.R. was spending more money on space annually than U.S. "...when we view Luna 16 and 17 in the context of trends in our program and theirs, then we must be concerned about our future in aeronautics and space—about our position of leadership that we have worked so hard to achieve." (CR, 11/30/70, S19002)

- Dr. Theos J. Thompson, member of AEC and professor of nuclear engineering at MIT, was killed in aircraft near Las Vegas. Dr. Thompson has received AEC's Ernest O. Lawrence Memorial Award for outstanding contributions to nuclear research in 1964. (NYT, 11/26/70, 78; Who's Who, 1970-1971)

November 26: Broken section of propeller spinner from *Spirit of St. Louis*— aircraft in which Charles A. Lindbergh made first solo crossing of Atlantic May 20, 1927— was donated to Smithsonian Institution by Stanley I. Vaughn of Columbus, Ohio. Truncated metal cone 508 mm (20 in) in diameter and weighing 1.5 kg (3¼ lb) had been replaced just before takeoff of transatlantic flight after separating from spinner during preliminary flight from San Diego, Calif., to Roosevelt Field, N.Y. Vaughn, then factory manager for

Curtiss Aeroplane and Motor Co. plant in Garden City, N.Y., had furnished new spinner for Lindbergh aircraft and kept broken one 43 yrs. (AP, *NYT*, 12/9/70, 46)

November 27: U.S.S.R. launched *Molniya I-16* communications satellite into orbit with 39 359-km (24 456.6-mi) apogee, 467-km (290.2-mi) perigee, 707-min period, and 65.4° inclination. (GSFC *SSR*, 11/30/70)

- Tass reported U.S.S.R.'s *Lunokhod I* unmanned lunar explorer had been transmitting satisfactorily during lunar night from its position on Sea of Rains at 38° 17' north latitude and 35° west longitude. Vehicle was positioned with reflector cover open to ensure reflector's operation for laser studies. Laser reflector had been built by France under cooperative agreement. (*Sov Rpt*, 12/4/70, 7)
- Work of scientists using NASA's *Oao II* (launched Dec. 7, 1968) and GSFC team to investigate UV radiation from stars was described in *Science*. They had mapped large areas of sky in UV, discovered larger amounts of UV radiation than previously expected, and derived detailed information that would lead to "better understanding of stellar genesis, evolution, and decay." (Hammond, *Science*, 11/27/70, 960-1)

November 28: Vertikal I high-altitude geophysical probe launched by U.S.S.R. carried recoverable payload containing experiments from six Communist countries to 487.6-km (303-mi) altitude. Equipment included x-ray cameras and x-ray spectrogeography experiment designed and built in Poland; x-ray spectrometers designed and made in U.S.S.R.; meteor particle experiment designed and fabricated by Hungary, Czechoslovakia, and U.S.S.R.; altophotometer and radiofrequency pulse probe designed and fabricated in East Germany; and solar radiation instruments provided by Bulgaria, East Germany, U.S.S.R., and Czechoslovakia. (*SBD*, 12/3/70, 131; *Sov Rpt*, 12/4/70, 5)

November 29: Observations of more than 100 000 photos of 21 satellites made from 19 cameras and tracking stations of Smithsonian Astrophysical Observatory had shown that satellites bobbed and dipped slightly in earth orbit, *New York Times* reported. Erratic motions were caused by variations in earth's gravitational field. Satellites were attracted and pulled downward more as they passed over high areas and attracted less over low areas. Scientists had concluded that if earth were round orbits would be perfectly elliptical. Instead, bulges and depressions covered earth in two bands—high band with neck across North Pole and ends stretching over Europe and West Africa on one side and Pacific Ocean on other; low band over South Pole with ends over North America and Asia. (*NYT*, 11/29/70, 24)

November 30: NASA's OAO-B Orbiting Astronomical Observatory failed to reach orbit after launch from ETR by Atlas-Centaur launch vehicle at 5:40 pm EST, apparently because nose-fairing failed to separate from vehicle. Additional weight of nose fairing prevented Centaur from reaching necessary orbital velocity and satellite impacted shortly after completion of Centaur burn. Failure investigation board would be established by LaRC.

Primary objective of OAO-B was to obtain moderate resolution spectrophotometric data in UV bands between 1100 and 4000 Å to investigate photometry of peculiar stars, law of interstellar reddening, magnitude and intensity of Lyman-alpha red shift for nearby galaxies, spectra of emission and reflection nebulas, and spectral energy distribution of normal stars, galaxies, and intergalactic media. OAO-B at 2123 kg (4680 lbs) was largest U.S. satellite to date and carried world's largest telescope—914-mm (36-in) Goddard experiment package—which could make detailed measurements and observations previously unobtainable in space astronomy. OAO-B was third in series of four observatories planned by NASA. *Oao II*, launched Dec. 7, 1968, was still operating satisfactorily. (NASA Release 70-174; NASA Proj Off)

- RAM C-III launched by NASA Sept. 30 was adjudged successful by NASA. Vehicle had achieved planned trajectory, velocity, and data return and had impacted within planned target area. During normally blacked-out period of reentry both VHF and X-band telemetry signals were improved by injection of fluid. Highly satisfactory data covering blackout period were transmitted after RAM C-III emerged from blackout. (NASA Proj Off)
- NASA discontinued use of Guaymas-Empalme station in Mexico for space observations because of changes in Apollo mission profile. Station would be operated by Mexican government for Mexican scientific programs. NASA and Mexican National Commission for Outer Space (CNEE), which had cooperated in operation of station, would continue cooperation in developing system approach to use of weather data from U.S. weather satellites acquired by automatic picture transmission (APT) cameras. They also would develop capabilities and applications for earth observations using advanced, airborne, remote-sensing instrumentation. Plans were being completed for new cooperative project for synoptic meteorological sounding rocket launchings. With release of Guaymas-Empalme station, built in 1961, NASA had closed three ground stations in tracking and data acquisition network, as well as retiring three tracking ships and four aircraft, in past year. (NASA Release 70-198)
- NAS panel headed by physicist Dr. Thomas F. Malone of Univ. of Connecticut warned that large-scale deployment of SST might increase incidence of skin cancer if exhaust from aircraft thinned protective layer that separated earth's atmosphere from solar rays. Warning was issued during meeting with DOT's Office of SST Development in Washington, D.C. (Randal, W *Star*, 12/1/70, A11)
- USAF announced award of additional $78 million to General Dynamics Corp. to cover cost of overruns on F-111 aircraft and said further overrun payments were forthcoming. USAF had announced award of $191.3 million to General Dynamics Corp. Aug. 25 as final payment for cost overruns on production models of F-111. (*WSJ*, 12/1/70)

During November: USAF total-force concept for 1970s was described in *Air Force and Space Digest* by Dr. Robert C. Seamans, Jr., Secretary of the Air Force: Increased numbers and total payload of Soviet ICBMs and Soviet deployment of ABM system could reduce effectiveness of U.S. land- and sea-based missiles. "We are deploying ABM protection for our missile fields and strengthening the penetration capability of our missiles with the deployment of MIRV. But . . .effectiveness of our missile forces will tend to vary, depending upon the length of time it takes us to respond to new Soviet capabilities." Dispersed manned bomber force with quick reaction would provide stability. "It might be possible to undermine the effectiveness of either missiles or bombers alone, but to counter both at the same time would be a vastly more difficult problem." B-1 bomber represented most economically feasible means to retain stabilizing capability. "Given the decade of leadtime involved, we must expedite the development of this aircraft." With Soviet fighters "becoming superior to our own," air-to-air combat was "primary concern in our development of the F-15, which will have superior capability for close-in, highly maneuverable combat" and "provide the effective weapon system necessary to defeat any enemy fighter. ...we are convinced that effective close air support will continue to be a vital mission...and that an aircraft especially tailored for that role is required. ...we are proceeding with the development of the A-X. To modernize our airlift capability, we are relying heavily on the C-5, which will produce a revolution in air mobility." In 1973, "with the C-5 in the inventory, projections indicate that we will be able to move an Army division with equipment and six fighter squadrons with support units to Europe in less than one week." (AF/SD, 11/70, 68-71)

- *Atlas of the Universe* by Patrick Moore was published by Rand McNally & Co. Volume contained almost 1500 maps, drawings, and charts, many provided through cooperation of NASA and U.S. Geological Survey. Book had foreword by Sir Bernard Lovell, Director of U.K.'s Jodrell Bank Experimental Station, and epilogue by Dr. Thomas O. Paine, former NASA Administrator. (Gordon, *W News*, 11/16/70)

- Three engine control assemblies were removed from XB-70 research aircraft on exhibition at Air Force Museum at Wright-Patterson AFB, Ohio, to be used for F-15 engine/inlet compatibility tests in 5-m (16-ft) wind tunnels at Arnold Engineering Development Center at Tullahoma, Tenn. Equipment would be on loan through June 1972. (AFSC Release 333.70)

December 1970

December 1: Francis B. Smith, NASA Assistant Administrator for University Affairs, began working with Commission on Government Procurement as Vice Chairman of Study Group on Major Systems Acquisition. During six months that Smith worked with group, Dr. Frank D. Hansing would be Acting Assistant Administrator for University Affairs. (NASA Ann, 12/17/70)

• James J. Harford, AIAA Executive Secretary, defended SST in *New York Times* article and urged Federal funding for its development: "...if I were a Senator or an editor I might think hard about approving the $290 million appropriation if I thought the money and the 20 000 engineers and technicians were transferable to 1971 technological programs in pollution control, mass transit, housing, or another neglected problem. To our disgrace, there are no programs which have advanced to the point at which they can use those funds and engineers in 1971. If the $290 million is voted down, no one benefits." (*NYT*, 12/1/70, 47)

• U.S. was continuing to survey Suez Canal area periodically by earth-orbiting satellites, *New York Times* sources said. Meanwhile U.S. was keeping two U-2 high-level reconnaissance aircraft at British base of Akrotiri in Cyprus after stopping their flights over Suez area. Sources thought flights had been ended because U.S. believed that Egypt had completed buildup of antiaircraft missiles and likelihood of renewed fighting between Israel and Egypt was diminishing. (Welles, *NYT*, 12/2/70, 7)

• Aviatrix Ruth Law Oliver, exhibition flyer who set speed record in 1916 by flying Curtiss biplane 1094 km (680 mi) from Chicago to Binghamton, N.Y., in 6 hrs 7 min, died at age 85. Mrs. Oliver had been co-owner of Ruth Law's Flying Circus and had become known for being first woman to loop-the-loop and fly at night and for piloting aircraft over Washington, D.C., in 1917 as part of Liberty Loan campaign during World War I. (Hoffman, *W Post*, 12/4/70, B18)

December 1-2: Principal investigators of candidate experiments for High Energy Astronomy Observatory (HEAO) met with NASA officials and study contractors at MSFC. Meetings were part of preliminary design and definition study phase of program to prepare scientific carrier for astronomy experiments that would study universe from earth orbit after 1974 launch. (MSFC Release 70-248)

December 1-3: Third annual Earth Resources Survey Program Review was held at MSC. A.M. Woll of Bureau of Indian Affairs said Indians looked to NASA for help in management of their natural resources. He described project in which geologists had located possible water or mineral source by examining feature 6.4 km (4 mi) in diameter on photo taken by *Apollo 9* astronauts to assist Indians at San Carlos Reservation in eastern Arizona to plan water and mineral resources. Photo had shown strange circular formation. Indians also hoped to locate minerals from photos taken from NASA aircraft.

NRL scientist N.W. Guinard described use of radar to detect and map oil spills. Tests could be accurate in seas experiencing high winds in which photos would not show oil traces. Radar also was more effective than photography at higher altitudes and could be used in all weather.

During meeting LaRC scientists proposed that Army Corps of Engineers, Federal Water Quality Administration, and NASA combine forces in technological assault on pollution in James River in Virginia. Wallops officials described plans for regional workshop in March with Smithsonian Institution and states of Maryland and Virginia to discuss need for regionalized environmental information system. (*H Chron*, 12/3/70; *H Post*, 12/4/70; Lannan, W *Star*, 12/20/70, 2; MSC PAO)

December 1-11: Meetings at MSFC reviewed design approaches to space shuttle main engines taken by contractors, discussed potential problems in designs, and reviewed efforts to see that they were proceeding satisfactorily under same basic design requirements. (MSFC Release 70-249; MSFC PIO)

December 2: U.S.SR. launched two Cosmos satellites. *Cosmos CCCLXXXI*, launched from Plesetsk, entered orbit with 1013-km (629.5-mi) apogee, 967-km (600.9-mi) perigee, 104.8-min period, and 74° inclination.

Cosmos CCCLXXXII, launched from Baykonur, entered orbit with 5040-km (3132-mi) apogee, 320-km (199-mi) perigee, 143-min period, and 51.5° inclination. By Dec. 7 spacecraft had changed orbit to 5072-km (3152-mi) apogee, 1615-km (1004-mi) perigee, 158.9-min period, and 51.5° inclination. Orbit changed again Dec. 8. By Dec. 31 it had stabilized at 5071-km (3151-mi) apogee, 2584-km (1605.6-mi) perigee, 171-min period, and 55.8° inclination. (*Spacewarn*, 12/15/70; *SF*, 6/71, 213-4; GSFC *SSR*, 12/31/70; *SBD*, 12/17/70, 190-1; 12/18/70, 203)

- Aerobee 150 sounding rocket was launched by NASA from WSMR carrying Princeton Univ. experiment to study stellar UV. Rocket and instruments functioned satisfactorily. (SR list)
- LeRC Director Bruce T. Lundin established review board to investigate failure of Atlas-Centaur booster during launch of OAO-B Nov. 30. H. Warren Plohr, Chief of LeRC Advanced Systems Div., was named chairman. (NASA Release 70-208)
- Dr. Robert H. Goddard Memorial Trophy was presented to Smithsonian Institution for permanent retention at National Space Club Christmas reception in Washington, D.C. Frederick C. Durant III, Assistant Director of Smithsonian's Air and Space Museum, received award from donor, Mrs. Esther Goddard, widow of rocket pioneer. (NSC *News Letter*, 1/8/71)
- NASA announced selection of LTV's Service Technology Corp. for award of $14-million, cost-plus-award-fee contract to provide maintenance and operations, engineering and construction, technical information, and supply and public affairs support services at MSC. One-year contract with provisions for two one-year extensions was effective April 1, 1971. (MSC Release 70-124)

- Aircraft arresting system of barrier net linked to energy absorbers to prevent transport aircraft from taxiing off runway was undergoing final tests at Edwards AFB, Calif., AP reported. Net developed by Aérazur Constructions Aéronautique of Paris and energy absorbers developed by All American Engineering Co. could be available to airports in one year. Installation sufficient to stop aircraft of DC-8 size would cost $500 000 to $600 000; tandem system durable enough to stop Boeing 747 at end of runway would cost about $800 000. Major airports could require 10 such devices—one at each end of runway. (*Huntsville News*, 12/2/70)
- U.K. Aviation Supply Minister Frederick Corfield said in House of Commons that U.K. would not join other European countries in producing airbus for short-range flights. European airbus program had been offered at $72-million cost. (Reuters, *W Post*, 12/3/70, A23)
- Urban Coalition draft study report on SST was quoted in *Wall Street Journal* by Albert R. Karr as saying SST was "environmental danger of unknown but enormous proportions." As Senate vote neared, Karr said, "whether those hazards are 'enormous' remains hotly at issue. The great SST debate hasn't answered this question. In fact, it hasn't even decided how the answers ought to be found." (*WSJ*, 12/2/70, 16)
- Senate approved nomination of William D. Ruckelshaus to head Environmental Protection Agency. (*CR*, 12/2/70, S19272)

December 2-3: NASA manned earth-orbital experiment program for next decade was discussed during meeting of more than 150 Government and industry scientists and engineers at MSFC. Purpose of conference was to discuss changes to planning document *Reference Earth Orbital Research and Applications Investigations*. NASA used book as primary source of experiment program data for continuing space station studies, research and applications module studies, and other manned space flight planning activities. General Dynamics Corp. had NASA contract to update book. (MSFC Release 70-252)

December 3: U.S.S.R. launched *Cosmos CCCLXXXIII* from Plesetsk into orbit with 323-km (200.7-mi) apogee, 180-km (111.9-mi) perigee, 89.5-min period, and 65.4° inclination. Satellite reentered Dec. 16. (GSFC *SSR*, 12/31/70; *SF*, 6/71, 213)
- NASA announced consolidation of all NASA life science activities under new position of Director of Life Sciences in OMSF and named M/G James W. Humphreys, Jr. (USAF, Ret.), Director of Space Medicine, to post. (NASA Release 70-206)
- Dr. Wernher von Braun, NASA Deputy Associate Administrator for Planning, spoke on future of national space program before AIAA in Washington, D.C. He said he thought if U.S. "can still afford foreign travel, still afford Scotch whiskey, can still afford to buy our wives Italian shoes, it is mainly due to the space program." Unprecedented growth in U.S. economy during first decade of space could have been accomplished only "by such programs as demanding as the landing of men on the Moon." Endeavor focused science and technology "into a direction that raises the entire plateau of knowledge and the plateau of engineering to enable you to do better things." (Program; *SBD*, 12/4/70, 133)

December 3

- MSFC announced award of $21 029 756 modification to Boeing Co. contract for work on Saturn V 1st stages and extension of contract performance period through March 31, 1973.

  MSFC also issued RFP on preliminary design of Research and Applications Module (RAM) that could be used with space station and space shuttle. Proposals were due Jan. 8, 1971. (MSFC releases 70-250, 70-251)

- NASA decision to search nationwide for suitable site for launching $6-billion space shuttle had stirred "anguished complaints from Florida politicians and other proponents of Cape Kennedy," *Wall Street Journal* said. But NASA experts had insisted shuttle's unique characteristics and estimated $200-million to $400-million cost of complex ground equipment made search essential. While Cape Kennedy was "leading launch-site candidate," other possibilities included FRC, WSMR, and Dugway Proving Ground in Utah. Essential requirement was flat area for landing shuttle, with several alternate landing locations in case of emergency. Cape Kennedy was not equipped for shuttle landings and lacked facilities to support plane-like flight through atmosphere. (Spivak, *WSJ*, 12/3/70, 36)

- Senate, by voice vote, passed H.R. 17755, $2.7-billion FY 1971 DOT appropriations bill, after voting 52 to 41 to strike $290 million for SST development. (*CR*, 12/3/70, S19327-95)

- President Nixon announced establishment of Aviation Advisory Commission and designation of Crocker Snow, Director of Aeronautics for Commonwealth of Massachusetts, as Chairman. Commission would make recommendations on long-term aviation requirements, including airport plans and use of surrounding lands, ground access, airways, air service, and aircraft compatible with plan to be prepared by Secretary of Transportation. Commission would report to President and Congress on or before Jan. 1, 1972. (*PD*, 12/7/70, 1619-20)

- H. W. Withington, director of SST development for Boeing Co., said in Seattle that total rejection of SST program could cost Seattle area about 4800 Boeing jobs. Washington state had 1.2 million persons working and 96 700 seeking unemployment benefits. Jobless figure in Seattle area was 12.1% of insured workers. Boeing's earnings had fallen from peak $83.9 million in 1967 to operating loss in 1969, although other factors had offset deficit. (AP, *B Sun*, 12/5/70)

- NAS published *The Life Sciences*, report of four-year study by Committee on Research in the Life Sciences: "From the best estimates we can make, in the current year (fiscal year 1970) appropriations for research, *per se*, are approximately 20 percent less than required to ensure that the nation's truly qualified life scientists are fully and usefully engaged." (NAS Release)

December 4: NASA announced modifications to 2nd stage (S-II-9) of Saturn V launch vehicle scheduled to boost Apollo 14 toward moon. Accumulator, compartment to suppress pogo-effect oscillations felt on previous Apollo flights, had been installed in liquid-oxygen line feeding the center engine and filled with helium gas to cushion pressures of fluid flowing through line. Three acceleration-actuated modules would be installed on center cross-beam structure as back-up to initiate cutoff of center engine if oscillations were excessive. J-2 engine propellant-utilization valve had been redesigned from mo-

tor driven to pneumatically actuated, to bypass much onboard stage electronic circuitry, simplify propellant utilization, and enhance stage's reliability. (NASA Release 70-207)
- Ground-test version of Saturn Workshop for Skylab program was shipped to Michoud Assembly Facility from McDonnell Douglas Astronautics Co. Huntington Beach, Calif. Workshop would arrive at Michoud Dec. 17 and undergo testing until Dec. 30, when it would be shipped by barge to MSC for further tests. (MSFC Release 70-246)
- Buddy secondary life support system (BSLSS) to be carried by Apollo 14 astronauts during surface EVA was described by MSC *Roundup*. If water cooling system in one of backpacks failed, astronauts could retrieve BSLSS from MET cart and attach the 2½-m (8½-ft) hoses to their portable life support systems (PLSS). Tether snapped to waist restraint straps of astronauts' spacesuits would prevent damage to hoses or spacesuits during return to LM. By sharing water supply between two crewmen, BLSS stretched time that emergency oxygen would last from 30 min to 60 min. (MSC *Roundup*, 12/4/70, 1; NASA Release, 12/13/70)
- *New York Times* editorial said Senate vote to delete SST funding from DOT appropriations bill was "major event in the new environmental politics. Citizens' organizations and individual conservationists collided head on with an entrenched economic interest group, strongly backed by the political and the propaganda resources of the Nixon Administration—and the conservationists triumphed." Public would be watching "to see that this useless, wasteful and potentially dangerous project is not artificially resuscitated with a tranfusion of tax money behind closed doors of a conference committee." (*NYT*, 12/4/70, 46)
- Senate vote to deny Federal funds for SST development [see Dec. 3] dealt "severe blow" to Fairchild Hiller Corp.'s growth plans over next decade, *Washington Post* said. Company was largest subcontractor to Boeing Co. for building SST prototypes and had projected revenues of more than $2.5 billion for next decade for building sections of 500 aircraft. (*W Post*, 12/4/70, C9)

December 5: First positive identification of amino acids of extraterrestrial origin was reported in *Nature* by ARC scientific team headed by Dr. Cyril A. Ponnamperuma. Dr. Ponnamperuma said discovery of abundance of amino acids in meteorite that fell near Murchison, Australia, Sept. 28, 1969, was probably first conclusive proof of extraterrestrial chemical evolution—chemical processes which preceded origin of life. Discovery was strong new evidence for theory of chemical evolution, suggested possible existence of life elsewhere in universe, and might provide new time sequence for origin of life on earth and elsewhere in universe. Important contributions to find had also been made by UCLA scientist Dr. Ian R. Kaplan and Univ. of Arizona scientist Dr. Carleton Moore. (*Nature*, 12/5/70, 923-6; NASA Release 70-205)
- President Nixon issued statement urging reversal of Senate disapproval of SST funding [see Dec. 3]: "I am well aware of the many concerns that have been voiced about the possible effects supersonic transports might have on the environment. I want to reassure the Congress that the two prototype aircraft will in no way affect the environment. As for possible later effects, we have an extensive

December 5: Dr. Cyril A. Ponnamperuma, heading a team of Ames Research Center Scientists, reported the first positive identification of amino acids formed outside the earth's environment. A meteorite that fell near Murchison, Australia, Sept. 28, 1969, was found to contain these constituents of living cells and was probable proof of extraterrestrial chemical evolution—chemical processes preceding the origin of life.

research project under way to insure against damage. Further progress on the part of the United States in the SST field will give this country a much stronger voice with regard to any long range effects on the environment than if we permit other nations to take over the entire field. . . . The SST is an airplane that will be built and flown. The issue is simply which nation will build them." (PD, 12/7/70, 1630)

- Test flights of Anglo-French supersonic transport Concorde had gone well and protests against its sonic boom had been "less than expected," AP article reported. Big hurdle would come in spring when

U.K. and France would decide whether to go into full production. Decision would depend on number of aircraft ordered by then. To date, 16 airlines had taken options on 74 Concordes, with first batch going to Pan American World Airways, Air France, and British Overseas Airways Corp. None had made firm commitment to buy and U.K. and France had assured foreign airlines they could revoke their options if Air France and BOAC backed out. AP said major airlines doubted Concorde would be profitable. It would carry 130 passengers, against Boeing 747's 490, and require more fuel than conventional jets. At possibly $52.8 million, its cost would be about five times that of conventional airliner. Concorde's only selling point was speed. It would halve flying time across Atlantic to 3½ hrs. (Pinder, *W Post*, 12/5/70, A13)

- Walker G. Bennett, newly appointed President of Augustana Hospital in Chicago, said at Chicago dinner that he hoped to reduce patient costs by 5% to 10% by equipment and techniques for space industry. Bennett was formally introduced by Dr. Wernher von Braun, NASA Deputy Associate Administrator for Planning, who later said sensors developed to monitor astronauts were being used to monitor patients at Augustana and other hospitals. (*C Trib*, 12/6/70)

December 5-6: Contact with French laser reflector on board U.S.S.R.'s *Lunokhod I* resting on lunar surface was made by Soviet Academy of Sciences' Crimean Astrophysical Observatory. Ground-based equipment sent signals to moon and recorded "clearly reflected signals," Tass reported. (*SBD*, 12/11/70, 170; *Sov Rpt*, 12/30/70)

December 7: Senate passed H.R. 19830, FY 1971 Independent Offices and HUD appropriations bill, which contained $3.269-billion NASA appropriation. As budgeted by NASA, appropriation allocated $678.7 million for research and program management, $25 million for construction of facilities, and $2.565 billion for R&D. Amendment proposed by Sen. Walter F. Mondale (D-Minn.) to eliminate funds for space shuttle and station was rejected. Bill was cleared for President's signature. (*CR*, 12/7/70, S19521-59; Conf Rpt 91-1345)

- DOT's SST development environmental impact statements and comments received by DOT on draft statement from other agencies were transmitted to Council on Environmental Quality by James M. Beggs, DOT Under Secretary of Transportation. Statement, required by National Environmental Policy Act of 1969, said development of two prototype SSTs "will not give rise to any long-range environmental problems." Possible adverse environmental consequences arising from fleet operation of SSTs represented "only possible effects on long-term productivity of environment." Such effects "remain in the realm of speculation at this time." Prototype development "would not involve an irreversible commitment of any environmental resources." Prototype would demonstrate performance, economics, safety, and environmental qualities of SST design before large-scale production of SSTs. Government research application to SST environmental factors included: LaRC R&D program to determine feasibility of tubular noise-suppression technique for SST engines, noise reduction and control at source program, system analysis of aircraft noise abatement, university research program to develop and expand noise research interest within educational institutions, theoretical and experimental studies of structure and turbulence levels of SSTs to reduce sideline noise and turbomachinery noise, LeRC, studies of noise reduction potential of suppressors and engine cycles, and work on control and evaluation of noise on receiving end, including human factors. Other SST environmental programs were: DOT-NASA-USAF high-altitude-radiation research program and NOAA air-transportation atmospheric weather-modification research program.

NASA and DOD gave unqualified support to draft impact statement. HUD said it would support prototype development on assurance that new airports would not be needed for SSTs. State Dept. felt "insufficient information exists now to negate all possible significant adverse environmental effects." Dept. of Commerce said there were still significant questions about SST effects on climate and atmosphere. Dept. of Interior endorsed impact statement after DOT had met its previous objections. (Text; DOT Release)

- Sen. Gaylord Nelson (D-Wis.) introduced S. 4565 to amend FAA Act of 1958, to prohibit operation within territorial jurisdiction of U.S. of any civil supersonic aircraft until and unless sonic boom and stratospheric pollution had been reduced to zero or effectual equivalent. (*CR*, 12/7/70, S19494-5)
- Successful heating of portion of upper atmosphere with high-powered radio beam was reported at American Geophysical Union meeting in San Francisco by Dr. William F. Utlaut, Deputy Director of NOAA Institute of Telecommunication Sciences at Boulder, Colo. Two-million-watt beam had raised temperatures by 30% over 4300-km (2700-mi) area 298 km (185 mi) above earth. Heated zone had expanded and emitted more infrared light than usual and radio echoes became unusually fuzzy. Dr. Utlaut said understanding of these phenomena could have bearing on quality of commercial and defense communications. (UPI, *NYT*, 12/10/70)
- Rep. Bertram L. Podell (D-N.Y.) announced resignation from House Committee on Science and Astronautics. He would take seat on House Committee on Interstate and Foreign Commerce. (NASA *LAR IX*/130)
- DOD reported two Soviet long-range reconnaissance aircraft had arrived in Cuba after 17-hr flight from U.S.S.R. Aircraft, modified Tu-95 bombers, were identified and shadowed by U.S. interceptors near Iceland during more than 11 000-km (7000-mi) flight. (*W Post*, 12/8/70, A6)

December 8: House, by vote of 213 to 174, tabled motion to instruct House conferees to agree to Senate amendment that would strike SST funds from H.R. 17755, FY 1971 DOT appropriations bill [see Dec. 3]. (*CR*, 12/8/70, H11306-24)

- Astronauts Alan B. Shepard, Jr., and Edgar D. Mitchell took four-hour walk-through of planned EVA activity in KSC training area in preparation for Apollo 14 lunar landing mission Jan. 31, 1971. (*Marshall Star*, 12/9/70, 1)
- *New York Times* editorial commented on discovery by ARC scientist Dr. Cyril A. Ponnamperuna of amino acids of extraterrestrial origin in meteorite [see Dec. 5]: Despite progress in discovery, "it should not be assumed that the mystery of the origin of life is near solution." Though it was becoming increasingly clear how cell's components could come into spontaneous being, "we still do not know

how they were put together to form the first functioning cell, the first living organism, the first organized spark of self-reproducing life." (*NYT*, 12/8/70)

- *Wall Street Journal* editorial said Congressmen were "squabbling" over location of future NASA space shuttle base: "There's a lot to be learned from the space shuttle, we have no doubt. The scientific results will be greatest, though, if NASA manages to locate the project where it can be managed most efficiently—and not merely where local Congressmen are most adept at gathering spoils." (*WSJ*, 12/8/70)
- U.S. Court of Appeals was studying suit to force disclosure of SST report prepared by panel headed by physicist Richard L. Garwin at President Nixon's request, Washington *Evening Star* said. Suit, filed by conservationist Sierra Club and Friends of the Earth, had been dismissed Aug. 21 by U.S. District Court Judge John H. Pratt on grounds report was covered by doctrine of executive privilege. (*W Star*, 12/8/70, A3)
- *Washington Post* editorial urged House to follow Senate in denying SST funding [see Dec. 3]: "For that was no idle expression of narrow interests or shallow sentiment on the Senate's part; it was...almost an explosion of public concern, as surprising to the backers of the move as it was apparently shocking to the supporters of the SST." What Senate vote said "was that somewhere out there, away from the corporate boardrooms and the offices of the Federal Aviation Administration and the corridors of Congress, a lot of people care more about the quality of life and the nation's urgent economic and social needs than they do about breakfasting in London and New York at roughly the same time on the same day." (*W Post*, 12/8/70, A20)
- U.S.S.R. announced at opening of Supreme Soviet in Moscow that Soviet government planned to spend equivalent of $19.9 billion on defense in 1971. Amount was similar to that announced for 1970, which was highest level in Soviet history. (Gwertzman, *NYT*, 12/9/70, 1)
- New York City Mayor John V. Lindsay addressed more than 300 mayors at National League of Cities Conference in Atlanta, Ga.: "The human potential of technology is boundless. We have only scratched the surface. For the sake of our cities and for the sake of the men and women in our aerospace. industry, we must learn more." Mayor Lindsay said AIAA and city of New York would sponsor first annual Urban Technology Conference in May 1971. Conference would focus on application of space techniques and talent to urban crisis. "From it we hope to learn how to convert the creativity that put a man on the Moon into an ally in the struggle to help men live in our cities." (*A&A*, 1/71, 9)

December 8-9: U.S.S.R.'s *Lunokhod I*, resting on lunar surface, resumed exploration of Sea of Rains after two weeks of inactivity during lunar night. On Dec. 9, one-hour communications session was held, solar panels were opened and positioned toward sun, and photos of rising sun and lunar surface were taken. (*SBD*, 12/11/70, 170; Tass, *Sov Rpt*, 12/30/70, 1)

December 9: NASA released text of Oct. 29 agreement with Soviet Academy of Sciences for joint efforts to design compatible space docking and rendezvous arrangements. Agreement had been confirmed by exchange of notes between Dr. George M. Low, Acting NASA Administrator, and Mstislav V. Keldysh, President of Soviet Academy. [See Oct. 29.] (NASA Release 70-210)

- House-Senate conferees on H.R. 17255, clean air bill, agreed to give Environmental Protection Agency power to limit emissions of pollutants from aircraft and to set standards banning use of aircraft fuels dangerous to public health. FAA would enforce standards. (*W Post*, 12/10/70, A2)
- In economic presentation to DOT, Air Transport Assn. said 12 major U.S. airlines would lose $123 million in 1970. Estimated loss for 1971 was $192 million at prevailing fare levels. ATA said at least 5.1% upward adjustment in fares would be needed to produce $100-million profit in 1971. (Samuelson, *W Post*, 12/11/70, D9)
- Dr. James F. Danielli, Univ. of Buffalo biologist, in Washington, D.C., press conference called for "watchdog agency" to steer scientists clear of experiments in germ warfare or genetic engineering that might change fundamental nature of man for the worst. Dr. Danielli was head of Buffalo Univ. team that synthesized single-celled organisms from component parts of other similar organisms [see Nov. 12]. (Randal, W *Star*, 12/9/70, A21)
- M/G Kenneth W. Schultz (USAF), Deputy for Minuteman, received USAF's Zuckert Award for Outstanding Proficiency in Management for having "cohesively bound together, effectively motivated and skillfully managed a 30,000 member Air-Force-contractor team," significantly increasing program's effectiveness, and saving millions of program dollars. (AFSC Release 304.70)
- Artem I. Mikoyan, codesigner of Soviet MIG aircraft, died in Moscow at age 65 after long illness. (*NYT*, 12/10/70, 1)
- Dr. Abram I. Alikhanov, director of U.S.S.R.'s Institute of Theoretical and Experimental Physics and designer of first Soviet atomic reactor, died in Moscow at age 66. (*NYT*, 12/10/70, 70)

December 10: U.S.S.R. launched *Cosmos CCCLXXXIV* from Plesetsk into orbit with 277-km (172.1-mi) apogee, 203-km (126.1-mi) perigee, 89.2-min period, and 72.8° inclination. Satellite reentered Dec. 22. (GSFC *SSR* 12/31/70; *SBD*, 12/17/70, 191)

- Nike-Tomahawk sounding rocket launched by NASA from Wallops Station carried West German payload to 257.5-km (160-mi) altitude. Payload, launched for West German Federal Ministry for Education and Science, consisted of leeches in Biosonde. Primary objective was to test under weightlessness a newly developed life-support system designed for use in satellites. Secondary objective was to obtain information on behavior of leeches when subjected to high-stress conditions of rocket launch.

  Weightlessness was maintained for 7½ min and data were obtained on operation of special oxygen-producing electrolytic cell in life-monitoring system. Data were also obtained on leeches' behavior by monitoring their movement. Payload was recovered by USNS *Range Recoverer* 158 km (85 nm) southeast of launch site.

  Launch, first in series of two, was conducted on cost-reimbursable basis in support of bioinstrumented tests for Univ. of Frankfurt, Germany. Second launch would be conducted Dec. 16. (WS Release 70-24)

December 10

- *Sert II* Space Electric Rocket Test, launched Feb. 3 to accumulate 4383 hrs of electric ion thruster operation, was adjudged unsuccessful by NASA. Thruster No. 1 had operated 3782 hrs and No. 2, 2011 hrs. Both had failed apparently because of electrical shorts in high-voltage system. Although endurance objectives were not met, secondary objectives were met and mission was considered significant contribution to advancement of ion system. (NASA Proj Off)
- House-Senate Conference Committee voted appropriation of $210 million for development of SST. (Hoffman, *W Post*, 12/11/70, 1)
- President Nixon commented on SST during Washington, D.C., press conference: "I have satisfied myself, after long deliberation...that the arguments with regard to the environment could be met, that this prototype should be built. What is involved here is not just 150,000 jobs which will be lost if we don't build it, not just the fact that billions of dollars in foreign exchange will be lost if we do not build it; but what is lost here is the fact that the United States of America which has been first in the world in commercial aviation from the time of the Wright Brothers decides not just to be second but not even to show." (*PD*, 12/14/70, 1652)
- House Committee on Science and Astronautics published *The National Space Program—Present and Future*, compilation of papers prepared for Subcommittee on NASA Oversight. Papers represented attitudes of Government, aerospace industry, and academic community on objectives of national space program during next decade and funding level needed to support program. Statements had been prepared for hearings, originally scheduled for September and October 1970, that had been canceled because of press of legislative business and "uncertain status of the NASA appropriation measure" for FY 1971 because of Presidential veto.

    Dr. George M. Low, Acting NASA Administrator, had urged that "new initiatives proposed for the 1970s represent not only a space program that will maintain our Nation's leadership, but also an opportunity to utilize resources which will otherwise remain idle or grossly underemployed. In this sense the 'real cost' to the society of NASA's programs for the 1970s is very low and the cost-benefit of these programs correspondingly high."

    Dr. Thomas O. Paine, Vice president and Group Executive of General Electric Co. and former NASA Administrator, said in statement that U.S.-U.S.S.R. discussion on common spacecraft docking had been "excellent step forward" in space programming. "The United States, Western Europe, Australia, Canada, and Japan should also work more closely together in the forefront of space science and technology, eliminating duplicative projects and obsolescent developments for purely nationalistic purposes. Space activities are inherently global; they challenge us to develop new international institutions as advanced as our technologies. Attractive new international arrangements should also be established for the developing nations to help them take full advantage for new application opportunities."

    Dr. Wernher von Braun, NASA Deputy Associate Administrator for Planning, urged vigorous space program with space shuttle as "cornerstone." Answers to question "Why go to the moon?" would not be found in budget books and cost-benefit analyses but "in the hearts and minds of men, who, responding to a mystical, metaphysical imperative, join hands and reach out to the stars."

Committee found "clear consensus...that a vigorous aeronautics and space effort is in the National interest, since it sets the pace for the country's technological advancement. There is also general agreement that a revitalized space program, given strong direction and adequate funding, is needed for the United States to retain its technological preeminence in the decades ahead." (Text)

- President Nixon issued Proclamation 4024 to Observe Dec. 17 as Wright Brothers Day: "The names of Orville and Wilbur Wright symbolize American ingenuity and courage." Their Dec. 17, 1903, flight was "forerunner of the aviation and space technology which today strengthens America's defense and contributes to better understanding throughout the world by promoting commerce and encouraging travel." (PD, 12/14/70, 1649)
- Findings of international team of scientists aboard NSF-sponsored 14th voyage of deep-sea drilling ship *Glomar Challenger* were discussed by Columbia Univ. geologist Dr. Dennis E. Hayes at New York press conference. Voyage between Lisbon and Puerto Rico had ended Dec. 1. Deep-sea drilling data had shown eastern margin of Atlantic Ocean off northwest African coast appeared to be younger than western margin off U.S. Discovery suggested that North Africa and North America were not joined millions of years ago, but had been drifting apart from what once was probably much smaller "proto-Atlantic" ocean. (Wilford, *NYT*, 12/11/70)
- NSF reported local government R&D expenditures had doubled from $20 million in 1966 to $40 million in 1969 but constituted less than 0.1% of total local government expenditures for all purposes during period. In 1968 state governments had overall expenditures of same level as local governments but R&D expenditures nearly five times as great. (NSF *Highlights*, 12/10/70, 1)

December 10-11: U.S.S.R.'s *Lunokhod I* completed nine-hour exploration mission, traveling 244 m (800 ft) over lunar surface. Vehicle traveled in and out of crater 16 m (52.5 ft) in diameter and 2 m (6.6 ft) deep, experiencing 27° list and 17° trim, and transmitted TV pictures of landscape and data on soil properties and operation of Lunokhod systems. (*SBD*, 12/14/70, 173; 12/15/70, 180; Tass, *Sov Rpt*, 12/30/70)

December 11: Noaa I (ITOS-A) National Oceanic and Atmospheric Administration meteorological satellite was successfully launched by NASA from WTR at 3:35 am PST by two-stage, long-tank, thrust-augmented Thor-Delta (DSV-3N-6) booster. Orbital parameters: apogee, 1472.2 km (914.8 mi); perigee, 1422.6 km (884 mi); period, 114.8 min; and inclination, 102°.

Primary objective was to place 306-kg (675-lb) spacecraft in sun-synchronous orbit with local equator-crossing time between 3:00 and 3:20 pm and conduct in-orbit engineering evaluation so that daytime and nighttime cloud-cover observations could be obtained regularly and dependably by both direct readout and onboard storage. Equipment included two AVCS, two APT camera systems, and two scanning radiometers to measure surface temperature in cloud-free areas and provide cloud-cover data in visible and infrared channels.

As secondary objective launch vehicle carried 2.3-kg (5-lb) *Cepe* Cylindrical Electrostatic Probe Experiment permanently attached to Delta 2nd stage, to provide information on electron density and temperature and on ion current during first two orbits. *Cepe* entered orbit with 1475.1-km (916.6-mi) apogee, 1425.4-km (885.7-mi) perigee, 114.9-min period, and 101.9° inclination.

Noaa I was first Tiros spacecraft funded by NOAA and second spacecraft in ITOS series. First, *Itos I* (Tiros-M), had been launched Jan. 23. (NASA Proj Off; GSFC *SSR*, 12/31/70; NOAA Release 70-6; *SBD*, 12/14/70, 175)

- *Ofo* Orbiting Frog Otolith, launched Nov. 9, was adjudged successful by NASA. All mission objectives—including maintenance of bullfrogs in space and collection of scientific and engineering data—were achieved. Mission, originally scheduled to last three to five days, had lasted until Nov. 15, when batteries ran down. (NASA Proj Off)
- NASA and Univ. of Rome's Aerospace Research Center dedicated Explorer XLII satellite, scheduled for launch Dec. 12, to Kenya. Dedication was in honor of Kenya's independence day Dec. 12 and in recognition of Kenya's cooperation with Italy in establishing San Marco launch platform off coast of Kenya. If successfully orbited, satellite would be named "Uhuru," Swahili word for freedom. (NASA Release 70-212)
- NAS-NRC Space Science Board released *Life Sciences in Space*, report of study to review NASA life sciences programs. Report recommended functional reorganization along disciplinary lines, central direction of all biological and medical programs, improvements in evaluation of all research and in criteria for selection of flight experiments, and formation of board of extramural scientists to provide continuing advice on overall program. (Text)
- U.N. General Assembly Political Committee voted 85 to 8 in favor of proposal urging speedy completion of treaty covering damages caused by objects launched into space. Group of Soviet Bloc states voted against measure on grounds that provisions for compensation clashed with Soviet law. Next round of meetings on treaty was scheduled for June in Geneva. (*NYT*, 12/12/70, C15)

December 12: *Uhuru (Explorer XLII)* (SAS-A) Small Astronomy Satellite was launched for NASA by Italy at 1:45 pm local time (5:45 am EST) from San Marco launch platform off coast of Kenya. Four-stage Scout launch vehicle boosted 142.9-kg (315-lb) satellite into orbit with 572-km (355.4-mi) apogee, 531-km (330.0-mi) perigee, 95.7-min period, and 3.04° inclination. Primary objective was to develop catalog of celestial x-ray sources by systematic scanning of sky in 2- to 20-kev range.

Satellite—first equippeed with sensitive instruments to detect high-energy x-ray sources in space—would pinpoint location of x-ray sources; transmit data on intensity, frequency, and time variation; and map diffuse x-ray background. NASA controllers would maneuver satellite from ground by energizing electromagnet in attitude control system. Electrical energy caused electromagnet to act like compass needle, aligning itself to north-south lines of earth's magnetic field and permitting controllers to point spacecraft to any direction.

December 12: Uhuru (Explorer XLII) *Small Astronomy Satellite was launched into an equatorial orbit by Italy from the San Marco launch platform off the coast of Kenya, East Africa, on Kenya's Independence Day. "Uhuru" was the Swahili word for "freedom." Boosted by a four-stage Scout vehicle to map x-ray sources both within and beyond the Milky Way Galaxy, the satellite was the first launched for the United States by another country. At left Mrs. Marjorie Townsend, project manager, discussed satellite performance with a colleague during preflight tests at Goddard Space Flight Center.*

During first day of operation *Uhuru* was expected to collect more data than had been obtained with sounding rockets in past eight years. Satellite's long observing time, rather than the few minutes for sounding rockets, would permit observation of x-ray sources from 30 to 50 times fainter than those observed thus far, probably increasing number of observed x-ray sources from 540 to several hundred.

Uhuru was first in series of three explorers in Small Astronomy Satellite program to survey sky above earth's atmosphere and identify x-ray, gamma-ray, UV, and infrared sources. It was first U.S. satellite launched by another country. Under NASA and Univ. of Rome agreement NASA provided booster and satellite; Italian team, trained at Wallops Station, conducted assembly, checkout, and launch services on cost-reimbursable basis. SAS-B would be launched in October 1971 and SAS-C in December 1972. (NASA Proj Off; NASA Release 70-203)

- France successfully launched *Peole* (Preliminaire Eole) experimental meteorological satellite on Diamant-B booster from Kourou, French

Guiana, into orbit with 749-km (465.4-mi) apogee, 635-km (394.6-mi) perigee, 98.6-min period, and 15.0° inclination. *Peole*—70-kg (154-lb) octahedron with 711-mm (28-in) diameter—was preliminary to Eole (IAS-A) International Applications Satellite and carried interrogating system to collect data from balloon system to be utilized by Eole. Launch, originally scheduled for Aug. 24, had been postponed because of vibration problem in Diamant-B booster. (GSFC *SSR*, 12/31/70; *SBD*, 12/15/70, 181; *Spacewarn*, 12/29/70, 1; *SF*, 6/71, 213, 5)

- U.S.S.R. launched *Cosmos CCCLXXXV* from Plesetsk into orbit with 984-km (611.4-mi) apogee, 980-km (608.9-mi) perigee, 104.7-min period, and 74.0° inclination. (GSFC *SSR*, 12/31/70; *SBD*, 12/15/70, 181)
- Nixon Administration was finding "no takers" in its efforts to replace Dr. Thomas O. Paine as NASA Administrator, *New York Times* said. Lack of NASA Administrator seemed "as much a reflection of the low level of attention that space commands among White House priorities as it is an indication of the unavailability of candidates." Since Dr. Paine's Sept. 15 retirement, U.S.S.R. had successfully launched 33 space payloads. (Lyons, *NYT*, 12/13/70, 50)

December 13: Photo of chromosomes in human blood could be analyzed in three minutes—one tenth of time required under manual system—by using computerized process developed by JPL team under Dr. Kenneth R. Castleman and Dr. Robert Nathan, NASA announced. System, combining space-photo analyzing techniques with automatic chromosome study methods, used closed circuit TV. Operator watched as automated microscope searched slide prepared from blood sample. Image of suitable group of chromosomes was automatically enlarged, photographed, and transferred to computer for measuring and classifying and then digital picture was transferred to photographic printer. Manual chromosome analysis, used to spot hereditary disorders, was time-consuming and expensive. JPL team hoped to develop small computer system facilitating wider use of new process. (NASA Release 70-211)

- DOD and Lockheed Aircraft Corp. were resolving their financial differences at cost to U.S. taxpayer of minimum of $600 million, *New York Times* said. By choosing to rescue Lockheed rather than letting it go bankrupt, DOD was also "setting a precedent for Government behavior toward other major military manufacturers that might flounder financially and need help." Exact portion of total of more than billion dollars involved that would be borne by taxpayer was unclear. Lockheed's principal creditors had lent Lockheed $30 million for its short-term needs and agreed to lend additional $250 million if Lockheed's negotiations with DOD ended favorably. Total of $280 million would ensure production of L-1011 trijet airbus. (Sheehan, *NYT*, 12/13/70, 1)

December 14: MSFC announced its engineers had used blindfolded jeep with TV camera, outsized station wagon, and other vehicles during tests on Flagstaff, Ariz., desert to determine accuracy of remote operation of LRV with simplest and least amount of equipment. Tests, in March and April and during summer, had proved that traverses of 30.5 km (19-mi) could be made with less than two percent

overall error by using directional gyroscope, odometer calibration, and sun compass alignment and updating. (MSFC Release 70-256)
- Senate received President Nixon's nomination of Col. James A. McDivitt (USAF), *Apollo 9* astronaut and Apollo Spacecraft Program Manager at MSC, to be brigadier general. (*CR*, 12/14/70, S20137)
- Lidar, new light detection and ranging device to detect air pollutants, was described in *New York Times* interview of meteorologist Ronald Collins of Stanford Research Institute. Device sent energy signal outward to bounce off air particle and return to receiver on ground. Returned signal was then translated from energy pulse into visual pattern seen on picture tube. Unlike radar, which used longer microwave, lidar used wavelengths measuring 0.6943μ, which picked up extremely small particles in concentration too thin to be seen with naked eye. It was even possible to track some gases in atmosphere with lidar. (Blakeslee, *NYT*, 12/14/70, 53)
- AAAS announced election of Dr. Glenn T. Seaborg, AEC Chairman, as President-elect. (AAAS Release)
- Last test model of USAF Minuteman III missile was launched from underground silo at ETR toward South Atlantic target, ending 28-mo development program. Missile was 17th Minuteman III launched from ETR since Aug. 16, 1968; 15 had been launched under simulated operational conditions at Vandenberg AFB, Calif. Tests at Vandenberg would continue. (UPI, W *Star*, 12/14/70, A6)

December 14-15: NASA launched series of five Nike-Apache sounding rockets from Wallops Station between 10:08 pm and 6:27 am EST carrying Geophysical Corp. of America payloads. First four rockets carried trimethylaluminum (TMA) experiments which created pale white clouds. Fifth carried sodium vapor experiment which created huge reddish-orange cloud visible for hundreds of kilometers along East Coast. Rockets reached altitudes between 195 km (121.2 mi) and 214.5 km (133.3 mi). Purpose of experiments was to compare vapor measurements of atmospheric winds and diffusion with east-west components of meteor shower winds. Data were obtained by photographing motion of vapor trails from five camera sites on ground. (WS Release 70-25; NASA Release 70-209; NASA Rpts SRL)

December 15: U.S.S.R.'s *Venus VII*, launched Aug. 17, reached atmosphere of Venus. At 8:02 am Moscow time (12:02 am EST) orbital module and descent package separated, Tass announced. After aerodynamic braking of descent package, parachute system was deployed, antennas unfolded, and transmission begun. Data was transmitted for 35 min during descent. Soviet scientists later reported surface of Venus had been calculated to have temperature of 773 K (500° C; 932°F) and pressure of 100 atmospheres. On Jan. 26, 1971, Tass announced *Venus VII* had landed and had continued transmitting data from surface for 23 min. Longest survival time of three previous Venus landing missions attempted by U.S.S.R. had been 94 min, recorded for *Venus IV* Oct. 18, 1967. (*Sov Rpt*, 12/20/70; 2/2/71; *SBD*, 12/21/70; 204; O'Toole, *W Post*, 12/16/70, A25)
- U.S.S.R. launched *Cosmos CCCLXXXVI* from Baykonur into orbit with 270-km (167.8-mi) apogee, 215-km (133.6-mi) perigee, 89.3-min period, and 64.9° inclination. Satellite reentered Dec. 28. (GSFC *SSR*, 12/31/70; *SBD*, 12/17/70, 191)

- NAS released *Venus: Strategy for Exploration,* report of NASA-supported study by scientific panel chaired by Harvard Univ. astronomer Dr. Richard M. Goody and Kitt Peak National Observatory astronomer Dr. Donald M. Hunten. Report urged unmanned scientific study of Venus with Planetary Explorer spacecraft, 385-kg (850-lb) "universal bus" capable of carrying entry probes, orbiters, balloons, and landers. Estimated cost of unmanned Venus probe program was $100-million total for first three missions and up to $25 million for each succeeding mission. Program would be undertaken in addition to planned NASA program of planetary exploration.

  Panel urged Venus exploration because information on cloud cover blanketing planet and planet's carbon dioxide atmosphere could contribute to understanding of general atmospheric systems, including earth's, and conflicting theories could be tested with relatively simple measurements. Determination of concentration of hydrogen atoms and compounds could reveal whether oceans once covered Venus and, if so, what happened to them. Panel urged that higher failure risks be accepted for Venus missions to lower equipment costs, that experimental operations be simplified, that landers be permitted to impact on Venus, and that NASA cooperate with space organizations of other nations in planning and carrying out investigations. (Text)
- Meeting at MSFC reviewed parallel studies of proposed chemical interorbital space shuttle by North American Rockwell Corp. and McDonnell Douglas Astronautics Co. Analyses of modifications required to adapt either 2nd or 3rd stage of Saturn V launch vehicle into interorbital shuttle between earth and lunar orbit were to lead to preliminary designs, determine orbital launch operations and procedures, and gather data on preliminary costs, logistics, and facilities. Discussions included results of analyses to date, definition of systems requirements, necessary system and subsystem trade-offs, and selection of preliminary configurations. (MSFC Release 70-262)
- House approved, by vote of 319 to 71, Conference Report on H.R. 17755, $2.5-billion FY 1971 DOT appropriation bill that included $210 million for SST development. (*CR,* 12/15/70, H11700-12)
- Dynalectron Corp. announced it had been awarded $3 805 000 NASA contract for maintaining and servicing lunar landing training vehicles and aircraft used by astronauts and management at MSC. Contract was one-year renewal of agreement under which Dynalectron had carried out management, maintenance, engineering, technical, and logistic functions in support of aircraft at MSC. (Dynalectron Release)
- State Dept. spokesman said USN would build $19-million post for satellite communications on British island of Diego-Garcia in Indian Ocean. As part of joint U.S.-U.K. project, island would also serve as air base for reconnaissance aircraft monitoring increased U.S.S.R,. naval presence in Indian Ocean. (Reuters, B *Sun,* 12/16/70, 6; DOD PIO)

December 16: U.S.S.R. launched *Cosmos CCCLXXXVII* from Plesetsk into orbit with 538-km (334.3-mi) apogee, 526-km (326.8-mi) perigee, 95.2-min period, and 74° inclination. (GSFC *SSR,* 12/31/70; *SBD,* 12/18/70, 200)

December 16

- Fifth anniversary of NASA's *Pioneer VI* interplanetary probe. Since launch Dec. 16, 1965, *Pioneer VI* had orbited sun six times, covering 5.3 billion km (3.3 billion mi); passed behind sun's far side (relative to earth) once, providing new data on solar atmosphere and pioneering solar weather reports from far side; made important contributions to defining exact orbits and mass of moon and planets; and performed experiment to check theory of relativity. *Pioneer VI* data had filled 11 700 km (7300 mi) of analog data tapes, recording 35 billion data bits in 9500 hrs of communications sessions during almost 45 000 hrs of flight. Its maximum distance from earth on sun's far side had been 290 million km (180 million mi).

  First of four Pioneer probes now in orbit, *Pioneer VI* was firmly stabilized in plane of earth's orbit by its 60-rpm spin and would need no further orientation maneuvers. Last maneuver had been made in 1966. One of two radio receivers was operating satisfactorily and solar cells, although damaged by massive solar flares, were still providing power. All other experiments except magnetometer were operating satisfactorily and a number of special experiments based on spacecraft's position in space had been conducted. (ARC Release 70-14)

- NASA launched Nike-Tomahawk sounding rocket from Wallops Station carrying live leeches in Biosonde experiment for West German Federal Ministry for Education and Science. Rocket reached 246.2-km (153-mi) altitude and good data were obtained. Launch was second in series of two to test new satellite life-support system and obtain information on leeches' behavior during flight. First launch had been conducted Dec. 10. (WS Release 70-26)

- Technology was now available to make SST almost as quiet as subsonic jet, Dr. Leo L. Beranek—world authority on acoustics and head of scientific panel appointed to advise DOT's Dept. of SST Development on noise—said in interview with *New York Times*. Engine design for noise level could be perfected by 1973 but whether engineers could cut "sideline" noise to target level of 108 epndb by then was not known. (Witkin, *NYT*, 12/17/70, 26)

- Secretary of Defense Melvin R. Laird issued statement saying U.S.S.R. might have begun to slow down SS-9 missile construction. Some U.S. strategic weapon analysts thought SS-9s and SS-11s "are approaching what might be called leveling-off phases." (B *Sun*, 12/17/70, A1)

- French aviation pioneer André Laurent-Eynac, France's first air minister, died in Paris at age 84. He had served as Under Secretary of State for Air in French government from 1921-1926 and in 1940. (*W Post*, 12/17/70, D6)

December 16-17: Progress reports on nuclear shuttle studies were given to MSFC officials by Lockheed Missiles & Space Co., North American Rockwell Corp., and McDonnell Douglas Corp. at MSFC meetings. (MSFC Release 70-260)

December 17: Dept. of Agriculture would start project to determine properties in moon dust that led to growth of plants bigger than those fertilized by potassium, phosporus, or nitrogen, *Washington Post* reported. USDA scientist Dr. Charles P. Walkinshaw had said moon dust "does something to plants that nothing else can duplicate." He had observed uninterrupted growth of almost 40 different species in moon dust soil. Liverworts, tobacco plants, and ferns had grown larger and to deeper green in moon dust than in other soil. Corn cells buried in moon dust in 1969 were still alive and had grown tough, stringy roots. Dr. Walkinshaw believed moon dust might act as supercatalyst, causing soil nutrients to be absorbed quickly through walls of plants cells. USDA Forestry Service and Chevron Chemical Co. planned to investigate how moon dust acted. (Brett, *W Post*, 12/17/70, G2)

- President Nixon signed H.R. 19830 as P.L. 91-556, Independent Offices and Dept. of Housing and Urban Development Appropriation Act, 1971, which included NASA appropriation of $3.269 billion. (*PD*, 12/21/70, 1705)
- Sixty-second anniversary of Wright brothers' first successful flight. At ceremonies in Kitty Hawk, N.C.—sponsored by NAA, Air Force Assn., and First Flight Society—portraits of aviation pioneers Igor I. Sikorsky and Wiley Post were added to Kitty Hawk's First Flight Shrine. Aviation Hall of Fame in Dayton, Ohio, honored Robert E. Gross, founder of Lockheed Aircraft Corp.; L/G Ira C. Eaker, Eighth Air Force Commander during World War II; Alexander P. de Seversky, military aviation expert; and Juan T. Trippe, founder of Pan American World Airways. At annual Wright Memorial Dinner in Washington, D.C., sponsored by Aero Club of Washington, Wright Brothers Memorial Trophy was presented to C. R. Smith, former President of American Airlines and Secretary of Commerce under President Lyndon B. Johnson. (NAA *News*, 1/71, 1; Varborough, *W Star*, 12/16/70, A18; Flatley, *W Star*; 12/18/70, C5)
- Nike-Apache sounding rocket was launched by NASA from WSMR carrying Univ. of Minnesota experiment to study atmospheric composition. Rocket and instruments functioned satisfactorily. (SR list)
- NASA had applied for patent on magnetic hammer that synthesized small quantities of clear and yellow diamonds from graphite, MSFC announced. Devised by MSFC engineers John R. Rasquin and Marvin F. Estes, process produced diamond grit of kind used by industry for cutting operations. Hammer, which might be used for making other crystals, had been developed for use in Saturn rocket production. (MSFC Release 70-265)

December 17-20: Apollo 11 Spacecraft Touring Exhibit visited Washington, D.C., during 50-state tour which began April 17. *Apollo 11* Astronauts Neil A. Armstrong and Michael Collins attended opening day ceremonies, during which Dr. George M. Low, Acting NASA Administrator, compared *Apollo 11* mission with first flight of Wright brothers in aircraft Dec. 17, 1903: Wright brothers' flight had lasted 12 secs; *Apollo 11*, 156 hrs. Wright brothers' aircraft weighed 274 kg (605 lbs); Apollo weighed 3 million kg (6½ million lbs) at liftoff. Wright brothers' aircraft flew 3 m per sec (10 fps); *Apollo 11*, 10 700 m per sec (35 000 fps). Wright brothers flew to 4.6-m (15-ft) altitude; *Apollo 11* reached about 389 000 km (242 000 mi). Wright brothers' aircraft used 1.5 kg (51 oz) of gasoline; *Apollo 11* burned 2.7 million kg (6 million lbs) of liquid hydrogen, liquid oxygen, and kerosene. Wright brothers' aircraft flew distance of 36.6 m (120 ft); *Apollo 11* flew about 1 530 000 km (950 000 mi). Five spectators watched Wright brothers' flight, while estimated 500 million persons—one sixth of the world's population—witnessed flight of

Apollo 11 July 19-20, 1969. (Program; NASA Hist Off; *CR*, 12/17/70, H11884-5)

December 18: U.S.S.R. launched two Cosmos satellites, from Plesetsk. *Cosmos CCCLXXXVIII* entered orbit with 499-km (310.1-mi) apogee, 269-km (167.2-mi) perigee, 92.2-min period, and 70.9° inclination and reentered May 10, 1971. *Cosmos CCCLXXXIX* entered orbit with 687-km (426.9-mi) apogee, 642-km (398.9-mi) perigee, 98-min period, and 81.1° inclination. (GSFC *SSR*, 12/31/70; 5/31/71; *SF*, 6/71, 213)

- Senate Committee on Government Operations' Permanent Subcommittee on Investigations published *TFX Contract Investigation* (Rpt. 91-1496). Report on history of DOD TFX program concluded: Sept. 1, 1961, decision by Secretary of Defense Robert S. McNamara to start program was wrong; November 1962 decision to choose second best TFX proposal at higher price was wrong; failure to heed warnings in February, July, and October 1964 of technical difficulties and to allow redesign of F-111B was wrong; order to start Project Icarus in October 1966 and to place personal management of TFX in Office of Secretary of Defense was poor management decision "made in desperation"; and decision to continue F-111A production line April-May 1967 was wrong.

  TFX program had been "failure" on which Federal Government would spend more than $7.8 billion to procure 500 aircraft, although original production schedule called for more than 1700 aircraft for less money. "Of the 500 planes we will have, less than 100 (the F-111F's) come reasonably close to meeting the original standards. Spending so great a sum for so few aircraft represents a fiscal blunder of the greatest magnitude. It is clear that vital financial resources were squandered in the attempt to make the TFX program produce satisfactory results." (Text)

- USAF awarded $41.2 million to Fairchild Hiller Corp. and $28.9 million to Northrop Corp. to construct prototypes of new AX twin-engine combat-support aircraft. Fairchild Hiller version would be larger. Prototypes would be flown in competitive tests after 26-mo development phase. Aircraft would have top speed of 640 km per hr (400 mph); carry up to 7300 kg (16 000 lbs) of arms, rockets, and ammunition; and spend up to four hours in direct support of frontline troops. (Beecher, *NYT*, 12/19/70, 1)

- Environmental scientist Dr. H. Peter Metzger filed suit in Federal Court in Denver, Colo., to block static nuclear rocket test scheduled for January in Nevada as experiment for possible future deep space probes. Metzger charged AEC-NASA Space Nuclear Propulsion Office had failed to comply with National Environmental Policy Act requirements that test be fully studied and reported to other agencies for comment. Defendants named were Dr. Glenn T. Seaborg, AEC Chairman; Milton Klein, Space Nuclear Propulsion Office Director; and Dr. George M. Low, Acting NASA Administrator. (*NYT*, 12/19/70, 14)

- Third round of SALT ended in Helsinki with joint U.S.-U.S.S.R. statement expressing determination of both delegations to pursue negotiations. Fourth round would open in Vienna March 15. (Roberts, *W Post*, 12/19/70, A15)

- DOT announced selection of Welton Becket and Associates to receive three-month $87 000 contract for master plan for U.S. first International Transportation Exposition to be held at Dulles International Airport May 27-June 4, 1972. (DOT Release 25370)

December 18-19: Apollo 14 spacecraft underwent second day of flight-readiness tests after all-night vigil by guards following anonymous telephone threat that KSC would be blown up. Deadline of 11:00 pm given by caller passed without incident. Tests were interrupted for several hours during night to replace faulty computer but engineers expected to complete simulated countdown and liftoff satisfactorily. (UPI, W *Star*, 12/20/70, A2)

December 19: Aerobee 170 sounding rocket was launched by NASA from WSMR carrying Cal Tech experiment to conduct x-ray studies. Rocket and instruments functioned satisfactorily. (SR list)

- Senate voted 48 to 43 against forcing end to debate on SST. Closure motion required two-thirds support to win. (CR, S20753-67, S20792-9)

December 21: Capt. Chester M. Lee (USN, Ret.), Apollo Mission Director, described plans and preparations for Apollo 14 mission at Washington, D.C., press briefing: Mission—scheduled for launch from KSC Jan. 31, 1971—would be cut from 10 days to 9 days to provide extra margin of safety by shortening lunar orbit and lunar surface stay time. "On this mission we will have done most of our lunar orbit photography and there is really no substantial reason to stay there. We would like to get home; there is no need to keep them there unless we have to. On Apollo 15. . .we intend to stay around because we can get some good data, but on Apollo 14 we don't have the set-up we will have on Apollo 15."

New equipment weighing 293 kg (645 lbs)—including 36-kg (79-lb) oxygen tank for 141 kg (310 lbs) of oxygen, 61-kg (135-lb) SM battery, and CM storage bags for 18 kg (40 lbs) potable water—had been added to supplement regular equipment in case of emergency.

Apollo 14 crew quarantine procedures were most strict to date to prevent possibility of crew illness before flight. ". . .for example, when they are moving from the crew quarters to the handball court. . .they will clear the hallways while they are going down there" by sounding a horn or bell.

"Kitty Hawk" had been chosen as name for CM to honor birthplace of aviation. LM would be named after Antares, brightest star in constellation Scorpio, which would be visible out LM window during descent to lunar surface. (Transcript; O'Toole, W *Post*, 12/22/70, A3)

- Grumman Corp. test pilot Robert Smyth flew USN F-14 Tomcat fighter, successor to F-4 Phantom and F-111B, on successful 10-min maiden flight from Grumman test facility at Calverton, N.Y. Aircraft, weighing 27 700 kg (61 000 lb) and capable of mach 2 speeds, was limited by bad weather to 460-m (1500-ft) altitude and low speeds. Flight was month ahead of schedule. USN planned to buy 722 F-14s at $11.5 million each. (Witkin, *NYT*, 12/22/70, 66; Getler, W *Post*, 12/23/70, A4)
- LeRC's Plum Brook Station began test series to ensure Skylab shroud would separate from Saturn V properly during launch scheduled for late 1972. Shroud, weighing 10 900 kg (24 000 lbs), was

December 21: *The U.S. Navy's F-14 Tomcat fighter made a successful 10-minute maiden flight from the Grumman test flight facility in Calverton, N.Y. Capable of mach 2 speeds, the F-14 would be successor to the F-4 Phantom and the F-111B.*

composed of quadrants separated by explosives lining seams. Each panel had to be ejected without touching Skylab vehicle in flight. Plum Brook vacuum chamber had been selected for shroud test because tank, 36.6 m (120 ft) tall and 30.5 m (100 ft) wide, could produce hard vacuum equivalent to 483-km (300-mi) altitude and easily house shroud 17 m (56 ft) high and 6.7 m (22 ft) wide. (NASA Release 70-214)

- NASA announced appointment of George W. Cherry, Deputy Associate Director of MIT's Charles Stark Draper Laboratory, as Director of Aeronautical Operating Systems Div., OART. Newly created division was responsible for research and technology programs contributing to solution of problems in navigation, control, reduction of operational hazards, and impacts of aviation on environment. (NASA Release 70-216)
- Intergovernmental group studying DOT-DOD-HEW military assistance for safety in traffic (MAST) program reported at least 61 lives possibly saved by helicopter airlift of accident victims from remote areas to hospitals since introduction of program July 15. About 65% of emergencies had occurred on highways. About 70% of 60 000 highway fatalities in 1969 had occurred in remote areas. (DOT Release 25170)

December 22: Boeing Co. announced completion of lunar roving vehicle (LRV) qualification unit, last in series of test units leading to manufacture of first LRV flight model. Fabrication of major components for three flight-model LRVs would begin soon, with first LRV delivered to NASA April 1, 1971, and carried to moon on Apollo 15 in July 1971. (Boeing Release)

- NASA announced selection of Hughes Aircraft Co. for $22-million, cost-plus-award-fee contract for new, more advanced series of Orbiting Solar Observatory (OSO) spacecraft. Hughes would deliver

three flight spacecraft, OSO I, J, and K, for missions in 1973, 1974 and 1975; provide ground support equipment; and perform launch services. (NASA Release 70-218)
- General Telephone & Electronics Corp. announced proposed plan to establish satellite communications system within U.S. If FCC approved, system would be operational within 24 mos. Hughes Aircraft Co. would lease eight radio channels required by General Telephone & Electronics in one of two 12-channel comsats that Hughes Aircraft had proposed for domestic use. Backup channels would be provided in second comsat for reliable short-term use during peak periods. Gross investment for system was estimated at nearly $27 million. (GT&E Release; Reuters, B *Sun*, 12/23/70)
- ComSatCorp reported third quarter net income of $4 271 000 (43 cents per share), up from $1 446 000 (14 cents per share) for third quarter of 1969. Increase was attributed to increased comsat traffic. (ComSatCorp Rpt to Shareholders)
- Appointment of *Apollo 8* Astronaut Frank Borman as Senior Vice President-Operations Group was announced by Eastern Airlines Chairman and Chief Executive Officer Floyd D. Hall. (EA Release)

December 23: International Workshop on Earth Resources Survey Systems at Univ. of Michigan May 3-14, 1971, would study use of satellites to survey earth resources and surface features, NASA announced. Workshop would be sponsored by NASA, Dept of Agriculture, NOAA, U.S. Geological Survey, Dept. of State, AID, and Naval Oceanographic Office. Invitations had been sent to administrators and experts from U.N. member states in accordance with President Nixon's Sept. 18 statement before U.N. General Assembly. Workshop would survey data already acquired and methods of collecting, processing, and analyzing earth resources data. Potential action for countries initiating earth resources programs would be discussed, including costs and opportunities for international cooperation and assistance. (NASA Release 70-215)
- USAF Sprint missile—a key weapon in Safeguard antimissile system—successfully completed first inflight test interception of ICBM over Pacific. Sprint, launched from Kwajalein Missile Range in Pacific, came within "kill range" of ICBM target nosecone launched from California 6800 km (4200 mi) away. First inflight test interception by Spartan missile had been made Aug. 28. (AP, *W Post*, 12/31/70, A3)
- Treasury Dept. graduated first class of 46 customs security officers to replace temporary sky marshals serving aboard U.S. international and domestic flights to combat skyjacking. Washington, D.C., ceremony marked transition of sky marshal program from multiagency to permanent single-agency status. Treasury Dept. would graduate about 50 officers a week until complement of 2100 was qualified for duty. (*Av Wk*, 1/4/71, 22)

December 25: U.S.S.R. launched *Molniya I-17* comsat from Plesetsk into orbit with 39 569-km (24 587.0-mi) apogee, 493-km (306.3-mi) perigee, 711.8-min period, and 64.7° inclination. Tass said satellite would transmit telephone, telegraph, radio, and TV communications to remote areas. (GSFC *SSR*, 12/31/70; AP, *W Post*, 12/26/70, A9; *SF*, 6/71, 213)
- JPL had announced availability of certain facilities to qualified investigators from academic community, *Science* reported. Included were

microscopes, spectrometers, wind tunnels, vacuum chamber, hypervelocity laboratory, and image-processing laboratory. Investigators were expected to help defray expenses. Some limited funding would be furnished by JPL. (*Science*, 12/25/70, 1389)

- Experiment to determine if ozone was trapped in solid carbon dioxide polar cap of Mars was described in *Science* by Univ. of Colorado and York Univ. of Toronto, Canada, scientists. Experiment showed that solid carbon dioxide was effective trap for ozone at temperatures as high as 156 K (-179°F). "Ultraviolet reflection-absorption spectra of ozone in solid carbon dioxide at 127°K [-231°F] indicate that the ozone observed over the polar cap of Mars may be trapped in solid carbon dioxide." (Broida, Lundell, *et al.*, *Science*, 12/25/70, 1402)

December 26-31: AAAS held 137th Annual Meeting in Chicago. Thirteenth Annual Meeting of Society for the History of Technology (SHOT) was held Dec. 26-30 in conjunction with AAAS meeting.

Dr. Edward H. Teller, physicist and atomic scientist of Univ. of California, advocated abolition of secrecy surrounding U.S. scientific research "so we can clearly understand what we are talking about" in growing debate over impact of science and technology on society. "Secrecy in science should be abolished so that the democratic process can be better able to work in making the decisions on how science is applied." (Wilford, *NYT*, 12/28/70, 1)

Dr. Bentley Glass, retiring AAAS president and Academic Vice President of Univ. of New York at Stoney Brook, said number of scientists and technologists in U.S. had doubled in decade to form 20% of professional labor force. Growth rate was much faster than that of population and could not surpass limit of about 25% of professional force. Dr. Glass believed that the greater the volume of research, the less likely any one project would produce truly original results. "It is in fact becoming more and more difficult, as scientific knowledge grows, to make a totally new and unexpected discovery or to break through the dogmas of established scientific views."

SHOT session on Perspectives of Apollo History, chaired by Dr. James Lee Cate of Univ. of Chicago with Dr. John M. Logsdon of George Washington Univ. as commentator, presented sample narratives: "Space Science and Automatic Spacecraft Behind Apollo" by R. Cargill Hall, JPL Historian; "The Saturn Family of Rockets" by Dr. John S. Beltz of Univ. of Alabama at Huntsville; and "The Apollo Command Spaceship" by Dr. Loyd S. Swenson, Jr., of Univ. of Houston. Narratives directed attention to problems of contemporary scholarship and to basic elements of Apollo. (Program; NASA Hist Off)

Jeffrey V. Odom, manager of National Bureau of Standards metric study, told AAAS seminar that report on metrication would be submitted to Congress in August 1971. Report was expected to include recommendations for gradual conversion to metric system in U.S., possibly over 10 yrs. *New York Times*, noting metric system proponents, emphasized that 90% of nations in world already used system and that U.K. expected to be almost completely converted by 1975, said NASA was first U.S. Government agency to convert to system in its scientific and technical publications. (Wilford, *NYT*, 12/29/70, 17)

Stewart L. Udall, former Secretary of the Interior, in AAAS speech accused science establishment, particularly NAS, of failing to exert moral and political leadership on issues of technology impact on man. He proposed that consumer advocate Ralph Nader organize team of young scientists to make "dispassionate and intensive study of the National Academy and the whole scientific enterprise in this country." (Wilford, *NYT*, 12/31/70, 3:6)

Dr. Glenn T. Seaborg, AEC Chairman and AAAS President-elect, left AAAS meeting before radical young scientists assumed podium to accuse him of "crime of science against the people." (Auerbach, *W Post*, 12/31/70, A4)

December 27: NASA announced Stanford Univ. scientists were measuring solar wind in 61-million-km (38-million-mi) space between sun-orbiting *Pioneer VI* (launched Dec. 16, 1965) and *Pioneer VIII* (launched Dec. 13, 1967). Purpose of experiment was to study influence of solar wind on earth's long-term weather cycles and obtain information on solar workings for particle physics. ARC-managed project would continue until May 1971; spacecraft would hold almost fixed position 100 million km (62 million mi) from earth until that time. Measurements were obtained by determining way radio signals between Stanford's 46-m (150-ft) antenna and Pioneers were slowed by interplanetary electrons. Stanford scientist Dr. Thomas Croft had said measurements were "probably the most accurate ever made of the density of free electrons in the solar wind." (NASA Release 70-217)

December 28: NASA held press briefing on preliminary results of *Uhuru (Explorer XLII)* Small Astronomy Satellite (launched Dec. 12). Mrs. Marjorie R. Townsend, SAS-A Project Manager at GSFC, said launch went smoothly and only few minor anomalies had occurred so far: "We are running at higher temperatures than we anticipated seeing at specific angles with respect to the sun. . . .Because of this it took us longer to despin than we had anticipated. The despin dumps a lot of excess [heat] into the spacecraft so we had to turn it on for a while and then turn it off. . .until we did get it into nominal operation. As of this morning SAS-A is working perfectly." (Transcript)

- Crippled flight of *Apollo 13* had been chosen top story of 1970 by AP member editors, AP announced. (*W Star*, 12/28/70)

December 29: European industrial consortium STAR (Satellites for Telecommunications, Applications and Research) was formed to compete for future ESRO contracts. Members were British Aircraft Corp., Contraves A. G. of Switzerland, CGE Fiar and Montedel-Montecatini Edison Electronica S.p.A. of Italy, SABCA of Belgium, L. M. Ericsson AB of Sweden, Fokker FFW of the Netherlands, Dornier-System GmbH of West Germany, and Thomson CSF of France. AEG-Telefunken of West Germany would contribute on comsat work. (*SF*, 3/71, 81)

- Dept. of Justice asked for legislation to force AT&T and other major communications firms out of ownership and management of ComSatCorp. It recommended overturn of FCC's "authorized user" decision (under which ComSatCorp could sell its services only to other communications firms) and earth station ownership decision (under which ComSatCorp could own only 50% of each earth station built, with communications firms owning other 50%). Recommendations were in letters from Assistant Attorney General Richard W. McLaren to Sen. Mike Gravel (D-Alaska), who had asked for investigations of links between ComSatCorp and other communications firms. (Aug, W *Star*, 1/8/71, B10)

- *Pravda* published year-end review of U.S.S.R. space activities by A. Dmitriyev. Soviet leadership was swinging full circle back to belief that Soviet approach was distinctly superior to that of U.S. Automatic spacecraft were cheaper than manned spacecraft but Soviet efforts were encompassing across-the-board capabilities—manned as well as unmanned. Main directions were systematic research in near-earth space using automatic vehicles and manned craft, moon and circumlunar space as testing ground for Soviet cosmonautics, and research of distant planets, primarily Venus, with aid of automatic devices. Recent Soviet accomplishments had paved way for significant breakthroughs in all of these directions. (*AF Mag*, 6/71, 54-9)
- DOT announced award of $1 353 000 grant to JPL for research and design of "people mover" automated transit system. People mover—to be constructed around and through Univ. of West Virginia at Morgantown—would consist of fully automated cars carrying passengers at up to 40 km per hr (25 mph). (DOT Release 25970)

December 30: USN F-14 Tomcat fighter aircraft crashed and burned after developing hydraulic problem during second flight from Grumman Corp. test flight facility at Calverton, N.Y. Aircraft had made maiden flight Dec. 21. Grumman Corp. officials and engineers said aircraft was not downed by major design errors. Both pilots ejected at 90 m (300 ft) and parachuted to safety. (Hoffman, *W Post*, 12/31/70, A1)

- DOD proposal to save Lockheed Aircraft Corp. from bankruptcy by adding $1.45 billion to Government payments on four Lockheed contracts was outlined by David M. Packard, Deputy Secretary of Defense, in letter to Sen. John C. Stennis (D-Miss.), Chairman of Senate Armed Forces Committee. (DOD Release 1057-70; Text)
- Dr. Carl McIntire, head of International Council of Christian Churches and Fundamentalist Bible Presbyterian Church, had announced purchase of property in city of Cape Canaveral, Fla., including Cape Kennedy Hilton Hotel, *Washington Post* reported. Property would be used for multimillion-dollar "Reformation Freedom Center and Christian Conference." Dr. McIntire had said site, 1.6-km (1-mi) strip from Banana River to Atlantic Ocean, "is alongside the rocket gantries and the moon shots" and would lend to God-in-space-conquest theme. *Post* said property had been valued speculatively at $25 million. (*W Post*, 12/30/70, A4)

December 31: NASA announced 1971 flight schedule, highlighted by two manned Apollo lunar landing missions—Apollo 14 Jan. 31 and Apollo 15 July 25—and by two unmanned Mariner Mars launches in May. Other launches tentatively scheduled for 1971 included three Intelsat comsats for ComSatCorp, NATO-B comsat for NATO, IMP-I Interplanetary Monitoring Platform, SOLRAD scientific satellite for

USN, ISIS-B cooperative U.S.-Canada scientific satellite, PAET Planetary Atmosphere Experiments Test, OSO-H Orbiting Solar Observatory, SSS-A Small Scientific Satellite, UK-4 cooperative U.S.-U.K. scientific satellite, AFCRL-A magnetic storm satellite for AFCRL, OAO-C Orbiting Astronomical Observatory, and CAS-A cooperative U.S.-France meteorological satellite.

NASA aeronautical program would include extensive testing of supercritical wing. (NASA Release 70-222)

- Dr. George M. Low, Acting NASA Administrator, would head six-man U.S. team to discuss possibilities for expanded U.S.-U.S.S.R. cooperation in space research at Moscow meetings during week of Jan. 18, 1971, NASA announced. Meetings would consider cooperation in scientific research by satellites, space meteorology, and space biology and medicine. Other members of team would be Dr. John E. Naugle, NASA Associate Administrator for Space Science and Applications; Arnold W. Frutkin, NASA Assistant Adminstrator for International Affairs; Arthur W. Johnson, Deputy Director of National Environmental Satellite Service; William A. Anders, NASC Executive Secretary and *Apollo 8* astronaut; and Robert F. Packard, Director of Office of Space-Atmospheric and Marine Science Affairs of Dept. of State. (NASA Release 70-223)
- Apollo 14 crew, following splashdown in Pacific Feb. 9, 1971, would be flown to MSC from Samoa instead of from Hawaii, reducing return time to MSC by five or six days, NASA announced. Crew would arrive in Houston in MQF about Feb. 12 and would remain under LRL quarantine until Feb. 26. (NASA Release 70-221)
- Senate confirmed nomination of Col. James A. McDivitt (USAF), *Apollo 9* astronaut and Apollo Spacecraft Program Manager at MSC, to be brigadier general. (*CR*, 1/2/71, S21852)
- By vote of 180 to 37, House passed H.J.R. 1421 making further continuing appropriations for FY 1971. Resolution would provide additional $51.7 million to DOT for SST development funding through March 30, 1971. (*CR*, 12/31/70, H12603-6; *NYT*, 1/1/71, 1)

During December: Discovery of two galaxies possibly only 10 million yrs old—one thousandth the age determined for galaxies under "big bang" theory that all material in universe had exploded in one place 10 billion yrs ago—was reported in *Astrophysical Journal* by Hale Observatory astronomers Dr. Wallace L. W. Sargent and Dr. Leonard Searle. Galaxies, about 100 light years in diameter against 100 000 for larger galaxies, contained only 1000 to 100 000 stars of short-lived blue type. Milky Way galaxy contained 10 billion stars, including those capable of living billions of years. If newly discovered galaxies were young, they would provide first evidence that galaxies evolved from infancy to old age and that galaxies could be created after "big bang." Dr. Sargent and Dr. Searle tentatively concluded that galaxies were "dense intergalactic clouds of neutral hydrogen in which the formation of massive stars is proceeding vigorously while the formation of low-mass stars is suppressed." (*Astrophysical Journal*, 12/70, L155-9)

- Results of international deliberations on definitive working arrangements for INTELSAT were described in *Astronautics & Aeronautics* magazine. Australia and Japan had presented Document 39, management plan detailing proposed functions of board of governors and assembly. Board would represent investors. Each member of board would have vote proportional in weight to his country's investment. System would reduce maximum voting percentage of U.S. from 52% to 40% and attempt to avoid veto power for either one nation or small group of major users. Assembly of delegates from each member country would decide by simple majority whether to raise subjects for action, which would then take two-thirds majority to pass. Permanent agreement proposed six-year transition period during which ComSatCorp would remain as systems manager. Three international consulting firms would report best form of management organization during transition period. INTELSAT guideline would be to contract out as many functions as possible to avoid bureaucracy detrimental to efficiency. About 80 nations would sign new agreement at Plenipotentiary Conference in spring 1971. (*A&A*, 12/70, 15-6)

- NSF published *National Patterns of R&D Resources: Funds and Manpower in the United States, 1953-71* (NSF 70-46). R&D expenditures were estimated at $27.8 billion for 1971, 3.7% increase from 1970, but actually no increase in constant dollars. Ratio of total R&D expenditure to GNP would drop from 3% in mid-60s to 2.7% in 1971 as R&D gains failed to keep pace with growth of economy. Federal R&D support rose 1% a year over past five years but in real terms support showed 3% annual decline. Over same period, non-Federal funding grew 10% annually in current dollars and 5% annually in real terms. Space and defense R&D efforts declined as emphasis shifted to other programs. DOD's intramural R&D spending declined, as did DOD basic research spending. (Text)

- NSF problems in stabilizing U.S. basic science research during FY 1971 "with budgetary squeezes on mission-oriented agencies such as NASA, and AEC, and Mansfield-amendment effects rippling through DOD," were described by NSF Director William D. McElroy in interview published by *Astronautics & Aeronautics* magazine: "We can identify positively right now the order of $70 million [in basic science research] that has been previously supported by other agencies. ...we don't have $70 million additional money to pick that up, so there is going to be a squeeze here." Though NSF FY 1971 budget would show $73-million increase over FY 1970 amount of $438 million, only $10 million could be earmarked for picking up programs cut by other Government agencies, (*A&A*, 12/70, 12-14)

During 1970: U.S. orbited 38 spacecraft in 28 launches—including *Apollo 13*'s CSM *Odyssey*, LM *Aquarius*, and S-IVB stage intentionally crashed onto moon. U.S.S.R. orbited 88 payloads in 81 launches; France, 2; Italy, 1 (for NASA on NASA launch vehicle); Japan, 1; and Communist China, 1. U.S. total included 21 orbited by DOD in 17 launches and 17 by NASA in 13 launches. NASA's total included 1 launched for NATO and 1 for DOD.

NASA's manned space flight program continued with launch of *Apollo 13*, but mission was cut short by an oxygen tank rupture. Three-man crew, taking emergency measures and using LM as "lifeboat," returned safely to earth after looping around moon. Analysis of Apollo system and modifications necessary to prevent or meet future emergencies and reevaluation of Apollo program caused rescheduling of Apollo 14 from 1970 to early 1971 and reduction in number of remaining Apollo missions from six to four.

Unmanned program included orbiting of *Intelsat-III F-6, Intelsat-III F-7,* and *Intelsat-III F-8* comsats for ComSatCorp; *Noaa I* meteorological satellite for NOAA and prototype meteorological satellite *Itos I;* NASA's *Nimbus IV* in support of world weather program; *Uhuru (Explorer XLII)* Small Astronomy Satellite (launched for NASA by Italy); *Sert II* Space Electric Rocket Test; and *Ofo* Orbiting Frog Otolith, carrying two bullfrogs. Launched as secondary payloads were *Oscar V* (U.S.-Australia) amateur radio-propagation satellite carried on *Itos I;* DOD's *Topo I* geodetic satellite, on *Nimbus IV; RM* Radiation/Meteoroid Satellite, on *Ofo;* and *Cepe* Cylindrical Electrostatic Probe Experiment, on *Noaa I.*

Almost 100 scientific sounding rockets were launched to altitudes between 160 and 1600 km (100 and 1000 mi) to study space physics and astronomy, including 19 during solar eclipse. One of the 19 detected unexpectedly high amount of neutral hydrogen streaming into earth's atmosphere. Meteorological sounding rockets included 46 Nike-Cajuns and more than 200 boosted Dart and Arcas rockets. About 60 high-altitude balloons—including one carrying Stratoscope II optical telescope—were launched to altitudes as high as 45 700 m (150 000 ft).

In joint NASA-USAF lifting-body program M2-F3, HL-10, and X-24A lifting-body vehicles made 22 flights. HL-10 reached mach 1.86 (1970 km per hr; 1224 mph), setting new lifting-body speed record. M2-F3 made its first glide flight and X-24A made its first supersonic flight, reaching 1223 km per hr (760 mph).

In aeronautics, YF-12A aircraft made 60 flights in joint NASA-USAF research program. Boeing 747 jumbo jet began commercial operations, and maiden flights were made by USN F-14A Tomcat air superiority fighter, Lockheed L-1011 airbus, and Douglas DC-10 airbus. C-5A cargo transport was delivered to Charleston AFB for operational use. SST development continued—amid much debate over costs, sonic boom, and environmental pollution—with Congress approving funding through March 1971. Full-scale hypersonic research engine was tested for first time in LaRC's 8-Foot High-Temperature Structures Tunnel at mach 7.4, and thick supercritical wing was flown on modified USN T-2-C aircraft. Other aeronautical research included studies on runway slipperiness, V/STOL avionics, aircraft noise, landing radar, and pilot warning indicators.

DOD space program included orbiting of last two Vela nuclear detection satellites, *Topo I* geodetic satellite (launched by NASA), and U.K.'s *Skynet B* military comsat (launched by NASA).

In joint NASA-AEC effort, SNAP-19 and SNAP-27 radioisotope generators completed one year of continuous operation on lunar surface, candidate fuel element for NERVA completed 10 hrs of electrical corrosion testing, and Peewee-2 experimental reactor was fabricated and delivered for testing in early 1971.

U.S.S.R.'s 88 payloads included 72 Cosmos satellites, 2 Lunas, 1 Zond, 1 Soyuz, 1 Venus, 2 Intercosmos, 4 Meteors, and 5 Molniya Is. Unmanned *Luna XVI* landed on moon and returned to earth with lunar samples. *Luna XVII* landed on moon and released *Lunokhod I* unmanned explorer that traveled across lunar surface and conducted experiments by ground command. *Venus VII* landed on planet Venus and transmitted data on Venusian surface and atmosphere. Only manned mission was *Soyuz IX*, which carried two cosmonauts on 18-day earth-orbital mission, a new space endurance record.

Two countries orbited their first satellites: Japan launched *Ohsumi* and Communist China launched *Chicom I*. France launched two satellites—*Dial*, carrying West German WIKA minicapsule, and *Peole* applications satellite. (Pres Rpt 71)

- In 1970 NASA reviewed priorities and redirected activities to conform to new national goals and budgetary restraints under Nixon Administration, which directed increased funding toward solution of nation's inflationary and environmental problems. Year was transition from dramatic successes of 1960s to new plans for 1970s and beyond, with balanced aeronautical and space program.

Major administrative event affecting NASA was resignation of Administrator, Dr. Thomas O. Paine, effective Sept. 15. Dr. Paine, who had been nominated NASA Administrator by President Nixon March 5, 1969, returned to General Electric Co. as a Vice President. Dr. George M. Low became Acting NASA Administrator. In March Dr. Wernher von Braun, MSFC Director, was sworn in as Deputy Associate Administrator for Planning, and in May NASA Planning Board was established under his direction for coordination of NASA planning.

NASA adjusted time-phasing of future programs to avoid commitments to excessive funding, but progress was made toward new goals and objectives announced in March. Design, development, and testing for Skylab program went forward and fabrication of flight hardware was begun. System characteristics for space shuttle, key element in program for late 1970s, were established and study contracts were let for vehicle and engine. Program definition studies for space station neared completion. ERTS spacecraft design was completed and aircraft flights tested remote-sensing devices for gathering earth resources data from space. Two flight spacecraft for Mariner Mars '71 missions were progressing toward final testing. Mission design for Grand Tour of outer planets later in decade was completed and conceptual design of key elements of spacecraft established. NASA continued aeronautical and space studies in conjunction with DOD, DOT, NOAA, and other agencies—including NASA-FAA V/STOL avionics program.

Scientific analysis of 56 kg (123 lbs) of lunar material from *Apollo 11* and *12* and data from Apollo experiments contributed significant knowledge on moon and possibly universe, including finding that moon was formed 4.5 billion yrs ago. GSFC research in optical methods for data processing experimented in applying lasers and coherent optics to analysis of large volumes of data from spacecraft.

During year NASA's research centers expanded work on applications of space technology to fields of medicine, engineering, agriculture, manufacturing, safety, and environmental sciences.

NASA FY 1971 appropriation of $3.269 billion was $64 million below $3.333 billion requested and $428 million below FY 1970 appropriation. Lowest budget since 1962 necessitated cutbacks and deferrals which NASA Administrator termed "austere." Cost of NASA space program in 1970 was $3.4 billion—1.7% of total Federal funding, or $17 for each person in U.S., against $400 per person for domestic program funding. Despite its challenging goals for the decade, NASA officials warned of increasing U.S.S.R. competition for space superiority. Observers expressed fears that cutbacks in aerospace industry jeopardized U.S. ability to sustain and capitalize on previous space successes and to maintain technological superiority. (A&A 70; Pres Rpt 71; House Com on Sci and Astronautics, *The National Space Program—Present and Future*, 12/10/70)

- In international cooperation program, NASA successfully launched U.K.'s *Skynet B* military comsat for DOD, *Nato I* military comsat for NATO, and *Oscar V* radio propagation satellite for U.S. and Australian amateur radio operators. Italian crew, trained by NASA, launched NASA's *Uhuru* on NASA launch vehicle.

  U.S.-U.S.S.R. agreement was reached on procedure for joint efforts to design compatible rendezvous and docking systems for manned spacecraft. Since European Space Conference had indicated that major participation in U.S. space transportation program would probably mean end to Europa III launcher program funding, ESC was assured U.S. would continue to make reimbursable launch services available for European payloads for peaceful purposes. Such participation was understood to mean at least 10% of developmental costs of space transportation system, or about $1 billion.

  During year NASA held briefings for U.N. Outer Space Committee and invited U.N. and specialized agencies to participate in International Workshop on Earth Resources Survey System in 1971. Discussions were held with Brazil, Canada, and Mexico on agreements to extend cooperation in earth resources area; experimenters from 28 countries signified intention to submit proposals to analyze data acquired by ERTS.

  Joint working group meetings were held in India and U.S. on satellite instructional TV experiment for ATS-F. NASA and AID began 18-mo experimental project with Korea in application of aerospace technology to Korean economic problems. NASA and Canadian government signed agreement to study application of augmentor-wing concept to V/STOL aircraft. Agreement was concluded with U.K. to test-fly NASA's XH-51 rigid-rotor helicopter in U.K. and to share data.

  Agreements for new projects were reached with Netherlands, U.K., and Italy. Netherlands agreement called for cooperative development and launch in 1974 of small astronomical satellite (ANS) with one U.S. and two Dutch experiments to investigate stellar UV and x-ray radiation. U.K. agreement was for 1973 launch of UK-R x-ray satellite, fifth in Ariel series. Italian agreement established principles for NASA launch on reimbursable basis of Italian Sirio satellite to study wave propagation and electron flows in magnetosphere. NASA accepted experiment from French National Laboratory for Space Astronomy to study stellar UV emissions from Skylab and accepted instrument developed by French National Laboratory for Stellar and Planetary Physics for investigating solar chromosphere structure from OSO-I, scheduled for 1973 launch. Assistance was accepted from U.K.'s Jodrell Bank Observatory in planning activities of radio science team for 1975 Project Viking Mars probe.

New sounding rocket agreements were signed with Australia, India, Spain, and Sweden, and negotiations were held with France for cooperative upper-atmosphere sounding rocket launchings at Kourou in French Guiana.

Lunar material returned by *Apollo 11* and *Apollo 12* was studied by 54 foreign principal investigators in 16 countries in addition to U.S. scientists. (*A&A* 70; Pres Rpt 71)

- Aerospace industry sales continued anticipated decline in 1969, to $24.9 billion from $26.1 billion—decrease of 4.6%. Commercial aerospace sales, primarily jet transports, increased 13.7% in 1970 from $4.342 million in 1969, to $4.910 million. Increase reflected delivery of wide-bodied jet transports. Space sales continued decline, to $3.606 million from $4.272 million in 1969, because of completion of Apollo program hardware phase and decline in military space expenditures. Aerospace industry remained Nation's largest manufacturing employer during 1970, with more than 1 million workers. Employment in aerospace industry declined from average of 1 347 000 workers in December 1969 to 1 067 000 in December 1970. Production workers dropped from 696 000 in December 1969 to 515 000 in December 1970, a 26% decrease. (*Aerospace*, 12/70, 3)
- *Apollo 11* and *12* lunar samples were seen by 41 million persons in U.S. and abroad during year. Samples were shown at 128 events covering all 50 states, with total attendance of some 12 million. Outside U.S., samples were shown to almost 30 million persons in 110 countries, including almost 14 million visitors to Expo '70 in Japan. NASA facilities across U.S. were inspected by 3 million persons, with heaviest influx at MSC (over 1 million) and KSC (1.2 million). (NASA Release 71-8)
- Scheduled domestic airlines experienced no passenger deaths within continental limits of U.S. during year. For all scheduled flights, airlines compiled record of 0.001 death per 100 million passenger-miles —lowest record in airline history. Aircraft hijacked to Cuba totaled 13 in 1970, 31 in 1969. (DOT Release 0971)
- Total scheduled airline traffic gained only 1% in 1970. Average growth rate in 1960s had been 16.6%. Airline industry incurred largest losses in 1970—$179 million. Losses caused rate of return on investment to fall to lowest level in history—1.5%. Domestic passenger traffic growth slowed almost to halt in 1970. Beginning in August, passenger traffic on domestic trunk carriers declined each month from level of same month in 1969. (*Air Transport 1971*, Annual Rpt of US scheduled airline industry)
- Operations at 335 FAA-staffed control towers in U.S. and possessions decreased 2% from level of 56 231 821 in 1969 to 55 280 498 in 1970. Operations had increased 2% in 1969 and 11% in 1968. Number of aircraft handled by FAA's air route traffic control centers declined 1% in 1970. In 1969 centers had recorded 11% increase following 1968 increase of 17%. (FAA Release 71-85)

Appendix A

SATELLITES, SPACE PROBES, AND MANNED SPACE FLIGHTS

A CHRONICLE FOR 1970

The following tabulation was compiled from open sources by Leonard C. Bruno of the Science and Technology Division of the Library of Congress. Sources included the United Nations Public Registry; the *Satellite Situation Report* issued by the Operations Control Center at Goddard Space Flight Center; public information releases of the Department of Defense, NASA, ESSA, and other agencies, as well as those of the Communications Satellite Corporation. Russian data are from the U.N. Public Registry, the *Satellite Situation Report*, translations from the Tass News Agency, statements in the Soviet press, and international news services reports. Data on satellites of other foreign nations are from the U.N. Public Registry, the *Satellite Situation Report*, governmental announcements, and international news services reports.

This tabulation lists payloads that have (a) orbited; (b) as probes, ascended to at least the 4000-mile altitude that traditionally has distinguished probes from sounding rockets, etc.; or (c) conveyed one or more human beings into space, whether orbit was attained or not. Furthermore, only flights that have succeeded—or at least can be shown by tracking data to have fulfilled our definition of satellite or probe or manned flight—are listed. Date of launch is referenced to local time at the launch site. An asterisk by the date marks dates that are one day earlier in this tabulation than in listings which are referenced to Greenwich Mean Time. A double asterisk by the date marks dates of Soviet launches which are a day later in this compilation than in listings which are referenced to Greenwich Mean Time.

World space activity increased slightly for the first time in four years. There was an increase in the total successful launches—114 against 110 in 1969—and an increase in total payloads orbited—130 against 124 in 1969. The difference between launches and payloads is of course accounted for by the multiple-payload launches (DOD, the principal user of this system in the past, made only 4 multiple launches in 1970, orbiting a total of 8 payloads; NASA also made 4 multiple launches totaling 8 payloads; the U.S.S.R., whose previous multiple-payload launch was in 1965, made one multiple launch orbiting 8 payloads.

Of the 1970 world total, the United States launched 28 boosters carrying 36 payloads (contrasting with 40 and 54 in 1969). Of these totals, DOD was responsible for 17 launches and 21 payloads. The 28 launches made the lowest U.S. total since 1960, and the 36 payloads are the fewest since 1961. Eight of NASA's 11 launches were non-NASA missions—*Intelsat-III F-6, Intelsat-III F-7, Intelsat-III F-8, Noaa I, Nato I, Skynet B, Oscar V,* and *Topo I*. The Soviet Union again dominated world totals, launching 88 payloads with a record 81 launches. It had launched 70 boosters in 1969.

The year was marked by the Soviet manned accomplishments of *Soyuz IX* and the unmanned exploits of *Luna XVI* and *Luna XVII*. The two-man crew of *Soyuz IX* set a new world endurance record for the longest manned space flight—nearly 18 days in orbit. The unmanned *Luna XVI* landed on the moon, mechanically retrieved bits of lunar soil, and returned its lunar payload to earth. *Luna XVII* deposited the first unmanned lunar rover, *Lunokhod I*, on the moon; it continued to function and return information nearly a year later. In addition, *Venus VII* transmitted the first information from that planet's surface and *Zond VIII* circled the moon and returned to earth.

Apollo 13 was the major event on the 1970 U.S. space calendar, but this third manned lunar landing attempt resulted in a mission failure. The lunar mission was aborted following a service module rupture and the three-man crew employed emergency procedures and returned to earth as quickly as possible. The U.S. also successfully completed a biological mission (*Ofo*), and initiated a new generation of improved meteorological satellites (*Itos I* and *Noaa I*). The NASA satellite *Explorer XLII*, launched by Italy near the end of the year, discovered several new extra-galaxy sources of intense energy.

Also during 1970, Japan became the fourth nation to orbit a satellite (*Ohsumi*) and the People's Republic of China became the fifth with its launch of *Chicom I*. France launched a satellite of its own (*Peole*) in addition to a French-German cooperative effort to orbit the *Dial/Mika-Wika satellites*. And Italy launched a NASA satellite from its mobile seaborne facility off the coast of Kenya.

As we have cautioned in previous years, the "Remarks" column of these appendixes is never complete, because of the inescapable lag behind each flight of the analysis and interpretation of the results.

Launch Date	Name, Country, International Designation, Vehicle	Payload Data	Apogee in Kilometers (and st mi)	Perigee in Kilometers (and st mi)	Period in Minutes	Inclination in Degrees	Remarks
Jan. 9	Cosmos CCCXVIII (U.S.S.R.) 1970-1A Not available	Total weight: Not available. Objective: "Continuation of Cosmos scientific satellite series." Payload: Not available.	276 (171.5)	203 (126.1)	89.2	64.9	Reentered 1/21/70.
Jan. 14	DOD spacecraft (United States) 1970-2A Titan IIIB-Agena D	Total weight: Not available. Objective: Develop space flight techniques and technology. Payload: Not available.	407.2 (253)	125.5 (78)	89.8	109.9	Reentered 2/1/70.
Jan. 14*	Intelsat-III F-6 (United States) 1970-3A DSV-3M	Total weight: 151.5 kg (334 lbs) in synchronous orbit; 293.9 kg (684 lbs) at liftoff. Objective: Place satellite and apogee motor into proper transfer orbit; provide tracking and telemetry and backup calculations through transfer orbit so satellite can be injected into synchronous orbit for commercial communications. Payload: 198-cm x 142-cm-dia (78-in x 56-in-dia) cylindrical satellite including mechanically despun receive or transmit antennas on top of spacecraft; outer cylindrical sleeve covered with 10 720 solar cells; 2 frequency translation mode repeaters with design capability for 1200 high-quality, 2-way voice or 4 TV channels, or a mix; attitude and control system, battery.	35 871.2 (22 289.3) After apogee motor firing. 35 821 (22 258.1)	289.5 (179.9) 35 795 (22 241.9)	634.5 1437.2	28.05 0.9	Launched by NASA long-tank, thrust-augmented, Thor-Delta into good transfer orbit for ComSatCorp. Apogee motor fired 1/16/70, stationing satellite in synchronous orbit at 24° west longitude. Replaced Intelsat III-F-2 over Atlantic. Carries communications traffic between U.S., Latin America, Europe, and Middle East. Spacecraft operating normally; began full-time commercial service 2/7/70. Still in orbit.
Jan. 15	Cosmos CCCXIX (U.S.S.R.) 1970-4A Not available	Total weight: Not available. Objective: "Continuation of Cosmos scientific satellite series." Payload: Not available.	1490 (925.8)	195 (121.2)	101.9	81.8	Reentered 7/1/70.
Jan. 16	Cosmos CCCXX (U.S.S.R.) 1970-5A Not available	Total weight: Not available. Objective: "Continuation of Cosmos scientific satellite series." Payload: Not available.	297 (184.6)	239 (148.5)	89.8	48.4	Reentered 2/10/70.
Jan. 20	Cosmos CCCXXI (U.S.S.R.) 1970-6A Not available	Total weight: Not available Objective: "Continuation of Cosmos scientific satellite series." Payload: Not available.	473 (293.9)	271 (168.4)	91.9	70.9	Reentered 3/23/70.

Launch Date	Name, Country, International Designation, Vehicle	Payload Data	Apogee in Kilometers (and st mi)	Perigee in Kilometers (and st mi)	Period in Minutes	Inclination in Degrees	Remarks
Jan. 21	Cosmos CCCXXII (U.S.S.R.) 1970-7A Not available	Total weight: Not available. Objective: "Continuation of Cosmos scientific satellite series." Payload: Not available.	311 (193.3)	195 (121.2)	89.5	65.4	Reentered 1/29/70.
Jan. 23	Itos I (United States) 1970-8A DSV-3N-6 and	Total weight: 306 kg (675 lbs). Objective: Flight qualify prototype spacecraft, thereby obtaining engineering data required for evaluation of single-momentum wheel-stabilization system for earth-oriented, stabilized platform; evaluate use of stabilized platform for operational methodology by performing cloud-cover observations in both direct readout and stored-modes of operation. Payload: Rectangular, box-shaped spacecraft with deployable 3-panel solar array. 3-axis stabilized, earth-oriented satellite carries 2 advanced vidicon camera subsystems (AVCS) and 2 automatic picture transmission (APT) camera subsystems which provide daytime coverage and 2 scanning radiometer (SR) subsystems (infrared) which furnish nighttime coverage. Solar panels measure 0.9-1.5 m (3-5 ft) each and total 4.6 sq m (48 sq ft). Spacecraft measures 4.3 m (14 ft) with panels deployed. Thermal control system. 4 antennas. 16 000 n-on-p solar cells produce 250 w avg. power.	1478.2 (918.7)	1432.3 (890.2)	115	101.9	Second-generation, operational meteorological satellite placed in circular, near-polar, sun-synchronous orbit by long-tank, thrust-augmented, Thor-Delta. Oscar V amateur radio satellite carried as secondary payload. First launch of Delta with 6 strap-ons. Spacecraft funded and developed by NASA. Provided first daily, global, day-night cloud-cover data; also furnishes both remote and local readout. Adjudged success and turned over to ESSA (which later became part of NOAA). Still in orbit.
	Oscar V 1970-8B	Total weight: 17.7 kg (39 lbs). Objective: Transmit low-power signals on 2 amateur bands to be used by radio amateurs world-wide for training in art of tracking and for experiments useful to NASA in radio propagation. Payload: Satellite carries 9.1 kg (20 lbs) of batteries and transmits at 29.45 mhz in 10-m band and 144.05 mhz in 2-m band.	1482 (920.9)	1434 (891.1)	115	102	Launched by NASA as secondary payload for Radio Amateur Satellite Corporation. Spacecraft built at University of Melbourne, Australia. Still in orbit.

Feb. 3*	Sert II (United States) 1970-9A Thorad-Agena D	Total weight: 1500 kg (3300 lbs) in orbit (of which 499 kg [1100 lbs] is spacecraft and SSU). Objective: Accumulate 6 mos (4383 hrs) of satisfactory operation on an electric ion thruster in space. Payload: 53-cm x 150-cm-dia (21-in x 59-in-dia) cylindrical spacecraft support unit (SSU), same size spacecraft support unit (SSU), which is attached to 6.6-m (21.5-ft) long and 1.5-m (5-ft)-dia Agena D 2nd stage. Two 1.5- x 5.8-m (5- x 15-ft) solar arrays attached to Agena D stage measure 17.4 sq m (187.5 sq ft) total and contain total of 33 300 solar cells for power total of 1471 w. Spacecraft contains two 1-kw ion thrusters each powered by 13.2 kg (29 lbs) of mercury propellant. Thermal control system.	1008.9 (626.9)	998.4 (620.4)	105.2	99.1	Launched into excellent orbit by long tank, thrust-augmented, Thor-Agena D. First orbital test of electron-bombardment ion engines. Spacecraft's solar arrays largest ever flown. No. 1 engine shut down 7/23/70 after successful operation for over 5 mos (3782 hrs). No. 2 engine shut down 10/17/70 after 2 ¾ mos operation (2011 hrs). Failure of each thruster caused by electrical short in high-voltage power supply system. Still in orbit.
Feb. 10	Cosmos CCCXXIII (U.S.S.R.) 1970-10A Not available	Total weight: Not available. Objective: "Continuation of Cosmos scientific satellite series." Payload: Not available.	314 (195.1)	200 (124.3)	89.6	65.3	Reentered 2/18/70.
Feb. 11	Ohsumi (Japan) 1970-11A Lambda 4S	Total weight: 38-kg (84-lb) satellite carries 10.9-kg (24-lb) instrumented payload. Objective: Launch scientific observation satellite into earth orbit. Payload: 46-cm (18-in) wide V-shaped satellite contains 295 mhz transmitter.	5136 (3191.4)	525 (326.2)	116.1	31.4	First Japanese orbital success after 4 previous failures. Japan became 4th nation to orbit earth satellite on own booster. Spacecraft and launch vehicle Japanese-built. Transmission of signals faded on 7th orbit. Still in orbit.
Feb. 11	DOD spacecraft (United States) 1970-12A Thor-Burner II	Total weight: Not available. Objective: Develop space-flight techniques and technology. Payload: Not available.	872.3 (542)	772.5 (480)	101.3	98.6	Still in orbit.
Feb. 19	Molniya I-13 (U.S.S.R.) 1970-13A Not available	Total weight: Not available. Objective: Develop and improve satellite and TV communications system. Payload: Not available.	39 309 (24 425.4)	335 (208.2)	703.4	65.3	Still in orbit.
Feb. 27	Cosmos CCCXXIV (U.S.S.R.) 1970-14A Not available	Total weight: Not available. Objective: "Continuation of Cosmos scientific satellite series." Payload: Not available.	465 (288.9)	264 (164)	91.9	71.0	Reentered 5/23/70.
Mar. 4	Cosmos CCCXXV (U.S.S.R.) 1970-15A Not available	Total weight: Not available. Objective: "Continuation of Cosmos scientific satellite series." Payload: Not available.	321 (199.5)	199 (123.7)	89.7	65.4	Reentered 3/12/70.

ASTRONAUTICS AND AERONAUTICS, 1970

Launch Date	Name, Country, International Designation, Vehicle	Payload Data	Apogee in Kilometers (and st mi)	Perigee in Kilometers (and st mi)	Period in Minutes	Inclination in Degrees	Remarks
Mar. 4	DOD spacecraft (United States) 1970-16A Thorad-Agena D	Total weight: Not available. Objective: Develop space-flight techniques and technology. Payload: Not available.	226.9 (141)	224.7 (139.7)	88.4	88.4	Dual launch with single booster. 16A reentered 3/26/70.
	and						
	DOD spacecraft 1970-16B	Total weight: Not available. Objective: Develop space-flight techniques and technology. Payload: Not available.	506.9 (315)	439.4 (273)	94.0	88.1	Still in orbit.
Mar. 10	Dial/Wika (France-W. Germany) 1970-17A Diamant-B	Total weight: 64.9 kg (143 lbs). Objective: Study equatorial particle (geocoronal) belt; measure light in wavelength of Lyman alpha ray, electron density. Payload: Not available.	1613 (1002.3)	308 (191.4)	104.3	5.4	Wika, German scientific satellite, launched by French Diamant-B from Kourou, French Guiana into approximate equatorial orbit. Operating successfully. Launched with Mika, French secondary payload. Still in orbit.
	and						
	Mika (France) 1970-17B	Total weight: 49.9 kg (110 lbs). Objective: Measure reactions of launch vehicle during flight. Payload: Satellite remained fixed to Diamant-B 3rd stage; 1-hr lifespan.	1656 (1029)	307 (190.8)	104.7	5.4	Mika, French technological satellite, launched as secondary payload; damaged by launch vibration between 2nd and 3rd stages and rendered inoperable. Entire orbital package (French Mika and German Wika) called Dial, "Diamant pour l'Allemagne." Still in orbit.
Mar. 13	Cosmos CCCXXVI (U.S.S.R.) 1970-18A Not available	Total weight: Not available. Objective: "Continuation of Cosmos scientific satellite series." Payload: Not available.	239 (148.5)	209 (129.9)	90.1	81.3	Reentered 3/21/70.
Mar. 17	Meteor III (U.S.S.R.) 1970-19A Not available	Total weight: Not available. Objective: Trace cloud cover, detect snow on both dark and daylight sides of earth, and record amount of heat energy radiated and reflected from atmosphere. Payload: Not available.	633 (393.3)	537 (333.7)	96.3	81.1	Meteor III meteorological satellite still in orbit.
Mar. 18	Cosmos CCCXXVII	Total weight: Not available.	823	268	95.5	70.9	Still in orbit.

ASTRONAUTICS AND AERONAUTICS, 1970

	(U.S.S.R.) 1970-20A Not available	Objective: "Continuation of Cosmos scientific satellite series." Payload: Not available.	(511.4)	(166.5)		
Mar. 20	Nato I (U.S.-NATO) 1970-21A DSV-3M	Total weight: 242.7 kg (535 lbs). Objective: Place communications satellite into proper transfer orbit so apogee motor can inject spacecraft into synchronous equatorial orbit. Payload: 81-cm x 137-cm-dia (32-in x 54-in-dia) cylindrical satellite constructed of 2 concentric cylinders with apogee motor within inner cylinder. Solar cells mounted on outside surface of outer cylinder. Despun antenna system and redundant X-band communications systems with power output of 3.5 w.	36 926.6 (22 950) After apogee motor firing, 36 619.2 (22 759)	290 (180.7) 34 421.3 (21 393)	653.0 1410	25.6 0.3	NATO military communications satellite launched into good transfer orbit by long-tank, thrust-augmented, Thor-Delta. Apogee kick motor placed spacecraft in synchronous orbit 3/23/70. Communications system failure on 10/25/70 necessitated turning on back-up system. Still in orbit and still operating on redundant system.
Mar. 27	Cosmos CCCXXVIII (U.S.S.R.) 1970-22A Not available	Total weight: Not available. Objective: "Continuation of Cosmos scientific satellite series." Payload: Not available.	316 (196.4)	206 (128)	89.7	72.8	Reentered 4/9/70.
Apr. 3	Cosmos CCCXXIX (U.S.S.R.) 1970-23A Not available	Total weight: Not available. Objective: "Continuation of Cosmos scientific satellite series." Payload: Not available.	241 (149.8)	210 (130.5)	88.9	81.4	Reentered 4/15/70.
Apr. 7	Cosmos CCCXXX (U.S.S.R.) 1970-24A Not available	Total weight: Not available. Objective: "Continuation of Cosmos scientific satellite series." Payload: Not available.	538 (334.3)	517 (321.3)	95.1	74	Still in orbit.
Apr. 8	Nimbus IV (United States) 1970-25A Thorad-Agena D and	Total weight: 675.2 kg (1488.6 lbs). Objective: Acquire sufficient number of global samples of atmospheric radiation measurements for purpose of comparing vertical temperature, water vapor, and ozone profiles and for comparing merits of several instrument approaches. Comparative data to be obtained from successful operation of at least 3 of 5 spectrometric experiments; satellite infrared spectrometer (SIRS), infrared interferometer spectrometer (IRIS), filter wedge spectrometer (FWS), selective chopper radiometer (SCR), or backscatter ultraviolet (BUV) spectrometer, or from successful operation of either FWS or SCR, plus either SIRS or IRIS. Payload: 305-cm (120-in)-high and 335-cm (132-in)-wide spacecraft consists of 3 major elements:142-cm (56-in) torus ring	1098 (682.4)	1087.2 (675.7)	107.1	99.9	Nimbus IV launched into nearly circular orbit by long-tank, thrust-augmented, Thor-Agena D; also carried Topo I as secondary payload. Spacecraft and sensors operating successfully except for FWS, which failed 6/7/70. Primary objectives achieved and flight a success. Still in orbit.

Launch Date	Name, Country, International Designation, Vehicle	Payload Data	Apogee in Kilometers (and st mi)	Perigee in Kilometers (and st mi)	Period in Minutes	Inclination in Degrees	Remarks
		forms base of spacecraft and houses major spacecraft electronics; smaller hexagon-shaped housing, connected to ring by truss, houses attitude stabilization and control system; and 2 0.9- x 1.5-m (3- x 5-ft) solar paddles. Active 3-axis stabilization. Solar cells provide 465 w of power and 8 nickel-cadmium batteries average 255 w.					
	Topo I 1970-25B	Total weight: 21.3 kg (47 lbs). Objective: Investigate new technique for accurate, real-time determination of positions on earth's surface. Payload: Rectangular box-shaped structure, 36 x 30 x 23 cm (14 x 12 x 9 in). Solar cell power supply, batteries, ranging transponder. Operating frequencies for command and data retrieval are 244.5 mhz and 449 mhz.	1092 (678.5)	1085 (674.2)	107	99.8	Launched as secondary payload by NASA for DOD (U.S. Army Topographic Command). Spacecraft uses modified SECOR equipment and is used through triangulation for ground tactical positioning. Spacecraft functioning normally and still in orbit.
Apr. 8	Cosmos CCCXXXI (U.S.S.R.) 1970-26A Not available	Total weight: Not available. Objective: "Continuation of Cosmos scientific satellite series." Payload: Not available.	323 (200.7)	203 (126.1)	89.6	64.9	Reentered 4/16/70.
Apr. 8	Vela XI (United States) 1970-27A Titan IIIC and	Total weight: 350 kg (770 lbs). Objective: Orbit nuclear detection sensors to monitor x-ray, gamma-ray, neutron, optical, electromagnetic-pulse, and airfluorescence emissions. Payload: 26-sided polyhedron with 127-cm (50-in) dia consisting of reaction wheel plus gas jets for stabilization, solar-cell array, 28 detectors.	110 739.5 (68 810.3)	110 091.8 (68 478.4)	6695	37.7	Dual launch of nuclear detection satellites; last in Vela series. Still in orbit.
	Vela XII 1970-27B	Total weight: 350 kg (770 lbs). Objective: Orbit nuclear detection sensors to monitor x-ray, gamma-ray, neutron, optical, electromagnetic-pulse, and airfluorescence emissions. Payload: 26-sided polyhedron with 127-cm (50-in) dia, consisting of reaction wheel	111 379.8 (69 281.8)	111 111.1 (69 041.2)	6699	32.9	Still in orbit.

ASTRONAUTICS AND AERONAUTICS, 1970

Date	Name	Details	Orbit data	Period	Remarks
Apr. 11	Cosmos CCCXXXII (U.S.S.R.) 1970-28A Not available	plus gas jets for stabilization, solar array, 28 detectors. Total weight: Not available. Objective: "Continuation of Cosmos scientific satellite series." Payload: Not available.	760 (472.2)	755 (469.1)	99.9	74	Still in orbit.
Apr. 11	Apollo 13 (United States) 1970-29A Saturn V	Total weight: 134 472.7 kg (296 463 lbs) (weight at initial earth orbit insertion, including S-IVB stage, instrument unit, spacecraft LM adaptor, LM, and CSM). Objective: Perform selenological inspection, survey, and sampling of materials in preselected region of Fra Mauro Formation; deploy and activate Apollo lunar surface experiments package (ALSEP); develop man's capability to work in lunar environment; obtain photographs of candidate exploration sites. Payload: 34.7-m (114-ft)-long S-IVB/IU/spacecraft LM adaptor/Block II command and service modules/lunar module; cameras; telemetry.	185.5 (115.3)	181.5 (112.8)			Apollo 13, third manned lunar landing mission, launched successfully carrying Astronauts James A. Lovell, Jr., John L. Swigert, Jr., and Fred W. Haise, Jr. Mission was aborted 56 hrs into flight toward moon because of SM oxygen tank rupture. First inflight mission failure in 23 manned U.S. flights. Crew followed LM "lifeboat" emergency plan, transferring to LM and using its systems and supplies to swing around moon and back to earth on fastest feasible course. Emergency procedures used to conserve electrical power, water, and oxygen. Crew returned to CM before reentry, jettisoned LM, and splashed down safely in Pacific 1:07 pm EST 4/17/70. Total flight time 142 hrs 54 min 41 secs.
Apr. 15	Cosmos CCCXXXIII (U.S.S.R.) 1970-30A Not available	Total weight: Not available. Objective: "Continuation of Cosmos scientific satellite series." Payload: Not available.	226 (140.4)	211 (131.1)	88.8	81.3	Reentered 4/28/70.
Apr. 15	DOD spacecraft (United States) 1970-31A Titan IIIB-Agena D	Total weight: Not available. Objective: Develop space-flight techniques and technology. Payload: Not available.	386.2 (240)	136.8 (85)	89.7	110.9	Reentered 5/6/70.
Apr. 22*	Intelsat-III F-7 (United States) 1970-32A DSV-3M	Total weight: 151.5 kg (334 lbs) in synchronous orbit; 293.5 kg (647 lbs) at liftoff. Objective: Place satellite and apogee motor into proper transfer orbit; provide tracking and telemetry and backup calculations through transfer orbit so satellite can be injected into synchronous orbit for commercial communications. Payload: 196-cm x 142-cm-dia (78-in x 56-in-dia) cylindrical satellite including mechanically despun receive/transmit and omnidirectional command and telemetry antennas on top of spacecraft; outer cylindrical sleeve covered with	19 895 (12 362.2) After apogee motor firing. 35 772.9 (22 233)	184 (114.3) 35 737.7 (22 211.1)	560 1408.1	0.3	NASA launch vehicle, long-tank thrust-augmented, Thor-Delta underperformed, and onboard hydrazine thrusters needed to push satellite into desired synchronous orbit 19° west longitude over Atlantic. Carries communications traffic between U.S., Europe, North Africa, and Middle East. Remaining fuel expected to maintain satellite in proper orbit 5 yrs. Began commercial service 5/8/70. Still in orbit.

Launch Date	Name, Country, International Designation, Vehicle	Payload Data	Apogee in Kilometers (and st mi)	Perigee in Kilometers (and st mi)	Period in Minutes	Inclination in Degrees	Remarks
		10 720 solar cells; 2 frequency translation mode repeaters with design capability for 1200 high-quality, 2-way voice or 4 TV channels, or a mix, attitude and control system; battery.					
Apr. 23	Cosmos CCCXXXIV (U.S.S.R.) 1970-33A Not available	Total weight: Not available. Objective: "Continuation of Cosmos scientific satellite series." Payload: Not available.	482 (299.5)	271 (168.4)	92	70.9	Reentered 8/9/70.
Apr. 24	Chicom I (People's Republic of China) 1970-34A Not available	Total weight: 172.8 kg (381 lbs). Objective: Provide conditions for development of research in astronomy, atmosphere, physics and other fields. Payload: Transmits on frequency of 20.009 mhz.	2387 (1483.2)	439 (272.8)	114	68.4	First Communist Chinese orbital success. Fifth nation to orbit earth satellite. Spacecraft broadcast telemetry data and revolutionary song "The East Is Red." Still in orbit.
Apr. 24	Cosmos CCCXXXV (U.S.S.R.) 1970-35A Not available	Total weight: Not available. Objective: "Continuation of Cosmos scientific satellite series." Payload: Not available.	398 (247.3)	249 (154.7)	91	48.4	Reentered 6/22/70.
Apr. 25	Cosmos CCCXXXVI (U.S.S.R.) 1970-36A Not available	Total weight: Not available. Objective: "Continuation of Cosmos scientific satellite series." Payload: Not available.	1489 (925.2)	1464 (909.7)	115.4	73.9	Eight satellites launched with single booster. Still in orbit.
	Cosmos CCCXXXVII 1970-36B	Total weight: Not available. Objective: "Continuation of Cosmos scientific satellite series." Payload: Not available.	1554 (965.6)	1469 (912.8)	116.2	74	Still in orbit.
	Cosmos CCCXXXVIII 1970-36C	Total weight: Not available. Objective: "Continuation of Cosmos scientific satellite series." Payload: Not available.	1517 (942.6)	1472 (914.7)	115.8	74	Still in orbit.
	Cosmos CCCXXXIX 1970-36D	Total weight: Not available. Objective: "Continuation of Cosmos scientific satellite series." Payload: Not available.	1468 (912.2)	1450 (901)	115	74	Still in orbit.

Date	Name	Description	Col4	Col5	Col6	Col7	Notes
	Cosmos CCCXL 1970-36E	Total weight: Not available. Objective: "Continuation of Cosmos scientific satellite series." Payload: Not available.	1468 (912.2)	1412 (877.4)	114.6	74	Still in orbit.
	Cosmos CCCXLI 1970-36F	Total weight: Not available. Objective: "Continuation of Cosmos scientific satellite series." Payload: Not available.	1470 (913.4)	1345 (835.7)	113.9	74	Still in orbit.
	Cosmos CCCXLII 1970-36G	Total weight: Not available. Objective: "Continuation of Cosmos scientific satellite series." Payload: Not available.	1470 (913.4)	1312 (815.2)	113.5	74	Still in orbit.
	Cosmos CCCXLIII 1970-36H	Total weight: Not available. Objective: "Continuation of Cosmos scientific satellite series." Payload: Not available.	1468 (912.2)	1379 (856.9)	114.2	74	Still in orbit.
Apr. 28	Meteor IV (U.S.S.R.) 1970-37A Not available	Total weight: Not available. Objective: Photograph cloud cover and snow and study atmospheric thermal energy radiated by earth. Payload: Not available.	743 (461.7)	636 (395.2)	98.2	81.2	Meteor IV meteorological satellite still in orbit.
May 12	Cosmos CCCXLIV (U.S.S.R.) 1970-38A Not available	Total weight: Not available. Objective: "Continuation of Cosmos scientific satellite series." Payload: Not available.	326 (202.6)	204 (126.8)	87.3	72	Reentered 5/20/70.
May 20	Cosmos CCCXLV (U.S.S.R.) 1970-39A Not available	Total weight: Not available. Objective: "Continuation of Cosmos scientific satellite series." Payload: Not available.	257 (159.7)	187 (116.2)	88.9	51.7	Reentered 5/28/70.
May 20	DOD spacecraft (United States) 1970-40A Thorad-Agena D and DOD spacecraft 1970-40B	Total weight: Not available. Objective: Develop space-flight techniques and technology. Payload: Not available. Total weight: Not available. Objective: Develop space-flight techniques and technology. Payload: Not available.	236.6 (147) 503.7 (313)	178.6 (111) 489.2 (304)	88.6 94.5	83 83.1	Dual payload launched with single booster. 40A reentered 6/17/70. Still in orbit.
June 2	Soyuz IX (U.S.S.R.) 1970-41A Not available	Total weight: Not available. Objective: Carry out extensive program of scientific and technical research and experiments in conditions of "solitary orbital flight."	249 (154.7)	236 (146.6)	89.3	51.6	Soyuz IX carrying Cosmonauts Andrian G. Nikolayev and Vitaly I. Sevastyanov set new world endurance record for longest manned space flight. After initial orbit, spacecraft made 2 orbital corrections to obtain extended-duration orbit. During record flight,

ASTRONAUTICS AND AERONAUTICS, 1970

Launch Date	Name, Country, International Designation, Vehicle	Payload Data	Apogee in Kilometers (and st mi)	Perigee in Kilometers (and st mi)	Period in Minutes	Inclination in Degrees	Remarks
June 10		Payload: 3-unit spacecraft with 2 crew cabins, TV cameras, 2 large wing-like solar panels.					crew performed biological, geological, and meteorological experiments and tested spacecraft's navigation and control system. Softlanded west of Karaganda 6/19/70 for total flight time of 17 days 16 hrs 59 min.
June 10	Cosmos CCCXLVI (U.S.S.R.) 1970-42A Not available	Total weight: Not available. Objective: "Continuation of Cosmos scientific satellite series." Payload: Not available.	351 (218.1)	206 (128)	90	51.8	Reentered 6/17/70.
June 12	Cosmos CCCXLVII (U.S.S.R.) 1970-43A Not available	Total weight: Not available. Objective: "Continuation of Cosmos scientific satellite series." Payload: Not available.	2005 (1245.9)	217 (134.8)	107.5	48.4	Still in orbit.
June 13	Cosmos CCCXLVIII (U.S.S.R.) 1970-44A Not available	Total weight: Not available. Objective: "Continuation of Cosmos scientific satellite series." Payload: Not available.	540 (335.5)	196 (121.8)	91.8	70.9	Reentered 7/25/70.
June 17	Cosmos CCCXLIX (U.S.S.R.) 1970-45A Not available	Total weight: Not available. Objective: "Continuation of Cosmos scientific satellite series." Payload: Not available.	350 (217.5)	203 (126.1)	89.8	64.4	Reentered 6/25/70.
June 19	DOD spacecraft (United States) 1970-46A Atlas-Agena D	Total weight: Not available. Objective: Develop space-flight techniques and technology. Payload: Not available.	35 840.1 (22 270)	35 791.8 (22 240)	1426.5	0.1	Still in orbit.
June 23	Meteor V (U.S.S.R.) 1970-47A Not available	Total weight: Not available. Objective: Photograph cloud cover and snow and study atmospheric thermal energy radiated by earth. Payload: Not available.	888 (551.8)	830 (515.7)	102	81.2	Meteor V meteorological satellite still in orbit.
June 25	DOD spacecraft (United States) 1970-48A Titan IIIB-Agena D	Total weight: Not available. Objective: Develop space-flight techniques and technology. Payload: Not available.	410.4 (255)	118.9 (73.9)	89.8	108.8	Reentered 7/6/70.

ASTRONAUTICS AND AERONAUTICS, 1970

Date	Spacecraft	Description	Orbit (km/mi)	Apogee/Perigee	Period	Incl.	Remarks
June 26	Molniya I-14 (U.S.S.R.) 1970-49A Not available	Total weight: Not available. Objective: Develop and improve satellite and TV communications system. Payload: Satellite with transmitter, command system, orientation system, orbit correction device, power supply.	39 233 (24 378.2)	468 (290.8)	704.5	65.4	Molniya I-14 to relay telephone, telegraph, and radio communications, as well as TV broadcasts to far north, Siberia, Central Asia, and Far East. Still in orbit.
June 26	Cosmos CCCL (U.S.S.R.) 1970-50A Not available	Total weight: Not available. Objective: "Continuation of Cosmos scientific satellite series." Payload: Not available.	249 (154.7)	200 (124.3)	89	51.7	Reentered 7/8/70.
June 27	Cosmos CCCLI (U.S.S.R.) 1970-51A Not available	Total weight: Not available. Objective: "Continuation of Cosmos scientific satellite series." Payload: Not available.	464 (288.3)	270 (167.8)	91.8	70.9	Reentered 10/13/70.
July 7	Cosmos CCCLII (U.S.S.R.) 1970-52A Not available	Total weight: Not available. Objective: "Continuation of Cosmos scientific satellite series." Payload: Not available.	292 (181.4)	205 (127.4)	89.4	51.7	Reentered 7/15/70.
July 9	Cosmos CCCLIII (U.S.S.R.) 1970-53A Not available	Total weight: Not available. Objective: "Continuation of Cosmos scientific satellite series." Payload: Not available.	304 (188.9)	205 (127.4)	89.6	65.4	Reentered 7/21/70.
July 23	DOD spacecraft (United States) 1970-54A Tital IIIB-Agena D	Total weight: Not available. Objective: Develop space-flight techniques and technology. Payload: Not available.	404 (251)	118.9 (73.9)	90.1	59.9	Reentered 8/19/70.
July 23	Intelsat-III F-8 (United States) 1970-55A DSV-3M	Total weight: 151.5 kg (334 lbs) in synchronous orbit; 293.5 kg (647 lbs) at liftoff. Objective: Place satellite and apogee motor into proper transfer orbit; provide tracking and telemetry and backup calculations through transfer orbit so satellite can be injected into synchronous orbit for commercial communications. Payload: 198-cm x 142-cm-dia (78-in. x 56-in-dia) cylindrical satellite including mechanically despun receive/transmit antennas located on top of spacecraft; outer cylindrical sleeve covered with 10 720 solar cells; 2 frequency translation mode repeaters with design capability for 1200 high-quality, 2-way voice or 4 TV channels, or a mix; attitude and control system; battery.	36 245.1 (22 526.5)	260.3 (161.8)	641	27.8	Launched successfully by long-tank, thrust-augmented, Thor-Delta for ComSatCorp. Comsat-controlled apogee motor began 27 sec burn at 27 hrs GET to place satellite into planned synchronous orbit at 128° east longitude over western Pacific, but motor cut off after burning 14.5 secs and contact was/lost. Repeated efforts to locate satellite unsuccessful. Last of Intelsat III series. Still in orbit.

Launch Date	Name, Country, International Designation, Vehicle	Payload Data	Apogee in Kilometers (and st mi)	Perigee in Kilometers (and st mi)	Period in Minutes	Inclination in Degrees	Remarks
July 29**	Cosmos CCCLIV (U.S.S.R.) 1970-56A Not available	Total weight: Not available. Objective: "Continuation of Cosmos scientific satellite series." Payload: Not available.	208 (129.3)	144 (89.5)		50	Reentered 7/29/70.
Aug. 7	Intercosmos III (U.S.S.R.) 1970-57A Not available	Total weight: Not available. Objective: Study effect of solar radiation on structure of upper atmosphere. Payload: Spacecraft equipment includes alpha-particle photometer, x-ray and optical photometer, and spectroheliograph and polarimeter.	1192 (740.7)	200 (124.3)	98.7	48.4	Intercosmos III carried instrumentation from Soviet Union and Czechoslovakia. Reentered 12/6/70.
Aug. 7	Cosmos CCCLV (U.S.S.R.) 1970-58A Not available	Total weight: Not available. Objective: "Continuation of Cosmos scientific satellite series." Payload: Not available.	321 (199.5)	198 (123)	89.6	65.3	Reentered 8/15/70.
Aug. 10	Cosmos CCCLVI (U.S.S.R.) 1970-59A Not available	Total weight: Not available. Objective: "Continuation of Cosmos scientific satellite series." Payload: Not available.	502 (311.9)	224 (139.2)	91.7	81.9	Reentered 10/2/70.
Aug. 17	Venus VII (U.S.S.R.) 1970-60A Not available	Total weight: 1177.5 kg (2596 lbs). Objective: Land on Venus, explore atmosphere in process of descent, and take measurements directly on surface. Payload: Not available.	Impacted on Venus				Venus VII, heavier and improved Venus spacecraft, landed instrument capsule on Venus 12/15/70.** First transmission of information from planet's surface. Capsule transmitted data on ambient temperature for 23 min. Showed temperature at surface approximately 728 K (850°F); atmospheric pressure about 90 times that of earth.
Aug. 18	DOD spacecraft (United States) 1970-61A Titan IIIB-Agena D	Total weight: Not available. Objective: Develop space-flight techniques and technology. Payload: Not available.	395.9 (246)	151.3 (94)	89.9	110.9	Reentered 9/3/70.
Aug. 19	Skynet B (U.S.-U.K.) 1970-62A DSV-3M	Total weight: 129.3 kg (285 lbs). Objective: Place into synchronous orbit over Indian Ocean a United Kingdom communications spacecraft as part of	37 426.5 (23 260.7)	264.9 (164.6)	Not available	26.02	After successful NASA launch into transfer orbit by long-tank, thrust-augmented, Thor-Delta, apogee motor fired 8/22/70 to place satellite into

ASTRONAUTICS AND AERONAUTICS, 1970

Date	Spacecraft	Description					Remarks
Aug. 19		Initial Defense Communications Satellite Program (Augmented) under U.S.-U.K. agreement. Payload: 81-cm x 137-cm-dia (32-in x 54-in dia) spacecraft constructed of 2 concentric cylinders with apogee motor within cylinder; solar cells mounted on outside surface of outer cylinder; despun antenna system, partially mounted within inner cylinder, on spacecraft bottom; high pressure hydrazine system used for spin-stabilization and positioning; command and telemetry system.					planned synchronous orbit at 39° east longitude. Approximately halfway through burn, all communications with spacecraft were lost. Still in orbit.
Aug. 19	Cosmos CCCLVII (U.S.S.R.) 1970-63A Not available	Total weight: Not available. Objective: "Continuation of Cosmos scientific satellite series." Payload: Not available.	466 (289.6)	271 (168.4)	91.9	70.9	Reentered 11/24/70.
Aug. 20	Cosmos CCCLVIII (U.S.S.R.) 1970-64A Not available	Total weight: Not available. Objective: "Continuation of Cosmos scientific satellite series." Payload: Not available.	538 (334.3)	515 (320)	95.1	74	Still in orbit.
Aug. 22	Cosmos CCCLIX (U.S.S.R.) 1970-65A Not available	Total weight: Not available. Objective: "Continuation of Cosmos scientific satellite series." Payload: Not available.	853 (530)	205 (127.4)	95.2	51.1	Reentered 11/6/70.
Aug. 26	DOD spacecraft (United States) 1970-66A Thorad-Agena D	Total weight: Not available. Objective: Develop space-flight techniques and technology. Payload: Not available.	503.7 (313)	484.4 (301)	94.4	74.9	Still in orbit.
Aug. 27	Oscar XIX (United States) 1970-67A Scout	Total weight: Not available. Objective: Provide navigation satellite. Payload: Not available.	1218.3 (757)	957.6 (595)	106.9	90	Oscar XIX navigation satellite part of Navy Navigation Satellite System for positioning fleet, ballistic-missile, attack carriers, oceanographic and miscellaneous vessels. Could locate objects in ocean within 90-m (300-ft) accuracy. Still in orbit.
Aug. 29	Cosmos CCCLX (U.S.S.R.) 1970-68A Not available	Total weight: Not available. Objective: "Continuation of Cosmos scientific satellite series." Payload: Not available.	286 (177.7)	207 (128.6)	89.4	64.9	Reentered 9/8/70.
Aug. 31*	DOD spacecraft (United States) 1970-69A Atlas-Agena D	Total weight: Not available. Objective: Develop space-flight techniques and technology. Payload: Not available.	39 831.3 (24 750)	31 913.3 (19 830)	1441.9	10.3	Still in orbit.

Launch Date	Name, Country, International Designation, Vehicle	Payload Data	Apogee in Kilometers (and st mi)	Perigee in Kilometers (and st mi)	Period in Minutes	Inclination in Degrees	Remarks
Sept. 3	DOD spacecraft (United States) 1970-70A Thor-Burner II	Total weight: Not available. Objective: Develop space-flight techniques and technology. Payload: Not available.	872 (542)	764 (475)	101.2	98.7	Still in orbit.
Sept. 8	Cosmos CCCLXI (U.S.S.R.) 1970-71A Not available	Total weight: Not available. Objective: "Continuation of Cosmos scientific satellite series." Payload: Not available.	338 (210)	195 (121.2)	89.8	72.8	Reentered 9/21/70.
Sept. 12	Luna XVI (U.S.S.R.) 1970-72A Not available	Total weight: 1876.1 kg (4136 lbs). Objective: Carry out scientific exploration of moon and near-moon space. Payload: 2-stage (ascent and descent) spacecraft; 4 legs; telephotometer; 2 mechanical arms.	Lunar orbit 104.6 (65)	14.5 (9)			Luna XVI first unmanned spacecraft to land on moon and return to earth with lunar samples. Spacecraft softlanded on moon's Sea of Fertility 1:18 am EDT 9/20/70. Earth-operated electric drill penetrated lunar surface to 35 cm (13.8 in). Samples obtained and transferred to container in return capsule and hermetically sealed. Spacecraft remained on moon 26 hrs 25 min; liftoff from moon 3:43 am EDT 9/21. Returned to earth 1:26 am EDT 9/24. Total flight time 11 days 16 hrs. Analysis of Sea of Fertility samples indicate roughly same relative abundance of 11 major elements as Apollo 12's Ocean of Storms samples; both differ significantly from Apollo 11's Sea of Tranquility samples. Luna XVI Sea of Fertility samples age estimated at 4.5 billion years.
Sept. 16	Cosmos CCCLXII (U.S.S.R.) 1970-73A Not available	Total weight: Not available. Objective: "Continuation of Cosmos scientific satellite series." Payload: Not available.	817 (507.7)	269 (167.2)	95.4	70.9	Still in orbit.
Sept. 17	Cosmos CCCLXIII (U.S.S.R.) 1970-74A Not available	Total weight: Not available. Objective: "Continuation of Cosmos scientific satellite series." Payload: Not available.	288 (179)	201 (124.9)	89.3	65	Reentered 9/29/70.
Sept. 22	Cosmos CCCLXIV	Total weight: Not available.	294	201	89.4	65.4	Reentered 10/2/70.

ASTRONAUTICS AND AERONAUTICS, 1970

Date	Satellite	Objective/Payload	Perigee (mi/km)	Apogee (mi/km)	Period (min)	Inclination (deg)	Remarks
	(U.S.S.R.) 1970-75A Not available	Objective: "Continuation of Cosmos scientific satellite series." Payload: Not available.	(182.7)	(124.9)			
Sept. 25	Cosmos CCCLXV (U.S.S.R.) 1970-76A Not available	Objective: "Continuation of Cosmos scientific satellite series." Payload: Not available.	210 (130.5)	144 (89.5)		49.5	Reentered 9/25/70.
Sept. 29	Molniya I-15 (U.S.S.R.) 1970-77A Not available	Total weight: Not available. Objective: Develop and improve satellite and TV communications system. Payload: Satellite with transmitter, command system, orientation system, orbit correction device, power supply.	39 300 (24 419.8)	480 (298.3)	706	65.5	Molniya I-15 to relay telephone, telegraph, and radio communications, as well as TV broadcasts to far north, Siberia, Central Asia, and Far East. Still in orbit.
Oct. 1	Cosmos CCCLXVI (U.S.S.R.) 1970-78A Not available	Total weight: Not available. Objective: "Continuation of Cosmos scientific satellite serries." Payload: Not available.	288 (179)	202 (125.5)	89.4	64.9	Reentered 10/13/70.
Oct. 3	Cosmos CCCLXVII (U.S.S.R.) 1970-79A Not available	Total weight: Not available. Objective: "Continuation of Cosmos scientific satellite series." Payload: Not available.	1026 (637.5)	919 (571)	104.5	65.2	Still in orbit.
Oct. 8	Cosmos CCCLXVIII (U.S.S.R.) 1970-80A Not available	Total weight: Not available. Objective: "Continuation of Cosmos scientific satellite series." Payload: Not available.	383 (238)	202 (125.5)	90.4	65	Reentered 10/14/70.
Oct. 8.	Cosmos CCCLXIX (U.S.S.R.) 1970-81A Not available	Total weight: Not available. Objective: "Continuation of Cosmos scientific satellite series." Payload: Not available.	467 (290.2)	261 (162.2)	91.8	70.9	Reentered 1/22/71.
Oct. 9	Cosmos CCCLXX (U.S.S.R.) 1970-82A Not available	Total weight: Not available. Objective: "Continuation of Cosmos scientific satellite series." Payload: Not available.	264 (164)	185 (115)	88.9	64.9	Reentered 10/22/70.
Oct. 12	Cosmos CCCLXXI (U.S.S.R.) 1970-83A Not available.	Total weight: Not available. Objective: "Continuation of Cosmos scientific satellite series." Payload: Not available.	757 (470.4)	750 (466)	99.8	74	Still in orbit.
Oct. 14	Intercosmos IV (U.S.S.R.) 1970-84A Not available	Total weight: Not available. Objective: Carry out studies of upper atmosphere phenomena and solar physics. Payload: Spacecraft equipment includes alpha-particle photometer, x-ray optical photometer, and spectroheliograph and polarimeter.	602 (374.1)	255 (158.5)	93.1	48.4	Intercosmos IV carried instrumentation from Soviet Union, East Germany, and Czechoslovakia. Reentered 1/17/71.

431

Launch Date	Name, Country, International Designation, Vehicle	Payload Data	Apogee in Kilometers (and st mi)	Perigee in Kilometers (and st mi)	Period in Minutes	Inclination in Degrees	Remarks
Oct. 15	Meteor VI (U.S.S.R.) 1970-85A Not available	Total weight: Not available. Objective: Photograph cloud cover and snow and study atmospheric thermal energy radiated by earth. Payload: Not available.	647 (402)	627 (389.6)	97.4	81.2	Meteor VI meteorological satellite still in orbit.
Oct. 16	Cosmos CCCLXXII (U.S.S.R.) 1970-86A Not available	Total weight: Not available. Objective: "Continuation of Cosmos scientific satellite series." Payload: Not available.	806 (500.8)	785 (487.8)	100.7	74	Still in orbit.
Oct. 20	Cosmos CCCLXXIII (U.S.S.R.) 1970-87A Not available	Total weight: Not available. Objective: "Continuation of Cosmos scientific satellite series." Payload: Not available.	562 (349.2)	459 (285.2)	94.8	62.9	Still in orbit.
Oct. 20	Zond VIII (U.S.S.R.) 1970-88A Not available	Total weight: Not available. Objective: Carry out physical research along flight path and near-moon space; take pictures of lunar surface; of earth and moon at different distances; check improved onboard systems, units, and construction of spacecraft. Payload: Not available.	Circumlunar flight				Zond VIII fourth successful Soviet circumlunar flight. Spacecraft circled moon within 1120 km (696 mi) and photographed moon in color and black and white. Splashed down in Indian Ocean 724 km (450 miles) southeast of Chagos Archipelago 10/27/70.
Oct. 23	Cosmos CCCLXXIV (U.S.S.R.) 1970-89A Not available	Total weight: Not available. Objective: "Continuation of Cosmos scientific satellite series." Payload: Not available.	2140 (1329.7)	521 (323.7)	112.2	62.9	Still in orbit.
Oct. 23	DOD spacecraft (United States) 1970-90A Titan IIIB-Agena D	Total weight: Not available. Objective: Develop space-flight techniques and technology. Payload: Not available.	399.1 (248)	130.4 (81)	89.8	111	Reentered 11/11/70.
Oct. 30	Cosmos CCCLXXV (U.S.S.R.) 1970-91A Not available	Total weight: Not available. Objective: "Continuation of Cosmos scientific satellite series." Payload: Not available.	2098 (1303.6)	525 (326.2)	111.8	62.8	Still in orbit.
Oct. 30	Cosmos CCCLXXVI (U.S.S.R.) 1970-92A Not available	Total weight: Not available. Objective: "Continuation of Cosmos scientific satellite series." Payload: Not available.	286 (177.7)	206 (128)	89.4	65.3	Reentered 11/12/70.

Date	Spacecraft	Description	Weight kg (lbs)	Apogee/Perigee km (mi)	Period (min)	Inclination (deg)	Remarks
Nov. 6	DOD spacecraft (United States) 1970-93A Titan IIIC	Total weight: Not available. Objective: Develop space-flight techniques and technology. Payload: Not available.	35 886.8 (22 299)	26 048.8 (16 186)	1197.1	7.8	Still in orbit.
Nov. 9	Ofo (United States) 1970-94A Scout	Total weight: 132.5 kg (292 lbs). Objective: Obtain information on functioning and adaptability in weightlessness of portion of inner ear which controls balance. Record under weightlessness and simulated gravity, instantaneous rate of firing from single vestibular units of each otolith from 2 microelectrodes for minimum of 3 days. Obtain data on acceleration of 0.0003 g or less at otolith during weightlessness and on water environment temperature. Payload: 119-cm x 76-cm-dia (47-in x 30-in-dia) truncated cone with spherical top affixed to octagonal platform. Two 9-cell battery packs deliver 28 volts; 5 antennas; 4 booms extend 198 cm (78 in) from side of spacecraft; yoyo despin assembly. 41.3-kg (91-lb) frog otolith experiment package (FOEP) contains 2 bullfrogs encased in pressure-tight 46-cm x 46-cm-dia (18-in x 18-in-dia) container.	528 (328)	299.7 (186.4)	92.8	37.4	Ofo (Orbiting Frog Otolith) met all research objectives and was considered success; 2 frogs monitored for 6 days. Frog's inner ear adapted readily to zero gravity of space after 3 days of disorientation. Indicates artificial gravity may be unnecessary for future long-duration manned flights. Still in orbit.
	and						
	RM 1970-94B	Total weight: 21 kg (46 lbs). Objective: Demonstrate feasibility and accuracy of advanced radiation dosimetry system in real-time mode in actual space radiation environment. Verify, in flight, proper operation of instrumentation (including sensor materials) to be used on long duration meteoroid flight experiments for future study of meteoroid population of solar system; will measure meteoroid impact flux, speed, and trajectory. Payload: 71-cm-high x 76-cm-dia (28-in-high x 30-in-dia) cylinder attached to Scout 4th stage (FW-4); 0.65 sq m (7 sq ft) of solar cells provide 25 w power; 2 nickel-cadmium batteries; spin-stabilized, total length (including FW-4) 170 cm (67 in).	495 (307.6)	292 (181.4)	92.4	37.4	RM (Radiation/Meteoroid) Satellite launched as secondary payload. Operated satisfactorily and returned good data. Reentered 2/7/71.
Nov. 10	Luna XVII (U.S.S.R.) 1970-95A Not available	Total weight: Not available. Objective: Develop new on-board systems and carry on further scientific exploration of moon and near-moon space.	Lunar orbit, 85 (53)		116	141	Luna XVII entered lunar orbit 11/15/70. Softlanded in moon's Sea of Rains 10:47 pm EST 11/16 and released Lunokhod I, self-propelled vehicle re-

Launch Date	Name, Country, International Designation, Vehicle	Payload Data	Apogee in Kilometers (and st mi)	Perigee in Kilometers (and st mi)	Period in Minutes	Inclination in Degrees	Remarks
		Payload: Spacecraft carried *Lunokhod 1*, self-propelled lunar vehicle, atop descent stage.					sembling large pot-bellied tub about size of small auto, with 8 spoked wheels powered by solar energy and batteries. Automatic lunar explorer—equipped with scientific apparatus, instruments, control systems, radio and TV—conducts operations during lunar days and hibernates during lunar nights. By 5/22/71, had logged 8458 meters (27 750 ft) during 7 lunar days and explored more than 400 000 sq meters (4 305 600 sq ft).
Nov. 11	Cosmos CCCLXXVII (U.S.S.R.) 1970-96A Not available	Total weight: Not available. Objective: "Continuation of Cosmos scientific satellite series." Payload: Not available.	286 (177.7)	204 (126.8)	89.3	65	Reentered 11/23/70.
Nov. 17	Cosmos CCCLXXVIII (U.S.S.R.) 1970-97A Not available	Total weight: Not available. Objective: "Continuation of Cosmos scientific satellite series." Payload: Not available.	1723 (1070.6)	234 (145.4)	104.6	74	Still in orbit.
Nov. 18	DOD spacecraft (United States) 1970-98A Thorad-Agena and DOD spacecraft 1970-98B	Total weight: Not available. Objective: Develop space-flight techniques and technology. Payload: Not available. Total weight: Not available. Objective: Develop space-flight techniques and technology. Payload: Not available.	225.9 (140.4) 510.9 (317.5)	177 (110) 486 (302)	88.5 94.5	82.9 83.1	Dual payload launched with single booster. 98A reentered 12/11/70. Still in orbit.
Nov. 24	Cosmos CCCLXXIX (U.S.S.R.) 1970-99A Not available	Total weight: Not available. Objective: "Continuation of Cosmos scientific satellite series." Payload: Not available.	14 026 (8715.4)	176 (109.4)	259.5	51.6	Still in orbit.
Nov. 24	Cosmos CCCLXXX (U.S.S.R.) 1970-100A Not available	Total weight: Not available. Objective: "Continuation of Cosmos scientific satellite series." Payload: Not available.	1491 (926.5)	198 (123)	101.7	81.9	Still in orbit.

Date	Name/Designation	Description	Weight kg (lbs)	Apogee/Perigee	Period (min)	Inclination	Remarks
Nov. 27	Molniya I-16 (U.S.S.R.) 1970-101A Not available	Total weight: Not available. Objective: Develop and improve satellite and TV communications system. Payload: Satellite with transmitter, command system, orientation system, orbit correction device, power supply.	39 359 (24 456.6)	467 (290.2)	707	65.4	Molniya I-16 to relay telephone, telegraph, and radio communications, as well as TV broadcasts to far north, Siberia, Central Asia, and Far East. Still in orbit.
Dec. 2	Cosmos CCCLXXXI (U.S.S.R.) 1970-102A Not available	Total weight: Not available. Objective: "Continuation of Cosmos scientific satellite series." Payload: Not available.	1013 (629.5)	967 (600.9)	104.8	74	Still in orbit.
Dec. 2	Cosmos CCCLXXXII (U.S.S.R.) 1970-103A Not available	Total weight: Not available. Objective: "Continuation of Cosmos scientific satellite series." Payload: Not available.	5071 (3151)	2584 (1605.6)	171	55.8	Still in orbit.
Dec. 3	Cosmos CCCLXXXIII (U.S.S.R.) 1970-104A Not available	Total weight: Not available. Objective: "Continuation of Cosmos scientific satellite series." Payload: Not available.	323 (200.7)	180 (111.9)	89.5	65.4	Reentered 12/16/70.
Dec. 10	Cosmos CCCLXXXIV (U.S.S.R.) 1970-105A Not available	Total weight: Not available. Objective: "Continuation of Cosmos scientific satellite series." Payload: Not available.	277 (172.1)	203 (126.1)	89.2	72.8	Reentered 12/22/70.
Dec. 11	Noaa I (United States) 1970-106A DSV-3N-6 and	Total weight: 306 kg (675 lbs). Objective: Place spacecraft in sun-synchronous orbit with local equator crossing time between 3:00 pm and 3:20 pm, and conduct in-orbit engineering evaluation so that daytime and nighttime cloudcover observations can be obtained regularly and dependably in both direct readout and stored modes of operation. Payload: 102- x 125- x 102-cm (40- x 49- x 40-in) rectangular, box-shaped spacecraft with 3-panel solar array. 3-axisstabilized, earth-oriented satellite carries 2 advanced vidicon camera subsystems (AVCS)‡ and 2 automatic picture transmission (APT) camera subsystems which provide daytime coverage, and 2 scanning radiometer (SR) subsystems (infrared) which furnish nighttime coverage. Solar panels measure 0.9 x 1.5 m (3 x 5 ft) and total 4.5 sq m (48 sq ft). Spacecraft measures 4.3 m (14 ft) with panels deployed. Thermal control system. 4 antennas. 10 000 n-on-p solar-cells produce 250 w average power.	1472.2 (914.8)	1422.6 (884)	114.8	101.95	Noaa I first operational spacecraft of Improved TIROS Operational System (ITOS) series. First managed by National Environmental Satellite Service of National Oceanic and Atmospheric Administration (NOAA). Spacecraft functioning normally and observes entire globe-twice daily by both stored and direct readout systems. Also carried Cepe as secondary payload. Still in orbit.

435

Launch Date	Name, Country, International Designation, Vehicle	Payload Data	Apogee in Kilometers (and st mi)	Perigee in Kilometers (and st mi)	Period in Minutes	Inclination in Degrees	Remarks
	Cepe 1970-106B	Total weight: 2.3 kg (5 lbs). Objective: Obtain information on electron density, temperature, and ion current in ionosphere during 2-orbit lifetime. Payload: Small cylindrical experiment package carries its own analog computer; permanently attached to 2nd stage of Delta launch vehicle.	1475.1 (916.6)	1425.4 (885.7)	114.9	101.9	Cepe (Cylindrical Electrostatic Probe Experiment) launched as secondary payload. Carried its own small analog computer and made its own inflight data analysis. Automatically adjusted its operation to collect only "optimum" data. During 2-orbit lifetime, obtained useful information which may aid in reducing future ground data processing. Still in orbit.
Dec. 12	Explorer XLII (Uhuru) (Italy-U.S.) 1970-107A Scout	Total weight: 142.9 kg (315 lbs). Objective: Develop catalog of celestial x-ray sources by systematic scanning of celestial sphere in 2-20 KEV energy range. Payload: 51-cm x 56-cm-dia (20-in x 22-in dia) cylindrical spacecraft; 4 solar paddles 25 x 135 cm (10 x 53 in) with solar cells on both sides, hinged to outer shell and measure 4 m (13 ft) tip-to-tip. 1 nickel-cadmium battery provides 27 w average power; thermal control system; spin stabilized; carries advanced x-ray instruments.	572 (355.4)	531 (330)	95.7	3.04	Explorer XLII launched into equatorial orbit from mobile seaborne facility off coast of Kenya as part of San Marco project. First launch of American spacecraft by crew of another country (Italy). Discovered several new extragalaxy sources of intense energy. In particular, discovered 3 new x-ray pulsars, identified as Cygnus X-1, Centaurus X-3, and Lupus X-1. Still in orbit.
Dec. 12	Cosmos CCCLXXXV (U.S.S.R.) 1970-108A Not available	Total weight: Not available. Objective: "Continuation of Cosmos scientific satellite series." Payload: Not available.	984 (611.4)	980 (608.9)	104.7	74	Still in orbit.
Dec. 12	Peole (France) 1970-109A Diamant-B	Total weight: 70 kg (154 lbs). Objective: Flight-qualify gravity-gradient stabilization system and tracking/localizer. Payload: 71 cm (28-in)-dia octahedron carries 44 laser reflectors and 2 advanced-design solar generators.	749 (465.4)	635 (394.6)	98.6	15	Peole experimental satellite preliminary to synchronous applications satellite. Entered lower than planned orbit, but functioning satisfactorily. Still in orbit.
Dec. 15	Cosmos CCCLXXXVI (U.S.S.R.) 1970-110A Not available	Total weight: Not available. Objective: "Continuation of Cosmos scientific satellite series." Payload: Not available.	270 (167.8)	215 (133.6)	89.3	64.9	Reentered 12/28/70.

Dec. 16	Cosmos CCCLXXXVII (U.S.S.R.) 1970-111A Not available	Total weight: Not available. Objective: "Continuation of Cosmos scientific satellite series." Payload: Not available.	538 (334.3)	526 (326.8)	95.2	74	Still in orbit.
Dec. 18	Cosmos CCCLXXXVIII (U.S.S.R.) 1970-112A Not available	Total weight: Not available. Objective: "Continuation of Cosmos scientific satellite series." Payload: Not available.	499 (310.1)	269 (167.2)	92.2	70.9	Reentered May 10, 1971.
Dec. 18	Cosmos CCCLXXXIX (U.S.S.R.) 1970-113A Not available	Total weight: Not available. Objective: "Continuation of Cosmos scientific satellite series." Payload: Not available.	687 (426.9)	642 (398.9)	98	81.1	Still in orbit.
Dec. 25	Molniya I-17 (U.S.S.R.) 1970-114A Not available	Total weight: Not available. Objective: Develop and improve satellite and TV communications system. Payload: Satellite with transmitter, command system, orientation system, orbit correction device, power supply.	39 569 (24 587)	493 (306.3)	711.8	64.7	Molniya I-17 to relay telephone, telegraph, and radio communications, as well as TV broadcasts to far north, Siberia, Central Asia, and Far East. Still in orbit.

*Local time at site; 1 day later by Greenwich time.
**Local time at site; 1 day earlier by Greenwich time.

Appendix B

CHRONOLOGY OF MAJOR NASA LAUNCHES, 1970

This chronology of major NASA launches in 1970 is intended to provide an accurate and ready historical reference, compiling and verifying information previously scattered in several sources. It includes launches of all rocket vehicles larger than sounding rockets launched either by NASA or under "NASA direction" (e.g., in 1970 NASA provided vehicles and launch facilities and launched ComSatCorp's three Intelsat III satellites, ESSA's *Noaa I,* NATO's *Nato I,* and U.K.'s *Skynet B,* as well as Australia's *Oscar V* and USA's *Topo I* as secondary payloads; also, under a NASA and University of Rome agreement, an Italian team launched a NASA satellite, *Explorer XLII,* from Italy's San Marco platform off the coast of Kenya, on NASA's Scout booster). NASA sounding rocket launches are published annually in *Goddard Projects Summary: Satellites and Sounding Rockets.*

An attempt has been made to classify performance of both the launch vehicle and the payload and to summarize total results in terms of primary mission. Three categories have been used for evaluating vehicle performance and mission results—successful (S), partially successful (P), and unsuccessful (U). A fourth category, unknown (Unk), has been added for payloads when vehicle malfunctions did not give the payload a chance to exercise its main experiments. These divisions are necessarily arbitrary; many of the results cannot be neatly categorized. Also they ignore the fact that a great deal is learned from missions that may have been classified as unsuccessful.

Date of launch is referenced to local time at the launch site. Open sources were used, verified when in doubt with the project offices in NASA Headquarters and with NASA Centers. For further information on each item, see Appendix A of this volume and the entries in the main chronology as referenced in the index. The information was compiled in May 1971 by Leonard C. Bruno of the Science and Technology Division of the Library of Congress.

Date	Name (NASA Code)	General Mission	Launch vehicle (site)	Performance Vehicle	Performance Payload	Performance Mission	Remarks
Jan. 14*	Intelsat-III F-6	Commercial communications satellite	Delta DSV-3M (ETR)	S	S	S	Launched by NASA into elliptical orbit. Transferred by ComSatCorp into synchronous orbit over Atlantic (24° west longitude). Replaced Intelsat-III F-2.
Jan. 23	Itos I (Tiros-M) and	Prototype meteorological satellite	Delta DSV-3-N-6 (WTR)	S	S	S	Second-generation, operational meteorological satellite placed in circular, near-polar, sun-synchronous orbit; provided first daily, global, day-night cloud-cover data. Spacecraft funded and developed by NASA; turned over to ESSA (which later became part of NOAA).
	Oscar V	Radio transmitter				S	Launched by NASA as secondary payload for Radio Amateur Satellite Corporation. Spacecraft built at University of Melbourne, Australia.
Feb. 3*	Sert II	Ion engine test satellite	Thorad-Agena D (WTR)	S	P	P	First orbital test of electron-bombardment ion engines. No. 1 engine shut down 7/23/70 after successful operation for over 5 mos (3782 hrs). No. 2 engine shut down 10/17/70 after 2 3/4 mos operation (2011 hrs). Spacecraft's solar arrays largest ever flown.
Mar. 20	Nato I (NATO-A)	Non-NASA mission; military communications satellite	Delta DSV-3M (ETR)	S	S	S	Launched by NASA for NATO into elliptical transfer orbit. Apogee kick motor placed spacecraft in synchronous orbit 3/23/70. Communications system failure on 10/25/70 necessitated use of redundant system.
Apr. 8	Nimbus IV (Nimbus-D) and	Experimental meteorological satellite	Thorad-Agena D (WTR)	S	S	S	Launched into nearly circular orbit. Spacecraft and sensors operating successfully except for filter wedge spectrometer (FWS), which failed 6/7/70. Primary objectives achieved and flight a success.
	Topo I (Topo-A)	U.S. Army geodetic satellite			S	S	Launched as secondary payload by NASA for DOD (U.S. Army Topographic Command). Secor-type satellite is used through triangulation for ground tactical positioning.
Apr. 11	Apollo 13 (AS-508 CSM-109, LM-7)	Apollo manned lunar landing and return	Saturn V (KSC)	S	P	U	Third manned lunar landing mission launched successfully, carrying Astronauts James A. Lovell, Jr., John L. Swigert, Jr., and Fred W. Haise, Jr. Mission was aborted 56 hrs into flight toward moon because of SM oxygen tank rupture. LM "Lifeboat" emergency plan employed. Crew returned to CM before reentry and splashed down in Pacific 1:07 pmEST 4/17/70. Total flight time 142 hrs 54 min 41 secs. First inflight mission failure in 23 manned U.S. flights.

Date	Name	Purpose	Vehicle				Remarks
Ap. 22*	Intelsat-III F-7	Commercial communications satellite	Delta DSV-3M (ETR)	P	S	S	NASA launch vehicle underperformed and onboard hydrazine thrusters needed to push satellite into desired synchronous orbit, 19° west longitude over Atlantic. Remaining fuel expected to maintain satellite in proper orbit 5 yrs. Began commercial service 5/8/70
July 23	Intelsat-III F-8	Commercial communications satellite	Delta DSV-3M (ETR)	S	U	U	Launched successfully by NASA into elliptical orbit. ComSatCorp-controlled apogee motor began 27-sec burn at 27 hrs GET to place satellite into planned synchronous orbit, 128° east longitude over western Pacific, but motor cut off after burning 14.5 secs and contact was lost. Last of Intelsat III series.
Aug. 19	Skynet B	Non-NASA mission; military communications satellite	Delta DSV-3M (ETR)	S	U	U	Launched successfully by NASA for U.K. into transfer orbit. Apogee motor fired 8/22/70 to place satellite into synchronous orbit at 39° east longitude. Approximately halfway through burn, all communications were lost.
Sept. 30	RAM C-III	Reentry probe	Scout (WS)	S	S	S	RAM C-III reentered at 7.6 km per sec (25 000 fps) to compare effectiveness of liquid electrophilic (Freon) with water in alleviating radio blackout during reentry.
Nov. 9	Ofo and	Biological experimentation in space	Scout (WS)	S	S	S	Ofo (Orbiting Frog \Otolith\) monitored 2 frogs in flight for 6 days. Frog's inner ear adapted to zero gravity after 3 days' disorientation. Indicated artificial gravity may be unnecessary for future long-duration manned flights.
	RM	Scientific satellite, radiation and meteoroid		S	S	S	RM (Radiation/Meteoroid) Satellite launched as secondary payload. Operated satisfactorily and returned good data.
Nov. 30	OAO-B	Scientific satellite, astronomy	Atlas-Centaur (ETR)	U	Unk	U	Launch vehicle nose-fairing failed to separate; additional weight prevented attainment of necessary orbital velocity and satellite impacted. Spacecraft carried world's largest telescope—914 mm (36 in).
Dec. 11	Noaa 1 (ITOS-A)	Meteorological satellite	Delta DSV-3N-6 (WTR)	S	S	S	First operational spacecraft of Improved TIROS Operational System (ITOS) series. The first managed by National Environmental Satellite Service of National Oceanic and Atmospheric Administration (NOAA). Spacecraft observes entire globe twice daily by both stored and direct readout systems.
	Cepe	Electrostatic probe experiment			S	S	Cepe (Cylindrical Electrostatic Probe Experiment) launched as secondary payload. Carried analog computer for inflight data analysis. Obtained useful information during 2-orbit lifetime, which may aid in reducing future ground data processing.

Date	NASA (NASA Code)	General Mission	Launch vehicle (site)	Performance			Remarks
				Vehicle	Payload	Mission	
Dec. 12	Explorer XLII (SAS-A)	Scientific satellite, astronomy	Scout (Kenya)	S	S	S	Launched into equatorial orbit from mobile seaborne facility off coast of Kenya as part of San Marco project. First launch of American spacecraft by crew of another country (Italy). Discovered several new extra-galaxy sources of intense energy.

*Time at launch site: 1 day later by Greenwich time.

Appendix C

CHRONOLOGY OF MANNED SPACE FLIGHT, 1970

This chronology contains basic information on all manned space flights during 1970 and, taken with Appendix C to the 1965, 1966, 1968, and 1969 editions of this publication, provides a summary record of manned exploration of the space environment through 1970. The information was compiled by Leonard C. Bruno of the Science and Technology Division of the Library of Congress.

The year 1970 marked a sharp downturn in manned space flight. Contrasting with 1969, in which four United States Apollo flights and five U.S.S.R. Soyuz flights put a record 23 men in space, 1970 saw only one Apollo flight and one Soyuz flight place a total of 5 men in space.

The three-man *Apollo 13* mission failed in its lunar landing attempt when a service module explosion crippled the ship on its way to the moon. The lunar mission was aborted and the crew immediately employed emergency procedures to effect a safe return to earth. Using the lunar module as a "lifeboat" and conserving all power and life support systems, the spacecraft circled the moon and returned safely to earth. But the third U.S. lunar landing mission was unsuccessful.

Soyuz IX, the only Soviet manned flight of the year, was highly successful. The two-man crew orbited the earth for nearly 18 days and set a new world record for the longest manned space flight. The crew performed biological, geological, and meteorological experiments. The previous manned space flight record of nearly 14 days in space was held by the U.S. *Gemini VII* flight of December 1965.

By the end of 1970, the United States had totaled 23 manned space flights—2 suborbital, 16 in earth orbit, 3 in lunar orbit, and 2 lunar landings—with a total of 26 different crewmen. Of the 26 American astronauts, 10 had participated in 2 flights each, 4 had flown three times, and 1 had flown four times. The Soviet Union had conducted 16 manned flights, all in earth orbit, with 22 cosmonauts. Four cosmonauts had flown twice each. Cumulative totals for manned spacecraft hours in flight had reached 2446 hours and 51 minutes for the United States and 1479 hours 7 minutes for the Soviet Union. Cumulative total man-hours in space were 6262 hours 42 minutes for the United States and 2548 hours 45 minutes for the U.S.S.R.

Data on U.S. flights are the latest available to date within NASA. Although minor details are subject to modification as data are refined, major aspects of all U.S. manned flights remain subject to direct observation by interested citizens of the world, with a significant portion of recent missions seen live on worldwide television.

ASTRONAUTICS AND AERONAUTICS, 1970

Date Launched	Date Recovered	Designation (NASA Code)	Crew	Weight in Kilograms (and in lbs)	Revolutions	Maximum distance from Earth in Kilometers (and in st mi)	Duration	Remarks
Apr. 11	Apr. 17	Apollo 13 (AS-508) CSM-109, LM-7	James A. Lovell, Jr. John L. Swigert, Jr. Fred W. Haise, Jr.	134 472.7 (296 463)	1 (of moon)		142 hrs 55 min	Third manned lunar landing mission. First in-flight mission failure in 23 manned U.S. flights. Spacecraft launched by Saturn V booster. After insertion into earth orbit, S-IVB initiated translunar injection. CSM separated from S-IVB and LM, transposed, docked with LM, and spacecraft separated from 3rd stage. S-IVB evasive maneuver accomplished stage set on lunar impact trajectory and impacted moon 8:09:40 pm EST 4/14/70. TLI maneuver placed manned spacecraft on hybrid lunar approach trajectory (non-free return). During lunar coast, crew reported undervoltage alarm on CSM main bus B at 55:55 GET. Pressure was rapidly lost in SM oxygen tank No. 2 and fuel cells 1 and 3 current dropped to zero. Decision made to abort mission. Crew activated LM, powered down CM, and used LM systems for power and life support. Midcourse maneuver performed by LM Descent Propulsion System (DPS) to place spacecraft in free-return trajectory around moon. During transearth coast, crew performed 2 additional midcourse maneuvers to shorten return time. Crew continued emergency procedures to deal with shortage of water to cool LM electronic systems, with decreasing temperatures in both cabins, and buildup of carbon monoxide in LM. Nearing reentry time, SM was jettisoned. Crew viewed and photographed damaged module, powered up CM, transferred, and jettisoned "lifeboat" LM. Reentry occurred as planned and CM splashed down in mid-Pacific 1:07:41 pm EST 4/17/70. Recovery by U.S.S. Iwo Jima. Total flight time 142 hrs 54 min 41 sec.
June 2	June 19	Soyuz IX	Andrian G. Nikolayev Vitaly I. Sevastyanov	Not available	287	267.1 (166)	424 hrs 59 min	Soyuz IX, carrying Cosmonauts Andrian G. Nikolayev and Vitaly I. Sevastyanov, set new world endurance record for longest manned space flight. After initial orbit, spacecraft made 2 orbital corrections to obtain extended duration orbit. During record flight, crew performed biological, geological, and meteorological experiments and tested spacecraft's navigation and

444

control system. Softlanded west of Karaganda 6/19/70. Long period of weightlessness adversely affected cosmonauts' coordination and crew experienced difficulty in readjusting to earth's gravity. Total flight time 17 days 16 hrs 59 min.

Appendix D

ABBREVIATIONS OF REFERENCES

Listed here are abbreviations for sources cited in the text. The list does not include all sources provided in the chronology, for some of the references cited are not abbreviated. Only references that appear in abbreviated form are listed below. Abbreviations used in the chronology entries themselves are cross-referenced in the Index.

A&A	AIAA's magazine, *Astronautics & Aeronautics*
A&A 1970	NASA's *Astronautics and Aeronautics, 1970* [this publication]
ABC	American Broadcasting Company
AEC Release	Atomic Energy Commission News Release
Aero Daily	*Aerospace Daily* newsletter
Aero Med	*Aerospace Medicine* magazine
Aero Tech	*Aerospace Technology* magazine (formerly *Technology Week*)
AF Mag	Air Force Association's *Air Force Magazine*
AF Mgmt	*Armed Forces Management* magazine
AFFTC Release	Air Force Flight Test Center News Release
AFHF Newsletter	*Air Force Historical Foundation Newsletter*
AFJ	*Armed Forces Journal* magazine
AFNS Release	Air Force News Service Release
AFOSR Release	Air Force Office of Scientific Research News Release
AFRPL Release	Air Force Rocket Propulsion Laboratory News Release
AFSC *Newsreview*	Air Force Systems Command's *Newsreview*
AFSC Release	Air Force Systems Command News Release
AF/SD	*Air Force and Space Digest* magazine
AFSSD Release	Air Force Space Systems Division News Release
AIA Release	Aerospace Industries Association News Release
AIAA *Facts*	American Institute of Aeronautics and Astronautics' *Facts*
AIAA *News*	American Institute of Aeronautics and Astronautics' *News*
AIAA Release	American Institute of Aeronautics and Astronautics News Release
AIP *Newsletter*	*American Institute of Physics Newsletter*
Amer Av	*American Aviation* magazine (formerly *Aerospace Technology*)
AP	Associated Press news service
ARC *Astrogram*	NASA Ames Research Center's *Astrogram*
Atlanta J/C	*Atlanta Journal and Constitution* newspaper
Av Daily	*Aviation Daily* newsletter
Av Wk	*Aviation Week & Space Technology* magazine
B News	*Birmingham News* newspaper
B Sun	*Baltimore Sun* newspaper
Bus Wk	*Business Week* magazine
C Daily News	*Chicago Daily News* newspaper

C Trib	*Chicago Tribune* newspaper
Can Press	Canadian Press news service
CBS	Columbia Broadcasting System
C&E News	*Chemical & Engineering News* magazine
ComSatCorp Release	Communications Satellite Corporation News Release
CQ	*Congressional Quarterly*
CR	*Congressional Record*
CSM	*Christian Science Monitor* newspaper
CTNS	Chicago Tribune News Service
DASA Release	Defense Atomic Support Agency News Release
D Post	*Denver Post* newspaper
DJ	Dow Jones news service
DOC PIO	Department of Commerce Public Information Office
DOD Release	Department of Defense News Release
DOT Release	Department of Transportation News Release
EOP Release	Executive Office of the President News Release
ERC Release	NASA Electronics Research Center News Release
ESSA Release	Environmental Science Services Administration News Release
FAA Release	Federal Aviation Administration News Release
FonF	*Facts on File*
FRC Release	NASA Flight Research Center News Release
FRC X-Press	NASA Flight Research Center's *FRC X-Press*
GE Forum	*General Electric Forum* magazine
Goddard News	NASA Goddard Space Flight Center's *Goddard News*
GSFC Release	NASA Goddard Space Flight Center News Release
GSFC SSR	NASA Goddard Space Flight Center's *Satellite Situation Report*
GT&E Release	General Telephone & Electronics News Release
H Chron	*Houston Chronicle* newspaper
H Post	*Houston Post* newspaper
JA	*Journal of Aircraft* magazine
JPL *Lab-Oratory*	Jet Propulsion Laboratory's *Lab-Oratory*
JPL Release	Jet Propulsion Laboratory News Release
JPRS	Department of Commerce Joint Publications Research Service
JSR	American Institute of Aeronautics and Astronautics' *Journal of Spacecraft and Rockets* magazine
KC Star	*Kansas City Star* newspaper
KC Times	*Kansas City Times* newspaper
KSC Release	NASA John F. Kennedy Space Center News Release
LA Her-Exam	Los Angeles *Herald-Examiner* newspaper
LA Times	*Los Angeles Times* newspaper
Langley Researcher	NASA Langley Research Center's *Langley Researcher*
LaRC Release	NASA Langley Research Center News Release
LATNS	Los Angeles Times News Service
LC *Info Bull*	Library of Congress *Information Bulletin*
LeRC Release	NASA Lewis Research Center News Release
Lewis News	NASA Lewis Research Center's *Lewis News*
M Her	*Miami Herald* newspaper
M News	*Miami News* newspaper
M Trib	*Minneapolis Tribune* newspaper
Marshall Star	NASA George C. Marshall Space Flight Center's *Marshall Star*
MJ	*Milwaukee Journal* newspaper
MSC Release	NASA Manned Spacecraft Center News Release
MSC *Roundup*	NASA Manned Spacecraft Center's *Space News Roundup*
MSFC Release	NASA George C. Marshall Space Flight Center News Release
N Hav Reg	*New Haven Register* newspaper
N News	*Newark News* newspaper
N Va Sun	*Northern Virginia Sun* newspaper
NAA News	*National Aeronautic Association News*
NAC Release	National Aviation Club News Release
NAE Release	National Academy of Engineering News Release
NANA	North American Newspaper Alliance

NAS Release	National Academy of Sciences News Release
NAS-NRC Release	National Academy of Sciences-National Research Council News Release
NAS-NRC-NAE News Rpt	National Academy of Sciences-National Research Council-National Academy of Engineering News Report
NASA Ann	NASA Announcement
NASA Hist Off	NASA Historical Office
NASA Hq WB	NASA Headquarters *Weekly Bulletin*
NASA Int Aff	NASA Office of International Affairs
NASA LAR IX/8	NASA *Legislative Activites Report*, Vol. IX, No. 8
NASA Proj Off	NASA project office (for project reported)
NASA Release	NASA Headquarters News Release
NASA Rpt SRL	NASA Report of Sounding Rocket Launching
NASA SP-4014	NASA Special Publication #4014
NASC Release	National Aeronautics and Space Council News Release
Natl Obs	*National Observer* newspaper
NBC	National Broadcasting Company
NGS Release	National Geographic Society News Release
NMI	NASA Management Instruction
NN	NASA Notice
NR Release	North American Rockwell Corp. News Release
NR *Skywriter*	North American Rockwell Corp. *Skywriter*
NSC Release	National Space Club News Release
NSF *Highlights*	National Science Foundation's *Science Resources Studies Highlights*
NSF Release	National Science Foundation News Release
NY *News*	*New York Daily News* newspaper
NYT 5:4	*New York Times* newspaper, section 5 page 4
NYTNS	New York Times News Service
O Sen	*Orlando Sentinel* newspaper
Oakland Trib	*Oakland Tribune* newspaper
Omaha W-H	*Omaha World-Herald* newspaper
OST Release	Office of Science and Technology News Release
P Bull	Philadelphia *Evening* and *Sunday Bulletin* newspaper
P Inq	*Philadelphia Inquirer* newspaper
PAO	Public Affairs Office
PD	National Archives and Records Service's *Weekly Compilation of Presidential Documents*
PIO	Public Information Office
PMR *Missile*	USN Pacific Missile Range's *Missile*
PMR *Release*	USN Pacific Missile Range News Release
Pres Rpt 71	*Aeronautics and Space Report of the President, Transmitted to the Congress January 1971* (report of activities during 1970)
SA	*Space Aeronautics* magazine
SBD	*Space Business Daily* newsletter
SAO Release	Smithsonian Astrophysical Observatory News Release
Sci Amer	*Scientific American* magazine
SciServ	Science Service news service
SD	*Space Digest* magazine
SD Union	*San Diego Union* newspaper
SF	*Spaceflight* magazine
SF Chron	*San Francisco Chronicle* newspaper
Sov Rpt	Center for Foreign\Technology's *Soviet Report* (translations)
SP	*Space Propulsion* newsletter
Spacewarn	IUWDS World Data Center A for Rockets and Satellites' *Spacewarn Bulletin*
SR	*Saturday Review* magazine
SR list	NASA compendium of sounding rocket launches
SSN	*Soviet Sciences in the News*, publication of Electro-Optical Systems, Inc.
St Louis G-D	*St. Louis Globe-Democrat* newspaper
St Louis P-D	*St. Louis Post-Dispatch* newspaper

Testimony	Congressional testimony, prepared statement
Text	Prepared report or speech text
Transcript	Official transcript of news conference or congressional hearing
UPI	United Press International news service
USGS Release	U.S. Geological Survey News Release
US News	*U.S. News & World Report* magazine
W News	*Washington Daily News* newspaper
W Post	*Washington Post* newspaper
W Star	Washington *Evening Star/Sunday Star* newspaper
WH Release	White House News Release
WJT	*World Journal Tribune* newspaper
WS Release	NASA Wallops Station News Release
WSJ	*Wall Street Journal* newspaper

INDEX AND LIST OF ABBREVIATIONS AND ACRONYMS

A

A-7D (Corsair II) (tactical fighter), 7
AAAS. See American Assn. for the Advancement of Science.
AAP (Apollo Applications Program). See Skylab.
Aarons, Leroy F., 174
AAS. See American Astronautical Society.
ABC. See American Broadcasting Co.
Abelson, Dr. Philip H., 153, 164, 185, 316
Abidjan, Ivory Coast, 82
ABM. See Antiballistic missile system.
ABMA. See Army Ballistic Missile Agency.
ABRES. See Advanced Ballistic Reentry System.
AC Electronics Div., General Motors Corp., 7, 29
Defense Research Laboratory, 181
Accident
 aircraft, 331, 379
  AH-56A (helicopter), 109
  Boeing 747, 278, 324, 349
  C-5A, 338
  F-14, 408
  F-111, 74, 100, 226, 244, 351
 automobile, 100
 spacecraft,
  Apollo 1 (AS-204), 232
  Apollo 13 (AS-508), 165
   cause, 121, 161
   damage, 125, 140, 181
   emergency procedures, 119, 121-125, 141, 143, 147, 148
   investigation, 125, 137, 142, 144, 155, 161, 167, 184, 232
   report, 161, 181, 191, 201-203, 204, 213, 217, 218-219
   Nixon, President Richard M., 134
   press comment, 131, 138, 140, 150-151, 164, 186, 232
   press conference, 127-128, 132-134, 143-144, 161
Acoustic grenade experiment, 14
ACS. See Attitude control system and Autonomous control subsystem.
Adams, John B., 210
Adams, Dr. Mac C., 21
ADC. See Aerospace Defense Command.
Adelaide, Australia, 200
Adenosine triphosphate (ATP), 366
Adey, Dr. W. Ross, 307
Advanced Ballistic Reentry System (ABRES), 93, 146, 198

Advanced technology experimental transport, 350
Advanced vidicon camera system (AVCS), 25, 394
Advisory Council on Management Improvement, 49
AE. See Atmosphere Explorer program.
AEC. See Atomic Energy Commission.
AEC-NASA Space Nuclear Propulsion Office, 116, 144, 402
AEG-Telefunken, 407
Aegir (undersea habitat), 196-197
Aero Club of Washington, 247, 366, 401
Aerobee (sounding rocket)
 150
  airglow experiment, 127
  solar astronomy, 77, 210, 377
  stellar data, 91, 190-191, 245
  sustainer, 41
  ultraviolet astronomy, 181, 190-191, 264, 384
  upper-atmosphere data, 154, 162, 170, 350
  x-ray astronomy, 377
 150MI, 27, 99-100
 150MII, 224-225
 170
  solar astronomy, 77, 102
  stellar data, 272, 353
  ultraviolet astronomy, 102, 272, 353
  water recovery system, 41
  x-ray astronomy, 45, 51, 169, 217, 311, 317
 350, 129, 264
Aeroflot, 263
Aerojet-General Corp., 33, 42, 51, 55, 58, 116, 220
Aerojet Liquid Rocket Co., 157, 328
Aeronautical Fixed Telecommunications Network (AFTN), 200
Aeronautics (see also Federal Aviation Administration), 32, 39, 169, 252, 375
 aircraft. See Aircraft.
 anniversary, 320, 346-347, 394, 401
 Aviation Advisory Commission, 386
 award, 19, 29-30, 92, 233, 240, 283, 305, 401
 cooperation, 28-29, 85-86, 180, 199, 259, 349, 385, 390, 411, 412, 413-414
 employment, 13, 109, 322, 366
 exposition, 259
 funds, 35-36, 92-93, 161-162
 general aviation. See General aviation.
 history, 13, 98, 110, 320-321, 352

military (see also U.S. Air Force, aircraft), 198, 310, 320, 321
NASA program. See National Aeronautics and Space Administration, programs.
noise abatement. See Noise, aircraft.
research (see also Aircraft, research; Lifting-body vehicle; and X-15, XB-70, YF-12, etc.), 28-29, 58, 60-62, 63-64, 73-74, 98, 99, 105, 109, 148, 166, 217, 219-220, 250, 310, 320-321, 340-342, 353, 359, 377, 411, 412, 413
statistics, 86, 167, 212, 367, 414
Aeronautics and Space Report of the President (to Congress), 39
Aerospace Corp., 199
Aerospace Defense Command (ADC) (USAF), 32, 42, 263
Aerospace Industries Assn. (AIA), 106, 321, 366
Aerospace industry, 321, 391, 412-413, 414
employment, 13, 27, 38, 98, 174-175, 221, 322, 325, 366, 386, 414
financial problems, 126-127, 237, 245, 322, 338
Aerospace Medical Assn., 152
Aerospace Research Pilot School, 16
Aerospace Safety Advisory Panel, 213, 219
AFCRL. See Air Force Cambridge Research Laboratories.
AFCRL-A (magnetic storm satellite), 409
AFETR. See Eastern Test Range.
Afghan Cartographic Institute, 330
Afghanistan, 330
Africa, 51, 68
AFSC. See Air Force Systems Command.
AFTN. See Aeronautical Fixed Telecommunications Network.
AFWTR. See Western Test Range.
Agency for International Development (AID), 98, 194, 330, 405, 413
Agnew, Vice President Spiro T., 3
Apollo 13 mission, 119, 135
awards by, 52, 88, 368
goodwill tour, 18
Stafford Field dedication, 151
Agnews State Hospital, Calif., 348
Agreement. See International cooperation; International cooperation, space; and Treaty.
Agriculture, Dept. of, 35-36, 98, 256, 400-401, 405
AH-56A (Cheyenne) (helicopter), 72, 109, 345
AHRU. See Attitude heading reference unit.
AIA. See Aerospace Industries Assn.
AIAA. See American Institute of Aeronautics and Astronautics.
AID. See Agency for International Development.
AIMS. See Air traffic control radar system.
Air cushion vehicle, 94
Air Force Assn., 29, 311, 401
Air Force Cambridge Research Laboratories (AFCRL), 82, 260, 320-321, 409

Air Force Materials Symposium, 174
Air Force Missile Development Center, 227
Air Force Museum, 102-103, 153, 166, 258, 342, 382
Air Force Office of Scientific Research, 22
Air Force Satellite Control Facility, 58, 105
Air Force School of Aerospace Medicine, 46
Air Force Scientific Advisory Board, 100
Air Force Systems Command (AFSC), 46, 69, 87, 89, 107-108, 142, 149, 192, 200, 208, 224, 227, 231, 283, 331, 377
Aeronautical Systems Div., 173
Electronic Systems Div., 68, 171
Air France, 278, 388-389
Air Line Pilots Assn., 259
Air pollution, 44-45, 47, 49, 264
aircraft, 21, 27, 28, 49, 102, 178, 228, 256, 260-261, 270, 286, 291, 293, 320, 366, 381, 389-390, 392, 393
automobile, 22-23, 28, 47, 260, 270
detection and measurement 99, 255, 262, 398
effects, 74-75
prevention of, cost, 10, 260
satellite use in monitoring, 72-73, 98, 221
Air traffic control radar system (AIMS), 331
Air Transport Assn. of America (ATA), 275, 392
Air transportation. See Air traffic control; Aircraft; Airlines; Airports; General aviation; and Supersonic transport.
Air University, Aerospace Studies Institute, 159
Air Weather Service, 171
Airborne Warning and Control System (AWACS), 228
Aircraft (see also individual aircraft, such as C-5A, F-111, X-15, XB-70), 148-149, 352, 401-402
accident, 74, 100, 109, 226, 244, 278, 324, 331, 338, 349, 351, 379, 408
air pollution. See Air pollution, aircraft.
arresting system, 385
award, 233, 240
bomber, 69, 93, 195, 221, 347, 381-382
cargo, 86-87, 243, 381-382
carrier, 15
channel wing, 91-92
collision study, 28-29, 43-44, 99, 204
cost, 149, 151, 179, 198-199, 350
exhibit, 258
fighter, 69, 74, 75-76, 93, 100, 151, 218, 262-263, 294, 381-382, 402, 403
foreign, 93, 98, 118, 198-199, 226, 260-261, 321, 347, 350, 364, 381-382, 385
general-aviation, 212
helicopter, 109, 110-111, 136, 198, 233-234, 283, 299, 310, 311, 312, 345, 413
hijacking. See Hijacking of aircraft.
hovercraft, 296
hypersonic, 220-221, 247, 411
industry, 13, 17, 32, 71-72, 126-127, 169,

174-175, 178, 237, 240, 320-321, 322, 326, 366, 386, 414
navigation, 44
noise. See Noise, aircraft.
private, 17, 22
race, 3
reconnaissance, 102, 164, 183, 231, 259, 290
record, 7, 175, 198, 367, 383
regulations, 133, 144, 172-173, 256, 331, 345, 370, 392
research (see also Aeronautics, research; and X-15, X-24, XB-70, YF-12, etc.), 35-36, 50, 58, 61, 63-64, 73-74, 82, 99-100, 166, 171, 180, 183, 200, 220-221, 225, 253, 320, 321, 359, 411
safety, 13, 15, 22, 28-29, 214-215, 310, 324, 353, 414
sonic boom. See Sonic boom.
statistics, 212, 414
STOL, 32, 63-64, 91-92, 118, 179, 348
supercritical wing, 58, 359, 377, 408-409, 411
supersonic. See Supersonic transport, Concorde, F-8, F-14A, F-111, Tu-144, X-15, XB-70, YF-12, etc.
tanker, 264
traffic, 414
traffic control. See Air traffic control.
training, 352
transport (see also Supersonic transport), 69, 110-111
    jet, 13, 21, 62, 73-74, 237, 242, 277, 303, 310, 324, 349, 370, 385, 387-389, 411
    STOL, 32, 63-64, 91-92, 118, 179, 349
    V/STOL, 36, 43, 63-64, 85, 166, 198-199, 204, 349-350, 411, 412, 413
    VTOL, 63, 148, 198-199, 321, 349-350
wake turbulence, 9, 62, 310, 389-390
wind-tunnel testing, 63, 109
Aircraft arresting system, 385
Aircraft carrier, 15
Airesearch Manufacturing Co. Div., Garrett Corp., 147
Airglow, 77, 96, 127, 311
Air traffic control (ATC), 92, 99, 109, 200, 335
    Air Traffic Control Advisory Committee (DOT), 76, 258
    FAA regulation, 172-173, 345, 353, 370
    NASA anticollision program, 28-29
    patent, 47
    radar use in, 68, 331
    R&D To Increase Airport and Airway System Capacity (FAA), 218
    satellite use in, 28, 47, 77, 85, 171, 241, 243, 252-253, 279-280
    statistics, 162
    supersonic transport (SST), 315
Airlines (see also Air pollution; Air traffic control; Noise, aircraft; and Supersonic transport)
    accident, 278
    aircraft, 13, 15, 21, 225, 237, 252, 278
    anniversary, 346-347
    cooperation, international, 15, 364
    fares, 392
    forecast, 167
    hijacking. See Hijacking of aircraft.
    safety, 13, 414
    services, 196, 208-209, 263, 308, 312
    statistics, 167, 414
    traffic growth, 414
Airlock, 103-105, 169, 246, 248, 372
Airports (see also Air pollution; Air traffic control; Noise, aircraft; and individual airports such as Washington National Airport), 386
    award, 113
    facilities, 218, 385
    funds for, 22, 178
    R&D Plan To Increase Airport and Airway System Capacity (DOT), 218
    regulation, 259, 345, 370
    statistics, 162
    supersonic transport (SST), 344-345, 389-390
Airship, 99
Akrotiri, Cyprus, 383
Alabama, 40
Alabama Space and Rocket Center, 102-103, 237, 344
Alabama Space Science Exhibit Commission, 93-94
Alabama, Univ. of, 204, 406
Alamogordo, N. Mex., 235
Alaska, 38-39, 196, 223, 242, 256
Alaska Airlines, 196
Alaska Rocket Range, 71, 73
Alaska, Univ. of, 40, 48
Albania, 151
Albany, N.Y., 139
Alberry, Milton W., 109
Albuquerque, N. Mex., 347
Aldrin, Col. Edwin E., Jr. (USAF), 277, 291, 346, 349
    awards and honors, 45, 52, 94, 155, 166, 234, 238, 256, 311, 368
Aleutian Islands, 103
Alexandria, Va., 355
Alfvén, Dr. Hannes O., 8, 348, 359
Alitalia, 196
All American Engineering Co., 385
Allen, Dr. James G., 308
Allen, Joe H., 7
Allen, William H., 159
Alley, Dr. Carroll O., 374
Allnutt, Robert F., 142, 144
Allott, Sen. Gordon L., 234
Alma, U.S.S.R., 346
Alouette I (Canadian satellite), 79, 247
Alouette II, 79, 247
ALSA. See Astronaut life support assembly.
ALSEP. See Apollo lunar surface experiments package.
Alsop, Stewart, 153
Alzar, Prof. Vital, 360
Amarillo, Tex., 346-347
Amelia Earhart Lives: A Trip Through Intrigue To Find America's First Lady of Mystery, 364

"American Advancement into Space" (AIAA meeting), 342
American Airlines, 305, 401
American Assn. for the Advancement of Science (AAAS), 185, 398, 406-407
American Astronautical Society (AAS), 106, 197, 308, 370
American Broadcasting Co. (ABC), 137-138
American Business Press, 41
American Cancer Society, 97
American Chemical Society, 59
American College of Cardiology, 68
American Council of Learned Societies, 173
American Council on Education, 96
American Federation of Technical Engineers, 274
American Geographical Society, 162-164
American Geophysical Union, 142-143, 390
American-Hayden Planetarium, 196
American Heart Assn., 68
American Helicopter Society, 198-199
American Helium Society, 99
American Horizons Foundation, 29
American Institute of Aeronautics and Astronautics (AIAA), 240, 283, 340-342, 370, 383, 385, 391
  Advanced Space Transportation Meeting, 41
  Aerospace Sciences Meeting, 18, 19-20
  Aircraft Design Award, 233
  awards, 19-20, 92, 106, 112, 233, 240
  Committee on International Cooperation in Space Flight, 149
  Conference on Test Effectiveness in the 70s, 107
  Earth Resources Conference, 72-73
  Launch Operations Meeting, 40
  officers, 21, 383
  Space Communications Award, 112
  Space Science Award, 19-20
  Technical Activities Committee, 149
  Technical Committee on Applications of Aerospace Technology to Society, 149
American Institute of Physics, 329
American Meteorological Society, 49
American Newspaper Publishers Assn., 146
American Nuclear Society, 154, 377
American Physical Society, 30, 152
American Rocket Society, 164
American Samoa, 125, 132, 135, 409
American Science and Engineering, Inc., 217, 317, 377
American Security Council, 347
American Society for Engineering Education, 204
American Society of Mechanical Engineers (ASME), 211
American Telephone & Telegraph Co. (AT&T), 28, 272, 290, 339, 407-408
Ames Research Center (ARC) (NASA), 61, 271, 295, 340-342, 348, 372
  Apollo 11 lunar rock analysis, 20
  Apollo 13 tank test, 191
  award, 166
  biomedical research, 40-41, 68, 258, 387, 390-391
  experiment, 154, 162, 362, 407
  meeting, 197, 295, 370
  Mission Analysis Div., 220-221
  particles and fields research, 33, 407
  personnel, 173-174, 250
  Polymer Research Laboratory, 284
  research, aircraft, 109, 320, 324
  solar eclipse observations, 79-80
  visitors, 346
Amino acids, 111, 387, 390-391
Anaheim, Calif., 197
Analysis of Apollo 8 Photography and Visual Observation (NASA SP-201), 166
Anchorage, Alaska, 196
Anders, L/C William A. (USAF, R.), 134, 305, 409
  awards and honors, 88-89
  Bauer, Louis H., Lecture, 152
Anderson, Sen. Clinton P., 39, 379
Anderson, Dr. John D., 16, 157, 170
Anderson, Orson L., 215
Anderson, Mrs. Pamela Brown, 308
Anderson, Robert, 57
Anderson, Rodney, 308
Anderson, Ronald, 274
Andoeya, Norway, 38, 82
Andrews AFB, Md., 224
Adrian, Tex., 276
Animal experiments, space, 257, 307, 324-325, 362, 367, 392, 395, 411
  tracking, 52, 57, 58, 59, 62, 107, 118
Annapolis, Md., 72
Anniversary, 49, 82-83, 92, 109, 217, 277, 345, 371, 395
  aeronautics, 320-321, 346-347, 394, 401
  atomic bomb, 235, 259, 260
  Government, 46, 49
  manned space flight, 126, 127, 165, 227-228, 236, 238-239, 243, 372-373
  NASA, 32, 165, 217, 227, 236, 238-239, 323, 395
  satellite, 32, 107, 112
  USAF, 307, 320-321
  U.S.S.R., 126, 127, 361
ANS (Netherlands satellite), 194
Antarctica, 196, 216
Antares (Apollo 14 LM). See Lunar module.
Antares (star), 403
Antenna, 65, 92, 100, 133, 180, 213
Antiballistic missile (ABM) system (see also Safeguard), 142, 154, 266, 381-382
  congressional consideration, 170-171, 200, 317, 331
  funds for, 200, 317, 331
  opposition to, 152, 185
  test, 276, 405
  U.S.S.R., 146, 152, 171, 266, 381-382
Anticollision system, aircraft, 28-29, 44, 99, 204
Antigua Tracking Station, 133, 210
Antivivisection, 307

Apollo (program), 3, 6, 32, 83-84, 108, 301, 414
  astronaut. See Astronaut.
  award, 155, 159, 166, 206, 283, 286, 311, 341-342, 376
  cost, 50
  criticism, 164, 290-291, 295-296, 340-341
  funds, 35-37, 87, 93, 161-162, 203, 284-285
  history, 406
  landing site, 294-295, 323
  launch, *Apollo 13* (AS-508), 119
  management, 251
  plans, 11-12, 42-44, 53-54, 71, 90-91, 101, 103, 115, 116, 229, 258, 263-264, 266-267, 278, 283, 284-285, 315, 323, 328, 331, 338, 386-387, 391, 403, 408-409, 410
  policy, 50, 147-148, 207-208, 211, 248, 295-296, 305, 309-310, 316-317, 330
  press comment (see also Apollo missions), 3-4, 86, 131, 287-288, 290-291, 294, 306
  progress, 186-187, 410-411
  tracking, 133, 284-285
Apollo (spacecraft), 50, 65-66, 196, 204, 218-219, 364, 410
  command and service module. See Command and service module.
  command module. See Command module.
  computer use, 164-165
  debris, 287
  equipment, 18, 29, 115-116, 181-182, 205, 208, 210, 212, 223, 258, 271, 284-285, 315, 338, 371, 403
  exhibit, 49, 209, 238, 401-402
  heat shield, 209
  landing system, 258
  launch, *Apollo 13* (AS-508), 119
  launch vehicle. See Saturn.
  life support system, 7, 387
  lunar module. See Lunar module.
  oxygen tank, 119, 121, 137, 140, 141, 161, 167, 181-182, 191, 201-203, 217, 218-219
  safety methods, 196, 204
  scientific instrument module, 249
  service module. See Service module.
  test, 403
Apollo 1 (AS-207) (spacecraft), 232
Apollo 6 (spacecraft), 28
Apollo 7 mission, 181
Apollo 8 mission, 88-89
Apollo 9 mission, 130, 186
Apollo 10 mission, 364
Apollo 11 (spacecraft), 49
Apollo 11 mission, 1, 65-66, 125-126, 190, 267, 289, 291, 293, 346, 349
  anniversary, 227-228, 238
  award, 45, 52, 94, 155, 166, 256, 283, 311, 368, 376
  commemorative stamp, 14-15
  exhibit, 401-402
  laser ranging retroflector experiment, 26, 210
  lunar samples, 4-6, 15, 18, 20, 51, 57, 59, 68-69, 76-77, 85, 103, 139, 142-143, 223, 234, 240, 244, 245, 250-251, 265, 270, 274, 289, 290, 295, 333, 353, 358-359, 369, 372, 412, 414
  photographs, 1
  preparations, 136
  press coverage, 1, 112
  results, 227-228, 240, 412
  Saturn V booster, 40
  significance, 32, 105-106, 240, 379
Apollo 11 Spacecraft Touring Exhibit, 401-402
Apollo 12 Lunar Sample Preliminary Examination Team, 76-77
Apollo 12 mission
  Apollo lunar surface experiments package, 329, 371-372, 374
  Apollo 12: Preliminary Science Report (NASA SP-235), 305
  awards and honors, 18, 41, 42, 82, 197, 368, 372, 376
  exhibit, 139
  Johnson, President Lyndon B., briefing, 15
  lightning hazard, 71, 250
  lunar landing, 2
  lunar samples, 7, 15, 51, 76-77, 85, 86, 91, 103, 125-126, 130, 142-143, 180-181, 184, 245, 295, 333, 353, 358-359, 369, 412, 414
  photographs, 176-177, 305
  press conference, 371-372
  results, 305, 371-372
  seismometer experiment, 244-245, 297
  suprathermal ion detector experiment (SIDE), 119-121
  Surveyor III spacecraft, 2, 7, 16, 54, 176-177, 178, 353-354
Apollo 13 mission, 233, 336
  accident, 119, 125, 137-139, 140-141, 143-144, 147-148, 154, 155, 165, 410
  investigation, 125, 137, 140, 144, 155, 167, 184, 232
  report, 147-148, 161, 181-182, 191, 201-203, 204, 213, 217, 218-219
  astronaut emblem, 25
  award, 133-134, 136-137, 249, 342, 368
  biological aspects, 101, 111, 115, 116
  celebrations, 136, 140, 162, 194
  countdown, 111, 116
  crew replacement, 111, 115, 116, 117-118, 119
  cuisine, 115
  emergency procedures, 121-125, 127-128, 130, 134, 165
  extravehicular activity, 62-63, 71
  launch, 119
  launch schedule, 7, 71
  lunar landing, 90-91
  lunar samples, 62-63, 71, 103, 125-126
  National Day of Prayer, 133, 139
  Nixon, President Richard M., 118, 128, 130, 133-137, 138, 139-140, 198
  photographs, 125, 140
  preparations, 62-63, 71, 101, 115

press comment, 46, 111, 116, 118, 126, 127, 128-129, 130-131, 136, 138-139, 140, 141, 145, 150, 165, 186, 232
press conference, 40, 62-63, 71, 90-91, 117-118, 127-128, 132-133, 134, 143-144, 161
public reaction, 121-122, 127, 128-132, 135-136, 150-151
quarantine, 51, 71
recovery, 125, 130, 132-134
splashdown, 119, 125, 135-136, 138
TV broadcasts, 71, 121
U.S.S.R. and, 126, 128, 130, 132, 136
Apollo 13 Mission Operations Team, 136-137, 368
Apollo 13 Review Board, 125, 137, 141, 144, 155, 161, 167, 184, 191, 201-203, 204, 213, 217, 218-219
Apollo 14 (spacecraft), 364, 403
Apollo 14 mission, 283
  buddy secondary life support system (BSLSS), 387
  camera, 115-116
  contract, 115-116
  crew, 167
  laser ranging retroflector, 210
  launch schedule, 7, 219, 233, 331, 403, 408, 410-411
  modularized equipment transport system (METS), 18
  preparations, 18, 205, 217, 233, 266, 283, 306, 319, 364, 386-387, 390, 403, 409
  press conference, 233, 338, 403
Apollo 15 (formerly 16) mission
  camera, 115-116
  cancellation of Apollo 15 mission, 284, 287, 295-296, 306, 309-310
  crew, 101, 285
  launch schedule, 219, 284-285, 323, 403, 408
  lunar communications relay units (LCRU), 315
  preparations, 323, 403
Apollo 16 (formerly 17) mission, 28, 42-43, 177, 205, 212, 284-285, 323
Apollo 17 (formerly 18) mission, 28, 42-43, 86, 177, 284-285, 323
Apollo 19 mission, 12, 266-267, 284-285, 287, 295-296, 309-310
Apollo Applications program (AAP). See Skylab.
Apollo lunar surface experiment package (ALSEP), 25, 37-38, 65-66, 125, 210, 212, 323, 329, 371-372
Apollo Telescope Mount (ATM), 104, 169, 209, 274, 296, 371
Applications Technology Satellite (ATS), 43, 161-162, 224, 284, 289, 367
APS. See auxiliary propulsion system.
APT. See Automatic picture transmission.
Aquanaut, 118-119, 226, 239, 245, 308
Aquarius (Apollo 13 LM). See Lunar module.
Arava (STOL aircraft), 118
ARC. See Ames Research Center.

ARCADE (French and U.S.S.R. space study of solar energy particles), 177
Arcas (sounding rocket), 14, 62, 71, 77, 133, 172, 208, 210, 257, 305, 309, 334, 338, 411
ARCS (Achievement Rewards for College Scientists) Foundation, 164
Arecibo (Puerto Rico) Ionospheric Observatory, 236
Arctic, 108, 127, 216
Argentina, 159, 252
Argentine Institute of Biochemical Research, 348
Argon, 333
Argonne National Laboratory, 26
Ariel I (U. K. satellite), 357
Ariel II, 357
Ariel III, 357
Aristarchus (lunar crater), 192, 228
Arizona, 311, 383
Arizona, Univ. of, 61, 69-70, 387
Arlington National Cemetery, 340
Arlington, Va., 224
Armalcolite (lunar material), 234
Armed Services Board of Contract Appeals, 335
Armenian Academy of Sciences, 97
Armstrong, Neil A., 1, 3, 102-103, 196
  Apollo 11 mission, 227-228, 238, 291, 401-402
  Apollo 13 mission, 132-133
  Apollo 13 Review Board, 144, 155
  appointment, 173-174, 223
  awards and honors, 23, 45, 52, 57, 94, 95, 166, 234, 267, 283, 368, 372, 376
  Museum, 153
  political aspirations, 8
  press conference, 132-133, 227-228
  U.S.S.R. visit, 185, 190, 192, 195, 366
Armstrong, Mrs. Neil A., 196
Armstrong, Neil A., Museum, 153
Armstrong, Stephen V., 291
Army Ballistic Missile Agency (ABMA), 103
Army Corps of Engineers, 230, 232, 384
Army Missile Command, 174, 226
Arnold Engineering Development Center, 89, 142, 382
Arnold, Henry H., Trophy, 311
Ascension Island, 336
Asia, 51
ASME. See American Society of Mechanical Engineers.
Assateague National Seashore Park, 111
Associated Aviation Underwriters, 9
Assessment of Space Communications Technology (House report), 39
Asteroid, 8, 209-210, 359
Astronaut (see also Cosmonaut; Extravehicular activity; Space biology), 292, 310, 340, 349, 401
  accident, 266
  Apollo mission. See Apollo missions (7, 9, 10, 11, 12, 13, 14).
  appointment, 173-174, 181, 223, 316, 405
  awards and honors, 8, 18, 41, 42, 45, 52, 57, 82, 88-89, 94, 133-134, 137, 155,

166, 196, 197, 249, 256, 267, 283, 311, 368, 372, 376
crew assignment, 111, 115, 116, 117-118, 167, 327
Expo '70 visit, 99
former, 6; 152
goodwill tour, 18, 51, 55, 59, 62, 68, 94, 99, 196, 314, 331, 336
hazards, 15, 20, 51, 54, 55, 71, 131, 137, 141, 191, 227, 360
memorial, 192
monument (proposed), 235
physiology, 46, 55-56, 57-58, 74, 90, 111, 115, 116, 117-118, 184, 319, 327, 387
political aspirations, 7, 8, 157
press comment, 2, 46, 126, 127, 128-129, 131, 136, 140
press conference, 40, 62-63, 90-91, 132-133, 134, 143-144, 212, 227, 292-293, 371-372
prisoner of war mission, 260
promotion, 198, 398
publication contract, 74, 314
quarantine, 15, 55, 403
reception, 157, 165, 194
record, 46
religion, 52-53
resignation, 211, 293
salary, 101
scientist-astronaut, 287
training, 20, 115, 264, 266, 352, 364, 371, 390
U.S.S.R. visit, 177, 185, 190, 192, 195, 366
women as, 46
Astronaut life support assembly (ALSA), 147
Astronautical Multilingual Dictionary, 158
Astronautics Engineer Award, 94
Astronomical Netherlands Satellite Program Authority, 194
Astronomy (see also individual observatories, planets, probes, sounding rockets, and satellites, such as Asteroid; Comet; Extraterrestrial life; Galaxy; *Mariner VI, Mariner VII, Oao II*; Pulsar; Quasar, Radioastronomy; Star; Telescope), 6, 33, 49, 116, 153, 201, 204-205
gamma ray, 396
infrared, 251
meeting, 4-6, 19-20, 267-268, 294-295, 370
NASA program, 11-12, 22, 26, 27, 36, 45, 51, 55, 60-62, 77, 79-82, 84, 101-102, 139, 159, 194, 203, 251, 256, 279, 284, 285-286, 286-287, 294-295, 297, 301, 340-342, 359, 361-362, 370, 379, 380-381, 383, 399, 400, 407, 410-411, 412-414
nomenclature, 267-268
solar, 30, 33, 43, 51-52, 60-62, 79, 84, 101-102, 129, 177, 210, 240, 306, 311, 320, 345, 377, 400, 407
stellar, 21, 27, 51, 55, 61-62, 91, 190-192, 198, 233, 245, 358, 380-381, 409, 411-412
ultraviolet, See Ultraviolet.
U.K., 21
U.S.S.R. program, 55, 266, 311-312, 358-359, 389, 399
x-ray. See X-ray.
ATA. See Air Transport Assn. of America.
ATC. See Air traffic control.
AT&T. See American Telephone & Telegraph Co.
Athena (missile), 146, 231, 256-257
H, 146
Atlanta, Ga., 219, 391
Atlantic City, N.J., 108-109
Atlantic Ocean, 45, 165, 198, 200, 308, 310, 312, 318, 379, 389
Atlas (booster)
E, 93
F, 93
Atlas-Agena (booster), 209, 278
Atlas-Centaur (booster), 83, 380, 384
Atlas of the Universe, 382
ATM. See Apollo Telescope Mount.
Atmosphere
stellar, 27
upper (earth), 12, 16, 111, 152, 154, 162, 170, 171, 189-190, 213, 216, 256, 305, 336-337, 375, 390, 399, 401, 414
Atmosphere Explorer (AE) program, 63
Atom, 176
Atomic bomb, 235, 258-259, 260
Atomic Energy Commission (AEC) (see also AEC-NASA Space Nuclear Propulsion Office; NERVA; SNAP, and Vela programs), 10, 67-68, 162, 235, 329, 335, 372
Argonne National Laboratory, 26
award, 94, 209, 277, 372
Brookhaven National Laboratory, 94
budget, 35-36, 235
camera, atomic, 30
contract, 12, 265
cooperation, 12, 75, 265, 309, 315, 329, 411
criticism, 235, 242-243
Div. of Radiation Protection Standards, 235
General Advisory Committee, 67, 250
iodine 123, 352-353
Los Alamos Scientific Laboratory, 209, 277
meeting, 263
nuclear reactor, 10, 265, 309, 411
nuclear rocket engine, 11, 36, 44, 64, 75, 116, 277, 329, 411
Pacific Northwest Laboratory, 30, 94, 228
personnel, 17, 224, 250, 277, 379, 398
R&D funds, 71-72, 98, 362
undersea nuclear probe, 228
Atomic Pioneers Award, 67-68
ATP. See Adenosine triphosphate.
Ats I (Applications Technology Satellite), 38-39, 165-166
Ats III, 39, 54, 79, 165-166

Ats V, 108, 346
ATS-F, 8, 42-43, 61, 115, 145, 146, 148, 156, 201, 224, 230, 273, 289, 318, 367, 413
ATS-G, 42-43, 61, 115, 145, 146, 148, 156, 201, 224, 230, 273, 289, 318
ATS Procurement Review Committee (PRC), 273, 289
Attitude control system (ACS), 129
Attitude heading reference unit (AHRU), 207
Atwood, J. L., 311
Auburn, Mass., 180
Auburn Rotary Club, 180
Auburn Univ., 204
Augustana Hospital, Chicago, 389
Aurora, 12, 16, 38, 48, 56, 69, 71, 73, 82, 99-100, 108, 126, 365-366, 369, 372
Aurorae (Esro I) (satellite), 214
Australia
 aircraft, 129, 157, 262-263
 Apollo 12 lunar samples, 51
 Apollo 13 mission, reaction to, 129
 communication via satellite, 96, 409-410
 communications station, 7
 international cooperation, space, 62, 77-78, 100, 150, 393, 413-414
  satellite, 24, 32, 150, 410-411, 413-414
  sounding rocket, 181, 185, 190-191, 414
 tracking station, 100, 199, 213
Australian National Univ., 295
Australian Overseas Telecommunications Commission, 74
Australis Oscar V. See *Oscar V.*
Autoland automatic landing system, 69
Automatic picture transmission (APT), 25, 67, 381, 394-395
Autonomous control subsystem (ACS), 194
Avco Research Laboratory, 156
Auxiliary propulsion system (APS), 119
AVCS. See Advanced vidicon camera system.
Aviation Advisory Commission, 386
Aviation Hall of Fame, 401
Aviation/Space Writers Assn., 15, 175
AWACS. See Airborne Warning and Control System.
Awards, 88-89, 256, 305
 civic, 57, 219
 Government, 18, 29, 55, 67, 94, 109, 113, 136-137, 175, 209, 245, 249, 269, 277, 368-369, 372, 379
  NASA. See National Aeronautics and Space Administration.
 institutions, 13, 23, 29, 313-314, 337
 military, 42, 109, 155, 227, 232, 392
 society
  achievement, 95, 126
  aeronautics, 19-20, 29, 92, 233, 240, 283, 305, 401
  astronautics, 19-20, 29, 37, 41, 45, 52, 57, 88-89, 94, 106, 112, 155, 159, 164, 166, 184, 206, 227, 271, 283, 286, 311, 341-342, 376, 384
  exploration, 41, 52
  science, 329, 348-349
 technology, 40, 87, 106, 197
AX (close support aircraft), 382, 402

B

B-1 (advanced strategic bomber), 69, 195, 221, 381-382
B-52 (Stratofortress), 93, 149
 H1-10 flights, 18, 27, 55, 66, 199
 M2-F3 flights, 178, 190, 357, 379
 X-24A flights, 94-95, 108, 145, 171-172, 206, 263, 273, 336, 375
B-58 (bomber), 93
B-70 (bomber), 366
BAC. See British Aircraft Corp.
Back contamination, 7, 15, 20, 51, 55-56, 71, 250-251
Bad Godesberg Institute for Aeronautical Medicine, 262
Baehr, Edward F., 317
Baker, Charles D., 28, 65, 153
Baker, Sen. Howard H., 44-45
Balbo, Gen. Italo (Italy), 110
Ball Brothers Research Corp., 296
Ball, Dr. Leslie W., 374
Balloon
 Atlantic Ocean crossing attempt, 308, 310
 record, 308, 312, 324
 research use, 73, 85, 101, 210, 230, 272, 286, 307, 336, 411
 tracking of, 85, 272
Balloon interrogation packages (BIP), 336
Baltimore, Md., 366
Bandermann, L. W., 345
Bank of America, 240
Bar Main, Alaska, 71, 75
Barnard, Dr. Christian N., 58
Barrett, Dr. Earl W., 74-75
Barrow, Dr. Thomas D., 277
Barstow, Calif., 108
Batavia, Ill., 225, 235
Battelle-Northwest Laboratories, 358
Bauer, Louis H., Lecture, 152
Baunsgaard, Hilmar, 128
Bay of Biscay, 287
Baykonur, U.S.S.R., 76
 launch
  Cosmos, 8, 198, 226, 248, 305, 315, 323, 325, 331, 332, 343, 345, 352, 366, 376, 384, 398
  Luna XVI, 299
  Luna XVII 364-365
  Soyuz IX, 189
  Venus VII, 266
Bayley, William H., 159
BBC. See British Broadcasting Co.
Beall, Rep. J. Glenn, 146
Bean, Capt. Alan L. (USN), 15, 18, 20, 37, 42, 46, 51, 99, 368, 372
Beaver, Okla., 276
Beech Aircraft Co., 161, 191, 201-203
Beggs, James M., 15, 86, 389
Beirut, Lebanon, 289-290, 306
Belfast (military transport aircraft), 69

Belgium, 51, 359
Bell Aerospace Systems, 55
Bell Telephone Laboratories, 110, 185, 269, 278
Bell, V. E., 285-286
Bellafore, Vito, 130
Bellcom, Inc., 1, 192
Beltz, Dr. John S., 406
Belyayev, Col. Pavel I. (U.S.S.R.), 9, 267
Bender, Dr. Peter L., 7
Bendix Corp., 212, 264-265
Bendix Field Engineering Corp., 320
Bendix Race, 212
Benn, Minister of Technology Anthony Wedgwood (U. K.), 151
Bennett (comet), 113, 116
Bennett, Dr. Ivan L., Jr., 89-90
Bennett, J. C., 113
Bennett, Walker G., 389
Beranek, Dr. Leo L., 400
Beregovoy, Georgy T., 160, 185, 190
Bermuda, 318
Berry, Dr. Charles A., 37, 46, 55-56, 89, 111, 116, 152, 186, 327
Bessborough, Lord Frederick E. N. P., 305
Bethesda, Md., 304
Bethpage, N.Y., 165
Bevan, William, Jr., 185
Beverly Shores, Ind., 305
Binaghi, Walter, 235, 265, 336
Binder, Robert H., 366
Binghampton, N.Y., 383
Biological warfare, 276-277
Biosatellite (program), 36
Biosatellite III, 307
Biosonde experiment, 392, 400
Bio-Space Technology Training Program, 257
Biosphere, 221
Biossat, Bruce, 145
BIP. See Balloon interrogation packages.
Bisplinghoff, Dr. Raymond L., 207, 265
Black Arrow (U.K. booster), 285
Black Brant IV (Canadian sounding rocket), 315, 318
Blagonravov, Dr. Anatoly A., 85, 128, 177, 335
Blair, Dr. William J., 94
Blake, Dr. Eugene Parson, 130
Blankenship, Charles, 209
Blount, Postmaster General Winton M., 14
Blue Book, Project, 3-4
Blue, James, 352-353
Blue Ribbon Action Commission (DOD), 274
Blue Ribbon Defense Panel, 223-224, 246, 257, 274
BOAC. See British Overseas Airways Corp.
Bochum Observatory (West Germany), 151, 299
Boeing 707 (jet passenger transport), 15, 73, 228, 279, 289, 300, 311
Boeing 727, 21, 293
Boeing 737, 21

Boeing 747, 113, 170-171, 277, 389
  certification, 13, 20
  christening, 15, 196
  cost, 303
  hijacking, 257, 289-290
  operational problems, 208-209, 278, 324, 349
  pollution control, 21
  record, 367
  regulations, 172-173, 370
  test, 13, 15, 62, 303
  transatlantic crossing, 13, 22, 411
Boeing 747B, 303
Boeing Co., 369
  Aerospace Group, 374
  Aerospace Systems Div., 54-55
  Airborne Warning and Control System, 228
  booster, Saturn V, 23, 266-267, 271, 386
  Commercial Airplane Div., 21
  contract, 7, 23, 179, 203, 228, 345, 386
  employment, 271, 386
  jet passenger transport. See Boeing 707, 727, 737, and 747.
  lunar roving vehicle, 29, 371, 404
  personnel, 21
  Southeast Div., 154
  space shuttle, 203
  supersonic transport, 279, 386, 387
Boeing Magazine, 321
Boeing Space Center, 353-354
Bogard, Dr. Donald D., 333
Bolam, Mrs. Guy, 364
Bolger, Philip H., 75, 328
Bollerud, B/G Jack (USAF), 194
Bolling AFB, Md., 303
Bolshaya Sovetskaya Entsklopediya, 291
Bombay, India, 233, 359
Bondi, Dr. Herman, 313
Bonestell, Chesley, 196
Bonn, West Germany, 86, 227, 320
Bonny (space monkey), 307
Boorstin, Dr. Daniel J., 145, 189
Boosted-Arcas (sounding rocket), 102
Boosted-Dart (sounding rocket), 411
Boothville, La., 264
Boreas (ESRO satellite), 49-50
Borman, Col. Frank (USAF, Ret.)
  Apollo 13 mission, 134
  appointment, 154, 260, 405
  awards and honors, 29, 88-89, 267
  docking, 335
  Gemini VII mission, 340
  resignation, 29
  Soyuz IX crew, message to, 203
  TV appearance, 96
  U. S. prisoners of war in Vietnam mission, 260
Born, Dr. Max, 4
Boston College, School of Management, 280
Boston, Mass., 13, 272, 296, 335
Boston Museum, 13
Boulder, Colo., 74-75, 205, 320, 351, 390
Boulding, Kenneth W., 291
Bourne, Mass., 12

Boyce, Dr. Peter, 101
Bradbury, Dr. Norris E., *209*, *277*
Bradley Air National Guard Base, Conn., 225
Braman, James D., 298
Brand, Vance D., 101
Brandt, John C., 306
Brandt, Chancellor Willy (West Germany), 118, 119
Brandt, Mrs. Willy, 118
Brans, Dr. Charles, 367
Brayton power conversion system, 334
Brazil, 85, 156, 252, 315, 318, 413
Breckenridge, William G., 185
Brevard County, Fla., 236
Brezhnev, Leonid I., 19, 127
Bridgetown, Barbados, 232
Bright, Capt. W. J. (U.K.), 3
Brighton, Malcolm, 308
Brighton, U.K., 267
Bristol, U.K., 292, 300
British Aircraft Corp. (BAC), 291, 292, 300, 407
British Broadcasting Co. (BBC), 129
British Interplanetary Society, 45, 357
British Overseas Airways Corp. (BOAC), 293, 300, 389
British Tourist Authority, 296
Britten-Norman Islander (light transport), 3
Broadwick, Georgia (Tiny), 352
Broecker, Dr. Wallace S., 216
Brookhaven National Laboratory, 94
Brooks AFB, Tex., 192
Brooks, Dr. George W., 325
Brooks, Dr. Harvey, 153
Brown, Dr. Allan H., 55
Brown Engineering Co., 113
Brown, Gen. George S. (USAF), 194, 283
Brown, Harold, 221
Brown, Herbert C., 55
Brown Univ., 90, 147, 189, 231
Browne, Secor D., 7, 40
Brussels, Belgium, 241, 243, 359
BSLSS. *See* Buddy secondary life support system.
Bucharest, Romania, 68
Buchsbaum, Dr. Solomon J., 89
Budapest, Hungary, 129
Buddy secondary life support system (BSLSS), 387
Budget, Bureau of, 174
Buffalo, N.Y., 368
Buffalo, Univ. of, 392
Buhl, Dr. David, 204, 262
Bulgaria, 380
Bulgarian Academy of Sciences, 200
Bullfrog experiment, 324-325
Bullpup-Cajun (sounding rocket), 16
Burakan Astrophysical Observatory, 97
Burch, Dean, 25
Bureau of Commercial Fisheries, 230
Bureau of Indian Affairs, 383
Bureau of Mines, 204
Bush, Dr. Vannevar, 67
Bushnell, Prof. David, 189
Buttenheim, Donald V., 41
Butler, T. Melvin, 326
Byrd, Adm. Richard E., 110

C

C-5 Review Board, 207, 373
C-5A (Galaxy) (military cargo transport), 214, 382, 411
  accident, 338
  contract, 9, 169-170, 214, 246, 270-271, 345
  cost, 149
  funds for, 19, 86-87, 129, 245, 331
  production problems, 16, 17, 19, 27, 72, 126-127
  record, 367
  static test, 16, 373
  test flights, 172-173, 225, 243
C-47 (research aircraft), 360
C-141 (military transport aircraft), 61, 136, 149
CAB. *See* Civil Aeronautics Board.
Cable and Wireless, Ltd., 74
Cable, undersea, 290
Caffrey, John, 96
Cahn, Robert, 30, 42
Cairns, Dr. Theodore L., 89
Cairo, United Arab Republic, 317
Caldera, President Rafael (Venezuela), 192
California, 39-40, 174-175, 283, 405
California Institute of Technology (Cal. Tech), 17, 29, 30, 59, 68-69, 221, 276, 294, 346
  Apollo 12 lunar rock samples, 142-143, 180-181
  award, 189
  Mariner VI, 157
  relativity, theory of, 201, 366-367
  solar research, 30, 101-102, 366-367
California, Univ. of, 276, 305, 337
  Berkeley, 1, 94, 108, 230, 307, 359
  Los Angeles (UCLA), 77, 194, 225, 328, 387
  San Diego, 19, 348, 359
Californium 252, 30
Calverton, N.Y., 403, 408
Cambridge Conference on Relativity, 198
Cambridge, Mass., 13, 21, 99, 169, 189
Cambridge Univ., 6
Camera, 2, 73
  Apollo 13, 71
  atomic, 30
  contract, 28, 115-116, 118
  Itos I, 24-25
  Nimbus IV, 113-114
  sounding rocket, 102
  Surveyor III, 178
  USAF, 200
Cameron, Dr. A. G. W., 143
Camp, Dennis W., 264
Campbell, P. A., 327
Canada, 118, 135, 242
  Apollo 12 lunar sample study, 51

cooperation, 71, 78, 79, 85, 148-149, 150, 242, 393, 408-409, 413
  Ministry of Transport, 272
  satellite, 32, 79, 150, 242, 247, 408-409, 413
  sounding rocket, 3, 12, 16, 40, 48, 56-57, 62, 71, 99, 108, 126, 133, 154, 162
Canadian Defence Research Establishment, 242
Canary Islands, 199, 320
Canberra, Australia, 74, 100
Cancer research, 97
Cannon, Sen. Howard W., 166
Cannon, Robert H., Jr., 40
Cape Canaveral, Fla., 103, 107, 408
Cape Kennedy, Fla., 45, 386
Capitol Records Co., 349
Caracas, Venezuela, 55
Carbon dioxide, 40, 49
Carbon monoxide, 110, 215
Caribbean, 200, 308
Caribbean Research Institute College, 210
Carnarvon, Australia, 100, 320
Carnegie Institution of Washington, 17, 164
  Geophysical Laboratory, 153
Carrillo, Mexico, 256-257
Carrington, Minister of Defence, Lord (U.K.), 241
Carswell AFB, Tex., 244, 331, 351
Carver, George Washington, 268
CAS-A (U.S.-France meteorological satellite), 408-409
Case, Sen. Clifford P., 217, 330
Case Western Reserve Univ., 189, 337
Casteau, Belgium, 294
Castleman, Dr. Kenneth R., 397
Castor (booster), 332
Castro, Premier Fidel (Cuba), 257
CAT. See Clear-air turbulence.
Cate, Dr. James Lee, 406
Catholic University of America, 94
CBS. See Columbia Broadcasting Co.
Centaur (booster), 214
Center for Study of Responsive Law, 324
Centre National d'Études Spatiales (CNES) (French National Center for Space Studies), 199
Centro Ricerche Aerospaziali (CRA) (Italian National Committee on Space Research), 275, 395
Cepe. See Cylindrical Electrostatic Probe Experiment.
Ceres (asteroid), 210
Cernan, Capt. Eugene A. (USN), 18, 125, 138, 198
Cernan, Mrs. Eugene A., 18
Cerro Tololo Inter-American Observatory, Chile, 209-210
Cessna 310 (light aircraft), 3
CETS. See Conférence Européenne sur les Télécommunications par Satellites.
CH-54 (helicopter), 182
Chaffee, Secretary of the Navy John H., 68
Chaffee, L/Cdr Roger B. (USN), 267, 340

Chalidze, V. N., 369-370
Chamberlain, J. W., 33
Chantilly, Va., 259
Chapman, Dr. Sydney, 205
Charleston AFB, S.C., 225, 243, 411
Charlesworth, Clifford E., 51
Charlotte Amalie, V.I., 110
Charpie, Robert A., 129
Charyk, Dr. Joseph V., 82
Cherington, Paul, 28
Cherry, George W., 404
Chesapeake, Va., 293-294
Chevron Chemical Co., 401
Chiang, Vice Premier, Ching-kuo (Nationalist China), 150
Chicago, Ill., 30, 162, 305, 338, 345, 383, 389, 406
Chicago, Univ. of, 26, 59, 176, 353, 406
Chico, Calif., 320
Chicom I (Communist China satellite), 147, 150-151, 152-153, 154, 156, 412
Chihuahua City, Mexico, 231
Chile, 199, 347
China, Communist, 136, 330
  aircraft, 307
  launch, *Chicom I*, 147, 150-151, 152-154, 176, 412
  missile program, 150-151, 175, 307, 360
  space program, 150-151, 152-154, 156, 165, 175, 257-258
China Lake, Calif., 105
China, Nationalist, 150
Chincoteague Coast Guard Station, 111
Chincoteague National Wildlife Refuge, 111
Chirnis, E. V., 327
Chou, Premier, En-lai (Communist China), 150-151
Chrysler Corp., 23, 180, 203
Chula Vista, Calif., 266
Chumikan (U.S.S.R. fishing vessel), 132
Churchill Research Range, Canada, 3, 12, 40, 48, 56, 62, 69, 71, 99, 108, 126, 154, 162, 170
Churchill, Sir Winston, 219
Cincinnati, Ohio, 352
City University of New York, 215
Civil Aeronautics Board (CAB), 7, 40, 308, 312
Civil Air Patrol, 267
Civil Aviation R&D Policy Study, 169
Civil Service Commission, 229
Civitan International, 219
Clark, Dr. John F., 128, 144
Clavius (moon), 228
Clean Waters Act, 47
Clear-air turbulence (CAT), 9
Cleveland, F.A., 240
Cleveland, Ohio, 27, 40, 208, 337
Clifton, Dr. H. Edward, 169
Clipper Young America (Boeing 747), 15
Cloud, 107, 116
Cloud, Dr. Preston E., Jr., 77, 306
CM. See Command module.
CNEE. See Mexican National Commission for Outer Space.

CNES. See Centre National d'Études Spatiales.
CNR. See Italian National Research Council.
Coal, 316
Coalition Against the SST, 239
Coast Guard Academy, 292
Coastal Zone Conference, 277
Cobble, Dr. James W., 94
Cochran, Jacqueline, 110
Cockerell, Sir Christopher, 296
Cocoa Beach, Fla., 40, 41, 149
Cohen, I. Bernard, 256
Cohen, Nathaniel B., 220
Cohn, Victor, 51, 111
Cold Lake Range, Canada, 334, 338
Coleman, Dr. James S., 89
College of Environmental Science, 49
College Park, Md., 178, 350
Collier, Robert J., Trophy, 166
Collins, Col. Michael (USAF, R.), 152, 401
  appointment, 6
  awards and honors, 45, 52, 94, 155, 166, 234, 238, 267, 311, 368
Collins, Ronald, 398
Colorado, 311
Colorado Springs, Colo., 150, 259
Colorado, Univ. of, 79, 100, 116, 233, 274, 291, 308, 311, 406
Columbia (Apollo 11 command module). See Command module.
Columbia Broadcasting Co., 138
Columbia Radiation Laboratories, 51
Columbia Univ., 6, 143, 216, 283, 372, 394
Columbus, Ohio, 347, 377, 379
Comet, 22, 27, 33, 116, 226, 344
Comité Spéciale de l'Année Géophysique Internationale (CSAGI) (Special Committee for the International Geophysical Year), 205
Command and service module (CSM), 71, 108, 119, 121, 123, 194, 213, 218-219, 245, 249, 278, 285, 323
  Apollo 13 (Odyssey), 410-411
Command module (CM), 7, 50, 213, 217, 406
  Apollo 13 (Odyssey), 90-91, 119-125
  Apollo 14 (Kitty Hawk), 403
Commerce, Dept. of, 239, 330, 390
Commission on Government Procurement, 383
Commissioned Officers Corps, 347
Committee for Human Rights (U. S. S. R.), 369-370
Committee for International Environmental Programs (IEPC), 214
Committee on Space Research (COSPAR), 176-177, 180-181, 185, 195
Communications satellite (see also individual satellites: *Echo II, Intelsat I, Intelsat-II F-2, Molniya I-13*, etc.), 162, 197, 346
  agreement, 74, 95, 96, 269, 409-410
  anniversary, 112
  benefits, 39, 65
  conference, 53, 96, 409-410
  contract, 74
  cooperation, international, 16, 43, 53, 65, 82-83, 96, 151, 214, 233, 241-242, 243, 245, 252-253, 257-258, 269, 327, 407, 409-410, 411, 413, 414
  earth station, 45, 65, 82-83, 223
  FCC regulation, 26, 28, 45-46, 99, 151, 249, 272, 290, 339-340, 405, 407-408
  launch, 410-413
    Intelsat-III F-6, 14
    Intelsat-III F-7, 144-145
    Intelsat-III F-8, 241-242
    Molniya I-13, 56
    Molniya I-14, 214
    Molniya I-15, 318
    Molniya I-16, 380
    Molniya I-17, 405
    Nato I (NATO A), 95
    plans, 408-409
    Skynet B, 269
  military, 16, 43-44, 95, 148-149, 198, 252, 269, 293-294, 346, 360, 411-413
  Nixon, President Richard M., 25, 28, 78-79
  use, 28, 43-44, 45, 59, 61, 171, 279-280, 375, 376
  U. S. policy, 25-26, 28, 39, 99, 290, 339-340, 405
Communications Satellite Act of 1962, 45-46, 65
Communications Satellite Corp. (ComSatCorp), 52
  Annual Meeting of Shareholders, 82-83
  contract, 74
  cooperation, 82-83, 409-410
  criticism, 151
  Early Bird. See *Intelsat I*.
  earth station, 45, 83, 223
  FCC regulation, 26, 28, 45-46, 99, 151, 249, 272, 290, 339-340, 405, 407-408
  INTELSAT, 83, 409-410
  Intelsat I (Early Bird), 112
  Intelsat-II F-2, 162
  Intelsat III series, 14, 45, 144-145, 241-242
  Intelsat-III F-4, 14
  Intelsat-III F-5, 14
  Intelsat-III F-6, 14, 45, 145, 411
  Intelsat-III F-7, 144-145, 162, 169, 411
  Intelsat-III F-8, 241-242, 411
  Intelsat IV series, 83
  personnel, 154, 233, 240
  Report to the President and the Congress, 142
  revenues, 32, 58, 136, 151, 236, 405
  satellite program, 290, 339-340, 408-409
  services, 26, 45, 74, 136, 151, 171, 339-340
Compañia Telefónica Nacional de España, 74
Compton, Karl Taylor, Award, 329
Computer, 11, 194, 250, 310, 341, 397
  Illiac IV, 172
  industry, 165
  NASA use, 7, 106, 165, 178, 180, 213, 310, 354-355

contract, 191, 225, 274, 353
Computer Program Abstracts, 6
Computer Sciences Corp., 75, 200, 353
Computing and Software, Inc., 131-132
ComSatCorp. *See* Communications Satellite Corp.
Conant, Dr. James B., 67
Concorde (U.K.-France supersonic transport), 170-171, 226, 260-261, 275, 326, 366, 388-389
  cost, 350, 364
  flights, 272, 277, 283, 291, 300, 359, 364
Condor (missile), 105
Cone, Clarence D., Jr., 97
Cone Crater (moon), 62-63
Conférence Européenne sur les Télécommunications par Satellites (CETS), 241
Conference on Experimental Tests of Gravitational Theories, 366-367
Conference on Human Environment, 255
Conference on international exploration of space (proposed), 74
Conference on Materials for Improved Fire Safety, 167
Conference on Space Shuttle Technology, 234, 246
Congo, 199
Congress, 10, 192-193
  Council on Environmental Quality report to, 1
  Defense, Dept. of, 150
  Joint Committee on Atomic Energy, 235
  Joint Committee on the Environment and Technology, 180
  Joint Economic Committee, Subcommittee on Economy in Government, 170-171, 207-208
  NASA's *Twenty-first Semiannual Report to*, 258
  National Science Foundation report to, 98
  President's messages
    Aeronautics and Space Report of the President, 39
    aircraft security, 302
    budget, 35, 36, 237
    Environmental Protection Agency, 229-230
    environmental quality program, 47, 262
    Independent Offices and HUD appropriations bill (including NASA), veto, 263
    marine resources and engineering development, 127
    National Oceanic and Atmospheric Administration, 229
    Office of Telecommunications Policy (proposed), 45-46
    State of the Union, 22-23
    technological change and its impact, 297-298
  Space Science and Technology Panel (President's Science Advisory Committee) report to, 87-88
  space program, 73-74, 86, 247, 248, 287, 316-317
  supersonic transport (SST), 387-388, 390, 411
Congress, House of Representatives, 67, 141, 146, 161-162, 165, 174, 203, 223, 247, 263, 264, 317, 331, 390, 392, 393
  bills introduced, 29, 41, 42, 92-93, 94, 264, 303
  bills passed, 145-146, 165, 170, 183, 200, 210, 248, 302, 331, 374, 377, 399, 409
  Committee on Appropriations, 373
    Subcommittee on Defense Appropriations, 67
    Subcommittee on DOD Research, Development, Test, and Evaluation, 152-153
    Subcommittee on Independent Offices, 167, 175
  Committee on Armed Services, 93, 95, 129
  Committee on Foreign Affairs, Subcommittee on National Security Policy and Scientific Development, 156-157
  Committee on Government Operations, Subcommittee on Government Activities, 28-29
  Committee on Interstate and Foreign Commerce, 390
  Committee on Judiciary, 29
  Committee on Merchant Marine and Fisheries, Subcommittee on Fisheries and Wildlife Conservation, 213
  Committee on Science and Astronautics, 53-54, 57-58, 60-62, 63-64, 65-66, 87, 95, 98, 99, 141, 145-146, 157, 178, 204, 277, 302, 309, 390
    Ad Hoc Subcommittee on Aerospace Museum Study, 234
    Special Ad Hoc Committee, 244
    Subcommittee on Advanced Research and Technology, 64-65
    Subcommittee on Manned Space Flight, 317
    Subcommittee on NASA Oversight, 50, 72-73, 317, 393-394
    Subcommittee on Science, Research, and Development, 132, 228, 239-240, 242, 249, 269-270, 337
    Research Management Advisory Panel, 64-65
    Subcommittee on Space Science and Applications, 39
  Committee on Ways and Means, 157, 310, 312
Congress, Senate, 166-167, 203, 217, 247, 257, 267, 317, 324, 392-393
  bills introduced, 39, 44, 74, 117, 234, 235, 265, 325, 335, 390
  bills passed, 113, 152, 166, 223, 226, 236, 257, 386, 389
  Committee on Aeronautical and Space Sciences, 92, 371
    Apollo 13 mission, 147-148, 218-219

international cooperation, space, 85-86
Low, Dr. George M., letter to, on *Luna XVI* and *XVII* missions, 379
NASA budget, 58, 66-67, 73-74, 77, 111-112, 145, 161-162
Committee on Appropriations, 67, 212, 310, 377
  Subcommittee on Independent Offices, 175
  Subcommittee on Transportation, 275, 286
Committee on Armed Services, 19, 67, 84-85, 86-87, 93, 171, 183, 306
Committee on Government Operations, Permanent Subcommittee on Investigations, 100, 113, 148, 402
Committee on Interior and Insular Affairs, 235
Committee on Labor and Public Welfare, NSF Subcommittee, 147
manned space flight, 330, 341
nominations approved and confirmed, 208, 214, 288-289, 385, 409
nominations submitted to, 28, 40, 99, 129, 185, 215, 240, 273, 361, 398
nominations withdrawn, 65
resolution, 117
supersonic transport (SST), 303-304, 387-388, 390, 393, 403
Connecticut, Univ. of, 381
Conrad, Capt. Charles, Jr. (USN)
  Apollo 12 mission
    briefing of President Lyndon B. Johnson, 15
    lunar landing, 2
    medical aspects, 37, 46
    TV account and photography, 197
  Apollo 13 splashdown, 138
  award, 18, 41, 42, 197, 368, 372, 376
  goodwill tour, 51, 68, 99
  international cooperation, space, 177
  space program, national, 164, 177
Constan, Dr. George N., 353
Constance, West Germany, 314, 326, 335, 340
Contract (see also under agencies, such as NASA, USAF)
  cost-plus-award-fee, 41, 44, 113, 115, 149, 154, 196, 233, 243, 265, 303, 320, 349, 353, 384, 404
  cost-plus-fixed fee, 54-55, 75, 116, 131-132, 171, 198, 200, 212, 227, 258, 315, 324, 349
  cost-plus-fixed-fee/award-fee, 194
  cost-plus-incentive-award-fee, 49, 87, 214
  cost-plus-incentive-fee 56, 58, 68, 208, 209, 228
  cost-reimbursement, 22, 108
  cost-sharing, 228
  fixed-price, 11, 55, 105, 157, 170, 179, 196, 204
  fixed-price-incentive-fee, 220
  study, 42, 55, 203, 208, 215, 227, 264, 292, 374

Contraves A. G., 407
Convair 990 (research aircraft), 101
Convocation on Challenge of Building Peace, 156
Cooby Creek, Australia, 100
Cook, Richard W., 215
Cooke, Lloyd M., 129
Cooper, Col. L. Gordon, Jr. (USAF, Ret.), 211
Cooper, Henry S. F., Jr., 2
Copernicus (lunar landing site), 323
Copernicus, Nicolaus, 354
Copernicus Quinquecentennial, 354
Cordtz, Dan, 322
Corfield, Frederick, 385
Corn blight sensing experiment, 290, 360
Cornell Aeronautical Laboratory, 75, 360
Cornell Univ., 75, 111, 152, 240, 287, 295, 330, 360
Cornfield, Minister of Aviation Supply Frederick V., 350
Corporation for Public Broadcasting (CPB), 54, 273
Corrigan, Douglas C., 110
Cortright, Edgar M., Jr., 125, 144, 181-182, 191; 201-203, 204, 218, 325-326, 349-350
Cosmic ray, 146-147, 152, 230, 286, 307
Cosmonaut, 195, 196, 330
  anniversary, 126, 127
  Apollo 13 mission message, 132
  astronauts, meeting with, 106, 185, 190, 192, 349
  awards and honors, 267
  death, 9
  Soyuz IX mission, 189-190, 201, 210, 225, 280, 327
  space cooperation, 185, 192, 346
  U.S. visit, 106, 336, 338, 344, 346, 349
  U.S.S.R. space program, 181, 340
  women as, 110, 340
Cosmos (U.S.S.R. satellite), 411
Cosmos CCXLIII, 326
Cosmos CCXLVIII, 42
Cosmos CCXLIX, 42
Cosmos CCLII, 42
Cosmos CCCXVI, 276
Cosmos CCCXVIII, 8
Cosmos CCCXIX, 15
Cosmos CCCXX, 16
Cosmos CCCXXI, 20
Cosmos CCCXXII, 21
Cosmos CCCXXIII, 46
Cosmos CCCXXIV, 66
Cosmos CCCXXV, 73
Cosmos CCCXXVI, 87
Cosmos CCCXXVII, 94
Cosmos CCCXXVIII, 102
Cosmos CCCXXIX, 109
Cosmos CCCXXX, 113
Cosmos CCCXXXI, 114
Cosmos CCCXXXII, 126
Cosmos CCCXXXIII, 130
Cosmos CCCXXXIV, 145
Cosmos CCCXXXV, 147
Cosmos CCCXXXVI, 149

Cosmos CCCXXXVII, 149
Cosmos CCCXXXVIII, 149
Cosmos CCCXXXIX, 149
Cosmos CCCXL, 149
Cosmos CCCXLI, 149
Cosmos CCCXLII, 149
Cosmos CCCXLIII, 149
Cosmos CCCXLIV, 170
Cosmos CCCXLV, 176
Cosmos CCCXLVI, 198
Cosmos CCCXLVII, 200
Cosmos CCCXLVIII, 201
Cosmos CCCXLIX, 206
Cosmos CCCL, 214
Cosmos CCCLI, 217
Cosmos CCCLII, 226
Cosmos CCCLIII, 229
Cosmos CCCLIV, 248
Cosmos CCCLV, 259
Cosmos CCCLV, 259
Cosmos CCCLVI, 261
Cosmos CCCLVII, 269
Cosmos CCCLVIII, 270
Cosmos CCCLIX, 272
Cosmos CCCLX, 277
Cosmos CCCLXI, 291-292
Cosmos CCCLXII, 304
Cosmos CCCLXIII, 305
Cosmos CCCLXIV, 311
Cosmos CCCLXV, 315, 320
Cosmos CCCLXVI, 323
Cosmos CCCLXVII, 325
Cosmos CCCLXVIII, 331
Cosmos CCCLXIX, 331
Cosmos CCCLXX, 332
Cosmos CCCLXXI, 334
Cosmos CCCLXXII, 337
Cosmos CCCLXXIII, 343, 345, 352, 357-358
Cosmos CCCLXXIV, 343, 345, 357
Cosmos CCCLXXV, 343, 352, 358
Cosmos CCCLXXVI, 352
Cosmos CCCLXXVII, 366
Cosmos CCCLXXVIII, 370
Cosmos CCCLXXIX, 376
Cosmos CCCLXXX, 376
Cosmos CCCLXXXI, 384
Cosmos CCCLXXXII, 384
Cosmos CCCLXXXIII, 385
Cosmos CCCLXXXIV, 392
Cosmos CCCLXXXV, 397
Cosmos CCCLXXXVI, 398
Cosmos CCCLXXXVII, 399
Cosmos CCCLXXXVIII, 402
Cosmos CCCLXXXIX, 402
COSPAR. See Committee on Space Research.
Cotton, Col. Joseph F. (USAF, Ret.), 342
Council for Advancement of Science Writing, 373
Council on Environmental Quality, 1, 30, 42, 170, 262, 304, 389
Council on Physical Fitness, 316
Cour-Palais, Burton G., 54
CPB. See Corporation for Public Broadcasting.

CRA. See Centro Ricerche Aerospaziali.
Crab Nebula, 152
Crain, Percy J., 326
Creutz, Edward C., 102
"Crew of Apollo 11" (painting), 274
Crewe, Dr. Albert V., 176
Crimean Astrophysical Observatory, 389
Croft, Dr. Thomas, 407
Cronkite, Walter, 13
Cross, Bert S., 116
Crossfield, A. Scott, 21
Cruikshank, Dale P., 69
CSM. See Command and service module.
CSAGI. See Comité Speciale de l'Année Géophysique Internationale.
Cuba, 15, 281, 390, 414
Culbertson, Philip E., 8
Cunningham, R. Walter, 20
Currie, James R., 352
Curtiss Aeroplane and Motor Co., 380
Curtiss-Wright Corp., 102, 272
CURV III (cable-controlled underwater recovery vehicle), 102
Custer Channel Wing Corp., 91-92
Cutler-Hammer, Inc., 220
CV-7A (Buffalo) (STOL aircraft), 179
Cyanoacetylenes, 325
Cygnus (constellation), 225
Cylindrical Electrostatic Probe Experiment (*Cepe*), 395, 411
Cyprus, 383
Czechoslovakia, 51, 159, 259, 380
 Academy of Sciences, 159

D

Da Nang, South Vietnam, 273
Daedalus, 268
Dalgarno, A., 333
Dallas, Tex., 48, 133
Daly, Gerald R., 293
Dana, William H., 27, 66, 178, 190, 357, 379
Daniel and Florence Guggenheim International Astronautics Award, 271
Danielli, Dr. James F., 368, 370, 392
Danilovich, Lt. Valentin I. (U.S.S.R.), 330
DAP I (dipetidyl aminopeptidase), 258
Dart (booster), 185
David, Dr. Edward E., Jr., 269, 271, 300, 303, 305, 345, 357
Davis, L/G Benjamin O., Jr. (USAF, Ret.), 310
Davos, Switzerland, 89
Davy (lunar landing site), 323
Day, LeRoy E., 107, 328
Day, Melvin S., 352
Dayton, Ohio, 166, 401
Dayton, Univ. of, 276
DC-6 (research aircraft), 54
DC-8 (jet transport), 73, 289, 300
DC-9 (jet transport), 21, 366
DC-10 (trijet transport), 237, 277, 370, 411
Dealy, John F., 145

Death ray, 27
Debus, Dr. Kurt H., 100, 236
Dedijer, S., 136
Deep Space Network (DSN) (NASA)), 16, 66, 207, 265
Deep Submergence Rescue Vehicle (DSRV), 27
Defense, Dept. of (DOD) (see also U.S. Air Force, U.S. Army, and U.S. Navy), 139, 150, 152, 154, 167, 175, 223-224, 232, 245-246, 330, 390
   Ad Hoc Advisory Group on Special Capabilities, 160
   aircraft. See Aircraft.
   antiwar demonstration against, 152
   budget, 35-36, 169, 223-224, 228, 317
   communications satellite system, 95-96, 269, 413
   computer programs, 6, 172
   contract, 9, 22, 56, 72, 103, 105, 184, 220, 223-224, 246, 273, 302, 397, 402, 408
   cooperation, 258, 265, 283, 311, 315, 350, 404
      NASA, 41, 53, 88, 172, 183, 265, 323, 375, 410-411, 412
      international, 95-96, 269, 411, 413
   employment, 362
   missile program, 35-36, 235, 276
   personnel, 187, 297, 300
   R&D, 67, 71-72, 98, 187; 228, 231, 239-240, 253, 362, 410
   reorganization study, 223-224, 246, 257, 274
   space program, 31, 35-36, 41, 53, 252, 303, 410-411
   weapon systems, 184, 245-246
DeFlorez Training Award, 92
Delta (booster) (see also Thor-Delta), 88, 215, 371
Delta inertial guidance system (DIGS), 215
Dempson, Peter, 247
Denisse, Jean-François, 305
Dennis, Stephen G., 297
DeNoyer, Dr. John M., 30, 213
Denver, Colo., 149, 208, 233, 274, 372, 402
Derbyshire, George A., 304
Descartes (lunar landing site), 294, 323, 341
Descent propulsion system (DPS), 122
Desirant, Maurice, 313
Detroit, Mich., 126, 212, 301
Deutch, Shea & Evans, 362
Dew Line Station, Alaska, 71, 75
Dial (French-West German satellite), 84, 412
Diamant-B (French booster), 84, 396-397
Dicke, Dr. Robert H., 129, 367
Dickey, Dr. Darrell J., 225
Dictionary of Scientific Biography, 173
Diego Garcia Island, 399
Diehl, William, 352
Dietz, Dr. Robert S., 296
DIGS. See Delta inertial guidance system.
Diode, 3

Disarmament, 82, 131, 133, 142, 230, 244, 248, 258, 266, 359, 371, 402
Disneyland, Calif., 346
Distinguished Civilian Service Award (DOD), 227
Distinguished Flying Cross (USAF), 29
Distinguished Service Medal (NASA), 29, 368
Distinguished Service Medal (USN), 42
District of Columbia, 51, 374
Djakarta, Indonesia, 94
Dmitriyev, A., 408
Docking
   Gemini VIII, 174
   Skylab adapter, 372
   U.S.-U.S.S.R. cooperation, 229, 334, 335-336, 346, 347, 349, 351, 358, 376, 391-392, 393, 409, 413
DOD. See Defense, Dept. of.
Donlan, Charles J., 342, 373
Donley, M/G Edwin I. (USA), 174
Donlan, Terry, 297
Doolittle, L/G James H. (USAF, Ret.), 175
Dornier-System GmbH, 407
Dosimeter, 362
DOT. See Transportation, Dept. of
Dover AFB, Del., 243
Downey, Calif., 292, 350
DPS. See Descent propulsion system.
Draley, Eugene C., 326
Draper, Dr. Charles Stark, 8, 87
Draper, Charles Stark, Laboratory, 8, 87, 125, 176, 404
Draper, Charles Stark, Research Center (Florida Institute of Technology), 8
Draper, Ronald F., 341
Drell, Dr. Sidney D., 90
Drummond, Dr. Andrew J., 101
Dryden, Dr. Hugh L., 267
DSN. See Deep Space Network.
DSRV-1 (Deep Submergence Rescue Vehicle), 27
Dublin, Ireland, 336
DuBridge, Dr. Lee A., 20, 47, 228, 244, 276-277, 346
   resignation, 267, 269, 271, 279
Dudley Observatory (Albany, N.Y.), 260
Duffy, Charles W., Jr., 21
Dugway Proving Ground, Utah, 386
Duke, L/C Charles M., Jr. (USAF), 111, 294
Dulles International Airport, 15, 259, 403
du Pont, E. I., de Nemours & Co., 89
Dupree, Prof. A. Hunter, 189
Durant, Frederick C., III, 384
Durham, N.H., 360
Duxbury, Thomas C., 185
Dyal, Palmer, 271
Dynalectron Corp., 399
Dynastat (airship), 99

E

Eagle (Apollo 11 LM). See Lunar module.
Eaker, L/G Ira C. (USAF, Ret.), 401
Earhart, Amelia, 110, 364

Early Bird (communications satellite). See *Intelsat I*.
Early Birds of Aviation, Inc., 352
Earth
  atmosphere. See Atmosphere, upper.
  Council on Environmental Quality, 1, 30, 42, 170, 262, 304, 389
  Environmental Advisory Council (DOT), 239
  environmental problems (see also Air pollution; Noise, aircraft; Pollution), 15-16, 44-45, 47, 49, 53, 77, 84, 86, 90, 106, 214, 215, 216, 231, 256, 262, 372
  gravity, 380
  land mass, 143, 296
  magnetic field, 7, 335
  magnetosphere, 20, 33, 413
  mapping, 215
  mass, 16
  National Environmental Policy Act of 1969, 1, 10, 75, 389, 402
  oceans. See Oceanography.
  origin, 143
  photographs, 96, 101, 165, 189-190, 196, 259, 290, 360, 383-384
  resources measurement (see also Earth Resources Technology Satellite), 72-73, 98, 146, 207, 208, 259, 372, 383-384, 385, 405, 412
Earth Resources Survey meetings
  AIAA conference, 72-73
  international workshop, 405, 413
  program review, 383-384
  U.N. committee review, 30
Earth Resources Survey Program Review, 383-384
Earth Resources Technology Satellite (ERTS), 78, 207, 284, 293, 412
  contract, 233, 264-265, 349
  funds, 36, 77, 161-162
  international cooperation, 30, 73, 177, 221, 280, 405, 413
East Hampton, N.Y., 308
Eastern Air Lines, Inc., 21, 154, 260, 293, 405
Eastern Test Range (ETR) (see also Cape Kennedy and Kennedy Space Center), launch, 172
  Atlas-Agena, 209, 278
  Atlas-Centaur, 380
  failure, 380, 384
  Intelsat-III F-6, 14
  Intelsat-III F-8, 241
  Long-tank Thor-Delta, 241
  Long-tank thrust-augmented Thor-Delta, 14, 95
  Minuteman III (missile), 398
  Nato I (NATO-A), 95-96
  OAO-B, 380-381
  Poseidon (missile), 206
  Titan IIIC, 114, 360
  unidentified satellite, 209, 278, 360, 376
  Vela XI, 114
  Vela XII, 114
Eastman Kodak Co., 90
EC-121 (Super Constellation), 233

Echo I (balloon satellite), 242
Echo II, 242
Eclipse, solar, 79-82, 102, 173, 243, 338, 411
Ecological Surveys from Space (NASA SP-230), 196
Economic Club, Detroit, 301
Economic Conversion Commission (proposed), 325
Ecuador, 199
EDP Technology, Inc., 75, 360
Edwards AFB, Calif., 16, 23, 109, 277, 367, 385
Edwards Test Station (USAF), Calif., 109
Efron, Leonard, 16
Eggers, Dr. Alfred J., Jr., 191, 220, 342
Eglin AFB, Fla., 194, 273, 375
Egypt. See United Arab Republic.
Ehricke, Dr. Kraft A., 56
Einstein, Dr. Albert, 377, 379
Eisele, Col. Donn F. (USAF), 181
Eisenhower, President Dwight D., 219
El Al Airlines, 289
El-Baz, Farouk, 1, 192
ELDO. See European Launcher Development Organization.
Electric propulsion, 38, 76, 173, 180, 243, 393, 411
Electronic Data Systems, Inc., 29
Electronics Research Center (ERC) (NASA), 108
  aeronautical research, 204, 341
  award, 369
  closing, 13, 20, 21, 43, 65, 219, 323-324
  pollution research, 73
  silicon carbide diode, 3
  transfer to DOT, 99, 105, 109, 169
Elgin National Industries, Helbros Watches Div., 115
Eliassen, Rolf, 250
Elizabeth II, Queen of Great Britain, 135
Ellington AFB, Tex., 140
Elliott, Dr. David G., 309
Elk tracking experiment, 52, 57, 58, 59, 62, 107, 118
Elms, James C., 99, 369
Elsztein, Pawel, 76
Emme, Dr. Eugene M., 189, 328, 342
"Emphasis Space" program, 164
Employment. See Aerospace industry, employment; and National Aeronautics and Space Administration, employment.
Engine (see also individual engines, such as F-1, H-1)
  aircraft, 373
  fire, 349
  jet, 19, 73, 228, 256, 293
  noise, 73, 279, 292, 366, 389-390, 400
  pollution, 102, 228, 256, 270, 286, 293, 304, 353
  supersonic transport, 266, 279, 389-390, 400
  turbofan, 68, 159
  turbojet, 266, 279
  turbulence, 62, 389-390

electric, 38, 51, 76, 173, 177, 180, 227, 243, 285, 293, 305, 309, 315, 338, 393, 411
nuclear (see also NERVA), 75, 227, 265, 277, 309, 374
rocket (see also NERVA), 25, 42, 66, 94-95, 109, 156, 164, 203, 217, 220, 223, 224-225
Engineering, 72
Engineers, 45, 65, 187, 253, 260, 331, 338-339, 362, 370, 383
Engle, L/C Joseph H. (USAF), 266
English Channel, 296
Eniwetok Atoll, 25
Environmental Advisory Council (DOT), 239
Environmental Data Serive, 230
Environmental Defense Fund, Inc., 256, 344
Environmental Protection Agency (EPA), 229-230, 235, 361, 385, 392
Environmental quality program, 1, 10, 30, 42, 47, 170-171, 262, 304, 389-390
Environmental Quality: The First Annual Report of the Council on Environmental Quality, 262
Environmental Science Services Administration (ESSA) (see also National Oceanic and Atmospheric Administration), 74-75, 165-166, 264
   award, 19-20
   budget, 35
   contract, 54, 155-156
   cooperation, 25, 77, 111, 336
   lightning research, 54, 250
   magnetometer, 7
   personnel, 347
   radiometer, 155-156
   Research Flight Facility, 54
   satellite, 2, 87, 203, 210, 243-244, 354
      launch, 23-25
   Solar Particle Alert Network (SPAN), 320
   Space Disburbance Laboratory, 320, 351
Eole (IAS-A) (International Applications Satellite), 396-397
Eole, Project, 85
EPA. See Environmental Protection Agency.
epndb: effective perceived noise in decibels.
Eppley Laboratory, 101
ERC. See Electronics Research Center.
Ericsson, L. M., AB, 407
ERTS. See Earth Resources Technology Satellite.
ERTS-A, 207, 213, 232, 259
ERTS-B, 207, 213, 232
ESC. See European Space Conference.
Escuintla, Guatemala, 311
Eshleman, V. R., 287
ESO. See European Space Organization.
ESRO. See European Space Research Organization.
ESSA. See Environmental Science Services Administration.
ESSA (satellite program), 107
Essa IX (meteorological satellite), 25
Essen, West Germany, 110
Estes, Marvin F., 401
ETR. See Eastern Test Range.
Etudes Techniques et Constructions Aérospatiales, 361
Europa I (booster), 241, 243
Europa II, 241, 243
Europa III, 243, 413
Europa IIIB, 241
Europe, 51, 112, 176, 407
European Broadcasting Union, 135
European Launcher Development Organization (ELDO), 88, 200, 241, 243, 252, 313
European Space Conference (ESC), 88, 227, 241, 243, 305, 313, 359-360, 413
European Space Organization (ESO), 241, 243
European Space Research Organization (ESRO), 32, 43, 77, 85, 88, 150, 193, 252, 314, 354, 407
   Aurorae (Esro I), 214
   Boreas (Esro IB), 49-50
   merger, 241, 243
EUV. See Ultraviolet, extreme.
EVA. See Extravehicular activity.
Evansville, Ind., 290
Evendale, Ohio, 142
Evins, Rep. Joe L., 247
Ewell, Dr. Raymond H., 368
EXAMETNET. See InterAmerican Meteorological Network.
Exceptional Bravery Medal (NASA), 368
Exceptional Scientific Achievement Medal (NASA), 369
Exceptional Service Medal (NASA), 29, 189, 369
Executives' Club, Chicago, 43
Exhibit, 99, 102-103, 111, 196, 209, 238, 258, 342, 346, 403
   lunar rock samples, 27, 57, 91, 102, 111, 126, 130, 139, 238, 244, 265, 270, 414
   spacecraft, 49, 99, 234, 401-402
Explorer (program), 162, 203
Explorer I (satellite), 32, 103, 111, 342
Explorer XXXI, 236
Explorer XXXIII, 236
Explorer XXXV, 271
Explorer XLII. See *Uhuru (Explorer XLII)*.
Export-Import Bank, 273
Expo '70, 91, 99, 414
Extraterrestrial life, 6, 60, 71, 88, 110, 204-205, 325, 359, 366, 370, 387, 390-391
Extravehicular activity (EVA), 370
   Apollo 11 mission, 187
   Apollo 13 mission, 62-63, 71
   Apollo 14 mission, 308, 338, 387, 390
   Apollo 15 mission, 285
Eynac, André Laurent, 400

F

F-4 (Phamtom II) (supersonic fighter and fighter-bomber aircraft), 151, 294, 403

F-4E, 129, 157, 262
F-5D (Skylancer) (training aircraft), 153
F-8 (supersonic carrier fighter), 359, 377
F-14 (Tomcat) (supersonic fighter aircraft), 151, 331, 403, 408
F-14A, 218, 411
F-14B, 68
F-15 (supersonic fighter aircraft), 69, 198-199, 382
F-102 (supersonic fighter-interceptor aircraft), 225
F-104 (Starfighter) (supersonic fighter aircraft), 173, 241
F-106 (research jet aircraft), 225, 292
F-111 (supersonic fighter), 10-11, 50, 56, 75-76, 99, 103, 113, 132, 149, 157, 244, 263, 273, 306, 317, 331, 381, 402
  accident, 74, 100, 226, 244, 351
F-111A, 331, 402
F-111B, 402, 403
F-111C, 129
F-111E, 300
F-111F, 402
FAA. See Federal Aviation Administration.
Faget, Dr. Maxime A., 342
FAI. See Fédération Aéronautique Internationale.
Fairbanks, Alaska, 38, 71, 73, 343, 365, 372
Fairbridge, Dr. Rhodes W., 143
Fairchild Camera and Instrument Corp., 116
Fairchild Hiller Corp., 42, 145, 146, 201, 224, 230, 273-274, 289, 318, 387
Faith 7 mission, 211
Fargette, Yves, 130
Farnborough Air Display, 300
Farnborough, U.K., 357
Farnham, Leon L., 318
Farrell, Richard, 3
FAS. See Federation of American Scientists.
FB-111 (supersonic bomber), 93, 244
FCC. See Federal Communications Commission.
Federal-aid Airport Program, 22
Federal Aviation Administration (FAA)
  air pollution, 293, 392
  air traffic control, 28-29, 68, 76, 77, 92, 171, 172-173, 200, 218, 258, 345, 353, 370, 414
  Air Traffic Control Advisory Committee, 76, 258
  aircraft certification, 13, 15, 20, 62, 144, 200
  airports, 22, 162, 218, 370
  award, 113, 175
  budget, 36
  Collision Prevention Advisory Group, 29
  cooperation, 28-29, 77, 350, 412
  landing system, 258, 335
  noise, aircraft, 133, 390
  organization, 107, 108-109, 204, 293
  personnel, 19, 99, 170, 293, 340
  regulations, 133, 200, 259, 278, 324, 331, 345
  statistics, 212, 414
  transport, supersonic (see also Supersonic transport), 107, 256, 390
  turbulence research, 310
Federal Communications Commission (FCC), 46
  ComSatCorp., 26, 290, 339-340, 407-408
  criticism of, 151
  policy, 99, 272, 290, 407-408
  requests to, 26, 28, 249, 339-340, 405
Federal Economy Act of 1970, 65
Federal Electric Corp., 223
Federal Funds for Research, Development, and Other Scientific Activities, Fiscal Years 1969, 1970, and 1971 (NSF 70-36), 322
Federal Highway Administration, 341
Federal National Accountants Assn., 173
Federal Support of Applied Research (NAE report), 221
Federal Support to Universities, Colleges, and Selected Nonprofit Institutions, Fiscal Year 1969 (NSF 70-27), 278-279
Federal Water Quality Administration, 384
Fédération Aéronautique Internationale (FAI), 108, 376
Federation of American Scientists (FAS), Strategic Weapons Committee, 347
Feller, William, 55
Fellgett, Dr. P. N., 21
Fendrick Gallery, Washington, D.C., 346
Feoktistov, Konstantin P., 106, 185, 267, 346
Ferguson, Gen. James (USAF), 16-17, 69, 149, 194, 283
Fermi, Enrico, Award, 209, 277
Fermi, Enrico, Institute, 353
Ferromagnetism, 348
Few, A. A., 179
Fiar, CGE, 407
Finger, Frederick G., 2
Fink, Daniel J., 224
Finland, 51
Fire protection, aircraft, 324
First Flight Shrine, 401
First Flight Society, 401
Fitch, Dr. Val L., 89
Flagstaff, Ariz., 54, 397
Flemming, Arthur S., Award, 57
Flight Research Center (FRC) (NASA), 153, 320, 348, 386
  award, 369
  contract, 131-132
  lifting-body vehicle flight
    HL-10, 18, 27, 55, 66, 199, 236
    M2-F3, 178, 190, 357, 379
    X-24A, 94-95, 108, 145, 171, 206, 247, 263, 273, 336, 347, 375
  space shuttle research, 334
  Space Shuttlecraft Symposium, 219-220
  supercritical-wing test, 58, 359, 377
Florida, 82
Florida Institute of Technology, 8

Florida, Univ. of, 189
Floyd Bennett Field, N.Y., 110
Fluorel, 9
FOBS. See Fractional Orbital Bombardment System.
Fokker FFW, 407
For the Benefit of All Mankind: A Survey of the Practical Returns from Space Investment (H. Rpt. 91-1446), 302
Ford, Rep. Gerald R., 141
Ford Trimotor (transport aircraft), 15
Foreign Service Wives Assn., 152
Forestry Service, 401
Fork River, Ontario, 307
Fort Bliss Oldtimers, 314
Fort Dix, N.J., 345
Fort Hood, Tex., 165
Fort Worth, Tex., 226
Fort Yukon, Alaska, 38
Foster, Dr. John S., Jr, 146, 152-153, 252, 311
Fra Mauro (lunar crater), 71, 90-91, 144, 147, 167, 210, 244-245, 295, 331, 364
Fractional Orbital Bombardment System (FOBS), 31, 68, 93, 320
France, 51, 159, 400
  aircraft. See Concorde.
  astronaut honored by, 249
  communications satellite, 412
  Concorde (U.K.-France supersonic transport), 170-171, 226, 260-261, 272, 275, 277, 283, 291, 300, 326, 350, 359, 364, 366, 388-389
  hydrogen bomb, 258
  international cooperation, space, 51, 84, 85, 150, 177, 199, 200, 287, 326, 354, 359, 409, 413-414
  laser reflector, 177, 374, 389
  launch, satellite, 84, 396-397, 410, 412
  satellite, 32, 85, 150, 177, 354, 396-397, 409, 410, 412, 413-414
  space program, 84, 85, 354, 410, 413
Frankfurt, Univ. of, 392
Franklin Institute, 313-314
Franklin, Kenneth L., 115
Fraser, Defence Minister Malcolm (Australia), 7, 74, 129
FRC. See Flight Research Center.
Fredericton, New Brunswick, 172
Free Life (balloon), 308
Freedom 7 (spacecraft), 165
Freedy, Dr. Amos, 194
French Guiana, 84, 199, 414
French Guiana National Space Study Center, 326
French National Laboratory for Space Astronomy, 413
French National Laboratory for Stellar and Planetary Physics, 413
Frey, Rep. Lou, Jr., 144
Frey, Royal D., 342
Friedheim, Jerry W., 103, 181
Friedman, Herbert, 159
Friends of the Earth, 239, 272, 391
Friendship Airport, Baltimore, 259
Frisbee, John L., 280

Frog experiment, space, 324-325, 362, 367, 395, 411
Frost, Col. Douglas H. (USAF), 7
Frutkin, Arnold W., 156, 193, 227, 243, 334, 335, 351, 409
Fubini, Dr. Eugene G., 157
Fuel cell, 198
Fulbright, Sen. J. W., 226, 303
Fuller, Buckminster, 191
Fulton, Fitzhugh L., Jr., 180
Fulton, Rep. James G., 42, 94, 277
Fundamentalist Bible Presbyterian Church, 408
Funkhouser, Dr. John G., 333
Furnas, Dr. Clifford C., 160

G

Gagarin, Col. Yuri A. (U.S.S.R.), 126, 127, 267
Gagarin, Mrs. Yuri, 195
Galaxy, 96-97, 110, 198, 325, 381, 409
Galaxy (star image measuring machine), 21
Gamma ray, 396
GAO. See General Accounting Office
Garden City, N.Y., 380
Gardner, Hy, 101
GARP. See Global Atmospheric Research Program.
Garrett Corp., Airesearch Manufacturing Co. Div., 147
Garwin, Richard L., 272, 391
Gast, Dr. Paul W., 181, 369
Gates, David M., 129
Gatland, Kenneth W., 357-358
Gavin, Thomas R., 341
Gazenko, Dr. Oleg G., 327, 328
GCA Corp., 234
GE. See General Electric Co.
GE4 (turbojet engine), 266, 279
Gemini (program), 186, 347
Gemini V mission, 211
Gemini VII mission, 340
Gemini VIII mission, 174
General Accounting Office (GAO), 224, 335
General aviation, 22, 28-29, 86, 113, 212
General Dynamics Corp., 328
  Aeronautical Systems Div., 273
  contract, 56, 103, 220, 227, 250, 273, 381, 384
  Convair Aerospace Div., 312, 314, 317
  Convair Div., 250
  Electro Dynamics Div., 312
  F-111, 56, 103, 113, 132, 157, 226, 273, 381
General Electric Co. (GE), 139, 157, 246, 393, 412
  Aircraft Engine Group, 19
  awards and honors, 19, 159
  B-1 (advanced strategic bomber), 195
  contract, 23, 42, 83-84, 87, 115, 145, 146, 148, 195, 196, 201, 223, 224, 230, 257, 264-265, 273-274, 289, 318
  Direct Energy Conversion Business Section, 198

laboratory, undersea, 110, 169, 308
Power Generation Group, 265
quiet jet engine, 74-75
Space Div., 196, 215, 224, 227
spacecraft, 115, 145, 146, 148, 201, 223, 224, 230, 273-274, 289, 318
turbojet engine, 266, 279
General Motors Corp., AC Electronics Div., 7, 29
General Services Administration, 255
General Telephone & Electronics Corp., 405
Genetic engineering, 392
Geneva, Switzerland, 277, 283, 364, 395
Gentry, Maj. Jerauld R. (USAF), 60, 88, 94, 145, 247, 273, 375
Geodetic satellite, 113-114, 241, 411
Geomagnetosphere, 331
George Washington Univ., 169, 221, 280, 406
Georgetown, Guyana, 128
Georgia, 82
Geos II (geodetic satellite), 212
Geostationary Operational Environmental Satellite, 354
Germ warfare, 392
German Federal Medical Council, 89
German Federal Republic. See Germany, West.
German measles (rubella), 111, 116, 117, 118, 119
Germantown, Md., 372
Germany, West, 15, 51, 159, 264, 291, 292
 aircraft, 85, 198, 241
 Brandt, Chancellor Willy, visit to U.S., 118
 Chicom I launch, reaction to, 150
 cooperation, 85, 317, 320
 cooperation, space, 51, 52, 84, 86, 91, 200, 292, 392, 400, 412
 space program, 32, 52, 84, 91, 200, 392, 400, 412
 sounding rockets, 392, 400
Gerry, Dr. Edward T., 156
Gervais, Maj. Joseph (USAF), 364
GET: ground elapsed time.
Getler, Michael, 186
ghz: gigahertz (one billion cycles per second).
Giaimo, Rep. Robert N., 303
Gibbs-Smith, Charles H., 296
Giberson, Walker E., 66
Gillilland, Whitney, 7
Gilruth, Dr. Robert R., 109, 135, 211, 334, 346, 347
Glass, Dr. Alexander J., 156
Glass, Dr. Bentley, 406
Glasser, L/G Otto J. (USAF), 93
Glenn, Col. John H., Jr. (USMC, Ret.), 15
Glennan, Dr. T. Keith, 157, 185, 214, 298
Global Atmospheric Research Program (GARP), 30, 36, 77, 85, 354
Glomar Challenger (deep-sea drilling ship), 394
Goddard, Mrs. Esther C., 256, 349, 384
Goddard, Dr. Frank E., 180

Goddard, Dr. Robert H., 92, 113, 127, 180, 256, 349
Goddard, Robert H., Award, 19
Goddard, Robert H., Memorial Award, 164
Goddard, Robert H., Memorial Trophy, 94, 384
Goddard Historical Essay Award, 94
Goddard Rocket and Space Museum, 92, 113, 234, 302
Goddard Space Flight Center (GSFC) (NASA), 128, 366, 412
 Advanced Development Div., 112
 Apollo 11 lunar rock sample, loss of, 223
 Apollo 12 lunar rock sample, 333
 ATS-F, 273-274
 ATS-G, 273-274
 award, 368
 Central Computing System, 353
 contract, 8, 353
 cooperation, 26, 212
 Delta (booster), 25
 Earth Resources Technology Laboratory, 203
 Earth Resources Technology Satellite (ERTS), 233, 293
 mapping, 212, 380
 OAO-B, 380
 personnel, 347
 satellite monitoring
  Nimbus IV, 343
  Oao II, 380
  Ogo II, 152
 Sert II test, 38
 solar research, 101-102
 sounding rocket experiments (see also Sounding rockets), 3, 9, 29
  astronomical, 129, 245, 311
  atmospheric data, 16, 56-57, 69, 84, 172, 208, 210, 213, 338, 360, 369
  electric fields, 71, 73
 Uhuru (*Explorer XLII*) (SAS-A), 407
Gofman, Dr. John W., 242
Gold, Dr. Thomas, 240, 287, 295, 330
Goldman, Eric F., 308
Goldstein, Dr. Richard M., 139, 286
Goldstone Tracking Station, 139, 213, 286
Goldwater, Sen. Barry M., 55, 226, 240-241
Golovin, Dr. Nicholas E., 267
Goody, Dr. Richard M., 399
Goodyear Aerospace Corp., 99, 204
Gordon, R., 328
Gordon, Capt. Richard F., Jr. (USN)
 Apollo 12 mission, 15, 46
 Apollo 13 mission, 125
 Apollo 15 mission, 101
 awards and honors, 18, 42, 372
 goodwill tour, 51, 99
Gorman, Harry H., 215
Gossick, M/G Lee V. (USAF), 173, 194, 283
Gould, Jack (John Ludlow), 138
Gownley, Hugh J., 366

Graduate Student Support and Manpower Resources in Graduate Science Education, Fall 1969 (NSF 70-40), 322
Grandy, Herbert W., 368
Gravel, Sen. Mike, 252, 311, 341, 408
Gravity, 59; 211, 366-367, 370, 380
Gray, Robert H., 201
Great Lameshur Bay, V.I., 226
Greece, 314
Green Bank, W. Va., 325
Green, Dr. James A., 266
Green River, Utah, 146, 231, 256
Greenberg, D. S., 345
Greene, Lawrence P., 250
Greenland, 200
Grenoble Univ., 348
Griffin, Gerald D., 137
Griffiss AFB, N.Y., 149
Grissom, L/C Virgil I. (USAF), 267, 340
Gross, Robert E., 401
Groves, George B., Jr., 325
Groves, L/G Leslie R. (USA, Ret.), 67, 232, 235-236
Grubb, H. Dale, 142, 162
Grumman Corp. 403, 408
 Grumman Aerospace Corp., 94, 130, 165, 179, 203, 223, 249
 Grumman Aircraft Engineering Corp., 151, 208
GSFC. See Goddard Space Flight Center.
Guatemala, 311
Guayaquil, Ecuador, 360
Guaymas-Empalme, Mexico, 381
Guinard, Norman W., 384
Gulf Coast, 207
Gulf General Atomic, 265
Gulton Industries, Data Systems Div., 155
Guyana, 27

H

Habib, Edmund J., 112
Hadley-Apennine (lunar landing site), 323
Hagerty, William A., 240
Haglund, Howard H., 103
Haise, Fred W., Jr., 364
 Apollo 13 mission, 25, 140-141, 147-148
  celebrations, 137, 140, 162, 194
  flight, 119-125
  medical examination, 101, 111, 115
  Nixon, President Richard M., 128, 133-134, 136-137, 139-140, 314
  plans, 62-63, 71, 101, 115
  press comment, 136
  press conference, 62-63, 90-91, 143
  splashdown, 125, 137-138
 awards and honors, 133-134, 137, 140, 368
 European goodwill tour, 314, 331
 press conference, 212
 tribute to, 128, 136, 137
Haise, Mrs. Fred W., Jr., 129, 137, 314
Halaby, Najeeb E., 41, 364
Hale, George E., 17
Hale Observatories, 17, 409

Hall, Edward N., 21
Hall, Floyd D., 154, 405
Hall, John H., 297
Hall, R. Cargill, 328, 406
Halpern, Rep. Seymour, 29
Hamilton Standard Div., United Aircraft Corp., 7
Hammer, magnetic, 401
Hammondsport, N.Y., 13
Handler, Dr. Philip, 239, 335
Hanessian, John, Jr., 221, 280
Hanford Atomic Works, 17
Hanford, Wash., 232
Hanscom, L.G., Field, Mass., 171
Hanzing, Dr. Frank D., 383
HAPPE. See High Altitude Particle Physics Experiment.
Hardy, George B., 334
Harford, James J., 383
Hargis, Calvin B., Jr., 198
Harmon International Aviation and Space Trophy, 88
Harmon Trophy, 212
Harnett, Daniel J., 285, 352
Harper, Charles W., 28, 63, 173-174
Harr, Dr. Karl G., Jr., 106, 321
Harrier (British V/STOL aircraft), 199
Harrington, Dr. Charles D., 213
Harris, Louis, 213
Harrisburg, Pa., 347
Harrison, Halstead, 369
Hartline, Ann, 226
Harvard College Observatory, 333
Harvard Univ. 147, 153, 154, 311, 329, 399
Harvey, Dr. Dodd L., 298
Harvey Mudd College, 189
Haskins, Dr. Caryl P., 164
Haughton, Daniel J., 72
Havana, Cuba, 15
Hawaii, 133, 137, 140, 223, 409
Hawaii, Univ. of, 295, 345, 348
Haydon, John M., 136
Hayes, Daniel F., 75
Hayes, Dr. Dennis E., 394
Hayes International Corp., 350
Haynes, Charles G., 98
Health, Education, and Welfare, Dept. of (HEW), 110, 352, 404
 Office of Education, 132
HEAO. See High Energy Astronomy Observatory.
HEAO-A (High Energy Astronomy Observatory), 361
HEAO-B, 361-362
Hearnes, Gov. Warren E., 238
Hearth, Donald P., 60, 73
Heat shield, 209
Heavy-lift helicopter (HLH), 310
Hechler, Rep. Ken, 94
Hedrick, B/G Walter R. (USAF), 144
Heenan, Cardinal John C., 135
Heflin, Dr. Woodford A., 159
Heidelberg, West Germany, 86, 289
Helgeson, Bob P., 17
Helicopter, 182, 333
 accident, 109

astronaut pickup, 125, 136
civil, 110-111, 312, 345
medical use, 233-234, 259, 283, 311, 404
military, 72, 198, 310, 345
NASA, 413
record, 198
U.S.S.R., 299
Helios-A (solar probe), 62
Helium, 96
Helsinki, Finland, 230, 266, 371, 402
Hendrie, Dr. Joseph M., 94
Henize, Dr. Karl G., 101
Henkin, Daniel Z., 175
Hercules, Inc., 25
Herr, Kenneth C., 1
Hess, Dr. Wilmot N., 20
HEW. See Health, Education, and Welfare, Dept. of.
Hewlett, William R., 90
Heyerdahl, Thor, 232
HH-53 (helicopter), 273
Hickel, Secretary of the Interior Walter J., 245
High Altitude Particle Physics Experiment (HAPPE), 230
High Energy Astronomy Observatory (HEAO), 47, 107, 152, 159, 179, 227, 284, 361-362, 383
Hijacking of aircraft, 304, 317, 414
  Congress, 310
  cost of, 302, 310
  Germany, West, 289, 291, 292
  hostages, 289-290, 292, 293, 300, 304, 306, 316, 317
  ICAO consideration of, 235, 265, 297, 310, 324, 336
  Israel, 289-290, 291
  Nixon, President Richard M., 235, 265, 291, 297, 300, 302, 304, 310, 336
  protective measure against, 291, 292, 293, 297, 302, 310, 330, 345, 405
  statistics, 304, 414
  Switzerland, 289, 292, 300, 316, 317
  treaty, 297, 300, 310, 324
  United Kingdom, 289, 291, 292, 293, 297, 300
  United Nations, 292, 302, 304
  U.S., 257, 289-290, 291, 292, 293, 297, 300, 302, 306, 317, 324, 345, 405, 414
Hill, Louis W., Space Transportation Award, 286, 341
Hines, William, 17-18, 52, 84, 232
Hirohito, Emperor (Japan), 364
Hiroshima, Japan, 259, 260
Hirschler, Otto, 226
HL-10 (lifting-body vehicle), 74, 411
  test flight
    award, 88-89
    powered, 18, 27, 55, 66, 199, 236
HLH. See Heavy-lift helicopter.
Hoag, Maj. Peter C. (USAF), 18, 55, 199, 236
Hodge, John D., 189
Hoffman, David, 366
Hoffman, Dr. Philip G., 109
Hoffman, Samuel K., 164

Holden, John C., 296
Holland, Brian, 3
Hollenbaugh, Roger C., 47
Holloman AFB, N. Mex., 208, 320
Hominy, Okla., 2
Honeywell, Inc., 208
Hong Kong, 157
Honolulu, Hawaii, 137, 139-140
  Kawaiahao Church, 139
Hooker, Dr. Stanley G., 286
Hoover, Cdr. George W. (USN, Ret.), 342
Hornig, Dr. Donald F., 90, 147
Hotz, Robert B., 11, 211
Houbolt, Dr. John C., 21
Houck, Dr. James R., 152
Houghton, Karl, 297
Houphouet-Boigny, President Felix (Ivory Coast), 82
Housing and Urban Development, Dept. of (HUD), 214-215, 390
  funding, 167, 170, 212, 226, 247, 248, 257, 263, 264, 303, 390, 401
Houston, Tex., 350
  Apollo 13 mission, 127
  astronauts at, 136-137, 141, 203, 231
  Lunar Receiving Laboratory, 51, 85, 178, 250-251, 319, 333, 409
  Lunar Science Institute, 2-3, 294
  meeting, 2-3, 4-6, 59, 164, 277, 283, 340-342
  National Space Hall of Fame, 6
Houston, Univ. of, 62, 71, 79, 109, 133, 333, 342, 406
Hovercraft, 296
Howard, John R., 33
Hoyle, Dr. Fred, 6
HST. See Hypersonic transport aircraft.
Hubbard Medal, 52
HUD. See Housing and Urban Development, Dept. of.
Hudson, M/G John B. (USAF), 227
Huey, William M., 340
Hughes Aircraft Co., 8, 26, 247, 404
Humble Oil & Refining Co., 277
Humphreys, M/G James W., Jr. (USAF, Ret.), 385
Humphreys, Lloyd E., 102
Hungary, 380
Hunten, Dr. Donald M., 399
Huntington Beach, Calif., 201, 292, 297, 387
Huntsville, Ala., 174, 189
  Alabama Space and Rocket Center, 93-94, 237, 344
  meeting, 154, 314, 370
Hurricane measurement, 264
Hussein, King (Jordan), 304-305
Hycon Manufacturing Co., 115
Hydrogen, liquid, 92
Hydrogen bomb, 258
Hypersonic aircraft, 220, 247, 411
Hypersonic transport aircraft (HST), 17

I

IAA. See International Academy of Astronautics.

IAEA. See International Atomic Energy Agency.
IAF. See International Astronautical Federation.
Iakovos, Archbishop, 139
IATA. See International Air Transport Assn.
IBM. See International Business Machines Corp.
ICAO. See International Civil Aviation Organization.
Icarus, 268
ICBM. See Intercontinental ballistic missile.
Iceland, 314, 390
ICSU. See International Council of Scientific Unions.
Idaho, 283
IEEE. See Institute of Electrical and Electronics Engineers.
IEPC. See Committee for International Environmental Programs.
IFR: instrument flight rules.
IGY. See International Geophysical Year.
ILC Industries, 194
Illinois State Fair, 265
Illinois, Univ. of, 172, 234
ILS. See Instrument landing system.
Imbrium Basin (moon), 341
IMP. See Interplanetary Monitoring Platform.
IMP-I (Interplanetary Monitoring Platform), 408
Impact of Changes in Federal Funding Patterns on Academic Institutions (NSF report 70-39), 337
IMU. See Inertial measuring unit.
Incirlik, Turkey, 297
India, 51, 85, 135
  Atomic Energy Dept., 239
  cooperation, space, 1, 85, 102, 233, 245, 327, 414
  rocket test, 18, 185
Indian Ocean, 83, 122, 187, 241, 269, 275, 342, 359, 399
Indian Space Research Organization (ISRO), 233, 245
Indianapolis, Ind., 347
Indians of North America, 383-384
Industrial Research, Inc., 305
Inertial measuring unit (IMU), 207
Infectious Disease in Manned Spaceflight: Probabilities and Countermeasures, 184
Information retrieval, 354-355
Infrared sensor, 231
Infrared telescope, 251
Infrared temperature profile radiometer (ITPR), 155-156
Inland Daily Press Assn., 338
Insomnia, 190
Institute for Space Studies, 170, 198
Institute for Strategic Studies, 289
Institute of Aeronautical Science (Japan), 244
Institute of Electrical and Electronics Engineers (IEEE), 40, 97

Institute of Telecommunication Sciences, 390
Institute of the United States of America, 199
Institute of Theoretical and Experimental Physics, U.S.S.R., 392
Instituto Nacional de Tecnica Aerospacial (INTA) (Spain), 207, 213
Instructional television (ITV), 233
Instrument landing system (ILS), 258
Instrument unit (IU), 119
INTA. See Instituto Nacional de Tecnica Aerospacial.
Intelsat. See International Telecommunications Satellite Consortium.
Intelsat (communications satellite), 409-410
Intelsat I (Early Bird), 112
Intelsat-II F-2, 162
Intelsat III series, 14, 45, 144-145, 241
Intelsat-III F-4, 14
Intelsat-III F-5, 14
Intelsat-III F-6, 14, 45, 145, 411
Intelsat-III F-7, 144-145, 162, 169, 411
Intelsat-III F-8, 241, 411
Intelsat IV series, 83
Inter-Academy Exchange program, 47
"Interactions of Technology and Society" (lecture), 191
Interagency Committee on Back Contamination, 7, 15, 51, 55
Interagency Conference on Weather Modification, 351
InterAmerican Meteorological Rocket Network (EXAMETNET), 252
Intercontinental ballistic missile (ICBM), 31, 69, 93, 150, 151, 230, 248, 266, 274, 276, 289, 303, 311, 358, 360, 381-382, 405
Intercosmos (U.S.S.R. comsat), 411
Intercosmos III, 259
Intercosmos IV, 336
Interior, Dept. of, 35-36, 59, 107, 110, 169, 204, 230, 274, 343, 345, 390, 407
  Earth Resources Observation Systems program, 259
International Academy of Astronautics (IAA), 159, 271
International Aeronautical Exposition (see also International Transposition Exhibition), 259
International Air Policy Statement, 210-211
International Air Transport Assn. (IATA), 211
International Aspects of Earth Resources Survey Satellite Programs, 221
International Assn. of Firefighters, 214
International Astronautical Federation (IAF), 328, 335-336, 340
International Astronautical Federation Congress, 314, 326-328
International Astronomical Union, 267
International Atomic Energy Agency (IAEA), 214, 263
  Fourteenth General Conference, 298, 312

International Business Machines Corp. (IBM), 191, 194, 274
International Civil Aviation Organization (ICAO), 310, 336
  Assembly, 235, 265
  Council, 235, 265, 298, 324
  Legal Committee, 336
International Conference on Plutonium and Other Actinides, Fourth, 329
International cooperation (see also Disarmament and Treaty), 118
  air transportation, 196, 210-211, 235
  aircraft (see also Concorde), 218, 241, 263, 385, 399
  astronomy, 267-268
  ecology, 214, 255
  meteorology, 2
  nuclear power, 225, 226, 307-308
  oceanography, 82, 103, 127, 179-180, 277, 345
  science and technology, 47, 106, 200, 307-308, 317, 320
  transportation, 15, 28
International cooperation, space (see also European Launcher Development Organization; European Space Research Organization; Global Atmospherics Research Program; International Telecommunications Satellite Consortium), 83, 85-86, 156, 258, 336, 354
  military, 7, 95-96, 149, 258, 269, 399, 411, 413
  satellite, 85, 221, 326-327
    communications. See Communications satellite, cooperation.
    DOD-U.K., 16, 44, 269, 399, 411, 413
    earth resources, 30, 72-73, 177, 221, 280, 383-384, 405, 413
    Europe, 95-96, 200, 252-253, 413
    France-Germany, West, 84, 200
    NASA
      -Australia, 24, 150, 411, 414
      -Canada, 79, 150, 242, 247, 409
      -Europe, 77, 85, 88, 150, 193-194
      -France, 85, 150, 409, 413-414
      -Germany, West, 91
      -India, 233, 245, 414
      -Italy, 88, 150, 275, 395, 407, 410, 413
      -Netherlands, 194, 413
      -U.K., 16, 150, 357, 409, 411, 413
    U.S.
      -NATO, 95-96, 408, 413
      -U.S.S.R., 212, 221
  Sounding rocket. See Sounding rocket, international programs.
  space research, 51, 78-79, 83, 85-86, 87-88, 112, 149, 176-177, 221, 243, 262, 326-328, 343
    Europe, 241, 354, 361, 380
    France-U.S.S.R., 177
    U.S.
      -Australia, 51, 62, 77-78
      -Brazil, 85
      -Canada, 51, 78
      -Europe, 44, 78, 193-194, 252-253, 255-256, 305, 313, 323, 354, 359-360
      -France, 51, 287, 413-414
      -Germany, West, 51, 52, 86, 292
      -Italy, 88
      -Japan, 51, 78
      -U.K., 51, 289, 292, 413
      -U.S.S.R., 32, 83, 85-86, 117, 156, 192, 197, 225, 229, 243, 289, 300, 334, 335-336, 343-344, 346, 347, 349, 351, 353, 358, 376, 391-392, 393, 409, 413
  space shuttle, 52, 88, 227, 292, 313, 323, 351, 361
  space station, 43, 56, 88, 193-194, 212, 314, 327, 346
  tracking, 199
    U.S.-Afghanistan, 330
    -Australia, 100, 199, 213
    -France, 199
    -Italy, 88
    -Mexico, 381
    -Spain, 207, 213
International Council of Christian Churches, 408
International Council of Scientific Unions (ICSU), 176
International Decade of Ocean Exploration, 127
International Federation of Airline Pilots, 292
International Geophysical Year (IGY), 205
International History of Astronautics Symposium, Fourth, 328
International Mineralogical Assn., Nomenclature Committee, 234
International Nomenclature Commission, 289
International Solar Energy Conference, 74-75
International Symposium on Space Rescue, Third, 327
International System of Units, 302
International Telecommunications Satellite Consortium (INTELSAT), 215
  conference, 53, 96, 409-410
  earth stations, 45, 65, 83, 223
  Interim Communications Committee, 130
  membership, 65, 184
  organization, 53, 241, 409-410
  satellite. See individual Intelsat satellites and series.
International Telephone & Telegraph World Communications, Inc., (ITT), 151
International Transportation Exhibition (see also International Aeronautical Exposition), 403
International Workshop on Earth Resources Survey Systems, 405, 413
Interplanetary Monitoring Platform (IMP), 43, 60
Interrogation, recording, and location system (IRLS), 52, 336
Intersales, Inc., 211

Intersociety Energy Conversion Engineering Conference, Fifth, 309
Intrepid (Apollo 12 LM). See Lunar module.
Introduction to the Solar Wind, 306
Iodine 123, 352
Iodine 131, 352
Ion, 95
Ion propulsion, 38, 51, 173, 243, 285, 305, 315, 338, 393
Ionosphere, 1, 48, 62, 71, 84, 102, 108, 133
Iowa, Univ. of, 105, 330, 360
Iraq, 306
Ireland, 314
Irish Sea, 261
IRLS. See Interrogation, recording, and location system.
Irwin, L/C James B. (USAF), 101, 285
ISIS (International Satellite for Ionospheric Studies), 43
Isis I, 79
ISIS-B, 409
Israel, 118, 232, 289-290, 291, 294, 383
Israel Aircraft Industries, 118
ISRO. See Indian Space Research Organization.
"Issues and Answers" (TV program), 91
Issues and Directions for Aeronautical Research and Development (House report), 98
Italian National Research Council (Consiglio Nazionale delle Ricerche) (CNR), 88
Italy, 51, 88, 150, 200, 275, 395-396, 407, 410, 413
Itek Corp., 28, 118
Ithaca, N.Y., 272
ITOS (program), 107, 394-395
Itos I (Tiros-M) (Improved Tiros Operational Satellite), 23-24, 87, 203, 354, 395, 411
ITPR. See Infrared temperature profile radiometer.
ITT. See International Telephone & Telegraph World Communications, Inc.
ITV. See Instructional televison.
IU. See Instrument unit.
Ivory Coast, 82

J

J-2 (rocket engine), 203, 386-387
Jackass Flats, Nev., 17
Jackson Hole, Wyo., 52, 107
Jackson, Nelson P., Aerospace Award, 94
Jackson, Roy P., 313
Jaffe, Dr. Leonard D., 176
Jamaica, N.Y., 110
James Madison (nuclear submarine), 245
James River, 384
Jamieson, Minister of Transportation Donald C. (Canada), 28
Japan, 151, 157, 187, 223, 364
 astronomy, 51
 international cooperation, space, 51, 78, 96, 354, 409-410
 launch, satellite, 48-49, 50, 412
  failure, 315-316
 lunar samples, 51, 414
 nuclear nonproliferation treaty, 40
 Pacific rocket tests protest, 103, 105
 science, 153
 space program, 59, 150, 197, 301, 354, 368, 410
Japanese Institute of Space and Aeronautical Science, 49
Jastrow, Dr. Robert, 170
Jaumotte, Prof. A., 328
Java, 195
Javelin (sounding rocket), 77, 108, 283, 328-329
Javits, Sen. Jacob K., 217, 330
Jefferies, John, Award, 342
Jefferson City, Mo., 238
Jet Propulsion Laboratory (JPL) (Cal Tech), 39-40, 159, 185, 328, 405-406
 anniversary, 109
 award, 126
 Civil Systems Project Office, 103
 Communications Research Section, 139
 computer, 341, 397
 Deep Space Network, 16, 66, 207, 265
 Goldstone Tracking Station, 139, 213, 286
 lunar research, 26, 176-177, 372
 Mariner program, 16, 33, 172, 279, 309, 408, 412
 Mariner Venus-Mercury Office, 66
 Mars (planet) research, 16, 172, 279
 Nuclear Power Sources Group, 309
 personnel, 77
 relativity, theory of, test, 157, 366-367, 377-379
 rocket motor, 25
 Thermoelectric Outer Planet Spacecraft, 180, 341
 transit system, 408
 Viking, Project, 255, 279
Jodrell Bank Experimental Station, 382, 413
Johannesburg, South Africa, 135
Johns Hopkins Hospital, 366
Johns Hopkins Univ., 89, 99-100, 185, 190-191, 366
Johnson, Arthur W., 409
Johnson, Caldwell C., 334
Johnson, David S., 354
Johnson, John A., 130
Johnson, President Lyndon B., 15, 401
Johnson, U. Alexis, 53, 72, 305
Johnson, Vincent L., 144, 313
Johnston, Richard S., 341
Johnston, S. Paul, 241
Johnston Island, 315
Joint Chiefs of Staff, 208, 223
Joint Institute for Laboratory Astrophysics, 7
Joint Research and Development Board, Committee on Atomic Energy, 67
Jones, R/A Don A. (USN), 347
Jones, Dr. Walton L., 63, 342
Jordan, 300, 304-305, 306, 316, 317

JPL. See Jet Propulsion Laboratory.
JR 100F (Japanese VTOL engine), 321
JT9D3 (jet engine), 278
JT9D3A, 278
Jupiter (planet), 12, 33, 45, 49, 60, 75, 84, 101, 251, 284, 341
Jupiter C (booster), 103
Justice, Dept. of, 407-408

K

Kabul, Afghanistan, 330
Kagoshima Space Center, 48
Kaiser Corp., 166
Kansas City, Mo., 165, 347
Kaplan, Dr. Ian R., 387
Kaplan, Dr. Joseph, 328
Kapp, Michael, 349
Kapustin Yar, U.S.S.R., 16, 147, 259, 336
Karaganda, Kazakhstan, 190
Karnik, K. S., 327
Karr, Albert R., 385
Karth, Rep. Joseph E., 72
Kasemir, Dr. Heinz W., 54
Kaufman, John W., 264
Kavanagh, Dr. Thomas C., 157
Kazakhstan, U.S.S.R., 299
Kazaryan, Prof. Raphael A., 97
KC-135 (jet research aircraft), 82, 275-276
KC-135A, 264
Keldysh, Prof. Mstislav V., 85, 219, 229, 314, 334, 335, 392
Kelley, Albert J., 280
Kelly AFB, Tex., 82
Kelly, Orr, 27
Kelly, Thomas J., 130
Kempner, L/G William E. (USAF, Ret.), 342
Kennedy, Sen. Edward M., 1, 117, 265
Kennedy, President John F., 164, 329
Kennedy, John F., International Airport, 13, 225
Kennedy Space Center (KSC) (NASA), 12
 accident, 100, 371
 Apollo/Saturn. See Apollo missions.
 astronauts at, 40, 62-63, 101, 116, 118-119, 205, 319, 390
 budget, 231
 employment, 236
 facilities, 329
 launch operations (see also Launch Complex 17, 34, 39, and 40; and Apollo missions), 25, 54, 71, 111, 118, 119, 125, 172, 331
 lightning study, 344
 personnel, 201, 231
 press conference, 40, 62-63, 117-118
 visits to, 414
  cosmonauts, 336
  Philip, Prince (U.K.), 51
Kent, Wash., 353
Kenya, 269, 275, 395
Kerwin, Cdr. Joseph P. (USN), 20
Keuper, Dr. Jerome P., 8
Key Biscayne, Fla., 51, 77
Khan, President Yahya (Pakistan), 151

Khodarev, Prof. Yu. K., 197
KIAS: knots indicated air speed.
Kiev Univ., 226
Kilby, Jack S. C., 55
Kilgore, Edwin C., 257
Killian, Dr. James R., Jr., 147
Kilpatrick, James J., 113
King, Elbert A., Jr., 333
King, Ambassador Spencer, 27
Kirillin, Vladimir A., 317
Kiruna, Sweden, 260
Kissel, Ralph R., 352
Kistiakowsky, Dr. George B., 147
Kitt Peak National Observatory, Ariz., 33, 110, 204, 333-334, 399
Kitty Hawk, N.C., 401, 403
Klass, L/C Joe (USAF), 364
Klein, Milton, 64, 75, 144, 402
Kleinknecht, Kenneth S., 51
Knothe, Dr. Adolf H., 40
Kodiak, Alaska, 38
Kohler, Dr. Foy D., 298
Komarov, Col. Vladimir M. (U.S.S.R.), 31, 267
Komarov, Mrs. Vladimir, 195
Kondratyev, Dr. K.Y., 326
Konovalov, B., 316
Kopp, Roger A., 306
Korea, 51, 194, 413
Kosygin, Premier Aleksey N. (U.S.S.R.), 130, 132, 176, 190
Kourou Space Center, French Guiana, 84, 396-397, 414
Kraft, Christopher C., Jr., 1, 21, 286, 341
Kranz, Eugene F., 57, 137
Kranzberg, Prof. Melvin, 189
Krieger, Robert L., 369
Kronauer, M/G Clifford J., Jr. (USAF), 194
Krupp, Fried., Co., 110
Krylov, Marshal Nikolay (U.S.S.R.), 374
KSC. See Kennedy Space Center.
Kumsan, South Korea, 273
Kurzweg, Dr. Hermann H., 64
Kwajalein Missile Range, 276, 405

L

L-1011 (Tristar) (jet transport), 237, 245, 370, 397, 411
La Guardia Airport, 225
Laird, Secretary of Defense Melvin R.
 award by, 227
 Communist China satellite launch, 150
 defense budget, 67
 DOD procurement policy, 246
 F-111, 75-76
 F-111C, 129
 MAST, Project, 259
 nuclear weapons, 142
 press conference, 75-76, 274
 U.S.S.R. missile program, 171, 230, 274, 358, 400
Lake Erie International Jetport Project, 377

Lakenheath RAF Station, U.K., 243
Lamont-Doherty Geological Observatory, 215, 283, 372
LaMoreaux, Philip E., 40
Landon, Alfred M., Lecture on Public Issues, 304
Langley Research Center (LaRC) (NASA), 97, 125, 348, 350
  award, 19-20
  cooperation, 384
  exhibit, 342
  Life Support Technology Laboratory, 195, 201
  Lifting Body Team, 369
  OAO-B, 380
  organization, 325-326
  personnel, 325-326
  radio communications symposium, 336
  Sert II, 338
  space station study, 266
  Space Systems Research Div., 181
  telescope, 79
  trailing vortex research, 320
  wind-tunnel research, 19, 113, 166, 411
Lannan, John, 307
Lapp, Dr. Ralph E., 58-59, 207
Laptev (U.S.S.R. ship), 256
LaRC. See Langley Research Center.
Larson, Dr. Edwin E., 274
Las Vegas, Nev., 126, 130, 175, 309, 379
Laser
  award, 337
  lunar experiments, 177
    Apollo 11, 26, 210, 240, 374
    Apollo 14, 210
    Lunokhod 1, 380, 389
  use, 142, 278, 320, 321, 376, 412
  military, 27, 97, 156
Laser ranging retroreflector (LRRR), 26, 210
Latham, Dr. Gary V., 6, 244, 262, 283, 286, 348, 361, 372
Latin America, 51, 108
Laughlin, Charles R., 47
Launch Complex 17, 371
Launch Complex 34, 172
Launch Complex 39, 172, 184
Launch Complex 40, 87
Lawrence, Ernest Orlando, Memorial Award, 94, 379
Lawrence Radiation Laboratory (Univ. of Calif.), 94, 242
Law's, Ruth, Flying Circus, 383
LCRU. See Lunar communications relay unit.
Leary, Frank, 69
Leavitt, William, 251, 354
Lederer, Jerome F., 9
Lee, Capt. Chester M. (USN, Ret.), 71, 403
Lee, Frederick B., 305
Leech experiment, 392, 400
Lefevre, Theo, 243, 305, 313, 359
Lefkowitz, Louis J., 225
Legion of Honor, 249
Lehrer, Max, 173
Leloir, Dr. Luis F., 348

Leningrad, U.S.S.R., 176, 181, 185, 195, 263
Leninist Subbotnik, 127
Leonov, L/C Aleksey A. (U.S.S.R.), 267, 330
Leonardo da Vinci, 296
LeRC. See Lewis Research Center.
Lerner, Max, 141
Les VI ((Lincoln Laboratory Experimental Satellite), 148
Lessing, Lawrence P., 279
Leussink, Minister of Science and Education Hans (West Germany), 320
Levin, Lewis, 102
Lewis, John S., 251
Lewis Research Center (LeRC) (NASA), 230, 348, 377
  Aerospace Safety Research and Data Institute, 219
  air pollution study, 27, 260
  Apollo 13 tank test, 191
  award, 305
  budget, 40
  contract, 257
  exhibit, 342
  noise abatement, 292, 390
  personnel, 21, 317, 377, 384
  propulsion conference, 373
  research, 92, 225, 257, 352-353
  Sert II, 38, 315
  Space Shuttle Technology Conference, 234
  Special Products Div., 148
Lewis, Richard S., 160
Lexington, Mass., 176
Ley, Willy, 267
Leyton, John, 315
Library of Congress
  Legislative Reference Service, 30, 68, 311
  Science Policy Div., 270
Lieurance, Newton A., 20
Life magazine astronaut contract, 74
Life raft, inflatable, 12-13
The Life Sciences (NAS study), 386
Life Sciences in Space (NAS-NRC report), 395
Lifting-body vehicle, 219-220, 318
  HL-10, 18, 27, 55, 66, 74, 88-89, 199, 236, 411
  M2-F3, 74, 178, 190, 357, 379, 411
  X-24A, 60, 74, 94-95, 108, 145, 171, 206, 247, 263, 273, 336, 347, 375, 411
Lightning, 54, 71, 179, 250, 276, 344
Lilly, William E., 37
Lima, Peru, 55, 135, 231, 259
Lincoln Laboratory, 176
Lincoln, Nebr., 164
Lindbergh, B/G Charles A. (USAF, Ret.), 131, 308, 379-380
Lindsay, Mayor John V., 135, 194, 286, 391
Lindstrom, Robert E., 20
Ling-Temco-Vought, Inc., 166, 297
  LTV Aerospace Corp., 30
  Missiles and Space Div., 285

Service Technology Corp., 29, 44, 384
Link Div., Singer-General Precision, Inc., 196, 303
Liquid-fuel rocket, 92
Lisbon, Portugal, 59, 394
Liston, Sally McReady, 342
Lithium hydroxide, 123
Lloyd's of London, 9
LM. See Lunar module.
Lockheed Aircraft Corp., 400
  Ah-56A (helicopter), 72, 109, 345
  C-5A (cargo transport), See C-5A.
  contract, 58, 86-87, 105, 196, 203-204, 270-271, 302, 345
  Electra, 231
  financial problems, 72, 126-127, 169-170, 183, 237, 241, 245, 302, 397, 408
  L-1011 (jet transport), 237, 245, 370, 397, 411
  T-33 (jet trainer), 352
Lockheed-California Co., 107, 370
Lockheed-Georgia Co., 9, 16, 17, 27
Lockheed Missiles & Space Co., 102, 171, 179, 214, 264, 324, 400
Lockheed Propulsion Co., 345
Loening, Dr. Grover, 286
Loftin, Laurence K., Jr., 325
Logan International Airport, 296
Logistics Management Institute, 139, 173
Logsdon, Dr. John M., 94
Lohman, Robert L., 197
LOI: lunar orbit insertion.
Loki-Dart (sounding rocket), 210, 305, 309
London Airport, 13
London, U.K., 22, 45, 102, 135, 226, 289, 296, 349, 357
London-Sydney Air Race, 3
Long Beach, Calif., 162, 277
Long-tank Thor-Delta (booster), 23-24, 144
Long-tank, thrust-augmented Thor-Delta (booster), 14, 95, 241, 269
Lop Nor, Communist China, 147
Lord, Douglas R., 323
Los Alamos, N. Mex., 232
Los Alamos (N. Mex.) Scientific Laboratory, 209, 277
Los Angeles, Calif., 40, 55, 68, 112, 240, 346, 359
Los Angeles International Airport, 162
Losey, Robert M., Award, 20
Lost City, Okla., 2, 8, 12, 14, 18, 146
Louisville, Ky., 240
Lovelace, Dr. W. Randolph, II, 267
Lovell, Sir Bernard, 382
Lovell, Capt. James A., Jr. (USN), 203, 316, 340
  Apollo 13 mission, 25, 136, 140
    flight, 119-125
    medical examination, 101, 111, 115
    Nixon, President Richard M., 118, 128, 133-134, 136-137, 139-140
    plans, 63, 71
    press conference, 40, 143-144
    telecast, 121
  awards and honors, 89, 137, 162, 194, 249, 267, 368
  goodwill tour, 314, 331
Lovell, Mrs. Marilyn, 129, 137
Low, Dr. George M., 313, 325, 402
  Apollo 13 mission, 134, 161
  appointment, 277, 303, 412
  awards and honors, 94, 159
  computer, 164-165
  international cooperation, space, 176, 334, 335, 391-392, 409
  Life Support Technology Laboratory (LaRC), dedication, 195
  Luna 16 mission, 309, 314
  lunar rock sample exhibit, 244
  NASA Aerospace Museum, 234
  National Air and Space Museum, 234
  press conference, 134, 161
  space program, 3, 38, 65, 164-165, 195, 305, 340-341, 368, 393
  U.S.S.R. space program, 379
Lowell Observatory, 101
LRL. See Lunar Receiving Laboratory.
LRRR. See Laser ranging retroreflector.
LRV. See Lunar roving vehicle.
LTTAT. See Long-tank thrust-augmented Thor.
LTV, Inc. See Ling-Temco-Vought, Inc.
Lubarsky, Dr. Bernard, 317
Lucerne, Switzerland, 331
Lucet, Ambassador Charles (France), 249
Luke AFB, Ariz., 283
Lumb, Dr. Dale R., 166
Luna (U.S.S.R. lunar probe), 411
Luna XV, 299
Luna SVI, 299, 309, 311-312, 314, 316, 318, 325, 349, 358, 365, 369, 375, 379, 411
Luna XVII, 364-365, 373, 374, 379, 411
Lunar communications relay units (LCRU), 258, 315
Lunar landing training vehicle (LLTV), 399
Lunar module (LM), 7, 18, 50, 94, 136, 217, 219, 315
  Apollo 11 (Eagle), 45
  Apollo 12 (Intrepid), 2, 119, 297
    moon landing, 54, 177, 276, 360-361
  Apollo 13 (Aquarius), 71, 91, 119, 125, 130, 143, 410
  Apollo 14 (Antares), 18, 205, 331, 387, 403
  contract, 7, 223
Lunar Orbiter (program), 192
Lunar Receiving Laboratory (LRL) (NASA), 85, 178, 319
  Apollo 14 astronauts, 409
  Lower Animal Test Team, 250
  lunar samples
    Apollo 11, 333
    Apollo 12, 7, 51
  Surveyor III spacecraft parts, 16
Lunar roving vehicle (LRV), 43, 285, 371
  design, 29, 181, 205
  test, 276, 397, 404
Lunar Science Conference, 2, 4-6, 20, 51
Lunar Science Institute, 2, 294-295

Lunar time system, 115
Lund, Univ. of (Sweden), 136
Lundin, Bruce T., 26, 230, 317, 384
Lunney, Glynn S., 137, 147, 148, 334, 342
Lunokhod I (Moonwalker I), 365, 372, 375, 380, 389, 391, 394, 411
Lydda Airport, Israel, 118
Lyman-alpha experiment, 116
Lyons, Richard D., 42, 290
Lyra (constellation), 225
Lysdale, Clarence A., 33

M

M2-F3 (lifting-body vehicle), 74, 178, 190, 357, 379, 411
MAC. See Military Airlift Command.
McBain, Donald C., 334
McBirney, Alexander R., 333
MCC: midcourse correction.
McCall, Robert, 349
McClellan, Sen. John L., 229
McCormack, James, 82, 142, 154
McCormack Rep. John W., 141
McCrosky, Dr. Richard E., 2, 8, 14
McCurdy, Richard C., 20, 220, 306
MacDill AFB, Fla., 262
McDivitt, B/G James A. (USAF), 140, 326, 398, 409
MacDonald, Dr. Gordon J. F., 30, 42
McDonald, Dr. J. Ken, 258
McDonald Observatory, Tex., 26, 374
McDonnell Douglas Corp., 159, 170, 181, 264, 277
  McDonnell Douglas Aircraft Co., 237, 366, 370
  McDonnell Douglas Astronautics Co.
    airlock, 246, 248, 372
    contract, 179, 217-218, 248, 285, 374
    Saturn Workshop, 292, 387
    space shuttle, 179, 285, 374, 399, 400
    space station simulator, 231, 297
McElroy, M.B., 333
McElroy, Dr. William D., 249, 329, 410
McGill Univ., 355
McGovern, Sen. George S., 325
McIlwan, Dr. Carl E., 19
McIntire, Dr. Carl, 408
McInturff, Raymond M., 2
McKee, Gen. William F. (USAF, Ret.), 19
McLaren, Richard W., 408
MacMillan, R/A Donald B. (USN, Ret.), 291
McMullen, Col. A. B. (USAF,Ret.), 113
McNamara, Secretary of Defense Robert S., 402
MacReady, Col. John A. (USAF, Ret.), 342
Macy, John W., Jr., 54
Madagascar, 199
Madrid, Spain, 207, 213
Magnetic field, 7, 102, 271, 274, 335
Magnetohydrodynamic (MHD) system, 309
Magnetohydrodynamics, 348

Magnetometer, 7, 271
Magnetoplasmadynamics, 293
Magnetosphere, 20, 33, 413
Magomezulu (witch doctor), 135
Magruder, William M., 107, 179
Mahnken, Conrad V. W., 169
Mahon, Joseph B., 61
Maier, Mayor Henry W., 208
Maisch, Carl F., 293
Malim, D. W., 361
Mallick, Donald L., 180
Malone, Dr. Thomas F., 381
Malta, 314
Man-powered flight, 186
Management, 280
Manhattan Project, 67, 235
Manke, John A., 108, 171, 206, 263, 336, 347
Manned Flight Awareness Workshop, 94
Manned Orbiting Laboratory (MOL), 11, 36
Manned space flight (see also Apollo program, *Apollo 8, 9, 10, 11, 12, 13,* and *14* missions; Gemini program; Astronaut; Cosmonaut; Manned Orbiting Laboratory; *Soyuz IV, V, VI, VII, VIII,* and *IX* missions; Space biology; and Space station), 10, 78, 171
  achievements, 3, 53, 105-106, 111-112, 133-134, 136-137, 138-139, 164-165, 166, 174, 204, 227-228, 238-239, 245, 301, 340-342, 368-369, 412
  advantages, 106, 372
  anniversary, 126, 127, 165, 227, 236, 238, 243, 372
  cooperation (see also Docking), 32, 41, 43, 83, 87-88, 156, 192, 225, 229, 243, 287, 313, 346, 349, 413
  criticism, 207-208, 217, 295-296, 329-330, 341, 360
  EVA. See Extravehicular activity.
  funding, 3, 10, 35-36, 39, 42-43, 66-67, 86, 92-93, 161-162, 248, 284-285, 287
  hazards, 15, 20, 51, 54, 55, 71, 130, 137, 142, 191, 227, 360
  long-duration, 83, 88, 104-105, 192, 203, 412
  lunar landing. See Moon landing, manned.
  military aspects, 11, 36, 105
  policy and plans
    U.S., 3, 7, 10, 11-12, 19, 35-36, 41, 43-44, 50, 66-67, 69, 71, 78-79, 87-88, 90, 91, 103-105, 147-148, 167, 181, 207-208, 218-219, 229, 233, 243, 251, 263-264, 284-285, 290-291, 293, 301, 303, 306, 319, 323, 328, 341-342, 385, 410
    U.S.S.R., 31-32, 68, 181, 192, 225, 229, 326, 411-412
  record, 108, 190, 203, 231, 401
Manned Space Flight Network (MSFN), 66, 210, 212, 320, 353
Manned Space Flight: Present and Future (House staff study), 50
Manned Spacecraft Center (MSC) (NASA),

55, 88-89, 111, 186, 262, 319, 334, 347, 387
ALSEP, 25, 37-38, 210, 212
Apollo spacecraft, 137, 155, 249
Apollo 11 Lunar Science Conference, 2, 4-6, 20
Apollo 12 Lunar Science Conference, 51
Apollo 13 mission, 127, 132-133, 134, 136, 137, 141, 143, 161, 191, 218-219
astronauts at, 20, 118, 141, 143, 283, 303, 352, 409
award, 40, 57, 92, 136-137, 211
Calibration Laboratory, 284
contract, 7, 29-30, 41, 44, 112-113, 115-116, 118, 147, 149, 170, 171, 183, 196, 208, 212, 223, 249, 285, 303, 315, 324, 384, 399
earth resources data, 98, 146, 231, 383-384
experiment (see also ALSEP), 318, 362
fire-resistant materials research, 9
Lunar and Earth Sciences Div., 181, 348, 369
lunar communications relay units (LCRU), 315
Lunar Receiving Laboratory. See Lunar Receiving Laboratory.
lunar rock sample exhibit, 414
lunar roving vehicle, 371
meeting, 30, 383-384
personnel, 1-2, 109, 399
press conference, 143, 161, 191
space shuttle, 165, 170, 198, 285, 324, 374
space station, 147, 183, 208, 368, 387
space tug, 112
spacecraft test, 84, 387
Surveyor III spacecraft parts, 16, 54
Mansfield, Sen. Michael J., 239
Mansur, Dr. George F., Jr., 288
Mao Tse-tung, 147, 150, 156
Mar, Dr. James W., 297
Marconi Space and Defence Systems, 361
Mare Foecunditatis. See Sea of Fertility (moon).
Mare Imbrium. See Sea of Rains (moon).
Mare Tranquillitatis. See Sea of Tranquility (moon).
Marietta, Ga., 27, 338
Marietta, Ohio, 139
Marine Game Fish Research Program, 230
Marine Minerals Technology Center, 230
Marine pollution, 22, 73, 127, 179, 255
Mariner II (Venus probe), 306
Mariner IV (Mars probe), 16
Mariner V (Venus probe), 286-287
Mariner VI (Mars probe), 1, 16, 60, 66, 79, 157, 170, 333
Mariner VII (Mars probe), 1, 16, 60, 66, 170, 172, 333
Mariner-Mars 1971 (program), 60-61, 66, 279, 408, 412
Mariner-Venus-Mercury 1973 (MVM-73) mission, 66, 247
Marius Hills (lunar landing site), 323, 341
Mark IV (eddy current proximity gauge), 330
Mark, Dr. Hans, 144
Marquardt Corp., 55
Mars (planet) (see also *Mariner IV*, *Mariner VI*, *Mariner VII*, and Viking Project), 49
atmosphere, 1, 101, 333-334
colonization, 170
ephemeris, 16
exploration, 146
benefits, 292-293
cost, 217
international cooperation, 177, 197
manned, 10, 69, 90, 192
plans, 11-12, 43, 60-61, 69, 75, 84, 255, 279, 370
spacecraft, 89, 279, 284
unmanned, 11-12, 16, 60-61, 75, 185, 284, 312, 374
gravity, 16
life on, 60, 279, 368, 370
mass, 16
moon, 172
photographs, 152
poles, 406
surface, 60
Marshall Space Flight Center (MSFC) (NASA), 334, 372
anniversary, 217
Apollo Telescope Mount, 169, 209, 274, 296, 371, 383
Astronautics Laboratory, 113
contract, 29, 42, 113, 170, 179, 181, 227, 248, 257, 274, 296
Saturn, 23, 28, 87, 223, 278, 285
cooperation, 174
employment, 95, 265, 286
High Energy Astronomy Observatory (HEAO), 107, 179, 227
launch vehicle. See Saturn.
lunar roving vehicle (LRV), 29, 181, 205, 276, 371, 397
Manufacturing Engineering Laboratory, 169
maxometer, 264
meeting, 94, 107, 218, 292, 313, 371, 383, 384, 385, 399
Multispectral Photo Laboratory, 284
patent, 232
personnel, 88, 95, 154, 353, 374
proximity gauge, 330
Quality and Reliability Assurance Laboratory, 113
Saturn Workshop (see also Skylab), 209, 285, 292, 310, 370
Space Flight Orientation Course, 267
space shuttle, 170, 179, 181, 313, 384, 399
space station, 218, 246, 248, 257, 367-368, 371-372
tachometer, 352
Technology Utilization Office, 258
trailing vortex research, 320
turbulence research, 9
visitors, 58, 344, 370

Marsten, Dr. Richard B., 61, 245
Martin Marietta Corp., 89, 191, 257, 273, 275-276, 283, 372
  Denver Div., 246
Martin, Minta, Lecture, 191
Martin, R., 333
Martin, Warren L., 157
Maryland, 384
Maryland, Univ. of, 12, 16, 26, 59, 198, 374
Maser, 337
Masling, Jack, 3
Mason, Dr. Brian H., 270
Mason-Rust Co., 29
Massachusetts, 21
Massachusetts Institute of Technology (MIT), 130, 147, 216, 269, 271, 276, 329, 379
  award, 342
  contract, 22, 47, 98
  Draper, Charles Stark, Laboratory, 8, 87, 125, 176
  experiment, 169, 190
  infrared research, 152, 251
  Instrumentation Laboratory, 8, 87
  Lincoln Laboratory, 171, 176
  pollution research, 73, 158, 255
  Radiation Laboratory, 320
  School of Engineering, 207, 265
Massachusetts Port Authority, 296
MAST. See Military assistance for safety in traffic.
Materials technology, 3, 9, 33, 64, 142, 167, 209, 214-215, 219, 220
Mathematics, 11
Matheson, Capt. James C. (USN), 198
Mathews, Charles W., 371
Mathias, Sen. Charles McC., 148
Mattingly, L/Cdr Thomas K., II (USN), 25, 71, 101, 111, 115, 116, 117, 119, 126
Max Planck Institute, 328
Maxometer, 264
May, Dr. Michael M., 94
Mazan, Walter L., 260
MDA. See Multiple docking adapter.
Mead, Dr. Margaret, 151-152
Mead, Dr. Sylvia E., 226
Medal of Freedom, 133, 136-137, 140
Meissner, Dr. Rolf, 295, 348
Melbourne, Australia, 74
Melbourne Univ., 24
Memphis, Tenn., 177
Mercury (planet), 12, 60, 139, 247, 284
Mercury (program), 164-165, 341
Meredith, J.F., 139
Meridian, Miss., 184
Merrill, Owen S., 309
Messerschmitt-Boelkow-Blohm, 292, 361
MET. See Mobile equipment transporter.
Meteor (U.S.S.R. meteorological satellite), 411
Meteor III, 92
Meteor IV, 154
Meteor V, 211
Meteor VI, 337

Meteorite, 2, 8, 12, 14, 18, 30, 146, 226, 387
Meteoroid, 177
Meteoroid shield, 33
Meteorological satellite (see also individual satellites, such as *Ats III, Ats V, Essa IX, Itos I, Meteor III, Meteor IV, Meteor V, Meteor VI, Nimbus III, Nimbus IV, Noaa I*)
  anniversary, 107
  benefits, 2, 78, 107
  contract, 223
  cooperation, 85, 243, 252, 327, 407
  Global Atmospheric Research Program, 30, 36, 77, 85, 354
  ITOS program, 107, 203, 354, 394-395, 411
  Nimbus program, 107, 113-114, 155-156, 223
  Synchronous Meteorological Satellite, 43, 85, 243, 354
  Tiros program, 23-24, 107
  U.S.S.R., 92, 154, 211, 337, 354, 411
  weather modification experiment, 351-352
Meteorology, 49, 165-166, 171, 176, 179, 255, 272, 336
Metric system, 302, 406-407
METS. See Modularized equipment transport system.
Metzger, Dr. H. Peter, 402
Mexican National Commission for Outer Space (Comision Nacional del Espacio Exterior (CNEE), 381
Mexico, 79, 82, 85, 381, 413
  Foreign Ministry, 257
MHD. See Magnetohydrodynamic system.
mhz: megahertz (one million cycles per second).
Miahuatlan, Mexico, 79
Miami, Fla., 166, 264
Miami, Univ. of, 298
Miatech, Gerald J., 33
Michelson-Morley science award, 337
Michelwait, Lowell P., 13
Michigan City, Ind., 290
Michigan, Univ. of, 12, 79, 127, 256, 264, 290, 336, 360, 375, 405
Michoud Assembly Facility (MFSC), 29, 37, 189, 353, 387
Midway Islands, 97, 237, 349
Mighty Mouse program, 344
MIKA (French minicapsule), 84
Mikoyan, Artem I., 392
Military Airlift Command (MAC) (USAF), 225, 273, 283
Military assistance for safety in traffic (MAST), 233-234, 259, 283, 311, 404
The Military Balance, 1970-71, 289
Milky Way (galaxy), 59, 110, 198, 325, 409
Miller, Rep. George P., 41, 92, 141, 178, 295, 309
Miller, Harold G., 92
Millionshchikov, Dr. M.D., 300
Milwaukee, Wis., 208
Minneapolis, Minn., 286, 307

Minnesota Mining and Manufacturing Co., 9, 116
Minnesota, Univ. of, 126, 170, 365, 372, 401
Minuteman (missile), 392
Minuteman I, 276
Minuteman III, 181, 289, 305, 398
MIRV. See Multiple independently targetable reentry vehicle.
Misawa Air Base, Japan, 273
Missile, 36, 72, 174, 192, 256-257, 258, 305, 321, 383
  air-to-surface, 105
  antiballistic (ABM), 59, 142, 146, 152, 154, 171, 185, 200, 266, 276, 311, 317, 331, 381, 405
  contract, 345
  foreign
    Communist China, 150-151, 175, 307, 360
    U.S.S.R., 31, 67, 93, 103, 132, 146, 167, 171, 200, 230, 232, 266, 274, 278, 289, 311, 320, 347, 349, 358, 360, 381, 400
  intercontinental ballistic (ICBM), 31, 69, 93, 150, 230, 248, 266, 274, 276, 289, 303, 311, 358, 360, 381, 405
  limitation of. See Strategic Arms Limitation Talks.
  multiple independently targetable reentry vehicle (MIRV), 84-85, 93, 152, 181, 381
  orbital, 31, 68, 93, 320
  short-range attack, 72, 93, 345
  submarine-launched, 93, 200, 206, 244, 289, 358, 381
  test, 103, 105, 146, 231, 244, 245, 257, 349, 398, 405
  tracking, 15, 93, 132, 245-246, 360
Mississippi Test Facility (MTF) (NASA), 37, 42, 43, 44, 86, 189, 207, 229, 251, 266, 274, 293, 318, 364
Mission Agency Support of Basic Research (House report), 64-65
MIT. See Massachusetts Institute of Technology.
Mitchell, Cdr. Edgar D. (USN), 167, 205, 264, 306, 390
Mitchell, Frederick J., 178
Mitchell, Harry, 65
Mitchell, James V., 334
Mitchell, Jesse L., 61
Mitrovich, M., 186
Mobile equipment transporter (MET) (see also Modularized equipment transport system, METS), 338, 387
Mobile quarantine facility (MQF), 409
Modena, Marco, 360
Modularized equipment transport system (METS) (see also Mobile equipment transporter, MET), 18
Mojave Ground Station, Calif., 108
MOL. See Manned Orbiting Laboratory.
Mollo Christensen, Dr. Erik L., 342
Molniya I (U.S.S.R. comsat), 411
Molniya I-13, 56
Molniya I-14, 214
Molniya I-15, 318
Molniya I-16, 380
Molniya I-17, 405
Molodezhnaya, Antarctica, 196
Mondale, Sen. Robert F., 166, 217, 226, 330, 389
Mondt, Jack F., 309
Monique (instrumented elk), 52, 57, 58, 59, 62, 107, 118
Monkey experiment, 307
Montana, 52
Monte Alto, Guatemala, 311
Montedel-Montecatini Edison Electronica S.p.A., 407
Montgomery, Rep. Gillespie V., 184
Montoya, Sen. Joseph M., 335
Montreal, Canada, 278, 355
Mooloolaba, Australia, 360
Moon (see also Apollo missions, Lunar Science Institute, Lunar Receiving Laboratory, Sea of Rains, etc.)
  age, 5, 20, 341, 414
  Apollo lunar surface experiment package (ALSEP), 25, 37-38, 66, 210, 212, 329, 371
  base, 197, 208
  colonization, 10
  communication on, 315
  communication with, 65
  contamination from, 7, 15, 20, 51, 55, 71, 250-251
  crater, 1, 59, 63, 76, 90-91, 192, 244, 267-268, 285, 294, 331, 365
  dust, 46, 54, 295, 316, 400-401
  exploration, 32, 88, 90, 287-288, 303, 305-306, 318, 342, 386
    equipment, 35-37, 112, 258, 275-276, 338, 389, 391, 394
    international cooperation, 32, 197, 287
    manned, 293, 318, 323
    training, 266, 283
  landing
    manned, 1, 2, 3, 7, 17, 23, 45, 49, 53-54, 62-63, 68, 71, 90-91, 112, 119, 138, 154, 164, 166, 167, 181, 227-228, 229, 238, 243, 323, 331, 338, 340, 346, 364, 372, 379, 406
    anniversary, 227, 238, 243, 372
    commemorative stamp, 14-15
    implications, 105-106
    plans, 386-387, 390, 403, 410
    U.S.S.R., 32
    unmanned, U.S.S.R., soft, 299-300, 364-365
  landing site, 71, 116, 147, 289, 294-295, 323, 341
  laser experiment, 26, 177, 210, 240, 374, 380, 389
  life on, 6, 71
  light, 58
  lunar landing training vehicle, 399
  lunar roving vehicle (see also *Lunokhod I*), 29, 43, 181, 205, 276, 371, 397, 404

Lunar Science Conference, 4-6, 9, 20, 51
magnetic field, 271, 274, 295
mapping, 115-116
MET, 18, 338, 387
nuclear explosion on, proposed, 6
origin, 341, 345
photographs, 1, 28, 166, 176-177, 185, 192, 294, 342, 364, 365, 391
probe
   Luna XV, 299-300
   Luna XVI, 299, 309, 311, 314, 316, 318, 325, 349, 358, 365, 369, 379, 411
   Luna XVII, 364-365, 373, 374, 379, 394, 411
quake, 244, 283, 295, 348, 364, 372
seismic experiment, 5-6, 58, 71, 119, 244, 262, 276, 295, 297, 308, 348, 360-361, 364, 372
surface, 1, 26, 166, 176-177, 208, 283, 285
composition, 59
cracks, 272
flare, 285
sample, 85, 103, 338, 379
   Apollo 11, 4-6, 15, 18, 20, 51, 57, 59, 68, 76, 85, 103, 139, 142-143, 223, 234, 240, 244, 245, 250, 265, 270, 274, 289, 290, 295, 333, 353, 358, 369, 372, 412, 414
   Apollo 12, 7, 15, 51, 76, 85, 86, 91, 103, 125-126, 130, 142-143, 180-181, 184, 245, 295, 333, 353, 358, 369, 372, 412, 414
   Apollo 13, 62-63, 71, 103, 125
   Apollo 14, 18
   exhibit, 27, 57, 85, 102, 111, 126, 130, 139, 237, 244, 265, 270, 414
   fraudulent claim, 15
   loss, 68
   Luna XVI, 299, 309, 311, 314, 316, 325, 349, 358, 365, 369, 372, 411
temperature, 58, 295, 372
volcanic activity, 1
water on, 274, 375
The Moon as Viewed by Lunar Orbiter (NASA SP-200), 192
Moore, Dr. Carleton, 387
Moore, Patrick, 382
Moore, Thomas W., 273
Moorer, Adm. Thomas H. (USN), 42, 208
Moreland, Grover, 290
Morse, Rep. F. Bradford, 29
Morton, Dr. Louis, 189
Moscow, 171, 181, 199, 287, 316, 369, 391, 392
   airline service, 263
   Apollo 13 mission reaction, 128
   astronaut visit, 195
   communications, 15, 97
   international space cooperation meetings, 117, 334, 346, 349, 351, 376, 407
   nuclear nonproliferation treaty ceremony, 75
   nuclear scientists visit, 307-308
   press conference, 349
   Soyuz IX mission, 189-190, 210
   scientists exchange, 317, 320
Mt. Elbrus, Caucasus, U.S.S.R., 55
Mt. Ranier, 343
Mount Wilson Observatory, 210
Mountain Home AFB, Idaho, 283
Mountain View, Calif., 295
Mozhaisky, Alexander F., 98
MPS-16 (radar equipment), 164
MQF. See Mobile quarantine facility.
MS-F1 (Japanese satellite), 315
MSC. See Manned Spacecraft Center.
MSFC. See Marshall Space Flight Center.
MSFN. See Manned Space Flight Network.
MTF. See Mississippi Test Facility.
Mu (Japanese booster), 50
Mu 4S1, 315
Mueller, Dr. George E., 7, 312, 328
Muhleman, Dr. Duane O., 157, 170
Mulligan, James H., Jr., 157
Multiple docking adapter (MDA), 104, 191, 246, 257, 347, 372
Multiple independently targetable reentry vehicle (MIRV), 84, 93, 152, 181, 381
Munich, West Germany, 292
Murase, Tsutomu, 333
Murchison, Australia, 387
Murmansk, U.S.S.R., 287
Muroc Dry Lake, Calif., 109
Murray Hill, N.J., 278
Murray, Peter Ross, 227
Murthy, H.G.S., 327
Muskie, Sen. Edmund S., 44
Myers, Boyd C., II, 323-324
Myers, Dale D., 7, 37, 57, 66, 180, 219, 227, 247, 284, 313, 323, 373
Myers, Philip F., 99
Mytton, R. G. V., 184

N

NAA. See National Aeronautic Assn.
NACA. See National Advisory Committee for Aeronautics.
NADC. See Naval Air Development Center.
Nader, Ralph, 22, 324, 407
NAE. See National Academy of Engineering.
Nance, Weldon B., 333
Nantucket Island, Mass., 82
Napolitana, Prof. Luigi G., 328
Narita, Japan, 321
NAS. See National Academy of Sciences.
NAS-NRC Space Science Board. See National Academy of Sciences and National Research Council.
NASA Aerospace Museum (proposed), 234
NASA Apollo Lunar Exploration Office, 192
NASA ATS Procurement Review Committee, 230
NASA Communications Network (NASCOM), 162
NASA Earth Observations Programs, 30

NASA Historical Advisory Committee, 189
NASA Lunar and Planetary Missions Board, 263, 284
NASA Management Advisory Panel, 20
NASA Motion Picture Film Depository, 255
NASA Office of Advanced Research and Technology (OART), 38, 47, 63-64, 201, 219, 223, 257, 313, 334, 342, 348, 404
NASA Office of Industry Affairs and Technology Utilization, 352
NASA Office of Manned Space Flight (OMSF), 9, 47, 75, 87, 107, 219, 263
NASA Office of Space Science and Applications (OSSA), 47, 60-61, 77, 212, 213, 219
  Science Steering Committee, Planetary Biology Subcommittee, 51
NASA Office of Technology Utilization, 6, 352
NASA Office of Tracking and Data Acquisition, 65-66
NASA Planning Board, 162, 412
NASA RECON (computer system), 310
NASA Research and Technology Advisory Committee, Subcommittee on Aircraft Operating Problems, 29
NASA Rocket Engine Test Site, Calif., 42
NASA Scientific and Technical Information Facility, 178
NASA Space Shuttle Task Group, 328
NASA Space Station Task Force, 197
NASC. See National Aeronautics and Space Council.
NASCOM. See NASA Communications Network.
Nasser, President Gamal Abdel (United Arab Republic), 317
Natal, Brazil, 315, 318
Nathan, Dr. Robert, 397
National Academy of Engineering (NAE), 48, 60, 87, 157, 221
National Academy of Sciences (NAS), 407
  annual meeting, 153
  Committee for International Environmental Programs (IEPC), 214
  Committee on Research in the Life Sciences, 386
  contract, 108
  Copernicus Quinquecentennial committee, 354
  exchange program, 47-48, 69-70
  funds, 147
  NASA Resident Research Associateship Program, 108
  space program, 51, 55, 60, 177, 239-240, 399
  Space Science Board, 1, 184, 263, 284, 304, 395
  supersonic transport (SST), 381
  Venus: Strategy for Exploration, 399
National Academy of Television Arts and Sciences, 197
National Advisory Committee for Aeronautics (NACA), 109
National Aeronautic Assn. (NAA), 166, 305, 401
National Aeronautics and Space Act of 1958, 94, 298, 314, 323
National Aeronautics and Space Administration (NASA) (see also NASA centers, programs, satellites, and related headings, such as Ames Research Center, Apollo program, *Itos I*; and Space program, national).
  accomplishments. See Space program, national.
  agreement. See International cooperation, space; and Treaty.
  anniversary, 32, 165, 217, 227, 236, 238-239, 323, 400
  astronaut. See Astronaut.
  awards and honors, 18, 19-20, 23, 37, 40, 41, 42, 45, 52, 92, 94, 95, 166, 184, 196-197, 206, 211, 219, 229, 249, 256, 269, 286, 305, 310, 311, 313, 341-342, 364, 368, 372, 376
    Distinguished Service Medal, 29, 368
    Exceptional Bravery Medal, 368
    Exceptional Scientific Achievement Medal, 369
    Exceptional Service Medal, 29, 369
    Group Achievement Award, 369
    Outstanding Leadership Medal, 369
  budget, FY 1971, 9, 11, 12, 19, 35-37, 39-40, 42-43, 47, 71-72, 91, 167, 207-208, 284-285, 303, 339, 350, 393-394, 412-413
    bills introduced, 39, 41, 42, 92-93
    bills signed, 224, 401
    bills vetoed, 263, 264, 303, 309, 373
    House consideration, 63-64, 65, 72-73
      appropriations, 167, 170, 175, 239, 247, 248, 264, 373, 377
      authorization, 41, 42, 53-54, 57-58, 60-62, 63-64, 65-66, 87, 92-93, 95, 99, 145-146, 203, 210
    Senate consideration, 85
      appropriations, 212, 217, 226, 257, 389
      authorization, 39, 58, 66, 73-74, 75, 77, 145, 161-162, 166, 203, 210, 223
  contract, 108, 273-274, 404
    aeronautics, 169, 179
    communications system, 8, 258, 315
    computer services, 7, 75, 191, 200, 225, 353, 374
    engine, 42, 55, 203, 223, 257
    facilities, 40-41, 325
    fuel cell, 198
    guidance and navigation, 7, 215
    instrumentation, 220
    launch vehicle, 28, 44, 55, 153, 157, 181, 203-204, 209, 214, 223, 275, 278, 292, 386
    life-support system, 7, 132, 147
    lunar science, 2-3, 28
    nuclear propulsion, 51, 116
    space equipment, 28, 115, 116, 118, 194, 196, 209, 212, 248, 257, 349
    space shuttle, 42, 55, 157, 170, 179, 181, 198, 203-204, 215, 217, 275,

285, 292, 324, 325, 374, 386, 400, 412
space station, 171, 183, 191, 194, 195, 208, 209, 219-220, 227, 250, 264, 278, 285, 303, 386
space tug, 112, 194
spacecraft, 7, 42, 49, 115, 177-178, 179, 194, 223, 233, 243, 283, 404
  controversy, 116, 145, 146, 201, 223, 229, 230, 273-274, 289, 318
  study, 179, 183, 195, 203-204, 208, 215, 227, 275
  support services, 7, 29-30, 41, 54-55, 75, 83-84, 87, 131-132, 149, 154, 194, 195, 200, 213, 223, 264, 384, 399
telescope, 296
tracking, 265, 320, 353
cooperation, 37, 52, 97, 98, 206-207, 209, 214, 265, 290, 359-360, 366, 410, 412
AEC, 12, 75, 265, 309, 329, 411
Agriculture, Dept. of, 405
AID, 405
Coast Guard, 110, 229
DOD, 41, 53, 88, 172, 183, 265, 323, 350, 410-412
ESSA (see also NOAA), 25, 77, 320
FAA, 29, 77, 350, 412
Geological Survey, 259, 382, 405
HEW, 110, 352
HUD, 214
Interior, Dept. of, 36, 59, 107, 110, 169, 259, 274, 343
NOAA (see also ESSA), 336, 394-395, 405, 411, 412
NSF, 110
Post Office Dept., 214
Smithsonian Institution, 59, 107, 110, 234, 270
State, Dept. of, 405
Transportation, Dept. of, 36, 169, 219, 250, 350, 389, 412
USAF (see also X-24A), 16, 61-62, 95, 105, 166, 174, 180, 214, 225, 259, 290, 323, 375, 382, 390, 409, 411
USN, 110, 169, 212, 405, 409
cooperation, international. See International cooperation, space; and Sounding rocket, international programs.
criticism, 3-4, 58, 229, 287, 295-296, 306, 307, 329-330
employment, 12, 19, 36-37, 39-40, 45, 65, 91, 95, 169, 175, 184, 226, 231, 233, 236, 243, 265, 271, 274, 286, 293, 331, 362
exhibit, 27, 49, 57, 99, 102, 111, 126, 130, 139, 196, 211, 244, 265, 270, 342, 346, 401, 414
facilities, 1, 11-12, 20, 21, 29-30, 37, 39-40, 43, 62, 65-66, 93, 99, 105, 109, 133, 145, 161, 169, 171-172, 189, 195, 203, 207, 219, 227, 284, 323-324, 354, 414
fire, 371

launch, 410-411, 413
  Apollo 13 (AS-508), 119
  cosmic ray experiment, 286, 307
  failure, 380, 384
  misfire, 9
  postponed, 9, 324
  RAM C-III (RAM C-C), 318
  satellite
    Intelsat-III F-6, 14, 411
    Intelsat-III F-7, 144, 411
    Intelsat-III F-8, 241, 411
    Itos I (Tiros-M), 24-25, 411
    Nato I (NATO-A), 95, 413
    Nimbus IV (Nimbus-D), 113-114, 411
    Noaa I (Itos-A), 394-395, 411
    Ofo (Orbiting Frog Otolith), 362, 411
    RM (Radiation/Meteoroid Satellite), 362, 411
  sounding rocket, 399, 411
    Aerobee 150, 77, 91, 127, 154, 162, 170, 181, 190, 210, 245, 264, 350, 377, 384
    Aerobee 150MI, 27, 99
    Aerobee 150MII, 224
    Aerobee 170, 41, 45, 51, 77, 169, 217, 272, 311, 317, 353
    Aerobee 350, 129, 264
    Arcas, 14, 62, 71, 77, 133, 172, 208, 305, 309, 334, 338, 360, 369
    Arcas, Boosted, 102
    Black Brant IV, 315, 318
    Bullpup Cajun, 16
    Javelin, 77, 108, 283, 328
    Loki-Dart, 305, 309
    Nike-Apache, 1, 12, 13-14, 48, 56, 69, 77, 95, 102, 133, 217, 256, 260, 276, 305, 311, 375, 398, 401
    Nike-Cajun, 3, 9, 12, 13-14, 29, 77, 96, 102, 210, 213, 234, 305, 309, 336
    Nike-Iroquois, 77
    Nike-Tomahawk, 12, 16, 38, 40, 48, 71, 73, 75, 77, 82, 84, 100, 126, 250, 331, 365, 369, 372, 392, 400
    Trailblazer, 376
    Viper Dart, 305, 309
management, 316-317
manpower. See Employment.
meeting, 336, 370, 371, 373, 377, 384, 399, 405
metric system, 302
organization, 50-51, 162, 325-326, 335, 348, 352, 385, 412
patents, 232, 340, 375, 401
personnel, 45, 65, 152, 165, 170, 173-174, 185, 214, 260, 265, 274, 277, 316, 383, 397, 398, 405, 409, 412
  appointment, 1, 7, 8, 20, 26, 28, 51, 73, 75, 77, 88, 98, 142, 162, 173-174, 201, 215, 220, 223, 257, 313, 317, 325, 353, 373, 374, 404
  death, 17, 65, 308
  resignation, 98, 154, 189, 246, 248, 251, 255, 265, 303, 412

retirement, 109
procurement, 285, 306
programs, 161, 295-296, 338-339, 409, 410-414
   aeronautics, 28-29, 36-37, 39, 42-43, 58, 60-62, 63-64, 73-74, 85, 93, 148, 161, 164, 173-174, 180, 204, 220, 221, 225, 248, 257, 284, 292, 320-321, 324-326, 341-342, 348, 350, 352, 359, 373, 377, 389-390, 409-413
   astronomy, 12, 22, 26, 27, 36, 45, 51, 55, 60-62, 78-79, 84, 101-102, 116, 139, 159, 194, 203, 251, 255, 279, 284, 285, 286-287, 294-295, 297, 301, 341, 359, 361-362, 370, 379-380, 381, 383, 399, 400, 407, 411, 412-413
   communications, 39, 61, 360
   computer, 6, 106, 164-165, 180, 191, 225, 274, 310, 353, 354-355, 397
   earth resources, 30, 36, 72-73, 77, 78, 161-162, 207, 232, 233, 264, 280, 284, 293, 349, 383-384, 405, 412, 413
   manned space flight, 75, 410
      achievements, 3, 32, 50, 53-54, 105-106, 111-112, 133-134, 136-137, 164-165, 166, 204, 227-228, 238, 245, 301, 340-342, 368-369, 412-413
      criticism, 207-208, 217, 295-296, 329-330, 340-341, 360
      funding, 3, 10, 35-36, 39, 42-43, 66, 67, 86-87, 92-93, 161-162, 248, 284-285, 287
      international cooperation (see also Docking), 32, 41, 44, 83, 88-89, 156, 192, 225, 243, 287, 349, 413-414
      plans, 3, 7, 10, 11-12, 19, 41, 43-44, 50, 66-67, 69, 71, 78, 87-88, 90, 91, 101, 103-105, 167, 207-208, 218-219, 229, 233, 251, 263, 284-285, 290, 293, 301, 303, 306, 319, 323, 327-328, 331, 340-341, 342, 385, 403, 409, 410, 412
      policy, 50, 105-106, 207-208, 248, 284-285, 287-288, 295-296, 329-330, 412
   meteorology, 108, 365, 411
   nuclear propulsion, 64, 75, 411
   sounding rocket, 36, 85, 233, 381, 411, 414
   space biology, 36, 57-58, 63, 74, 88, 90, 110, 178, 186-187, 257, 266, 307, 319, 324-325, 328, 361, 366, 367, 385, 395, 403, 411
   space rescue, 327-328
   space science, 32, 35, 36, 295-296, 306, 340-342, 409
   space shuttle, 7, 18, 37, 43, 50-51, 52, 54, 67, 105, 198, 217, 219, 285, 301, 323, 367, 372-373, 386, 393, 412
      criticism, 217, 330
      design, 12, 78, 107, 179, 203, 246, 328, 374, 412
      engine, 42, 55, 74, 204, 215, 275, 313, 328, 384, 400
      funds, 35-36, 37, 43-44, 166, 207-208, 217, 340-341, 389
      international cooperation, 41, 52, 88, 227, 292, 313, 323, 328, 351, 361
      meeting, 89, 220, 234, 246, 384, 400
      test, 182-183, 197, 301
   space station, 10, 11-12, 29, 37, 39, 50, 52, 55, 62, 87-88, 118, 172, 181, 191, 194, 196, 209, 227, 232, 246, 251, 257, 265, 285, 295, 303, 307, 310, 361, 372-373, 385, 386, 412
      criticism, 217, 295-296, 330
      design, 67, 104, 169, 183, 208, 218, 264, 273, 292, 328, 367-368, 370, 372, 412
      funds, 35-36, 166, 207-208, 217, 229, 284, 389
      international cooperation, 43, 55, 88, 192-193, 212, 225, 292, 295, 305, 327, 346, 351
      meeting, 89, 267, 295, 371, 385
      plans, 10, 284-285, 293, 301, 323
      test, 201, 231, 266, 275-276, 387
   space tug, 50, 112, 174, 181, 194, 248, 301, 307, 323
   technology utilization. See Technology utilization, space.
   tracking and data acquisition (see also Tracking), 65-66, 162, 199, 207, 210, 213, 284, 381
   test
      aircraft, 58, 359, 377, 411
      ion engine, 38, 51, 173, 243, 285, 315, 338, 393
      launch vehicle, 25, 44, 85, 157, 251, 318
      lifting-body vehicle, 18, 27, 55, 66, 74, 88-89, 94, 108, 145, 171, 178, 190, 199, 206, 220, 236, 247, 263, 273, 336, 347, 357, 375, 379, 411
      lunar roving vehicle, 276, 397, 404
      nuclear, 64
      spacecraft, 41, 74, 76, 84, 87-88, 89, 161, 165, 173, 197, 201, 205, 220, 276, 297, 306, 334, 387, 403, 404
      X-15. See X-15.
   universities, 36, 47, 66, 79, 108, 112, 144, 204, 216, 231, 234, 256, 322
National Aeronautics and Space Council (NASC), 94, 152, 227, 305, 314, 409
National Air and Space Museum, 55, 94, 234, 240, 349, 379-380, 384
National Air Pollution Control Administration, 260
National Archives and Records Service, 255
National Assn. of Home Builders, 214
National Audio-visual Center, 255
National Aviation Club, 113, 166
National Bison Range, 52
National Broadcasting Co. (NBC), 137

National Bureau of Standards, 169, 406
National Center for Atmospheric Research (NCAR), 79, 230, 286, 306, 336, 350
National Civil Service League, 95
National Commission on Productivity, 298
National Cosmonauts Day, 126, 127
National Data Buoy Development Project, 229, 230, 330
National Day of Prayer, 133, 138, 139
National Defense Research Committee, 67
National Elk Refuge, 52
National Environmental Policy Act of 1969, 1, 10, 75, 389, 402
National Environmental Satellite Center, 230
National Exhibits, Inc., 211
National Geographic Society, 52, 155
National Goals Research Staff, 237
National Highway Safety Bureau, 341
National Hurricane Center, 166, 264
National Industrial Pollution Control Council, 116
National Institutes of Health (NIH), 68
National Institutes of Research and Advanced Studies (NIRAS) (proposed), 132, 337
The National Institutes of Research and Advanced Studies: A Recommendation for Centralization of Federal Science Responsibilities, 132
National Laboratory for Environmental Sciences (proposed), 29
National League of Cities Conference, 391
National Ocean Survey, 347
National Oceanic and Atmospheric Administration (NOAA) (see also Environmental Science Services Administration), 229, 317, 324, 325, 330, 347, 351, 354, 390, 394, 405, 411, 412
National Oceanographic Data Center, 230
National Oceanographic Instrumentation Center, 230
National Operational Meteorological Satellite System (NOMSS), 25, 244
National Patterns of R&D Resources: Funds and Manpower in the United States, 1953-71 (NSF 70-46), 410
National Portrait Gallery, 274
National Press Club, 157, 239
National Radio Astronomy Observatory, 110, 204, 262, 325
National Research Council (NRC), 177
  Space Science Board, 1, 184, 263, 284, 304, 395
National Science Board, 57, 129
National Science Foundation (NSF), 45, 110, 132, 147, 187, 230, 260, 277, 303, 335, 357, 376, 394, 410
  annual report to Congress, 53
  cooperation, 110
  Federal Support for Research and Development at Universities and Colleges and Selected Nonprofit Institutions, FY 1968, 98

Federal Funds for Research, Development, and Other Scientific Activities, Fiscal Years 1968, 1969, and 1970, 71-72
  funds, 35-36, 44, 117, 167, 223, 244, 247, 257
  grants, 101, 173, 221, 329, 394
  Impact Changes in Federal Funding Patterns on Academic Institutions, 337
  National Patterns of R&D Resources: Funds and Manpower in the United States, 1953-71, 410
  personnel, 67, 102, 265, 267-268
  Research and Development in Industry, 1968: Funds, 1968; Scientists & Engineers, January 1969, 253
National Science Medal, 55
National Scientific Balloon Flight Station, 101
National Sea Grant Program, 230
National security, 39, 240, 311
National Severe Storms Forecast Center, 165
National Space Club, 94, 270, 349, 384
National Space Hall of Fame, 6
National Space Program—Present and Future (House document), 393
National Telemetry Conference, 166
National Telemetry Man of the Year Award, 166
National Transportation Safety Board, 278, 324, 349
NATO. See North Atlantic Treaty Organization.
Nato I (NATO-A) (military comsat), 95, 198, 346, 360, 413
NATO-B, 408
Naugle, Dr. John E., 60, 409
Naval Air Development Center (NADC), 321
Naval Air Reserve, 110
Naval Oceanographic Office, 405
Naval Radio Receiving Station, 293-294
Naval Research Laboratory (NRL), 91, 102, 197-198, 264, 272, 377
Navigation satellite, 43-44, 77, 275, 327
Navy League, 177
NBC. See National Broadcasting Co.
NBS. See Neutral Buoyancy Simulator.
NCAR. See National Center for Atmospheric Research.
Nebraska Wesleyan Univ., 164
Neel, Louis E., 348
Neilon, John J., 201
Nein, Max E., 314
Nellis AFB, Nev., 100
Nelson, Clifford H., 325
Nelson, Sen. Gaylord, 390
Neptune (planet), 33, 84
Neptunium, 164
NERVA. See Nuclear engine for rocket vehicle application.
Netherlands, 149, 194, 291, 413
Neumann, Gerhard, 19
Neutral Buoyancy Simulator (NBS), 370
Nevada, 402

New Brunswick, Univ. of, 172
New Delhi, India, 233, 245, 376
New Hampshire, Univ. of, 250, 360
New Horizons in Science, Eighth Annual Briefing, 373
New London, Conn., 292
New Mexico State Univ., 172, 330
New Orleans, La., 37, 68, 189, 266, 271
New York, N.Y., 272
 air services and traffic, 13, 22, 198, 263, 286, 345
 Apollo 13 mission reaction 135
 astronauts in, 196, 238
 exhibit, 196
 meeting, 146, 162, 198, 302, 329, 391
 national day of prayer, 139
 solar eclipse, 82
 train services, 335
New York State Medal for Valor, 29
New York State Museum, 139
New York State Supreme Court Appellate Div., 360
New York, Univ. of at Stoney Brook, 406
Newark Airport, N.J., 345
Newell, Dr. Homer E., 227
Newfoundland, 310, 312, 324
Newkirk, Dr. Gordon A., Jr., 79
Newport News, Va., 108
Newmann, Gerhard, 159
Niagara Falls, N.Y., 200
Nicks, Oran W., 26, 63, 73, 313
NIH. See National Institutes of Health.
Nike (booster), 41
Nike-Apache (sounding rocket)
 airglow experiment, 311
 auroral data, 56, 69
 cloud study, 260
 electron measurement, 95
 ionospheric experiments, 1, 48, 95, 102, 133
 magnetic field measurement, 102
 solar eclipse data, 77
 upper atmosphere data, 12, 13, 256, 305, 309, 375, 398, 401
 x-ray astronomy, 217, 276
Nike-Cajun (sounding rocket), 326, 411
 grenade experiment, 3, 9, 12, 29, 38, 102, 210, 213
 solar eclipse data, 77
 trimethylaluminum experiment, 13-14
 upper-atmosphere data, 96, 210, 213, 234, 305, 309, 336-337
Nike-Iroquois (sounding rocket), 77
Nike-Tomahawk (sounding rocket)
 auroral data, 12, 16, 38, 40, 48, 71, 73, 75, 82, 100, 126, 365-366, 372
 electric fields measurement, 71, 73, 75, 331
 energetic particle data, 250
 geomagnetosphere data, 331
 instrumentation test, 392, 400
 ionospheric experiment, 48, 84
 leech experiment, 92, 400
 solar eclipse data, 77
 upper-atmosphere data, 77

Nikolayev, Andrian G., 196
 awards and honors, 267, 271
 Soyuz IX mission, 189-190, 201, 211, 225, 327
 visit to U.S., 336, 338, 340, 344, 346, 349
Nikolayeva-Tereshkova, Maj. Valentina (U.S.S.R.), 110, 196, 267
Nimbus (program), 107, 113-114, 156, 223
Nimbus III (meteorological satellite), 52, 57, 59, 62, 107, 114, 210, 305, 309
Nimbus IV, 113-114, 118, 210, 239, 305, 309, 334, 338, 343, 365, 369, 411
Nimbus E, 156, 223
Nimbus F, 223
NIRAS. See National Institutes of Research and Advanced Studies.
Nitrogen, 333-334
Nixon, President Richard M., 235, 277
 Advisory Council on Management Improvement, 49
 Aeronautics and Space Report, 39
 Apollo 13 mission, 136-137
  astronauts
   awards, 137, 140
   communications with, 118, 133-135
   tribute, 133-134, 135, 137
   White House dinner, 198
  briefing, 128, 134
  messages regarding, 130, 135, 137
  National Day of Prayer, 133, 138, 139
 appointments and nominations by, 7, 28, 30, 40, 65, 89-90, 99, 102, 129, 185, 198, 215, 240, 250, 260, 265, 271, 273, 288, 298, 316, 345-346, 361, 386, 398
 astronaut goodwill tour, 51
 astronaut promotion, 198, 398
 astronauts, visit with, 139-140
 Aviation Advisory Commission, 386
 awards by, 55, 67, 136-137, 229
 bills signed, 75, 116, 178, 224, 244, 288, 314, 330, 401
 budget, 35-36, 44, 45, 53-54, 87, 109, 237
 communications satellite, 25-26, 28, 78
 Communications Satellite Act of 1962, report, 65
 Defense, Dept. of, 223-224, 246
 Electronics Research Center (ERC), 13, 99
 environmental quality program, 1, 10, 30, 47, 262
 Federal Economy Act of 1970, 65
 hijacking of aircraft, 235, 265, 291, 297, 300, 302, 304, 310, 336
 International Air Policy Statement, 210-211
 lunar landing anniversary statement, 238, 243
 marine resources and engineering development report, 127
 NASA appropriations, 257, 263, 264, 309-310, 373, 401, 412-413
 National Environmental Policy Act, 1, 10
 oceans policy, 179-180
 Paine, Dr. Thomas O., resignation, 246

The Physical Sciences, National Science Board report, 57
reorganization plans, 45-46, 229-230, 235, 316, 317
reports transmitted to Congress, 39, 57, 258, 262
resignations accepted by, 246, 267, 269
SALT message, 133
science, 53, 57, 302, 357
space program, national, 10, 35, 77-78, 82, 83, 84, 86, 90, 91, 134, 148, 150, 237, 246-247, 248, 290-291, 316-317, 397
State of the Union message, 22-23, 27
supersonic transport, 272, 387-388, 393
technology, 297-298, 357
Turbotrain demonstration experiment, 335
U.N. anniversary statement, 345
Vietnam war, 157
violence, 304
Wright Brothers Day, 394
Nixon, Mrs. Richard M., 137, 139, 198
Nizer, Louis, 314
NOA: new obligational authority (in budget).
NOAA. See National Oceanic and Atmospheric Administration.
Noaa I (ITOS-A) (National Oceanic and Atmospheric Administration meteorological satellite), 394-395, 411
Nobel, Alfred B., Prize in Chemistry, 348
Nobel, Alfred B., Prize in Physics, 4, 348, 359
Noerdlingen Crater, West Germany, 264, 266
Noise, aircraft, 29, 73-74, 133, 170-171, 178, 225, 239, 256, 261, 279, 291, 296, 304, 311, 344-345, 353, 366, 388, 389, 400, 411
Nome, Alaska, 38
NOMSS. See National Operational Meteorological Satellite System.
Noonan, Fred, 364
NORAD. See North American Air Defense Command.
Normyle, William J., 19, 246
North Africa, 394
North American Air Defense Command (NORAD), 147, 150
North American Rockwell Corp (NR)
Aerospace Div., 233
aircraft, 195, 221, 359, 377
Apollo spacecraft, 106, 125, 137, 201-203, 213, 249
contract, 153, 170, 179, 195, 213, 221, 249, 278, 374
cosmonaut visit, 106
employment, 98, 174-175, 221, 271
personnel, 7, 57, 116, 311
polyurethane foam, 27
Rocketdyne Div., 42, 55, 157, 164, 223
rotational machine, 266
Saturn V, 153, 179, 271, 278
Space Div., 56, 108, 170, 179, 183, 199, 208, 264, 292, 295

space shuttle, 7, 170, 179, 374, 399
space station, 183, 266
North Atlantic Treaty Organization (NATO), 44, 95, 293, 300, 358, 360, 408, 410, 413
North Dakota, 181
North Pole, 291, 380
Northern Research and Engineering Corp., 228
Northrop Corp., 233, 352
Aircraft Div., 313
Norway, 38, 51, 82, 291
Novosibirsk, U.S.S.R., 185, 195
Now It Can Be Told, 232
NR. See North American Rockwell Corp.
NRC. See National Research Council.
NRL. See Naval Research Laboratory.
NSF. See National Science Foundation.
Nuclear accelerator, 225, 235
Nuclear detection satellite (see also Vela), 114
Nuclear engine for rocket vehicle application (NERVA), 12, 36, 43, 64, 75, 116, 277, 329, 411
Nuclear nonproliferation treaty, 40, 75
Nuclear power, 10, 67, 69, 75, 225, 235, 265, 277, 309, 329, 411
Nuclear propulsion (see also Nuclear engine for rocket vehicle application), 10, 31, 64, 75, 277, 309
Nuclear reactor, 10, 265, 309, 411
Nuclear rocket test, 402
Nuclear submarine, 244, 245
Nuclear test ban treaty, 44, 186
Nuclear weapons, 235, 371
Nuñez, A. Carlos, 130

O

Oahu, Hawaii, 196
Oak Ridge, Tenn., 232
OAO. See Orbiting Astronomical Observatory.
Oao II (Orbiting Astronomical Observatory), 22, 380-381
OAO-B, 380-381, 384
OAO-C, 409
OART. See NASA Office of Advanced Research and Technology.
Oban, Scotland, 283, 291
Oberth, Prof. Hermann, 92
Ocean of Storms (Oceanus Procellarum) (moon), 76, 143, 177, 180, 271, 333, 369
Ocean Station Papa, Canada, 272
Ocean Station Vessel Meteorological Program, 330
Oceanographic Instrumentation Center (DOD), 330
Oceanography (see also Aegir [underwater habitat]; Aquanaut; Marine pollution; and Project Tektite I and II)
cooperation, 110, 179-180
international aspects, 82, 179-180, 283-284, 371
research, 99, 110, 118-119, 216, 226, 228, 239, 272, 308, 394

U. S. program, 110, 127, 158, 179, 216, 229-230, 317, 324, 325, 330, 347
Oceanus Procellarum. See Ocean of Storms (moon).
O'Connor, M/G Edmund F. (USAF), 227
Odessa, Tex., 230
Odom, Jeffrey V., 406
Odyssey (Apollo 13 CM). See Command module.
Office of Science and Technology (OST) (President's), 20, 87-88, 132, 213, 228, 276-277, 300, 357
   director
      appointment, 302, 303, 345-346
      resignation, 267, 269, 279
Office of Scientific Research and Development, 67
Office of Technology Assessment (proposed), 234
Office of Telecommunications Policy, 45-46, 215, 288-289
Ofo (Orbiting Frog Otolith), 324-325, 362, 367, 395, 411
Ogo I (Orbiting Geophysical Observatory), 236
Ogo II, 152
Ogo III, 236
Ogo V, 116
O'Hair, Mrs. Madalyn Murray, 52-53, 113
O'Hare International Airport, 162
Ohio Historical Society, 153
Ohio State Univ., 131
Ohsumi (Lambda 45-5) (Japanese satellite), 48-49, 50, 316, 412
OIL. See Orbiting International Laboratory.
O'Keefe, Dr. John A., 143, 195, 333
O'Leary, Dr. Brian T., 287
Oliver, Ruth Law, 383
Olsen, John P., 153
Olsen, Dr. Walter T., 21
Omar Khayyam, 268
OMSF. See NASA Office of Manned Space Flight.
O'Neill, L/G John W. (USAF), 100, 192
Operation Overflight, 164
Oppenheimer, Dr. J. Robert, 268
Orbital workshop. See Skylab.
Orbiting Astronomical Observatory (OAO), 77, 225, 380
Orbiting International Laboratory (OIL), 327
Orbiting Solar Observatory (OSO), 43, 77, 404-405
Oregon, 283
Oregon, Univ. of, 333
Osaka, Japan, 99
Osborn, G. H., 328
Oscar V (Oscar A) (U.S.-Australia amateur radio-propagation satellite), 24, 411, 413
Oscar XVIII (navigation satellite), 275
Oscar XIX, 275
OSO. See Orbiting Solar Observatory.
Oso II (Orbiting Solar Observatory), 276
Oso III, 236
Oso IV, 236
Oso V, 79
Oso VI, 79
OSO-H, 409
OSO-I, 51-52, 405, 413
OSO-J, 405
OSO-K, 405
OSSA. See NASA Office of Space Science and Applications.
OST. See Office of Science and Technology (President's).
Oswald, Lee Harvey, 164
OTDA. See NASA Office of Tracking and Data Acquisition.
O'Toole, Thomas, 10, 94, 229, 316
"Our World in Space—Yesterday, Today and Tomorrow" (lecture), 349
Outstanding Leadership Medal (NASA), 369
OV I-20 (Orbiting Vehicle Research Satellite), 250
OV I-21, 250
Owen, Thomas B., 102
Oxygen, 119, 121, 216, 218-219
Oxygen, liquid, 92, 137
Ozone, 96, 172, 208, 334, 338, 369, 406

P

Pacific Missile Range (PMR), 172, 208, 233
   Space Fair, 338
Pacific Northwest Laboratory (AEC), 30, 94, 228
Pacific Ocean, 82, 360, 380
   Apollo 13, 119, 125, 133
   Apollo 14, 331
   Ats I, 165
   Intelsat-III F-8, 241
   stellar x-ray test, 315
   U.S. missile test, 276, 405
   U.S.S.R. missile and rocket test, 97, 103, 105, 237-238, 272, 349
Packard, David M., 72, 183, 195, 408
Packard, Robert F., 409
PAET (Planetary Atmosphere Experiments Test), 409
Pageos I (Passive Geodetic Earth-Orbiting Satellite), 242
Paine, Dr. Thomas O., 55, 98, 117, 150, 178, 223, 382
   aeronautics, 58, 85
   Apollo 11 mission, 238
   Apollo 13 mission, 119, 125, 127-128, 134, 135, 147, 161, 213, 217, 218-219, 232
   Apollo 14 mission, 233, 284
   Apollo 15 mission, 284-285
   Apollo 16 mission, 284-285
   Apollo 17 mission, 284-285
   Apollo 18 mission, 285
   Apollo 19 mission, 284
   appointment, 265
   appointments by, 26, 28, 88, 289
   ATS-F and ATS-G contract, 201, 224, 230, 289

awards and honors, 45, 229, 303, 368
budget, 11-12, 36-37, 39-40, 47, 58, 175, 284-285, 303
Communist China satellite launch, 150
Electronics Research Center (ERC), 324
Goddard Rocket and Space Museum, 92
international cooperation, space, 62, 78-79, 85-86, 193-194, 300, 335, 393
Jet Propulsion Laboratory (JPL), 39
Lewis Research Center (LeRC), 40
Lunar Science Institute dedication, 2-3
nuclear propulsion, 154
resignation, 246, 248, 251, 255, 277, 303, 412
safety, 269
space program, national, 11-12, 78-79, 111-112, 164, 226, 284-285, 301, 303
space shuttle, 183, 284-285, 301, 303
space station, 193, 284-285, 301, 303
space transportation system (STS), 53, 301
U.S.S.R. space program, 226, 301, 303
White Sands, N. Mex., Test Facility, 62
Pakistan, 151, 330
Palestine, Tex., 101, 230
Pallas (asteroid), 210
Palmdale, Calif., 370
Palo Alto, Calif., 102, 107
Pan American World Airways, Inc. (Pan Am), 13, 15, 22, 41, 198, 236, 257, 263, 349, 364, 389
Panofsky, Wolfgang K. H., 55
Papanastassiou, Dr. D. A., 143
Papers of Robert H. Goddard, 256
Parachute, retrorocket, 103
Paris, France, 128, 193, 400
Parker, Robert A., 101
Parkin, Curtis W., 271
Parsons, Ralph M., Co., 325
Particles, energetic, 315, 318
Pasadena, Calif., 39
Patent, 12, 47, 232, 240, 293, 340, 375, 401
Patrick AFB, Fla., 41, 58, 108, 236, 319
Paul, Hans G., 206
Paul VI, Pope, 128
Paul, Prof. Rodman W., 189
Pearson, Lester, 304
Peer, William, 274
Peewee-2 (nuclear reactor), 411
Peking, Communist China, 150
Pendray, G. Edward, 256
Pendray, G. Edward, Award, 20
Pennsylvania, 18
Pennsylvania Highway Dept., 330
Pennsylvania, Univ. of, 55
Pentagon, 42
Peole (French satellite), 396-397, 412
People mover, 408
Perchard, Robert J., 12
Percy, Sen. Charles H., 74
Perkins, Courtland D., 252
Perkins, Kendall, 159
Perry, Robert L., 189
Perseus (constellation), 21
Peru, 231, 259
Pěsek, Prof. R., 159

Peshawar, Pakistan, 330
Petersen, Richard H., 220
Peterson, Rudolph A., 240
Petrone, Dr. Rocco A., 147-148, 217, 219, 323, 371
Petrov, Prof. Boris N., 19, 171, 355
Petrov, Georgy I., 312
Pettis, Rep. Jerry L., 157, 167
PFLP. See Popular Front for Liberation of Palestine.
Philadelphia, Pa., 18, 173, 224, 313, 321, 347
Philco-Ford Corp., 105, 243, 265
Philip, Prince (U.K.), 51
Philippines, 157
Philips, Ronald, 195
Phillips, L/G Samuel C. (USAF), 43-44, 311
Phobos (Martian moon), 172
Phoenix, Ariz., 101
Photography (see also Moon, photographs; Earth, photographs), 118, 196-197, 200, 346
Photopolarimeter, 45
The Physical Sciences (National Science Board report), 57
Pickering, Dr. William H., 66, 103, 126
Picturephone, 375
Pierce, William, 274
Pimentel, Dr. Goerge C., 1
Pinkel, I. Irving, 283
Pioneer (program), 66, 284
Pioneer VI (interplanetary probe), 49, 400, 407
Pioneer VII, 49
Pioneer VIII, 49, 407
Pioneer IX, 49, 166
Pioneer F, 45, 49, 66
Pioneer G, 45, 49, 66
Piper Cub (light aircraft), 17
Piper Twin Comanche B (light aircraft), 3
Piper, William T., 17
Pittsburgh, Pa., 18, 347
Pittsburgh, Univ. of, 100, 375
Pitzer, Dr. Kenneth S., 214
PLACE. See Position-location and aircraft communications experiment.
Pleasanton, Kans., 2
Plesetsk, U.S.S.R.
Cosmos launch, 15, 20, 21, 94, 102, 109, 113, 126, 206, 229, 291, 304, 311, 331, 334, 337, 352, 370, 376, 384, 385, 392, 397, 399, 402
Meteor IV launch, 154
Meteor VI launch, 337
Molniya I-13 launch, 56
Molniya I-17 launch, 405
Plohr, H. Warren, 384
Plowshare, Project, 235
PLSS. See Portable life support system.
Plum Brook Station (LeRC), 240, 403
Pluto (planet), 33, 84
Plutonium, 30, 125, 329
PMR. See Pacific Missile Range.
pndb: perceived noise in decibels.
Podell, Rep. Bertram L., 390
Point Barrow, Alaska, 9, 29, 38

Point Mugu, Calif., 338
Poland, 57, 209, 380
Polaris (missile), 180, 358
Polhamus, Edward C., 113
Pollution (see also Air pollution), 10, 35, 42, 44-45, 47, 49, 64, 72-73, 86, 99, 106, 116, 166, 178, 216, 221, 228, 244, 306, 361, 411
Pompidou, Georges J., 130
Ponnamperuma, Dr. Cyril A., 387
Pope, Alan Y., 21
Popular Front for Liberation of Palestine (PFLP), 289, 293, 297, 300, 304-305, 306, 316
Portable life support system (PLSS), 7, 387
Porter, Dwight J., 298
Porter, Rufus, 277
Portugal, 21
Poseidon (missile), 206, 244, 245, 256, 258, 289
Position-location and aircraft communications experiment (PLACE), 367
Post Office Dept., 214
Post, Wiley, 110, 401
Powers, Francis Gary, 164
Prague, Czechoslovakia, 15
Prairie Network, 2, 14
Pratt & Whitney Div., United Aircraft Corp., 42, 55, 68, 118, 157, 198, 278, 293
Pratt, Judge John H., 272, 391
Pratt, Kans., 276
PRC. See ATS Procurement Review Committee.
Presidential Awards Dinner, 310
Presidential Mangement Improvement Award, 310
Presidential Medal of Freedom, 136-137, 368
President's Commission on the Assassination of President Kennedy, 164
President's Council of Economic Advisers, 109
President's Council on Environmental Quality. See Council on Environmental Quality.
President's Safety Award, 229, 269
President's Science Advisory Committee, 287
  Space Science and Technology Panel, 87, 89-90
President's Space Task Group, 1, 112-113
President's Task Force on Science Policy, 158
Press Comment
  aerospace technology funding, 50, 211
  Apollo program, 3-4, 86, 130-131, 287, 290, 294, 306
  Apollo 13 mission
    accident, 130-131, 135-136, 138, 140, 186
    foreign, 128-129, 135-136, 150
    U.S., 46, 111, 116, 118, 126, 127, 129, 135-136, 138, 140, 141-142, 145, 165, 232
  atomic bomb anniversary, 235, 260
  Blue Ribbon Defense Panel, 257
  Boeing 747 (jet passenger transport), 21
  C-5A (cargo aircraft), 270-271
  Cape Kennedy, 45
  communications satellite, 28, 375-376
  Communist China, 165, 176
  Communist China satellite, 136, 150, 153-155
  David, Dr. Edward E., Jr., 271
  Dubridge, Dr. Lee A., 279
  Electronics Research Center (ERC) (NASA), 105
  energy shortage, 316
  environment, 15-16
  Federal budget, 44, 109
  Grand Tour, 84
  Groves, L/G Leslie R. (USA, Ret.), 235-236
  hijacking of aircraft, 300, 304
  hydrogen bomb, 258
  Illinois, Univ. of, antiwar demonstration, 172
  international cooperation, space, 83, 255-256, 287, 343-344, 353, 358
  Lockheed Aircraft Corp., 397
  Lovell, Capt. James A., Jr. (USN), 46
  Luna XVI, 314, 318, 375
  Luna XVII, 373, 374
  lunar landing anniversary, 238-239
  lunar rock samples, 184
  Lunar Science Conference, 9
  Mars, manned exploration, 10, 50, 69, 89
  metric system, 406-407
  NASA Administrator, 255, 397
  1960's decade, 32-33
  nuclear test ban treaty, 186
  Paine, Dr. Thomas O., resignation, 251, 255, 397
  relativity theory, 377-379
  R&D, 44, 109, 157
  solar eclipse, 82
  space program, national, 9, 10, 17, 19, 50, 69, 82-84, 86, 90-91, 150, 232, 238-239, 251, 256-257, 271, 287-288, 290, 294
  space shuttle, 50, 58, 91, 391
  space station, 10, 83, 318
  State of the Union message, 27
  supersonic transport (SST), 17-18, 113, 272, 326, 387, 391
  Tektite II, 118
  television via satellite, 257-258
  UFOs, 3-4
  USAF satellite, 278
  U.S., U.S.S.R. exchange of scientists, 225
  U.S.S.R. space program, 193
  Venus VII, 273
Press Conference
  antiballistic missile (ABM) system, 266
  Apollo 12 mission, 371-372
  Apollo 13 mission, 40, 62-63, 71, 90-91, 117-118, 127-128, 132-133, 134, 143, 161
  Apollo 14 mission, 233, 338, 403
  F-111 (supersonic fighter), 75-76

germ warfare, 392
Glomar Challenger voyage, 394
international cooperation, space, 212, 335
Jet Propulsion Laboratory (JPL), 39-40
laser, military use, 156
Lewis Research Center (LeRC), 40
Lovell, Capt. James A., Jr. (USN), 40
Luna XVI mission, 349
manned space flight, 212, 227-228, 238, 330
meteorite, 12, 14
moon landing, manned, 227-228, 238
NASA FY 1971 budget, 36-37
Paine, Dr. Thomas O., on resignation, 248
radiation, infrared, 152
Skylab, 104-105, 218
Soyuz (U.S.S.R. spacecraft), 196, 212
Soyuz IX mission, 212
space program, national, 11-12, 78-79, 88, 103-105, 284-285, 292, 323
space shuttle, 228
SS-11 (U.S.S.R. ICBM), 274
Staran IV, 204
supersonic transport (SST), 42, 179, 239, 393
Uhuru (Explorer XLII), 407
U.S.S.R. missile program 230, 275
U.S.S.R. space program, 229, 340, 349, 357-358
Press, Frank, 129
Pretoria, South Africa, 113
Primrose Lake, Canada, 338
Princeton Univ., 22, 27, 55, 89, 101, 192, 384
  School of Engineering and Applied Sciences, 252
Prinn, Ronald G., 251
Priorities in Applied Research: An Initial Appraisal (NAE report), 221
Prisoners of War mission, 260
Probe (see also individual probes, such as *Mariner VI, Mariner VII, Venus IV, Venus V, Venus VI,* and *Venus VII*)
  interplanetary, 8, 32, 33, 36, 43, 49, 60, 78, 84, 85, 180, 284, 301, 341, 343, 383, 412
  Jupiter, 12, 45, 49, 60, 75, 84, 284, 341
  lunar. See *Luna XV, Luna XVI, Luna XVII, Surveyor VI, Zond V, Zond VI, Zond VII,* and *Zond VIII.*
  Mars, 11-12, 16, 43, 60-61, 62, 69, 75, 89, 172, 185, 247, 284, 301, 370, 377-379, 412
  Mercury, 12, 60, 247, 284
  Neptune, 84
  Pluto, 84
  Saturn, 84
  sun, 60, 361-362
  Uranus, 84
  U.S.S.R., 266, 286-287, 289, 299, 364-365, 379, 398
  Venus, 12, 247, 284, 286-287, 301, 379, 398

Professional Air Traffic Controllers Organization, 315
Propulsion, 373
Propulsion Specialist Conference, 6th, 206
Prouty, Sen. Winston L., 44
Providence, R.I., 90
Proximity gauge, 330
Proxmire, Sen. William, 217, 226, 267, 330
Puerto Rico, 257, 259, 394
Puerto Rico Trench, 212
Pulsar, 159
Puppi, Prof. Giampietro, 88
Purdue Univ., 8, 55, 94, 290, 360
Purser, Paul E., 109

Q

Quadra (weather ship), 272
Quaide, Dr. William L., 20
Quarantine. 7, 15, 16, 55, 71
Quasar, 96-97
Quebec, Canada, 247

R

RA II (papyrus boat), 232
Radar
  air traffic control, 28, 68, 171, 341
  aircraft, 94
  contract, 227
  mapping use, 236
  marine use, 341
  military use, 68, 232
  tracking, 42, 133, 139, 197-198
  U.S.S.R., 146, 164, 232
  weather use, 171, 272
Radiation (see also Ultraviolet and X-ray)
  contract, 227
  cosmic, 152, 227
  effects, 49, 227, 242-243, 304, 327, 390
  gamma, 325
  infrared, 152, 251
  lunar, 325
  measurement, 44, 239, 362
  protection from, 235
Radiation/Meteoroid Satellite (*RM*), 325, 362, 411
Radio Amateur Satellite Corp., 24
Radio Corporation of America (RCA), 116, 315, 349
  Astro-Electronics Div., 258
  Defense Electronics Products Div., 173
Radio interference, 180
Radio signal, 170
Radioastronomy, 251, 377-379
Radioisotope thermoelectric generator (RTG), 329-341
Radioisotopes, 10
Radiometer, 114, 220
Radiotelescope, 262
Rae, Prof. John B., 189
RAF. See Royal Air Force.
Rakiety Sondujace Atmosfere, 76
Raleigh, N.C., 373

RAM. See Research and Applications Module.
RAM C-I (Radio Attenuation Measurement space craft), 318
RAM C-II, 318
RAM C-III (RAM-C-C), 318, 381
Ramaty, Dr. Reuven, 152
Ramdohr, Paul, 289
RAND Corp., 189
Rand McNally & Co., 382
R&D. See Research and development.
R&D Plan To Increase Airport System Capacity, 218
Rankine cycle, 334
Rasquin, John R., 401
Rauschenberg, Robert, 346
Raybestos-Manhattan, Inc., 9
Raytheon Co., 7
RB-57F (reconnaissance aircraft), 183, 259, 290
RCA. See Radio Corp. of America.
RCA Canada, Ltd., 247
RCA Global Communications, Inc., 38-39
RCS. See Reaction control system.
RDT&E (research, development, test, and engineering). See Research and development.
Rea, Dr. Donald G., 77
Reaction control system, 125
Reading Univ., Scotland, 21
RECON (remote control) system, 178
Reconnaissance satellite, 278
Record, 111-112, 346
   aircraft, 7, 175, 367, 383
   helicopter, intercity, 198
   spacecraft, 108, 190, 203, 231, 401
Recruit (rocket), 332
Reed, David, 2
Reed, Sylvanus Albert, Award, 19
Reentry vehicle, 182-183, 198, 241, 333, 381
Rees, Dr. Eberhard F. M., 28, 174, 204, 265
Reference Earth Orbital Research and Applications (NASA planning guidebook), 385
Refset L-3203-6, 9
Regina, Saskatchewan, 307
Rego, J. Dominguez, 293
Regueiro, Enrique G., 108
Reiger, Siegfried H., 233
Reinhold, Dr. Karl, 313
Relativity, theory of, 157, 170, 201, 377-379
Rendezvous, 347, 351, 391-392, 413
Report to the President and the Secretary of Defense on the Department of Defense, 223
Republic Steel Corp., 377
Rescue of astronaut. See Space rescue.
Research and Applications Module (RAM), 386
Research and development (R&D), 136, 269-270, 272, 280, 394
   aeronautics (see also Aeronautics, research), 98, 99, 109, 250, 320, 321, 412
   benefits, 167, 258
   employment, 187, 362
   Federal support, 44, 45, 47, 63-65, 67, 71-72, 98, 153, 157-158, 187, 207-208, 221, 228, 239-240, 253, 265, 278-279, 322, 357, 362, 376, 410
   funds, 109, 410
      AEC, 72, 98, 362
      Agriculture, Dept. of, 98
      DOD, 72, 98, 228, 239, 322, 362, 410
      HEW, 322
      NASA, 47, 72, 92-93, 98, 161-162, 226, 239-240, 246, 322, 362, 373
Research and Development in Industry, 1968: Funds, 1968; Scientists & Engineers, January 1969 (NSF 70-29), 253
Research and Development in State Government Agencies, Fiscal Years 1967 & 1968 (NSF 70-22), 187
Research and Technology: Objective and Plan (NASA N-7029204), 230
Resident Research Associateship Program, 108
Resources for Scientific Activities at Universities and Colleges, 1969 (NSF 70-16), 187
Reuss, Rep. Henry S., 174
RF: radio frequency.
RFP: request for proposals.
Rhein Main Air Base, West Germany, 243
Rice Univ., 3, 179, 251
Richardson, Herbert W., 271
Richland, Wash., 18, 30, 228
al-Rifai, Premier Abdel Moneim (Jordan), 304-305
Rimae Parry (moon), 192
Ringwood, Dr. Alfred E., 295
Ripley, Anthony, 235
Rippon, Minister of Technology Geoffrey (U.K.), 226
Ritchie-Calder, Lord, 33
RM (Radiation/Meteoroid Satellite), 325, 362, 411
Roach, Dr. Franklin E., 358
Roberts, Chalmers M., 258
Roberts, Dr. Walter Orr, 49
Robertson, Reuben B., III, 324
Robinson, Dr. George S., Jr., 355
Rochester, Univ. of, 90, 376
Rockefeller, John D., III, 273
Rockefeller, Gov. Nelson A., 139
Rockefeller Univ., 329
Rocketdyne Div., North American Rockwell Corp., 42, 55
Rockwell, Willard F., Jr., 57, 98, 116
Rogers Dry Lake, Calif., 95
Rogers, Secretary of State William P., 291, 292
Rohini (Indian rocket), 18, 185
Rohr Corp., 266
Rolls-Royce, Ltd., Bristol Engine Div., 286
Romania, 151

Rome Air Development Center, N.Y., 148, 200
Rome, Italy, 196
Rome, Univ. of, Centro Ricerche Aerospaziali, 275, 395
Roncho Los Amigos Hospital, Downey, Calif., 350
Roosa, Maj. Stuart A. (USAF), 167, 264, 364
Roosevelt Field, N.Y., 379
Roosevelt, President Franklin D., 308
Rosamond, Calif., 18, 27
ROSEAU (Radio Observation par Satellite Excentrique à Automatisme Unique) (French satellite), 177
Rosenberg, Dr. Paul, 159
Ross, Dr. Robert S., 99
Roswell, N. Mex., 13, 92
Rothberg, Donald, 242
Royal Air Force (RAF), 69
Royal Aircraft Establishment, 357
Royal Institute of Technology (Sweden), 8
Royal Library, Copenhagen, 296
RTG. See Radioisotopic thermoelectric generator.
Rubey, Dr. William W., 3
Ruckelshaus, William D., 361, 385
Rulis, Raymond J., 315
Rumsey, H., Jr., 286
Ruppenthal, Dr. Karl M., 239
Rush, Ambassador Kenneth, 86
Russell, Carl D., 240
Rutgers Univ. Center of Transportation, 270
Ryan, Gen. John D. (USAF), 50, 93

S

S-IC. See Saturn V (booster), stage, 1st.
S-II. See Saturn V, stage, 2nd.
S-IVB. See Saturn V, stage, 3rd.
SABCA, 407
SAC. See Strategic Air Command.
Saenger-Bredt, Irene, 328.
Safeguard (antiballistic missile system), 59, 142, 200, 276, 317, 405
Safi, Morocco, 232
Sagan, Dr. Carl, 111
Sagittarius (constellation), 59, 233
Saint (satellite inspection and interception satellite), 42
St. Louis. Mo., 246, 347, 372
St. Peter's Basilica, Rome, 128
St. Petersburg, Fla., 139
St. Severin, France, 30
Sakharov, Andrey D., 369
Salas, Gabriel, 360
SALT. See Strategic Arms Limitation Talks.
Salt Lake City, Utah, 86, 243
SAM-2 (U.S.S.R. antiaircraft missile), 232
SAM-3, 232
Samos (satellite), 267
SAMSO. See Space and Missile Systems Organization.
SAMTEC. See Space and Missile Test Center.
Samuelson, Robert J., 151
San Antonio, Tex., 46, 97, 192
San Carlos Reservation, Ariz., 383
San Carlos Yautepec, Mexico, 79
San Clemente, Calif., 1, 7, 140, 185, 215, 246, 250, 288-289
San Diego, Calif., 27, 206, 359, 379
San Francisco, Calif., 96, 346, 390
San Jose, Calif., 348
San Juan, Puerto Rico, 257
San Marco (Italian launch site), Indian Ocean, 275, 395
Sandbridge, Va., 79
Sandia Corp., 21
Sandia Laboratories, 89
Santa Ana, Calif., 162
Santa Barbara, Calif., 181, 371
Santa Fe, N. Mex., 329
Santa Ninfa, Sicily, 130
SAO. See Smithsonian Astrophysical Observatory.
Sarabhai, Vikram, 239
Sargent, Gov. Francis W., 13
Sargent, Dr. Frederick, II, 49
Sargent, Adm. T. R., III (USCG), 229
Sargent, Dr. Wallace L. W., 409
SAS. See Small Astronomy Satellite.
Satellite infrared spectrometer (SIRS), 32
Satellite Photometric Observatory, mobile (NASA), 79
Satellite Tracking and Data Acquisition Network (STADAN), 199
Sato, Prime Minister Eisaku (Japan), 259
Saturn (planet), 33, 84
Saturn (booster), 406
Saturn I (booster), 217
Saturn IB (booster), 37, 50, 172, 189, 217, 285
Saturn V (booster), 100, 169, 189, 217
  capability, 31, 103-104
  contract, 28, 108, 153, 223, 278, 386
  launch, AS-508, 119
  program, 3, 11-12, 37, 43, 50, 65, 88, 92, 107, 172, 229, 248, 266-267, 271, 285, 301, 318, 364, 379
  stage
    1st (S-IC), 40, 266-267, 271, 386
      test, 44, 267, 318
    2nd (S-II), 27, 92, 108, 153, 271, 278, 330, 386-387, 399
      test, 86, 157, 251
    3rd (S-IVB), 92, 119, 399, 410
Saturn Workshop (spacecraft) (see also Skylab), 50, 67, 208-209, 285, 292, 310, 370, 387
Saunders-Roe, Ltd., 296
SCAD. See Subsonic Cruise Armed Decoy.
Schaefer, Herbert, 328
Schaeffer, Dr. Oliver A., 333
Scheer, Julian, 12, 68
Schilling, David C., Flight Trophy, 29
Schirra, Capt. Walter M., Jr. (USN, Ret.), 138
Schmitt, Harrison H., 101

Schneiderman, Dan, 279
Schreiber, Edward, 215
Schriever, Gen. Bernard A. (USAF, Ret.), 49
Schroter's Valley (moon), 192
Schultz, M/G Kenneth W. (USAF), 392
Schuman, Elmer, 250
Schwartz, Gunther, 8
Schwartz, Harry, 11, 273
Schwarzchild, Dr. Martin, 192
Schweickart, Russell L., 370
Science (see also National Academy of Sciences), 172-173, 226, 291
   award, 3, 94, 329, 348
   benefits, 13, 53, 162-164, 385
   Communist China, 156
   criticism, 162, 238
   Government support of, 47, 71, 153, 207-208, 228, 231, 238, 239-240, 249, 278-279, 322, 329, 337, 357, 410-411
   human needs, 53, 158, 244, 303, 329, 373, 374
   national policy and goals, 53, 57-58, 153, 158, 228, 231, 239-240, 249, 269-270, 298-299, 329, 338-339, 357, 406
   The Physical Sciences, report of National Science Board, 57
   President's Task Force on Science Policy, 158
   President's Science Advisory Committee, 87, 89, 90, 287
   U.S.S.R., 27, 153, 225, 298-299
Science and Technology, Dept. of (proposed), 335
Science and Technology: Tools for Progress (report), 158
Science Museum, London, 102
Science Research Council (SRC) (U.K.), 357
Scientific American, 277
Scientific and Technical Personnel in the Federal Government, 1968 (NSF-24), 187
Scientific instrument module (SIM), 249
Scientific satellite, 383, 409
Scientist-astronaut, 287
Scientists, 63-65, 187, 253, 257, 260, 329, 338-339, 362, 370, 395, 406-407
Scorpius (constellation), 233
Scott, Mrs. Blanche S., 13
Scott, Col. David R. (USAF), 101, 125, 285
Scout (booster), 107, 194, 275, 295, 318, 357, 362
Scoville, Dr. Herbert, Jr., 347
Scribner's, Charles, Sons, 173
Scripps Institution of Oceanography, 8
Sea of Fertility (Mare Foecundatis) (moon), 299, 316, 369
Sea of Rains (Mare Imbrium) (moon), 90-91, 365, 380, 391, 394
Sea of Tranquility (Mare Tranquillitatis) (moon), 4, 26, 143, 228, 267, 333, 369, 374
Seaborg, Dr. Glenn T., 10, 94, 162, 224, 263, 277, 298, 312, 329, 398
Seal Beach, Calif., 271

Seamans, Secretary of the Air Force Robert C., Jr.
   Airborne Warning and Control System (AWACS), 228
   awards, by, 155
   B-1, 195
   budget, 95
   C-5A, 86
   F-111, 10-11, 306
   missile program, 84, 311, 381-382
   space transportation system (STS), 53
   supersonic bomber, 195
   turbofan engine contract, 68
   USAF science policy, 242
   U.S.S.R. missile and space program, 311
Searle, Dr. Leonard, 409
Seattle, Wash., 185, 259, 266, 349, 386
SECOR (Sequential Collation of Range program), 114
Securities and Exchange Commission (SEC), 169
Sedov, Prof. Leonid I., 328
Seek Storm, Project, 171
Seismometer experiment, lunar
   Apollo 12, 5-6, 58, 71, 119, 244-245, 262, 276, 295, 297, 348, 360-361, 372
   Apollo 14, 308, 364
Seitz, Dr. Frederick, 329
Selenia-Industrie Elettroniche Associate, S. p. A., 361
Self-test and repair (STAR) computer, 341
Selfridge AFB, Mich., 292
Senegal, 184
Serpens (constellation), 51
Serpukhov, U.S.S.R., 225
SERT I (Space Electric Rocket Test), 38
Sert II, 38, 51, 76, 173, 177, 180, 243, 285, 315, 338, 393, 411,
Service module (SM), 50, 213, 278, 285
   Apollo 13, 119, 125, 140, 167
   Apollo 14, 323, 403
Service Technology Corp., 29, 44, 384
SESP. See space experiments support program.
Sessler, Dr. Andrew M., 94
SEV. See Space escape vehicle.
Sevastyanov, Vitaly I., 189, 201, 211, 225, 271, 327, 336, 338, 344, 346, 349
Seventh Air Force, 283
Seventh Day Adventists, Central State Conference, 139
Seversky, Alexander P. de, 401
Seychelles (islands), 187
Seyfert galaxies, 192
Shaffer, John H., 15, 113
Shanghai, Communist China, 165
Shatalov, Vladimir A., 132, 267
Sheldon, Dr. Charles S., II, 31, 68, 270
Shell Oil Co., 20, 220
Shepard, Capt. Alan B., Jr., (USN), 165, 167, 205, 231, 264, 306, 390
Sheyma Island, Aleutians, 273
Shook, Edward M., 311
Short Brothers & Harland, Ltd., 69
Short, Dr. Nicholas M., 223

Short-range attack missile (SRAM), 72, 93, 345
SHOT. See Society for the History of Technology.
Shuang-ch'eng-tsu, China, 147
SIDE. See Suprathermal ion detector experiment.
Siemens Aktiengesellschaft, 361
Sierra Club, 239, 272, 303, 391
Sikorsky Aircraft Corp., 198
Sikorsky, Igor I., 401
Sikorsky S-65 (helicopter), 198
Silver, Dr. Leon T., 294
Silver Quill Award, 41
Silverstein, Dr. Abe, 377
SIM. See Scientific instrument module.
Simons, Howard, 129
Singer, Dr. S. Fred, 143, 345
Singer-General Precision, Inc., Link Div., 196, 303
Sinus Medii (Central Bay) (moon), 26
Sirio (Italian satellite), 88, 413
SIRS. See Satellite infrared spectrometer.
Sjoberg, Sigurd A., 1, 137
Skylab (formerly Apollo Applications) program, 169, 293, 303, 330, 341, 351, 413
   contract, 7, 83-84, 118, 147, 191, 194, 196, 208, 209, 220, 257, 296, 303
   design, 218, 246, 292, 328, 370, 371, 372, 412
   funds, 35
   international cooperation, 225
   NASA program, 39, 42-43, 50-51, 57-58, 62, 67, 88, 104-105, 115, 172, 218, 226-227, 229, 251, 284-285, 301, 310, 361
   press conference, 103-105, 218
   test, 157, 201, 275-276, 292, 387, 412
Skylark SL-971 (sounding rocket), 233
Skynet A (IDCSP-A) (U.K. comsat), 16, 43-44, 269
Skynet B, 269, 411, 413
Slayton, Donald K., 40, 118, 293, 294
Slichter, Dr. Charles P., 90
Slidell, La., 207
SM. See Service module.
Small Astronomy Satellite (SAS), 60, 275, 395, 407, 411, 414
Small Scientific Satellite (SSS), 275
Smelt, Dr. Ronald, 21
Smith, Bradford A., 172
Smith, C. R., 305, 401
Smith, Francis B., 47, 383
Smith, Gerard C., 133
Smith, Keith M., 3
Smith, Kenneth M., 99, 170
Smith, Dr. Lloyd H., 90
Smith, Dr. Malcom, Jr., 115
Smith, Sen. Margaret Chase, 139
Smith MK 29 (automatic landing system), 69
Smith, Gov. Preston, 98
Smith, Sen. Ralph T., 371
Smith, Ross, 3
Smith, Adm. Willard J. (USCG, Ret.), 260
Smithsonian Associates, 349

Smithsonian Astrophysical Observatory (SAO) (Cambridge, Mass.), 2, 8, 14, 18, 59, 286, 380
   Center for Study of Short-Lived Phenomena, 262
Smithsonian Institution, 107, 110, 152, 165, 236, 274
   Astrophysical Laboratory, 146
   exhibit
      aircraft, 103
      lunar rock sample, 244
      spacecraft, 234
   Hall of Electricity, 197-198
   Museum of History and Technology, 189
   Museum of Natural History, 270
   NASA historical artifacts agreement, 234
   National Air and Space Museum, 55, 94, 234, 236, 240-241, 349, 379, 384
SMS. See Synchronous Meteorological Satellite.
Smylie, Robert E., 18
Smyth, Henry DeWolf, 185
Smyth, Robert, 403
SNAP-3A (nuclear generator), 217
SNAP-8, 51
SNAP-19, 411
SNAP-27, 125, 374, 411
Snow, Crocker, 386
Snyder, Dr. Conway M., 372
Snyder, Dr. Lewis E., 262
Société Anonyme de Télécommunications, 361
Société Nationale Industrielle Aérospatiale, 361
Society for the History of Technology (SHOT), 406-407
Society of Manufacturing Engineers, Interprofessional Cooperation Award, 126
Sodium experiment, 13, 398
Solar cell, 231
Solar corona, 44, 61, 79, 306
Solar energy, 44, 74-75, 101-102, 177
Solar flare, 77
Solar Particle Alert Network (SPAN), 320
Solar physics, 306, 370
Solar wind, 33, 306, 372, 407
Solid propellant, 31
Solid Waste Disposal Act, 47
SOLRAD (solar radiation satellite), 408-409
Sonar, 68
Sonett, Dr. Charles P., 295, 372
Sonic boom, 2, 29, 133, 239, 261, 283, 291, 359, 388, 390, 411
Soter, Dr. Steven, 297
Sounding rocket (see also individual sounding rockets: Aerobee, Arcas, Black Brant IV, Boosted Arcas, Boosted Dart, Javelin, Nike-Apache, Nike-Cajun, Nike-Iroquois, Nike-Tomahawk, Trailblazer, Viper-Dart), 36, 233, 411, 413-414
   international programs, 85, 381, 413-414
      NASA-Australia, 181, 185, 191, 414
      -Brazil, 252, 315, 318

—Canada, 3, 12, 16, 40, 48, 56, 62, 71, 99, 108, 126, 133, 154, 162, 334, 338
—France, 414
—Germany, West, 392, 400
—India, 1, 84, 95, 102, 414
—Norway, 38, 82, 369
—Pakistan, 102
—Spain, 414
—Sweden, 260, 414
U.K.-Australia, 233
South Africa, 51, 199
South Carolina, 54
South Pole, 380
South Rogers Lake Bed, Calif., 60
Southampton Univ., Man Powered Aircraft Group, 186
Southeast Asia, 209
Southern California, Univ. of, 23
Southwick, Thomas P., 260
Southwind (USCG cutter), 287
Sovetskaya Entsklopediya, 98
Soviet Academy of Sciences, 156, 277, 314, 327
   Luna XVI lunar rock samples, 299, 316
   scientist exchange, 47-48, 69-70
   space cooperation, 85-86, 117, 192, 229, 334, 335, 349, 351, 355
Soviet Institute for Space Research, 312
Soviet Institute of Biomedical Problems, 327
Soyuz VI mission, 108, 326
Soyuz VII mission, 108, 326
Soyuz VIII, mission, 108
Soyuz IX mission, 189-190, 192, 201, 203, 212, 225, 231, 271, 280, 326-327, 412
Soyuz IX (spacecraft), 196
Spaatz, Gen. Carl (USAF, Ret.), 307
Space Age Management, 166
Space and Missiles Systems Organization (SAMSO) (USAF), 43, 46, 87, 107-108, 146, 199, 231, 250, 311, 377
Space and Missile Test Center (SAMTEC) (USAF), 108
Space biology
   animal experiments, 257, 307, 324-325, 362, 367, 392, 395, 400, 411
   atmosphere, artificial, 131
   award, 37, 342
   back contamination, 7, 15, 20, 51, 55, 71
   environment, effects, 37-38, 47, 55-56, 57-58, 189-190, 203, 210, 211, 212, 262, 266, 280, 301-302, 308, 327
   extraterrestrial life, 6, 60, 71, 88, 110, 205, 325, 359, 366, 370, 387, 390-391
   international cooperation, 117, 327, 411
   Life Sciences in Space (NAS-NRC report), 395
   life support system, 7, 40-41, 74, 132, 147, 201, 297, 387, 400
   lunar dust experiment, 400-401
   medical benefits, 68, 152, 366, 397
   NASA program, 36, 57-58, 63, 74, 87-88, 90, 110, 178, 186-187, 257, 266, 307, 319, 324-325, 328, 361, 366, 367, 385, 395, 403, 411
   nutrition, 40-41, 115, 327, 361
   radiation effects, 227, 242-243, 304, 327, 390
   weightlessness, effects, 46, 57-58, 63, 110, 189-190, 203, 210, 262, 275-276, 280, 327, 362, 392
   women as astronauts, 46, 110
Space Congress, Seventh, 149
Space Development Committee (Japan), 368
Space Disturbance Laboratory, 320, 351
Space escape vehicle (SEV), 199
Space experiments support program (SESP), 231
Space law, 355
Space medicine. See Space biology.
Space, military use, 192
   China, Communist, 152-153
   communications, 16, 43-44, 95-96, 148-149, 198, 252, 269, 293-294, 346, 360, 413
   navigation, 43-44
   reconnaissance, 43-44, 252, 411
   U.K., 269, 411, 413
   U.S., 41, 42, 43-44, 95-96, 105, 252, 303, 323, 360, 371, 376, 410-411
   U.S.S.R., 30-31, 42, 226, 303, 358
Space Orbiter Shuttlecraft, 165, 181, 198
Space, peaceful use of, 19, 326
Space program, national (see also individual programs, such as Apollo program and National Aeronautics and Space Administration, budget), 314
   achievements, 31-32, 50, 111-112, 147-148, 166, 246-247, 269, 368-369, 410-411
   manned space flight, 3, 53-54, 105-106, 111-112, 133-134, 136-137, 138-139, 164, 166, 204, 227-228, 238, 245, 301, 341-342, 368-369, 412
   benefits. See Space results.
   objectives, 11, 28, 31-32, 77-78, 87-88, 145, 147-148, 192, 207-208, 231, 301, 303, 393-394
   policy, 10, 11-12, 17, 32, 35, 90, 111, 141, 211, 233, 294, 295-296, 298, 360, 397
   post-Apollo, 11-12, 35, 50, 77-78, 87-88, 103-105, 213-214, 218-219, 246-247, 248, 256, 263-264, 279, 284, 301, 303, 309-310, 316-317, 341, 359, 368, 393-394, 399
   press comment, 9, 10, 17, 19, 50, 69, 82, 83, 84, 86, 90, 91, 150, 232, 238-239, 251, 255-256, 271, 287-288, 290, 294
   press conference, 11-12, 78, 88, 103-105, 284-285, 292, 323
   significance, 111,112, 238-239, 248, 338-339, 349-350, 385, 393-394
   U.S.S.R. vs. U.S. See Space race.
Space race, 31-32, 91, 154-155, 156, 330, 413
   booster, 31, 270, 379
   funds, 31, 270, 379
   lunar exploration, 225, 303, 375, 379
   manned space flight, 379
   military, 303

planetary flights, 379
space station, 91, 225, 303, 379
Space rescue, 44, 327-328
Space results (see also Earth; Moon; Mars; Venus; individual probes, satellites, and sounding rockets), 106, 146
  agriculture, 326
  aircraft, 17-18, 84
  astronomy, 1, 21, 76, 79, 86, 102, 112, 240, 245, 361
  communications, 39, 65, 375-376
  earth sciences, 15-16, 73, 96, 101, 196, 240, 326, 391
  economic benefits, 302
  education, 112, 144, 375-376
  engineering, 204, 302, 330
  For the Benefit of All Mankind: A Survey of the Practical Returns from Space Investment (H. Rpt. 91-1446), 302
  geology, 326
  international relations, 358
  materials technology, 27, 92, 142, 192, 302, 324
  medicine, 58, 92, 152, 192, 302, 366, 389
  meteorology, 84, 276, 327
  military, 192
  navigation, 326, 327
  oceanography, 326
  science and technology, 111-112, 166, 393-394
  social science, 11
  transportation, 41, 302, 326, 375
Space Science Board (NAS-NRC). See National Academy of Sciences and National Research Council.
Space shuttle (see also Space Transportation system), 7, 106, 174, 180, 192, 247, 316-317, 318, 341
  contract, 42, 55, 157, 170, 179, 181, 198, 203-204, 215, 217, 275, 285, 292, 324, 325, 374, 386, 399, 400, 412
  cost, 217
  criticism, 217, 330
  design, 12, 78, 107, 179, 203-204, 246, 313, 328, 374, 384, 400
  funds, 31, 36-37, 43, 166, 207-208, 217, 341, 389
  international cooperation, 41, 52, 88, 227, 292, 313, 323, 327-328, 351, 361
  launch site, 386, 391
  meeting, 88, 219-220, 234, 246, 384, 400
  military, 105, 323
  NASA program, 17-18, 36-37, 54, 67, 74, 78, 88, 165, 181, 198, 207-208, 219-220, 248, 284-285, 301-302, 323, 327-328, 334, 393-394, 412-413.
  press comment, 50, 58-59, 91, 391
  press conference, 227-228
  reusable, 12, 37, 41, 42, 50, 55, 84, 88, 107, 170, 174, 183, 203-204, 217-218, 227-228, 284-285, 301, 367
  test, 182-183, 197, 301-302
Space Shuttle Technology Conference, 234

Space Shuttlecraft Symposium, 219-220
Space Station (see Manned Orbiting Laboratory; Saturn Workshop; Skylab), 181, 197, 240, 307, 314, 329
  contract, 171, 183, 208, 227, 265, 266, 386
  criticism, 217, 295-296, 330
  design, 67, 104-105, 169, 183, 208, 218-219, 264, 273, 292, 328, 367, 370, 372, 412
  funds, 35, 166, 207-208, 217, 229, 284, 389
  international cooperation, 43, 56, 88-89, 192-193, 212, 225, 292, 295, 305, 327, 346, 351
  meeting, 89, 267, 295, 371, 385
  military, 11, 36, 105
  NASA program, 3, 10, 11-12, 29, 37, 43-44, 50, 52, 66-67, 78-79, 87-88, 89, 103-105, 197, 232, 248, 251, 263-264, 267, 284-285, 293, 301, 303, 323, 361, 385, 412-413
  patent, 232, 340
  press comment, 10, 83, 318
  radiation hazards, 227
  test, 201, 231, 266, 275-276, 387
  use, 87-88, 267, 295
  U.S.S.R., 19, 91, 160, 181, 193, 203, 225, 229, 280, 312, 342, 355, 379
Space station simulator, 297
Space Station Utilization Conference, 267, 295
Space Task Group, 36-37, 43, 53
Space transportation system (STS) (see also Space shuttle; Space tug), 53, 58-59, 88, 217, 305
Space tug, 50, 112, 174, 182-183, 194, 248, 301, 307, 323
Spacecraft (see also individual spacecraft, such as Apollo, Luna, Lunar Orbiter, Mariner, Surveyor), 11-12, 52, 88, 227, 234, 244, 250, 283, 284, 285, 309, 323, 410-411
  accident. See Accident, spacecraft.
  airlock, 104-105, 169, 246, 248, 372
  command and service module. See Command and service module.
  command module. See Command module.
  contract controversy, 115, 145, 146, 201, 224, 229, 230, 273-274, 289, 318
  debris, 276, 287, 297, 306, 395
  design, 11-12, 38, 47, 49, 67, 78, 104-105, 112, 170, 177-178, 179, 180, 183, 203-204, 218-219, 223, 264, 292, 328, 342, 365, 383, 399
  electrical systems, 209
  equipment, 25, 73-74, 114, 115, 116, 118, 169, 185, 218-219, 279, 296, 340, 380-381, 394-395
  escape system, 12-13, 199
  exhibit, 49, 99, 234, 401
  extravehicular equipment, 18, 29, 338
  hazards, 54, 71, 130-131, 141-142, 227
  heating, 209
  instrumentation, 25, 73, 76, 79, 114, 155,

185, 209, 232, 247, 314, 362, 365, 367, 374, 380, 395
life support system, 40-41, 74, 121, 123, 132, 147, 201, 218-219, 361, 387, 392
lunar module. See Lunar module.
multiple docking adapter, 104-105, 191, 246, 257, 347, 372
propulsion. See Engine and individual launch vehicles, such as Saturn.
record, 108, 190, 203, 231, 401
reentry, 182-183, 198, 301, 318, 332-333, 381
reusable (see also Space shuttle), 36-37, 209, 217
scientific instrument module, 249
service module. See Service module.
stabilization system, 87
test, 38, 74, 76, 84, 87, 89, 161, 165, 173, 198, 201, 203, 219-220, 275-276, 297, 306, 334, 387, 403-404.
Spacecraft debris, 40, 42, 175, 276, 287, 297, 306, 395
Spacecraft support unit (SSU), 38
Spacemedic Award, 37
Spacesuit, 194
SPACO Inc., 113
Spain, 51, 207, 213, 414
SPAN. See Solar Particle Alert Network.
Spartan (missile), 276, 405
Spaulding Distinguished Series lecture, 360
Spectrometer, 114, 365, 380
SPED. See Supersonic Planetary Entry Decelerator.
Sperry, Lawrence, Award, 342
Sperry Rand Corp.
 Sperry Gyroscope Div., 272
 Univac Div., 225
Spirit of St. Louis (aircraft), 379
Spokane, Wash., 180
Spong, Sen. William B., 235
Sprince, Richard H., 245
Springfield, Ill., 265
Sprint (missile), 405
SPS: service propulsion system.
Sputnik I, 151
SRAM. See Short-range attack missile.
SRC. See Science Research Council (U.K.).
SRET (Satellite de Récherches et d'Environment Technique) (French satellite), 177
SRN 1 (hovercraft), 296
SS-9 (U.S.S.R. ICBM), 103, 167, 230, 266, 311, 320, 400
SS-11, 230, 274, 400
SS-13, 230
S.S. Manhattan (ice breaker), 108
SSS. See Small Scientific Satellite.
SSS-A (Small Scientific Satellite), 409
SST. See Supersonic transport.
The SST: The Issues of Environmental Compatability, 311
SSU. See Spacecraft support unit.
Staats, Elmer B., 224

STADAN. See Satellite Tracking and Data Acquisition Network.
Stafford Field, Okla., 151
Stafford, Col. Thomas P. (USAF), 18, 151
Stafford, Mrs. Thomas P., 18
Stanford Research Institute, 398
Stanford Univ., 68, 191, 214, 239, 407
 Linear Accelerator Center, 55
Star, 21, 26, 51, 55, 61, 91, 191-192, 198, 233, 245, 358, 380, 381, 409, 414
STAR (Satellites for Telecommunications, Applications and Research) (consortium), 407
STAR. See Self-test and repair computer.
Staran IV (anticollision system), 204
Stardust Hotel, Las Vegas, Nev., 126, 130
Starr, Dr. Chauncey, 157
State Dept. of, 6, 53, 252, 305, 390, 399, 405, 409
State Univ. of New York, 333
 Buffalo, 368
STB. See Subsystem test bed.
Steele, John L., 166
Stehling, Kurt R., 221
Stennis, Sen. John C., 275, 286, 306, 408
Sternberg Astronomical Institute, 346
STERO (French experiment for measurement of solar radio emissions), 177
Stever, H. Guyford, 129
Stevenson, Sen. Adlai E., III, 371
Stewart, Francis M., 197
Stewart, Prof. Homer J., 84
Stewart, L/G James T. (USAF), 194
Stockholm, Sweden, 8
Stokes, Mayor Carl, 208
STOL (short takeoff and landing) aircraft, 32, 63-64, 91-92, 118, 179, 347
Stommel, Dr. Henry, 216
Stonewall, Tex., 15
Strangway, David, 274
Strategic Air Command (SAC) (USAF), 93
Strategic Arms Limitation Talks (SALT), 82, 131, 133, 142, 230, 244, 248, 258, 266, 359, 371, 402
Stratoscope II (balloon-borne telescope), 101, 189, 192, 411
Streptococcus mitis, 178
SST. See Space transportation system.
Study of Critical Environmental Problems, 255
Submarine, missile-carrying, 206, 244, 245-246, 289, 358
Subsonic Cruise Armed Decoy (SCAD), 93
Subsystem test bed (STB), 84
Suez Canal, 232, 294, 383
Suffield, Alberta, 242
Sullivan, Walter S., 30, 238, 376
Sun, 30, 33, 43, 52, 61, 79, 83, 101-102, 177, 210, 240, 306, 311, 320, 345, 372, 400, 407
Super Guppy (cargo aircraft), 372
Supercritical wing, 58, 359, 377, 409, 411
Supersonic Planetary Entry Decelerator (SPED), 332
Supersonic transport (SST) (see also Concorde and Tu-144), 17

administration, 107
cost, 170, 174, 350, 411
criticism, 17-18, 113, 286, 326
design and development, 180, 266, 279, 389-390
economic aspects, 32, 178, 179, 252, 275, 326, 366
foreign, 260-261, 275, 291, 359, 364, 366
funds, 36, 113, 183, 377, 383, 386, 391, 393, 399, 403, 409, 411
hazards, 42, 178, 216, 239, 252, 255, 256, 272, 286, 291, 303-304, 311, 320-321, 345, 381, 385, 387-388, 389-390, 400, 411
navigation, 43-44, 47, 315
Nixon, President Richard M., 272, 387-388, 393
press comment, 17-18, 113, 272, 326, 387, 390-391
press conference, 42, 179, 239, 393, 400
sonic boom, 2, 239, 261, 283, 291, 388, 390, 411
weather forecasting for, 2

Supersonic Transport Development Program, 107
Suprathermal ion detector experiment (SIDE), 119
Surveyor III (spacecraft), 2, 7, 16, 54, 177, 178, 353
Surveyor VI (lunar probe), 26
Surveyor VII (lunar probe), 294
Sweden, 260, 291, 414
Swedish-American of the Year, 256
Swenson, Dr. Loyd S., Jr., 342, 406
Swerdling, Melvin, 309
Swigert, John L., Jr.
  Apollo 13 mission, 147-148
    celebrations for, 136, 140, 162, 194
    flight, 119-125
    Nixon, President Richard M., 127, 133, 136-137, 139-140, 314
    press comment, 126, 136
    press conference, 143
    splashdown, 125
    training, 115, 116-118
  awards and honors, 133-134, 137, 140-141, 368
  goodwill tour, 314, 331
  income tax, 136
  press conference, 292
  space program, national, 292-293
Swiss Transport Museum, 331
Swissair, 289, 300
Switzerland, 51, 291, 292, 314
Sydney, Australia, 92, 129
Symington, Sen. W. Stuart, 307
Symposium on Environmental Aspects of Nuclear Power Stations, 263
Synchronous Meteorological Satellite (SMS), 43, 85, 243, 354
Syria, 306
Syvertson, Clarence A., 250
Szmant, Alina, 226

T

T2-C (jet trainer), 377, 411
T-33 (jet trainer), 352
T-38 (jet trainer), 20, 352
T-39 (jet trainer), 214
Tachometer, heart-rate, 352
TACRV. See Tracked air-cushion research vehicle.
Tacsat I (tactical communications satellite), 148
Tafunda Airport, American Samoa, 136
Tago-Sato-Kosaka comet, 22, 27
Taiwan, 157
Talkeetna, Alaska, 223
Tama (U.S.S.R. oceanographic vessel), 342
Tamplin, Dr. Arthur, 242
Tanzania, 82
Teague, Rep. Olin E., 244, 317
Technology, 159-160, 191, 211, 226, 228, 233-234, 237, 244, 249, 292, 298, 338-339, 357, 372, 373-374, 379
Technology Assessment: Annotated Bibliography and Inventory of Congressional Organization for Science and Technology, 234,
Technology utilization, space, 106, 111-112, 180, 258, 412-413
  agriculture, 106, 290, 359-360, 412
  air transportation, 63-64, 214, 279-280, 330, 341
  award, 106
  communications, 39, 78, 106, 346
  earth resources, 39, 78, 98, 106, 196, 259, 280, 383-384, 412
  economic progress, 87-88, 164
  international aspects, 194-195, 327
  materials technology, 167, 214, 324, 330
  medicine, 106, 194-195, 348, 350, 352, 366, 389, 397, 412
  meteorology, 78, 264, 343, 351-352
  military communications, 43-44
  pollution control, 43-44, 64, 72-73, 98, 106, 196, 221, 255, 260, 412
  science, 51-52, 57-58, 59, 61-62, 87-88, 111-112, 295
  social progress, 35, 37, 72-73, 77, 84, 106, 149, 164, 280, 340, 391
  systems engineering, 325, 412
  transportation, 106, 180, 334
  urban planning, 106, 183, 259
Tektite (rock), 195, 239, 333
Tektite I (underwater research project), 110, 169
Tektite II, 110, 118, 226, 239, 308
Tel Aviv, Israel, 118
Telecommunications, 39, 45-46, 200, 215, 288
Telemetry, 164-165, 166
Telescope
  infrared, 61, 251
  optical, 17, 55, 191
  radioastronomy, 110
  reflection, 17, 101

space, 381
U.S.S.R., 55, 216
white-light, 79
Television
  Apollo 13, 119-121, 127, 137-138
  color, 258
  educational, 38-39, 54, 61, 156-157, 233, 243, 413
  Luna XVII, 364-365
  Lunokhod I, 394
  Mariner-Mars mission, 279
  via satellite, 14, 32, 38-39, 54, 61, 89, 156-157, 233, 245, 257, 414
Television relay using small terminals (TRUST), 367
Teller, Dr. Edward H., 406
Teokaekara, Dr. Matthew P., 101
Tereshkova-Nikolayeva, Valentina V. See Nikolayev-Tereshkova.
TERLS. See Thumba Equatorial Rocket Launching Station.
The Terrestrial Environment: Solid-Earth and Ocean Physics (NASA CR-1579), 158
Tetrenault, Norman, 360
Texas, 98, 311
Texas Committee for the Study of Land Use and Environmental Control, 146
Texas Instruments, Inc., 55
Texas Medical Center, Houston, 350
Texas, Univ. of, 26, 48, 133
TFX. See F-111.
Thailand, 157
Thant, U, U.N. Secretary General, 238, 297, 302, 304
Thermal reactor, 260
Thermoelectric Outer Planet Spacecraft (TOPS), 180, 341
Thermonuclear fusion, 277, 376
Thibodaux, Joseph G., Jr., 206
Thiokol Chemical Corp., 24
Thomas, David D., 19, 99
Thompson, Milton O., 219
Thompson, Robert F., 51
Thompson, Dr. Theos J., 379
Thompson Trophy Race, 212
Thomson CSF, 407
Thor (booster first stage), 113-114, 315
Thor-Agena (booster), 175, 231, 273 D, 113-114, 371
Thor-Burner II (booster), 49, 286
Thor-Delta (booster) (see also Delta), 33, 394
  long-tank, 23-24, 144, 241
  long-tank, thrust-augmented, 14, 95, 269
Thorad-Agena (booster), 38, 73
Thrust vector control system (TVC), 46
Thumba Equatorial Rocket Launching Station (TERLS), 1, 18, 84, 95, 102, 185
Tidbinbilla, Australia, 100, 213
Timofeyev, Sergey, 176
Tipton, Stuart G., 275
Tiros I (Television Infrared Observation Satellite), 24, 107
Tiros Operational Satellite (TOS), 23-25
Titan IIIB (booster), 220
Titan IIIB-Agena, 242, 267, 345

Titan IIIB-Agena D, 14, 130
Titan IIIC, 46, 58, 61-62, 114, 220, 360
Titan IIID, 214, 220
Titan-Centaur (booster), 61-62
Titov, LJC Gherman S. (U.S.S.R.), 267
TLI: translunar injection.
TMA. See Trimethylaluminum.
Tobias Mayer Dome (moon), 192
Tokyo, Japan, 128, 321, 344, 364
Tokyo Observatory, 344
Topo I (geodetic satellite), 114, 411
TOPS. See Thermoelectric Outer Planet Spacecraft.
Toronto, Univ. of, 274
Torrejon Air Base, Spain, 243
TOS. See Tiros Operational Satellite.
Toulouse-Blagnac Aerodrome, 359
Toward a Science Policy for the United States (House report), 337
Toward Balanced Growth: Quantity with Quality (National Goals Research Staff report), 237
Townes, Dr. Charles H., 1, 90, 337
Townsend, Dr. John W., 347
Townsend, Mrs. Marjorie R., 407
Tracked air-cushion research vehicle (TACRV), 94
Tracking
  aircraft, 210, 381-382
  animal, 52, 57, 58-59, 62, 107, 118
  Apollo 11 mission, 66, 210, 381
  budget, 66, 93
  contract, 105, 236
  cooperation, international, 88, 100, 199, 207, 213, 381
  deep space (DSN), 16, 66, 100, 207, 213, 265
  military, 105
  MSFN, 66, 210, 212, 320, 353
  NASCOM, 162
  ship, 210, 212, 381
  STADAN, 199
  station, 199, 210, 380
    Afghanistan, 330
    Antigua, 210
    Australia, 100, 199, 213
    Chile, 199
    Ecuador, 199
    England, 199
    Madagascar, 199
    Mexico, 381
    South Africa, 199
    Spain, 207, 213
    U.S.S.R., 30-31
Trailblazer II (sounding rocket), 376
Trailing vortex, 320
Train, Russell E., 30, 42
Tranquilite (lunar mineral), 289
Tranquility Base (moon), 76
Trans World Airlines, Inc. (TWA), 346, 364
Transit IV-A (navigation satellite), 217
Transportation, Dept. of (DOT) (see also Federal Aviation Administration), 314, 330
  air cushion vehicle, 94
  air traffic control, 36

Air Traffic Control Advisory Committee, 76, 258
aircraft security, 302, 310, 345
automated transit system, 408
award, 113
budget, 183, 377, 390, 399, 409
contract, 94
cooperation, 15, 36, 169, 219, 283, 350, 389-390, 404, 412
Environmental Advisory Council, 239
exhibit, 259
facilities, 99, 105, 109, 169, 323-324
noise abatement, 353, 389-390, 400
personnel, 40, 65, 153, 169, 189, 250, 260, 271, 297, 340-366
R&D, 99, 105, 109, 250, 341
supersonic transport, 107, 174, 178, 179, 183, 239, 303-304, 377, 381, 389-390, 399, 400, 409
Transportation Systems Center, 99, 105, 169, 189, 219, 341
VTOL aircraft, 350
Transportation Development Center. See Transportation Systems Center (DOT).
Transportation Systems Center (DOT), 99, 105, 169, 189, 219, 341
Treasury Dept., 275, 405
Treaty, 40, 44, 75, 186, 283, 294, 371, 395
Tribus, Dr. Myron, 239
Trimethylaluminum (TMA), 398
Trippe, Juan T., 401
Trubshaw, Brian, 300
Trudeau, Prime Minister Pierre E. (Canada), 135
True, Dr. Ranate, 226
Truly, Richard H., 310
TRUST. See Television relay using small terminals.
Trust Territories Upper Air Observation Program, 330
Truszynski, Gerald M., 66, 74, 176
TRW Inc., 55, 181, 220, 264
TRW Systems Group, 49
Tsien Hsueh-shen, 153
Tsiolkovsky, Konstantine, 92
Tu-16 (Badger) (U.S.S.R. bomber), 307
Tu-95 (U.S.S.R. bomber), 390
Tu-144 (U.S.S.R. supersonic transport), 359, 366, 367
Tuebingen, West Germany, 264
Tullahoma, Tenn., 89, 142, 382
Turbotrain, 335
Turbulence, 9, 62, 310, 389-390
Turcat, André, 359
Turgarinov, Aleksey, 316
Turkevich, Dr. Anthony, 59
Turkey, 20
Turner, Dr. Barry E., 325
Turner, Robert E., 264
Turner, Col. Roscoe (USAF), 110, 212
Turner, Roscoe, Aeronautical Corp., 212
TVC. See Thrust vector control system.
Tverdokhelebov, A.N., 370
TWA. See Trans World Airlines, Inc.
20th Tactical Fighter Wing, 300
Tyche (lunar crater), 59, 294

U

U-2 (reconnaissance aircraft), 164, 383
U.A.R. See United Arab Republic.
Uchinora Space Center, Japan, 48, 315
Udall, Stewart L., 407
Ueda, Dr. Takao, 197
UFO. See Unidentified flying object.
UHF: ultrahigh frequency.
Uhl, Edward G., 145
Uhuru (Explorer XLII) (SAS-A) (Small Astronomy Satellite), 395, 407, 411, 413
U.K. See United Kingdom.
UK-4. (U.K. scientific satellite), 357, 409
UK-5, 357
UK-R (U.K. x-ray satellite), 413
Ultraviolet (UV), 16, 27, 100, 102, 181, 191, 194, 225, 251, 272, 304, 333-334, 353, 377, 380-381, 384, 396, 413
extreme (EUV), 264
U.N. See United Nations.
UNESCO. See United Nations Educational, Scientific, and Cultural Organization.
Unidentified flying object (UFO), 3-4
Unidentified satellite, 14, 49, 73, 175, 212, 242, 267, 278, 286, 345, 360, 371, 376
Union Carbide Corp., Linde Div., 200
United Action for Animals, Inc., 307
United Air Lines, 334
United Aircraft Corp.
  Hamilton Standard Div., 7
  Pratt & Whitney Div., 42, 55, 68, 118, 157, 198, 278
  Sikorsky Aircraft Div., 21
United Arab Republic (U.A.R.), 317, 383
United Auto Workers Union, 239
United Kingdom (U.K.), 151
  air traffic control, 62
  aircraft, 241, 385, 414
    Concorde, 170-171, 226, 260-261, 272, 275, 283, 291, 300, 326, 350, 359, 364, 366, 388-389
    V/STOL, 85
  booster, 200, 285
  cooperation, defense, 44, 241, 269, 399, 411
  cooperation, space, 16, 44, 51, 149, 150, 199, 200, 233, 287, 292, 354, 357, 359, 408-409, 411, 413
  hijacking of aircraft, 289-290, 291, 292, 293, 297, 300
  House of Commons, 350, 385
  launch
    satellite, 200, 269, 285, 411, 413
    sounding rocket, 233
  lunar sample study, 51
  metric system, 406
  nuclear weapons, 258-259
  satellite, 16, 32, 44, 354, 357, 408-409, 411, 413
  science and technology, 21, 382
United Nations (U.N.), 2, 263, 284
  Apollo 11 lunar sample exhibits, 238
  Committee on the Peaceful Uses of Outer Space, 30, 128, 413

Committee on the Peaceful Uses of the Seabed and Ocean Floor, 82, 277
Conference on Human Environment, 255
General Assembly, 302, 345, 371, 395, 405
Political Committee, 395
hijacking of aircraft, 304
Secretary General, 238, 292, 297, 304
Security Council, 293
United Nations Educational, Scientific, and Cultural Organization (UNESCO), 110
United States (U.S) (see also appropriate agencies and Congress), 49, 199
  award, 55, 57, 67, 87, 94, 113, 137, 166, 209, 269, 277, 348
  budget, 10, 35-37, 44, 65, 208, 237
  communications, 7, 25-26, 28, 45-46, 78, 96, 99, 112, 151, 290, 405
  computer industry, 164-165
  defense, 16-17, 27, 44, 67, 75, 93, 97, 139, 142, 171, 172, 173, 185, 223-224, 246, 252, 257-258, 286, 289, 358
  disarmament, 82, 131, 133, 142, 244, 248, 258, 266, 359, 371, 402
  economy, 32,
  education, 214, 215-216
  employment, 17, 325, 362, 370
  Expo '70, 91
  housing, 263
  international cooperation, 16, 27, 47-48, 53, 75, 82-83, 96, 106, 179, 221, 225, 242, 307-308
  international cooperation, space. See International cooperation, space.
  meteorology, 2, 49, 127, 169, 171, 351-352
  metric system, 406
  nuclear nonproliferation treaty. See Nuclear nonproliferation treaty.
  oceanography, 99, 110, 118-119, 127, 179, 216, 226, 228, 229-230, 239, 308, 317, 324, 325, 330, 347, 394
  pollution abatement, 1, 10, 16, 20, 22-23, 27, 30, 35, 42, 44-45, 47, 74-75, 77, 99, 106, 116-117, 132, 171, 176, 231, 237, 255, 256, 260, 262, 286, 296, 304, 326, 366
  research and development, 12-13, 17, 21, 44, 45, 71, 98, 99, 109, 136, 152-153, 157-158, 228, 242, 249, 253, 265, 316, 329-330, 362, 376, 394, 410
  science and technology, 53, 86, 89-90, 102, 129, 162-164, 187, 237, 260, 266-267, 269-270, 277, 370
    Government support of, 47, 53, 71, 136, 147, 152-153, 157-158, 207-208, 227-228, 231, 238, 239-240, 249, 279, 322, 329-330, 337, 338, 357, 410
    human needs, 244, 298, 303, 357, 373-374
    national policy and goals, 132, 136, 158, 207-208, 227-228, 231, 234, 237, 239-240, 249, 271, 329-330, 335, 337, 338-339, 345-346, 370

Nixon, President Richard M., 53, 57, 297-298, 302, 357
space program. See Space program, national.
State of the Union Message, 22, 27
transportation (see also Supersonic transport), 15, 22, 28, 105, 107, 108-109, 158, 197, 210-211, 312, 335
Vietnam war. See Vietnam war.
United States and Soviet Progress in Space: How Do the Nations Compare? (report), 30-31
United States Space Science Program (NAS-NRC report), 177
Universities, 110, 187, 194, 269-270, 322
  Federal support, 45, 47, 98, 153, 172, 187, 215-216, 231, 238, 278-279, 322, 337
  grants to, 47, 131, 215-216
  military research, 172, 176, 231
  NASA program, 36, 47, 66, 79, 108, 111-112, 144, 204, 216, 230, 234, 256, 322
Universities Space Research Assn., 2
University Corp. for Atmospheric Research, 49
Upper Heyford, England, 300
Upper Volta, 199
Uranus (planet), 33, 84, 101, 189
Urban Coalition, 385
Urban Technology Conference, 391
Urbana, Ill., 172
Urey, Dr. Harold C., 287, 305
Ursa Major (constellation), 225
U.S. Air Force (USAF) (see also individual bases, centers, and commands, such as Air Force Systems Command, Arnold Engineering Development Center, Edwards AFB)
  aircraft (see also individual aircraft, such as, B-1, C-5A, C-130, F-111, helicopter, X-15, XB-70), 7, 9, 16, 26-27, 82, 93, 95, 129, 142, 149, 173, 174, 208, 225, 264, 300, 331, 381
  anniversary, 307, 320
  award, 175, 227, 229
  booster (see also names of boosters, such as Atlas-Agena, Thor-Agena, Thor-Burner II, Titan IIIB), 46, 58, 61-62
  budget, 93, 95
  camera, 200
  communications satellite, 16, 43-44, 269, 411, 413
  contract, 9, 22, 56, 58, 68, 72, 103, 105, 195, 199, 220, 221, 227, 228, 231, 250, 270-271, 273, 345, 377, 381, 402
  cooperation, 16, 43-44, 53, 61-62, 68, 95, 105, 166, 172, 174, 180, 183, 214, 225, 259, 283, 290, 323, 329, 375, 382, 390, 408-409, 411
  launch
    reentry vehicle, 198
    satellite, 14, 49, 73, 114, 130, 175, 209, 212-213, 242, 267, 275, 278, 286, 305, 345, 360, 371, 411

lifting-body vehicle, 74, 336, 347, 375, 411
missile program, 84, 93, 146, 181, 214, 231, 248, 256, 276, 289, 305, 311, 346, 381, 392, 398, 405
MOL, 36
navigation satellite, 43-44
personnel, 173, 194, 227, 283, 297
research, 174, 242, 301, 320
satellite, 221, 278, 376
space program (see also Defense, Dept. of), 174
Vietnam war, 273
U.S. Army (USA), 73, 114, 226, 308
Aeronautical Laboratory, 109
Meritorious Civilian Service Award for Bravery, 109
U.S. Army Office of Military History, 189
U.S. Bureau of Public Roads, 180
U.S. Coast and Geodetic Survey, 230
U.S. Coast Guard (USCG), 110, 229, 287, 312, 324
U.S. Comptroller General, 145, 224
U.S. Conference of Mayors, 208
U.S. Congressional Medal of Honor Society, 184
U.S. Court of Appeals, 391
U.S. Geological Survey, 183, 259, 382, 405
U.S. Lake Survey, 230
U.S. Marine Corps (USMC). 198
U.S. National Committee of International Scientific Radio Union, 139
U.S. Navy (USN), 110, 230, 272, 342
aircraft, 58, 68, 151, 218, 377, 408, 411
aircraft carrier, 15
award, 42
computer, 106
contract, 68, 72, 345
control center, 293-294
cooperation, 110, 169, 212, 248, 275, 337, 408-409
deep submergence rescue vehicle, 27
launch, 275
missile, 105, 180, 206, 244, 245-246, 248, 256, 258, 289
personnel, 197-198
Project Tektite, 110, 118-119, 169, 226, 239, 308
research, 321
satellite, 275, 408-409
tracking ship, 212
U.S. Patent Office, 47
U.S. Public Health Service, 352
U.S. Smelting, Refining and Mining Co., 153
U.S. Supreme Court, 113
U.S. Weather Bureau, 2, 46, 230
USA. See U.S. Army.
USAF. See U.S. Air Force.
USAF Institute of Technology, 276
USAF Office of Aerospace Research, 224
USAF Weather Service, 259
USCG. See U.S. Coast Guard.
USMC. See U.S. Marine Corps.
USNS *Range Recoverer*, 333, 392
USNS *Vanguard*, 212
U.S.S. *Hornet*, 15
U.S.S. *Iwo Jima*, 125, 133, 136
U.S.S.R. (Union of Soviet Socialist Republics) (see also Soviet Academy of Sciences, etc.), 165, 176, 245-246, 256, 291, 301, 330, 392
aircraft, 93, 98, 154, 307, 347, 359, 366, 367, 372, 381-382, 390
accident, 330
airlines, 263
anniversary, 126, 127, 361
antiballistic missile system, 146, 171, 266, 311, 381
Apollo 13 mission
reaction to, 126, 128, 130, 132, 137
rescue aid offer, 130, 132
astronaut visit to, 176, 185, 190, 192, 195, 366
astronomy, 55, 286-287
award, 267, 271
booster, 31
budget, 391
Committee for Human Rights, 369-370
communications satellite, 15, 56, 214, 318, 380, 405, 411
cooperation, 47-48, 69-70, 106, 225, 263, 283-284, 307-308, 317, 320
cooperation, space, 32, 69-70, 83, 85-86, 106, 117, 156, 160, 176-177, 192, 197, 221, 225, 229, 243, 287, 300, 334, 335, 343-344, 346, 347, 349, 351, 353, 358, 376, 380, 389, 391-392, 393, 409, 413
cosmonaut. See Cosmonaut.
disarmament (see also U.S.S.R., SALT talks), 82, 245, 258, 266, 359, 371, 402
Expo '70 exhibit, 91
Institute of the United States of America, 199
laser, 156, 380, 389
launch, 410-412
probe
Luna XVI, 299
Luna XVII, 364
Venus VII, 266, 398
Vertikal I, 380
satellite
Cosmos, 8, 15, 16, 20, 21, 42, 46, 66, 73, 87, 94, 102, 109, 113, 114, 126, 130, 145, 147, 149, 170, 176, 198, 200, 201, 206, 214, 217, 226, 229, 248, 259, 261, 269, 270, 272, 277, 291-292, 304, 305, 311, 315, 320, 323, 325, 331, 332, 334, 337, 343, 345, 352, 366, 370, 376, 384, 385, 392, 397, 398, 399, 402
Intercosmos III, 259
Intercosmos IV, 336
Meteor III, 92
Meteor IV, 154
Meteor V, 211
Meteor VI, 337
Molniya I-13, 56
Molniya I-14, 214
Molniya I-15, 318

Molniya I-16, 380
Molniya I-17, 405
Soyuz IX, 189-190, 201, 203
Zond VIII, 342-343
launch site, 76
Lenin centennial, 127
Lunar XVI lunar samples, 299, 309, 311-312, 314, 316, 325, 349, 358, 365, 369, 372, 411
Lunokhod I, 365, 372, 380, 389, 391, 394, 411
meteorological satellite, 92, 154, 211, 337, 354, 411
missile and rocket program, 31-32, 67-68, 93, 103, 132, 142, 146, 167, 171, 200, 230, 232, 266, 274, 278, 289, 311, 320, 347, 349, 360, 361, 381-382, 400
nuclear nonproliferation treaty, 75, 186
nuclear weapons on seabed treaty, 283-284, 371
probe, 266, 286-287, 289, 299, 364-365, 379, 380, 398, 411-412
radar, 146
rocket test, 97, 103, 105, 272
SALT talks, 82, 131, 133, 142, 230, 244, 248, 258, 266, 359, 371, 402
science and technology, 27, 153, 225-226, 298-299
space program, 19, 30-31, 59, 68, 114, 126, 153-154, 160, 181, 187, 193, 196, 211, 229, 252, 270, 271, 280, 294, 298-299, 309, 311-312, 314, 318, 326-327, 330, 340, 346, 349, 355, 358, 359, 379, 397, 408, 410, 411-412
 military use, 30-31, 42, 95, 226, 303, 358
space station, 19, 91, 160, 181, 193, 203, 225, 229, 280, 312, 342, 355, 379
spacecraft. See U.S.S.R., launch; and individual spacecraft, such as *Luna XV, Molniya I-13, Soyuz IV, Zond VIII.*
spacecraft debris, 287, 395
spacecraft record, 108
submarine, missile-carrying, 358
supersonic transport, 359, 366-367
U-2 incident, 164
weapons, 82, 93, 103, 142, 171, 186, 259, 280-281, 347, 358, 361, 374-375
Utah, 386
Utah Launch Complex, 146
Utlaut, Dr. William F., 390
UV. See Ultraviolet.

V

Valley Forge Space Technology Center, 318
VAM-17 (booster), 99
VAM-20 (booster), 27, 224
Van Allen, Dr. James A., 105, 330, 360
Van Allen radiation belt, 103
Van Nuys, Calif., 175

Vance, Dale L., 7
Vancouver, B.C., 272
Vancouver (weather ship), 272
Vandenberg AFB, Calif. (see also Western Test Range)
 Advanced Ballistic Reentry System, 198
 contract, 231
 Minuteman I, 276
 satellite launch vehicle
  Thor-Agena, 73, 175, 273
  Thor-Burner II, 49, 286
  Thorad-Agena, 371
  Titan IIIB-Agena, 242, 267
  Titan IIIB-Agena D, 14, 130, 212
Vanderwalker, John G., 169
Vasa Medal, 256
VASI. See Visual-approach slope indicator.
Vaughn, Stanley I., 379-380
VC-10 (jet transport), 293, 300
Vecchietti, George J., 45, 352
Vegesack (German ship), 40
Vela (nuclear test detection satellite), 44, 114, 221, 411
Venezuela, 192-193
Venice, Italy, 313
Venus (planet)
 atmosphere, 286, 399
 exploration of, 3, 11-12, 247, 284, 311-312, 374, 398, 408
 map, 236
 probe, 247, 266, 273, 379, 398, 412
 radius, 286
 surface, 139, 236
 temperature, 398
Venus IV (U.S.S.R. interplanetary probe), 266, 286-287, 398
Venus V, 266
Venus VI, 266
Venus VII, 266, 273, 289, 398, 412
Venus: Strategy for Exploration (NAS report), 399
Vermiyle, William V., Medal, 313
Verona Test Site, N.Y., 149
Versatile upper stage (VUS) (launch vehicle), 275
Vesper, Howard G., 250
Vesta (asteroid), 210
Vertikal I (U.S.S.R. geophysical probe), 380
VFR: visual flight rules.
VHF: very high frequency.
Vienna, Austria, 68, 82, 131, 133, 230, 248, 312, 402
Vietnam, North, 151
Vietnam war, 1, 3, 152, 157, 238, 260, 273, 279
Viking Molecular Analysis Science Team, 255
Viking, Project
 contract, 283
 experiments, 255, 279, 341
 plans, 11-12, 43, 61, 284, 413
 test, 89, 412
Vinogradov, Aleksander, 358-359
Viper Dart (sounding rocket), 305, 309
Virgin Islands, 110, 210

Virginia, 82, 384
Virginia, Univ. of, 204, 257, 262
Virginia Beach, Va., 351
Visual-approach slope indicator (VASI), 335
Vlachý, J., 159
Vogel, Lawrence W., 98
Volcker, Paul A., 275
Volpe, Secretary of Transportation John A., 108-109, 258, 324
   air transport security, 204, 293, 310, 353
   appointments by, 153, 260, 340, 366
   contract award, 94, 228
   exhibit, 259
   international cooperation, 15, 28
   pollution control, 28, 270, 353
   R&D, 15, 99, 341
   SST, 107
von Braun, Dr. Wernher, 94, 118, 162, 226, 314, 389
   appointment, 28, 88, 412
   awards and honors, 219, 286
   press conference, 103-105
   R&D, 269-270
   space program, national, 91, 103-105, 146, 226, 385, 393
von Dohnanyi, Klaus, 86, 326
von Kármán, Dr. Theodore, 159, 267
von Kármán Lecture, 252, 342
Voskhod II mission, 9, 330
Vostok I (U.S.S.R. spacecraft), 126
Vostok VI mission, 196
V/STOL (vertical or short takeoff and landing) aircraft, 36, 43, 63-64, 85, 166, 198-199, 204, 350, 411-412, 413
VTOL (vertical takeoff and landing) aircraft, 63, 148, 198-199, 321, 350
VUS. See Versatile upper stage.

W

Wackenhut Services, Inc., 149
Waetjen, Richard M., 108
Waite, Roy, 352
Walker, Eric A., 87, 159
Walker, Joseph A., 267
Walkinshaw, Dr. Charles P., 400-401
Waller, Richard A., 169
Wallops Island, Va., 102
Wallops Station (NASA), 212, 384, 396
   award, 369
   Bio-Space Technology Training Program, 257
   contract, 200
   exhibit, lunar sample, 111
   launch
      Ofo (Orbiting Frog Otolith), 362
      RAM C-III (RAM C-C), 318
      RM (Radiation/Meteoroid Satellite), 362
      sounding rocket
         Aerobee 150, 77, 264
         Aerobee 170, 41, 77, 102
         Aerobee 350, 264
         Arcas, 77, 257, 305
         Bullpup-Cajun, 16

         Javelin, 77, 283, 328
         Loki-Dart, 305
         Nike-Apache, 13, 77, 256, 305, 398
         Nike-Cajun, 3, 9, 13, 77, 210, 213, 234, 305
         Nike-Iroquois, 77
         Nike-Tomahawk, 77, 250, 331, 392, 400
         Trailblazer II, 376-377
         Viper Dart, 305
   SPED (Supersonic Planetary Entry Decelerator), 332
Walsh, John, 215
Walton Beckett and Associates, 403
Wan Hu, 268
Wapakoneta, Ohio, 153
Warning system, emergency, 240
Warren AFB, Wyo., 317
Warsaw, Poland, 57, 76
Washburn, Bradford, Award, 13
Washington, 386
Washington Airways, Inc., 312
Washington, D.C., 44, 47, 82, 108, 129, 139, 198, 207, 224, 232, 234, 238, 259, 262, 272, 274, 312, 315, 345, 346, 349, 383
   astronauts in, 6, 152, 155, 157, 166
   awards presented in, 41, 94, 155, 166, 249, 310, 368, 384
   cosmonauts in, 336, 338
   exhibit, 103, 346
   meetings, 15, 21, 28, 37, 41, 99, 134, 139, 152, 157, 166, 270, 292, 300, 309, 311, 349-350, 352, 360, 381
   nuclear nonproliferation treaty signed, 75
   press conference, 42, 179, 239, 277, 284, 330, 392, 403
Washington Heart Assn., 37
Washington, Univ. of, 185
Wasserburg, Dr. Gerald J., 30, 59, 181
Waterman, Dr. Alan T., 268
Waterman, Walter, 352
Watkins, Capt. Augustine (Pan Am), 257
Wayne State Univ., 156
WC-130 (weather reconnaissance aircraft), 171
Weather forecasting, 2, 107
Weather modification, 49, 351-352, 390
Weatherford, Okla., 151
Webb, James E., 152, 165, 166
Weber, Dr. Joseph, 59, 198
Webster, William, 250
Weightlessness, effects, 46, 58, 63, 110, 190, 203, 210, 262, 275-276, 280, 327, 362, 392
Weiss, Dr. Harold S., 131
Welch, John J., Jr., 297
Wente, Van A., 310
West, Julian M., 325
West Virginia, Univ. of, 408
Western Governor's Conference, 86
Western Test Range (WTR), 23, 38, 107, 113, 194, 198, 203, 275, 394
Western Union Telegraph Co., 249
Wetherington, Ryndal L., 197

Wheelchair, motorized, 350
Whipple, John A., 185
Whitcomb, Dr. Richard T., Jr., 19, 350, 359
White, L/C Edward H., II (USAF), 192, 267
White, Edward H., II, Memorial Museum, 192
White House, 3, 211, 237, 266, 297
  appointments, 347
  astronauts, 51, 133, 198
  awards presented at, 55, 67, 229
  press conference, 134
  space program, 21, 28, 255-256, 317, 397
  visits to, 13, 118, 128, 198
White, Mayor Kevin H., 296
White Sands Missile Range (WSMR), N. Mex., 386
  contract, 227
  launch
    Aerobee 150, 77
      airglow, 127
      atmospheric data, 350
      solar astronomy, 210, 376-377
      stellar data, 91, 190-191, 245, 384
      ultraviolet astronomy, 190-191, 224, 225, 384
      x-ray astronomy, 377
    Aerobee 150MI, 27
    Aerobee 170
      ultraviolet astronomy, 272, 353
      x-ray astronomy, 45, 169, 311, 317
    Aerobee 350
      solar astronomy, 129
      stellar data, 129
      x-ray astronomy, 129
    Athena missile, 146, 231
    Nike-Apache
      airglow, 311
      atmospheric data, 401
      x-ray astronomy, 276
    Nike-Cajun, 96
  Space Orbiter Shuttlecraft drop test, 165, 198
White Sands, N. Mex., Test Facility, 41, 62
White, Gen. Thomas D., Space Trophy, 155
Whitehead, Dr. Clay T., 215, 288
Whiteman AFB, Mo., 317
Whitworth College, 180
Wichita, Kans., 347
Wiesner, Dr. Jerome B., 147, 329
WIKA (West German minicapsule), 84, 412
Wild, Dr. Jack W., 328
Wilford, John N., 74, 137, 236
Williams College, 158, 255
Williams, Dr. Donald J., 351
Willis, Charles F., Jr., 196
Willkie, Wendell L., 191
Wilson, Rep. Charles H., 264
Wilson, George C., 27, 167, 262
Wilson, Prime Minister Harold, 130
Wilson, M/G Louis L., Jr. (USAF), 194
Wind tunnel, 63, 109
Windecker Research, Inc., 99, 170
Windler, Milton L., 137

Windsor Locks, Conn., 225
Wing, aircraft, 10-11, 17, 19, 32, 58, 62, 359, 377, 409, 412
Winslow, Ariz., 347
Winter, Frank H., 94
Winter, Lumen Martin, 25
Winzen Research Corp., 286, 307
Wisconsin, 157
Wisconsin, Univ. of, 22, 191, 224
Wiser, F. C., 364
Withington, H. W., 386
Wolfe, George M., 298
Wolfle, Dael L., 185
Woll, Arthur M., 383
Wolper, David L., 314
Wolper Productions, Inc., 314
Women as astronauts, 46, 110
Wong, S. Kuen, 16
Wong, Wilson, 297
Wood, Carlos C., 21
Wood, Dr. John A., 59
Woomera Rocket Range, Australia, 7, 100, 181, 191, 233
Worcester Polytechnic Institute, 40
Worden, Maj. Alfred M. (USAF), 101, 285
Working Group on Lunar Nomenclature, 268
World Academy of Art and Science, 162
World Citizen Award, 219
World War II, 320, 321
World Weather Watch, 2
World Wildlife Fund, 372
Wright brothers, 98, 352
Wright Brothers Day, 394
Wright Brothers Lecture, 240
Wright Brothers Memorial Trophy, 305, 401
Wright Memorial Dinner, 401
Wright, Orville, 394
Wright-Patterson AFB, Ohio, 103, 153, 173, 194, 227, 258, 342, 382
Wright, Robert A., 237
Wright, Dr. Theodore P., 272
Wright, Wilbur, 394
WSMR. See White Sands Missile Range.
WTR. See Western Test Range.
Wuenscher, Hans F., 232, 340
Wyatt, Dr. DeMarquis D., 162
Wyld, James H., Propulsion Award, 206

X

X-15 (rocket research aircraft), 102-103, 220
X-24A (lifting-body vehicle), 74, 411
  test flight
    glide, 60
    powered, 94-95, 108, 145, 171, 206, 247, 263, 273, 336, 347, 375
XB-70 (supersonic bomber), 180, 233, 342, 382
XC-142 (V/STOL aircraft), 166, 199
XH-51 (helicopter), 413
XLR-11 (rocket engine), 66, 95
X-ray, 159

experiment, 45, 129, 191, 194, 276, 311, 317, 336, 361-362, 377
source, 51, 52, 169, 217, 233, 357, 395, 513

Y

Yagotin, U.S.S.R., 226
Yarborough, Charles, 62
Yates, Rep. Sidney R., 170
Yegerov, Col. Gen. Nikita V. (U.S.S.R.), 165
Yeliseyev, Aleksey, 181
Yerevan, Armenia, 97
Yeshiva Univ., 143
Yevdokimov, Dr. Valery, 197
YF-12 (reconnaissance aircraft), 180
YF-12A, 411
Yost, Ambassador to U.N. Charles W., 297

Young, Capt. John W. (USN), 198
Young Communist League, 181

Z

Zaehringer, Dr. Josef, 289, 333
Zelenchuk Observatory, 216
Zerqua, Jordan, 289
Zia Co., 41
Ziegler, Ronald L., 297
Zond (spacecraft), 379
Zond V (U.S.S.R. space probe), 343
Zond VI, 343
Zond VII, 343
Zond VIII, 342-343, 346, 359, 411
Zuckert Award for Outstanding Proficiency in Management, 392
Zvezdny Gorodok (Star City) (U.S.S.R.), 212, 346

NASA HISTORICAL PUBLICATIONS

HISTORIES

- Robert L. Rosholt, *An Administrative History of NASA, 1958-1963*, NASA SP-4101, 1966, GPO, $4.00.*
- Loyd S. Swenson, James M. Grimwood, and Charles C. Alexander, *This New Ocean: A History of Project Mercury*, NASA SP-4201, 1966, GPO, $5.50.
- Constance McL. Green and Milton Lomask, *Vanguard—A History*, NASA SP-4202, 1970; also Washington: Smithsonian Institution Press, 1971, $12.50.
- Alfred Rosenthal, *Venture Into Space: Early Years of Goddard Space Flight Center*, NASA SP-4301, 1968, GPO, $2.50.
- Edwin P. Hartman, *Adventures in Research: A History of the Ames Research Center, 1940-1965*, NASA SP-4302, 1970, GPO, $4.75.

HISTORICAL STUDIES

- Eugene M. Emme (ed.), *History of Rocket Technology*, Detroit: Wayne State University, 1964, $8.50.
- Mae Mills Link, *Space Medicine in Project Mercury*, NASA SP-4003, 1965, NTIS, $6.00.**
- *Historical Sketch of NASA*, NASA EP-29, 1965 and 1966, NTIS, $6.00.
- Katherine M. Dickson (Library of Congress), *History of Aeronautics and Astronautics: A Preliminary Bibliography*, NASA HHR-29, NTIS, $6.00.
- William R. Corliss, *NASA Sounding Rockets, 1958-1968*. NASA SP-4401, 1971, GPO, $1.75.
- Eugene M. Emme (ed.), *Statements by Presidents of the United States on International Cooperation in Space*, Senate Committee on Aeronautical and Space Sciences, Sen. Doc. 92-40, 1971, GPO, $0.55.

CHRONOLOGIES

- *Aeronautics and Astronautics: An American Chronology of Science and Technology in the Exploration of Space, 1915-1960*, compiled by E. M. Emme, Washington: NASA, 1961, NTIS, $6.00.
- *Aeronautical and Astronautical Events of 1961*, published by the House Committee on Science and Astronautics, 1962, NTIS, $6.00.
- *Astronautical and Aeronautical Events of 1962*, published by the House Committee on Science and Astronautics, 1963, NTIS, $6.00.
- *Astronautics and Aeronautics, 1963*, NASA SP-4004, 1964, NTIS, $6.00.
- *Astronautics and Aeronautics, 1964*, NASA SP-4005, 1965, NTIS, $6.00.
- *Astronautics and Aeronautics, 1965*, NASA SP-4006, 1966, NTIS, $6.00.
- *Astronautics and Aeronautics, 1966*, NASA SP-4007, 1967, NTIS, $6.00.
- *Astronautics and Aeronautics, 1967*, NASA SP-4008, 1968, GPO, $2.25.
- *Astronautics and Aeronautics, 1968*, NASA SP-4010, 1969, GPO, $2.00
- *Astronautics and Aeronautics, 1969*, NASA SP-4014, 1970, GPO, $2.25.
- James M. Grimwood, *Project Mercury: A Chronology*, NASA SP-4001, 1963, NTIS, $6.00.
- James M. Grimwood and Barton C. Hacker, with Peter J. Vorzimmer, *Project Gemini Technology and Operations: A Chronology*, NASA SP-4002, 1969, GPO, $2.75.
- Ivan D. Ertel and Mary Louise Morse, *The Apollo Spacecraft: A Chronology*, Vol. I, *Through November 7, 1962*, NASA SP-4009, 1969, GPO, $2.50.
- Mary Louise Morse and Jean Kernahan Bays, *The Apollo Spacecraft: A Chronology*, Vol. II, *November 8, 1962-September 30, 1964*, NASA SP-4009 (1972).
- R. Cargill Hall, *Project Ranger: A Chronology*, JPL/HR-2, 1971, NTIS, $6.00.

---

* GPO: Titles can be ordered from the Superintendent of Documents, Government Printing Office, Washington, D.C. 20402.
**NTIS: Titles can be ordered from National Technical Information Service, Springfield, Va. 22151

www.ingramcontent.com/pod-product-compliance
Lightning Source LLC
Chambersburg PA
CBHW081714170526
45167CB00009B/3575